The Atom

"The Building Blocks of Matter"

Edited by Paul F. Kisak

Contents

Chapter 1

The Atom

An **atom** is the smallest constituent unit of ordinary matter that has the properties of a chemical element.[1] Every solid, liquid, gas, and plasma is made up of neutral or ionized atoms. Atoms are very small; typical sizes are around 100 pm (a ten-billionth of a meter, in the short scale).[2] However, atoms do not have well defined boundaries, and there are different ways to define their size which give different but close values.

Atoms are small enough that classical physics give noticeably incorrect results. Through the development of physics, atomic models have incorporated quantum principles to better explain and predict the behavior.

Every atom is composed of a nucleus and one or more electrons bound to the nucleus. The nucleus is made of one or more protons and typically a similar number of neutrons (none in hydrogen-1). Protons and neutrons are called nucleons. Over 99.94% of the atom's mass is in the nucleus. The protons have a positive electric charge, the electrons have a negative electric charge, and the neutrons have no electric charge. If the number of protons and electrons are equal, that atom is electrically neutral. If an atom has more or fewer electrons than protons, then it has an overall negative or positive charge, respectively, and it is called an ion.

Electrons of an atom are attracted to the protons in an atomic nucleus by this electromagnetic force. The protons and neutrons in the nucleus are attracted to each other by a different force, the nuclear force, which is usually stronger than the electromagnetic force repelling the positively charged protons from one another. Under certain circumstances the repelling electromagnetic force becomes stronger than the nuclear force, and nucleons can be ejected from the nucleus, leaving behind a different element: nuclear decay resulting in nuclear transmutation.

The number of protons in the nucleus defines to what chemical element the atom belongs: for example, all copper atoms contain 29 protons. The number of neutrons defines the isotope of the element.[3] The number of electrons influences the magnetic properties of an atom. Atoms can attach to one or more other atoms by chemical bonds to form chemical compounds such as molecules. The ability of atoms to associate and dissociate is responsible for most of the physical changes observed in nature, and is the subject of the discipline of chemistry.

Not all the matter of the universe is composed of atoms. Dark matter comprises more of the Universe than matter, and is composed not of atoms, but of particles of a currently unknown type.

1.1 History of atomic theory

Main article: Atomic theory

1.1.1 Atoms in philosophy

Main article: Atomism

The idea that matter is made up of discrete units is a very old one, appearing in many ancient cultures such as Greece and India. The word "atom", in fact, was coined by ancient Greek philosophers. However, these ideas were founded in philosophical and theological reasoning rather than evidence and experimentation. As a result, their views on what atoms look like and how they behave were incorrect. They also could not convince everybody, so atomism was but one of a number of competing theories on the nature of matter. It was not until the 19th century that the idea was embraced and refined by scientists, when the blossoming science of chemistry produced discoveries that only the concept of atoms could explain.

1.1.2 First evidence-based theory

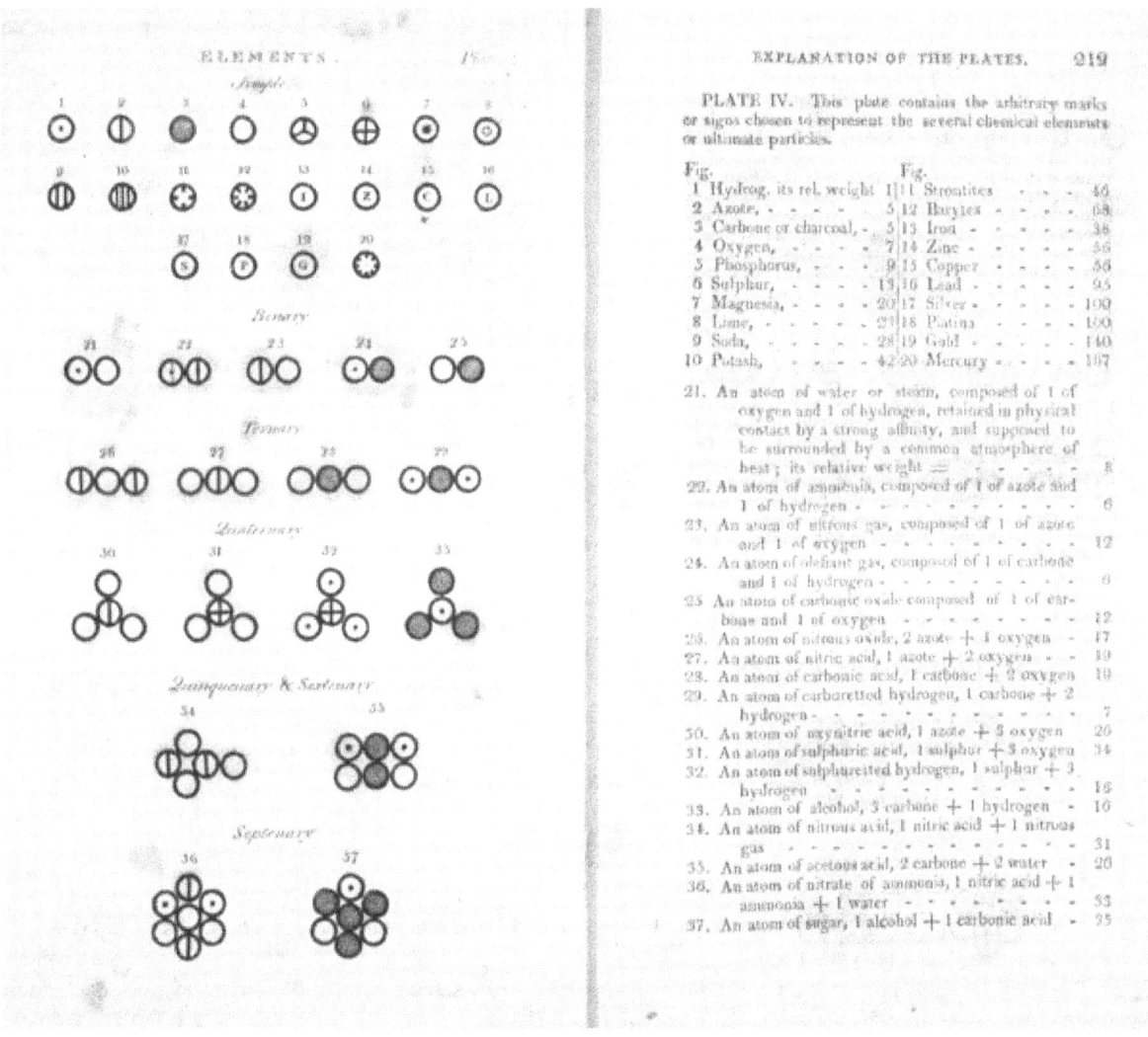

Various atoms and molecules as depicted in John Dalton's A New System of Chemical Philosophy *(1808).*

In the early 1800s, John Dalton used the concept of atoms to explain why elements always react in ratios of small whole numbers (the law of multiple proportions). For instance, there are two types of tin oxide: one is 88.1% tin and 11.9% oxygen and the other is 78.7% tin and 21.3% oxygen (tin(II) oxide and tin dioxide respectively). This means that 100g of tin will combine either with 13.5g or 27g of oxygen. 13.5 and 27 form a ratio of 1:2, a ratio of small whole numbers. This common pattern in chemistry suggested to Dalton that elements react in whole number multiples of discrete units—in other words, atoms. In the case of tin oxides, one tin atom will combine with either one or two oxygen atoms.[4]

Dalton also believed atomic theory could explain why water absorbs different gases in different proportions. For example, he found that water absorbs carbon dioxide far better than it absorbs nitrogen.[5] Dalton hypothesized this was due to the differences in mass and complexity of the gases' respective particles. Indeed, carbon dioxide molecules (CO_2) are heavier and larger than nitrogen molecules (N_2).

1.1.3 Brownian motion

In 1827, botanist Robert Brown used a microscope to look at dust grains floating in water and discovered that they moved about erratically, a phenomenon that became known as "Brownian motion". This was thought to be caused by water molecules knocking the grains about. In 1905 Albert Einstein produced the first mathematical analysis of the motion.[6][7][8] French physicist Jean Perrin used Einstein's work to experimentally determine the mass and dimensions of atoms, thereby conclusively verifying Dalton's atomic theory.[9]

1.1.4 Discovery of the electron

The physicist J. J. Thomson measured the mass of cathode rays, showing they were made of particles, but were around 1800 times lighter than the lightest atom, hydrogen. Therefore, they were not atoms, but a new particle, the first *subatomic* particle to be discovered, which he originally called "*corpuscle*" but was later named *electron*, after particles postulated by George Johnstone Stoney in 1874. He also showed they were identical to particles given off by photoelectric and radioactive materials.[10] It was quickly recognized that they are the particles that carry electric currents in metal wires, and carry the negative electric charge within atoms. Thomson was given the 1906 Nobel Prize in Physics for this work. Thus he overturned the belief that atoms are the indivisible, ultimate particles of matter.[11] Thomson also incorrectly postulated that the low mass, negatively charged electrons were distributed throughout the atom in a uniform sea of positive charge. This became known as the plum pudding model.

1.1.5 Discovery of the nucleus

Main article: Geiger-Marsden experiment

In 1909, Hans Geiger and Ernest Marsden, under the direction of Ernest Rutherford, bombarded a metal foil with alpha particles to observe how they scattered. They expected all the alpha particles to pass straight through with little deflection, because Thomson's model said that the charges in the atom are so diffuse that their electric fields could not affect the alpha particles much. However, Geiger and Marsden spotted alpha particles being deflected by angles greater than 90°, which was supposed to be impossible according to Thomson's model. To explain this, Rutherford proposed that the positive charge of the atom is concentrated in a tiny nucleus at the center of the atom.[12]

1.1.6 Discovery of isotopes

While experimenting with the products of radioactive decay, in 1913 radiochemist Frederick Soddy discovered that there appeared to be more than one type of atom at each position on the periodic table.[13] The term isotope was coined by Margaret Todd as a suitable name for different atoms that belong to the same element. J.J. Thomson created a technique for separating atom types through his work on ionized gases, which subsequently led to the discovery of stable isotopes.[14]

1.1.7 Bohr model

Main article: Bohr model

In 1913 the physicist Niels Bohr proposed a model in which the electrons of an atom were assumed to orbit the nucleus but could only do so in a finite set of orbits, and could jump between these orbits only in discrete changes of energy corresponding to absorption or radiation of a photon.[15] This quantization was used to explain why the electrons orbits are stable (given that normally, charges in acceleration, including circular motion, lose kinetic energy which is emitted as electromagnetic radiation, see *synchrotron radiation*) and why elements absorb and emit electromagnetic radiation in discrete spectra.[16]

Later in the same year Henry Moseley provided additional experimental evidence in favor of Niels Bohr's theory. These results refined Ernest Rutherford's and Antonius Van den Broek's model, which proposed that the atom contains in its nucleus a number of positive nuclear charges that is equal to its (atomic) number in the periodic table. Until these experiments, atomic number was not known to be a physical and experimental quantity. That it is equal to the atomic nuclear charge remains the accepted atomic model today.[17]

1.1.8 Chemical bonding explained

Chemical bonds between atoms were now explained, by Gilbert Newton Lewis in 1916, as the interactions between their constituent electrons.[18] As the chemical properties of the elements were known to largely repeat themselves according to the periodic law,[19] in 1919 the American chemist Irving Langmuir suggested that this could be explained if the electrons in an atom were connected or clustered in some manner. Groups of electrons were thought to occupy a set of electron shells about the nucleus.[20]

1.1.9 Further developments in quantum physics

The Stern–Gerlach experiment of 1922 provided further evidence of the quantum nature of the atom. When a beam of silver atoms was passed through a specially shaped magnetic field, the beam was split based on the direction of an atom's angular momentum, or spin. As this direction is random, the beam could be expected to spread into a line. Instead, the beam was split into two parts, depending on whether the atomic spin was oriented up or down.[21]

In 1924, Louis de Broglie proposed that all particles behave to an extent like waves. In 1926, Erwin Schrödinger used this idea to develop a mathematical model of the atom that described the electrons as three-dimensional waveforms rather than point particles. A consequence of using waveforms to describe particles is that it is mathematically impossible to obtain precise values for both the position and momentum of a particle at the same time; this became known as the uncertainty principle, formulated by Werner Heisenberg in 1926. In this concept, for a given accuracy in measuring a position one could only obtain a range of probable values for momentum, and vice versa. This model was able to explain observations of atomic behavior that previous models could not, such as certain structural and spectral patterns of atoms larger than hydrogen. Thus, the planetary model of the atom was discarded in favor of one that described atomic orbital zones around the nucleus where a given electron is most likely to be observed.[22][23]

1.1.10 Discovery of the neutron

The development of the mass spectrometer allowed the mass of atoms to be measured with increased accuracy. The device uses a magnet to bend the trajectory of a beam of ions, and the amount of deflection is determined by the ratio of an atom's mass to its charge. The chemist Francis William Aston used this instrument to show that isotopes had different masses. The atomic mass of these isotopes varied by integer amounts, called the whole number rule.[24] The explanation for these different isotopes awaited the discovery of the neutron, an uncharged particle with a mass similar to the proton, by the physicist James Chadwick in 1932. Isotopes were then explained as elements with the same number of protons, but different numbers of neutrons within the nucleus.[25]

1.1.11 Fission, high-energy physics and condensed matter

In 1938, the German chemist Otto Hahn, a student of Rutherford, directed neutrons onto uranium atoms expecting to get transuranium elements. Instead, his chemical experiments showed barium as a product.[26] A year later, Lise Meitner and her nephew Otto Frisch verified that Hahn's result were the first experimental *nuclear fission*.[27][28] In 1944, Hahn received the Nobel prize in chemistry. Despite Hahn's efforts, the contributions of Meitner and Frisch were not recognized.[29]

In the 1950s, the development of improved particle accelerators and particle detectors allowed scientists to study the impacts of atoms moving at high energies.[30] Neutrons and protons were found to be hadrons, or composites of smaller particles called quarks. The standard model of particle physics was developed that so far has successfully explained the properties of the nucleus in terms of these sub-atomic particles and the forces that govern their interactions.[31]

1.2 Structure

1.2.1 Subatomic particles

Main article: Subatomic particle

Though the word *atom* originally denoted a particle that cannot be cut into smaller particles, in modern scientific usage the atom is composed of various subatomic particles. The constituent particles of an atom are the electron, the proton and the neutron; all three are fermions. However, the hydrogen-1 atom has no neutrons and the hydron ion has no electrons.

The electron is by far the least massive of these particles at 9.11×10^{-31} kg, with a negative electrical charge and a size that is too small to be measured using available techniques.[32] It is the lightest particle with a positive rest mass measured. Under ordinary conditions, electrons are bound to the positively charged nucleus by the attraction created from opposite electric charges. If an atom has more or fewer electrons than its atomic number, then it becomes respectively negatively or positively charged as a whole; a charged atom is called an ion. Electrons have been known since the late 19th century, mostly thanks to J.J. Thomson; see history of subatomic physics for details.

Protons have a positive charge and a mass 1,836 times that of the electron, at 1.6726×10^{-27} kg. The number of protons in an atom is called its atomic number. Ernest Rutherford (1919) observed that nitrogen under alpha-particle bombardment ejects what appeared to be hydrogen nuclei. By 1920 he had accepted that the hydrogen nucleus is a distinct particle within the atom and named it proton.

Neutrons have no electrical charge and have a free mass of 1,839 times the mass of the electron,[33] or 1.6929×10^{-27} kg, the heaviest of the three constituent particles, but it can be reduced by the nuclear binding energy. Neutrons and protons (collectively known as nucleons) have comparable dimensions—on the order of 2.5×10^{-15} m—although the 'surface' of these particles is not sharply defined.[34] The neutron was discovered in 1932 by the English physicist James Chadwick.

In the Standard Model of physics, electrons are truly elementary particles with no internal structure. However, both protons and neutrons are composite particles composed of elementary particles called quarks. There are two types of quarks in atoms, each having a fractional electric charge. Protons are composed of two up quarks (each with charge $+^2/_3$) and one down quark (with a charge of $-^1/_3$). Neutrons consist of one up quark and two down quarks. This distinction accounts for the difference in mass and charge between the two particles.[35][36]

The quarks are held together by the strong interaction (or strong force), which is mediated by gluons. The protons and neutrons, in turn, are held to each other in the nucleus by the nuclear force, which is a residuum of the strong force that has somewhat different range-properties (see the article on the nuclear force for more). The gluon is a member of the family of gauge bosons, which are elementary particles that mediate physical forces.[35][36]

1.2.2 Nucleus

Main article: Atomic nucleus

All the bound protons and neutrons in an atom make up a tiny atomic nucleus, and are collectively called nucleons. The radius of a nucleus is approximately equal to $1.07 \sqrt[3]{A}$ fm, where A is the total number of nucleons.[37] This is much smaller than the radius of the atom, which is on the order of 10^5 fm. The nucleons are bound together by a short-ranged attractive potential called the residual strong force. At distances smaller than 2.5 fm this force is much more powerful than the electrostatic force that causes positively charged protons to repel each other.[38]

Atoms of the same element have the same number of protons, called the atomic number. Within a single element, the number of neutrons may vary, determining the isotope of that element. The total number of protons and neutrons determine the nuclide. The number of neutrons relative to the protons determines the stability of the nucleus, with certain isotopes undergoing radioactive decay.[39]

The proton, the electron, and the neutron are classified as fermions. Fermions obey the Pauli exclusion principle which prohibits *identical* fermions, such as multiple protons, from occupying the same quantum state at the same time. Thus, every proton in the nucleus must occupy a quantum state different from all other protons, and the same applies to all neutrons of the nucleus and to all electrons of the electron cloud. However, a proton and a neutron are allowed to occupy the same quantum state.[40]

For atoms with low atomic numbers, a nucleus that has more neutrons than protons tends to drop to a lower energy state through radioactive decay so that the neutron–proton ratio is closer to one. However, as the atomic number increases, a higher proportion of neutrons is required to offset the mutual repulsion of the protons. Thus, there are no stable nuclei with equal proton and neutron numbers above atomic number $Z = 20$ (calcium) and as Z increases, the neutron–proton ratio of stable isotopes increases.[40] The stable isotope with the highest proton–neutron ratio is lead-208 (about 1.5).

The number of protons and neutrons in the atomic nucleus can be modified, although this can require very high energies because of the strong force. Nuclear fusion occurs when multiple atomic particles join to form a heavier nucleus, such as through the energetic collision of two nuclei. For example, at the core of the Sun protons require energies of 3–10 keV to overcome their mutual repulsion—the coulomb barrier—and fuse together into a single nucleus.[41] Nuclear fission is the opposite process, causing a nucleus to split into two smaller nuclei—usually through radioactive decay. The nucleus can also be modified through bombardment by high energy subatomic particles or photons. If this modifies the number of protons in a nucleus, the atom changes to a different chemical element.[42][43]

If the mass of the nucleus following a fusion reaction is less than the sum of the masses of the separate particles, then the difference between these two values can be emitted as a type of usable energy (such as a gamma ray, or the kinetic energy of a beta particle), as described by Albert Einstein's mass–energy equivalence formula, $E = mc^2$, where m is the mass loss and c is the speed of light. This deficit is part of the binding energy of the new nucleus, and it is the non-recoverable loss of the energy that causes the fused particles to remain together in a state that requires this energy to separate.[44]

The fusion of two nuclei that create larger nuclei with lower atomic numbers than iron and nickel—a total nucleon number of about 60—is usually an exothermic process that releases more energy than is required to bring them together.[45] It is this energy-releasing process that makes nuclear fusion in stars a self-sustaining reaction. For heavier nuclei, the binding energy per nucleon in the nucleus begins to decrease. That means fusion processes producing nuclei that have atomic numbers higher than about 26, and atomic masses higher than about 60, is an endothermic process. These more massive nuclei can not undergo an energy-producing fusion reaction that can sustain the hydrostatic equilibrium of a star.[40]

1.2.3   Electron cloud

Main articles: Atomic orbital and Electron configuration
 The electrons in an atom are attracted to the protons in the nucleus by the electromagnetic force. This force binds the electrons inside an electrostatic potential well surrounding the smaller nucleus, which means that an external source of energy is needed for the electron to escape. The closer an electron is to the nucleus, the greater the attractive force. Hence electrons bound near the center of the potential well require more energy to escape than those at greater separations.

Electrons, like other particles, have properties of both a particle and a wave. The electron cloud is a region inside the potential well where each electron forms a type of three-dimensional standing wave—a wave form that does not move relative to the nucleus. This behavior is defined by an atomic orbital, a mathematical function that characterises the probability that an electron appears to be at a particular location when its position is measured.[46] Only a discrete (or quantized) set of these orbitals exist around the nucleus, as other possible wave patterns rapidly decay into a more stable form.[47] Orbitals can have one or more ring or node structures, and they differ from each other in size, shape and orientation.[48]

Each atomic orbital corresponds to a particular energy level of the electron. The electron can change its state to a higher energy level by absorbing a photon with sufficient energy to boost it into the new quantum state. Likewise, through spontaneous emission, an electron in a higher energy state can drop to a lower energy state while radiating the excess energy as a photon. These characteristic energy values, defined by the differences in the energies of the quantum states, are responsible for atomic spectral lines.[47]

The amount of energy needed to remove or add an electron—the electron binding energy—is far less than the binding energy of nucleons. For example, it requires only 13.6 eV to strip a ground-state electron from a hydrogen atom,[49] compared to 2.23 *million* eV for splitting a deuterium nucleus.[50] Atoms are electrically neutral if they have an equal number of protons and electrons. Atoms that have either a deficit or a surplus of electrons are called ions. Electrons that are farthest from the nucleus may be transferred to other nearby atoms or shared between atoms. By this mechanism, atoms are able to bond into molecules and other types of chemical compounds like ionic and covalent network crystals.[51]

1.3 Properties

1.3.1 Nuclear properties

Main articles: Isotope, Stable isotope, List of nuclides and List of elements by stability of isotopes

By definition, any two atoms with an identical number of *protons* in their nuclei belong to the same chemical element. Atoms with equal numbers of protons but a different number of *neutrons* are different isotopes of the same element. For example, all hydrogen atoms admit exactly one proton, but isotopes exist with no neutrons (hydrogen-1, by far the most common form,[52] also called protium), one neutron (deuterium), two neutrons (tritium) and more than two neutrons. The known elements form a set of atomic numbers, from the single proton element hydrogen up to the 118-proton element ununoctium.[53] All known isotopes of elements with atomic numbers greater than 82 are radioactive.[54][55]

About 339 nuclides occur naturally on Earth,[56] of which 254 (about 75%) have not been observed to decay, and are referred to as "stable isotopes". However, only 90 of these nuclides are stable to all decay, even in theory. Another 164 (bringing the total to 254) have not been observed to decay, even though in theory it is energetically possible. These are also formally classified as "stable". An additional 34 radioactive nuclides have half-lives longer than 80 million years, and are long-lived enough to be present from the birth of the solar system. This collection of 288 nuclides are known as primordial nuclides. Finally, an additional 51 short-lived nuclides are known to occur naturally, as daughter products of primordial nuclide decay (such as radium from uranium), or else as products of natural energetic processes on Earth, such as cosmic ray bombardment (for example, carbon-14).[57][note 1]

For 80 of the chemical elements, at least one stable isotope exists. As a rule, there is only a handful of stable isotopes for each of these elements, the average being 3.2 stable isotopes per element. Twenty-six elements have only a single stable isotope, while the largest number of stable isotopes observed for any element is ten, for the element tin. Elements 43, 61, and all elements numbered 83 or higher have no stable isotopes.[58]

Stability of isotopes is affected by the ratio of protons to neutrons, and also by the presence of certain "magic numbers" of neutrons or protons that represent closed and filled quantum shells. These quantum shells correspond to a set of energy levels within the shell model of the nucleus; filled shells, such as the filled shell of 50 protons for tin, confers unusual stability on the nuclide. Of the 254 known stable nuclides, only four have both an odd number of protons *and* odd number of neutrons: hydrogen-2 (deuterium), lithium-6, boron-10 and nitrogen-14. Also, only four naturally occurring, radioactive odd–odd nuclides have a half-life over a billion years: potassium-40, vanadium-50, lanthanum-138 and tantalum-180m. Most odd–odd nuclei are highly unstable with respect to beta decay, because the decay products are even–even, and are therefore more strongly bound, due to nuclear pairing effects.[58]

1.3.2 Mass

Main articles: Atomic mass and mass number

The large majority of an atom's mass comes from the protons and neutrons that make it up. The total number of these particles (called "nucleons") in a given atom is called the mass number. It is a positive integer and dimensionless (instead of having dimension of mass), because it expresses a count. An example of use of a mass number is "carbon-12," which has 12 nucleons (six protons and six neutrons).

The actual mass of an atom at rest is often expressed using the unified atomic mass unit (u), also called dalton (Da). This unit is defined as a twelfth of the mass of a free neutral atom of carbon-12, which is approximately 1.66×10^{-27} kg.[59] Hydrogen-1 (the lightest isotope of hydrogen which is also the nuclide with the lowest mass) has an atomic weight of 1.007825 u.[60] The value of this number is called the atomic mass. A given atom has an atomic mass approximately equal (within 1%) to its mass number times the atomic mass unit (for example the mass of a nitrogen-14 is roughly 14 u). However, this number will not be exactly an integer except in the case of carbon-12 (see below).[61] The heaviest stable atom is lead-208,[54] with a mass of 207.9766521 u.[62]

As even the most massive atoms are far too light to work with directly, chemists instead use the unit of moles. One mole of atoms of any element always has the same number of atoms (about 6.022×10^{23}). This number was chosen so that if an element has an atomic mass of 1 u, a mole of atoms of that element has a mass close to one gram. Because of the definition of the unified atomic mass unit, each carbon-12 atom has an atomic mass of exactly 12 u, and so a mole of carbon-12 atoms weighs exactly 0.012 kg.[59]

1.3.3   Shape and size

Main article: Atomic radius

Atoms lack a well-defined outer boundary, so their dimensions are usually described in terms of an atomic radius. This is a measure of the distance out to which the electron cloud extends from the nucleus.[2] However, this assumes the atom to exhibit a spherical shape, which is only obeyed for atoms in vacuum or free space. Atomic radii may be derived from the distances between two nuclei when the two atoms are joined in a chemical bond. The radius varies with the location of an atom on the atomic chart, the type of chemical bond, the number of neighboring atoms (coordination number) and a quantum mechanical property known as spin.[63] On the periodic table of the elements, atom size tends to increase when moving down columns, but decrease when moving across rows (left to right).[64] Consequently, the smallest atom is helium with a radius of 32 pm, while one of the largest is caesium at 225 pm.[65]

When subjected to external forces, like electrical fields, the shape of an atom may deviate from spherical symmetry. The deformation depends on the field magnitude and the orbital type of outer shell electrons, as shown by group-theoretical considerations. Aspherical deviations might be elicited for instance in crystals, where large crystal-electrical fields may occur at low-symmetry lattice sites. Significant ellipsoidal deformations have recently been shown to occur for sulfur ions[66] and chalcogen ions[67] in pyrite-type compounds.

Atomic dimensions are thousands of times smaller than the wavelengths of light (400–700 nm) so they cannot be viewed using an optical microscope. However, individual atoms can be observed using a scanning tunneling microscope. To visualize the minuteness of the atom, consider that a typical human hair is about 1 million carbon atoms in width.[68] A single drop of water contains about 2 sextillion (2×10^{21}) atoms of oxygen, and twice the number of hydrogen atoms.[69] A single carat diamond with a mass of 2×10^{-4} kg contains about 10 sextillion (10^{22}) atoms of carbon.[note 2] If an apple were magnified to the size of the Earth, then the atoms in the apple would be approximately the size of the original apple.[70]

1.3.4   Radioactive decay

Main article: Radioactive decay
Every element has one or more isotopes that have unstable nuclei that are subject to radioactive decay, causing the nucleus to emit particles or electromagnetic radiation. Radioactivity can occur when the radius of a nucleus is large compared with the radius of the strong force, which only acts over distances on the order of 1 fm.[71]

The most common forms of radioactive decay are:[72][73]

- Alpha decay: this process is caused when the nucleus emits an alpha particle, which is a helium nucleus consisting of two protons and two neutrons. The result of the emission is a new element with a lower atomic number.

- Beta decay (and electron capture): these processes are regulated by the weak force, and result from a transformation of a neutron into a proton, or a proton into a neutron. The neutron to proton transition is accompanied by the emission of an electron and an antineutrino, while proton to neutron transition (except in electron capture) causes the emission of a positron and a neutrino. The electron or positron emissions are called beta particles. Beta decay either increases or decreases the atomic number of the nucleus by one. Electron capture is more common than positron emission, because it requires less energy. In this type of decay, an electron is absorbed by the nucleus, rather than a positron emitted from the nucleus. A neutrino is still emitted in this process, and a proton changes to a neutron.

- Gamma decay: this process results from a change in the energy level of the nucleus to a lower state, resulting in the emission of electromagnetic radiation. The excited state of a nucleus which results in gamma emission usually occurs following the emission of an alpha or a beta particle. Thus, gamma decay usually follows alpha or beta decay.

Other more rare types of radioactive decay include ejection of neutrons or protons or clusters of nucleons from a nucleus, or more than one beta particle. An analog of gamma emission which allows excited nuclei to lose energy in a different way, is internal conversion— a process that produces high-speed electrons that are not beta rays, followed by production of high-energy photons that are not gamma rays. A few large nuclei explode into two or more charged fragments of varying masses plus several neutrons, in a decay called spontaneous nuclear fission.

Each radioactive isotope has a characteristic decay time period—the half-life—that is determined by the amount of time needed for half of a sample to decay. This is an exponential decay process that steadily decreases the proportion of the remaining isotope by 50% every half-life. Hence after two half-lives have passed only 25% of the isotope is present, and so forth.[71]

1.3.5 Magnetic moment

Main articles: Electron magnetic moment and Nuclear magnetic moment

Elementary particles possess an intrinsic quantum mechanical property known as spin. This is analogous to the angular momentum of an object that is spinning around its center of mass, although strictly speaking these particles are believed to be point-like and cannot be said to be rotating. Spin is measured in units of the reduced Planck constant ($\hbar$), with electrons, protons and neutrons all having spin $\frac{1}{2}$ $\hbar$, or "spin-$\frac{1}{2}$". In an atom, electrons in motion around the nucleus possess orbital angular momentum in addition to their spin, while the nucleus itself possesses angular momentum due to its nuclear spin.[74]

The magnetic field produced by an atom—its magnetic moment—is determined by these various forms of angular momentum, just as a rotating charged object classically produces a magnetic field. However, the most dominant contribution comes from electron spin. Due to the nature of electrons to obey the Pauli exclusion principle, in which no two electrons may be found in the same quantum state, bound electrons pair up with each other, with one member of each pair in a spin up state and the other in the opposite, spin down state. Thus these spins cancel each other out, reducing the total magnetic dipole moment to zero in some atoms with even number of electrons.[75]

In ferromagnetic elements such as iron, cobalt and nickel, an odd number of electrons leads to an unpaired electron and a net overall magnetic moment. The orbitals of neighboring atoms overlap and a lower energy state is achieved when the spins of unpaired electrons are aligned with each other, a spontaneous process known as an exchange interaction. When the magnetic moments of ferromagnetic atoms are lined up, the material can produce a measurable macroscopic field. Paramagnetic materials have atoms with magnetic moments that line up in random directions when no magnetic field is present, but the magnetic moments of the individual atoms line up in the presence of a field.[75][76]

The nucleus of an atom will have no spin when it has even numbers of both neutrons and protons, but for other cases of odd numbers, the nucleus may have a spin. Normally nuclei with spin are aligned in random directions because of thermal equilibrium. However, for certain elements (such as xenon-129) it is possible to polarize a significant proportion of the nuclear spin states so that they are aligned in the same direction—a condition called hyperpolarization. This has important applications in magnetic resonance imaging.[77][78]

1.3.6 Energy levels

The potential energy of an electron in an atom is negative, its dependence of its position reaches the minimum (the most absolute value) inside the nucleus, and vanishes when the distance from the nucleus goes to infinity, roughly in an inverse proportion to the distance. In the quantum-mechanical model, a bound electron can only occupy a set of states centered on the nucleus, and each state corresponds to a specific energy level; see time-independent Schrödinger equation for theoretical explanation. An energy level can be measured by the amount of energy needed to unbind the electron from the atom, and is usually given in units of electronvolts (eV). The lowest energy state of a bound electron is called the ground state, i.e. stationary state, while an electron transition to a higher level results in an excited state.[79] The electron's energy raises when n increases because the (average) distance to the nucleus increases. Dependence of the energy on ℓ is caused not by electrostatic potential of the nucleus, but by interaction between electrons.

For an electron to transition between two different states, e.g. grounded state to first excited level (ionization), it must absorb or emit a photon at an energy matching the difference in the potential energy of those levels, according to Niels Bohr model, what can be precisely calculated by the Schrödinger equation. Electrons jump between orbitals in a particle-like fashion. For example, if a single photon strikes the electrons, only a single electron changes states in response to the photon; see Electron properties.

The energy of an emitted photon is proportional to its frequency, so these specific energy levels appear as distinct bands in the electromagnetic spectrum.[80] Each element has a characteristic spectrum that can depend on the nuclear charge, subshells filled by electrons, the electromagnetic interactions between the electrons and other factors.[81]

When a continuous spectrum of energy is passed through a gas or plasma, some of the photons are absorbed by atoms, causing electrons to change their energy level. Those excited electrons that remain bound to their atom spontaneously emit this energy as a photon, traveling in a random direction, and so drop back to lower energy levels. Thus the atoms behave like a filter that forms a series of dark absorption bands in the energy output. (An observer viewing the atoms from a view that does not include the continuous spectrum in the background, instead sees a series of emission lines from the photons emitted by the atoms.) Spectroscopic measurements of the strength and width of atomic spectral lines allow the composition and physical properties of a substance to be determined.[82]

Close examination of the spectral lines reveals that some display a fine structure splitting. This occurs because of spin–orbit coupling, which is an interaction between the spin and motion of the outermost electron.[83] When an atom is in an external magnetic field, spectral lines become split into three or more components; a phenomenon called the Zeeman effect. This is caused by the interaction of the magnetic field with the magnetic moment of the atom and its electrons. Some atoms can have multiple electron configurations with the same energy level, which thus appear as a single spectral line. The interaction of the magnetic field with the atom shifts these electron configurations to slightly different energy levels, resulting in multiple spectral lines.[84] The presence of an external electric field can cause a comparable splitting and shifting of spectral lines by modifying the electron energy levels, a phenomenon called the Stark effect.[85]

If a bound electron is in an excited state, an interacting photon with the proper energy can cause stimulated emission of a photon with a matching energy level. For this to occur, the electron must drop to a lower energy state that has an energy difference matching the energy of the interacting photon. The emitted photon and the interacting photon then move off in parallel and with matching phases. That is, the wave patterns of the two photons are synchronized. This physical property is used to make lasers, which can emit a coherent beam of light energy in a narrow frequency band.[86]

1.3.7 Valence and bonding behavior

Main articles: Valence (chemistry) and Chemical bond

Valency is the combining power of an element. It is equal to number of hydrogen atoms that atom can combine or displace in forming compounds.[87] The outermost electron shell of an atom in its uncombined state is known as the valence shell, and the electrons in that shell are called valence electrons. The number of valence electrons determines the bonding behavior with other atoms. Atoms tend to chemically react with each other in a manner that fills (or empties) their outer valence shells.[88] For example, a transfer of a single electron between atoms is a useful approximation for bonds that form between atoms with one-electron more than a filled shell, and others that are one-electron short of a full shell, such as occurs in the compound sodium chloride and other chemical ionic salts. However, many elements display multiple valences, or tendencies to share differing numbers of electrons in different compounds. Thus, chemical bonding between these elements takes many forms of electron-sharing that are more than simple electron transfers. Examples include the element carbon and the organic compounds.[89]

The chemical elements are often displayed in a periodic table that is laid out to display recurring chemical properties, and elements with the same number of valence electrons form a group that is aligned in the same column of the table. (The horizontal rows correspond to the filling of a quantum shell of electrons.) The elements at the far right of the table have their outer shell completely filled with electrons, which results in chemically inert elements known as the noble gases.[90][91]

1.3.8 States

Main articles: State of matter and Phase (matter)
Quantities of atoms are found in different states of matter that depend on the physical conditions, such as temperature and pressure. By varying the conditions, materials can transition between solids, liquids, gases and plasmas.[92] Within a state, a material can also exist in different allotropes. An example of this is solid carbon, which can exist as graphite or diamond.[93] Gaseous allotropes exist as well, such as dioxygen and ozone.

At temperatures close to absolute zero, atoms can form a Bose–Einstein condensate, at which point quantum mechanical effects, which are normally only observed at the atomic scale, become apparent on a macroscopic scale.[94][95] This super-cooled collection of atoms then behaves as a single super atom, which may allow fundamental checks of quantum mechanical behavior.[96]

1.4 Identification

The scanning tunneling microscope is a device for viewing surfaces at the atomic level. It uses the quantum tunneling phenomenon, which allows particles to pass through a barrier that would normally be insurmountable. Electrons tunnel through the vacuum between two planar metal electrodes, on each of which is an adsorbed atom, providing a tunneling-current density that can be measured. Scanning one atom (taken as the tip) as it moves past the other (the sample) permits plotting of tip displacement versus lateral separation for a constant current. The calculation shows the extent to which scanning-tunneling-microscope images of an individual atom are visible. It confirms that for low bias, the microscope images the space-averaged dimensions of the electron orbitals across closely packed energy levels—the Fermi level local density of states.[97][98]

An atom can be ionized by removing one of its electrons. The electric charge causes the trajectory of an atom to bend when it passes through a magnetic field. The radius by which the trajectory of a moving ion is turned by the magnetic field is determined by the mass of the atom. The mass spectrometer uses this principle to measure the mass-to-charge ratio of ions. If a sample contains multiple isotopes, the mass spectrometer can determine the proportion of each isotope in the sample by measuring the intensity of the different beams of ions. Techniques to vaporize atoms include inductively coupled plasma atomic emission spectroscopy and inductively coupled plasma mass spectrometry, both of which use a plasma to vaporize samples for analysis.[99]

A more area-selective method is electron energy loss spectroscopy, which measures the energy loss of an electron beam within a transmission electron microscope when it interacts with a portion of a sample. The atom-probe tomograph has sub-nanometer resolution in 3-D and can chemically identify individual atoms using time-of-flight mass spectrometry.[100]

Spectra of excited states can be used to analyze the atomic composition of distant stars. Specific light wavelengths contained in the observed light from stars can be separated out and related to the quantized transitions in free gas atoms. These colors can be replicated using a gas-discharge lamp containing the same element.[101] Helium was discovered in this way in the spectrum of the Sun 23 years before it was found on Earth.[102]

1.5 Origin and current state

Atoms form about 4% of the total energy density of the observable Universe, with an average density of about 0.25 atoms/m^3.[103] Within a galaxy such as the Milky Way, atoms have a much higher concentration, with the density of matter in the interstellar medium (ISM) ranging from 10^5 to 10^9 atoms/m^3.[104] The Sun is believed to be inside the Local Bubble, a region of highly ionized gas, so the density in the solar neighborhood is only about 10^3 atoms/m^3.[105] Stars form from dense clouds in the ISM, and the evolutionary processes of stars result in the steady enrichment of the ISM with elements more massive than hydrogen and helium. Up to 95% of the Milky Way's atoms are concentrated inside stars and the total mass of atoms forms about 10% of the mass of the galaxy.[106] (The remainder of the mass is an unknown dark matter.)[107]

1.5.1 Formation

Electrons are thought to exist in the Universe since early stages of the Big Bang. Atomic nuclei forms in nucleosynthesis reactions. In about three minutes Big Bang nucleosynthesis produced most of the helium, lithium, and deuterium in the Universe, and perhaps some of the beryllium and boron.[108][109][110]

Ubiquitousness and stability of atoms relies on their binding energy, which means that an atom has a lower energy than an unbound system of the nucleus and electrons. Where the temperature is much higher than ionization potential, the matter exists in the form of plasma—a gas of positively charged ions (possibly, bare nuclei) and electrons. When the temperature drops below the ionization potential, atoms become statistically favorable. Atoms (complete with bound electrons) became to dominate over charged particles 380,000 years after the Big Bang—an epoch called recombination, when the expanding Universe cooled enough to allow electrons to become attached to nuclei.[111]

Since the Big Bang, which produced no carbon or heavier elements, atomic nuclei have been combined in stars through the process of nuclear fusion to produce more of the element helium, and (via the triple alpha process) the sequence of elements from carbon up to iron;[112] see stellar nucleosynthesis for details.

Isotopes such as lithium-6, as well as some beryllium and boron are generated in space through cosmic ray spallation.[113] This occurs when a high-energy proton strikes an atomic nucleus, causing large numbers of nucleons to be ejected.

Elements heavier than iron were produced in supernovae through the r-process and in AGB stars through the s-process, both of which involve the capture of neutrons by atomic nuclei.[114] Elements such as lead formed largely through the radioactive decay of heavier elements.[115]

1.5.2 Earth

Most of the atoms that make up the Earth and its inhabitants were present in their current form in the nebula that collapsed out of a molecular cloud to form the Solar System. The rest are the result of radioactive decay, and their relative proportion can be used to determine the age of the Earth through radiometric dating.[116][117] Most of the helium in the crust of the Earth (about 99% of the helium from gas wells, as shown by its lower abundance of helium-3) is a product of alpha decay.[118]

There are a few trace atoms on Earth that were not present at the beginning (i.e., not "primordial"), nor are results of radioactive decay. Carbon-14 is continuously generated by cosmic rays in the atmosphere.[119] Some atoms on Earth have been artificially generated either deliberately or as by-products of nuclear reactors or explosions.[120][121] Of the transuranic elements—those with atomic numbers greater than 92—only plutonium and neptunium occur naturally on Earth.[122][123] Transuranic elements have radioactive lifetimes shorter than the current age of the Earth[124] and thus identifiable quantities of these elements have long since decayed, with the exception of traces of plutonium-244 possibly deposited by cosmic dust.[125] Natural deposits of plutonium and neptunium are produced by neutron capture in uranium ore.[126]

The Earth contains approximately 1.33×10^{50} atoms.[127] Although small numbers of independent atoms of noble gases exist, such as argon, neon, and helium, 99% of the atmosphere is bound in the form of molecules, including carbon dioxide and diatomic oxygen and nitrogen. At the surface of the Earth, an overwhelming majority of atoms combine to form various compounds, including water, salt, silicates and oxides. Atoms can also combine to create materials that do not consist of discrete molecules, including crystals and liquid or solid metals.[128][129] This atomic matter forms networked arrangements that lack the particular type of small-scale interrupted order associated with molecular matter.[130]

1.5.3 Rare and theoretical forms

Superheavy elements

Main article: Transuranium element

While isotopes with atomic numbers higher than lead (82) are known to be radioactive, an "island of stability" has been proposed for some elements with atomic numbers above 103. These superheavy elements may have a nucleus that is relatively stable against radioactive decay.[131] The most likely candidate for a stable superheavy atom, unbihexium, has 126 protons and 184 neutrons.[132]

Exotic matter

Main article: Exotic matter

Each particle of matter has a corresponding antimatter particle with the opposite electrical charge. Thus, the positron is a positively charged antielectron and the antiproton is a negatively charged equivalent of a proton. When a matter and corresponding antimatter particle meet, they annihilate each other. Because of this, along with an imbalance between the number of matter and antimatter particles, the latter are rare in the universe. The first causes of this imbalance are not yet fully understood, although theories of baryogenesis may offer an explanation. As a result, no antimatter atoms have been discovered in nature.[133][134] However, in 1996 the antimatter counterpart of the hydrogen atom (antihydrogen) was synthesized at the CERN laboratory in Geneva.[135][136]

Other exotic atoms have been created by replacing one of the protons, neutrons or electrons with other particles that have the same charge. For example, an electron can be replaced by a more massive muon, forming a muonic atom. These types of atoms can be used to test the fundamental predictions of physics.[137][138][139]

1.6 See also

- History of quantum mechanics
- Infinite divisibility
- List of basic chemistry topics
- Timeline of atomic and subatomic physics
- Vector model of the atom
- Nuclear model
- Radioactive isotope

1.7 Notes

[1] For more recent updates see Interactive Chart of Nuclides (Brookhaven National Laboratory).

[2] A carat is 200 milligrams. By definition, carbon-12 has 0.012 kg per mole. The Avogadro constant defines 6×10^{23} atoms per mole.

1.8 References

[1] "Atom". *Compendium of Chemical Terminology (IUPAC Gold Book)* (2nd ed.). IUPAC. Retrieved 2015-04-25.

[2] Ghosh, D. C.; Biswas, R. (2002). "Theoretical calculation of Absolute Radii of Atoms and Ions. Part 1. The Atomic Radii". *Int. J. Mol. Sci.* **3**: 87–113. doi:10.3390/i3020087.

[3] Leigh, G. J., ed. (1990). *International Union of Pure and Applied Chemistry, Commission on the Nomenclature of Inorganic Chemistry, Nomenclature of Organic Chemistry – Recommendations 1990*. Oxford: Blackwell Scientific Publications. p. 35. ISBN 0-08-022369-9. An atom is the smallest unit quantity of an element that is capable of existence whether alone or in chemical combination with other atoms of the same or other elements.

[4] Andrew G. van Melsen (1952). *From Atomos to Atom*. Mineola, N.Y.: Dover Publications. ISBN 0-486-49584-1.

[5] Dalton, John. "On the Absorption of Gases by Water and Other Liquids", in *Memoirs of the Literary and Philosophical Society of Manchester*. 1803. Retrieved on August 29, 2007.

[6] Einstein, Albert (1905). "Über die von der molekularkinetischen Theorie der Wärme geforderte Bewegung von in ruhenden Flüssigkeiten suspendierten Teilchen" (PDF). *Annalen der Physik* (in German) **322** (8): 549–560. Bibcode:1905AnP...322..549E. doi:10.1002/andp.19053220806. Retrieved 4 February 2007.

[7] Mazo, Robert M. (2002). *Brownian Motion: Fluctuations, Dynamics, and Applications*. Oxford University Press. pp. 1–7. ISBN 0-19-851567-7. OCLC 48753074.

[8] Lee, Y.K.; Hoon, K. (1995). "Brownian Motion". Imperial College. Archived from the original on 18 December 2007. Retrieved 18 December 2007.

[9] Patterson, G. (2007). "Jean Perrin and the triumph of the atomic doctrine". *Endeavour* **31** (2): 50–53. doi: PMID 17602746.

[10] Thomson, J. J. (August 1901). "On bodies smaller than atoms". *The Popular Science Monthly* (Bonnier Corp.): 323–335. Retrieved 2009-06-21.

[11] "J.J. Thomson". Nobel Foundation. 1906. Retrieved 20 December 2007.

[12] Rutherford, E. (1911). "The Scattering of α and β Particles by Matter and the Structure of the Atom" (PDF). *Philosophical Magazine* **21** (125): 669–88. doi:10.1080/14786440508637080.

[13] "Frederick Soddy, The Nobel Prize in Chemistry 1921". Nobel Foundation. Retrieved 18 January 2008.

[14] Thomson, Joseph John (1913). "Rays of positive electricity". *Proceedings of the Royal Society*. A**89**(607): 1–20. Bibcode:1913RSPSA..89...1T. doi:10.1098/rspa.1913.0057.

[15] Stern, David P. (16 May 2005). "The Atomic Nucleus and Bohr's Early Model of the Atom". NASA/Goddard Space Flight Center. Retrieved 20 December 2007.

[16] Bohr, Niels (11 December 1922). "Niels Bohr, The Nobel Prize in Physics 1922, Nobel Lecture". Nobel Foundation. Retrieved 16 February 2008.

[17] Pais, Abraham (1986). *Inward Bound: Of Matter and Forces in the Physical World*. New York: Oxford University Press. pp. 228–230. ISBN 0-19-851971-0.

[18] Lewis, Gilbert N. (1916). "The Atom and the Molecule". *Journal of the American Chemical Society* **38** (4): 762–786. doi:10.1021/ja02261a002.

[19] Scerri, Eric R. (2007). *The periodic table: its story and its significance*. Oxford University Press US. pp. 205–226. ISBN 0-19-530573-6.

[20] Langmuir, Irving (1919). "The Arrangement of Electrons in Atoms and Molecules". *Journal of the American Chemical Society* **41** (6): 868–934. doi:10.1021/ja02227a002.

[21] Scully, Marlan O.; Lamb, Willis E.; Barut, Asim (1987). "On the theory of the Stern-Gerlach apparatus". *Foundations of Physics* **17** (6): 575–583. Bibcode:1987FoPh...17..575S. doi:10.1007/BF01882788.

[22] Brown, Kevin (2007). "The Hydrogen Atom". MathPages. Retrieved 21 December 2007.

[23] Harrison, David M. (2000). "The Development of Quantum Mechanics". University of Toronto. Archived from the original on 25 December 2007. Retrieved 21 December 2007.

[24] Aston, Francis W. (1920). "The constitution of atmospheric neon". *Philosophical Magazine* **39**(6): 449–55.doi:10.1080/1478644 0408636058.

[25] Chadwick, James (12 December 1935). "Nobel Lecture: The Neutron and Its Properties". Nobel Foundation. Retrieved 21 December 2007.

[26] "Otto Hahn, Lise Meitner and Fritz Strassmann". *Chemical Achievers: The Human Face of the Chemical Sciences*. Chemical Heritage Foundation. Archived from the original on 24 October 2009. Retrieved 15 September 2009.

[27] Meitner, Lise; Frisch, Otto Robert (1939). "Disintegration of uranium by neutrons: a new type of nuclear reaction". *Nature* **143** (3615): 239–240. Bibcode:1939Natur.143..239M. doi:10.1038/143239a0.

[28] Schroeder, M. "Lise Meitner – Zur 125. Wiederkehr Ihres Geburtstages" (in German). Retrieved 4 June 2009.

[29] Crawford, E.; Sime, Ruth Lewin; Walker, Mark (1997). "A Nobel tale of postwar injustice". *Physics Today* **50** (9): 26–32. Bibcode:1997PhT....50i..26C. doi:10.1063/1.881933.

[30] Kullander, Sven (28 August 2001). "Accelerators and Nobel Laureates". Nobel Foundation. Retrieved 31 January 2008.

[31] "The Nobel Prize in Physics 1990". Nobel Foundation. 17 October 1990. Retrieved 31 January 2008.

[32] Demtröder, Wolfgang (2002). *Atoms, Molecules and Photons: An Introduction to Atomic- Molecular- and Quantum Physics* (1st ed.). Springer. pp. 39–42. ISBN 3-540-20631-0. OCLC 181435713.

[33] Woan, Graham (2000). *The Cambridge Handbook of Physics*. Cambridge University Press. p. 8. ISBN 0-521-57507-9. OCLC 224032426.

[34] MacGregor, Malcolm H. (1992). *The Enigmatic Electron*. Oxford University Press. pp. 33–37. ISBN 0-19-521833-7. OCLC 223372888.

[35] Particle Data Group (2002). "The Particle Adventure". Lawrence Berkeley Laboratory. Archived from the original on 4 January 2007. Retrieved 3 January 2007.

[36] Schombert, James (18 April 2006). "Elementary Particles". University of Oregon. Retrieved 3 January 2007.

[37] Jevremovic, Tatjana (2005). *Nuclear Principles in Engineering*. Springer. p. 63. ISBN 0-387-23284-2. OCLC 228384008.

[38] Pfeffer, Jeremy I.; Nir, Shlomo (2000). *Modern Physics: An Introductory Text*. Imperial College Press. pp. 330–336. ISBN 1-86094-250-4. OCLC 45900880.

[39] Wenner, Jennifer M. (10 October 2007). "How Does Radioactive Decay Work?". Carleton College. Retrieved 9 January 2008.

[40] Raymond, David (7 April 2006). "Nuclear Binding Energies". New Mexico Tech. Archived from the original on 11 December 2006. Retrieved 3 January 2007.

[41] Mihos, Chris (23 July 2002). "Overcoming the Coulomb Barrier". Case Western Reserve University. Retrieved 13 February 2008.

[42] Staff (30 March 2007). "ABC's of Nuclear Science". Lawrence Berkeley National Laboratory. Archived from the original on 5 December 2006. Retrieved 3 January 2007.

[43] Makhijani, Arjun; Saleska, Scott (2 March 2001). "Basics of Nuclear Physics and Fission". Institute for Energy and Environmental Research. Archived from the original on 16 January 2007. Retrieved 3 January 2007.

[44] Shultis, J. Kenneth; Faw, Richard E. (2002). *Fundamentals of Nuclear Science and Engineering*. CRC Press. pp. 10–17. ISBN 0-8247-0834-2. OCLC 123346507.

[45] Fewell, M. P. (1995). "The atomic nuclide with the highest mean binding energy". *American Journal of Physics* **63** (7): 653–658. Bibcode:1995AmJPh..63..653F. doi:10.1119/1.17828.

[46] Mulliken, Robert S. (1967). "Spectroscopy, Molecular Orbitals, and Chemical Bonding". *Science* **157**(3784): 13–24.Bibcode: 1967Sci...157...13M. doi:10.1126/science.157.3784.13. PMID 5338306.

[47] Brucat, Philip J. (2008). "The Quantum Atom". University of Florida. Archived from the original on 7 December 2006. Retrieved 4 January 2007.

[48] Manthey, David (2001). "Atomic Orbitals". Orbital Central. Archived from the original on 10 January 2008. Retrieved 21 January 2008.

[49] Herter, Terry (2006). "Lecture 8: The Hydrogen Atom". Cornell University. Retrieved 14 February 2008.

[50] Bell, R. E.; Elliott, L. G. (1950). "Gamma-Rays from the Reaction $H^1(n,\gamma)D^2$ and the Binding Energy of the Deuteron". *Physical Review* **79** (2): 282–285. Bibcode:1950PhRv...79..282B. doi:10.1103/PhysRev.79.282.

[51] Smirnov, Boris M. (2003). *Physics of Atoms and Ions*. Springer. pp. 249–272. ISBN 0-387-95550-X.

[52] Matis, Howard S. (9 August 2000). "The Isotopes of Hydrogen". *Guide to the Nuclear Wall Chart*. Lawrence Berkeley National Lab. Archived from the original on 18 December 2007. Retrieved 21 December 2007.

[53] Weiss, Rick (17 October 2006). "Scientists Announce Creation of Atomic Element, the Heaviest Yet". Washington Post. Retrieved 21 December 2007.

[54] Sills, Alan D. (2003). *Earth Science the Easy Way*. Barron's Educational Series. pp. 131–134. ISBN 0-7641-2146-4. OCLC 51543743.

[55] Dumé, Belle (23 April 2003). "Bismuth breaks half-life record for alpha decay". Physics World. Archived from the original on 14 December 2007. Retrieved 21 December 2007.

[56] Lindsay, Don (30 July 2000). "Radioactives Missing From The Earth". Don Lindsay Archive. Archived from the original on 28 April 2007. Retrieved 23 May 2007.

[57] Tuli, Jagdish K. (April 2005). "Nuclear Wallet Cards". National Nuclear Data Center, Brookhaven National Laboratory. Retrieved 16 April 2011.

[58] CRC Handbook (2002).

[59] Mills, Ian; Cvitaš, Tomislav; Homann, Klaus; Kallay, Nikola; Kuchitsu, Kozo (1993). *Quantities, Units and Symbols in Physical Chemistry* (PDF) (2nd ed.). Oxford: International Union of Pure and Applied Chemistry, Commission on Physiochemical Symbols Terminology and Units, Blackwell Scientific Publications. p. 70. ISBN 0-632-03583-8. OCLC 27011505.

[60] Chieh, Chung (22 January 2001). "Nuclide Stability". University of Waterloo. Retrieved 4 January 2007.

[61] "Atomic Weights and Isotopic Compositions for All Elements". National Institute of Standards and Technology. Archived from the original on 31 December 2006. Retrieved 4 January 2007.

[62] Audi, G.; Wapstra, A.H.; Thibault, C. (2003). "The Ame2003 atomic mass evaluation (II)" (PDF). *Nuclear Physics A* **729** (1): 337–676. Bibcode:2003NuPhA.729..337A. doi:10.1016/j.nuclphysa.2003.11.003.

[63] Shannon, R. D. (1976). "Revised effective ionic radii and systematic studies of interatomic distances in halides and chalco-genides". *Acta Crystallographica A* **32** (5): 751–767. Bibcode:1976AcCrA..32..751S. doi:10.1107/S0567739476001551.

[64] Dong, Judy (1998). "Diameter of an Atom". The Physics Factbook. Archived from the original on 4 November 2007. Retrieved 19 November 2007.

[65] Zumdahl, Steven S. (2002). *Introductory Chemistry: A Foundation* (5th ed.). Houghton Mifflin. ISBN 0-618-34342-3. OCLC 173081482. Archived from the original on 4 March 2008. Retrieved 5 February 2008.

[66] Birkholz, M.; Rudert, R. (2008). "Interatomic distances in pyrite-structure disulfides – a case for ellipsoidal modeling of sulfur ions]" (PDF). *phys. stat. sol. b* **245**: 1858–1864. Bibcode:2008PSSBR.245.1858B. doi:10.1002/pssb.200879532.

[67] Birkholz, M. (2014). "Modeling the Shape of Ions in Pyrite-Type Crystals". *Crystals* **4**: 390–403. doi:10.3390/cryst4030390.

[68] Staff (2007). "Small Miracles: Harnessing nanotechnology". Oregon State University. Retrieved 7 January 2007.—describes the width of a human hair as 10^5 nm and 10 carbon atoms as spanning 1 nm.

[69] Padilla, Michael J.; Miaoulis, Ioannis; Cyr, Martha (2002). *Prentice Hall Science Explorer: Chemical Building Blocks*. Upper Saddle River, New Jersey USA: Prentice-Hall, Inc. p. 32.ISBN0-13-054091-9.OCLC47925884. There are 2,000,000,000,000,000,000,000(that's 2 sextillion) atoms of oxygen in one drop of water—and twice as many atoms of hydrogen.

[70] Feynman, Richard (1995). *Six Easy Pieces*. The Penguin Group. p. 5. ISBN 978-0-14-027666-4. OCLC 40499574.

[71] "Radioactivity". Splung.com. Archived from the original on 4 December 2007. Retrieved 19 December 2007.

[72] L'Annunziata, Michael F. (2003). *Handbook of Radioactivity Analysis*. Academic Press. pp. 3–56. ISBN 0-12-436603-1. OCLC 16212955.

[73] Firestone, Richard B. (22 May 2000). "Radioactive Decay Modes". Berkeley Laboratory. Retrieved 7 January 2007.

[74] Hornak, J. P. (2006). "Chapter 3: Spin Physics". *The Basics of NMR*. Rochester Institute of Technology. Archived from the original on 3 February 2007. Retrieved 7 January 2007.

[75] Schroeder, Paul A. (25 February 2000). "Magnetic Properties". University of Georgia. Archived from the original on 29 April 2007. Retrieved 7 January 2007.

[76] Goebel, Greg (1 September 2007). "[4.3] Magnetic Properties of the Atom". *Elementary Quantum Physics*. In The Public Domain website. Retrieved 7 January 2007.

[77] Yarris, Lynn (Spring 1997). "Talking Pictures". *Berkeley Lab Research Review*. Archived from the original on 13 January 2008. Retrieved 9 January 2008.

[78] Liang, Z.-P.; Haacke, E. M. (1999). Webster, J. G., ed. *Encyclopedia of Electrical and Electronics Engineering: Magnetic Resonance Imaging*. vol. 2. John Wiley & Sons. pp. 412–426. ISBN 0-471-13946-7.

[79] Zeghbroeck, Bart J. Van (1998). "Energy levels". Shippensburg University. Archived from the original on 15 January 2005. Retrieved 23 December 2007.

[80] Fowles, Grant R. (1989). *Introduction to Modern Optics*. Courier Dover Publications. pp. 227–233. ISBN 0-486-65957-7. OCLC 18834711.

[81] Martin, W. C.; Wiese, W. L. (May 2007). "Atomic Spectroscopy: A Compendium of Basic Ideas, Notation, Data, and Formulas". National Institute of Standards and Technology. Archived from the original on 8 February 2007. Retrieved 8 January 2007.

[82] "Atomic Emission Spectra — Origin of Spectral Lines". Avogadro Web Site. Retrieved 10 August 2006.

[83] Fitzpatrick, Richard (16 February 2007). "Fine structure". University of Texas at Austin. Retrieved 14 February 2008.

[84] Weiss, Michael (2001). "The Zeeman Effect". University of California-Riverside. Archived from the original on 2 February 2008. Retrieved 6 February 2008.

[85] Beyer, H. F.; Shevelko, V. P. (2003). *Introduction to the Physics of Highly Charged Ions*. CRC Press. pp. 232–236. ISBN 0-7503-0481-2. OCLC 47150433.

[86] Watkins, Thayer. "Coherence in Stimulated Emission". San José State University. Archived from the original on 12 January 2008. Retrieved 23 December 2007.

[87] oxford dictionary – valency

[88] Reusch, William (16 July 2007). "Virtual Textbook of Organic Chemistry". Michigan State University. Retrieved 11 January 2008.

[89] "Covalent bonding – Single bonds". chemguide. 2000.

[90] Husted, Robert et al. (11 December 2003). "Periodic Table of the Elements". Los Alamos National Laboratory. Archived from the original on 10 January 2008. Retrieved 11 January 2008.

[91] Baum, Rudy (2003). "It's Elemental: The Periodic Table". Chemical & Engineering News. Retrieved 11 January 2008.

[92] Goodstein, David L. (2002). *States of Matter*. Courier Dover Publications. pp. 436–438. ISBN 0-13-843557-X.

[93] Brazhkin, Vadim V. (2006). "Metastable phases, phase transformations, and phase diagrams in physics and chemistry". *Physics-Uspekhi* **49** (7): 719–24. Bibcode:2006PhyU...49..719B. doi:10.1070/PU2006v049n07ABEH006013.

[94] Myers, Richard (2003). *The Basics of Chemistry*. Greenwood Press. p. 85. ISBN 0-313-31664-3. OCLC 50164580.

[95] Staff (9 October 2001). "Bose-Einstein Condensate: A New Form of Matter". National Institute of Standards and Technology. Archived from the original on 3 January 2008. Retrieved 16 January 2008.

[96] Colton, Imogen; Fyffe, Jeanette (3 February 1999). "Super Atoms from Bose-Einstein Condensation". The University of Melbourne. Archived from the original on 29 August 2007. Retrieved 6 February 2008.

[97] Jacox, Marilyn; Gadzuk, J. William (November 1997). "Scanning Tunneling Microscope". National Institute of Standards and Technology. Archived from the original on 7 January 2008. Retrieved 11 January 2008.

[98] "The Nobel Prize in Physics 1986". The Nobel Foundation. Retrieved 11 January 2008.—in particular, see the Nobel lecture by G. Binnig and H. Rohrer.

[99] Jakubowski, N.; Moens, Luc; Vanhaecke, Frank (1998). "Sector field mass spectrometers in ICP-MS". *Spectrochimica Acta Part B: Atomic Spectroscopy* **53** (13): 1739–63. Bibcode:1998AcSpe..53.1739J. doi:10.1016/S0584-8547(98)00222-5.

[100] Müller, Erwin W.; Panitz, John A.; McLane, S. Brooks (1968). "The Atom-Probe Field Ion Microscope". *Review of Scientific Instruments* **39** (1): 83–86. Bibcode:1968RScI...39...83M. doi:10.1063/1.1683116.

[101] Lochner, Jim; Gibb, Meredith; Newman, Phil (30 April 2007). "What Do Spectra Tell Us?". NASA/Goddard Space Flight Center. Archived from the original on 16 January 2008. Retrieved 3 January 2008.

[102] Winter, Mark (2007). "Helium". WebElements. Archived from the original on 30 December 2007. Retrieved 3 January 2008.

[103] Hinshaw, Gary (10 February 2006). "What is the Universe Made Of?". NASA/WMAP. Archived from the original on 31 December 2007. Retrieved 7 January 2008.

[104] Choppin, Gregory R.; Liljenzin, Jan-Olov; Rydberg, Jan (2001). *Radiochemistry and Nuclear Chemistry*. Elsevier. p. 441. ISBN 0-7506-7463-6. OCLC 162592180.

[105] Davidsen, Arthur F. (1993). "Far-Ultraviolet Astronomy on the Astro-1 Space Shuttle Mission". *Science* **259** (5093): 327–34. Bibcode:1993Sci...259..327D. doi:10.1126/science.259.5093.327. PMID 17832344.

[106] Lequeux, James (2005). *The Interstellar Medium*. Springer. p. 4. ISBN 3-540-21326-0. OCLC 133157789.

[107] Smith, Nigel (6 January 2000). "The search for dark matter". Physics World. Archived from the original on 16 February 2008. Retrieved 14 February 2008.

[108] Croswell, Ken (1991). "Boron, bumps and the Big Bang: Was matter spread evenly when the Universe began? Perhaps not; the clues lie in the creation of the lighter elements such as boron and beryllium". *New Scientist* (1794): 42. Archived from the original on 7 February 2008. Retrieved 14 January 2008.

[109] Copi, Craig J.; Schramm, DN; Turner, MS (1995). "Big-Bang Nucleosynthesis and the Baryon Density of the Universe". *Science* **267** (5195): 192–99. arXiv:astro-ph/9407006. Bibcode:1995Sci...267..192C. doi:10.1126/science.7809624. PMID 7809624.

[110] Hinshaw, Gary (15 December 2005). "Tests of the Big Bang: The Light Elements". NASA/WMAP. Archived from the original on 17 January 2008. Retrieved 13 January 2008.

[111] Abbott, Brian (30 May 2007). "Microwave (WMAP) All-Sky Survey". Hayden Planetarium. Retrieved 13 January 2008.

[112] Hoyle, F. (1946). "The synthesis of the elements from hydrogen". *Monthly Notices of the Royal Astronomical Society* **106**: 343–83. Bibcode:1946MNRAS.106..343H. doi:10.1093/mnras/106.5.343.

[113] Knauth, D. C.; Knauth, D. C.; Lambert, David L.; Crane, P. (2000). "Newly synthesized lithium in the interstellar medium". *Nature* **405** (6787): 656–58. doi:10.1038/35015028. PMID 10864316.

[114] Mashnik, Stepan G. (2000). "On Solar System and Cosmic Rays Nucleosynthesis and Spallation Processes". arXiv:astro-ph/0008382 [astro-ph].

[115] Kansas Geological Survey (4 May 2005). "Age of the Earth". University of Kansas. Retrieved 14 January 2008.

[116] Manuel 2001, pp. 407–430, 511–519.

[117] Dalrymple, G. Brent (2001). "The age of the Earth in the twentieth century: a problem (mostly) solved". *Geological Society, London, Special Publications* **190** (1): 205–21. Bibcode:2001GSLSP.190..205D. doi:10.1144/GSL.SP.2001.190.01.14. Retrieved 14 January 2008.

[118] Anderson, Don L.; Foulger, G. R.; Meibom, Anders (2 September 2006). "Helium: Fundamental models". MantlePlumes.org. Archived from the original on 8 February 2007. Retrieved 14 January 2007.

[119] Pennicott, Katie (10 May 2001). "Carbon clock could show the wrong time". PhysicsWeb. Archived from the original on 15 December 2007. Retrieved 14 January 2008.

[120] Yarris, Lynn (27 July 2001). "New Superheavy Elements 118 and 116 Discovered at Berkeley Lab". Berkeley Lab. Archived from the original on 9 January 2008. Retrieved 14 January 2008.

[121] Diamond, H et al. (1960). "Heavy Isotope Abundances in Mike Thermonuclear Device". *Physical Review* **119** (6): 2000–04. Bibcode:1960PhRv..119.2000D. doi:10.1103/PhysRev.119.2000.

[122] Poston Sr., John W. (23 March 1998). "Do transuranic elements such as plutonium ever occur naturally?". Scientific American.

[123] Keller, C. (1973). "Natural occurrence of lanthanides, actinides, and superheavy elements". *Chemiker Zeitung* **97** (10): 522–30. OSTI 4353086.

[124] Zaider, Marco; Rossi, Harald H. (2001). *Radiation Science for Physicians and Public Health Workers.* Springer. p. 17. ISBN 0-306-46403-9. OCLC 44110319.

[125] Manuel 2001, pp. 407–430,511–519.

[126] "Oklo Fossil Reactors". Curtin University of Technology. Archived from the original on 18 December 2007. Retrieved 15 January 2008.

[127] Weisenberger, Drew. "How many atoms are there in the world?". Jefferson Lab. Retrieved 16 January 2008.

[128] Pidwirny, Michael. "Fundamentals of Physical Geography". University of British Columbia Okanagan. Archived from the original on 21 January 2008. Retrieved 16 January 2008.

[129] Anderson, Don L. (2002). "The inner inner core of Earth". *Proceedings of the National Academy of Sciences* **99** (22): 13966–68. Bibcode:2002PNAS...9913966A. doi:10.1073/pnas.232565899. PMC 137819. PMID 12391308.

[130] Pauling, Linus (1960). *The Nature of the Chemical Bond.* Cornell University Press. pp. 5–10. ISBN 0-8014-0333-2. OCLC 17518275.

[131] Anonymous (2 October 2001). "Second postcard from the island of stability". *CERN Courier.* Archived from the original on 3 February 2008. Retrieved 14 January 2008.

[132] Jacoby, Mitch (2006). "As-yet-unsynthesized superheavy atom should form a stable diatomic molecule with fluorine". *Chemical & Engineering News* **84** (10): 19. doi:10.1021/cen-v084n010.p019a.

[133] Koppes, Steve (1 March 1999). "Fermilab Physicists Find New Matter-Antimatter Asymmetry". University of Chicago. Retrieved 14 January 2008.

[134] Cromie, William J. (16 August 2001). "A lifetime of trillionths of a second: Scientists explore antimatter". Harvard University Gazette. Retrieved 14 January 2008.

[135] Hijmans, Tom W. (2002). "Particle physics: Cold antihydrogen". *Nature* **419** (6906): 439–40. Bibcode:2002Natur.419..439H. doi:10.1038/419439a. PMID 12368837.

[136] Staff (30 October 2002). "Researchers 'look inside' antimatter". BBC News. Retrieved 14 January 2008.

[137] Barrett, Roger (1990). "The Strange World of the Exotic Atom". *New Scientist* (1728): 77–115. Archived from the original on 21 December 2007. Retrieved 4 January 2008.

[138] Indelicato, Paul (2004). "Exotic Atoms". *Physica Scripta* **T112** (1): 20–26. arXiv:physics/0409058.Bibcode:2004PhST..112... doi:10.1238/Physica.Topical.112a00020. 20I.

[139] Ripin, Barrett H. (July 1998). "Recent Experiments on Exotic Atoms". American Physical Society. Retrieved 15 February 2008.

1.9   Sources

- Manuel, Oliver (2001). *Origin of Elements in the Solar System: Implications of Post-1957 Observations.* Springer. ISBN 0-306-46562-0. OCLC 228374906.

1.10   Further reading

- Dalton, J. (1808). *A New System of Chemical Philosophy, Part 1.* London and Manchester: S. Russell.

- Gangopadhyaya, Mrinalkanti (1981). *Indian Atomism: History and Sources.* Atlantic Highlands, New Jersey: Humanities Press. ISBN 0-391-02177-X. OCLC 10916778.

- Harrison, Edward Robert (2003). *Masks of the Universe: Changing Ideas on the Nature of the Cosmos.* Cambridge University Press. ISBN 0-521-77351-2. OCLC 50441595.

- Iannone, A. Pablo (2001). *Dictionary of World Philosophy.* Routledge. ISBN 0-415-17995-5. OCLC 44541769.

- King, Richard (1999). *Indian philosophy: an introduction to Hindu and Buddhist thought.* Edinburgh University Press. ISBN 0-7486-0954-7.

- Levere, Trevor, H. (2001). *Transforming Matter – A History of Chemistry for Alchemy to the Buckyball.* The Johns Hopkins University Press. ISBN 0-8018-6610-3.

- Liddell, Henry George; Scott, Robert. "A Greek-English Lexicon". Perseus Digital Library.

- Liddell, Henry George; Scott, Robert. "ἄτομος". *A Greek-English Lexicon.* Perseus Digital Library. Retrieved 21 June 2010.

- McEvilley, Thomas (2002). *The shape of ancient thought: comparative studies in Greek and Indian philosophies.* Allworth Press. ISBN 1-58115-203-5.

- Moran, Bruce T. (2005). *Distilling Knowledge: Alchemy, Chemistry, and the Scientific Revolution.* Harvard University Press. ISBN 0-674-01495-2.

- Ponomarev, Leonid Ivanovich (1993). *The Quantum Dice.* CRC Press. ISBN 0-7503-0251-8. OCLC 26853108.

- Roscoe, Henry Enfield (1895). *John Dalton and the Rise of Modern Chemistry.* Century science series. New York: Macmillan. Retrieved 3 April 2011.

- Siegfried, Robert (2002). *From Elements to Atoms: A History of Chemical Composition.* DIANE. ISBN 0-87169-924-9. OCLC 186607849.

- Teresi, Dick (2003). *Lost Discoveries: The Ancient Roots of Modern Science.* Simon & Schuster. pp. 213–214. ISBN 0-7432-4379-X.

- Various (2002). Lide, David R., ed. *Handbook of Chemistry & Physics* (88th ed.). CRC. ISBN 0-8493-0486-5. OCLC 179976746. Archived from the original on 23 May 2008. Retrieved 23 May 2008.

- Wurtz, Charles Adolphe (1881). *The Atomic Theory.* New York: D. Appleton and company. ISBN 0-559-43636-X.

1.11 External links

- "Quantum Mechanics and the Structure of Atoms" on YouTube

- Freudenrich, Craig C. "How Atoms Work". How Stuff Works. Archived from the original on 8 January 2007. Retrieved 9 January 2007.

- "The Atom". *Free High School Science Texts: Physics*. Wikibooks. Retrieved 10 July 2010.

- Anonymous (2007). "The atom". Science aid+. Retrieved 10 July 2010.—a guide to the atom for teens.

- Anonymous (3 January 2006). "Atoms and Atomic Structure". BBC. Archived from the original on 2 January 2007. Retrieved 11 January 2007.

- Various (3 January 2006). "Physics 2000, Table of Contents". University of Colorado. Archived from the original on 14 January 2008. Retrieved 11 January 2008.

- Various (3 February 2006). "What does an atom look like?". University of Karlsruhe. Retrieved 12 May 2008.

THOMSON

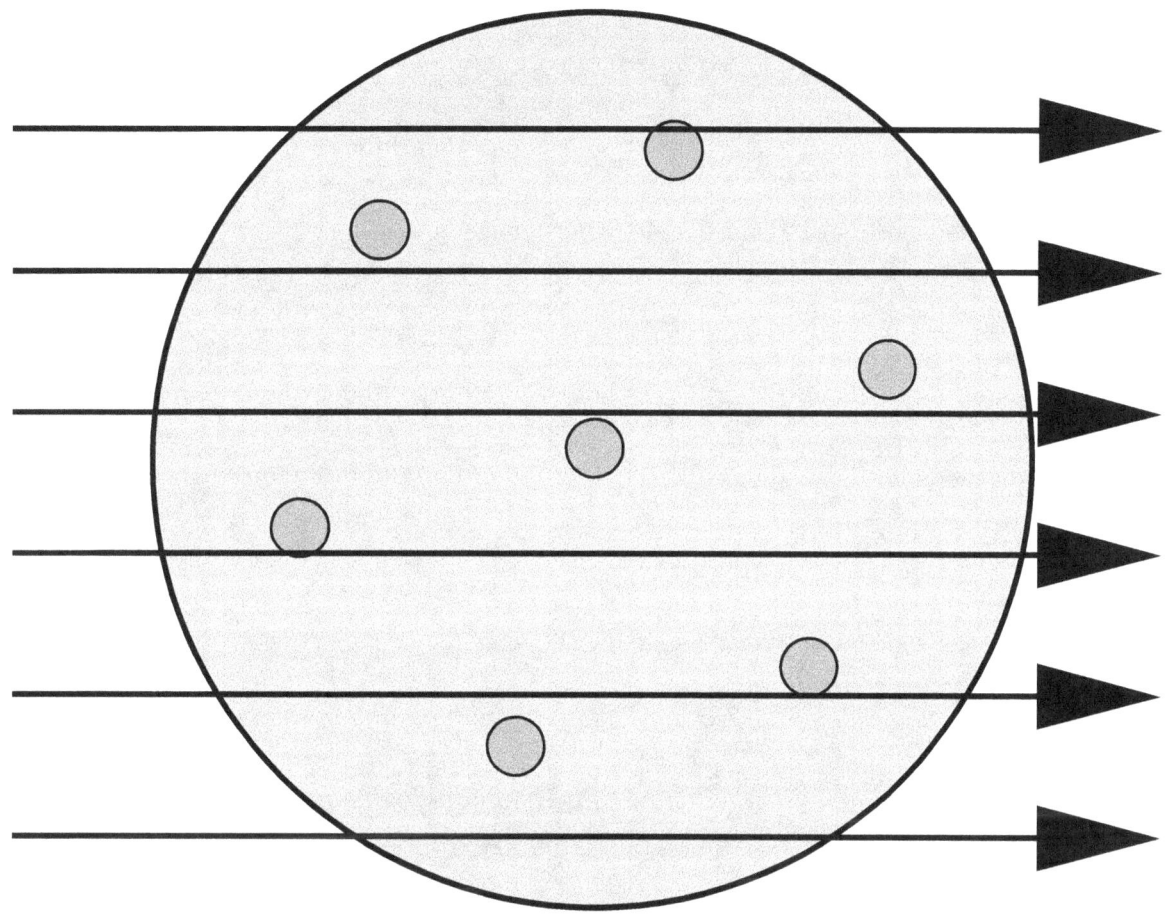

RUTHERFORD

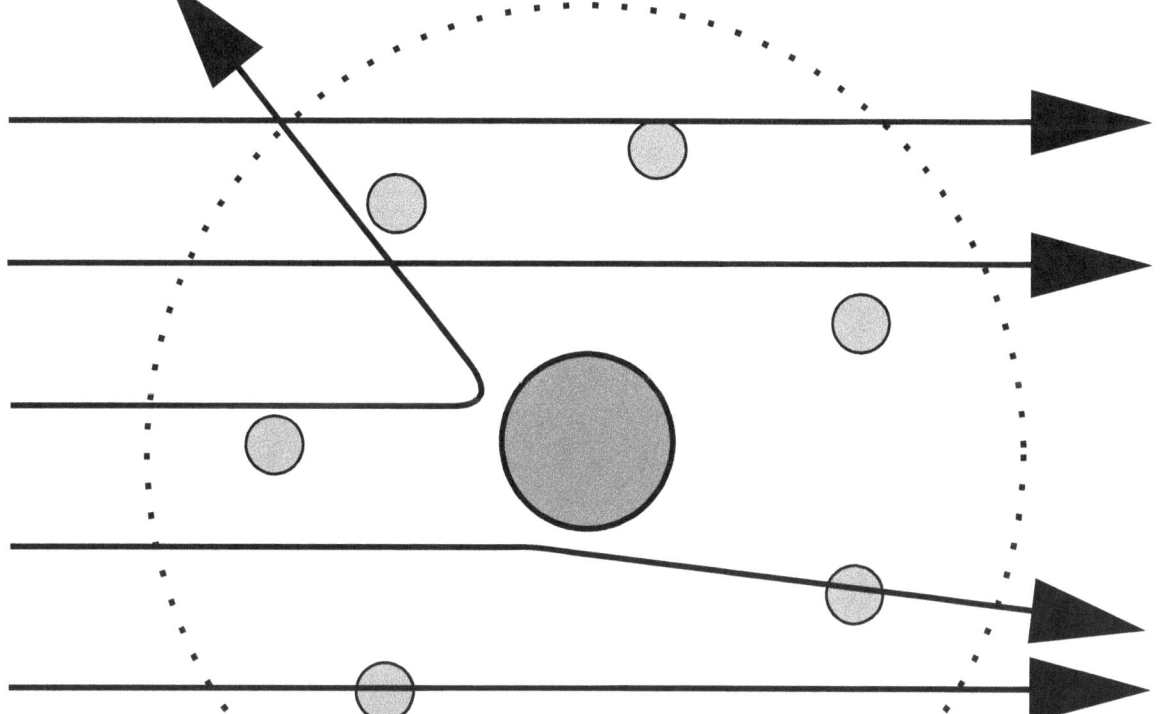

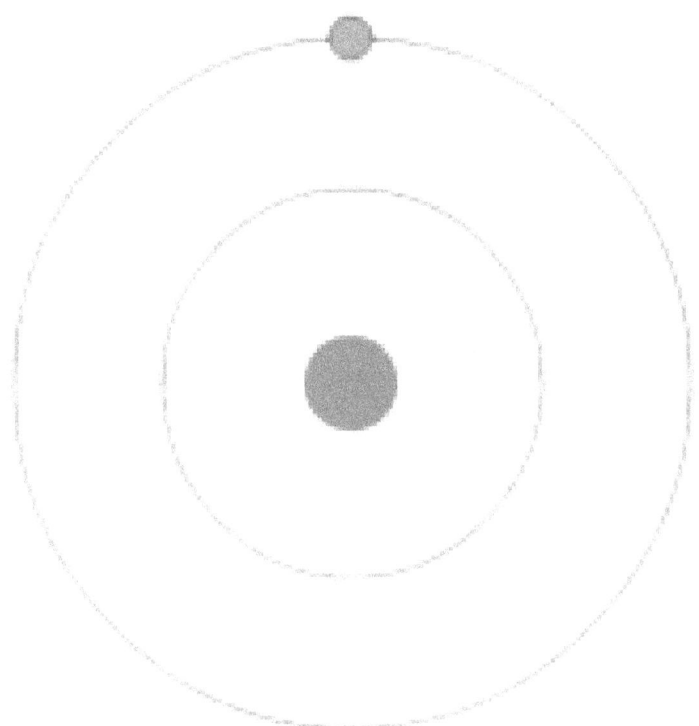

The Bohr model of the atom, with an electron making instantaneous "quantum leaps" from one orbit to another. This model is obsolete.

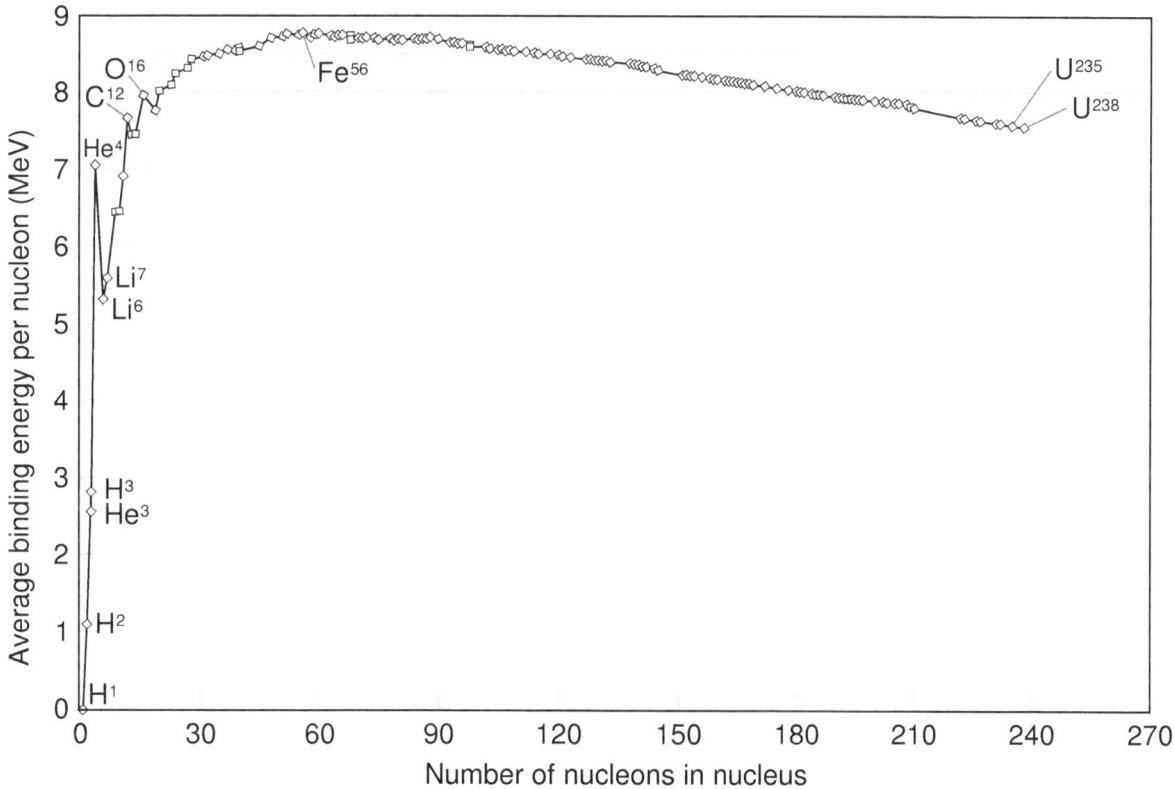

The binding energy needed for a nucleon to escape the nucleus, for various isotopes

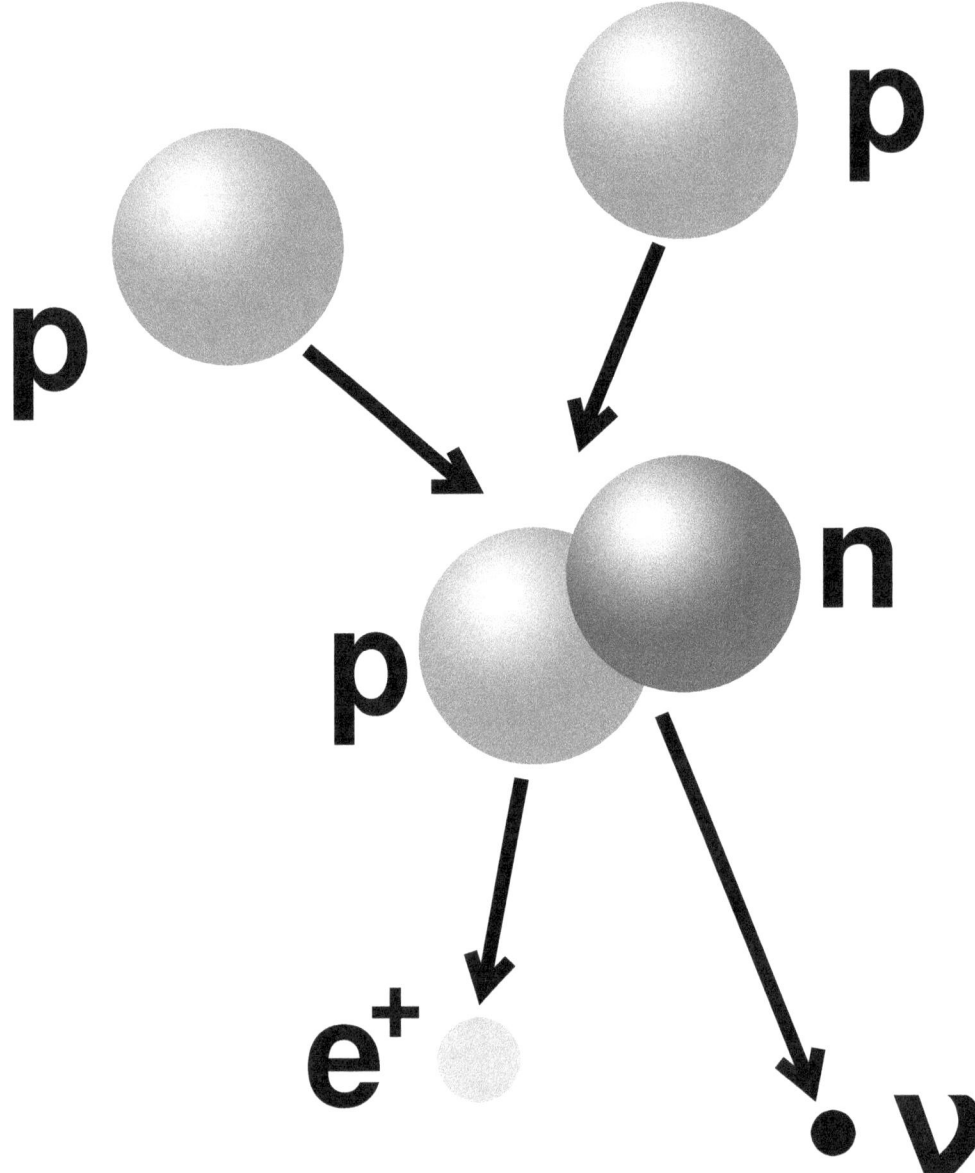

Illustration of a nuclear fusion process that forms a deuterium nucleus, consisting of a proton and a neutron, from two protons. A positron (e⁺)—an antimatter electron—is emitted along with an electron neutrino.

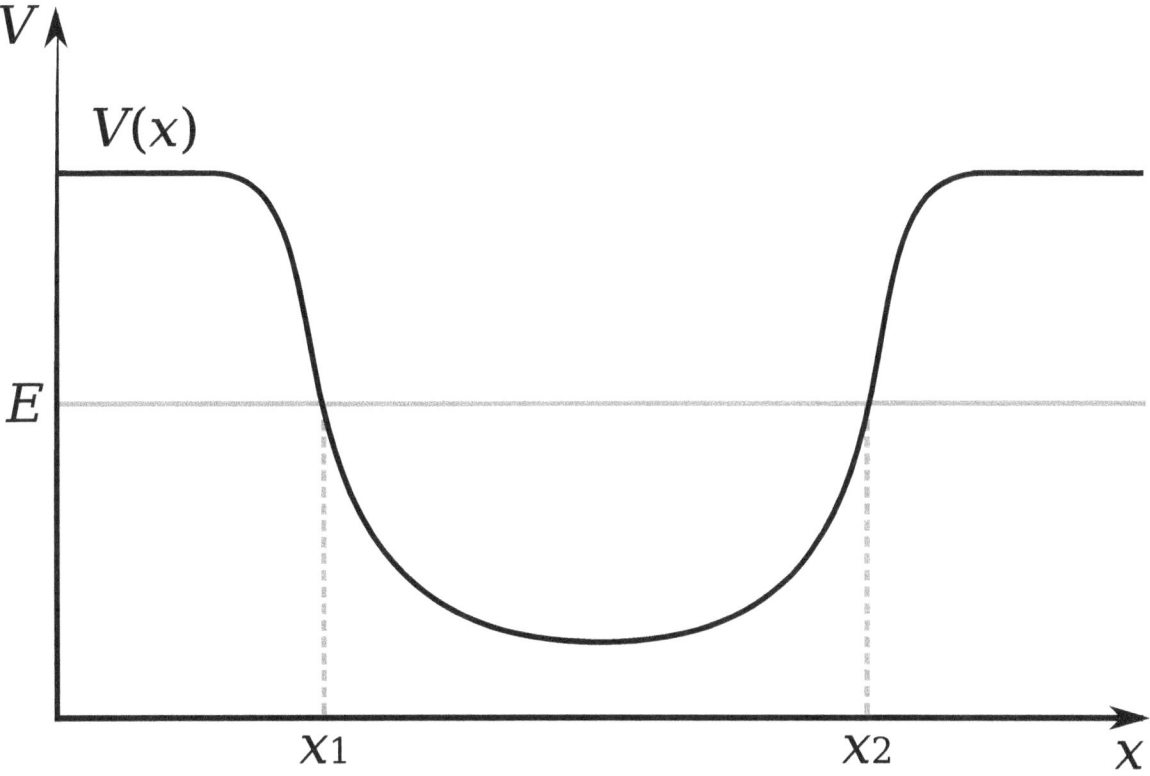

A potential well, showing, according to classical mechanics, the minimum energy V(x) needed to reach each position x. Classically, a particle with energy E is constrained to a range of positions between x_1 and x_2.

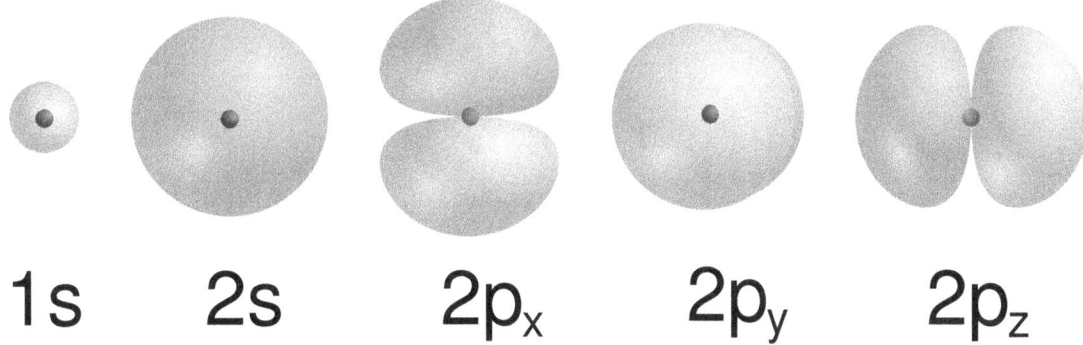

Wave functions of the first five atomic orbitals. The three 2p orbitals each display a single angular node that has an orientation and a minimum at the center.

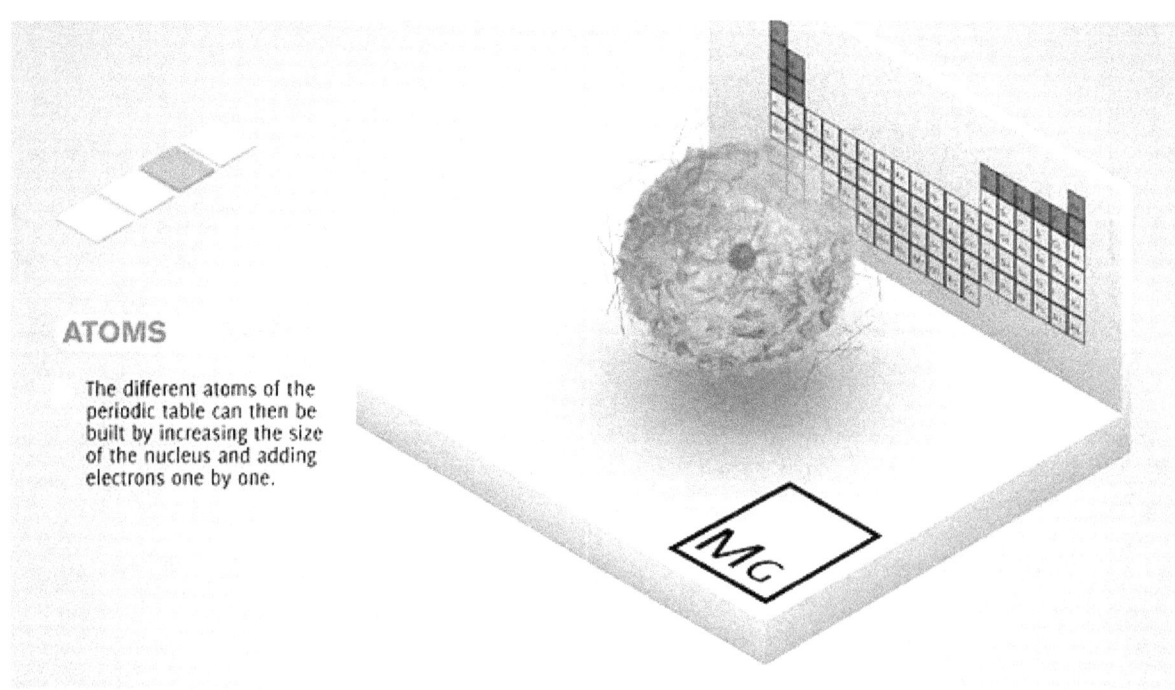

ATOMS

The different atoms of the periodic table can then be built by increasing the size of the nucleus and adding electrons one by one.

How atoms are constructed from electron orbitals and link to the periodic table

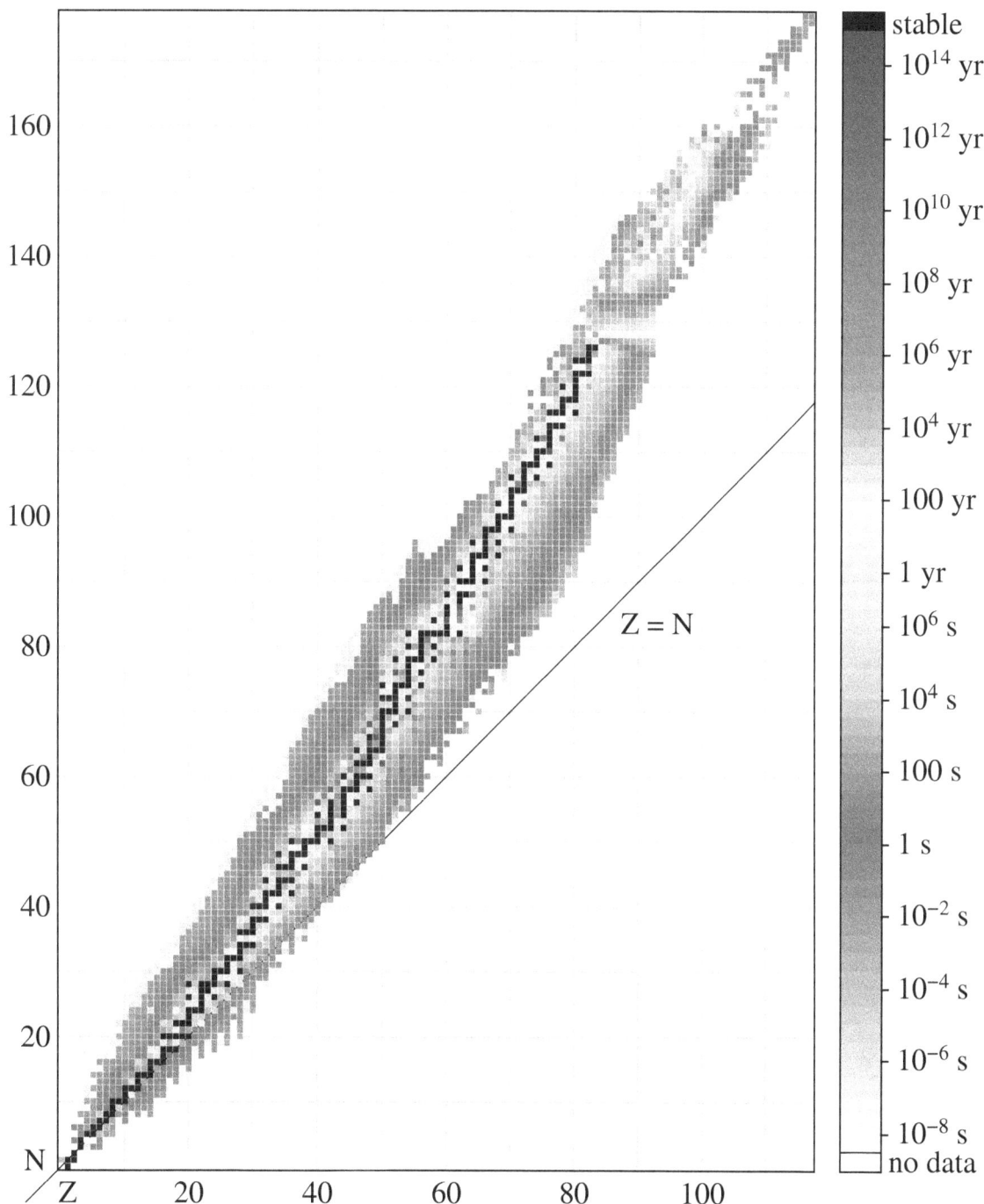

This diagram shows the half-life (T½) of various isotopes with Z protons and N neutrons.

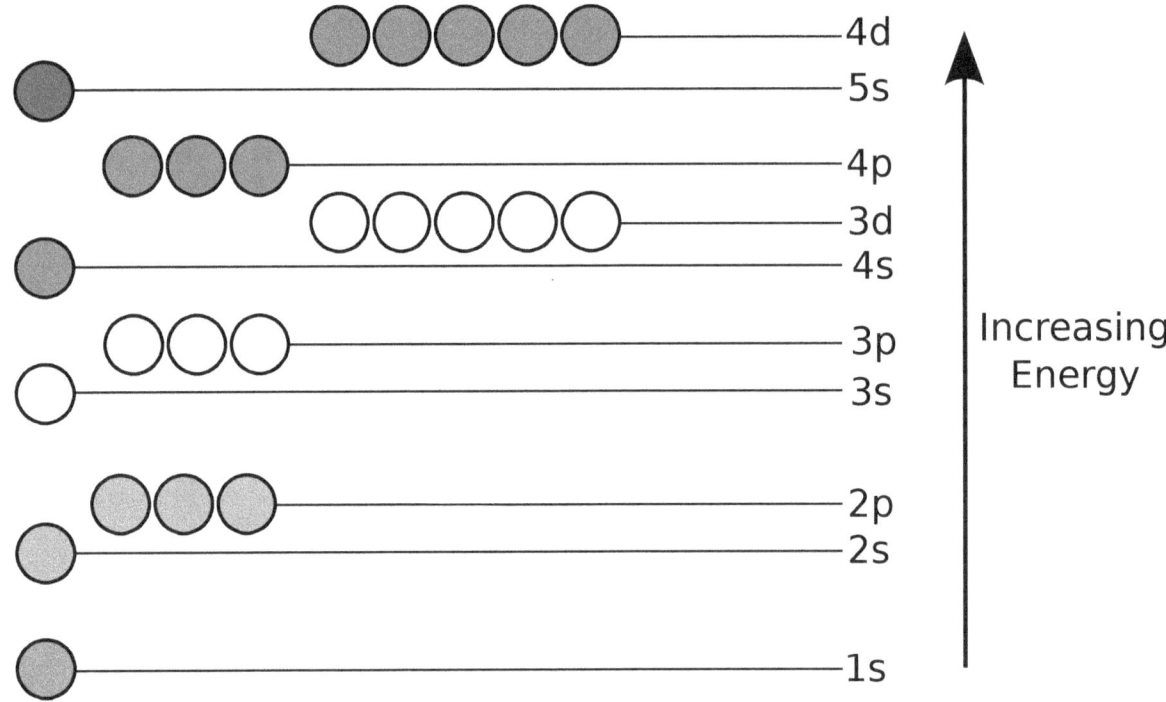

These electron's energy levels (not to scale) are sufficient for ground states of atoms up to cadmium ($5s^2\ 4d^{10}$) inclusively. Do not forget that even the top of the diagram is lower than an unbound electron state.

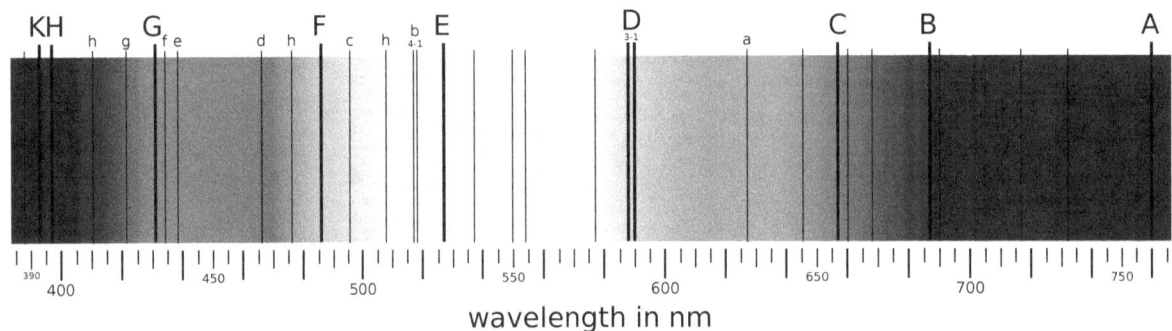

An example of absorption lines in a spectrum

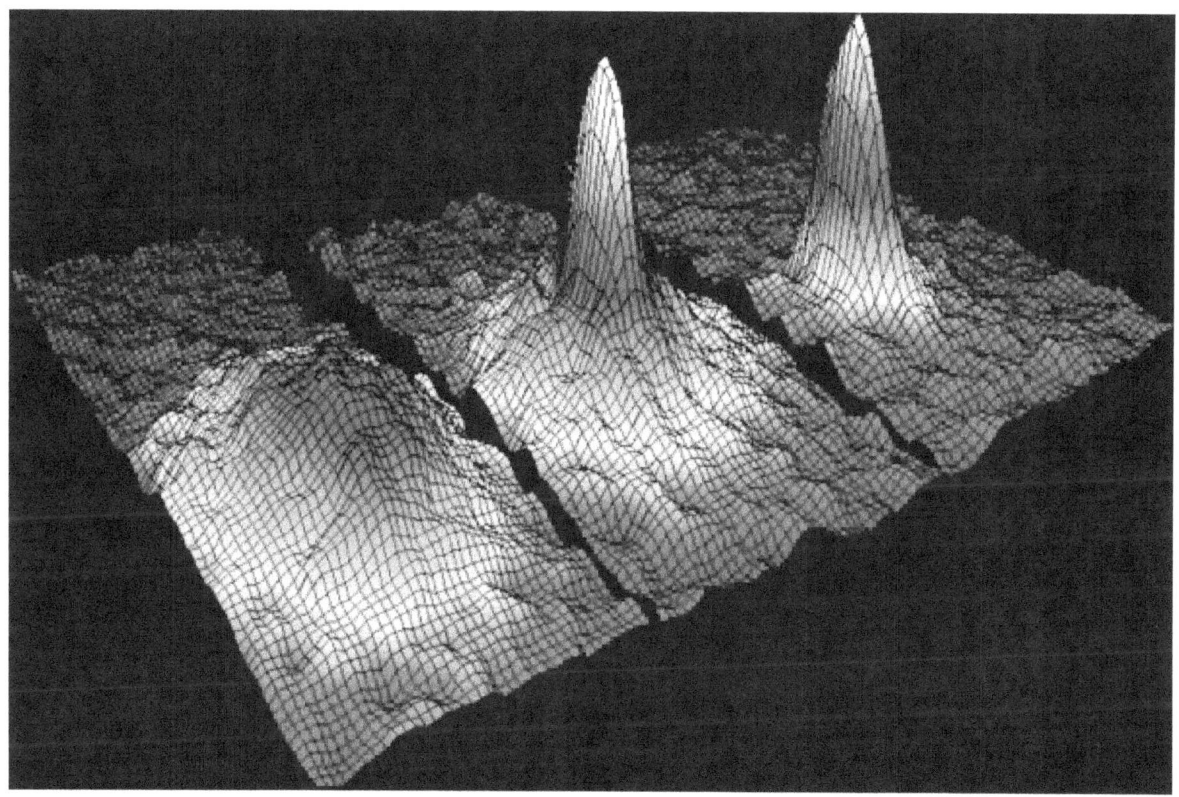

Snapshots illustrating the formation of a Bose–Einstein condensate

Scanning tunneling microscope image showing the individual atoms making up this gold (100) surface. The surface atoms deviate from the bulk crystal structure and arrange in columns several atoms wide with pits between them (See surface reconstruction).

Chapter 2

Atomic theory

"Atomic model" redirects here. For the unrelated term in mathematical logic, see Atomic model (mathematical logic). This article is about the historical models of the atom. For a history of the study of how atoms combine to form molecules, see History of molecular theory.

In chemistry and physics, **atomic theory** is a scientific theory of the nature of matter, which states that matter is composed of discrete units called atoms. It began as a philosophical concept in ancient Greece and entered the scientific mainstream in the early 19th century when discoveries in the field of chemistry showed that matter did indeed behave as if it were made up of atoms.

The word *atom* comes from the Ancient Greek adjective *atomos*, meaning "uncuttable".[1] 19th century chemists began using the term in connection with the growing number of irreducible chemical elements. While seemingly apropos, around the turn of the 20th century, through various experiments with electromagnetism and radioactivity, physicists discovered that the so-called "uncuttable atom" was actually a conglomerate of various subatomic particles (chiefly, electrons, protons and neutrons) which can exist separately from each other. In fact, in certain extreme environments, such as neutron stars, extreme temperature and pressure prevents atoms from existing at all. Since atoms were found to be divisible, physicists later invented the term "elementary particles" to describe the "uncuttable", though not indestructible, parts of an atom. The field of science which studies subatomic particles is particle physics, and it is in this field that physicists hope to discover the true fundamental nature of matter.

2.1 History

2.1.1 Philosophical atomism

Main article: Atomism

The idea that matter is made up of discrete units is a very old one, appearing in many ancient cultures such as Greece and India. However, these ideas were founded in philosophical and theological reasoning rather than evidence and experimentation. Because of this, they could not convince everybody, so atomism was but one of a number of competing theories on the nature of matter. It was not until the 19th century that the idea was embraced and refined by scientists, as the blossoming science of chemistry produced discoveries that could easily be explained using the concept of atoms.

2.1.2 Dalton

Near the end of the 18th century, two laws about chemical reactions emerged without referring to the notion of an atomic theory. The first was the law of conservation of mass, formulated by Antoine Lavoisier in 1789, which states that the total

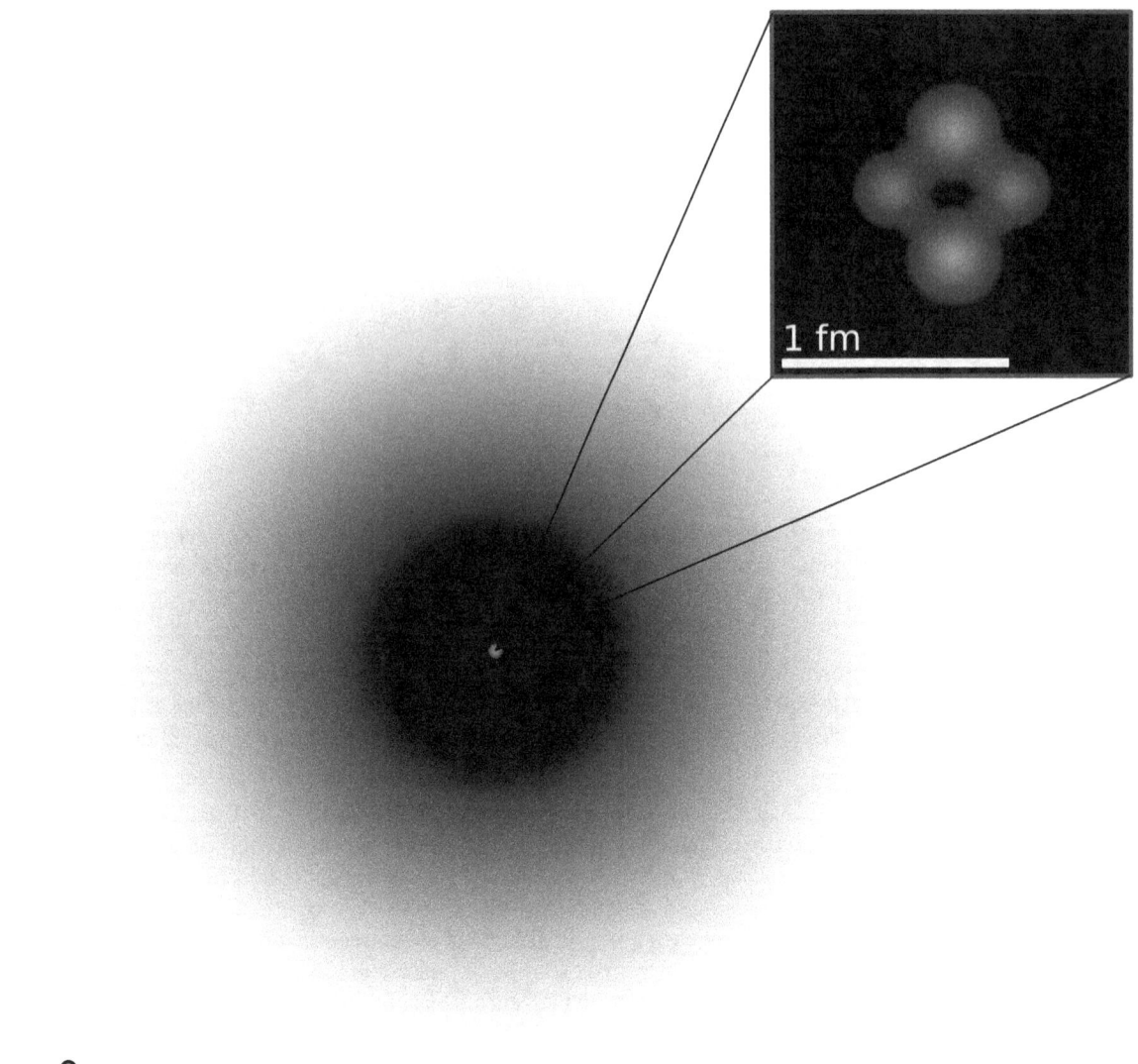

$$1 \text{ Å} = 100,000 \text{ fm}$$

The current theoretical model of the atom involves a dense nucleus surrounded by a probabilistic "cloud" of electrons

mass in a chemical reaction remains constant (that is, the reactants have the same mass as the products).[2] The second was the law of definite proportions. First proven by the French chemist Joseph Louis Proust in 1799,[3] this law states that if a compound is broken down into its constituent elements, then the masses of the constituents will always have the same proportions, regardless of the quantity or source of the original substance.

John Dalton studied and expanded upon this previous work and developed the law of multiple proportions: if two elements can be combined to form a number of possible compounds, then the ratios of the masses of the second element which combine with a fixed mass of the first element will be ratios of small whole numbers. For example: Proust had studied tin oxides and found that their masses were either 88.1% tin and 11.9% oxygen or 78.7% tin and 21.3% oxygen (these were tin(II) oxide and tin dioxide respectively). Dalton noted from these percentages that 100g of tin will combine either with 13.5g or 27g of oxygen; 13.5 and 27 form a ratio of 1:2. Dalton found that an atomic theory of matter could elegantly explain this common pattern in chemistry. In the case of Proust's tin oxides, one tin atom will combine with either one or two oxygen atoms.[4]

Dalton also believed atomic theory could explain why water absorbed different gases in different proportions - for example, he found that water absorbed carbon dioxide far better than it absorbed nitrogen.[5] Dalton hypothesized this was due to the differences in mass and complexity of the gases' respective particles. Indeed, carbon dioxide molecules (CO_2) are heavier and larger than nitrogen molecules (N_2).

Dalton proposed that each chemical element is composed of atoms of a single, unique type, and though they cannot be altered or destroyed by chemical means, they can combine to form more complex structures (chemical compounds). This marked the first truly scientific theory of the atom, since Dalton reached his conclusions by experimentation and examination of the results in an empirical fashion.

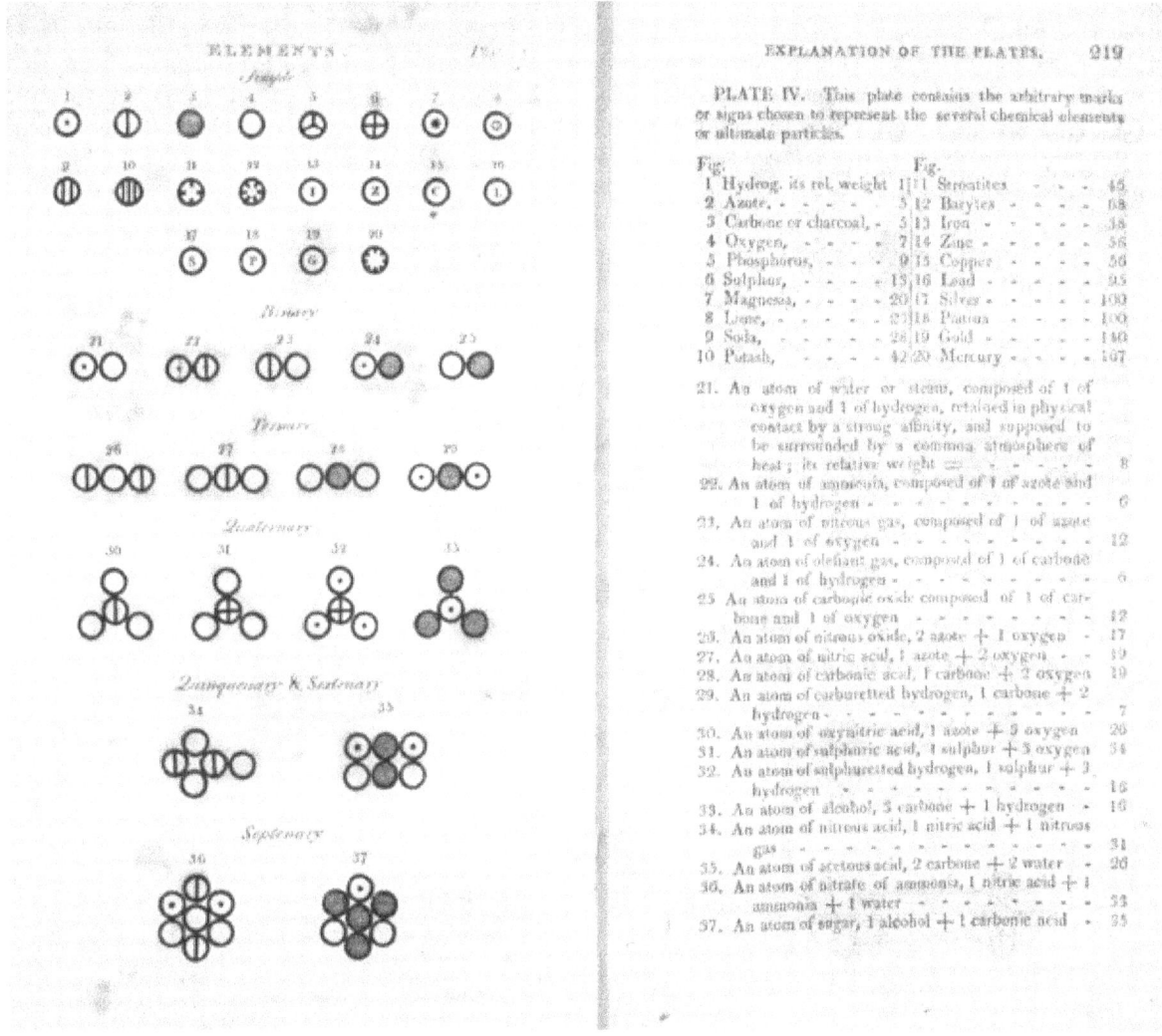

Various atoms and molecules as depicted in John Dalton's A New System of Chemical Philosophy *(1808).*

In 1803 Dalton orally presented his first list of relative atomic weights for a number of substances. This paper was published in 1805, but he did not discuss there exactly how he obtained these figures.[5] The method was first revealed in 1807 by his acquaintance Thomas Thomson, in the third edition of Thomson's textbook, *A System of Chemistry*. Finally, Dalton published a full account in his own textbook, *A New System of Chemical Philosophy*, 1808 and 1810.

Dalton estimated the atomic weights according to the mass ratios in which they combined, with the hydrogen atom taken as unity. However, Dalton did not conceive that with some elements atoms exist in molecules — e.g. pure oxygen exists as O_2. He also mistakenly believed that the simplest compound between any two elements is always one atom of each (so he thought water was HO, not H_2O).[6] This, in addition to the crudity of his equipment, flawed his results. For instance, in 1803 he believed that oxygen atoms were 5.5 times heavier than hydrogen atoms, because in water he measured 5.5

grams of oxygen for every 1 gram of hydrogen and believed the formula for water was HO. Adopting better data, in 1806 he concluded that the atomic weight of oxygen must actually be 7 rather than 5.5, and he retained this weight for the rest of his life. Others at this time had already concluded that the oxygen atom must weigh 8 relative to hydrogen equals 1, if one assumes Dalton's formula for the water molecule (HO), or 16 if one assumes the modern water formula.[7]

2.1.3 Avogadro

The flaw in Dalton's theory was corrected in principle in 1811 by Amedeo Avogadro. Avogadro had proposed that equal volumes of any two gases, at equal temperature and pressure, contain equal numbers of molecules (in other words, the mass of a gas's particles does not affect the volume that it occupies).[8] Avogadro's law allowed him to deduce the diatomic nature of numerous gases by studying the volumes at which they reacted. For instance: since two liters of hydrogen will react with just one liter of oxygen to produce two liters of water vapor (at constant pressure and temperature), it meant a single oxygen molecule splits in two in order to form two particles of water. Thus, Avogadro was able to offer more accurate estimates of the atomic mass of oxygen and various other elements, and made a clear distinction between molecules and atoms.

2.1.4 Brownian Motion

In 1827, the British botanist Robert Brown observed that dust particles inside pollen grains floating in water constantly jiggled about for no apparent reason. In 1905, Albert Einstein theorized that this Brownian motion was caused by the water molecules continuously knocking the grains about, and developed a hypothetical mathematical model to describe it.[9] This model was validated experimentally in 1908 by French physicist Jean Perrin, thus providing additional validation for particle theory (and by extension atomic theory).

2.1.5 Discovery of subatomic particles

Main articles: Electron and Plum pudding model

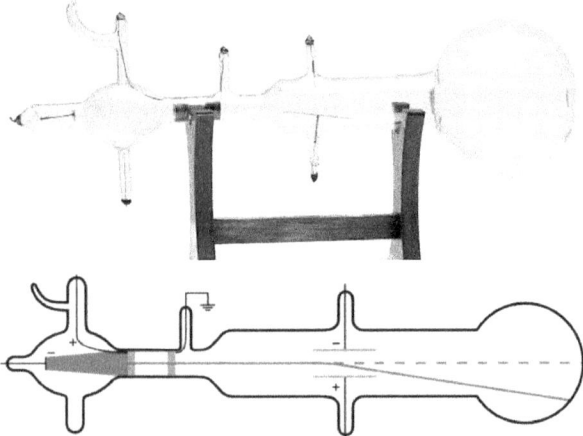

The cathode rays (blue) were emitted from the cathode, sharpened to a beam by the slits, then deflected as they passed between the two electrified plates.

Atoms were thought to be the smallest possible division of matter until 1897 when J.J. Thomson discovered the electron through his work on cathode rays.[10]

A Crookes tube is a sealed glass container in which two electrodes are separated by a vacuum. When a voltage is applied across the electrodes, cathode rays are generated, creating a glowing patch where they strike the glass at the opposite end of the tube. Through experimentation, Thomson discovered that the rays could be deflected by an electric field (in addition to magnetic fields, which was already known). He concluded that these rays, rather than being a form of light, were composed of very light negatively charged particles he called "corpuscles" (they would later be renamed electrons by other scientists). He measured the mass-to-charge ratio and discovered it was 1800 times smaller than that of hydrogen, the smallest atom. These corpuscles were a particle unlike any other previously known.

Thomson suggested that atoms were divisible, and that the corpuscles were their building blocks.[11] To explain the overall neutral charge of the atom, he proposed that the corpuscles were distributed in a uniform sea of positive charge; this was the plum pudding model[12] as the electrons were embedded in the positive charge like plums in a plum pudding (although in Thomson's model they were not stationary).

2.1.6 Discovery of the nucleus

Main article: Rutherford model

Thomson's plum pudding model was disproved in 1909 by one of his former students, Ernest Rutherford, who discovered

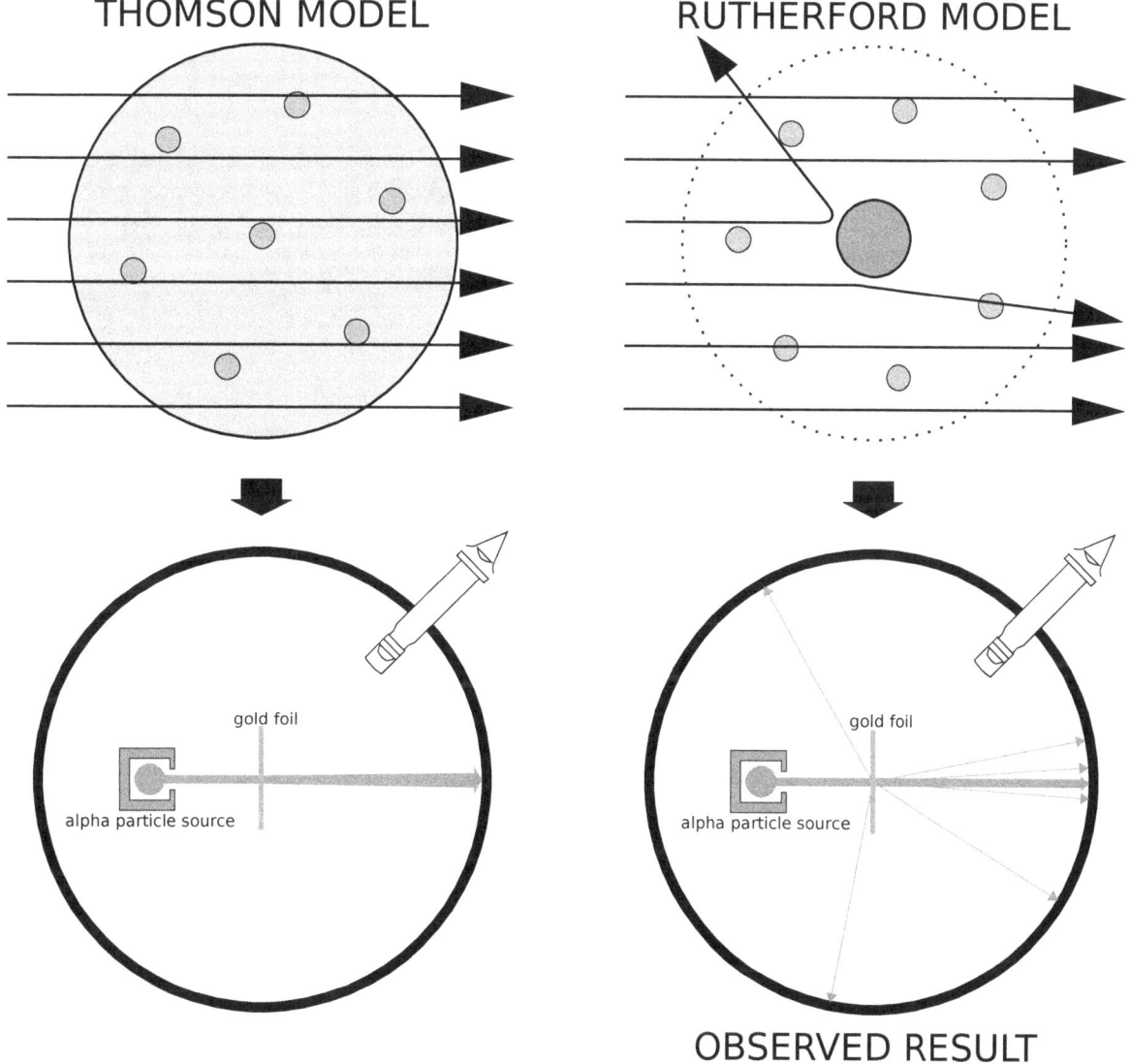

The Geiger-Marsden experiment

Left: *Expected results: alpha particles passing through the plum pudding model of the atom with negligible deflection.*

Right: *Observed results: a small portion of the particles were deflected by the concentrated positive charge of the nucleus.*

that most of the mass and positive charge of an atom is concentrated in a very small fraction of its volume, which he assumed to be at the very center.

In the Geiger–Marsden experiment, Hans Geiger and Ernest Marsden (colleagues of Rutherford working at his behest) shot alpha particles at thin sheets of metal and measured their deflection through the use of a fluorescent screen.[13] Given the very small mass of the electrons, the high momentum of the alpha particles, and the low concentration of the positive charge of the plum pudding model, the experimenters expected all the alpha particles to pass through the metal foil without significant deflection. To their astonishment, a small fraction of the alpha particles experienced heavy deflection. Rutherford concluded that the positive charge of the atom must be concentrated in a very tiny volume to produce an electric field sufficiently intense to deflect the alpha particles so strongly.

This led Rutherford to propose a planetary model in which a cloud of electrons surrounded a small, compact nucleus of positive charge. Only such a concentration of charge could produce the electric field strong enough to cause the heavy deflection.[14]

2.1.7   First steps toward a quantum physical model of the atom

Main article: Bohr model

The planetary model of the atom had two significant shortcomings. The first is that, unlike planets orbiting a sun, electrons are charged particles. An accelerating electric charge is known to emit electromagnetic waves according to the Larmor formula in classical electromagnetism. An orbiting charge should steadily lose energy and spiral toward the nucleus, colliding with it in a small fraction of a second. The second problem was that the planetary model could not explain the highly peaked emission and absorption spectra of atoms that were observed.

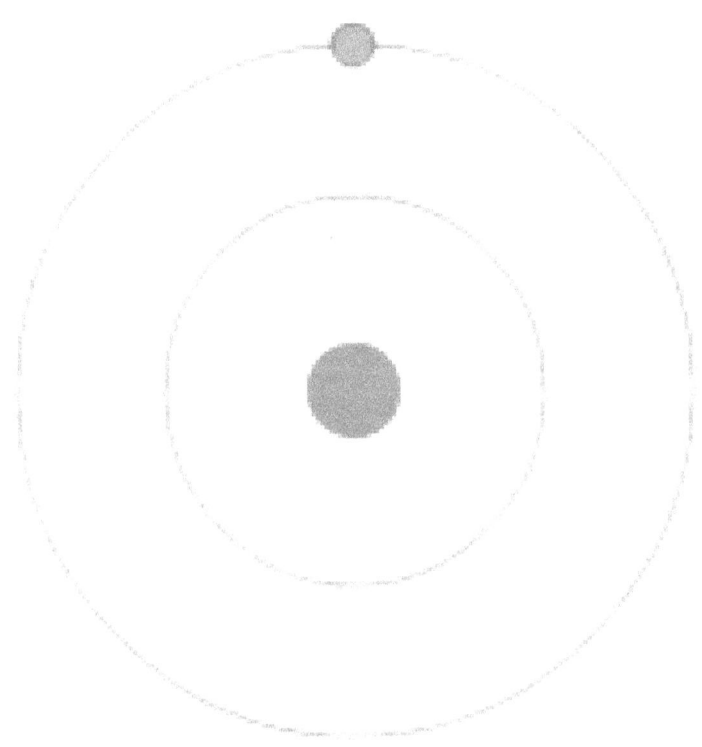

The Bohr model of the atom

Quantum theory revolutionized physics at the beginning of the 20th century, when Max Planck and Albert Einstein postulated that light energy is emitted or absorbed in discrete amounts known as quanta (singular, *quantum*). In 1913, Niels Bohr incorporated this idea into his Bohr model of the atom, in which an electron could only orbit the nucleus in particular circular orbits with fixed angular momentum and energy, its distance from the nucleus (i.e., their radii) being proportional to its energy.[15] Under this model an electron could not spiral into the nucleus because it could not lose energy in a continuous manner; instead, it could only make instantaneous "quantum leaps" between the fixed energy levels.[15] When this occurred, light was emitted or absorbed at a frequency proportional to the change in energy (hence the absorption and emission of light in discrete spectra).[15]

Bohr's model was not perfect. It could only predict the spectral lines of hydrogen; it couldn't predict those of multielectron atoms. Worse still, as spectrographic technology improved, additional spectral lines in hydrogen were observed which Bohr's model couldn't explain. In 1916, Arnold Sommerfeld added elliptical orbits to the Bohr model to explain the extra emission lines, but this made the model very difficult to use, and it still couldn't explain more complex atoms.

2.1.8   Discovery of isotopes

Main article: Isotope

While experimenting with the products of radioactive decay, in 1913 radiochemist Frederick Soddy discovered that there appeared to be more than one element at each position on the periodic table.[16] The term isotope was coined by Margaret Todd as a suitable name for these elements.

That same year, J.J. Thomson conducted an experiment in which he channeled a stream of neon ions through magnetic and electric fields, striking a photographic plate at the other end. He observed two glowing patches on the plate, which suggested two different deflection trajectories. Thomson concluded this was because some of the neon ions had a different mass.[17] The nature of this differing mass would later be explained by the discovery of neutrons in 1932.

2.1.9   Discovery of nuclear particles

Main article: Atomic nucleus

In 1917 Rutherford bombarded nitrogen gas with alpha particles and observed hydrogen nuclei being emitted from the gas (Rutherford recognized these, because he had previously obtained them bombarding hydrogen with alpha particles, and observing hydrogen nuclei in the products). Rutherford concluded that the hydrogen nuclei emerged from the nuclei of the nitrogen atoms themselves (in effect, he had split a nitrogen).[18]

From his own work and the work of his students Bohr and Henry Moseley, Rutherford knew that the positive charge of any atom could always be equated to that of an integer number of hydrogen nuclei. This, coupled with the atomic mass of many elements being roughly equivalent to an integer number of hydrogen atoms - then assumed to be the lightest particles - led him to conclude that hydrogen nuclei were singular particles and a basic constituent of all atomic nuclei. He named such particles protons. Further experimentation by Rutherford found that the nuclear mass of most atoms exceeded that of the protons it possessed; he speculated that this surplus mass was composed of hitherto unknown neutrally charged particles, which were tentatively dubbed "neutrons".

In 1928, Walter Bothe observed that beryllium emitted a highly penetrating, electrically neutral radiation when bombarded with alpha particles. It was later discovered that this radiation could knock hydrogen atoms out of paraffin wax. Initially it was thought to be high-energy gamma radiation, since gamma radiation had a similar effect on electrons in metals, but James Chadwick found that the ionization effect was too strong for it to be due to electromagnetic radiation, so long as energy and momentum were conserved in the interaction. In 1932, Chadwick exposed various elements, such as hydrogen and nitrogen, to the mysterious "beryllium radiation", and by measuring the energies of the recoiling charged particles, he deduced that the radiation was actually composed of electrically neutral particles which could not be massless like the gamma ray, but instead were required to have a mass similar to that of a proton. Chadwick now claimed these particles as Rutherford's neutrons.[19] For his discovery of the neutron, Chadwick received the Nobel Prize in 1935.

2.1.10   Quantum physical models of the atom

Main article: Atomic orbital
In 1924, Louis de Broglie proposed that all moving particles — particularly subatomic particles such as electrons —

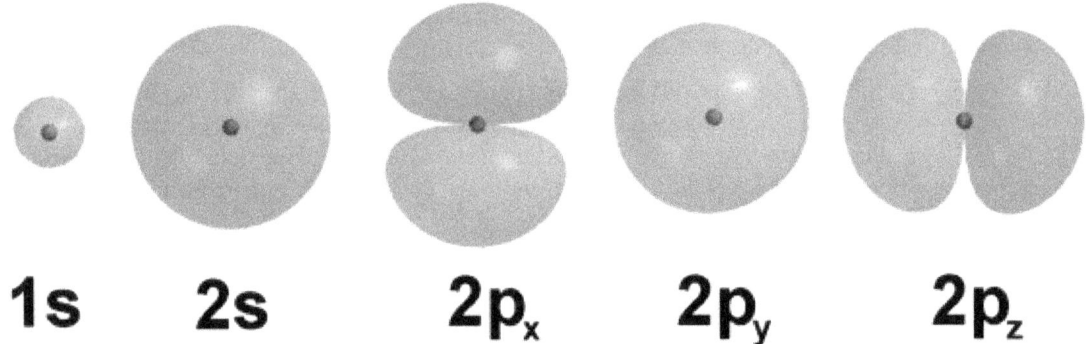

The five filled atomic orbitals of a neon atom separated and arranged in order of increasing energy from left to right, with the last three orbitals being equal in energy. Each orbital holds up to two electrons, which most probably exist in the zones represented by the colored bubbles. Each electron is equally present in both orbital zones, shown here by color only to highlight the different wave phase.

exhibit a degree of wave-like behavior. Erwin Schrödinger, fascinated by this idea, explored whether or not the movement of an electron in an atom could be better explained as a wave rather than as a particle. Schrödinger's equation, published in 1926,[20] describes an electron as a wavefunction instead of as a point particle. This approach elegantly predicted many of the spectral phenomena that Bohr's model failed to explain. Although this concept was mathematically convenient, it was difficult to visualize, and faced opposition.[21] One of its critics, Max Born, proposed instead that Schrödinger's wavefunction described not the electron but rather all its possible states, and thus could be used to calculate the probability of finding an electron at any given location around the nucleus.[22] This reconciled the two opposing theories of particle versus wave electrons and the idea of wave–particle duality was introduced. This theory stated that the electron may exhibit the properties of both a wave and a particle. For example, it can be refracted like a wave, and has mass like a particle.[23]

A consequence of describing electrons as waveforms is that it is mathematically impossible to simultaneously derive the position and momentum of an electron. This became known as the Heisenberg uncertainty principle after the theoretical physicist Werner Heisenberg, who first described it and published it in 1927.[24] This invalidated Bohr's model, with its neat, clearly defined circular orbits. The modern model of the atom describes the positions of electrons in an atom in terms of probabilities. An electron can potentially be found at any distance from the nucleus, but, depending on its energy level, exists more frequently in certain regions around the nucleus than others; this pattern is referred to as its atomic orbital. The orbitals come in a variety of shapes-sphere, dumbbell, torus, etc.-with the nucleus in the middle.[25]

2.2   See also

- History of the molecule

- Discoveries of the chemical elements

- Introduction to quantum mechanics

- Kinetic theory

- Atomism

- *The Physical Principles of the Quantum Theory*

2.3   Notes

[1] Berryman, Sylvia, "Ancient Atomism", *The Stanford Encyclopedia of Philosophy* (Fall 2008 Edition), Edward N. Zalta (ed.), http://plato.stanford.edu/archives/fall2008/entries/atomism-ancient/

[2] Weisstein, Eric W. "Lavoisier, Antoine (1743-1794)". scienceworld.wolfram.com. Retrieved 2009-08-01.

[3] Proust, Joseph Louis. "Researches on Copper", excerpted from *Ann. chim.* 32, 26-54 (1799) [as translated and reproduced in Henry M. Leicester and Herbert S. Klickstein, *A Source Book in Chemistry*, 1400–1900 (Cambridge, Massachusetts: Harvard, 1952)]. Retrieved on August 29, 2007.

[4] Andrew G. van Melsen (1952). *From Atomos to Atom*. Mineola, N.Y.: Dover Publications. ISBN 0-486-49584-1.

[5] Dalton, John. "On the Absorption of Gases by Water and Other Liquids", in *Memoirs of the Literary and Philosophical Society of Manchester*. 1803. Retrieved on August 29, 2007.

[6] Johnson, Chris. "Avogadro - his contribution to chemistry". Archived from the original on 27 June 2009. Retrieved 2009-08-01.

[7] Alan J. Rocke (1984). *Chemical Atomism in the Nineteenth Century*. Columbus: Ohio State University Press.

[8] Avogadro, Amedeo (1811). "Essay on a Manner of Determining the Relative Masses of the Elementary Molecules of Bodies, and the Proportions in Which They Enter into These Compounds". *Journal de Physique* **73**: 58–76.

[9] Einstein, A. (1905). "Über die von der molekularkinetischen Theorie der Wärme geforderte Bewegung von in ruhenden Flüssigkeiten suspendierten Teilchen". *Annalen der Physik* **322** (8): 549. Bibcode:1905AnP...322..549E. doi:10.1002/andp.06.

[10] Thomson, J.J. (1897). "Cathode rays" ([facsimile from Stephen Wright, Classical Scientific Papers, Physics (Mills and Boon, 1964)]). *Philosophical Magazine* **44** (269): 293. doi:10.1080/14786449708621070.

[11] Whittaker, E. T. (1951), *A history of the theories of aether and electricity. Vol 1*, Nelson, London

[12] Thomson, J.J. (1904). "On the Structure of the Atom: an Investigation of the Stability and Periods of Oscillation of a number of Corpuscles arranged at equal intervals around the Circumference of a Circle; with Application of the Results to the Theory of Atomic Structure". *Philosophical Magazine* **7** (39): 237. doi:10.1080/14786440409463107.

[13] Geiger, H (1910). "The Scattering of the α-Particles by Matter". *Proceedings of the Royal Society* **A 83**: 492–504.

[14] Rutherford, Ernest (1911). "The Scattering of α and β Particles by Matter and the Structure of the Atom" (PDF). *Philosophical Magazine* **21** (4): 669. Bibcode:2012PMag...92..379R. doi:10.1080/14786435.2011.617037.

[15] Bohr, Niels (1913). "On the constitution of atoms and molecules" (PDF). *Philosophical Magazine* **26** (153): 86441308634993.

[16] "Frederick Soddy, The Nobel Prize in Chemistry 1921". Nobel Foundation. Retrieved 2008-01-18.

[17] Thomson, J.J. (1913). "Rays of positive electricity". *Proceedings of the Royal Society* **A 89** (607): 1–20. Bibcode:....1T. doi:10.1098/rspa.1913.0057. [as excerpted in Henry A. Boorse & Lloyd Motz, *The World of the Atom*, Vol. 1 (New York: Basic Books, 1966)]. Retrieved on August 29, 2007.

[18] Rutherford, Ernest (1919). "Collisions of alpha Particles with Light Atoms. IV. An Anomalous Effect in Nitrogen". *Philosophical Magazine* **37** (222): 581. doi:10.1080/14786440608635919.

[19] Chadwick, James (1932). "Possible Existence of a Neutron" (PDF). *Nature* **129** (3252): 312. Bibcode:1932Natur.129Q.312C. doi:10.1038/129312a0.

[20] Schrödinger, Erwin (1926). "Quantisation as an Eigenvalue Problem". *Annalen der Physik* **81** (18): 109–139. Bibco6..109S. doi:10.1002/andp.19263861802.

[21] Mahanti, Subodh. "Erwin Schrödinger: The Founder of Quantum Wave Mechanics". Retrieved 2009-08-01.

[22] Mahanti, Subodh. "Max Born: Founder of Lattice Dynamics". Retrieved 2009-08-01.

[23] Greiner, Walter. "Quantum Mechanics: An Introduction". Retrieved 2010-06-14.

[24] Heisenberg, W. (1927). "Über den anschaulichen Inhalt der quantentheoretischen Kinematik und Mechanik". *Zeitschrift für Physik* (in German) **43** (3–4): 172–198. Bibcode:1927ZPhy...43..172H. doi:10.1007/BF01397280.

[25] Milton Orchin, Roger Macomber, Allan Pinhas, R. Wilson. "The Vocabulary and Concepts of Organic Chemistry, Second Edition," (PDF). Retrieved 2010-06-14.

2.4 Further reading

- Bernard Pullman (1998) *The Atom in the History of Human Thought*, trans. by Axel Reisinger. Oxford Univ. Press.

- Eric Scerri (2007) *The Periodic Table, Its Story and Its Significance*, Oxford University Press, New York.

- Charles Adolphe Wurtz (1881) *The Atomic Theory*, D. Appleton and Company, New York.

2.5 External links

- Atomism by S. Mark Cohen.

- Atomic Theory - detailed information on atomic theory with respect to electrons and electricity.

Chapter 3

Atomic nucleus

The **nucleus** is the small, dense region consisting of protons and neutrons at the center of an atom. The atomic nucleus was discovered in 1911 by Ernest Rutherford based on the 1909 Geiger–Marsden gold foil experiment. After the discovery of the neutron in 1932, models for a nucleus composed of protons and neutrons were quickly developed by Dmitri Ivanenko[1] and Werner Heisenberg.[2][3][4][5][6] Almost all of the mass of an atom is located in the nucleus, with a very small contribution from the electron cloud. Protons and neutrons are bound together to form a nucleus by the nuclear force.

The diameter of the nucleus is in the range of 1.75 fm (1.75×10^{-15} m) for hydrogen (the diameter of a single proton)[7] to about 15 fm for the heaviest atoms, such as uranium. These dimensions are much smaller than the diameter of the atom itself (nucleus + electron cloud), by a factor of about 23,000 (uranium) to about 145,000 (hydrogen).

The branch of physics concerned with the study and understanding of the atomic nucleus, including its composition and the forces which bind it together, is called nuclear physics.

3.1 Introduction

3.1.1 History

Main article: Rutherford model

The nucleus was discovered in 1911, as a result of Ernest Rutherford's efforts to test Thomson's "plum pudding model" of the atom.[8] The electron had already been discovered earlier by J.J. Thomson himself. Knowing that atoms are neutral, Thomson postulated that there must be a positive charge as well. In his plum pudding model, Thomson stated that an atom consisted of negative electrons randomly scattered within a sphere of positive charge. Ernest Rutherford later devised an experiment, performed by Hans Geiger and Ernest Marsden under Rutherford's direction, that involved the deflection of alpha particles directed at a thin sheet of metal foil. He reasoned that if Thomson's model were correct, the positively charged alpha nuclei would easily pass through the foil with very little deviation in their paths as the foil should act in a manner as to be neutrally charged if the negative and positive charges are so intimately mixed as to make it appear neutral. To his surprise, many of the particles were deflected at very large angles. Because the mass of alpha particles is about 8000 times that of an electron, it became apparent that a very strong force must be present if it could deflect the massive and fast moving helium nuclei. He realized that the plum pudding model could not be accurate and that the deflections of the alpha particles could only be explained if the positive and negatives charges were in fact separated from each other and that the mass of the atom was a concentrated point of positive charge and mass. Thus, the idea of a nuclear atom with a dense center of positive charge and mass became justified.

A model of the atomic nucleus showing it as a compact bundle of the two types of nucleons: protons (red) and neutrons (blue). In this diagram, protons and neutrons look like little balls stuck together, but an actual nucleus (as understood by modern nuclear physics) cannot be explained like this, but only by using quantum mechanics. In a nucleus which occupies a certain energy level (for example, the ground state), each nucleon can be said to occupy a range of locations.

3.1.2 Etymology

The term **nucleus** is from the Latin word *nucleus*, a diminutive of *nux* ("nut"), meaning the kernel (i.e., the "small nut") inside a watery type of fruit (like a peach). In 1844, Michael Faraday used the term to refer to the "central point of an atom". The modern atomic meaning was proposed by Ernest Rutherford in 1912.[9] The adoption of the term "nucleus" to atomic theory, however, was not immediate. In 1916, for example, Gilbert N. Lewis stated, in his famous article *The Atom and the Molecule*, that "the atom is composed of the *kernel* and an outer atom or *shell*"[10]

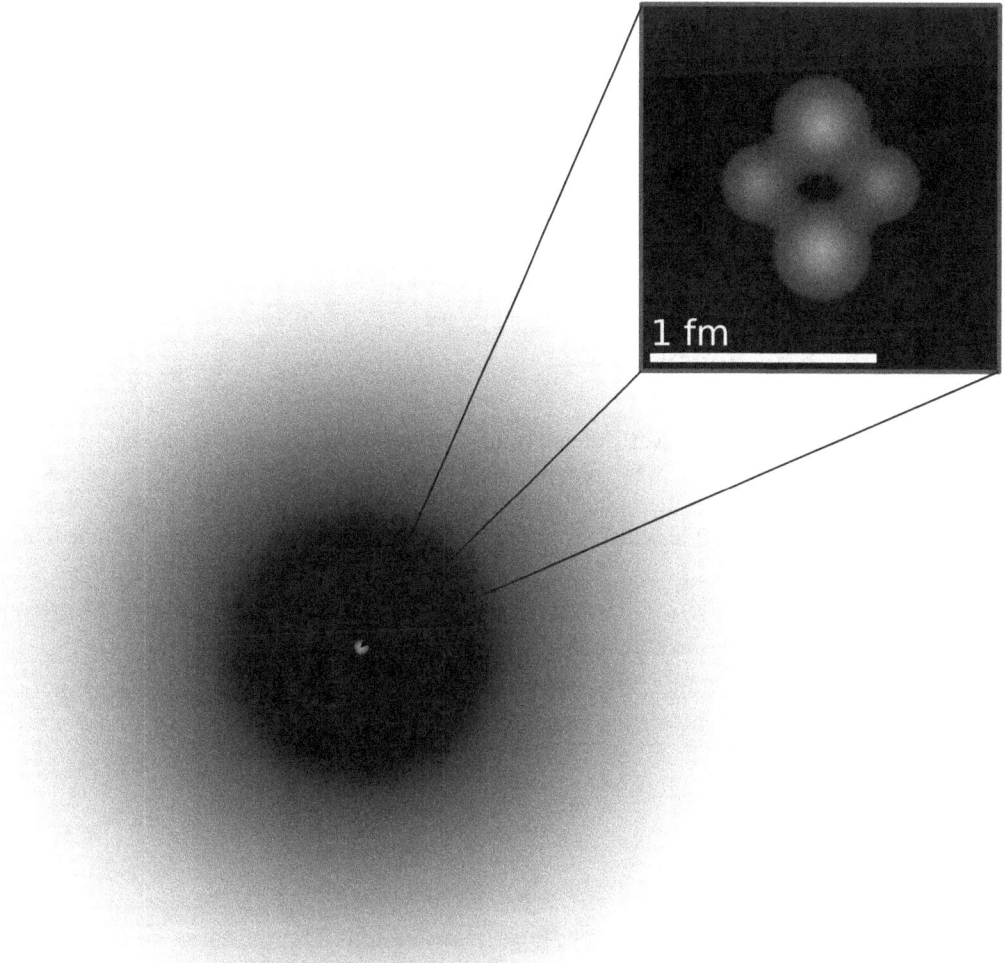

1 Å = 100,000 fm

*A figurative depiction of the helium−4 atom with the electron cloud in shades of gray. In the nucleus, the two protons and two neutrons are depicted in red and blue. This depiction shows the particles as separate, whereas in an actual helium atom, the protons are superimposed in space and most likely found at the very center of the nucleus, and the same is true of the two neutrons. Thus, all four particles are most likely found in exactly the same space, at the central point. Classical images of separate particles fail to model known charge distributions in very small nuclei. A more accurate image is that the spatial distribution of nucleons in helium's nucleus, although on a far smaller scale, is much closer to the helium **electron cloud** shown here, than to the fanciful nucleus image.*

3.1.3   Nuclear makeup

Fundamental particles, called quarks, are held in association by the nuclear strong force in certain stable combinations of hadrons called baryons, that manifest themselves as the neutrons and the protons of the nucleus. The nuclear strong force extends far enough from each baryon so as to bind the neutrons and protons together against the repulsive force of the positively charged protons. The nuclear strong force has a very short range and essentially drops to zero just beyond the edge of the nucleus. The collective action of the positively charged nucleus is to hold the electrically negative charged electrons in their orbits about the nucleus. The collection of negatively charged electrons orbiting the nucleus display an affinity for certain configurations and numbers of electrons that make their orbits stable. Which chemical element an atom represents is determined by the number of protons in the nucleus; the atom will have an equal number of electrons orbiting that nucleus. Individual chemical elements can create more stable electron configurations by combining to share their electrons. It is that sharing of electrons to create stable electronic orbits about the nucleus that appears to us as the chemistry of our macro world.

While protons define the entire charge of a nucleus and, hence, its chemical identity, neutrons are electrically neutral, but contribute to the mass of a nucleus to nearly the same extent as the protons. Neutrons explain the phenomenon of isotopes – varieties of the same chemical element which differ only in their atomic mass, not their chemical action.

3.2   Protons and neutrons

Protons and neutrons are fermions, with different values of the strong isospin quantum number, so two protons and two neutrons can share the same space wave function since they are not identical quantum entities. They sometimes are viewed as two different quantum states of the same particle, the *nucleon*.[11][12] Two fermions, such as two protons, or two neutrons, or a proton + neutron (the deuteron) can exhibit bosonic behavior when they become loosely bound in pairs.

In the rare case of a hypernucleus, a third baryon called a hyperon, with a different value of the strangeness quantum number can also share the wave function. However, the latter type of nuclei are extremely unstable and are not found on Earth except in high energy physics experiments.

The neutron has a positively charged core of radius ≈ 0.3 fm surrounded by a compensating negative charge of radius between 0.3 fm and 2 fm. The proton has an approximately exponentially decaying positive charge distribution with a mean square radius of about 0.8 fm.[13]

3.3   Forces

Nuclei are bound together by the residual strong force (nuclear force). The residual strong force is a minor residuum of the strong interaction which binds quarks together to form protons and neutrons. This force is much weaker *between* neutrons and protons because it is mostly neutralized within them, in the same way that electromagnetic forces *between* neutral atoms (such as van der Waals forces that act between two inert gas atoms) are much weaker than the electromagnetic forces that hold the parts of the atoms internally together (for example, the forces that hold the electrons in an inert gas atom bound to its nucleus).

The nuclear force is highly attractive at the distance of typical nucleon separation, and this overwhelms the repulsion between protons which is due to the electromagnetic force, thus allowing nuclei to exist. However, because the residual strong force has a limited range because it decays quickly with distance (see Yukawa potential), only nuclei smaller than a certain size can be completely stable. The largest known completely stable (e.g., stable to alpha, beta, and gamma decay) nucleus is lead-208 which contains a total of 208 nucleons (126 neutrons and 82 protons). Nuclei larger than this maximal size of 208 particles are unstable and (as a trend) become increasingly short-lived with larger size, as the number of neutrons and protons which compose them increases beyond this number. However, bismuth-209 is also stable to beta decay and has the longest half-life to alpha decay of any known isotope, estimated at a billion times longer than the age of the universe.

The residual strong force is effective over a very short range (usually only a few fermis; roughly one or two nucleon diameters) and causes an attraction between any pair of nucleons. For example, between protons and neutrons to form [NP] deuteron, and also between protons and protons, and neutrons and neutrons.

3.4   Halo nuclei and strong force range limits

The effective absolute limit of the range of the strong force is represented by halo nuclei such as lithium-11 or boron-14, in which dineutrons, or other collections of neutrons, orbit at distances of about ten fermis (roughly similar to the 8 fermi radius of the nucleus of uranium-238). These nuclei are not maximally dense. Halo nuclei form at the extreme edges of the chart of the nuclides—the neutron drip line and proton drip line—and are all unstable with short half-lives, measured in milliseconds; for example, lithium-11 has a half-life of 8.8 milliseconds.

Halos in effect represent an excited state with nucleons in an outer quantum shell which has unfilled energy levels "below" it (both in terms of radius and energy). The halo may be made of either neutrons [NN, NNN] or protons [PP, PPP]. Nuclei which have a single neutron halo include ^{11}Be and ^{19}C. A two-neutron halo is exhibited by ^{6}He, ^{11}Li, ^{17}B, ^{19}B and ^{22}C. Two-neutron halo nuclei break into three fragments, never two, and are called *Borromean nuclei* because of this behavior (referring to a system of three interlocked rings in which breaking any ring frees both of the others). ^{8}He and ^{14}Be both exhibit a four-neutron halo. Nuclei which have a proton halo include ^{8}B and ^{26}P. A two-proton halo is exhibited by ^{17}Ne and ^{27}S. Proton halos are expected to be more rare and unstable than the neutron examples, because of the repulsive electromagnetic forces of the excess proton(s).

3.5   Nuclear models

Although the standard model of physics is widely believed to completely describe the composition and behavior of the nucleus, generating predictions from theory is much more difficult than for most other areas of particle physics. This is due to two reasons:

- In principle, the physics within a nuclei can be derived entirely from quantum chromodynamics (QCD). In practice however, current computational and mathematical approaches for solving QCD in low-energy systems such as the nuclei are extremely limited. This is due to the phase transition that occurs between high-energy quark matter and low-energy hadronic matter, which renders perturbative techniques unusuable, making it difficult to construct an accurate QCD-derived model of the forces between nucleons. Current approaches are limited to either phenomenological models such as the Argonne v18 potential or chiral effective field theory.[14]

- Even if the nuclear force is well constrained, a significant amount of computational power is required to accurately compute the properties of nuclei *ab initio*. Developments in many-body theory have made this possible for many low mass and relatively stable nuclei, but further improvements in both computational power and mathematical approaches are required before heavy nuclei or highly unstable nuclei can be tackled.

Historically, experiments have been compared to relatively crude models that are necessarily imperfect. None of these models can completely explain experimental data on nuclear structure.[15]

The nuclear radius (R) is considered to be one of the basic quantities that any model must predict. For stable nuclei (not halo nuclei or other unstable distorted nuclei) the nuclear radius is roughly proportional to the cube root of the mass number (A) of the nucleus, and particularly in nuclei containing many nucleons, as they arrange in more spherical configurations:

The stable nucleus has approximately a constant density and therefore the nuclear radius R can be approximated by the following formula,

$$R = r_0 A^{1/3}$$

where A = Atomic mass number (the number of protons Z, plus the number of neutrons N) and $r_0 = 1.25$ fm $= 1.25 \times 10^{-15}$ m. In this equation, the constant r_0 varies by 0.2 fm, depending on the nucleus in question, but this is less than 20% change from a constant.[16]

In other words, packing protons and neutrons in the nucleus gives *approximately* the same total size result as packing hard spheres of a constant size (like marbles) into a tight spherical or almost spherical bag (some stable nuclei are not quite spherical, but are known to be prolate).

3.5.1 Liquid drop model

Main article: Semi-empirical mass formula

Early models of the nucleus viewed the nucleus as a rotating liquid drop. In this model, the trade-off of long-range electromagnetic forces and relatively short-range nuclear forces, together cause behavior which resembled surface tension forces in liquid drops of different sizes. This formula is successful at explaining many important phenomena of nuclei, such as their changing amounts of binding energy as their size and composition changes (see semi-empirical mass formula), but it does not explain the special stability which occurs when nuclei have special "magic numbers" of protons or neutrons.

The terms in the semi-empirical mass formula, which can be used to approximate the binding energy of many nuclei, are considered as the sum of five types of energies (see below). Then the picture of a nucleus as a drop of incompressible liquid roughly accounts for the observed variation of binding energy of the nucleus:

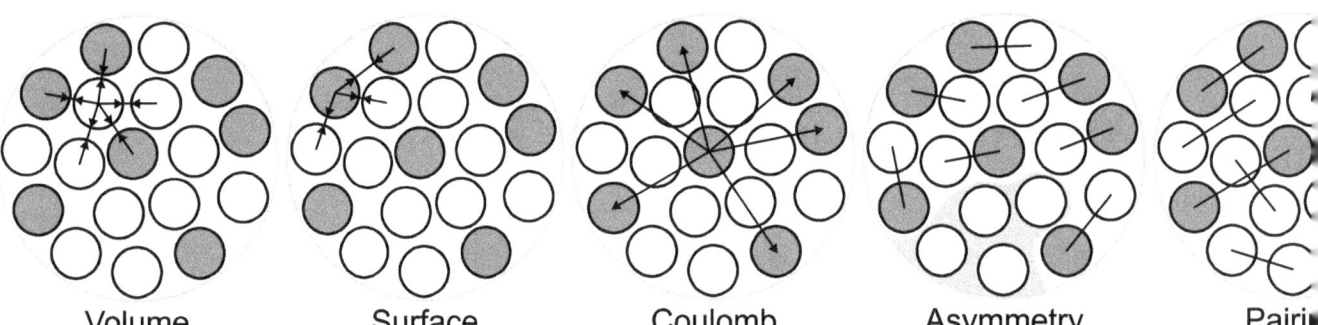

Volume   Surface   Coulomb   Asymmetry   Pairi

Volume energy. When an assembly of nucleons of the same size is packed together into the smallest volume, each interior nucleon has a certain number of other nucleons in contact with it. So, this nuclear energy is proportional to the volume.

Surface energy. A nucleon at the surface of a nucleus interacts with fewer other nucleons than one in the interior of the nucleus and hence its binding energy is less. This surface energy term takes that into account and is therefore negative and is proportional to the surface area.

Coulomb Energy. The electric repulsion between each pair of protons in a nucleus contributes toward decreasing its binding energy.

Asymmetry energy (also called Pauli Energy). An energy associated with the Pauli exclusion principle. Were it not for the Coulomb energy, the most stable form of nuclear matter would have the same number of neutrons as protons, since unequal numbers of neutrons and protons imply filling higher energy levels for one type of particle, while leaving lower energy levels vacant for the other type.

Pairing energy. An energy which is a correction term that arises from the tendency of proton pairs and neutron pairs to occur. An even number of particles is more stable than an odd number.

3.5.2 Shell models and other quantum models

Main article: Nuclear shell model

A number of models for the nucleus have also been proposed in which nucleons occupy orbitals, much like the atomic orbitals in atomic physics theory. These wave models imagine nucleons to be either sizeless point particles in potential wells, or else probability waves as in the "optical model", frictionlessly orbiting at high speed in potential wells.

In the above models, the nucleons may occupy orbitals in pairs, due to being fermions, which allows to explain even/odd Z and N effects well-known from experiments. The exact nature and capacity of nuclear shells differs from those of electrons in atomic orbitals, primarily because the potential well in which the nucleons move (especially in larger nuclei) is quite different from the central electromagnetic potential well which binds electrons in atoms. Some resemblance to atomic orbital models may be seen in a small atomic nucleus like that of helium-4, in which the two protons and two neutrons separately occupy 1s orbitals analogous to the 1s orbital for the two electrons in the helium atom, and achieve unusual stability for the same reason. Nuclei with 5 nucleons are all extremely unstable and short-lived, yet, helium-3, with 3 nucleons, is very stable even with lack of a closed 1s orbital shell. Another nucleus with 3 nucleons, the triton hydrogen-3 is unstable and will decay into helium-3 when isolated. Weak nuclear stability with 2 nucleons {NP} in the 1s orbital is found in the deuteron hydrogen-2, with only one nucleon in each of the proton and neutron potential wells. While each nucleon is a fermion, the {NP} deuteron is a boson and thus does not follow Pauli Exclusion for close packing within shells. Lithium-6 with 6 nucleons is highly stable without a closed second 1p shell orbital. For light nuclei with total nucleon numbers 1 to 6 only those with 5 do not show some evidence of stability. Observations of beta-stability of light nuclei outside closed shells indicate that nuclear stability is much more complex than simple closure of shell orbitals with magic numbers of protons and neutrons.

For larger nuclei, the shells occupied by nucleons begin to differ significantly from electron shells, but nevertheless, present nuclear theory does predict the magic numbers of filled nuclear shells for both protons and neutrons. The closure of the stable shells predicts unusually stable configurations, analogous to the noble group of nearly-inert gases in chemistry. An example is the stability of the closed shell of 50 protons, which allows tin to have 10 stable isotopes, more than any other element. Similarly, the distance from shell-closure explains the unusual instability of isotopes which have far from stable numbers of these particles, such as the radioactive elements 43 (technetium) and 61 (promethium), each of which is preceded and followed by 17 or more stable elements.

There are however problems with the shell model when an attempt is made to account for nuclear properties well away from closed shells. This has led to complex *post hoc* distortions of the shape of the potential well to fit experimental data, but the question remains whether these mathematical manipulations actually correspond to the spatial deformations in real nuclei. Problems with the shell model have led some to propose realistic two-body and three-body nuclear force effects involving nucleon clusters and then build the nucleus on this basis. Two such cluster models are the Close-Packed Spheron Model of Linus Pauling and the 2D Ising Model of MacGregor.[15]

3.5.3   Consistency between models

Main article: Nuclear structure

As with the case of superfluid liquid helium, atomic nuclei are an example of a state in which both (1) "ordinary" particle physical rules for volume and (2) non-intuitive quantum mechanical rules for a wave-like nature apply. In superfluid helium, the helium atoms have volume, and essentially "touch" each other, yet at the same time exhibit strange bulk properties, consistent with a Bose–Einstein condensation. The latter reveals that they also have a wave-like nature and do not exhibit standard fluid properties, such as friction. For nuclei made of hadrons which are fermions, the same type of condensation does not occur, yet nevertheless, many nuclear properties can only be explained similarly by a combination of properties of particles with volume, in addition to the frictionless motion characteristic of the wave-like behavior of objects trapped in Erwin Schrödinger's quantum orbitals.

3.6   See also

- Giant resonance

- List of particles

- Nuclear medicine

- Radioactivity

- Semi-empirical mass formula

3.7   References

[1] Iwanenko, D.D., The neutron hypothesis, Nature **129** (1932) 798.

[2] Heisenberg, W. (1932). "Über den Bau der Atomkerne. I". *Z. Phys.* **77**: 1–11. Bibcode:1932ZPhy...77....1H.doi:10.1007/33.

[3] Heisenberg, W. (1932). "Über den Bau der Atomkerne. II". *Z. Phys.* **78** (3–4): 156–164. Bibcode:1932ZPhy...78..156H. doi:10.1007/BF01337585.

[4] Heisenberg, W. (1933). "Über den Bau der Atomkerne. III". *Z. Phys.* **80** (9–10): 587–596. Bibcode:1933ZPhy...80..587H. doi:10.1007/BF01335696.

[5] Miller A. I. *Early Quantum Electrodynamics: A Sourcebook*, Cambridge University Press, Cambridge, 1995, ISBN 0521568919, pp. 84–88.

[6] Bernard Fernandez and Georges Ripka (2012). "Nuclear Theory After the Discovery of the Neutron". *Unravelling the Mystery of the Atomic Nucleus: A Sixty Year Journey 1896 — 1956*. Springer. p. 263. ISBN 9781461441809. Retrieved 15 February 2013.

[7] Geoff Brumfiel (July 7, 2010). "The proton shrinks in size". *Nature*. doi:10.1038/news.2010.337.

[8] *Rutgers University*. "The Rutherford Experiment". physics.rutgers.edu. Retrieved February 26, 2013.

[9] D. Harper. "Nucleus". *Online Etymology Dictionary*. Retrieved 2010-03-06.

[10] G.N. Lewis (1916). "The Atom and the Molecule". *Journal of the American Chemical Society* **38** (4): 4. doi:10.1021/ja02202.

[11] A.G. Sitenko, V.K. Tartakovskiĭ (1997). *Theory of Nucleus: Nuclear Structure and Nuclear Interaction*. Kluwer Academic. p. 3. ISBN 0-7923-4423-5.

[12] M.A. Srednicki (2007). *Quantum Field Theory*. Cambridge University Press. pp. 522–523. ISBN 978-0-521-86449-7.

[13] J.-L. Basdevant, J. Rich, M. Spiro (2005). *Fundamentals in Nuclear Physics*. Springer. p. 155. ISBN 0-387-01672-4.

[14] Machleidt, R.; Entem, D.R. (2011). "Chiral effective field theory and nuclear forces".*Physics Reports***503**(1): 1–75.arXiv:19v1. doi:10.1016/j.physrep.2011.02.001.

[15] N.D. Cook (2010). *Models of the Atomic Nucleus* (2nd ed.). Springer. p. 57 ff. ISBN 978-3-642-14736-4.

[16] K.S. Krane (1987). *Introductory Nuclear Physics*. Wiley-VCH. ISBN 0-471-80553-X.

3.8   External links

- The Nucleus – a chapter from an online textbook

- The LIVEChart of Nuclides – IAEA in Java or HTML

- Article on the "nuclear shell model," giving nuclear shell filling for the various elements. Accessed Sept. 16, 2009.

- Timeline: Subatomic Concepts, Nuclear Science & Technology.

Chapter 4

Subatomic particle

In the physical sciences, **subatomic particles** are particles much smaller than atoms.[1] There are two types of sub-atomic particles: elementary particles, which according to current theories are not made of other particles; and *composite* particles.[2] Particle physics and nuclear physics study these particles and how they interact.[3]

In particle physics, the concept of a particle is one of several concepts inherited from classical physics. But it also reflects the modern understanding that at the quantum scale matter and energy behave very differently from what much of everyday experience would lead us to expect.

The idea of a particle underwent serious rethinking when experiments showed that light could behave like a stream of particles (called photons) as well as exhibit wave-like properties. This led to the new concept of wave–particle duality to reflect that quantum-scale "particles" behave like both particles and waves (also known as wavicles). Another new concept, the uncertainty principle, states that some of their properties taken together, such as their simultaneous position and momentum, cannot be measured exactly.[4] In more recent times, wave–particle duality has been shown to apply not only to photons but to increasingly massive particles as well.[5]

Interactions of particles in the framework of quantum field theory are understood as creation and annihilation of *quanta* of corresponding fundamental interactions. This blends particle physics with field theory.

4.1 Classification

4.1.1 By statistics

Main article: Spin–statistics theorem
Any subatomic particle, like any particle in the 3-dimensional space that obeys laws of quantum mechanics, can be either a boson (an integer spin) or a fermion (a half-integer spin).

4.1.2 By composition

The elementary particles of the Standard Model include:[6]

- Six "flavors" of quarks: up, down, bottom, top, strange, and charm;

- Six types of leptons: electron, electron neutrino, muon, muon neutrino, tau, tau neutrino;

- Twelve gauge bosons (force carriers): the photon of electromagnetism, the three W and Z bosons of the weak force, and the eight gluons of the strong force;

- The Higgs boson.

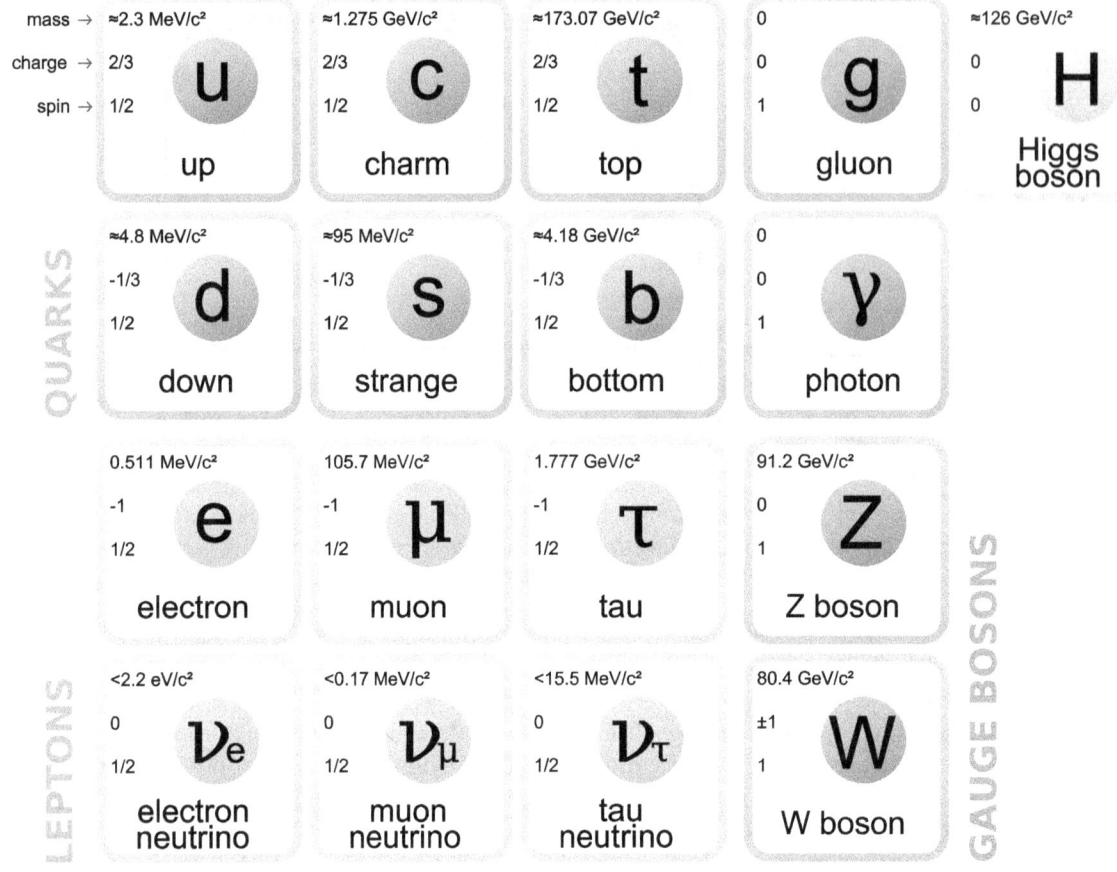

The Standard Model classification of particles

Various extensions of the Standard Model predict the existence of an elementary graviton particle and many other elementary particles.

Composite subatomic particles (such as protons or atomic nuclei) are bound states of two or more elementary particles. For example, a proton is made of two up quarks and one down quark, while the atomic nucleus of helium-4 is composed of two protons and two neutrons. Composite particles include all hadrons: these include baryons (such as protons and neutrons) and mesons (such as pions and kaons).

4.1.3 By mass

In special relativity, the energy of a particle at rest equals its mass times the speed of light squared ($E = mc^2$). That is, mass can be expressed in terms of energy and vice versa. If a particle has a frame of reference where it lies at rest, then it has a positive rest mass and is referred to as *massive*.

All composite particles are massive. Baryons (meaning "heavy") tend to have greater mass than mesons (meaning "intermediate"), which in turn tend to be heavier than leptons (meaning "lightweight"), but the heaviest lepton (the tau particle) is heavier than the two lightest flavours of baryons (nucleons). It is also certain that any particle with an electric charge is massive.

All massless particles (particles whose invariant mass is zero) are elementary. These include the photon and gluon, although the latter cannot be isolated.

The question of the masses of neutrinos is uncertain.

4.2   Other properties

Through the work of Albert Einstein, Louis de Broglie, and many others, current scientific theory holds that *all* particles also have a wave nature.[7] This has been verified not only for elementary particles but also for compound particles like atoms and even molecules. In fact, according to traditional formulations of non-relativistic quantum mechanics, wave–particle duality applies to all objects, even macroscopic ones; although the wave properties of macroscopic objects cannot be detected due to their small wavelengths.[8]

Interactions between particles have been scrutinized for many centuries, and a few simple laws underpin how particles behave in collisions and interactions. The most fundamental of these are the laws of conservation of energy and conservation of momentum, which let us make calculations of particle interactions on scales of magnitude that range from stars to quarks.[9] These are the prerequisite basics of Newtonian mechanics, a series of statements and equations in *Philosophiae Naturalis Principia Mathematica*, originally published in 1687.

4.3   Dividing an atom

The negatively charged electron has a mass equal to $1/1836$ of that of a hydrogen atom. The remainder of the hydrogen atom's mass comes from the positively charged proton. The atomic number of an element is the number of protons in its nucleus. Neutrons are neutral particles having a mass slightly greater than that of the proton. Different isotopes of the same element contain the same number of protons but differing numbers of neutrons. The mass number of an isotope is the total number of nucleons (neutrons and protons collectively).

Chemistry concerns itself with how electron sharing binds atoms into structures such as crystals and molecules. Nuclear physics deals with how protons and neutrons arrange themselves in nuclei. The study of subatomic particles, atoms and molecules, and their structure and interactions, requires quantum mechanics. Analyzing processes that change the numbers and types of particles requires quantum field theory. The study of subatomic particles *per se* is called particle physics. The term *high-energy physics* is nearly synonymous to "particle physics" since creation of particles requires high energies: it occurs only as a result of cosmic rays, or in particle accelerators. Particle phenomenology systematizes the knowledge about subatomic particles obtained from these experiments.

4.4   History

Main articles: History of subatomic physics and Timeline of particle discoveries

The term "*subatomic* particle" is largely a retronym of 1960s made to distinguish a big number of baryons and mesons (that comprise hadrons) from particles that are now thought to be truly elementary. Before that hadrons were usually classified as "elementary" because their composition was unknown.

A list of important discoveries follows:

4.5   See also

- *Atom: Journey Across the Subatomic Cosmos* (book)

- *Atom: An Odyssey from the Big Bang to Life on Earth...and Beyond* (book)

- CPT invariance

- Dark Matter

- Hot spot effect in subatomic physics

- List of fictional elements, materials, isotopes and atomic particles

- List of particles

- Poincaré symmetry

- Ylem

4.6 References

[1] "Subatomic particles". NTD. Retrieved 5 June 2012.

[2] Bolonkin, Alexander (2011).*Universe, Human Immortality and Future Human Evaluation*. Elsevier. p. 25.ISBN978012415816.

[3] Fritzsch, Harald (2005). *Elementary Particles*. World Scientific. pp. 11–20. ISBN 978-981-256-141-1.

[4] Heisenberg, W. (1927), "Über den anschaulichen Inhalt der quantentheoretischen Kinematik und Mechanik", *Zeitschrift für Physik* (in German) **43** (3–4): 172–198, Bibcode:1927ZPhy...43..172H, doi:10.1007/BF01397280.

[5] Arndt, Markus; Nairz, Olaf; Vos-Andreae, Julian; Keller, Claudia; Van Der Zouw, Gerbrand; Zeilinger, Anton (2000). "Wave-particle duality of C60 molecules". *Nature* **401** (6754): 680–682. Bibcode:1999Natur.401..680A. doi:10.1038/44348. PMID 18494170.

[6] Cottingham, W. N.; Greenwood, D. A. (2007). *An introduction to the standard model of particle physics*. Cambridge University Press. p. 1. ISBN 978-0-521-85249-4.

[7] Walter Greiner (2001). *Quantum Mechanics: An Introduction*. Springer. p. 29. ISBN 3-540-67458-6.

[8] R. Eisberg & R. Resnick (1985). *Quantum Physics of Atoms, Molecules, Solids, Nuclei, and Particles* (2nd ed.). John Wiley & Sons. pp. 59–60. ISBN 0-471-87373-X. For both large and small wavelengths, both matter and radiation have both particle and wave aspects. [...] But the wave aspects of their motion become more difficult to observe as their wavelengths become shorter. [...] For ordinary macroscopic particles the mass is so large that the momentum is always sufficiently large to make the de Broglie wavelength small enough to be beyond the range of experimental detection, and classical mechanics reigns supreme.

[9] Isaac Newton (1687). Newton's Laws of Motion (*Philosophiae Naturalis Principia Mathematica*)

[10] Klemperer, Otto (1959). *Electron Physics: The Physics of the Free Electron*. Academic Press.

[11] Some sources such as The Strange Quark indicate 1947.

[12] http://press.web.cern.ch/press-releases/2014/06/cern-experiments-report-new-higgs-boson-measurements

4.7 Further reading

General readers

- Feynman, R.P. & Weinberg, S. (1987). *Elementary Particles and the Laws of Physics: The 1986 Dirac Memorial Lectures*. Cambridge Univ. Press.

- Brian Greene (1999). *The Elegant Universe*. W.W. Norton & Company. ISBN 0-393-05858-1.

- Oerter, Robert (2006). *The Theory of Almost Everything: The Standard Model, the Unsung Triumph of Modern Physics*. Plume.

- Schumm, Bruce A. (2004). *Deep Down Things: The Breathtaking Beauty of Particle Physics*. Johns Hopkins University Press. ISBN 0-8018-7971-X.

- Martinus Veltman (2003). *Facts and Mysteries in Elementary Particle Physics*. World Scientific. ISBN 981-238-149-X.

Textbooks

- Coughlan, G. D., J. E. Dodd, and B. M. Gripaios (2006). *The Ideas of Particle Physics: An Introduction for Scientists*, 3rd ed. Cambridge Univ. Press. An undergraduate text for those not majoring in physics.

- Griffiths, David J. (1987). *Introduction to Elementary Particles*. Wiley, John & Sons, Inc. ISBN 0-471-60386-4.

- Kane, Gordon L. (1987). *Modern Elementary Particle Physics*. Perseus Books. ISBN 0-201-11749-5.

4.8   External links

- particleadventure.org: The Standard Model.

- cpepweb.org: Particle chart.

- University of California: Particle Data Group.

- Annotated Physics Encyclopædia: Quantum Field Theory.

- Jose Galvez: Chapter 1 Electrodynamics (pdf).

Chapter 5

Neutron

This article is about the subatomic particle. For other uses, see Neutron (disambiguation).

The **neutron** is a subatomic particle, symbol n or n0, with no net electric charge and a mass slightly larger than that of a proton. Protons and neutrons, each with mass approximately one atomic mass unit, constitute the nucleus of an atom, and they are collectively referred to as nucleons.[4] Their properties and interactions are described by nuclear physics.

The nucleus consists of Z protons, where Z is called the atomic number, and N neutrons, where N is the neutron number. The atomic number defines the chemical properties of the atom, and the neutron number determines the isotope or nuclide.[5] The terms isotope and nuclide are often used synonymously, but they refer to chemical and nuclear properties, respectively. The atomic mass number, symbol A, equals Z+N. For example, carbon has atomic number 6, and its abundant carbon-12 isotope has 6 neutrons, whereas its rare carbon-13 isotope has 7 neutrons. Some elements occur in nature with only one stable isotope, such as fluorine (see stable nuclide). Other elements occur as many stable isotopes, such as tin with ten stable isotopes. Even though it is not a chemical element, the neutron is included in the table of nuclides.[6]

Within the nucleus, protons and neutrons are bound together through the nuclear force, and neutrons are required for the stability of nuclei. Neutrons are produced copiously in nuclear fission and fusion. They are a primary contributor to the nucleosynthesis of chemical elements within stars through fission, fusion, and neutron capture processes.

The neutron is essential to the production of nuclear power. In the decade after the neutron was discovered in 1932,[7] neutrons were used to effect many different types of nuclear transmutations. With the discovery of nuclear fission in 1938,[8] it was quickly realized that, if a fission event produced neutrons, each of these neutrons might cause further fission events, etc., in a cascade known as a nuclear chain reaction.[5] These events and findings led to the first self-sustaining nuclear reactor (Chicago Pile-1, 1942) and the first nuclear weapon (Trinity, 1945).

Free neutrons, or individual neutrons free of the nucleus, are effectively a form of ionizing radiation, and as such, are a biological hazard, depending upon dose.[5] A small natural "neutron background" flux of free neutrons exists on Earth, caused by cosmic ray muons, and by the natural radioactivity of spontaneously fissionable elements in the Earth's crust.[9] Dedicated neutron sources like neutron generators, research reactors and spallation sources produce free neutrons for use in irradiation and in neutron scattering experiments.

5.1 Description

Neutrons and protons are both nucleons, which are attracted and bound together by the nuclear force to form atomic nuclei. The nucleus of the most common isotope of the hydrogen atom (with the chemical symbol "H") is a lone proton. The nuclei of the heavy hydrogen isotopes deuterium and tritium contain one proton bound to one and two neutrons, respectively. All other types of atomic nuclei are composed of two or more protons and various numbers of neutrons. The most common nuclide of the common chemical element lead, ^{208}Pb has 82 protons and 126 neutrons, for example.

The free neutron has a mass of about 1.675×10^{-27} kg (equivalent to 939.6 MeV/c^2, or 1.0087 u).[3] The neutron has a mean square radius of about 0.8×10^{-15} m, or 0.8 fm,[10] and it is a spin-½ fermion.[11] The neutron has a magnetic moment with a negative value, because its orientation is opposite to the neutron's spin.[12] The neutron's magnetic moment causes its motion to be influenced by magnetic fields. Although the neutron has no net electric charge, it does have a slight distribution of charge within it. With its positive electric charge, the proton is directly influenced by electric fields, whereas the response of the neutron to this force is much weaker.

A free neutron is unstable, decaying to a proton, electron and antineutrino with a mean lifetime of just under 15 minutes (881.5±1.5 s). This radioactive decay, known as beta decay,[13] is possible since the mass of the neutron is slightly greater than the proton. The free proton is stable. Neutrons or protons bound in a nucleus can be stable or unstable, however, depending on the nuclide. Beta decay, in which neutrons decay to protons, or vice versa, is governed by the weak force, and it requires the emission or absorption of electrons and neutrinos, or their antiparticles.

Protons and neutrons behave almost identically under the influence of the nuclear force within the nucleus. The concept of isospin, in which the proton and neutron are viewed as two quantum states of the same particle, is used to model the interactions of nucleons by the nuclear or weak forces. Because of the strength of the nuclear force at short distances, the binding energy of nucleons is more than seven orders of magnitude larger than the electromagnetic energy binding electrons in atoms. Nuclear reactions (such as nuclear fission) therefore have an energy density that is more than ten million times that of chemical reactions. Because of the mass–energy equivalence, nuclear binding energies add or subtract from the mass of nuclei. Ultimately, the ability of the nuclear force to store energy arising from the electromagnetic repulsion of nuclear components is the basis for most of the energy that makes nuclear reactors or bombs possible. In nuclear fission, the absorption of a neutron by a heavy nuclide (e.g., uranium-235) causes the nuclide to become unstable and break into light nuclides and additional neutrons. The positively charged light nuclides then repel, releasing electromagnetic potential energy.

The neutron is classified as a hadron, since it is composed of quarks, and as a baryon, since it is composed of three quarks.[14] The finite size of the neutron and its magnetic moment indicate the neutron is a composite, rather than elementary, particle. The neutron consists of two down quarks with charge $-\frac{1}{3}\,e$ and one up quark with charge $+\frac{2}{3}\,e$, although this simple model belies the complexities of the Standard Model for nuclei.[15] The masses of the three quarks sum to only about 12 MeV/c^2, whereas the neutron's mass is about 940 MeV/c^2, for example.[15] Like the proton, the quarks of the neutron are held together by the strong force, mediated by gluons.[16] The nuclear force results from secondary effects of the more fundamental strong force.

5.2 Discovery

Main article: Discovery of the neutron

The story of the discovery of the neutron and its properties is central to the extraordinary developments in atomic physics that occurred in the first half of the 20th century, leading ultimately to the atomic bomb in 1945. In the 1911 Rutherford model, the atom consisted of a small positively charged massive nucleus surrounded by a much larger cloud of negatively charged electrons. In 1920 Rutherford suggested the nucleus consisted of positive protons and neutrally-charged particles, suggested to be a proton and an electron bound in some way.[17] Electrons were assumed to reside within the nucleus because it was known that beta radiation consisted of electrons emitted from the nucleus.[17] Rutherford called these uncharged particles *neutrons*, by the Latin root for *neutralis* (neuter) and the Greek suffix *-on* (a suffix used in the names of subatomic particles, i.e. *electron* and *proton*).[18][19] References to the word *neutron* in connection with the atom can be found in the literature as early as 1899, however.[20]

Throughout the 1920s, physicists assumed that the atomic nucleus was composed of protons and "nuclear electrons"[21][22] but there were obvious problems. It was difficult to reconcile the proton–electron model for nuclei with the Heisenberg uncertainty relation of quantum mechanics.[23][24] The Klein paradox,[25] discovered by Oskar Klein in 1928, presented further quantum mechanical objections to the notion of an electron confined within a nucleus.[23] Observed properties of atoms and molecules were inconsistent with the nuclear spin expected from proton–electron hypothesis. Since both protons and electrons carry an intrinsic spin of ½ $\hbar$, there is no way to arrange an odd number of spins ±½ $\hbar$ to give a spin integer multiple of $\hbar$. Nuclei with integer spin are common, e.g., ^{14}N.

In 1931, Walther Bothe and Herbert Becker found that if alpha particle radiation from polonium fell on beryllium, boron, or lithium, an unusually penetrating radiation was produced. The radiation was not influenced by an electric field, so Bothe and Becker assumed it was gamma radiation.[26][27] The following year Irène Joliot-Curie and Frédéric Joliot in Paris showed that if this "gamma" radiation fell on paraffin, or any other hydrogen-containing compound, it ejected protons of very high energy.[28] Neither Rutherford nor James Chadwick at the Cavendish Laboratory in Cambridge were convinced by the gamma ray interpretation.[21] Chadwick quickly performed a series of experiments that showed that the new radiation consisted of uncharged particles with about the same mass as the proton.[7][29][30] These particles were neutrons. Chadwick won the Nobel Prize in Physics for this discovery in 1935.[2]

Models for atomic nucleus consisting of protons and neutrons were quickly developed by Werner Heisenberg[31][32][33] and others.[34][35] The proton–neutron model explained the puzzle of nuclear spins. The origins of beta radiation were explained by Enrico Fermi in 1934 by the process of beta decay, in which the neutron decays to a proton by *creating* an electron and a (as yet undiscovered) neutrino.[36] In 1935 Chadwick and his doctoral student Maurice Goldhaber, reported the first accurate measurement of the mass of the neutron.[37][38]

By 1934, Fermi had bombarded heavier elements with neutrons to induce radioactivity in elements of high atomic number. In 1938, Fermi received the Nobel Prize in Physics *"for his demonstrations of the existence of new radioactive elements produced by neutron irradiation, and for his related discovery of nuclear reactions brought about by slow neutrons"*.[39] In 1938 Otto Hahn, Lise Meitner, and Fritz Strassmann discovered nuclear fission, or the fractionation of uranium nuclei into light elements, induced by neutron bombardment.[40][41][42] In 1945 Hahn received the 1944 Nobel Prize in Chemistry *"for his discovery of the fission of heavy atomic nuclei."* [43][44][45] The discovery of nuclear fission would lead to the development of nuclear power and the atomic bomb by the end of World War II.

5.3 Beta decay and the stability of the nucleus

Under the Standard Model of particle physics, the only possible decay mode for the neutron that conserves baryon number is for one of the neutron's quarks to change flavour via the weak interaction. The decay of one of the neutron's down quarks into a lighter up quark can be achieved by the emission of a W boson. By this process, the Standard Model description of beta decay, the neutron decays into a proton (which contains one down and two up quarks), an electron, and an electron antineutrino.

Since interacting protons have a mutual electromagnetic repulsion that is stronger than their attractive nuclear interaction, neutrons are a necessary constituent of any atomic nucleus that contains more than one proton (see diproton and neutron–proton ratio).[46] Neutrons bind with protons and one another in the nucleus via the nuclear force, effectively moderating the repulsive forces between the protons and stabilizing the nucleus.

See also: Beta-decay stable isobars and Neutron emission

5.3.1 Free neutron decay

Outside the nucleus, free neutrons are unstable and have a mean lifetime of 881.5±1.5 s (about 14 minutes, 42 seconds); therefore the half-life for this process (which differs from the mean lifetime by a factor of ln(2) = 0.693) is 611.0±1.0 s (about 10 minutes, 11 seconds).[13] Beta decay of the neutron, described above, can be denoted by the radioactive decay:[47]

n0 → p+ + e− + ν
e

where p+, e−, and ν
e denote the proton, electron and electron antineutrino, respectively. For the free neutron the decay energy for this process (based on the masses of the neutron, proton, and electron) is 0.782343 MeV. The maximal energy of the beta decay electron (in the process wherein the neutrino receives a vanishingly small amount of kinetic energy) has been measured at 0.782 ± .013 MeV.[48] The latter number is not well-enough measured to determine the comparatively tiny rest mass

of the neutrino (which must in theory be subtracted from the maximal electron kinetic energy) as well as neutrino mass is constrained by many other methods.

A small fraction (about one in 1000) of free neutrons decay with the same products, but add an extra particle in the form of an emitted gamma ray:

n0 → p+ + e− + ν
e + γ

This gamma ray may be thought of as a sort of "internal bremsstrahlung" that arises as the emitted beta particle interacts with the charge of the proton in an electromagnetic way. Internal bremsstrahlung gamma ray production is also a minor feature of beta decays of bound neutrons (as discussed below).

A very small minority of neutron decays (about four per million) are so-called "two-body (neutron) decays", in which a proton, electron and antineutrino are produced as usual, but the electron fails to gain the 13.6 eV necessary energy to escape the proton, and therefore simply remains bound to it, as a neutral hydrogen atom (one of the "two bodies"). In this type of free neutron decay, in essence all of the neutron decay energy is carried off by the antineutrino (the other "body").

The transformation of a free proton to a neutron (plus a positron and a neutrino) is energetically impossible, since a free neutron has a greater mass than a free proton.

5.3.2   Bound neutron decay

Main article: Atomic nucleus

While a free neutron has a half life of about 10.2 min, most neutrons within nuclei are stable. According to the nuclear shell model, the protons and neutrons of a nuclide are a quantum mechanical system organized into discrete energy levels with unique quantum numbers. For a neutron to decay, the resulting proton requires an available state at lower energy than the initial neutron state. In stable nuclei the possible lower energy states are all filled, meaning they are each occupied by two protons with spin up and spin down. The Pauli exclusion principle therefore disallows the decay of a neutron to a proton within stable nuclei. The situation is similar to electrons of an atom, where electrons have distinct atomic orbitals and are prevented from decaying to lower energy states, with the emission of a photon, by the exclusion principle.

Neutrons in unstable nuclei can decay by beta decay as described above. In this case, an energetically allowed quantum state is available for the proton resulting from the decay. One example of this decay is carbon-14 (6 protons, 8 neutrons) that decays to nitrogen-14 (7 protons, 7 neutrons) with a half-life of about 5,730 years.

Inside a nucleus, a proton can transform into a neutron via inverse beta decay, if an energetically allowed quantum state is available for the neutron. This transformation occurs by emission of an antielectron (also called positron) and an electron neutrino:

p+ → n0 + e+ + ν
e

The transformation of a proton to a neutron inside of a nucleus is also possible through electron capture:

p+ + e− → n0 + ν
e

Positron capture by neutrons in nuclei that contain an excess of neutrons is also possible, but is hindered because positrons are repelled by the positive nucleus, and quickly annihilate when they encounter electrons.

5.3.3   Competition of beta decay types

Three types of beta decay in competition are illustrated by the single isotope copper-64 (29 protons, 35 neutrons), which has a half-life of about 12.7 hours. This isotope has one unpaired proton and one unpaired neutron, so either the proton or the neutron can decay. This particular nuclide (though not all nuclides in this situation) is almost equally likely to decay through proton decay by positron emission (18%) or electron capture (43%), as through neutron decay by electron emission (39%).

5.4 Intrinsic properties

5.4.1 Electric charge

The total electric charge of the neutron is $0\ e$. This zero value has been tested experimentally, and the present experimental limit for the charge of the neutron is $-2(8) \times 10^{-22}\ e$,[49] or $-3(13) \times 10^{-41}$ C. This value is consistent with zero, given the experimental uncertainties (indicated in parentheses). By comparison, the charge of the proton is, of course, $+1\ e$.

5.4.2 Electric dipole moment

Main article: Neutron electric dipole moment

The Standard Model of particle physics predicts a tiny separation of positive and negative charge within the neutron leading to a permanent electric dipole moment.[50] The predicted value is, however, well below the current sensitivity of experiments. From several unsolved puzzles in particle physics, it is clear that the Standard Model is not the final and full description of all particles and their interactions. New theories going beyond the Standard Model generally lead to much larger predictions for the electric dipole moment of the neutron. Currently, there are at least four experiments trying to measure for the first time a finite neutron electric dipole moment, including:

- Cryogenic neutron EDM experiment being set up at the Institut Laue–Langevin[51]

- nEDM experiment under construction at the new UCN source at the Paul Scherrer Institute[52]

- nEDM experiment being envisaged at the Spallation Neutron Source[53]

- nEDM experiment being built at the Institut Laue–Langevin[54]

5.4.3 Magnetic moment

Main article: Neutron magnetic moment

Even though the neutron is a neutral particle, the magnetic moment of a neutron is not zero. Since the neutron is a neutral particle, it is not affected by electric fields, but with its magnetic moment it is affected by magnetic fields. The magnetic moment of the neutron is an indication of its quark substructure and internal charge distribution.[55] The value for the neutron's magnetic moment was first directly measured by Luis Alvarez and Felix Bloch at Berkeley, California in 1940,[56] using an extension of the magnetic resonance methods developed by Rabi. Alvarez and Bloch determined the magnetic moment of the neutron to be $\mu_n = -1.93(2)\ \mu N$, where μN is the nuclear magneton.

5.4.4 Structure and geometry of charge distribution

An article published in 2007 featuring a model-independent analysis concluded that the neutron has a negatively charged exterior, a positively charged middle, and a negative core.[57] In a simplified classical view, the negative "skin" of the neutron assists it to be attracted to the protons with which it interacts in the nucleus. (However, the main attraction between neutrons and protons is via the nuclear force, which does not involve charge.)

The simplified classical view of the neutron's charge distribution also "explains" the fact that the neutron magnetic dipole points in the opposite direction from its spin angular momentum vector (as compared to the proton). This gives the neutron, in effect, a magnetic moment which resembles a negatively charged particle. This can be reconciled classically with a neutral neutron composed of a charge distribution in which the negative sub-parts of the neutron have a larger average radius of distribution, and therefore contribute more to the particle's magnetic dipole moment, than do the positive parts that are, on average, nearer the core.

5.4.5 Mass

The mass of a neutron cannot be directly determined by mass spectrometry due to lack of electric charge. However, since the mass of protons and deuterons can be measured by mass spectrometry, the mass of a neutron can be deduced by subtracting proton mass from deuteron mass, with the difference being the mass of the neutron plus the binding energy of deuterium (expressed as a positive emitted energy). The latter can be directly measured by measuring the energy (B_d) of the single 0.7822 MeV gamma photon emitted when neutrons are captured by protons (this is exothermic and happens with zero-energy neutrons), plus the small recoil kinetic energy (E_{rd}) of the deuteron (about 0.06% of the total energy).

$$m_n = m_d - m_p + B_d - E_{rd}$$

The energy of the gamma ray can be measured to high precision by X-ray diffraction techniques, as was first done by Bell and Elliot in 1948. The best modern (1986) values for neutron mass by this technique are provided by Greene, et al.[58] These give a neutron mass of:

$$m_{\text{neutron}} = 1.008644904(14) \text{ u}$$

The value for the neutron mass in MeV is less accurately known, due to less accuracy in the known conversion of u to MeV:[59]

$$m_{\text{neutron}} = 939.56563(28) \text{ MeV}/c^2.$$

Another method to determine the mass of a neutron starts from the beta decay of the neutron, when the momenta of the resulting proton and electron are measured.

5.4.6 Anti-neutron

Main article: Antineutron

The antineutron is the antiparticle of the neutron. It was discovered by Bruce Cork in the year 1956, a year after the antiproton was discovered. CPT-symmetry puts strong constraints on the relative properties of particles and antiparticles, so studying antineutrons yields provide stringent tests on CPT-symmetry. The fractional difference in the masses of the neutron and antineutron is $(9\pm6)\times10^{-5}$. Since the difference is only about two standard deviations away from zero, this does not give any convincing evidence of CPT-violation.[13]

5.5 Neutron compounds

5.5.1 Dineutrons and tetraneutrons

Main articles: Dineutron and Tetraneutron

The existence of stable clusters of 4 neutrons, or tetraneutrons, has been hypothesised by a team led by Francisco-Miguel Marqués at the CNRS Laboratory for Nuclear Physics based on observations of the disintegration of beryllium−14 nuclei. This is particularly interesting because current theory suggests that these clusters should not be stable.

The dineutron is another hypothetical particle. In 2012, Artemis Spyrou from Michigan State University and coworkers reported that they observed, for the first time, the dineutron emission in the decay of ^{16}Be. The dineutron character is evidenced by a small emission angle between the two neutrons. The authors measured the two-neutron separation energy to be 1.35(10) MeV, in good agreement with shell model calculations, using standard interactions for this mass region.[60]

5.5.2 Neutronium and neutron stars

Main articles: Neutronium and Neutron star

At extremely high pressures and temperatures, nucleons and electrons are believed to collapse into bulk neutronic matter, called neutronium. This is presumed to happen in neutron stars.

The extreme pressure inside a neutron star may deform the neutrons into a cubic symmetry, allowing tighter packing of neutrons.[61]

5.6 Detection

Main article: Neutron detection

The common means of detecting a charged particle by looking for a track of ionization (such as in a cloud chamber) does not work for neutrons directly. Neutrons that elastically scatter off atoms can create an ionization track that is detectable, but the experiments are not as simple to carry out; other means for detecting neutrons, consisting of allowing them to interact with atomic nuclei, are more commonly used. The commonly used methods to detect neutrons can therefore be categorized according to the nuclear processes relied upon, mainly neutron capture or elastic scattering. A good discussion on neutron detection is found in chapter 14 of the book *Radiation Detection and Measurement* by Glenn F. Knoll (John Wiley & Sons, 1979).

5.6.1 Neutron detection by neutron capture

A common method for detecting neutrons involves converting the energy released from neutron capture reactions into electrical signals. Certain nuclides have a high neutron capture cross section, which is the probability of absorbing a neutron. Upon neutron capture, the compound nucleus emits more easily detectable radiation, for example an alpha particle, which is then detected. The nuclides 3He, 6Li, 10B, 233U, 235U, 237Np and 239Pu are useful for this purpose.

5.6.2 Neutron detection by elastic scattering

Neutrons can elastically scatter off nuclei, causing the struck nucleus to recoil. Kinematically, a neutron can transfer more energy to light nuclei such as hydrogen or helium than to heavier nuclei. Detectors relying on elastic scattering are called fast neutron detectors. Recoiling nuclei can ionize and excite further atoms through collisions. Charge and/or scintillation light produced in this way can be collected to produce a detected signal. A major challenge in fast neutron detection is discerning such signals from erroneous signals produced by gamma radiation in the same detector.

Fast neutron detectors have the advantage of not requiring a moderator, and therefore being capable of measuring the neutron's energy, time of arrival, and in certain cases direction of incidence.

5.7 Sources and production

Main articles: Neutron source, neutron generator and research reactor

Free neutrons are unstable, although they have the longest half-life of any unstable sub-atomic particle by several orders of magnitude. Their half-life is still only about 10 minutes, however, so they can be obtained only from sources that produce them freshly.

Natural neutron background. A small natural background flux of free neutrons exists everywhere on Earth. In the atmosphere and deep into the ocean, the "neutron background" is caused by muons produced by cosmic ray interaction with the atmosphere. These high energy muons are capable of penetration to considerable depths in water and soil. There, in striking atomic nuclei, among other reactions they induce spallation reactions in which a neutron is liberated from the nucleus. Within the Earth's crust a second source is neutrons produced primarily by spontaneous fission of uranium and thorium present in crustal minerals. The neutron background is not strong enough to be a biological hazard, but it is of importance to very high resolution particle detectors that are looking for very rare events, such as (hypothesized) interactions that might be caused by particles of dark matter.[9] Recent research has shown that even thunderstorms can produce neutrons with energies of up to several tens of MeV.[62]

Even stronger neutron background radiation is produced at the surface of Mars, where the atmosphere is thick enough to generate neutrons from cosmic ray muon production and neutron-spallation, but not thick enough to provide significant protection from the neutrons produced. These neutrons not only produce a Martian surface neutron radiation hazard from direct downward-going neutron radiation but may also produce a significant hazard from reflection of neutrons from the Martian surface, which will produce reflected neutron radiation penetrating upward into a Martian craft or habitat from the floor.[63]

Sources of neutrons for research. These include certain types of radioactive decay (spontaneous fission and neutron emission), and from certain nuclear reactions. Convenient nuclear reactions include tabletop reactions such as natural alpha and gamma bombardment of certain nuclides, often beryllium or deuterium, and induced nuclear fission, such as occurs in nuclear reactors. In addition, high-energy nuclear reactions (such as occur in cosmic radiation showers or accelerator collisions) also produce neutrons from disintigration of target nuclei. Small (tabletop) particle accelerators optimized to produce free neutrons in this way, are called neutron generators.

In practice, the most commonly used small laboratory sources of neutrons use radioactive decay to power neutron production. One noted neutron-producing radioisotope, californium−252 decays (half-life 2.65 years) by spontaneous fission 3% of the time with production of 3.7 neutrons per fission, and is used alone as a neutron source from this process. Nuclear reaction sources (that involve two materials) powered by radioisotopes use an alpha decay source plus a beryllium target, or else a source of high-energy gamma radiation from a source that undergoes beta decay followed by gamma decay, which produces photoneutrons on interaction of the high energy gamma ray with ordinary stable beryllium, or else with the deuterium in heavy water. A popular source of the latter type is radioactive antimony-124 plus beryllium, a system with a half-life of 60.9 days, which can be constructed from natural antimony (which is 42.8% stable antimony-123) by activating it with neutrons in a nuclear reactor, then transported to where the neutron source is needed.[64]

Nuclear fission reactors naturally produce free neutrons; their role is to sustain the energy-producing chain reaction. The intense neutron radiation can also be used to produce various radioisotopes through the process of neutron activation, which is a type of neutron capture.

Experimental nuclear fusion reactors produce free neutrons as a waste product. However, it is these neutrons that possess most of the energy, and converting that energy to a useful form has proved a difficult engineering challenge. Fusion reactors that generate neutrons are likely to create radioactive waste, but the waste is composed of neutron-activated lighter isotopes, which have relatively short (50–100 years) decay periods as compared to typical half-lives of 10,000 years for fission waste, which is long due primarily to the long half-life of alpha-emitting transuranic actinides.[65]

5.7.1   Neutron beams and modification of beams after production

Free neutron beams are obtained from neutron sources by neutron transport. For access to intense neutron sources, researchers must go to a specialist neutron facility that operates a research reactor or a spallation source.

The neutron's lack of total electric charge makes it difficult to steer or accelerate them. Charged particles can be accelerated, decelerated, or deflected by electric or magnetic fields. These methods have little effect on neutrons. However, some effects may be attained by use of inhomogeneous magnetic fields because of the neutron's magnetic moment. Neutrons can be controlled by methods that include moderation, reflection, and velocity selection. Thermal neutrons can be polarized by transmission through magnetic materials in a method analogous to the Faraday effect for photons. Cold neutrons of wavelengths of 6–7 angstroms can be produced in beams of a high degree of polarization, by use of magnetic mirrors and magnetized interference filters.[66]

5.8 Applications

The neutron plays an important role in many nuclear reactions. For example, neutron capture often results in neutron activation, inducing radioactivity. In particular, knowledge of neutrons and their behavior has been important in the development of nuclear reactors and nuclear weapons. The fissioning of elements like uranium-235 and plutonium-239 is caused by their absorption of neutrons.

Cold, *thermal* and *hot* neutron radiation is commonly employed in neutron scattering facilities, where the radiation is used in a similar way one uses X-rays for the analysis of condensed matter. Neutrons are complementary to the latter in terms of atomic contrasts by different scattering cross sections; sensitivity to magnetism; energy range for inelastic neutron spectroscopy; and deep penetration into matter.

The development of "neutron lenses" based on total internal reflection within hollow glass capillary tubes or by reflection from dimpled aluminum plates has driven ongoing research into neutron microscopy and neutron/gamma ray tomography.[67][68][69]

A major use of neutrons is to excite delayed and prompt gamma rays from elements in materials. This forms the basis of neutron activation analysis (NAA) and prompt gamma neutron activation analysis (PGNAA). NAA is most often used to analyze small samples of materials in a nuclear reactor whilst PGNAA is most often used to analyze subterranean rocks around bore holes and industrial bulk materials on conveyor belts.

Another use of neutron emitters is the detection of light nuclei, in particular the hydrogen found in water molecules. When a fast neutron collides with a light nucleus, it loses a large fraction of its energy. By measuring the rate at which slow neutrons return to the probe after reflecting off of hydrogen nuclei, a neutron probe may determine the water content in soil.

5.9 Medical therapies

Main articles: Fast neutron therapy and Neutron capture therapy of cancer

Because neutron radiation is both penetrating and ionizing, it can be exploited for medical treatments. Neutron radiation can have the unfortunate side-effect of leaving the affected area radioactive, however. Neutron tomography is therefore not a viable medical application.

Fast neutron therapy utilizes high energy neutrons typically greater than 20 MeV to treat cancer. Radiation therapy of cancers is based upon the biological response of cells to ionizing radiation. If radiation is delivered in small sessions to damage cancerous areas, normal tissue will have time to repair itself, while tumor cells often cannot.[70] Neutron radiation can deliver energy to a cancerous region at a rate an order of magnitude larger than gamma radiation[71]

Beams of low energy neutrons are used in boron capture therapy to treat cancer. In boron capture therapy, the patient is given a drug that contains boron and that preferentially accumulates in the tumor to be targeted. The tumor is then bombarded with very low energy neutrons (although often higher than thermal energy) which are captured by the boron-10 isotope in the boron, which produces an excited state of boron-11 that then decays to produce lithium-7 and an alpha particle that have sufficient energy to kill the malignant cell, but insufficient range to damage nearby cells. For such a therapy to be applied to the treatment of cancer, a neutron source having an intensity of the order of billion (10^9) neutrons per second per cm^2 is preferred. Such fluxes require a research nuclear reactor.

5.10 Protection

Exposure to free neutrons can be hazardous, since the interaction of neutrons with molecules in the body can cause disruption to molecules and atoms, and can also cause reactions that give rise to other forms of radiation (such as protons). The normal precautions of radiation protection apply: Avoid exposure, stay as far from the source as possible, and keep exposure time to a minimum. Some particular thought must be given to how to protect from neutron exposure, however. For other types of radiation, e.g. alpha particles, beta particles, or gamma rays, material of a high atomic number and with high density make for good shielding; frequently, lead is used. However, this approach will not work with neutrons, since the absorption of neutrons does not increase straightforwardly with atomic number, as it does with alpha, beta, and gamma radiation. Instead one needs to look at the particular interactions neutrons have with matter (see the section on detection above). For example, hydrogen-rich materials are often used to shield against neutrons, since ordinary hydrogen both scatters and slows neutrons. This often means that simple concrete blocks or even paraffin-loaded plastic blocks afford better protection from neutrons than do far more dense materials. After slowing, neutrons may then be absorbed with an isotope that has high affinity for slow neutrons without causing secondary capture radiation, such as lithium-6.

Hydrogen-rich ordinary water affects neutron absorption in nuclear fission reactors: Usually, neutrons are so strongly absorbed by normal water that fuel enrichment with fissionable isotope is required. The deuterium in heavy water has a very much lower absorption affinity for neutrons than does protium (normal light hydrogen). Deuterium is, therefore, used in CANDU-type reactors, in order to slow (moderate) neutron velocity, to increase the probability of nuclear fission compared to neutron capture.

5.11 Neutron temperature

Main article: Neutron temperature

5.11.1 Thermal neutrons

A *thermal neutron* is a free neutron that is Boltzmann distributed with kT = 0.0253 eV (4.0×10^{-21} J) at room temperature. This gives characteristic (not average, or median) speed of 2.2 km/s. The name 'thermal' comes from their energy being that of the room temperature gas or material they are permeating. (see *kinetic theory* for energies and speeds of molecules). After a number of collisions (often in the range of 10–20) with nuclei, neutrons arrive at this energy level, provided that they are not absorbed.

In many substances, thermal neutron reactions show a much larger effective cross-section than reactions involving faster neutrons, and thermal neutrons can therefore be absorbed more readily (i.e., with higher probability) by any atomic nuclei that they collide with, creating a heavier — and often unstable — isotope of the chemical element as a result.

Most fission reactors use a neutron moderator to slow down, or *thermalize* the neutrons that are emitted by nuclear fission so that they are more easily captured, causing further fission. Others, called fast breeder reactors, use fission energy neutrons directly.

5.11.2 Cold neutrons

Cold neutrons are thermal neutrons that have been equilibrated in a very cold substance such as liquid deuterium. Such a *cold source* is placed in the moderator of a research reactor or spallation source. Cold neutrons are particularly valuable for neutron scattering experiments.

5.11.3 Ultracold neutrons

Ultracold neutrons are produced by inelastically scattering cold neutrons in substances with a temperature of a few kelvins, such as solid deuterium or superfluid helium. An alternative production method is the mechanical deceleration of cold neutrons.

5.11.4 Fission energy neutrons

Main article: nuclear fission

A *fast neutron* is a free neutron with a kinetic energy level close to 1 MeV (1.6×10^{-13} J), hence a speed of ~14000 km/s (~ 5% of the speed of light). They are named *fission energy* or *fast* neutrons to distinguish them from lower-energy thermal neutrons, and high-energy neutrons produced in cosmic showers or accelerators. Fast neutrons are produced by nuclear processes such as nuclear fission. Neutrons produced in fission, as noted above, have a Maxwell–Boltzmann distribution of kinetic energies from 0 to ~14 MeV, a mean energy of 2 MeV (for U-235 fission neutrons), and a mode of only 0.75 MeV, which means that more than half of them do not qualify as fast (and thus have almost no chance of initiating fission in fertile materials, such as U-238 and Th-232).

Fast neutrons can be made into thermal neutrons via a process called moderation. This is done with a neutron moderator. In reactors, typically heavy water, light water, or graphite are used to moderate neutrons.

5.11.5 Fusion neutrons

For more details on this topic, see Nuclear fusion § Criteria and candidates for terrestrial reactions.

D–T (deuterium–tritium) fusion is the fusion reaction that produces the most energetic neutrons, with 14.1 MeV of kinetic energy and traveling at 17% of the speed of light. D–T fusion is also the easiest fusion reaction to ignite, reaching near-peak rates even when the deuterium and tritium nuclei have only a thousandth as much kinetic energy as the 14.1 MeV that will be produced.

14.1 MeV neutrons have about 10 times as much energy as fission neutrons, and are very effective at fissioning even non-fissile heavy nuclei, and these high-energy fissions produce more neutrons on average than fissions by lower-energy neutrons. This makes D–T fusion neutron sources such as proposed tokamak power reactors useful for transmutation of transuranic waste. 14.1 MeV neutrons can also produce neutrons by knocking them loose from nuclei.

On the other hand, these very high energy neutrons are less likely to simply be captured without causing fission or spallation. For these reasons, nuclear weapon design extensively utilizes D–T fusion 14.1 MeV neutrons to cause more fission. Fusion neutrons are able to cause fission in ordinarily non-fissile materials, such as depleted uranium (uranium-238), and these materials have been used in the jackets of thermonuclear weapons. Fusion neutrons also can cause fission in substances that are unsuitable or difficult to make into primary fission bombs, such as reactor grade plutonium. This physical fact thus causes ordinary non-weapons grade materials to become of concern in certain nuclear proliferation discussions and treaties.

Other fusion reactions produce much less energetic neutrons. D–D fusion produces a 2.45 MeV neutron and helium-3 half of the time, and produces tritium and a proton but no neutron the other half of the time. D–^{3}He fusion produces no neutron.

5.11.6 Intermediate-energy neutrons

A fission energy neutron that has slowed down but not yet reached thermal energies is called an epithermal neutron.

Cross sections for both capture and fission reactions often have multiple resonance peaks at specific energies in the epithermal energy range. These are of less significance in a fast neutron reactor, where most neutrons are absorbed before slowing down to this range, or in a well-moderated thermal reactor, where epithermal neutrons interact mostly with moderator nuclei, not with either fissile or fertile actinide nuclides. However, in a partially moderated reactor with more interactions of epithermal neutrons with heavy metal nuclei, there are greater possibilities for transient changes in reactivity that might make reactor control more difficult.

Ratios of capture reactions to fission reactions are also worse (more captures without fission) in most nuclear fuels such as plutonium-239, making epithermal-spectrum reactors using these fuels less desirable, as captures not only waste the one neutron captured but also usually result in a nuclide that is not fissile with thermal or epithermal neutrons, though still fissionable with fast neutrons. The exception is uranium-233 of the thorium cycle, which has good capture-fission ratios at all neutron energies.

5.11.7 High-energy neutrons

These neutrons have much more energy than fission energy neutrons and are generated as secondary particles by particle accelerators or in the atmosphere from cosmic rays. They can have energies as high as tens of joules per neutron. These neutrons are extremely efficient at ionization and far more likely to cause cell death than X-rays or protons.[72][73]

5.12   See also

- Ionizing radiation
- Isotope
- List of particles
- Neutronium
- Neutron magnetic moment
- Neutron radiation and the Sievert radiation scale
- Nuclear reaction
- Thermal reactor
- Nucleosynthesis
  - Neutron capture nucleosynthesis
  - R-process
  - S-process

5.12.1   Neutron sources

- Neutron generator
- Neutron sources

5.12.2   Processes involving neutrons

- Neutron bomb
- Neutron diffraction
- Neutron flux
- Neutron transport

5.13 References

[1] Ernest Rutherford. Chemed.chem.purdue.edu. Retrieved on 2012-08-16.

[2] 1935 Nobel Prize in Physics. Nobelprize.org. Retrieved on 2012-08-16.

[3] Mohr, P.J.; Taylor, B.N. and Newell, D.B. (2011), "The 2010 CODATA Recommended Values of the Fundamental Physical Constants" (Web Version 6.0). The database was developed by J. Baker, M. Douma, and S. Kotochigova. (2011-06-02). National Institute of Standards and Technology, Gaithersburg, Maryland 20899.

[4] Thomas, A.W.; Weise, W. (2001), *The Structure of the Nucleon*, Wiley-WCH, Berlin, ISBN 3-527-40297-7

[5] Glasstone, Samuel; Dolan, Philip J., eds. (1977), *The Effects of Nuclear Weapons, Third Edition*, U.S. Dept. of Defense and Energy Research and Development Administration, U.S. Government Printing Office, ISBN 1-60322-016-X

[6] Nudat 2. Nndc.bnl.gov. Retrieved on 2010-12-04.

[7] Chadwick, James (1932). "Possible Existence of a Neutron". *Nature* **129** (3252): 312. Bibcode:1932Natur.129Q.312C. doi:10.1038/129312a0.

[8] O. Hahn and F. Strassmann (1939). "Über den Nachweis und das Verhalten der bei der Bestrahlung des Urans mittels Neutronen entstehenden Erdalkalimetalle ("On the detection and characteristics of the alkaline earth metals formed by irradiation of uranium with neutrons")". *Naturwissenschaften* **27** (1): 11–15. Bibcode:1939NW.....27...11H. doi:10.1007/BF01488241.. The authors were identified as being at the Kaiser-Wilhelm-Institut für Chemie, Berlin-Dahlem. Received 22 December 1938.

[9] M. J. Carson et al. (2004). "Neutron background in large-scale xenon detectors for dark matter searches". *Astroparticle Physics* **21** (6): 667–687. doi:10.1016/j.astropartphys.2004.05.001.

[10] Povh, B.; Rith, K.; Scholz, C.; Zetsche, F. (2002). *Particles and Nuclei: An Introduction to the Physical Concepts*. Berlin: Springer-Verlag. p. 73. ISBN 978-3-540-43823-6.

[11] J.-L. Basdevant, J. Rich, M. Spiro (2005). *Fundamentals in Nuclear Physics*. Springer. p. 155. ISBN 0-387-01672-4.

[12] Paul Allen Tipler, Ralph A. Llewellyn (2002). *Modern Physics* (4 ed.). Macmillan. p. 310. ISBN 0-7167-4345-0.

[13] Nakamura, K (2010). "Review of Particle Physics". *Journal of Physics G: Nuclear and Particle Physics* **37** (7A): 075021. Bibcode:2010JPhG...37g5021N. doi:10.1088/0954-3899/37/7A/075021. PDF with 2011 partial update for the 2012 edition The exact value of the mean lifetime is still uncertain, due to conflicting results from experiments. The Particle Data Group reports values up to six seconds apart (more than four standard deviations), commenting that "our 2006, 2008, and 2010 Reviews stayed with 885.7±0.8 s; but we noted that in light of SEREBROV 05 our value should be regarded as suspect until further experiments clarified matters. Since our 2010 Review, PICHLMAIER 10 has obtained a mean life of 880.7±1.8 s, closer to the value of SEREBROV 05 than to our average. And SEREBROV 10B[...] claims their values should be lowered by about 6 s, which would bring them into line with the two lower values. However, those reevaluations have not received an enthusiastic response from the experimenters in question; and in any case the Particle Data Group would have to await published changes (by those experimenters) of published values. At this point, we can think of nothing better to do than to average the seven best but discordant measurements, getting 881.5±1.5s. Note that the error includes a scale factor of 2.7. This is a jump of 4.2 old (and 2.8 new) standard deviations. This state of affairs is a particularly unhappy one, because the value is so important. We again call upon the experimenters to clear this up."

[14] R.K. Adair (1989). *The Great Design: Particles, Fields, and Creation*. Oxford University Press. p. 214.

[15] Cho, Adiran (2 April 2010). "Mass of the Common Quark Finally Nailed Down". *http://news.sciencemag.org".* *American Association for the Advancement of Science. Retrieved 27 September 2014.*

[16] W.N. Cottingham, D.A. Greenwood (1986).*An Introduction to Nuclear Physics.*Cambridge University Press.ISBN97805216334.

[17] E. Rutherford (1920). "Nuclear Constitution of Atoms".*Proceedings of the Royal Society A***97**(686): 374–400.Bibcode:1920RSPR. doi:10.1098/rspa.1920.0040.

[18] "Wolfgang Pauli". *Sources in the History of Mathematics and Physical Sciences.* Sources in the History of Mathematics and Physical Sciences **6**: 105–144. 1985. doi:10.1007/978-3-540-78801-0_3. ISBN 978-3-540-13609-5. |chapter= ignored (help)

[19] Hendry, John, ed. (1984), *Cambridge Physics in the Thirties*, Adam Hilger Ltd, Bristol, ISBN 0852747616

[20]N. Feather (1960)."A history of neutrons and nuclei. Part 1".*Contemporary Physics***1**(3): 191–203.doi:10.1080/00107516008611.

[21]Brown, Laurie M. (1978). "The idea of the neutrino".*Physics Today***31**(9): 23.Bibcode:1978PhT....31i..23B.doi:10.1063/195181.

[22] Friedlander G., Kennedy J.W. and Miller J.M. (1964) *Nuclear and Radiochemistry* (2nd edition), Wiley, pp. 22–23 and 38–39

[23] Stuewer, Roger H. (1985). "Niels Bohr and Nuclear Physics". In French, A. P.; Kennedy, P. J. *Niels Bohr: A Centenary Volume.* Harvard University Press. pp. 197–220. ISBN 0674624165.

[24] Pais, Abraham (1986). *Inward Bound.* Oxford: Oxford University Press. p. 299. ISBN 0198519974.

[25] Klein, O. (1929). "Die Reflexion von Elektronen an einem Potentialsprung nach der relativistischen Dynamik von Dirac". *Zeitschrift für Physik* **53** (3–4): 157–165. Bibcode:1929ZPhy...53..157K. doi:10.1007/BF01339716.

[26] Bothe, W.; Becker, H. (1930). "Künstliche Erregung von Kern-γ-Strahlen" [Artificial excitation of nuclear γ-radiation]. *Zeitschrift für Physik* **66** (5–6): 289–306. Bibcode:1930ZPhy...66..289B. doi:10.1007/BF01390908.

[27] Becker, H.; Bothe, W. (1932). "Die in Bor und Beryllium erregten γ-Strahlen" [Γ-rays excited in boron and beryllium]. *Zeitschrift für Physik* **76** (7–8): 421–438. Bibcode:1932ZPhy...76..421B. doi:10.1007/BF01336726.

[28] Joliot-Curie, Irène and Joliot, Frédéric (1932). "Émission de protons de grande vitesse par les substances hydrogénées sous l'influence des rayons γ très pénétrants" [Emission of high-speed protons by hydrogenated substances under the influence of very penetrating γ-rays]. *Comptes Rendus* **194**: 273.

[29] "Atop the Physics Wave: Rutherford Back in Cambridge, 1919–1937". *Rutherford's Nuclear World.* American Institute of Physics. 2011–2014. Retrieved 19 August 2014.

[30] Chadwick, J. (1933). "Bakerian Lecture. The Neutron". *Proceedings of the Royal Society A: Mathematical, Physical and Engineering Sciences* **142** (846): 1–25. Bibcode:1933RSPSA.142....1C. doi:10.1098/rspa.1933.0152.

[31] Heisenberg, W. (1932). "Über den Bau der Atomkerne. I". *Z. Phys.* **77**: 1–11. doi:10.1007/BF01342433.

[32] Heisenberg, W. (1932). "Über den Bau der Atomkerne. II". *Z. Phys.* **78** (3–4): 156–164. doi:10.1007/BF01337585.

[33] Heisenberg, W. (1933). "Über den Bau der Atomkerne. III". *Z. Phys.* **80** (9–10): 587–596. doi:10.1007/BF01335696.

[34] Iwanenko, D.D., The neutron hypothesis, Nature **129** (1932) 798.

[35] Miller A. I. *Early Quantum Electrodynamics: A Sourcebook*, Cambridge University Press, Cambridge, 1995, ISBN 0521568919, pp. 84–88.

[36] Wilson, Fred L. (1968). "Fermi's Theory of Beta Decay". *Am. J. Phys.* **36** (12): 1150–1160. Bibcode:1968AmJPh..36.1150W. doi:10.1119/1.1974382.

[37] Chadwick, J.; Goldhaber, M. (1934). "A nuclear photo-effect: disintegration of the diplon by gamma rays". *Nature* **134**: 237–238. doi:10.1038/134237a0.

[38] Chadwick, J.; Goldhaber, M. (1935). "A nuclear photoelectric effect" (PDF). *Proc. R. Soc. Lond* **151**: 479–493. doi:10.12.

[39] Cooper, Dan (1999). *Enrico Fermi: And the Revolutions in Modern physics.* New York: Oxford University Press. ISBN 0-19-511762-X. OCLC 39508200.

[40] Hahn, O. (1958). "The Discovery of Fission". *Scientific American* **198** (2): 76. doi:10.1038/scientificamerican0258-76.

[41] Rife, Patricia (1999). *Lise Meitner and the dawn of the nuclear age.* Basel, Switzerland: Birkhäuser. ISBN 0-8176-3732-X.

[42] Hahn, O.; Strassmann, F. (10 February 1939). "Proof of the Formation of Active Isotopes of Barium from Uranium and Thorium Irradiated with Neutrons; Proof of the Existence of More Active Fragments Produced by Uranium Fission". *Die Naturwissenschaften* **27**: 89–95.

[43] "The Nobel Prize in Chemistry 1944". Nobel Foundation. Retrieved 2007-12-17.

[44] Bernstein, Jeremy (2001). *Hitler's uranium club: the secret recordings at Farm Hall.* New York: Copernicus. p. 281. ISBN 0-387-95089-3.

[45] "The Nobel Prize in Chemistry 1944: Presentation Speech". Nobel Foundation. Retrieved 2008-01-03.

[46] Sir James Chadwick's Discovery of Neutrons. ANS Nuclear Cafe. Retrieved on 2012-08-16.

[47] Particle Data Group Summary Data Table on Baryons. lbl.gov (2007). Retrieved on 2012-08-16.

[48] Basic Ideas and Concepts in Nuclear Physics: An Introductory Approach, Third Edition K. Heyde Taylor & Francis 2004. Print ISBN 978-0-7503-0980-6. eBook ISBN 978-1-4200-5494-1. DOI: 10.1201/9781420054941.ch5. full text

[49] Olive, K.A. et al. (2014). "Review of Particle Physics". *Chin. Phys. C* **38**: 090001. doi:10.1088/1674-1137/38/9/090001.

[50] "Pear-shaped particles probe big-bang mystery" (Press release). University of Sussex. 20 February 2006. Retrieved 2009-12-14.

[51] A cryogenic experiment to search for the EDM of the neutron. Hepwww.rl.ac.uk. Retrieved on 2012-08-16.

[52] Search for the neutron electric dipole moment: nEDM. Nedm.web.psi.ch (2001-09-12). Retrieved on 2012-08-16.

[53] SNS Neutron EDM Experiment. P25ext.lanl.gov. Retrieved on 2012-08-16.

[54] Measurement of the Neutron Electric Dipole Moment. Nrd.pnpi.spb.ru. Retrieved on 2012-08-16.

[55] Gell, Y.; Lichtenberg, D. B. (1969). "Quark model and the magnetic moments of proton and neutron". *Il Nuovo Cimento A*. Series 10 **61**: 27–40. Bibcode:1969NCimA..61...27G. doi:10.1007/BF02760010.

[56] Alvarez, L. W; Bloch, F. (1940). "A quantitative determination of the neutron magnetic moment in absolute nuclear magnetons". *Physical Review* **57**: 111–122. doi:10.1103/physrev.57.111.

[57] Miller, G.A. (2007). "Charge Densities of the Neutron and Proton".*Physical Review Letters***99**(11): 112001.Bibcode:2007PhM. doi:10.1103/PhysRevLett.99.112001.

[58] Greene, GL et al. (1986). "New determination of the deuteron binding energy and the neutron mass". *Phys. Rev. Lett.* **56**: 819–822. Bibcode:1986PhRvL..56..819G. doi:10.1103/PhysRevLett.56.819.

[59] Byrne, J. *Neutrons, Nuclei, and Matter*, Dover Publications, Mineola, New York, 2011, ISBN 0486482383, pp. 18–19

[60] Spyrou, A. et al. (2012). "First Observation of Ground State Dineutron Decay: 16Be". *Physical Review Letters* **108** (10): 102501. Bibcode:2012PhRvL.108j2501S. doi:10.1103/PhysRevLett.108.102501. PMID 22463404.

[61] Llanes-Estrada, Felipe J.; Moreno Navarro, Gaspar (2011). "Cubic neutrons". arXiv:1108.1859v1 [nucl-th].

[62] Köhn, C., Ebert, U., Calculation of beams of positrons, neutrons and protons associated with terrestrial gamma-ray flashes, J. Geophys. Res. Atmos. (2015), vol. 23, doi:10.1002/2014JD022229

[63] Clowdsley, MS; Wilson, JW; Kim, MH; Singleterry, RC; Tripathi, RK; Heinbockel, JH; Badavi, FF; Shinn, JL (2001). "Neutron Environments on the Martian Surface" (PDF). *Physica Medica* **17** (Suppl 1): 94–6. PMID 11770546.

[64] Byrne, J. *Neutrons, Nuclei, and Matter*, Dover Publications, Mineola, New York, 2011, ISBN 0486482383, pp. 32–33.

[65] Science/Nature | Q&A: Nuclear fusion reactor. BBC News (2006-02-06). Retrieved on 2010-12-04.

[66] Byrne, J. *Neutrons, Nuclei, and Matter*, Dover Publications, Mineola, New York, 2011, ISBN 0486482383, p. 453.

[67] Kumakhov, M. A.; Sharov, V. A. (1992). "A neutron lens". *Nature* **357** (6377): 390–391. Bibcode:1992Natur.357..390K. doi:10.1038/357390a0.

[68] Physorg.com, "New Way of 'Seeing': A 'Neutron Microscope'". Physorg.com (2004-07-30). Retrieved on 2012-08-16.

[69] "NASA Develops a Nugget to Search for Life in Space". NASA.gov (2007-11-30). Retrieved on 2012-08-16.

[70] Hall EJ. Radiobiology for the Radiologist. Lippincott Williams & Wilkins; 5th edition (2000)

[71] Johns HE and Cunningham JR. The Physics of Radiology. Charles C Thomas 3rd edition 1978

[72] Tami Freeman (May 23, 2008). "Facing up to secondary neutrons". Medical Physics Web. Retrieved 2011-02-08.

[73] Heilbronn, L.; Nakamura, T; Iwata, Y; Kurosawa, T; Iwase, H; Townsend, LW (2005). "Expand+Overview of secondary neutron production relevant to shielding in space". *Radiation Protection Dosimetry* **116** (1–4): 140–143. doi:10.1093/rpd/nci033. PMID 16604615.

5.14   Further reading

- Annotated bibliography for neutrons from the Alsos Digital Library for Nuclear Issues

- Abraham Pais, *Inward Bound*, Oxford: Oxford University Press, 1986. ISBN 0198519974.

- Sin-Itiro Tomonaga, *The Story of Spin*, The University of Chicago Press, 1997

- Herwig Schopper, *Weak interactions and nuclear beta decay*, Publisher, North-Holland Pub. Co., 1966.

5.15   External links

- neutron properties at Particle Data Group, Lawrence Berkeley National Laboratory in Berkeley, CA. (pdgLive)

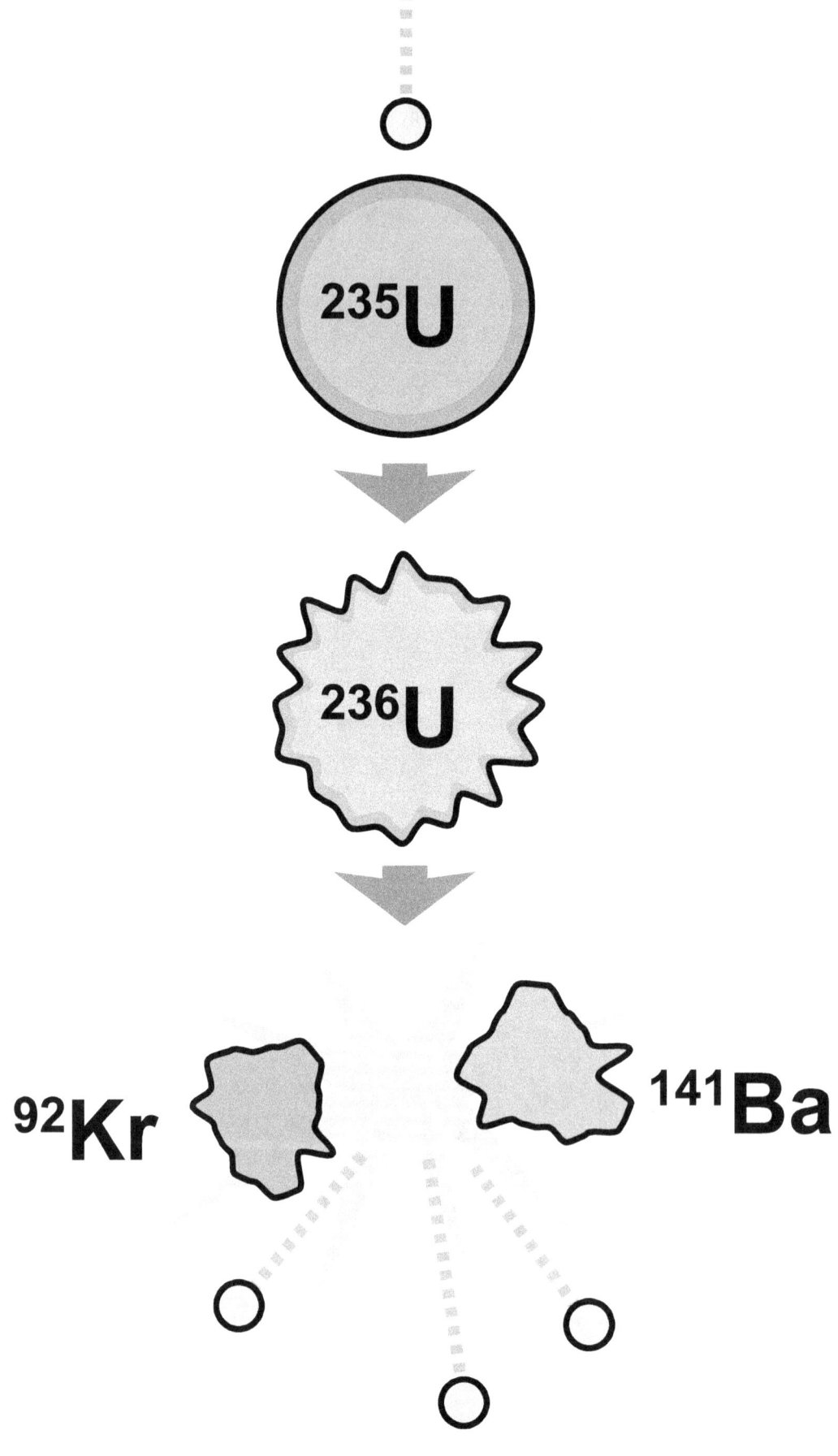

Nuclear fission caused by absorption of a neutron by uranium-235. The heavy nuclide fragments into lighter components and additional

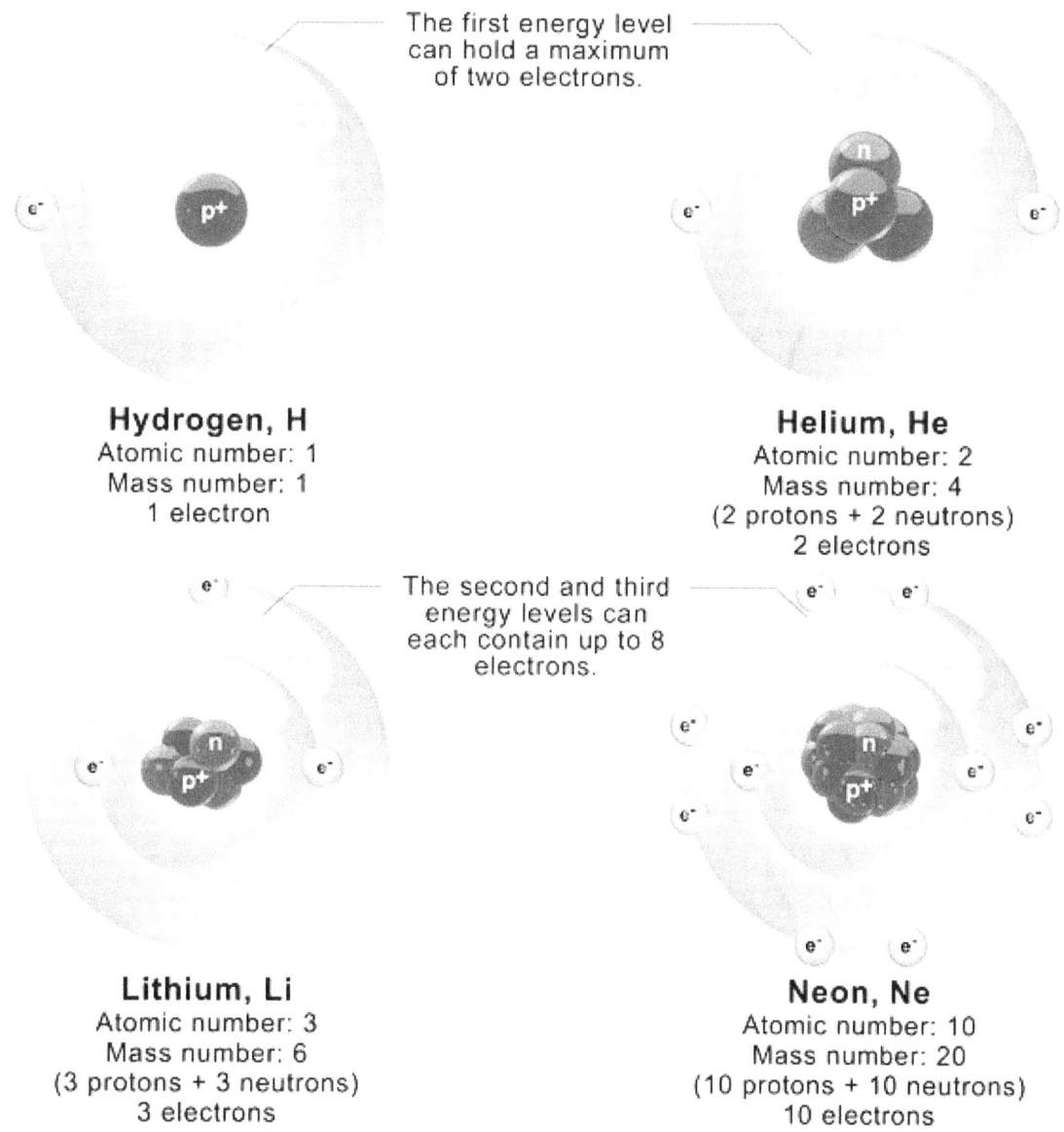

The first energy level can hold a maximum of two electrons.

The second and third energy levels can each contain up to 8 electrons.

Hydrogen, H
Atomic number: 1
Mass number: 1
1 electron

Helium, He
Atomic number: 2
Mass number: 4
(2 protons + 2 neutrons)
2 electrons

Lithium, Li
Atomic number: 3
Mass number: 6
(3 protons + 3 neutrons)
3 electrons

Neon, Ne
Atomic number: 10
Mass number: 20
(10 protons + 10 neutrons)
10 electrons

Models depicting the nucleus and electron energy levels in hydrogen, helium, lithium, and neon atoms. In reality, the diameter of the nucleus is about 100,000 times smaller than the diameter of the atom.

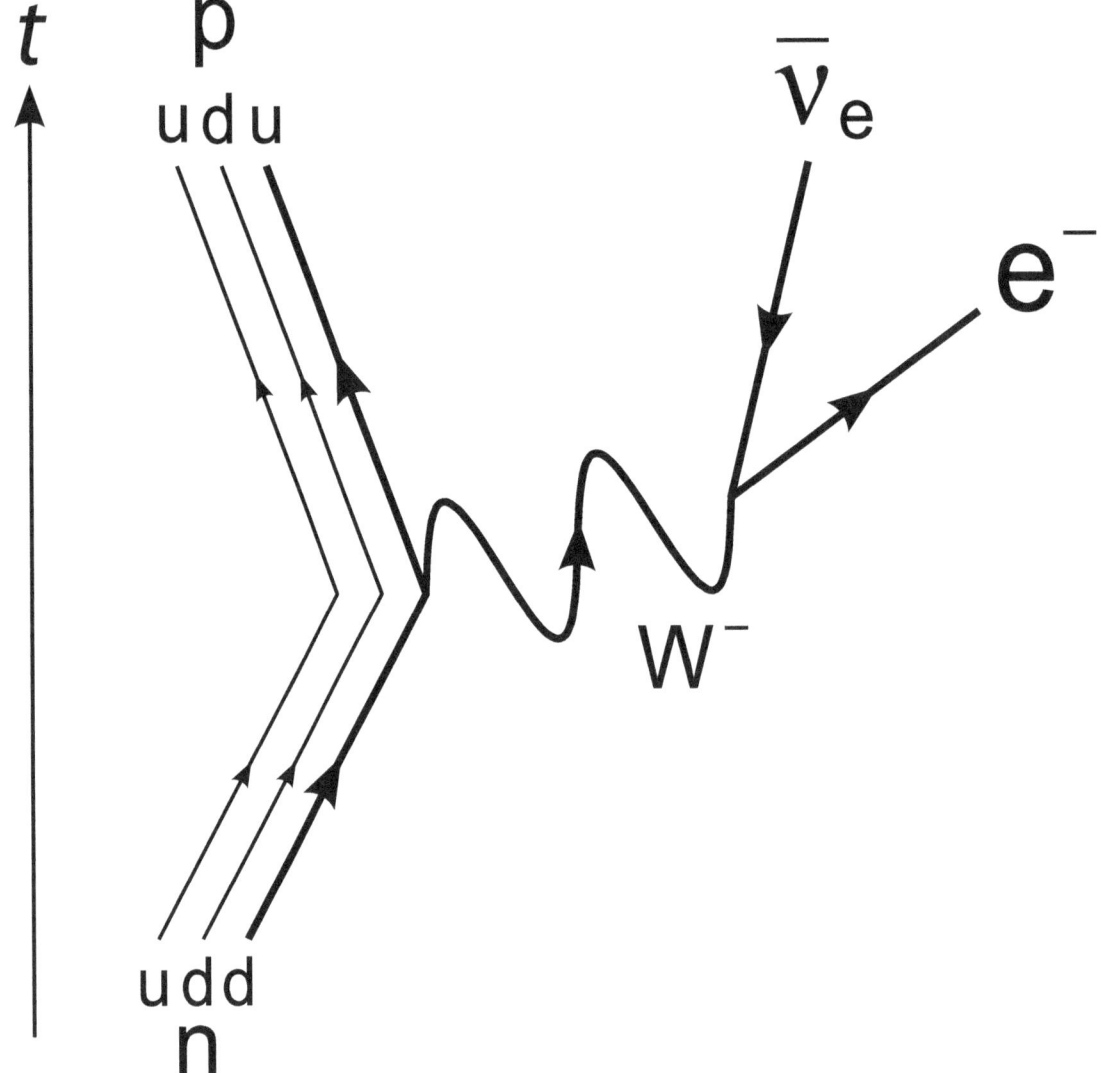

The Feynman diagram for beta decay of a neutron into a proton, electron, and electron antineutrino via an intermediate heavy W boson

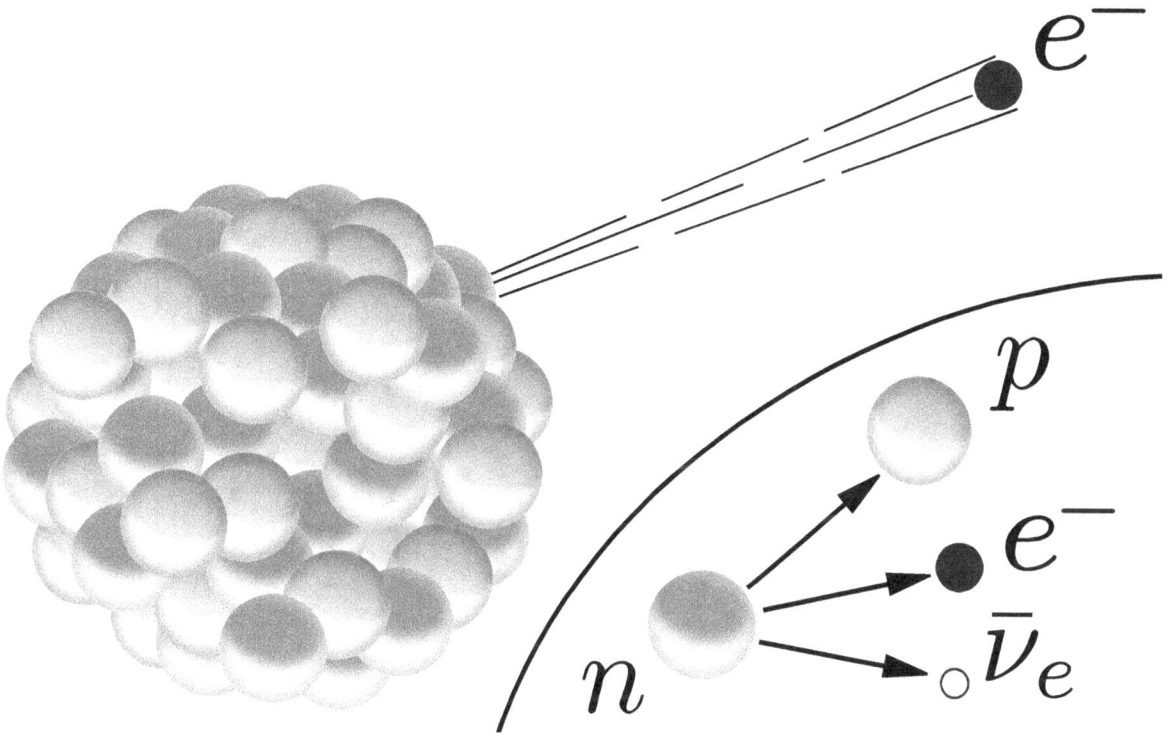

A schematic of the nucleus of an atom indicating β– radiation, the emission of a fast electron from the nucleus (the accompanying antineutrino is omitted). In the Rutherford model for the nucleus, red spheres were protons with positive charge and blue spheres were protons tightly bound to an electron with no net charge.

*The **inset** shows beta decay of a free neutron as it is understood today; an electron and antineutrino are created in this process.*

Institut Laue–Langevin (ILL) in Grenoble, France – a major neutron research facility.

Example of Cold
Neutron Source

Cold neutron source providing neutrons at about the temperature of liquid hydrogen

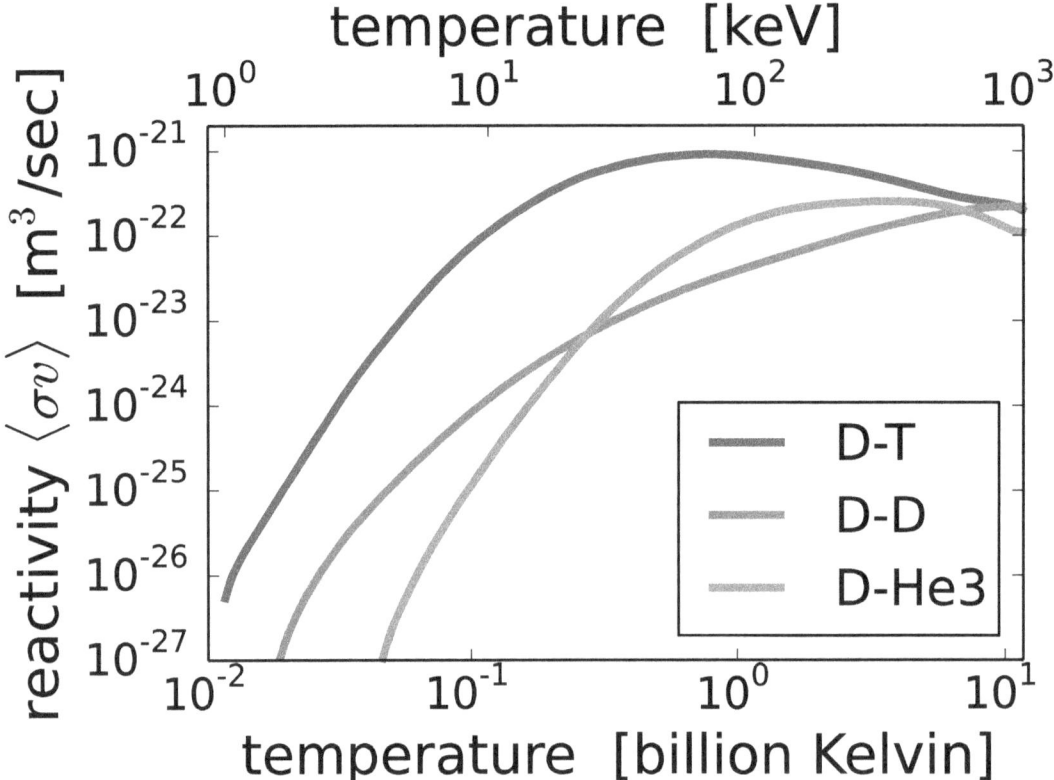

The fusion reaction rate increases rapidly with temperature until it maximizes and then gradually drops off. The DT rate peaks at a lower temperature (about 70 keV, or 800 million kelvins) and at a higher value than other reactions commonly considered for fusion energy.

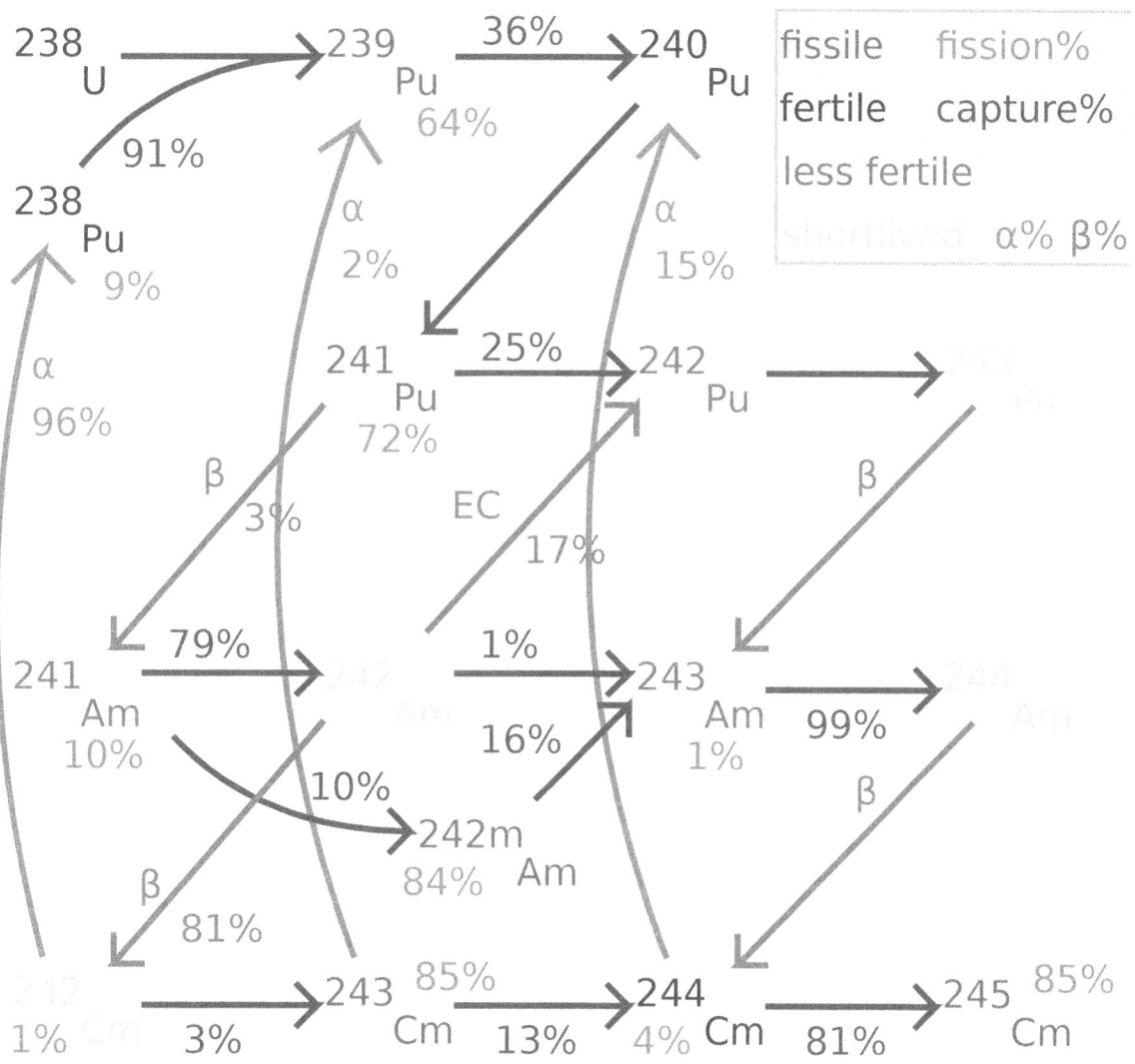

Transmutation flow in light water reactor, which is a thermal-spectrum reactor

Chapter 6

Proton

This article is about the proton as a subatomic particle. For other uses, see Proton (disambiguation).

The **proton** is a subatomic particle, symbol p or p+, with a positive electric charge of +1e elementary charge and mass slightly less than that of a neutron. Protons and neutrons, each with mass approximately one atomic mass unit, are collectively referred to as "nucleons". One or more protons are present in the nucleus of an atom. The number of protons in the nucleus is referred to as its atomic number. Since each element has a unique number of protons, each element has its own unique atomic number. The word *proton* is Greek for "first", and this name was given to the hydrogen nucleus by Ernest Rutherford in 1920. In previous years Rutherford had discovered that the hydrogen nucleus (known to be the lightest nucleus) could be extracted from the nuclei of nitrogen by collision. The proton was therefore a candidate to be a fundamental particle and a building block of nitrogen and all other heavier atomic nuclei.

In the modern Standard Model of particle physics, the proton is a hadron, and like the neutron, the other nucleon (particle present in atomic nuclei), is composed of three quarks. Although the proton was originally considered a fundamental particle, it is composed of three valence quarks: two up quarks and one down quark. The rest masses of the quarks contribute only about 1% of the proton's mass, however.[2] The remainder of the proton mass is due to the kinetic energy of the quarks and to the energy of the gluon fields that bind the quarks together. Because the proton is not a fundamental particle, it possesses a physical size; the radius of the proton is about 0.84–0.87 fm.[3]

At sufficiently low temperatures, free protons will bind to electrons. However, the character of such bound protons does not change, and they remain protons. A fast proton moving through matter will slow by interactions with electrons and nuclei, until it is captured by the electron cloud of an atom. The result is a protonated atom, which is a chemical compound of hydrogen. In vacuum, when free electrons are present, a sufficiently slow proton may pick up a single free electron, becoming a neutral hydrogen atom, which is chemically a free radical. Such "free hydrogen atoms" tend to react chemically with many other types of atoms at sufficiently low energies. When free hydrogen atoms react with each other, they form neutral hydrogen molecules (H_2), which are the most common molecular component of molecular clouds in interstellar space. Such molecules of hydrogen on Earth may then serve (among many other uses) as a convenient source of protons for accelerators (as used in proton therapy) and other hadron particle physics experiments that require protons to accelerate, with the most powerful and noted example being the Large Hadron Collider.

6.1 Description

Protons are spin-½ fermions and are composed of three valence quarks,[4] making them baryons (a sub-type of hadrons). The two up quarks and one down quark of the proton are held together by the strong force, mediated by gluons.[5]:21–22 A modern perspective has the proton composed of the valence quarks (up, up, down), the gluons, and transitory pairs of sea quarks. The proton has an approximately exponentially decaying positive charge distribution with a mean square radius of about 0.8 fm.[6]

Protons and neutrons are both nucleons, which may be bound together by the nuclear force to form atomic nuclei. The nucleus of the most common isotope of the hydrogen atom (with the chemical symbol "H") is a lone proton. The nuclei of the heavy hydrogen isotopes deuterium and tritium contain one proton bound to one and two neutrons, respectively. All other types of atomic nuclei are composed of two or more protons and various numbers of neutrons.

6.2 History

The concept of a hydrogen-like particle as a constituent of other atoms was developed over a long period. As early as 1815, William Prout proposed that all atoms are composed of hydrogen atoms (which he called "protyles"), based on a simplistic interpretation of early values of atomic weights (see Prout's hypothesis), which was disproved when more accurate values were measured.[7]:39–42

In 1886, Eugen Goldstein discovered canal rays (also known as anode rays) and showed that they were positively charged particles (ions) produced from gases. However, since particles from different gases had different values of charge-to-mass ratio (e/m), they could not be identified with a single particle, unlike the negative electrons discovered by J. J. Thomson.

Following the discovery of the atomic nucleus by Ernest Rutherford in 1911, Antonius van den Broek proposed that the place of each element in the periodic table (its atomic number) is equal to its nuclear charge. This was confirmed experimentally by Henry Moseley in 1913 using X-ray spectra.

In 1917, (in experiments reported in 1919) Rutherford proved that the hydrogen nucleus is present in other nuclei, a result usually described as the discovery of the proton.[8] Rutherford had earlier learned to produce hydrogen nuclei as a type of radiation produced as a product of the impact of alpha particles on nitrogen gas, and recognize them by their unique penetration signature in air and their appearance in scintillation detectors. These experiments were begun when Rutherford had noticed that, when alpha particles were shot into air (mostly nitrogen), his scintillation detectors showed the signatures of typical hydrogen nuclei as a product. After experimentation Rutherford traced the reaction to the nitrogen in air, and found that when alphas were produced into pure nitrogen gas, the effect was larger. Rutherford determined that this hydrogen could have come only from the nitrogen, and therefore nitrogen must contain hydrogen nuclei. One hydrogen nucleus was being knocked off by the impact of the alpha particle, producing oxygen-17 in the process. This was the first reported nuclear reaction, $^{14}N + \alpha \rightarrow {}^{17}O + p$. (This reaction would later be observed happening directly in a cloud chamber in 1925).

Rutherford knew hydrogen to be the simplest and lightest element and was influenced by Prout's hypothesis that hydrogen was the building block of all elements. Discovery that the hydrogen nucleus is present in all other nuclei as an elementary particle, led Rutherford to give the hydrogen nucleus a special name as a particle, since he suspected that hydrogen, the lightest element, contained only one of these particles. He named this new fundamental building block of the nucleus the *proton,* after the neuter singular of the Greek word for "first", πρῶτον. However, Rutherford also had in mind the word *protyle* as used by Prout. Rutherford spoke at the British Association for the Advancement of Science at its Cardiff meeting beginning 24 August 1920.[9] Rutherford was asked by Oliver Lodge for a new name for the positive hydrogen nucleus to avoid confusion with the neutral hydrogen atom. He initially suggested both *proton* and *prouton* (after Prout).[10] Rutherford later reported that the meeting had accepted his suggestion that the hydrogen nucleus be named the "proton", following Prout's word "protyle".[11] The first use of the word "proton" in the scientific literature appeared in 1920.[12]

6.3 Stability

Main article: Proton decay

The free proton (a proton not bound to nucleons or electrons) is a stable particle that has not been observed to break down spontaneously to other particles. Free protons are found naturally in a number of situations in which energics or temperatures are high enough to separate them from electrons, for which they have some affinity. Free protons exist in plasmas in which temperatures are too high to allow them to combine with electrons. Free protons of high energy and velocity make up 90% of cosmic rays, which propagate in vacuum for interstellar distances. Free protons are emitted directly from atomic nuclei in some rare types of radioactive decay. Protons also result (along with electrons and antineutrinos) from the radioactive decay of free neutrons, which are unstable.

The spontaneous decay of free protons has never been observed, and the proton is therefore considered a stable particle. However, some grand unified theories of particle physics predict that proton decay should take place with lifetimes of the order of 10^{36} years, and experimental searches have established lower bounds on the mean lifetime of the proton for various assumed decay products.[13][14][15]

Experiments at the Super-Kamiokande detector in Japan gave lower limits for proton mean lifetime of 6.6×10^{33} years for decay to an antimuon and a neutral pion, and 8.2×10^{33} years for decay to a positron and a neutral pion.[16] Another experiment at the Sudbury Neutrino Observatory in Canada searched for gamma rays resulting from residual nuclei resulting from the decay of a proton from oxygen-16. This experiment was designed to detect decay to any product, and established a lower limit to the proton lifetime of 2.1×10^{29} years.[17]

However, protons are known to transform into neutrons through the process of electron capture (also called inverse beta decay). For free protons, this process does not occur spontaneously but only when energy is supplied. The equation is:

$$p+ + e- \rightarrow n + \nu$$
$$e$$

The process is reversible; neutrons can convert back to protons through beta decay, a common form of radioactive decay. In fact, a free neutron decays this way, with a mean lifetime of about 15 minutes.

6.4   Quarks and the mass of the proton

In quantum chromodynamics, the modern theory of the nuclear force, most of the mass of the proton and the neutron is explained by special relativity. The mass of the proton is about 80–100 times greater than the sum of the rest masses of the quarks that make it up, while the gluons have zero rest mass. The extra energy of the quarks and gluons in a region within a proton, as compared to the rest energy of the quarks alone in the QCD vacuum, accounts for almost 99% of the mass. The rest mass of the proton is, thus, the invariant mass of the system of moving quarks and gluons that make up the particle, and, in such systems, even the energy of massless particles is still measured as part of the rest mass of the system.

Two terms are used in referring to the mass of the quarks that make up protons: *current quark mass* refers to the mass of a quark by itself, while *constituent quark mass* refers to the current quark mass plus the mass of the gluon particle field surrounding the quark.[18]:285–286 [19]:150–151 These masses typically have very different values. As noted, most of a proton's mass comes from the gluons that bind the current quarks together, rather than from the quarks themselves. While gluons are inherently massless, they possess energy—to be more specific, quantum chromodynamics binding energy (QCBE)—and it is this that contributes so greatly to the overall mass of the proton (see mass in special relativity). A proton has a mass of approximately $938 \, \text{MeV/c}^2$, of which the rest mass of its three valence quarks contributes only about $9.4 \, \text{MeV/c}^2$; much of the remainder can be attributed to the gluons' QCBE.[20][21][22]

The internal dynamics of the proton are complicated, because they are determined by the quarks' exchanging gluons, and interacting with various vacuum condensates. Lattice QCD provides a way of calculating the mass of the proton directly from the theory to any accuracy, in principle. The most recent calculations[23][24] claim that the mass is determined to better than 4% accuracy, even to 1% accuracy (see Figure S5 in Dürr *et al.*[24]). These claims are still controversial, because the calculations cannot yet be done with quarks as light as they are in the real world. This means that the predictions are found by a process of extrapolation, which can introduce systematic errors.[25] It is hard to tell whether these errors are controlled properly, because the quantities that are compared to experiment are the masses of the hadrons, which are known in advance.

These recent calculations are performed by massive supercomputers, and, as noted by Boffi and Pasquini: "a detailed description of the nucleon structure is still missing because ... long-distance behavior requires a nonperturbative and/or numerical treatment..."[26] More conceptual approaches to the structure of the proton are: the topological soliton approach originally due to Tony Skyrme and the more accurate AdS/QCD approach that extends it to include a string theory of gluons,[27] various QCD-inspired models like the bag model and the constituent quark model, which were popular in the 1980s, and the SVZ sum rules, which allow for rough approximate mass calculations.[28] These methods do not have the same accuracy as the more brute-force lattice QCD methods, at least not yet.

6.5   Charge radius

Main article: Charge radius

The internationally accepted value of the proton's charge radius is 0.8768 fm (see orders of magnitude for comparison to other sizes). This value is based on measurements involving a proton and an electron.

However, since 5 July 2010, an international research team has been able to make measurements involving an exotic atom made of a proton and a negatively charged muon. After a long and careful analysis of those measurements, the team concluded that the root-mean-square charge radius of a proton is "0.84184(67) fm, which differs by 5.0 standard deviations from the CODATA value of 0.8768(69) fm".[29] In January 2013, an updated value for the charge radius of a proton—0.84087(39) fm—was published. The precision was improved by 1.7 times, but the difference with CODATA value persisted at 7σ significance.[30]

The international research team that obtained this result at the Paul Scherrer Institut (PSI) in Villigen (Switzerland) includes scientists from the Max Planck Institute of Quantum Optics (MPQ) in Garching, the Ludwig-Maximilians-Universität (LMU) Munich and the Institut für Strahlwerkzeuge (IFWS) of the Universität Stuttgart (both from Germany), and the University of Coimbra, Portugal.[31][32] They are now attempting to explain the discrepancy, and re-examining the results of both previous high-precision measurements and complicated calculations. If no errors are found in the measurements or calculations, it could be necessary to re-examine the world's most precise and best-tested fundamental theory: quantum electrodynamics.[31] The proton radius remains a puzzle as of early 2015.[33]

6.6   Interaction of free protons with ordinary matter

Main article: Proton therapy

Although protons have affinity for oppositely charged electrons, free protons must lose sufficient velocity (and kinetic energy) in order to become closely associated and bound to electrons, since this is a relatively low-energy interaction. High energy protons, in traversing ordinary matter, lose energy by collisions with atomic nuclei, and by ionization of atoms (removing electrons) until they are slowed sufficiently to be captured by the electron cloud in a normal atom.

However, in such an association with an electron, the character of the bound proton is not changed, and it remains a proton. The attraction of low-energy free protons to any electrons present in normal matter (such as the electrons in normal atoms) causes free protons to stop and to form a new chemical bond with an atom. Such a bond happens at any sufficiently "cold" temperature (i.e., comparable to temperatures at the surface of the Sun) and with any type of atom. Thus, in interaction with any type of normal (non-plasma) matter, low-velocity free protons are attracted to electrons in any atom or molecule with which they come in contact, causing the proton and molecule to combine. Such molecules are then said to be "protonated", and chemically they often, as a result, become so-called Bronsted acids.

6.7   Proton in chemistry

6.7.1   Atomic number

In chemistry, the number of protons in the nucleus of an atom is known as the atomic number, which determines the chemical element to which the atom belongs. For example, the atomic number of chlorine is 17; this means that each chlorine atom has 17 protons and that all atoms with 17 protons are chlorine atoms. The chemical properties of each atom are determined by the number of (negatively charged) electrons, which for neutral atoms is equal to the number of (positive) protons so that the total charge is zero. For example, a neutral chlorine atom has 17 protons and 17 electrons, whereas a Cl^- anion has 17 protons and 18 electrons for a total charge of −1.

All atoms of a given element are not necessarily identical, however, as the number of neutrons may vary to form different isotopes, and energy levels may differ forming different nuclear isomers. For example, there are two stable isotopes of

chlorine: 35
17Cl with 35 − 17 = 18 neutrons and 37
17Cl with 37 − 17 = 20 neutrons.

6.7.2 Hydrogen ion

See also: Hydron (chemistry)
In chemistry, the term proton refers to the hydrogen ion, H+
. Since the atomic number of hydrogen is 1, a hydrogen ion has no electrons and corresponds to a bare nucleus, consisting of a proton (and 0 neutrons for the most abundant isotope *protium* 1
1H). The proton is a "bare charge" with only about **1/64,000** of the radius of a hydrogen atom, and so is extremely reactive chemically. The free proton, thus, has an extremely short lifetime in chemical systems such as liquids and it reacts immediately with the electron cloud of any available molecule. In aqueous solution, it forms the hydronium ion, H_3O^+, which in turn is further solvated by water molecules in clusters such as $[H_5O_2]^+$ and $[H_9O_4]^+$.[34]

The transfer of H+
in an acid–base reaction is usually referred to as "proton transfer". The acid is referred to as a proton donor and the base as a proton acceptor. Likewise, biochemical terms such as proton pump and proton channel refer to the movement of hydrated H+
ions.

The ion produced by removing the electron from a deuterium atom is known as a deuteron, not a proton. Likewise, removing an electron from a tritium atom produces a triton.

6.7.3 Proton nuclear magnetic resonance (NMR)

Also in chemistry, the term "proton NMR" refers to the observation of hydrogen-1 nuclei in (mostly organic) molecules by nuclear magnetic resonance. This method uses the spin of the proton, which has the value one-half. The name refers to examination of protons as they occur in protium (hydrogen-1 atoms) in compounds, and does not imply that free protons exist in the compound being studied.

6.8 Human exposure

Main article: Effect of spaceflight on the human body

The Apollo Lunar Surface Experiments Packages (ALSEP) determined that more than 95% of the particles in the solar wind are electrons and protons, in approximately equal numbers.[35][36]

> Because the Solar Wind Spectrometer made continuous measurements, it was possible to measure how the Earth's magnetic field affects arriving solar wind particles. For about two-thirds of each orbit, the Moon is outside of the Earth's magnetic field. At these times, a typical proton density was 10 to 20 per cubic centimeter, with most protons having velocities between 400 and 650 kilometers per second. For about five days of each month, the Moon is inside the Earth's geomagnetic tail, and typically no solar wind particles were detectable. For the remainder of each lunar orbit, the Moon is in a transitional region known as the magnetosheath, where the Earth's magnetic field affects the solar wind but does not completely exclude it. In this region, the particle flux is reduced, with typical proton velocities of 250 to 450 kilometers per second. During the lunar night, the spectrometer was shielded from the solar wind by the Moon and no solar wind particles were measured.[35]

Protons also occur in from extrasolar origin in space, from galactic cosmic rays, where they make up about 90% of the total particle flux. These protons often have higher energy than solar wind protons, but their intensity is far more uniform and less variable than protons coming from the Sun, the production of which is heavily affected by solar proton events such as coronal mass ejections.

Research has been performed on the dose-rate effects of protons, as typically found in space travel, on human health.[36][37] To be more specific, there are hopes to identify what specific chromosomes are damaged, and to define the damage, during cancer development from proton exposure.[36] Another study looks into determining "the effects of exposure to proton irradiation on neurochemical and behavioral endpoints, including dopaminergic functioning, amphetamine-induced conditioned taste aversion learning, and spatial learning and memory as measured by the Morris water maze.[37] Electrical charging of a spacecraft due to interplanetary proton bombardment has also been proposed for study.[38] There are many more studies that pertain to space travel, including galactic cosmic rays and their possible health effects, and solar proton event exposure.

The American Biostack and Soviet Biorack space travel experiments have demonstrated the severity of molecular damage induced by heavy ions on micro organisms including Artemia cysts.[39]

6.9 Antiproton

Main article: Antiproton

CPT-symmetry puts strong constraints on the relative properties of particles and antiparticles and, therefore, is open to stringent tests. For example, the charges of the proton and antiproton must sum to exactly zero. This equality has been tested to one part in 10^8. The equality of their masses has also been tested to better than one part in 10^8. By holding antiprotons in a Penning trap, the equality of the charge to mass ratio of the proton and the antiproton has been tested to one part in 6×10^9.[40] The magnetic moment of the antiproton has been measured with error of 8×10^{-3} nuclear Bohr magnetons, and is found to be equal and opposite to that of the proton.

6.10 See also

- Fermion field

- Hydrogen

- Hydron (chemistry)

- List of particles

- Proton-proton chain reaction

- Quark model

- Proton spin crisis

6.11 References

[1] Mohr, P.J.; Taylor, B.N. and Newell, D.B. (2011), "The 2010 CODATA Recommended Values of the Fundamental Physical Constants", National Institute of Standards and Technology, Gaithersburg, MD, US.

[2] Cho, Adiran (2 April 2010). "Mass of the Common Quark Finally Nailed Down". *http://news.sciencemag.org"*. *American Association for the Advancement of Science. Retrieved 27 September 2014.*

[3] "Proton size puzzle reinforced!". Paul Shearer Institute. 25 January 2013.

[4] Adair, R.K. (1989). *The Great Design: Particles, Fields, and Creation*. Oxford University Press. p. 214.

[5] Cottingham, W.N.; Greenwood, D.A. (1986). *An Introduction to Nuclear Physics*. Cambridge University Press. ISBN 978034.

[6] Basdevant, J.-L.; Rich, J.; M. Spiro (2005). *Fundamentals in Nuclear Physics*. Springer. p. 155. ISBN 0-387-01672-4.

[7] Department of Chemistry and Biochemistry UCLA Eric R. Scerri Lecturer. *The Periodic Table : Its Story and Its Significance: Its Story and Its Significance*. Oxford University Press. ISBN 978-0-19-534567-4.

[8] Petrucci, R.H.; Harwood, W.S.; Herring, F.G. (2002). *General Chemistry* (8th ed.). p. 41.

[9] See meeting report and announcement

[10] Romer A (1997). "Proton or prouton? Rutherford and the depths of the atom".*Amer. J. Phys.***65**(8): 707.Bibcode:1997Am07R. doi:10.1119/1.18640.

[11] Rutherford reported acceptance by the *British Association* in a footnote to Masson, O. (1921). "XXIV.The constitution of atoms". *Philosophical Magazine Series 6* **41** (242): 281. doi:10.1080/14786442108636219.

[12] Pais, A. (1986) *Inward Bound*, Oxford Press, ISBN 0198519974, p. 296. Pais believed the first science literature use of the word *proton* occurs in "Physics at the British Association". *Nature* **106** (2663): 357. 1920. doi:10.1038/106357a0.

[13] Buccella, F.; Miele, G.; Rosa, L.; Santorelli, P.; Tuzi, T. (1989). "An upper limit for the proton lifetime in SO(10)". *Physics Letters B* **233**: 178. doi:10.1016/0370-2693(89)90637-0.

[14] Lee, D. G.; Mohapatra, R.; Parida, M.; Rani, M. (1995). "Predictions for the proton lifetime in minimal nonsupersymmetric SO(10) models: An update". *Physical Review D* **51**: 229. arXiv:hep-ph/9404238. doi:10.1103/PhysRevD.51.229.

[15] "Proton lifetime is longer than 1034 years". Kamioka Observatory. November 2009.

[16] Nishino, H.; Clark, S.; Abe, K.; Hayato, Y.; Iida, T.; Ikeda, M.; Kameda, J.; Kobayashi, K.; Koshio, Y.; Miura, M.; Moriyama, S.; Nakahata, M.; Nakayama, S.; Obayashi, Y.; Ogawa, H.; Sekiya, H.; Shiozawa, M.; Suzuki, Y.; Takeda, A.; Takenaga, Y.; Takeuchi, Y.; Ueno, K.; Ueshima, K.; Watanabe, H.; Yamada, S.; Hazama, S.; Higuchi, I.; Ishihara, C.; Kajita, T. et al. (2009). "Search for Proton Decay via p→e$^+$π^0 and p→μ$^+$π^0 in a Large Water Cherenkov Detector". *Physical Review Letters* **102** (14). arXiv:0903.0676. Bibcode:2009PhRvL.102n1801N. doi:10.1103/PhysRevLett.102.141801.

[17] Ahmed, S.; Anthony, A.; Beier, E.; Bellerive, A.; Biller, S.; Boger, J.; Boulay, M.; Bowler, M.; Bowles, T.; Brice, S.; Bullard, T.; Chan, Y.; Chen, M.; Chen, X.; Cleveland, B.; Cox, G.; Dai, X.; Dalnoki-Veress, F.; Doe, P.; Dosanjh, R.; Doucas, G.; Dragowsky, M.; Duba, C.; Duncan, F.; Dunford, M.; Dunmore, J.; Earle, E.; Elliott, S.; Evans, H. et al. (2004). "Constraints on Nucleon Decay via Invisible Modes from the Sudbury Neutrino Observatory". *Physical Review Letters* **92** (10). arXiv:hep-ex/0310030. Bibcode:2004PhRvL..92j2004A. doi:10.1103/PhysRevLett.92.102004. PMID 15089201.

[18] Watson, A. (2004). *The Quantum Quark*. Cambridge University Press. pp. 285–286. ISBN 0-521-82907-0.

[19] Timothy Paul Smith (2003). *Hidden Worlds: Hunting for Quarks in Ordinary Matter*. Princeton University Press. ISBN 0-691-05773-7.

[20] Weise, W.; Green, A.M. (1984). *Quarks and Nuclei*. World Scientific. pp. 65–66. ISBN 9971-966-61-1.

[21] Ball, Philip (Nov 20, 2008). "Nuclear masses calculated from scratch". Nature. doi:10.1038/news.2008.1246. Retrieved Aug 27, 2014.

[22] Reynolds, Mark (Apr 2009). "Calculating the Mass of a Proton". *CNRS international magazine* (CNRS) (13). ISSN 2270-5317. Retrieved Aug 27, 2014.

[23] See this news report and links

[24] Durr, S.; Fodor, Z.; Frison, J.; Hoelbling, C.; Hoffmann, R.; Katz, S. D.; Krieg, S.; Kurth, T.; Lellouch, L.; Lippert, T.; Szabo, K. K.; Vulvert, G. (2008). "Ab Initio Determination of Light Hadron Masses". *Science* **322** (5905): 1224–7. arXiv:0906.3599. doi:10.1126/science.1163233. PMID 19023076.

[25] Perdrisat, C. F.; Punjabi, V.; Vanderhaeghen, M. (2007). "Nucleon electromagnetic form factors". *Progress in Particle and Nuclear Physics* **59** (2): 694. arXiv:hep-ph/0612014. Bibcode:2007PrPNP..59..694P. doi:10.1016/j.ppnp.2007.05.001.

[26] Boffi, Sigfrido; Pasquini, Barbara (2007). "Generalized parton distributions and the structure of the nucleon". *Rivista del Nuovo Cimento* **30**. arXiv:0711.2625. Bibcode:2007NCimR..30..387B. doi:10.1393/ncr/i2007-10025-7.

[27] Joshua, Erlich (December 2008). "Recent Results in AdS/QCD". *Proceedings, 8th Conference on Quark Confinement and the Hadron Spectrum, September 1–6, 2008, Mainz, Germany*. arXiv:0812.4976.

[28] Pietro, Colangelo; Alex, Khodjamirian (October 2000). "QCD Sum Rules, a Modern Perspective". In Shifman, M. *At the Frontier of Particle Physics / Handbook of QCD*. World Scientific. arXiv:hep-ph/0010175.

[29] Pohl, R.; Antognini, A.; Nez, F. O.; Amaro, F. D.; Biraben, F. O.; Cardoso, J. O. M. R.; Covita, D. S.; Dax, A.; Dhawan, S.; Fernandes, L. M. P.; Giesen, A.; Graf, T.; Hänsch, T. W.; Indelicato, P.; Julien, L.; Kao, C. Y.; Knowles, P.; Le Bigot, E. O.; Liu, Y. W.; Lopes, J. A. M.; Ludhova, L.; Monteiro, C. M. B.; Mulhauser, F. O.; Nebel, T.; Rabinowitz, P.; Dos Santos, J. M. F.; Schaller, L. A.; Schuhmann, K.; Schwob, C. et al. (2010). "The size of the proton". *Nature* **466** (7303): 213–6. doi:10.1038/nature09250. PMID 20613837.

[30] Antognini, A.; Nez, F.; Schuhmann, K.; Amaro, F. D.; Biraben, F.; Cardoso, J. M. R.; Covita, D. S.; Dax, A.; Dhawan, S.; Diepold, M.; Fernandes, L. M. P.; Giesen, A.; Gouvea, A. L.; Graf, T.; Hänsch, T. W.; Indelicato, P.; Julien, L.; Kao, C. -Y.; Knowles, P.; Kottmann, F.; Le Bigot, E. -O.; Liu, Y. -W.; Lopes, J. A. M.; Ludhova, L.; Monteiro, C. M. B.; Mulhauser, F.; Nebel, T.; Rabinowitz, P.; Dos Santos, J. M. F.; Schaller, L. A. (2013). "Proton Structure from the Measurement of 2S-2P Transition Frequencies of Muonic Hydrogen". *Science* **339** (6118): 417–420. doi:10.1126/science.1230016. PMID 23349284.

[31] Researchers Observes Unexpectedly Small Proton Radius in a Precision Experiment. *Azonano.* July 9, 2010

[32] "The Proton Just Got Smaller". *Photonics.Com.* 12 July 2010. Retrieved 2010-07-19.

[33] Carlson, Carl E. (February 19, 2015), *The Proton Radius Puzzle*, arXiv:1502.05314

[34] Headrick, J.M.; Diken, E.G.; Walters, R. S.; Hammer, N. I.; Christie, R.A.; Cui, J.; Myshakin, E.M.; Duncan, M.A.; Johnson, M.A.; Jordan, K.D. (2005). "Spectral Signatures of Hydrated Proton Vibrations in Water Clusters". *Science* **308** (5729): 1765–69. Bibcode:2005Sci...308.1765H. doi:10.1126/science.1113094. PMID 15961665.

[35] "Apollo 11 Mission". Lunar and Planetary Institute. 2009. Retrieved 2009-06-12.

[36] "Space Travel and Cancer Linked? Stony Brook Researcher Secures NASA Grant to Study Effects of Space Radiation". Brookhaven National Laboratory. 12 December 2007. Retrieved 2009-06-12.

[37] Shukitt-Hale, B.; Szprengiel, A.; Pluhar, J.; Rabin, B.M.; Joseph, J.A. "The effects of proton exposure on neurochemistry and behavior". Elsevier/COSPAR. Retrieved 2009-06-12.

[38] Green, N.W.; Frederickson, A.R. "A Study of Spacecraft Charging due to Exposure to Interplanetary Protons" (PDF). Jet Propulsion Laboratory. Retrieved 2009-06-12.

[39] Planel, H. (2004). *Space and life: an introduction to space biology and medicine*. CRC Press. pp. 135–138. ISBN 0-415-31759-2.

[40] Gabrielse, G. (2006). "Antiproton mass measurements". *International Journal of Mass Spectrometry* **251** (2–3): 273–280. Bibcode:2006IJMSp.251..273G. doi:10.1016/j.ijms.2006.02.013.

6.12   External links

- Particle Data Group

- Large Hadron Collider

- Eaves, Laurence; Copeland, Ed; Padilla, Antonio (Tony) (2010). "The shrinking proton". *Sixty Symbols*. Brady Haran for the University of Nottingham.

Ernest Rutherford at the first Solvay Conference, 1911

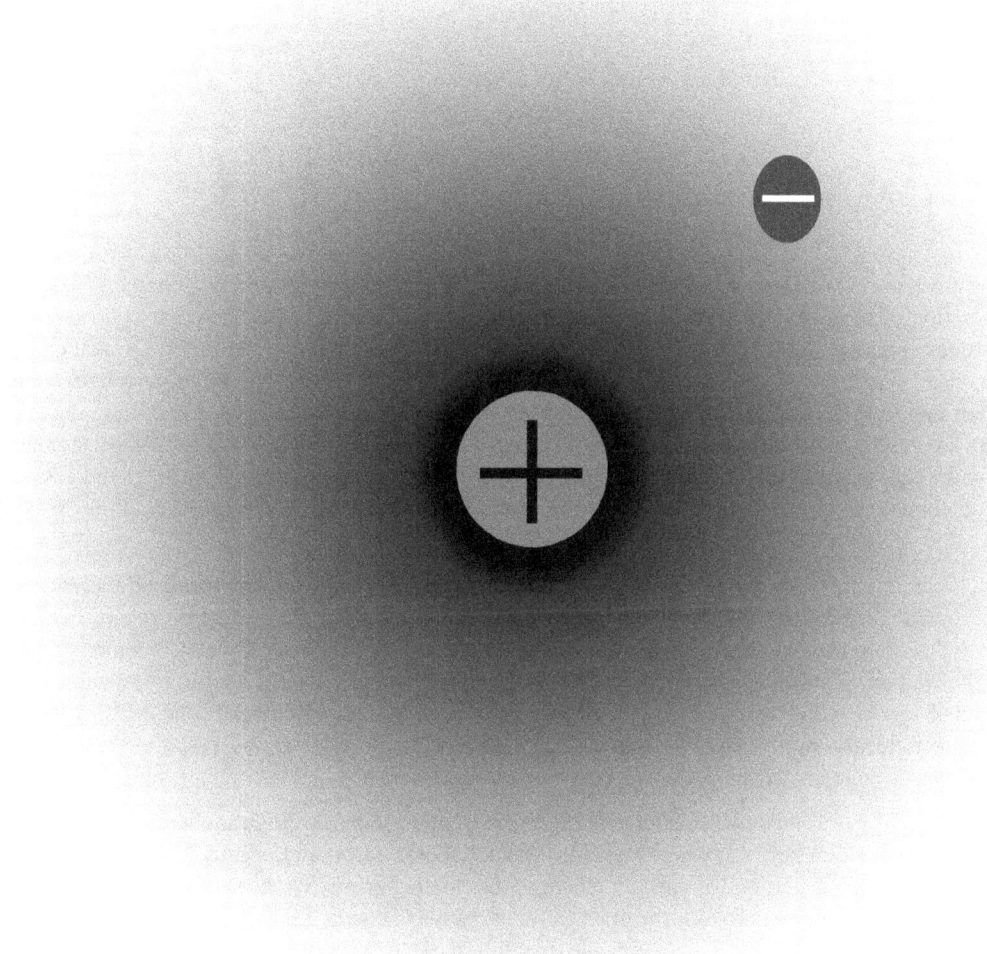

*Protium, the most common isotope of hydrogen, consists of one proton and one electron (it has no neutrons). The term "hydrogen ion"
(H+
) implies that that H-atom has lost its one electron, causing only a proton to remain. Thus, in chemistry, the terms "proton" and "hydrogen
ion" (for the protium isotope) are used synonymously*

Chapter 7

Electron

For other uses, see Electron (disambiguation).

The **electron** is a subatomic particle, symbol e− or β−, with a negative elementary electric charge.[7] Electrons belong to the first generation of the lepton particle family,[8] and are generally thought to be elementary particles because they have no known components or substructure.[1] The electron has a mass that is approximately 1/1836 that of the proton.[9] Quantum mechanical properties of the electron include an intrinsic angular momentum (spin) of a half-integer value in units of $\hbar$, which means that it is a fermion. Being fermions, no two electrons can occupy the same quantum state, in accordance with the Pauli exclusion principle.[8] Like all matter, electrons have properties of both particles and waves, and so can collide with other particles and can be diffracted like light. The wave properties of electrons are easier to observe with experiments than those of other particles like neutrons and protons because electrons have a lower mass and hence a higher De Broglie wavelength for typical energies.

Many physical phenomena involve electrons in an essential role, such as electricity, magnetism, and thermal conductivity, and they also participate in gravitational, electromagnetic and weak interactions.[10] An electron generates an electric field surrounding it. An electron moving relative to an observer generates a magnetic field. External magnetic fields deflect an electron. Electrons radiate or absorb energy in the form of photons when accelerated. Laboratory instruments are capable of containing and observing individual electrons as well as electron plasma using electromagnetic fields, whereas dedicated telescopes can detect electron plasma in outer space. Electrons have many applications, including electronics, welding, cathode ray tubes, electron microscopes, radiation therapy, lasers, gaseous ionization detectors and particle accelerators.

Interactions involving electrons and other subatomic particles are of interest in fields such as chemistry and nuclear physics. The Coulomb force interaction between positive protons inside atomic nuclei and negative electrons composes atoms. Ionization or changes in the proportions of particles changes the binding energy of the system. The exchange or sharing of the electrons between two or more atoms is the main cause of chemical bonding.[11] British natural philosopher Richard Laming first hypothesized the concept of an indivisible quantity of electric charge to explain the chemical properties of atoms in 1838;[3] Irish physicist George Johnstone Stoney named this charge 'electron' in 1891, and J. J. Thomson and his team of British physicists identified it as a particle in 1897.[5][12][13] Electrons can also participate in nuclear reactions, such as nucleosynthesis in stars, where they are known as beta particles. Electrons may be created through beta decay of radioactive isotopes and in high-energy collisions, for instance when cosmic rays enter the atmosphere. The antiparticle of the electron is called the positron; it is identical to the electron except that it carries electrical and other charges of the opposite sign. When an electron collides with a positron, both particles may be totally annihilated, producing gamma ray photons.

7.1   History

See also: History of electromagnetism

The ancient Greeks noticed that amber attracted small objects when rubbed with fur. Along with lightning, this phenomenon is one of humanity's earliest recorded experiences with electricity. [14] In his 1600 treatise *De Magnete*, the English scientist William Gilbert coined the New Latin term *electricus*, to refer to this property of attracting small objects after being rubbed. [15] Both *electric* and *electricity* are derived from the Latin *ēlectrum* (also the root of the alloy of the same name), which came from the Greek word for amber, ἤλεκτρον (*ēlektron*).

In the early 1700s, Francis Hauksbee and French chemist Charles François de Fay independently discovered what they believed were two kinds of frictional electricity—one generated from rubbing glass, the other from rubbing resin. From this, Du Fay theorized that electricity consists of two electrical fluids, *vitreous* and *resinous*, that are separated by friction, and that neutralize each other when combined.[16] A decade later Benjamin Franklin proposed that electricity was not from different types of electrical fluid, but the same electrical fluid under different pressures. He gave them the modern charge nomenclature of positive and negative respectively.[17] Franklin thought of the charge carrier as being positive, but he did not correctly identify which situation was a surplus of the charge carrier, and which situation was a deficit.[18]

Between 1838 and 1851, British natural philosopher Richard Laming developed the idea that an atom is composed of a core of matter surrounded by subatomic particles that had unit electric charges.[2] Beginning in 1846, German physicist William Weber theorized that electricity was composed of positively and negatively charged fluids, and their interaction was governed by the inverse square law. After studying the phenomenon of electrolysis in 1874, Irish physicist George Johnstone Stoney suggested that there existed a "single definite quantity of electricity", the charge of a monovalent ion. He was able to estimate the value of this elementary charge e by means of Faraday's laws of electrolysis.[19] However, Stoney believed these charges were permanently attached to atoms and could not be removed. In 1881, German physicist Hermann von Helmholtz argued that both positive and negative charges were divided into elementary parts, each of which "behaves like atoms of electricity".[3]

Stoney initially coined the term *electrolion* in 1881. Ten years later, he switched to *electron* to describe these elementary charges, writing in 1894: "... an estimate was made of the actual amount of this most remarkable fundamental unit of electricity, for which I have since ventured to suggest the name *electron*". A 1906 proposal to use *electrion* failed because Hendrik Lorentz preferred *electron*.[20][21] The word *electron* is a combination of the words *electric* and *ion*.[22] The suffix *-on* which is now used to designate other subatomic particles, such as a proton or neutron, is in turn derived from electron.[23][24]

7.1.1 Discovery

The German physicist Johann Wilhelm Hittorf studied electrical conductivity in rarefied gases: in 1869, he discovered a glow emitted from the cathode that increased in size with decrease in gas pressure. In 1876, the German physicist Eugen Goldstein showed that the rays from this glow cast a shadow, and he dubbed the rays cathode rays.[26] During the 1870s, the English chemist and physicist Sir William Crookes developed the first cathode ray tube to have a high vacuum inside.[27] He then showed that the luminescence rays appearing within the tube carried energy and moved from the cathode to the anode. Furthermore, by applying a magnetic field, he was able to deflect the rays, thereby demonstrating that the beam behaved as though it were negatively charged.[28][29] In 1879, he proposed that these properties could be explained by what he termed 'radiant matter'. He suggested that this was a fourth state of matter, consisting of negatively charged molecules that were being projected with high velocity from the cathode.[30]

The German-born British physicist Arthur Schuster expanded upon Crookes' experiments by placing metal plates parallel to the cathode rays and applying an electric potential between the plates. The field deflected the rays toward the positively charged plate, providing further evidence that the rays carried negative charge. By measuring the amount of deflection for a given level of current, in 1890 Schuster was able to estimate the charge-to-mass ratio of the ray components. However, this produced a value that was more than a thousand times greater than what was expected, so little credence was given to his calculations at the time.[28][31]

In 1892 Hendrik Lorentz suggested that the mass of these particles (electrons) could be a consequence of their electric charge.[32]

In 1896, the British physicist J. J. Thomson, with his colleagues John S. Townsend and H. A. Wilson,[12] performed experiments indicating that cathode rays really were unique particles, rather than waves, atoms or molecules as was believed earlier.[5] Thomson made good estimates of both the charge e and the mass m, finding that cathode ray particles, which he called "corpuscles," had perhaps one thousandth of the mass of the least massive ion known: hydrogen.[5][13] He

A beam of electrons deflected in a circle by a magnetic field[25]

showed that their charge to mass ratio, e/m, was independent of cathode material. He further showed that the negatively charged particles produced by radioactive materials, by heated materials and by illuminated materials were universal.[5][33] The name electron was again proposed for these particles by the Irish physicist George F. Fitzgerald, and the name has since gained universal acceptance.[28]

While studying naturally fluorescing minerals in 1896, the French physicist Henri Becquerel discovered that they emitted radiation without any exposure to an external energy source. These radioactive materials became the subject of much interest by scientists, including the New Zealand physicist Ernest Rutherford who discovered they emitted particles. He designated these particles alpha and beta, on the basis of their ability to penetrate matter.[34] In 1900, Becquerel showed that the beta rays emitted by radium could be deflected by an electric field, and that their mass-to-charge ratio was the same as for cathode rays.[35] This evidence strengthened the view that electrons existed as components of atoms.[36][37]

The electron's charge was more carefully measured by the American physicists Robert Millikan and Harvey Fletcher in their oil-drop experiment of 1909, the results of which were published in 1911. This experiment used an electric field to prevent a charged droplet of oil from falling as a result of gravity. This device could measure the electric charge from as few as 1–150 ions with an error margin of less than 0.3%. Comparable experiments had been done earlier by Thomson's team,[5] using clouds of charged water droplets generated by electrolysis,[12] and in 1911 by Abram Ioffe, who independently obtained the same result as Millikan using charged microparticles of metals, then published his results in 1913.[38] However, oil drops were more stable than water drops because of their slower evaporation rate, and thus more suited to precise experimentation over longer periods of time.[39]

Around the beginning of the twentieth century, it was found that under certain conditions a fast-moving charged particle caused a condensation of supersaturated water vapor along its path. In 1911, Charles Wilson used this principle to devise his cloud chamber so he could photograph the tracks of charged particles, such as fast-moving electrons.[40]

7.1.2 Atomic theory

See also: The proton–electron model of the nucleus

 By 1914, experiments by physicists Ernest Rutherford, Henry Moseley, James Franck and Gustav Hertz had largely established the structure of an atom as a dense nucleus of positive charge surrounded by lower-mass electrons.[41] In 1913, Danish physicist Niels Bohr postulated that electrons resided in quantized energy states, with the energy determined by the angular momentum of the electron's orbits about the nucleus. The electrons could move between these states, or orbits, by the emission or absorption of photons at specific frequencies. By means of these quantized orbits, he accurately explained the spectral lines of the hydrogen atom.[42] However, Bohr's model failed to account for the relative intensities of the spectral lines and it was unsuccessful in explaining the spectra of more complex atoms.[41]

Chemical bonds between atoms were explained by Gilbert Newton Lewis, who in 1916 proposed that a covalent bond between two atoms is maintained by a pair of electrons shared between them.[43] Later, in 1927, Walter Heitler and Fritz London gave the full explanation of the electron-pair formation and chemical bonding in terms of quantum mechanics.[44] In 1919, the American chemist Irving Langmuir elaborated on the Lewis' static model of the atom and suggested that all electrons were distributed in successive "concentric (nearly) spherical shells, all of equal thickness".[45] The shells were, in turn, divided by him in a number of cells each containing one pair of electrons. With this model Langmuir was able to qualitatively explain the chemical properties of all elements in the periodic table,[44] which were known to largely repeat themselves according to the periodic law.[46]

In 1924, Austrian physicist Wolfgang Pauli observed that the shell-like structure of the atom could be explained by a set of four parameters that defined every quantum energy state, as long as each state was inhabited by no more than a single electron. (This prohibition against more than one electron occupying the same quantum energy state became known as the Pauli exclusion principle.)[47] The physical mechanism to explain the fourth parameter, which had two distinct possible values, was provided by the Dutch physicists Samuel Goudsmit and George Uhlenbeck. In 1925, Goudsmit and Uhlenbeck suggested that an electron, in addition to the angular momentum of its orbit, possesses an intrinsic angular momentum and magnetic dipole moment.[41][48] The intrinsic angular momentum became known as spin, and explained the previously mysterious splitting of spectral lines observed with a high-resolution spectrograph; this phenomenon is known as fine structure splitting.[49]

7.1.3 Quantum mechanics

See also: History of quantum mechanics

In his 1924 dissertation *Recherches sur la théorie des quanta* (Research on Quantum Theory), French physicist Louis de Broglie hypothesized that all matter possesses a de Broglie wave similar to light.[50] That is, under the appropriate conditions, electrons and other matter would show properties of either particles or waves. The corpuscular properties of a particle are demonstrated when it is shown to have a localized position in space along its trajectory at any given moment.[51] Wave-like nature is observed, for example, when a beam of light is passed through parallel slits and creates interference patterns. In 1927, the interference effect was found in a beam of electrons by English physicist George Paget Thomson with a thin metal film and by American physicists Clinton Davisson and Lester Germer using a crystal of nickel.[52]

De Broglie's prediction of a wave nature for electrons led Erwin Schrödinger to postulate a wave equation for electrons moving under the influence of the nucleus in the atom. In 1926, this equation, the Schrödinger equation, successfully described how electron waves propagated.[53] Rather than yielding a solution that determined the location of an electron over time, this wave equation also could be used to predict the probability of finding an electron near a position, especially a position near where the electron was bound in space, for which the electron wave equations did not change in time. This approach led to a second formulation of quantum mechanics (the first being by Heisenberg in 1925), and solutions of Schrödinger's equation, like Heisenberg's, provided derivations of the energy states of an electron in a hydrogen atom that were equivalent to those that had been derived first by Bohr in 1913, and that were known to reproduce the hydrogen spectrum.[54] Once spin and the interaction between multiple electrons were considered, quantum mechanics later made it possible to predict the configuration of electrons in atoms with higher atomic numbers than hydrogen.[55]

In 1928, building on Wolfgang Pauli's work, Paul Dirac produced a model of the electron – the Dirac equation, consistent with relativity theory, by applying relativistic and symmetry considerations to the hamiltonian formulation of the quantum mechanics of the electro-magnetic field.[56] To resolve some problems within his relativistic equation, in 1930 Dirac developed a model of the vacuum as an infinite sea of particles having negative energy, which was dubbed the Dirac sea. This led him to predict the existence of a positron, the antimatter counterpart of the electron.[57] This particle was discovered in 1932 by Carl Anderson, who proposed calling standard electrons *negatrons*, and using *electron* as a generic term to describe both the positively and negatively charged variants.

In 1947 Willis Lamb, working in collaboration with graduate student Robert Retherford, found that certain quantum states of hydrogen atom, which should have the same energy, were shifted in relation to each other, the difference being the Lamb shift. About the same time, Polykarp Kusch, working with Henry M. Foley, discovered the magnetic moment of the electron is slightly larger than predicted by Dirac's theory. This small difference was later called anomalous magnetic dipole moment of the electron. This difference was later explained by the theory of quantum electrodynamics, developed by Sin-Itiro Tomonaga, Julian Schwinger and Richard Feynman in the late 1940s.[58]

7.1.4 Particle accelerators

With the development of the particle accelerator during the first half of the twentieth century, physicists began to delve deeper into the properties of subatomic particles.[59] The first successful attempt to accelerate electrons using electromagnetic induction was made in 1942 by Donald Kerst. His initial betatron reached energies of 2.3 MeV, while subsequent betatrons achieved 300 MeV. In 1947, synchrotron radiation was discovered with a 70 MeV electron synchrotron at General Electric. This radiation was caused by the acceleration of electrons, moving near the speed of light, through a magnetic field.[60]

With a beam energy of 1.5 GeV, the first high-energy particle collider was ADONE, which began operations in 1968.[61] This device accelerated electrons and positrons in opposite directions, effectively doubling the energy of their collision when compared to striking a static target with an electron.[62] The Large Electron–Positron Collider (LEP) at CERN, which was operational from 1989 to 2000, achieved collision energies of 209 GeV and made important measurements for the Standard Model of particle physics.[63][64]

7.1.5 Confinement of individual electrons

Individual electrons can now be easily confined in ultra small (L = 20 nm, W = 20 nm) CMOS transistors operated at cryogenic temperature over a range of −269 °C (4 K) to about −258 °C (15 K).[65] The electron wavefunction spreads in a semiconductor lattice and negligibly interacts with the valence band electrons, so it can be treated in the single particle formalism, by replacing its mass with the effective mass tensor.

7.2 Characteristics

7.2.1 Classification

In the Standard Model of particle physics, electrons belong to the group of subatomic particles called leptons, which are believed to be fundamental or elementary particles. Electrons have the lowest mass of any charged lepton (or electrically charged particle of any type) and belong to the first-generation of fundamental particles.[66] The second and third generation contain charged leptons, the muon and the tau, which are identical to the electron in charge, spin and interactions, but are more massive. Leptons differ from the other basic constituent of matter, the quarks, by their lack of strong interaction. All members of the lepton group are fermions, because they all have half-odd integer spin; the electron has spin $1/2$.[67]

7.2.2 Fundamental properties

The invariant mass of an electron is approximately 9.109×10^{-31} kilograms,[68] or 5.489×10^{-4} atomic mass units. On the basis of Einstein's principle of mass–energy equivalence, this mass corresponds to a rest energy of 0.511 MeV. The ratio between the mass of a proton and that of an electron is about 1836.[9][69] Astronomical measurements show that the proton-to-electron mass ratio has held the same value for at least half the age of the universe, as is predicted by the Standard Model.[70]

Electrons have an electric charge of -1.602×10^{-19} coulomb,[68] which is used as a standard unit of charge for sub-atomic particles, and is also called the elementary charge. This elementary charge has a relative standard uncertainty of 2.2×10^{-8}.[68] Within the limits of experimental accuracy, the electron charge is identical to the charge of a proton, but with the opposite sign.[71] As the symbol e is used for the elementary charge, the electron is commonly symbolized by e−, where the minus sign indicates the negative charge. The positron is symbolized by e+ because it has the same properties as the electron but with a positive rather than negative charge.[67][68]

The electron has an intrinsic angular momentum or spin of $1/2$.[68] This property is usually stated by referring to the electron as a spin-$1/2$ particle.[67] For such particles the spin magnitude is $\sqrt{3}/2\ \hbar$.[note 3] while the result of the measurement of a projection of the spin on any axis can only be $\pm \hbar/2$. In addition to spin, the electron has an intrinsic magnetic moment along its spin axis.[68] It is approximately equal to one Bohr magneton,[72][note 4] which is a physical constant equal to $9.27400915(23) \times 10^{-24}$ joules per tesla.[68] The orientation of the spin with respect to the momentum of the electron defines the property of elementary particles known as helicity.[73]

The electron has no known substructure.[1][74] and it is assumed to be a point particle with a point charge and no spatial extent.[8] In classical physics, the angular momentum and magnetic moment of an object depend upon its physical dimensions. Hence, the concept of a dimensionless electron possessing these properties might seem paradoxical and inconsistent to experimental observations in Penning traps which point to finite non-zero radius of the electron. A possible explanation of this paradoxical situation is given below in the "Virtual particles" subsection by taking into consideration the Foldy-Wouthuysen transformation. The issue of the radius of the electron is a challenging problem of the modern theoretical physics. The admission of the hypothesis of a finite radius of the electron is incompatible to the premises of the theory of relativity. On the other hand, a point-like electron (zero radius) generates serious mathematical difficulties due to the self-energy of the electron tending to infinity.[75] These aspects have been analyzed in detail by Dmitri Ivanenko and Arseny Sokolov.

Observation of a single electron in a Penning trap shows the upper limit of the particle's radius is 10^{-22} meters.[76] There *is* a physical constant called the "classical electron radius", with the much larger value of 2.8179×10^{-15} m, greater than the radius of the proton. However, the terminology comes from a simplistic calculation that ignores the effects of quantum mechanics; in reality, the so-called classical electron radius has little to do with the true fundamental structure of the electron.[77][note 5]

There are elementary particles that spontaneously decay into less massive particles. An example is the muon, which decays into an electron, a neutrino and an antineutrino, with a mean lifetime of 2.2×10^{-6} seconds. However, the electron is thought to be stable on theoretical grounds: the electron is the least massive particle with non-zero electric charge, so its decay would violate charge conservation.[78] The experimental lower bound for the electron's mean lifetime is 4.6×10^{26} years, at a 90% confidence level.[79][80]

7.2.3   Quantum properties

As with all particles, electrons can act as waves. This is called the wave–particle duality and can be demonstrated using the double-slit experiment.

The wave-like nature of the electron allows it to pass through two parallel slits simultaneously, rather than just one slit as would be the case for a classical particle. In quantum mechanics, the wave-like property of one particle can be described mathematically as a complex-valued function, the wave function, commonly denoted by the Greek letter psi (ψ). When the absolute value of this function is squared, it gives the probability that a particle will be observed near a location—a probability density.[81]:162–218

Electrons are identical particles because they cannot be distinguished from each other by their intrinsic physical properties. In quantum mechanics, this means that a pair of interacting electrons must be able to swap positions without an observable change to the state of the system. The wave function of fermions, including electrons, is antisymmetric, meaning that it changes sign when two electrons are swapped; that is, $\psi(r_1, r_2) = -\psi(r_2, r_1)$, where the variables r_1 and r_2 correspond to the first and second electrons, respectively. Since the absolute value is not changed by a sign swap, this corresponds to equal probabilities. Bosons, such as the photon, have symmetric wave functions instead.[81]:162–218

In the case of antisymmetry, solutions of the wave equation for interacting electrons result in a zero probability that each pair will occupy the same location or state. This is responsible for the Pauli exclusion principle, which precludes any two electrons from occupying the same quantum state. This principle explains many of the properties of electrons. For example, it causes groups of bound electrons to occupy different orbitals in an atom, rather than all overlapping each other in the same orbit.[81]:162–218

7.2.4   Virtual particles

Main article: Virtual particle

In a simplified picture, every photon spends some time as a combination of a virtual electron plus its antiparticle, the virtual positron, which rapidly annihilate each other shortly thereafter.[82] The combination of the energy variation needed to create these particles, and the time during which they exist, fall under the threshold of detectability expressed by the Heisenberg uncertainty relation, $\Delta E \cdot \Delta t \geq \hbar$. In effect, the energy needed to create these virtual particles, ΔE, can be "borrowed" from the vacuum for a period of time, Δt, so that their product is no more than the reduced Planck constant, $\hbar \approx 6.6 \times 10^{-16}$ eV·s. Thus, for a virtual electron, Δt is at most 1.3×10^{-21} s.[83]

While an electron–positron virtual pair is in existence, the coulomb force from the ambient electric field surrounding an electron causes a created positron to be attracted to the original electron, while a created electron experiences a repulsion. This causes what is called vacuum polarization. In effect, the vacuum behaves like a medium having a dielectric permittivity more than unity. Thus the effective charge of an electron is actually smaller than its true value, and the charge decreases with increasing distance from the electron.[84][85] This polarization was confirmed experimentally in 1997 using the Japanese TRISTAN particle accelerator.[86] Virtual particles cause a comparable shielding effect for the mass of the electron.[87]

The interaction with virtual particles also explains the small (about 0.1%) deviation of the intrinsic magnetic moment of the electron from the Bohr magneton (the anomalous magnetic moment).[72][88] The extraordinarily precise agreement of this predicted difference with the experimentally determined value is viewed as one of the great achievements of quantum electrodynamics.[89]

The apparent paradox (mentioned above in the properties subsection) of a point particle electron having intrinsic angular momentum and magnetic moment can be explained by the formation of virtual photons in the electric field generated by the electron. These photons cause the electron to shift about in a jittery fashion (known as zitterbewegung),[90] which results in a net circular motion with precession. This motion produces both the spin and the magnetic moment of the electron.[8][91] In atoms, this creation of virtual photons explains the Lamb shift observed in spectral lines.[84]

7.2.5   Interaction

An electron generates an electric field that exerts an attractive force on a particle with a positive charge, such as the proton, and a repulsive force on a particle with a negative charge. The strength of this force is determined by Coulomb's inverse square law.[92] When an electron is in motion, it generates a magnetic field.[81]:140 The Ampère-Maxwell law relates the magnetic field to the mass motion of electrons (the current) with respect to an observer. This property of induction supplies the magnetic field that drives an electric motor.[93] The electromagnetic field of an arbitrary moving charged particle is expressed by the Liénard–Wiechert potentials, which are valid even when the particle's speed is close to that of light (relativistic).

When an electron is moving through a magnetic field, it is subject to the Lorentz force that acts perpendicularly to the plane defined by the magnetic field and the electron velocity. This centripetal force causes the electron to follow a helical trajectory through the field at a radius called the gyroradius. The acceleration from this curving motion induces the electron to radiate energy in the form of synchrotron radiation.[81]:160[94][note 6] The energy emission in turn causes a recoil of the electron, known as the Abraham–Lorentz–Dirac Force, which creates a friction that slows the electron. This force is caused by a back-reaction of the electron's own field upon itself.[95]

Photons mediate electromagnetic interactions between particles in quantum electrodynamics. An isolated electron at a constant velocity cannot emit or absorb a real photon; doing so would violate conservation of energy and momentum.

Instead, virtual photons can transfer momentum between two charged particles. This exchange of virtual photons, for example, generates the Coulomb force.[96] Energy emission can occur when a moving electron is deflected by a charged particle, such as a proton. The acceleration of the electron results in the emission of Bremsstrahlung radiation.[97]

An inelastic collision between a photon (light) and a solitary (free) electron is called Compton scattering. This collision results in a transfer of momentum and energy between the particles, which modifies the wavelength of the photon by an amount called the Compton shift.[note 7] The maximum magnitude of this wavelength shift is $h/m_e c$, which is known as the Compton wavelength.[98] For an electron, it has a value of 2.43×10^{-12} m.[68] When the wavelength of the light is long (for instance, the wavelength of the visible light is 0.4–0.7 μm) the wavelength shift becomes negligible. Such interaction between the light and free electrons is called Thomson scattering or Linear Thomson scattering.[99]

The relative strength of the electromagnetic interaction between two charged particles, such as an electron and a proton, is given by the fine-structure constant. This value is a dimensionless quantity formed by the ratio of two energies: the electrostatic energy of attraction (or repulsion) at a separation of one Compton wavelength, and the rest energy of the charge. It is given by $\alpha \approx 7.297353 \times 10^{-3}$, which is approximately equal to $1/137$.[68]

When electrons and positrons collide, they annihilate each other, giving rise to two or more gamma ray photons. If the electron and positron have negligible momentum, a positronium atom can form before annihilation results in two or three gamma ray photons totalling 1.022 MeV.[100][101] On the other hand, high-energy photons may transform into an electron and a positron by a process called pair production, but only in the presence of a nearby charged particle, such as a nucleus.[102][103]

In the theory of electroweak interaction, the left-handed component of electron's wavefunction forms a weak isospin doublet with the electron neutrino. This means that during weak interactions, electron neutrinos behave like electrons. Either member of this doublet can undergo a charged current interaction by emitting or absorbing a W and be converted into the other member. Charge is conserved during this reaction because the W boson also carries a charge, canceling out any net change during the transmutation. Charged current interactions are responsible for the phenomenon of beta decay in a radioactive atom. Both the electron and electron neutrino can undergo a neutral current interaction via a Z0 exchange, and this is responsible for neutrino-electron elastic scattering.[104]

7.2.6   Atoms and molecules

Main article: Atom

 An electron can be *bound* to the nucleus of an atom by the attractive Coulomb force. A system of one or more electrons bound to a nucleus is called an atom. If the number of electrons is different from the nucleus' electrical charge, such an atom is called an ion. The wave-like behavior of a bound electron is described by a function called an atomic orbital. Each orbital has its own set of quantum numbers such as energy, angular momentum and projection of angular momentum, and only a discrete set of these orbitals exist around the nucleus. According to the Pauli exclusion principle each orbital can be occupied by up to two electrons, which must differ in their spin quantum number.

Electrons can transfer between different orbitals by the emission or absorption of photons with an energy that matches the difference in potential.[105] Other methods of orbital transfer include collisions with particles, such as electrons, and the Auger effect.[106] To escape the atom, the energy of the electron must be increased above its binding energy to the atom. This occurs, for example, with the photoelectric effect, where an incident photon exceeding the atom's ionization energy is absorbed by the electron.[107]

The orbital angular momentum of electrons is quantized. Because the electron is charged, it produces an orbital magnetic moment that is proportional to the angular momentum. The net magnetic moment of an atom is equal to the vector sum of orbital and spin magnetic moments of all electrons and the nucleus. The magnetic moment of the nucleus is negligible compared with that of the electrons. The magnetic moments of the electrons that occupy the same orbital (so called, paired electrons) cancel each other out.[108]

The chemical bond between atoms occurs as a result of electromagnetic interactions, as described by the laws of quantum mechanics.[109] The strongest bonds are formed by the sharing or transfer of electrons between atoms, allowing the formation of molecules.[11] Within a molecule, electrons move under the influence of several nuclei, and occupy molecular orbitals; much as they can occupy atomic orbitals in isolated atoms.[110] A fundamental factor in these molecular structures is the existence of electron pairs. These are electrons with opposed spins, allowing them to occupy the same molecular orbital without violating the Pauli exclusion principle (much like in atoms). Different molecular orbitals have different spatial distribution of the electron density. For instance, in bonded pairs (i.e. in the pairs that actually bind atoms together) electrons can be found with the maximal probability in a relatively small volume between the nuclei. On the contrary, in non-bonded pairs electrons are distributed in a large volume around nuclei.[111]

7.2.7 Conductivity

If a body has more or fewer electrons than are required to balance the positive charge of the nuclei, then that object has a net electric charge. When there is an excess of electrons, the object is said to be negatively charged. When there are fewer electrons than the number of protons in nuclei, the object is said to be positively charged. When the number of electrons and the number of protons are equal, their charges cancel each other and the object is said to be electrically neutral. A macroscopic body can develop an electric charge through rubbing, by the triboelectric effect.[115]

Independent electrons moving in vacuum are termed *free* electrons. Electrons in metals also behave as if they were free. In reality the particles that are commonly termed electrons in metals and other solids are quasi-electrons—quasiparticles, which have the same electrical charge, spin and magnetic moment as real electrons but may have a different mass.[116] When free electrons—both in vacuum and metals—move, they produce a net flow of charge called an electric current, which generates a magnetic field. Likewise a current can be created by a changing magnetic field. These interactions are described mathematically by Maxwell's equations.[117]

At a given temperature, each material has an electrical conductivity that determines the value of electric current when an electric potential is applied. Examples of good conductors include metals such as copper and gold, whereas glass and Teflon are poor conductors. In any dielectric material, the electrons remain bound to their respective atoms and the material behaves as an insulator. Most semiconductors have a variable level of conductivity that lies between the extremes of conduction and insulation.[118] On the other hand, metals have an electronic band structure containing partially filled electronic bands. The presence of such bands allows electrons in metals to behave as if they were free or delocalized electrons. These electrons are not associated with specific atoms, so when an electric field is applied, they are free to move like a gas (called Fermi gas)[119] through the material much like free electrons.

Because of collisions between electrons and atoms, the drift velocity of electrons in a conductor is on the order of millimeters per second. However, the speed at which a change of current at one point in the material causes changes in currents in other parts of the material, the velocity of propagation, is typically about 75% of light speed.[120] This occurs because electrical signals propagate as a wave, with the velocity dependent on the dielectric constant of the material.[121]

Metals make relatively good conductors of heat, primarily because the delocalized electrons are free to transport thermal energy between atoms. However, unlike electrical conductivity, the thermal conductivity of a metal is nearly independent of temperature. This is expressed mathematically by the Wiedemann–Franz law,[119] which states that the ratio of thermal conductivity to the electrical conductivity is proportional to the temperature. The thermal disorder in the metallic lattice increases the electrical resistivity of the material, producing a temperature dependence for electric current.[122]

When cooled below a point called the critical temperature, materials can undergo a phase transition in which they lose all resistivity to electric current, in a process known as superconductivity. In BCS theory, this behavior is modeled by pairs of electrons entering a quantum state known as a Bose–Einstein condensate. These Cooper pairs have their motion coupled to nearby matter via lattice vibrations called phonons, thereby avoiding the collisions with atoms that normally create electrical resistance.[123] (Cooper pairs have a radius of roughly 100 nm, so they can overlap each other.)[124] However, the mechanism by which higher temperature superconductors operate remains uncertain.

Electrons inside conducting solids, which are quasi-particles themselves, when tightly confined at temperatures close to absolute zero, behave as though they had split into three other quasiparticles: spinons, Orbitons and holons.[125][126] The former carries spin and magnetic moment, the next carries its orbital location while the latter electrical charge.

7.2.8 Motion and energy

According to Einstein's theory of special relativity, as an electron's speed approaches the speed of light, from an observer's point of view its relativistic mass increases, thereby making it more and more difficult to accelerate it from within the observer's frame of reference. The speed of an electron can approach, but never reach, the speed of light in a vacuum, *c*. However, when relativistic electrons—that is, electrons moving at a speed close to *c*—are injected into a dielectric medium such as water, where the local speed of light is significantly less than *c*, the electrons temporarily travel faster than light in the medium. As they interact with the medium, they generate a faint light called Cherenkov radiation.[127]

The effects of special relativity are based on a quantity known as the Lorentz factor, defined as $\gamma = 1/\sqrt{1 - v^2/c^2}$ where *v* is the speed of the particle. The kinetic energy K_e of an electron moving with velocity *v* is:

$$K_e = (\gamma - 1)m_e c^2,$$

where m_e is the mass of electron. For example, the Stanford linear accelerator can accelerate an electron to roughly 51 GeV.[128] Since an electron behaves as a wave, at a given velocity it has a characteristic de Broglie wavelength. This is given by $\lambda_e = h/p$ where *h* is the Planck constant and *p* is the momentum.[50] For the 51 GeV electron above, the wavelength is about 2.4×10^{-17} m, small enough to explore structures well below the size of an atomic nucleus.[129]

7.3   Formation

The Big Bang theory is the most widely accepted scientific theory to explain the early stages in the evolution of the Universe.[130] For the first millisecond of the Big Bang, the temperatures were over 10 billion Kelvin and photons had mean energies over a million electronvolts. These photons were sufficiently energetic that they could react with each other to form pairs of electrons and positrons. Likewise, positron-electron pairs annihilated each other and emitted energetic photons:

$$\gamma + \gamma \leftrightarrow e+ + e-$$

An equilibrium between electrons, positrons and photons was maintained during this phase of the evolution of the Universe. After 15 seconds had passed, however, the temperature of the universe dropped below the threshold where electron-positron formation could occur. Most of the surviving electrons and positrons annihilated each other, releasing gamma radiation that briefly reheated the universe.[131]

For reasons that remain uncertain, during the process of leptogenesis there was an excess in the number of electrons over positrons.[132] Hence, about one electron in every billion survived the annihilation process. This excess matched the excess of protons over antiprotons, in a condition known as baryon asymmetry, resulting in a net charge of zero for the universe.[133][134] The surviving protons and neutrons began to participate in reactions with each other—in the process known as nucleosynthesis, forming isotopes of hydrogen and helium, with trace amounts of lithium. This process peaked after about five minutes.[135] Any leftover neutrons underwent negative beta decay with a half-life of about a thousand seconds, releasing a proton and electron in the process,

$$n \rightarrow p + e- + \nu_e$$

For about the next 300000–400000 years, the excess electrons remained too energetic to bind with atomic nuclei.[136] What followed is a period known as recombination, when neutral atoms were formed and the expanding universe became transparent to radiation.[137]

Roughly one million years after the big bang, the first generation of stars began to form.[137] Within a star, stellar nucleosynthesis results in the production of positrons from the fusion of atomic nuclei. These antimatter particles immediately annihilate with electrons, releasing gamma rays. The net result is a steady reduction in the number of electrons, and a matching increase in the number of neutrons. However, the process of stellar evolution can result in the synthesis of radioactive isotopes. Selected isotopes can subsequently undergo negative beta decay, emitting an electron and antineutrino from the nucleus.[138] An example is the cobalt-60 (^{60}Co) isotope, which decays to form nickel-60 (60Ni).[139]

At the end of its lifetime, a star with more than about 20 solar masses can undergo gravitational collapse to form a black hole.[140] According to classical physics, these massive stellar objects exert a gravitational attraction that is strong enough to prevent anything, even electromagnetic radiation, from escaping past the Schwarzschild radius. However, quantum mechanical effects are believed to potentially allow the emission of Hawking radiation at this distance. Electrons (and positrons) are thought to be created at the event horizon of these stellar remnants.

When pairs of virtual particles (such as an electron and positron) are created in the vicinity of the event horizon, the random spatial distribution of these particles may permit one of them to appear on the exterior; this process is called quantum tunnelling. The gravitational potential of the black hole can then supply the energy that transforms this virtual particle into a real particle, allowing it to radiate away into space.[141] In exchange, the other member of the pair is given negative energy, which results in a net loss of mass-energy by the black hole. The rate of Hawking radiation increases with decreasing mass, eventually causing the black hole to evaporate away until, finally, it explodes.[142]

Cosmic rays are particles traveling through space with high energies. Energy events as high as 3.0×10^{20} eV have been recorded.[143] When these particles collide with nucleons in the Earth's atmosphere, a shower of particles is generated, including pions.[144] More than half of the cosmic radiation observed from the Earth's surface consists of muons. The particle called a muon is a lepton produced in the upper atmosphere by the decay of a pion.

$$\pi- \rightarrow \mu- + \nu$$
$$\mu$$

A muon, in turn, can decay to form an electron or positron.[145]

$$\mu- \rightarrow e- + \nu$$
$$e + \nu$$
$$\mu$$

7.4 Observation

Remote observation of electrons requires detection of their radiated energy. For example, in high-energy environments such as the corona of a star, free electrons form a plasma that radiates energy due to Bremsstrahlung radiation. Electron gas can undergo plasma oscillation, which is waves caused by synchronized variations in electron density, and these produce energy emissions that can be detected by using radio telescopes.[147]

The frequency of a photon is proportional to its energy. As a bound electron transitions between different energy levels of an atom, it absorbs or emits photons at characteristic frequencies. For instance, when atoms are irradiated by a source with a broad spectrum, distinct absorption lines appear in the spectrum of transmitted radiation. Each element or molecule displays a characteristic set of spectral lines, such as the hydrogen spectral series. Spectroscopic measurements of the strength and width of these lines allow the composition and physical properties of a substance to be determined.[148][149]

In laboratory conditions, the interactions of individual electrons can be observed by means of particle detectors, which allow measurement of specific properties such as energy, spin and charge.[107] The development of the Paul trap and Penning trap allows charged particles to be contained within a small region for long durations. This enables precise measurements of the particle properties. For example, in one instance a Penning trap was used to contain a single electron for a period of 10 months.[150] The magnetic moment of the electron was measured to a precision of eleven digits, which, in 1980, was a greater accuracy than for any other physical constant.[151]

The first video images of an electron's energy distribution were captured by a team at Lund University in Sweden, February 2008. The scientists used extremely short flashes of light, called attosecond pulses, which allowed an electron's motion to be observed for the first time.[152][153]

The distribution of the electrons in solid materials can be visualized by angle-resolved photoemission spectroscopy (ARPES). This technique employs the photoelectric effect to measure the reciprocal space—a mathematical representation of periodic structures that is used to infer the original structure. ARPES can be used to determine the direction, speed and scattering of electrons within the material.[154]

7.5 Plasma applications

7.5.1 Particle beams

Electron beams are used in welding.[156] They allow energy densities up to 10^7 W·cm^{-2} across a narrow focus diameter of 0.1–1.3 mm and usually require no filler material. This welding technique must be performed in a vacuum to prevent the electrons from interacting with the gas before reaching their target, and it can be used to join conductive materials that would otherwise be considered unsuitable for welding.[157][158]

Electron-beam lithography (EBL) is a method of etching semiconductors at resolutions smaller than a micrometer.[159] This technique is limited by high costs, slow performance, the need to operate the beam in the vacuum and the tendency of the electrons to scatter in solids. The last problem limits the resolution to about 10 nm. For this reason, EBL is primarily used for the production of small numbers of specialized integrated circuits.[160]

Electron beam processing is used to irradiate materials in order to change their physical properties or sterilize medical and food products.[161] Electron beams fluidise or quasi-melt glasses without significant increase of temperature on intensive irradiation: e.g. intensive electron radiation causes a many orders of magnitude decrease of viscosity and stepwise decrease of its activation energy.[162]

Linear particle accelerators generate electron beams for treatment of superficial tumors in radiation therapy. Electron therapy can treat such skin lesions as basal-cell carcinomas because an electron beam only penetrates to a limited depth before being absorbed, typically up to 5 cm for electron energies in the range 5–20 MeV. An electron beam can be used to supplement the treatment of areas that have been irradiated by X-rays.[163][164]

Particle accelerators use electric fields to propel electrons and their antiparticles to high energies. These particles emit synchrotron radiation as they pass through magnetic fields. The dependency of the intensity of this radiation upon spin polarizes the electron beam—a process known as the Sokolov–Ternov effect.[note 8] Polarized electron beams can be useful for various experiments. Synchrotron radiation can also cool the electron beams to reduce the momentum spread of the particles. Electron and positron beams are collided upon the particles' accelerating to the required energies; particle detectors observe the resulting energy emissions, which particle physics studies .[165]

7.5.2 Imaging

Low-energy electron diffraction (LEED) is a method of bombarding a crystalline material with a collimated beam of electrons and then observing the resulting diffraction patterns to determine the structure of the material. The required energy of the electrons is typically in the range 20–200 eV.[166] The reflection high-energy electron diffraction (RHEED) technique uses the reflection of a beam of electrons fired at various low angles to characterize the surface of crystalline materials. The beam energy is typically in the range 8–20 keV and the angle of incidence is 1–4°.[167][168]

The electron microscope directs a focused beam of electrons at a specimen. Some electrons change their properties, such as movement direction, angle, and relative phase and energy as the beam interacts with the material. Microscopists can record these changes in the electron beam to produce atomically resolved images of the material.[169] In blue light, conventional optical microscopes have a diffraction-limited resolution of about 200 nm.[170] By comparison, electron microscopes are limited by the de Broglie wavelength of the electron. This wavelength, for example, is equal to 0.0037 nm for electrons accelerated across a 100,000-volt potential.[171] The Transmission Electron Aberration-Corrected Microscope is capable of sub-0.05 nm resolution, which is more than enough to resolve individual atoms.[172] This capability makes the electron microscope a useful laboratory instrument for high resolution imaging. However, electron microscopes are expensive instruments that are costly to maintain.

Two main types of electron microscopes exist: transmission and scanning. Transmission electron microscopes function like overhead projectors, with a beam of electrons passing through a slice of material then being projected by lenses on a photographic slide or a charge-coupled device. Scanning electron microscopes rasteri a finely focused electron beam, as in a TV set, across the studied sample to produce the image. Magnifications range from 100× to 1,000,000× or higher for both microscope types. The scanning tunneling microscope uses quantum tunneling of electrons from a sharp metal tip into the studied material and can produce atomically resolved images of its surface.[173][174][175]

7.5.3   Other applications

In the free-electron laser (FEL), a relativistic electron beam passes through a pair of undulators that contain arrays of dipole magnets whose fields point in alternating directions. The electrons emit synchrotron radiation that coherently interacts with the same electrons to strongly amplify the radiation field at the resonance frequency. FEL can emit a coherent high-brilliance electromagnetic radiation with a wide range of frequencies, from microwaves to soft X-rays. These devices may find manufacturing, communication and various medical applications, such as soft tissue surgery.[176]

Electrons are important in cathode ray tubes, which have been extensively used as display devices in laboratory instruments, computer monitors and television sets.[177] In a photomultiplier tube, every photon striking the photocathode initiates an avalanche of electrons that produces a detectable current pulse.[178] Vacuum tubes use the flow of electrons to manipulate electrical signals, and they played a critical role in the development of electronics technology. However, they have been largely supplanted by solid-state devices such as the transistor.[179]

7.6   See also

- Anyon

- Electride

- Electron bubble

- Exoelectron emission

- *g*-factor

- Periodic systems of small molecules

- Spintronics

- Stern–Gerlach experiment

- Townsend discharge

- Zeeman effect

- List of particles

- Lepton

7.7   Notes

[1] The fractional version's denominator is the inverse of the decimal value (along with its relative standard uncertainty of 4.2×10^{-13} u).

[2] The electron's charge is the negative of elementary charge, which has a positive value for the proton.

[3] This magnitude is obtained from the spin quantum number as

$$S = \sqrt{s(s+1)} \cdot \frac{h}{2\pi}$$
$$= \frac{\sqrt{3}}{2}\hbar$$

for quantum number $s = {}^1\!/_2$.
See: Gupta, M.C. (2001). *Atomic and Molecular Spectroscopy*. New Age Publishers. p. 81. ISBN 81-224-1300-5.

[4] Bohr magneton:

$$\mu_{\mathrm{B}} = \frac{e\hbar}{2m_{\mathrm{e}}}.$$

[5] The classical electron radius is derived as follows. Assume that the electron's charge is spread uniformly throughout a spherical volume. Since one part of the sphere would repel the other parts, the sphere contains electrostatic potential energy. This energy is assumed to equal the electron's rest energy, defined by special relativity ($E = mc^2$).
From electrostatics theory, the potential energy of a sphere with radius r and charge e is given by:

$$E_{\mathrm{p}} = \frac{e^2}{8\pi\varepsilon_0 r},$$

where ε_0 is the vacuum permittivity. For an electron with rest mass m_0, the rest energy is equal to:

$$E_{\mathrm{p}} = m_0 c^2,$$

where c is the speed of light in a vacuum. Setting them equal and solving for r gives the classical electron radius.
See: Haken, H.; Wolf, H.C.; Brewer, W.D. (2005). *The Physics of Atoms and Quanta: Introduction to Experiments and Theory.* Springer. p. 70. ISBN 3-540-67274-5.

[6] Radiation from non-relativistic electrons is sometimes termed cyclotron radiation.

[7] The change in wavelength, $\Delta\lambda$, depends on the angle of the recoil, θ, as follows,

$$\Delta\lambda = \frac{h}{m_{\mathrm{e}}c}(1 - \cos\theta),$$

where c is the speed of light in a vacuum and m_{e} is the electron mass. See Zombeck (2007: 393, 396).

[8] The polarization of an electron beam means that the spins of all electrons point into one direction. In other words, the projections of the spins of all electrons onto their momentum vector have the same sign.

7.8   References

[1] Eichten, E.J.; Peskin, M.E.; Peskin, M. (1983). "New Tests for Quark and Lepton Substructure". *Physical Review Letters* **50** (11): 811–814. Bibcode:1983PhRvL..50..811E. doi:10.1103/PhysRevLett.50.811.

[2] Farrar, W.V. (1969). "Richard Laming and the Coal-Gas Industry, with His Views on the Structure of Matter". *Annals of Science* **25** (3): 243–254. doi:10.1080/00033796900200141.

[3] Arabatzis, T. (2006). *Representing Electrons: A Biographical Approach to Theoretical Entities.* University of Chicago Press. pp. 70–74. ISBN 0-226-02421-0.

[4] Buchwald, J.Z.; Warwick, A. (2001). *Histories of the Electron: The Birth of Microphysics.* MIT Press. pp. 195–203. ISBN 0-262-52424-4.

[5] Thomson, J.J. (1897). "Cathode Rays". *Philosophical Magazine* **44** (269): 293. doi:10.1080/14786449708621070.

[6] P.J. Mohr, B.N. Taylor, and D.B. Newell (2011), "The 2010 CODATA Recommended Values of the Fundamental Physical Constants" (Web Version 6.0). This database was developed by J. Baker, M. Douma, and S. Kotochigova. Available: http://physics.nist.gov/constants [Thursday, 02-Jun-2011 21:00:12 EDT]. National Institute of Standards and Technology, Gaithersburg, MD 20899.

[7] "JERRY COFF". Retrieved 10 September 2010.

[8] Curtis, L.J. (2003). *Atomic Structure and Lifetimes: A Conceptual Approach.* Cambridge University Press. p. 74. ISBN 0-521-53635-9.

[9] "CODATA value: proton-electron mass ratio". *2006 CODATA recommended values.* National Institute of Standards and Technology. Retrieved 2009-07-18.

[10] Anastopoulos, C. (2008). *Particle Or Wave: The Evolution of the Concept of Matter in Modern Physics.* Princeton University Press. pp. 236–237. ISBN 0-691-13512-6.

[11] Pauling, L.C. (1960). *The Nature of the Chemical Bond and the Structure of Molecules and Crystals: an introduction to modern structural chemistry* (3rd ed.). Cornell University Press. pp. 4–10. ISBN 0-8014-0333-2.

[12] Dahl (1997:122–185).

[13] Wilson, R. (1997). *Astronomy Through the Ages: The Story of the Human Attempt to Understand the Universe.* CRC Press. p. 138. ISBN 0-7484-0748-0.

[14] Shipley, J.T. (1945). *Dictionary of Word Origins.* The Philosophical Library. p. 133. ISBN 0-88029-751-4.

[15] Baigrie, B. (2006). *Electricity and Magnetism: A Historical Perspective.* Greenwood Press. pp. 7–8. ISBN 0-313-33358-0.

[16] Keithley, J.F. (1999). *The Story of Electrical and Magnetic Measurements: From 500 B.C. to the 1940s.* IEEE Press. pp. 15, 20. ISBN 0-7803-1193-0.

[17] "Benjamin Franklin (1706–1790)". *Eric Weisstein's World of Biography.* Wolfram Research. Retrieved 2010-12-16.

[18] Myers, R.L. (2006). *The Basics of Physics.* Greenwood Publishing Group. p. 242. ISBN 0-313-32857-9.

[19] Barrow, J.D. (1983). "Natural Units Before Planck". *Quarterly Journal of the Royal Astronomical Society* **24**: 24–26. Bibcode:

[20] Sōgo Okamura (1994). *History of Electron Tubes.* IOS Press. p. 11. ISBN 978-90-5199-145-1. Retrieved 29 May 2015. In 1881, Stoney named this electromagnetic 'electrolion'. It came to be called 'electron' from 1891. [...] In 1906, the suggestion to call cathode ray particles 'electrions' was brought up but through the opinion of Lorentz of Holland 'electrons' came to be widely used.

[21] Stoney, G.J. (1894). "Of the "Electron," or Atom of Electricity". *Philosophical Magazine* **38**(5): 418–420. doi:10.1080/1478643.

[22] "electron, n.2". OED Online. March 2013. Oxford University Press. Accessed 12 April 2013

[23] Soukhanov, A.H. ed. (1986). *Word Mysteries & Histories.* Houghton Mifflin Company. p. 73. ISBN 0-395-40265-4.

[24] Guralnik, D.B. ed. (1970). *Webster's New World Dictionary.* Prentice Hall. p. 450.

[25] Born, M.; Blin-Stoyle, R.J.; Radcliffe, J.M. (1989). *Atomic Physics.* Courier Dover. p. 26. ISBN 0-486-65984-4.

[26] Dahl (1997:55–58).

[27] DeKosky, R.K. (1983). "William Crookes and the quest for absolute vacuum in the 1870s". *Annals of Science* **40** (1): 1–18. doi:10.1080/00033798300200101.

[28] Leicester, H.M. (1971). *The Historical Background of Chemistry.* Courier Dover. pp. 221–222. ISBN 0-486-61053-5.

[29] Dahl (1997:64–78).

[30] Zeeman, P.; Zeeman, P. (1907). "Sir William Crookes, F.R.S". *Nature* **77** (1984): 1–3. Bibcode:1907Natur..77....1C. doi:10.1038/077001a0.

[31] Dahl (1997:99).

[32] Frank Wilczek: "Happy Birthday, Electron" *Scientific American*, June 2012.

[33] Thomson, J.J. (1906). "Nobel Lecture: Carriers of Negative Electricity" (PDF). The Nobel Foundation. Retrieved 2008-08-25.

[34] Trenn, T.J. (1976). "Rutherford on the Alpha-Beta-Gamma Classification of Radioactive Rays". *Isis* **67**(1): 61–75. doi:10.1086/5. JSTOR 231134.

[35] Becquerel, H. (1900). "Déviation du Rayonnement du Radium dans un Champ Électrique". *Comptes rendus de l'Académie des sciences* (in French) **130**: 809–815.

[36] Buchwald and Warwick (2001:90–91).

[37] Myers, W.G. (1976). "Becquerel's Discovery of Radioactivity in 1896". *Journal of Nuclear Medicine* **17** (7): 579–582. PMID 775027.

[38] Kikoin, I.K.; Sominskiĭ, I.S. (1961). "Abram Fedorovich Ioffe (on his eightieth birthday)". *Soviet Physics Uspekhi* **3** (5): 798–809. Bibcode:1961SvPhU...3..798K. doi:10.1070/PU1961v003n05ABEH005812. Original publication in Russian: Кикоин, И.К.; Соминский, М.С. (1960). "Академик А.Ф. Иоффе" (PDF). *Успехи Физических Наук* **72** (10): 303–321.

[39] Millikan, R.A. (1911). "The Isolation of an Ion, a Precision Measurement of its Charge, and the Correction of Stokes' Law". *Physical Review* **32** (2): 349–397. Bibcode:1911PhRvI..32..349M. doi:10.1103/PhysRevSeriesI.32.349.

[40] Das Gupta, N.N.; Ghosh, S.K. (1999). "A Report on the Wilson Cloud Chamber and Its Applications in Physics". *Reviews of Modern Physics* **18** (2): 225–290. Bibcode:1946RvMP...18..225G. doi:10.1103/RevModPhys.18.225.

[41] Smirnov, B.M. (2003). *Physics of Atoms and Ions*. Springer. pp. 14–21. ISBN 0-387-95550-X.

[42] Bohr, N. (1922). "Nobel Lecture: The Structure of the Atom" (PDF). The Nobel Foundation. Retrieved 2008-12-03.

[43] Lewis, G.N. (1916). "The Atom and the Molecule". *Journal of the American Chemical Society* **38** (4): 762–786. doi:10.10212.

[44] Arabatzis, T.; Gavroglu, K. (1997). "The chemists' electron". *European Journal of Physics* **18**(3): 150–163. Bibcode:1997EJPh0A. doi:10.1088/0143-0807/18/3/005.

[45] Langmuir, I. (1919). "The Arrangement of Electrons in Atoms and Molecules". *Journal of the American Chemical Society* **41** (6): 868–934. doi:10.1021/ja02227a002.

[46] Scerri, E.R. (2007). *The Periodic Table*. Oxford University Press. pp. 205–226. ISBN 0-19-530573-6.

[47] Massimi, M. (2005). *Pauli's Exclusion Principle, The Origin and Validation of a Scientific Principle*. Cambridge University Press. pp. 7–8. ISBN 0-521-83911-4.

[48] Uhlenbeck, G.E.; Goudsmit, S. (1925). "Ersetzung der Hypothese vom unmechanischen Zwang durch eine Forderung bezüglich des inneren Verhaltens jedes einzelnen Elektrons". *Die Naturwissenschaften* (in German) **13** (47): 953. Bibcode:1925NW....E.doi:10.1007/BF01558878.

[49] Pauli, W. (1923). "Über die Gesetzmäßigkeiten des anomalen Zeemaneffektes". *Zeitschrift für Physik* (in German) **16** (1): 155–164. Bibcode:1923ZPhy...16..155P. doi:10.1007/BF01327386.

[50] de Broglie, L. (1929). "Nobel Lecture: The Wave Nature of the Electron" (PDF). The Nobel Foundation. Retrieved 2008-08-30.

[51] Falkenburg, B. (2007). *Particle Metaphysics: A Critical Account of Subatomic Reality*. Springer. p. 85. ISBN 3-540-33731-8.

[52] Davisson, C. (1937). "Nobel Lecture: The Discovery of Electron Waves" (PDF). The Nobel Foundation. Retrieved 2008-08-30.

[53] Schrödinger, E. (1926). "Quantisierung als Eigenwertproblem". *Annalen der Physik* (in German) **385**(13): 437–490. 385..437S. doi:10.1002/andp.19263851302.

[54] Rigden, J.S. (2003). *Hydrogen*. Harvard University Press. pp. 59–86. ISBN 0-674-01252-6.

[55] Reed, B.C. (2007). *Quantum Mechanics*. Jones & Bartlett Publishers. pp. 275–350. ISBN 0-7637-4451-4.

[56] Dirac, P.A.M. (1928). "The Quantum Theory of the Electron". *Proceedings of the Royal Society A* **117** (778): 610–624. Bibcode:1928RSPSA.117..610D. doi:10.1098/rspa.1928.0023.

[57] Dirac, P.A.M. (1933). "Nobel Lecture: Theory of Electrons and Positrons" (PDF). The Nobel Foundation. Retrieved 2008-11-01.

[58] "The Nobel Prize in Physics 1965". The Nobel Foundation. Retrieved 2008-11-04.

[59] Panofsky, W.K.H. (1997). "The Evolution of Particle Accelerators & Colliders" (PDF). *Beam Line* (Stanford University) **27** (1): 36–44. Retrieved 2008-09-15.

[60] Elder, F.R. et al. (1947). "Radiation from Electrons in a Synchrotron". *Physical Review* **71**(11): 829–830. Bibcode:1947PhRv...E. doi:10.1103/PhysRev.71.829.5.

[61] Hoddeson, L. et al. (1997). *The Rise of the Standard Model: Particle Physics in the 1960s and 1970s*. Cambridge University Press. pp. 25–26. ISBN 0-521-57816-7.

[62] Bernardini, C. (2004). "AdA: The First Electron–Positron Collider". *Physics in Perspective* **6**(2): 156–183. Bibcode:2004PhP.56B. doi:10.1007/s00016-003-0202-y.

[63] "Testing the Standard Model: The LEP experiments". CERN. 2008. Retrieved 2008-09-15.

[64] "LEP reaps a final harvest". *CERN Courier* **40** (10). 2000.

[65] Prati, E.; De Michielis, M.; Belli, M.; Cocco, S.; Fanciulli, M.; Kotekar-Patil, D.; Ruoff, M.; Kern, D. P.; Wharam, D. A.; Verduijn, J.; Tettamanzi, G. C.; Rogge, S.; Roche, B.; Wacquez, R.; Jehl, X.; Vinet, M.; Sanquer, M. (2012). "Few electron limit of n-type metal oxide semiconductor single electron transistors". *Nanotechnology* **23** (21): 215204. doi:10.1088/0957-4484/23/21/215204. PMID 22552118.

[66] Frampton, P.H.; Hung, P.Q.; Sher, Marc (2000). "Quarks and Leptons Beyond the Third Generation". *Physics Reports* **330** (5–6): 263–348. arXiv:hep-ph/9903387. Bibcode:2000PhR...330..263F. doi:10.1016/S0370-1573(99)00095-2.

[67] Raith, W.; Mulvey, T. (2001). *Constituents of Matter: Atoms, Molecules, Nuclei and Particles*. CRC Press. pp. 777–781. ISBN 0-8493-1202-7.

[68] The original source for CODATA is Mohr, P.J.; Taylor, B.N.; Newell, D.B. (2006). "CODATA recommended values of the fundamental physical constants". *Reviews of Modern Physics* **80** (2): 633–730. arXiv:0801.0028. Bibcode:2008RvMP...80..633M. doi:10.1103/RevModPhys.80.633.

   Individual physical constants from the CODATA are available at: "The NIST Reference on Constants, Units and Uncertainty". National Institute of Standards and Technology. Retrieved 2009-01-15.

[69] Zombeck, M.V. (2007). *Handbook of Space Astronomy and Astrophysics* (3rd ed.). Cambridge University Press. p. 14. ISBN 0-521-78242-2.

[70] Murphy, M.T. et al. (2008). "Strong Limit on a Variable Proton-to-Electron Mass Ratio from Molecules in the Distant Universe". *Science* **320** (5883): 1611–1613. arXiv:0806.3081. Bibcode:2008Sci...320.1611M. doi:10.1126/science.1156352. PMID 18566280.

[71] Zorn, J.C.; Chamberlain, G.E.; Hughes, V.W. (1963). "Experimental Limits for the Electron-Proton Charge Difference and for the Charge of the Neutron". *Physical Review* **129** (6): 2566–2576. Bibcode:1963PhRv..129.2566Z. doi:10.1103/PhysRev.16.

[72] Odom, B. et al. (2006). "New Measurement of the Electron Magnetic Moment Using a One-Electron Quantum Cyclotron". *Physical Review Letters* **97** (3): 030801. Bibcode:2006PhRvL..97c0801O. doi:10.1103/PhysRevLett.97.030801. PMID16990.

[73] Anastopoulos, C. (2008). *Particle Or Wave: The Evolution of the Concept of Matter in Modern Physics*. Princeton University Press. pp. 261–262. ISBN 0-691-13512-6.

[74] Gabrielse, G. et al. (2006). "New Determination of the Fine Structure Constant from the Electron *g* Value and QED". *Physical Review Letters* **97** (3): 030802(1–4). Bibcode:2006PhRvL..97c0802G. doi:10.1103/PhysRevLett.97.030802.

[75] Eduard Shpolsky, Atomic physics (Atomnaia fizika),second edition, 1951

[76] Dehmelt, H. (1988). "A Single Atomic Particle Forever Floating at Rest in Free Space: New Value for Electron Radius". *Physica Scripta* **T22**: 102–10. Bibcode:1988PhST...22..102D. doi:10.1088/0031-8949/1988/T22/016.

[77] Meschede, D. (2004). *Optics, light and lasers: The Practical Approach to Modern Aspects of Photonics and Laser Physics*. Wiley-VCH. p. 168. ISBN 3-527-40364-7.

[78] Steinberg, R.I. et al. (1999). "Experimental test of charge conservation and the stability of the electron". *Physical Review D* **61** (2): 2582–2586. Bibcode:1975PhRvD..12.2582S. doi:10.1103/PhysRevD.12.2582.

[79] J. Beringer (Particle Data Group) et al. (2012). "Review of Particle Physics: [electron properties]" (PDF). *Physical Review D* **86** (1): 010001. Bibcode:2012PhRvD..86a0001B. doi:10.1103/PhysRevD.86.010001.

[80] Back, H. O. et al. (2002). "Search for electron decay mode e → γ + ν with prototype of Borexino detector". *Physics Letters B* **525**: 29–40. Bibcode:2002PhLB..525...29B. doi:10.1016/S0370-2693(01)01440-X.

[81] Munowitz, M. (2005). *Knowing, The Nature of Physical Law*. Oxford University Press. ISBN 0-19-516737-6.

[82] Kane, G. (October 9, 2006). "Are virtual particles really constantly popping in and out of existence? Or are they merely a mathematical bookkeeping device for quantum mechanics?". Scientific American. Retrieved 2008-09-19.

[83] Taylor, J. (1989). "Gauge Theories in Particle Physics". In Davies, Paul. *The New Physics*. Cambridge University Press. p. 464. ISBN 0-521-43831-4.

[84] Genz, H. (2001). *Nothingness: The Science of Empty Space*. Da Capo Press. pp. 241–243, 245–247. ISBN 0-7382-0610-5.

[85] Gribbin, J. (January 25, 1997). "More to electrons than meets the eye". *New Scientist*. Retrieved 2008-09-17.

[86] Levine, I. et al. (1997). "Measurement of the Electromagnetic Coupling at Large Momentum Transfer". *Physical Review Letters* **78** (3): 424–427. Bibcode:1997PhRvL..78..424L. doi:10.1103/PhysRevLett.78.424.

[87] Murayama, H. (March 10–17, 2006). *Supersymmetry Breaking Made Easy, Viable and Generic*. Proceedings of the XLIInd Rencontres de Moriond on Electroweak Interactions and Unified Theories (La Thuile, Italy). arXiv:0709.3041.—lists a 9% mass difference for an electron that is the size of the Planck distance.

[88] Schwinger, J. (1948). "On Quantum-Electrodynamics and the Magnetic Moment of the Electron". *Physical Review* **73** (4): 416–417. Bibcode:1948PhRv...73..416S. doi:10.1103/PhysRev.73.416.

[89] Huang, K. (2007). *Fundamental Forces of Nature: The Story of Gauge Fields*. World Scientific. pp. 123–125. ISBN 981-270-645-3.

[90] Foldy, L.L.; Wouthuysen, S. (1950). "On the Dirac Theory of Spin 1/2 Particles and Its Non-Relativistic Limit". *Physical Review* **78**: 29–36. Bibcode:1950PhRv...78...29F. doi:10.1103/PhysRev.78.29.

[91] Sidharth, B.G. (2008). "Revisiting Zitterbewegung". *International Journal of Theoretical Physics* **48**(2): 497–506.arXiv:0806.05. Bibcode:2009IJTP...48..497S. doi:10.1007/s10773-008-9825-8.

[92] Elliott, R.S. (1978). "The History of Electromagnetics as Hertz Would Have Known It". *IEEE Transactions on Microwave Theory and Techniques* **36** (5): 806–823. Bibcode:1988ITMTT..36..806E. doi:10.1109/22.3600.

[93] Crowell, B. (2000). *Electricity and Magnetism*. Light and Matter. pp. 129–152. ISBN 0-9704670-4-4.

[94] Mahadevan, R.; Narayan, R.; Yi, I. (1996). "Harmony in Electrons: Cyclotron and Synchrotron Emission by Thermal Electrons in a Magnetic Field". *The Astrophysical Journal* **465**: 327–337. arXiv:astro-ph/9601073. Bibcode:1996ApJ...465..327M. doi:10.1086/177422.

[95] Rohrlich, F. (1999). "The Self-Force and Radiation Reaction". *American Journal of Physics* **68**(12): 1109–1112.Bibcod1109R. doi:10.1119/1.1286430.

[96] Georgi, H. (1989). "Grand Unified Theories". In Davies, Paul. *The New Physics*. Cambridge University Press. p. 427. ISBN 0-521-43831-4.

[97] Blumenthal, G.J.; Gould, R. (1970). "Bremsstrahlung, Synchrotron Radiation, and Compton Scattering of High-Energy Electrons Traversing Dilute Gases". *Reviews of Modern Physics* **42** (2): 237–270. Bibcode:1970RvMP...42..237B. doi:10.11037.

[98] Staff (2008). "The Nobel Prize in Physics 1927". The Nobel Foundation. Retrieved 2008-09-28.

[99] Chen, S.-Y.; Maksimchuk, A.; Umstadter, D. (1998). "Experimental observation of relativistic nonlinear Thomson scattering". *Nature* **396** (6712): 653–655. arXiv:physics/9810036. Bibcode:1998Natur.396..653C. doi:10.1038/25303.

[100] Beringer, R.; Montgomery, C.G. (1942). "The Angular Distribution of Positron Annihilation Radiation". *Physical Review* **61** (5–6): 222–224. Bibcode:1942PhRv...61..222B. doi:10.1103/PhysRev.61.222.

[101] Buffa, A. (2000). *College Physics* (4th ed.). Prentice Hall. p. 888. ISBN 0-13-082444-5.

[102] Eichler, J. (2005). "Electron–positron pair production in relativistic ion–atom collisions". *Physics Letters A* **347** (1–3): 67–72. Bibcode:2005PhLA..347...67E. doi:10.1016/j.physleta.2005.06.105.

[103] Hubbell, J.H. (2006). "Electron positron pair production by photons: A historical overview". *Radiation Physics and Chemistry* **75** (6): 614–623. Bibcode:2006RaPC...75..614H. doi:10.1016/j.radphyschem.2005.10.008.

[104] Quigg, C. (June 4–30, 2000). *The Electroweak Theory*. TASI 2000: Flavor Physics for the Millennium (Boulder, Colorado): 80. arXiv:hep-ph/0204104.

[105] Mulliken, R.S. (1967). "Spectroscopy, Molecular Orbitals, and Chemical Bonding". *Science* **157**(3784): 13–24.Bibcode:1967SM. doi:10.1126/science.157.3784.13. PMID 5338306.

[106] Burhop, E.H.S. (1952). *The Auger Effect and Other Radiationless Transitions*. Cambridge University Press. pp. 2–3. ISBN 0-88275-966-3.

[107] Grupen, C. (2000). "Physics of Particle Detection". *AIP Conference Proceedings* **536**: 3–34. arXiv:physics/9906063.do756.

[108] Jiles, D. (1998). *Introduction to Magnetism and Magnetic Materials*. CRC Press. pp. 280–287. ISBN 0-412-79860-3.

[109] Löwdin, P.O.; Erkki Brändas, E.; Kryachko, E.S. (2003). *Fundamental World of Quantum Chemistry: A Tribute to the Memory of Per- Olov Löwdin*. Springer. pp. 393–394. ISBN 1-4020-1290-X.

[110] McQuarrie, D.A.; Simon, J.D. (1997). *Physical Chemistry: A Molecular Approach*. University Science Books. pp. 325–361. ISBN 0-935702-99-7.

[111] Daudel, R. et al. (1973). "The Electron Pair in Chemistry". *Canadian Journal of Chemistry* **52** (8): 1310–1320. doi:10.1139/v74-201.

[112] Rakov, V.A.; Uman, M.A. (2007). *Lightning: Physics and Effects*. Cambridge University Press. p. 4. ISBN 0-521-03541-4.

[113] Freeman, G.R.; March, N.H. (1999). "Triboelectricity and some associated phenomena". *Materials Science and Technology* **15** (12): 1454–1458. doi:10.1179/026708399101505464.

[114] Forward, K.M.; Lacks, D.J.; Sankaran, R.M. (2009). "Methodology for studying particle–particle triboelectrification in granular materials". *Journal of Electrostatics* **67** (2–3): 178–183. doi:10.1016/j.elstat.2008.12.002.

[115] Weinberg, S. (2003). *The Discovery of Subatomic Particles*. Cambridge University Press. pp. 15–16. ISBN 0-521-82351-X.

[116] Lou, L.-F. (2003). *Introduction to phonons and electrons*. World Scientific. pp. 162, 164. ISBN 978-981-238-461-4.

[117] Guru, B.S.; Hızıroğlu, H.R. (2004). *Electromagnetic Field Theory*. Cambridge University Press. pp. 138, 276. ISBN 0-521-83016-8.

[118] Achuthan, M.K.; Bhat, K.N. (2007). *Fundamentals of Semiconductor Devices*. Tata McGraw-Hill. pp. 49–67. ISBN 0-07-061220-X.

[119] Ziman, J.M. (2001). *Electrons and Phonons: The Theory of Transport Phenomena in Solids*. Oxford University Press. p. 260. ISBN 0-19-850779-8.

[120] Main, P. (June 12, 1993). "When electrons go with the flow: Remove the obstacles that create electrical resistance, and you get ballistic electrons and a quantum surprise". *New Scientist* **1887**: 30. Retrieved 2008-10-09.

[121] Blackwell, G.R. (2000). *The Electronic Packaging Handbook*. CRC Press. pp. 6.39–6.40. ISBN 0-8493-8591-1.

[122] Durrant, A. (2000). *Quantum Physics of Matter: The Physical World*. CRC Press. pp. 43, 71–78. ISBN 0-7503-0721-8.

[123] Staff (2008). "The Nobel Prize in Physics 1972". The Nobel Foundation. Retrieved 2008-10-13.

[124] Kadin, A.M. (2007). "Spatial Structure of the Cooper Pair". *Journal of Superconductivity and Novel Magnetism* **20** (4): 285–292. arXiv:cond-mat/0510279. doi:10.1007/s10948-006-0198-z.

[125] "Discovery About Behavior Of Building Block Of Nature Could Lead To Computer Revolution". *ScienceDaily*. July 31, 2009. Retrieved 2009-08-01.

[126] Jompol, Y. et al. (2009). "Probing Spin-Charge Separation in a Tomonaga-Luttinger Liquid". *Science* **325** (5940): 597–601. arXiv:1002.2782. Bibcode:2009Sci...325..597J. doi:10.1126/science.1171769. PMID 19644117.

[127] Staff (2008). "The Nobel Prize in Physics 1958, for the discovery and the interpretation of the Cherenkov effect". The Nobel Foundation. Retrieved 2008-09-25.

[128] Staff (August 26, 2008). "Special Relativity". Stanford Linear Accelerator Center. Retrieved 2008-09-25.

[129] Adams, S. (2000). *Frontiers: Twentieth Century Physics*. CRC Press. p. 215. ISBN 0-7484-0840-1.

[130] Lurquin, P.F. (2003). *The Origins of Life and the Universe*. Columbia University Press. p. 2. ISBN 0-231-12655-7.

[131] Silk, J. (2000). *The Big Bang: The Creation and Evolution of the Universe* (3rd ed.). Macmillan. pp. 110–112, 134–137. ISBN 0-8050-7256-X.

[132] Christianto, V. (2007). "Thirty Unsolved Problems in the Physics of Elementary Particles" (PDF). *Progress in Physics* **4**: 112–114.

[133] Kolb, E.W.; Wolfram, Stephen (1980). "The Development of Baryon Asymmetry in the Early Universe". *Physics Letters B* **91** (2): 217–221. Bibcode:1980PhLB...91..217K. doi:10.1016/0370-2693(80)90435-9.

[134] Sather, E. (Spring–Summer 1996). "The Mystery of Matter Asymmetry" (PDF). *Beam Line*. University of Stanford. Retrieved 2008-11-01.

[135] Burles, S.; Nollett, K.M.; Turner, M.S. (1999). "Big-Bang Nucleosynthesis: Linking Inner Space and Outer Space". arXiv:astro-ph/9903300 [astro-ph].

[136] Boesgaard, A.M.; Steigman, G. (1985). "Big bang nucleosynthesis – Theories and observations". *Annual Review of Astronomy and Astrophysics* **23** (2): 319–378. Bibcode:1985ARA&A..23..319B. doi:10.1146/annurev.aa.23.090185.001535.

[137] Barkana, R. (2006). "The First Stars in the Universe and Cosmic Reionization". *Science* **313** (5789): 931–934. arXiv:astro-ph/0608450. Bibcode:2006Sci...313..931B. doi:10.1126/science.1125644. PMID 16917052.

[138] Burbidge, E.M. et al. (1957). "Synthesis of Elements in Stars". *Reviews of Modern Physics* **29** (4): 548–647. Bibcode:1947B. doi:10.1103/RevModPhys.29.547.

[139] Rodberg, L.S.; Weisskopf, V. (1957). "Fall of Parity: Recent Discoveries Related to Symmetry of Laws of Nature". *Science* **125** (3249): 627–633. Bibcode:1957Sci...125..627R. doi:10.1126/science.125.3249.627. PMID 17810563.

[140] Fryer, C.L. (1999). "Mass Limits For Black Hole Formation". *The Astrophysical Journal* **522** (1): 413–418. arXiv:astro-ph/9902315. Bibcode:1999ApJ...522..413F. doi:10.1086/307647.

[141] Parikh, M.K.; Wilczek, F. (2000). "Hawking Radiation As Tunneling". *Physical Review Letters* **85** (24): 5042–5045. arXiv:hep-th/9907001. Bibcode:2000PhRvL..85.5042P. doi:10.1103/PhysRevLett.85.5042. PMID 11102182.

[142] Hawking, S.W. (1974). "Black hole explosions?". *Nature* **248** (5443): 30–31. Bibcode:1974Natur.248...30H. doi:10.1038/24803a0.

[143] Halzen, F.; Hooper, D. (2002). "High-energy neutrino astronomy: the cosmic ray connection". *Reports on Progress in Physics* **66** (7): 1025–1078. arXiv:astro-ph/0204527. Bibcode:2002astro.ph..4527H. doi:10.1088/0034-4885/65/7/201.

[144] Ziegler, J.F. (1998). "Terrestrial cosmic ray intensities". *IBM Journal of Research and Development* **42** (1): 117–139. doi:10.1147/17.

[145] Sutton, C. (August 4, 1990). "Muons, pions and other strange particles". *New Scientist*. Retrieved 2008-08-28.

[146] Wolpert, S. (July 24, 2008). "Scientists solve 30-year-old aurora borealis mystery". University of California. Retrieved 2008-10-11.

[147] Gurnett, D.A.; Anderson, R. (1976). "Electron Plasma Oscillations Associated with Type III Radio Bursts". *Science* **194** (4270): 1159–1162. Bibcode:1976Sci...194.1159G. doi:10.1126/science.194.4270.1159. PMID 17790910.

[148] Martin, W.C.; Wiese, W.L. (2007). "Atomic Spectroscopy: A Compendium of Basic Ideas, Notation, Data, and Formulas". National Institute of Standards and Technology. Retrieved 2007-01-08.

[149] Fowles, G.R. (1989). *Introduction to Modern Optics*. Courier Dover. pp. 227–233. ISBN 0-486-65957-7.

[150] Staff (2008). "The Nobel Prize in Physics 1989". The Nobel Foundation. Retrieved 2008-09-24.

[151] Ekstrom, P.; Wineland, David (1980). "The isolated Electron" (PDF). *Scientific American* **243** (2): 91–101. doi:10.1038/sn0880-104. Retrieved 2008-09-24.

[152] Mauritsson, J. "Electron filmed for the first time ever" (PDF). Lund University. Archived from the original (PDF) on March 25, 2009. Retrieved 2008-09-17.

[153] Mauritsson, J. et al. (2008). "Coherent Electron Scattering Captured by an Attosecond Quantum Stroboscope". *Physical Review Letters* **100** (7): 073003. arXiv:0708.1060. Bibcode:2008PhRvL.100g3003M. doi:10.1103/PhysRevLett.100.073003. PMID 18352546.

[154] Damascelli, A. (2004). "Probing the Electronic Structure of Complex Systems by ARPES". *Physica Scripta* **T109**: 61–74. arXiv:cond-mat/0307085. Bibcode:2004PhST..109...61D. doi:10.1238/Physica.Topical.109a00061.

[155] Staff (April 4, 1975). "Image # L-1975-02972". Langley Research Center, NASA. Retrieved 2008-09-20.

[156] Elmer, J. (March 3, 2008). "Standardizing the Art of Electron-Beam Welding". Lawrence Livermore National Laboratory. Retrieved 2008-10-16.

[157] Schultz, H. (1993). *Electron Beam Welding*. Woodhead Publishing. pp. 2–3. ISBN 1-85573-050-2.

[158] Benedict, G.F. (1987). *Nontraditional Manufacturing Processes*. Manufacturing engineering and materials processing **19**. CRC Press. p. 273. ISBN 0-8247-7352-7.

[159] Ozdemir, F.S. (June 25–27, 1979). *Electron beam lithography*. *Proceedings of the 16th Conference on Design automation* (San Diego, CA, USA: IEEE Press): 383–391. Retrieved 2008-10-16.

[160] Madou, M.J. (2002). *Fundamentals of Microfabrication: the Science of Miniaturization* (2nd ed.). CRC Press. pp. 53–54. ISBN 0-8493-0826-7.

[161] Jongen, Y.; Herer, A. (May 2–5, 1996). *Electron Beam Scanning in Industrial Applications. APS/AAPT Joint Meeting* (American Physical Society). Bibcode:1996APS..MAY.H9902J.

[162] Mobus G. et al. (2010). Journal of Nuclear Materials, v. 396, 264–271, doi:10.1016/j.jnucmat.2009.11.020

[163] Beddar, A.S.; Domanovic, Mary Ann; Kubu, Mary Lou; Ellis, Rod J.; Sibata, Claudio H.; Kinsella, Timothy J. (2001). "Mobile linear accelerators for intraoperative radiation therapy". *AORN Journal* **74** (5): 700. doi:10.1016/S0001-2092(06)61769-9.

[164] Gazda, M.J.; Coia, L.R. (June 1, 2007). "Principles of Radiation Therapy" (PDF). Retrieved 2013-10-31.

[165] Chao, A.W.; Tigner, M. (1999). *Handbook of Accelerator Physics and Engineering*. World Scientific. pp. 155, 188. ISBN 981-02-3500-3.

[166] Oura, K. et al. (2003). *Surface Science: An Introduction*. Springer. pp. 1–45. ISBN 3-540-00545-5.

[167] Ichimiya, A.; Cohen, P.I. (2004). *Reflection High-energy Electron Diffraction*. Cambridge University Press. p. 1. ISBN 0-521-45373-9.

[168] Heppell, T.A. (1967). "A combined low energy and reflection high energy electron diffraction apparatus". *Journal of Scientific Instruments* **44** (9): 686–688. Bibcode:1967JScI...44..686H. doi:10.1088/0950-7671/44/9/311.

[169] McMullan, D. (1993). "Scanning Electron Microscopy: 1928–1965". University of Cambridge. Retrieved 2009-03-23.

[170] Slayter, H.S. (1992). *Light and electron microscopy*. Cambridge University Press. p. 1. ISBN 0-521-33948-0.

[171] Cember, H. (1996). *Introduction to Health Physics*. McGraw-Hill Professional. pp. 42–43. ISBN 0-07-105461-8.

[172] Erni, R. et al. (2009). "Atomic-Resolution Imaging with a Sub-50-pm Electron Probe". *Physical Review Letters* **102** (9): 096101. Bibcode:2009PhRvL.102i6101E. doi:10.1103/PhysRevLett.102.096101. PMID 19392535.

[173] Bozzola, J.J.; Russell, L.D. (1999). *Electron Microscopy: Principles and Techniques for Biologists*. Jones & Bartlett Publishers. pp. 12, 197–199. ISBN 0-7637-0192-0.

[174] Flegler, S.L.; Heckman Jr., J.W.; Klomparens, K.L. (1995). *Scanning and Transmission Electron Microscopy: An Introduction* (Reprint ed.). Oxford University Press. pp. 43–45. ISBN 0-19-510751-9.

[175] Bozzola, J.J.; Russell, L.D. (1999). *Electron Microscopy: Principles and Techniques for Biologists* (2nd ed.). Jones & Bartlett Publishers. p. 9. ISBN 0-7637-0192-0.

[176] Freund, H.P.; Antonsen, T. (1996). *Principles of Free-Electron Lasers*. Springer. pp. 1–30. ISBN 0-412-72540-1.

[177] Kitzmiller, J.W. (1995). *Television Picture Tubes and Other Cathode-Ray Tubes: Industry and Trade Summary*. DIANE Publishing. pp. 3–5. ISBN 0-7881-2100-6.

[178] Sclater, N. (1999). *Electronic Technology Handbook*. McGraw-Hill Professional. pp. 227–228. ISBN 0-07-058048-0.

[179] Staff (2008). "The History of the Integrated Circuit". The Nobel Foundation. Retrieved 2008-10-18.

7.9   External links

- "The Discovery of the Electron". American Institute of Physics, Center for History of Physics.

- "Particle Data Group". University of California.

- Bock, R.K.; Vasilescu, A. (1998). *The Particle Detector BriefBook* (14th ed.). Springer. ISBN 3-540-64120-3.

- Copeland, Ed. "Spherical Electron". *Sixty Symbols*. Brady Haran for the University of Nottingham.

Robert Millikan

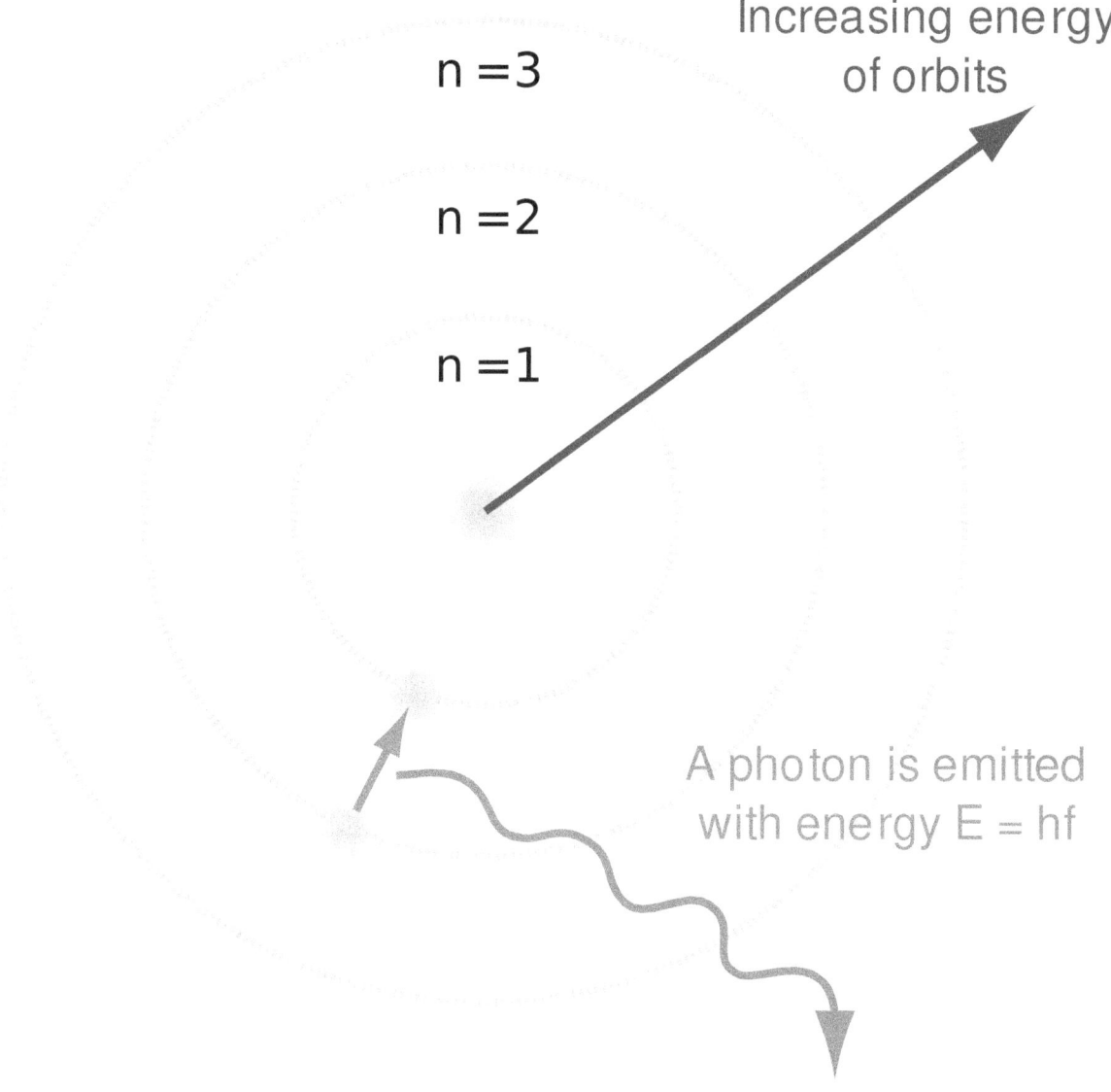

n =3

Increasing energy
of orbits

n =2

n =1

A photon is emitted
with energy E = hf

The Bohr model of the atom, showing states of electron with energy quantized by the number n. An electron dropping to a lower orbit emits a photon equal to the energy difference between the orbits.

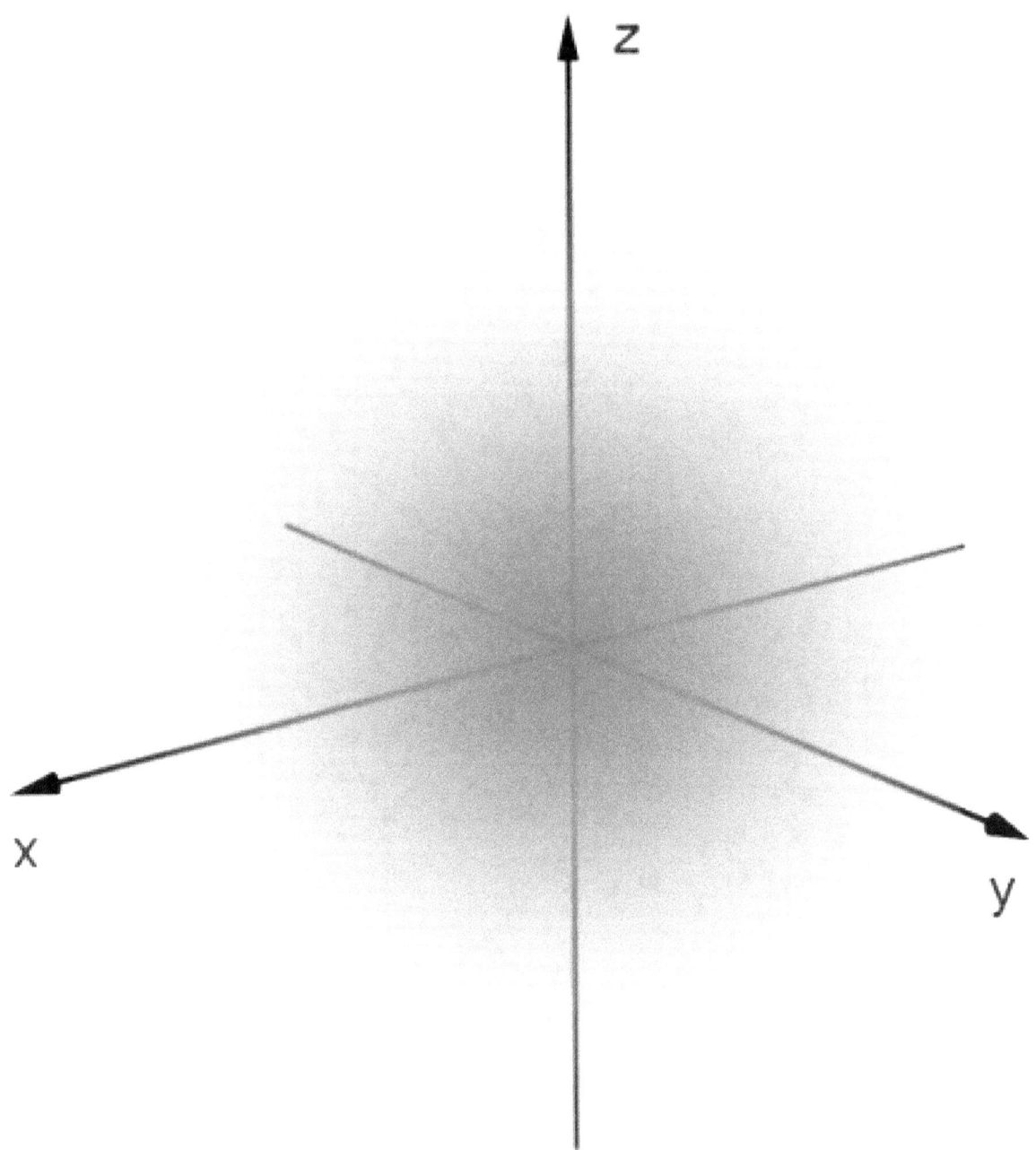

Orbital s ($\ell = 0$, $m_\ell = 0$)

In quantum mechanics, the behavior of an electron in an atom is described by an orbital, which is a probability distribution rather than an orbit. In the figure, the shading indicates the relative probability to "find" the electron, having the energy corresponding to the given quantum numbers, at that point.

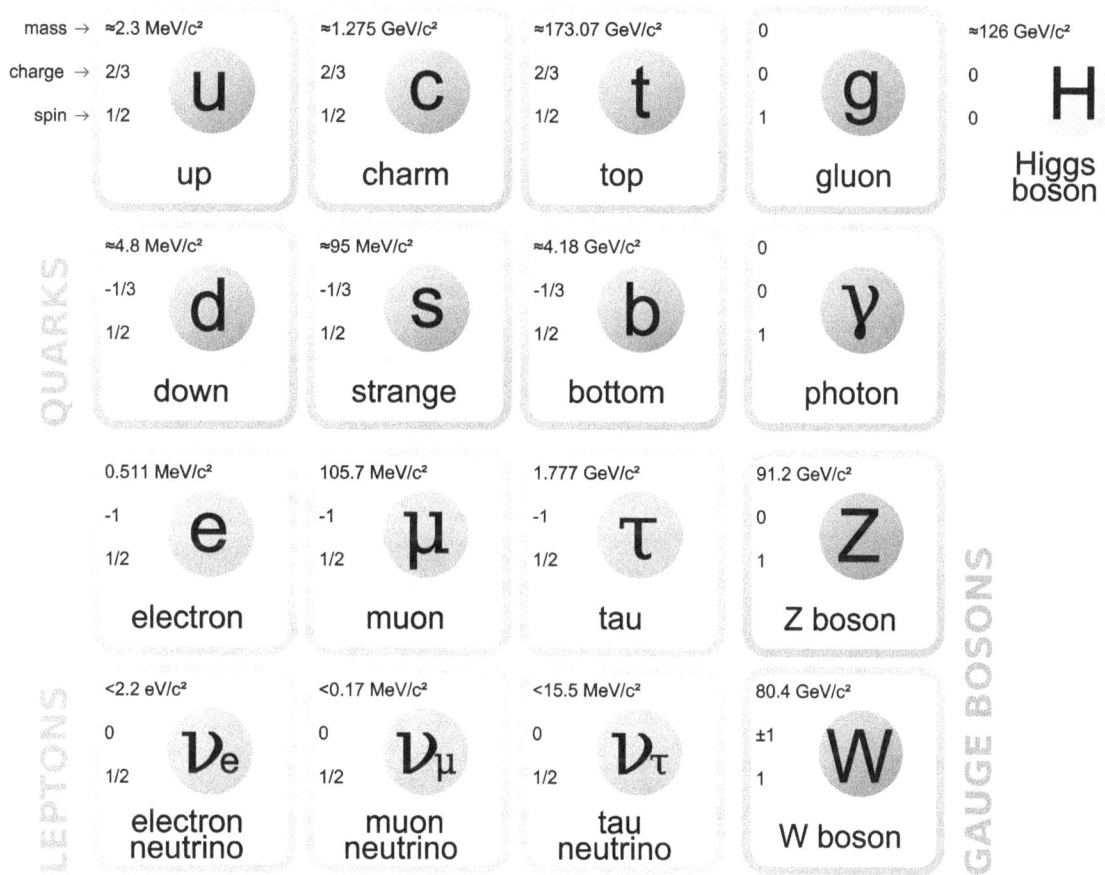

Standard Model of elementary particles. The electron (symbol e) is on the left.

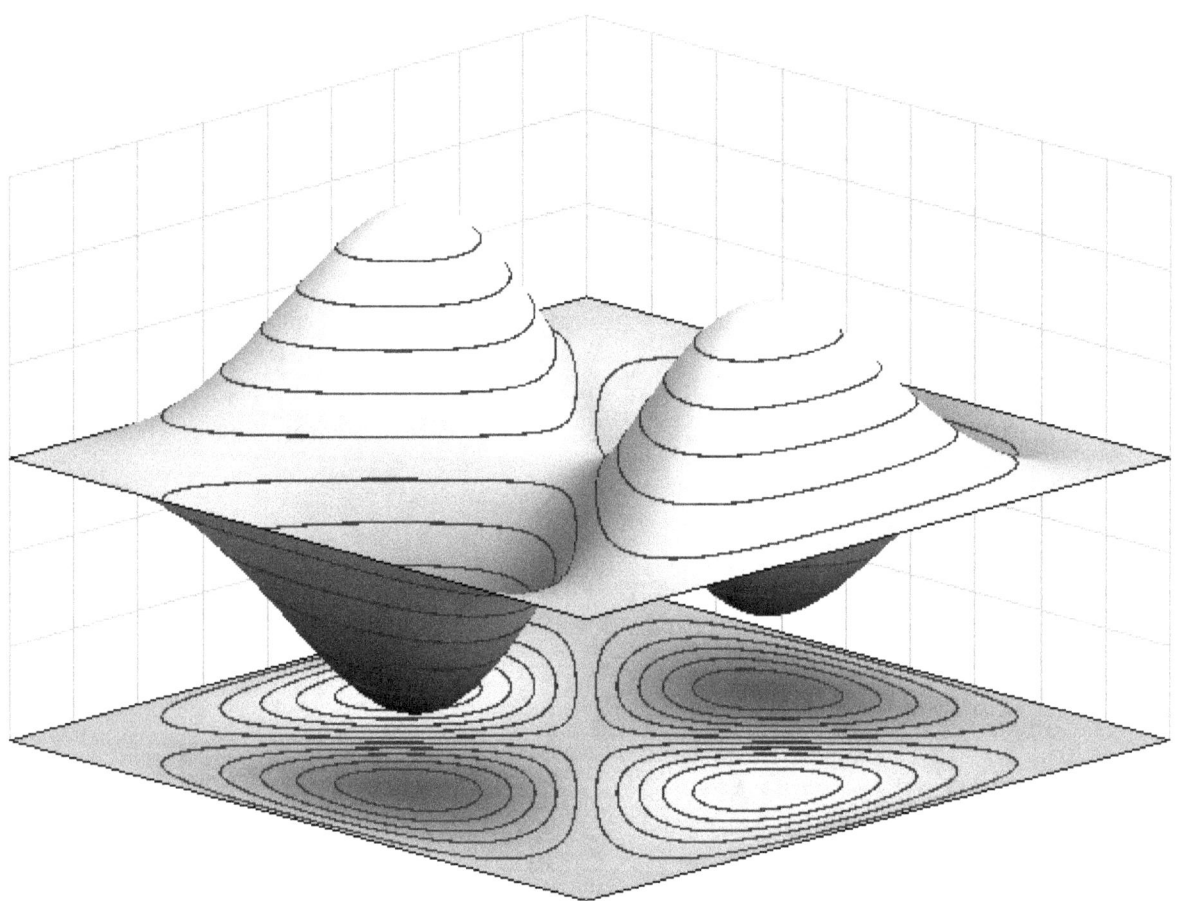

Example of an antisymmetric wave function for a quantum state of two identical fermions in a 1-dimensional box. If the particles swap position, the wave function inverts its sign.

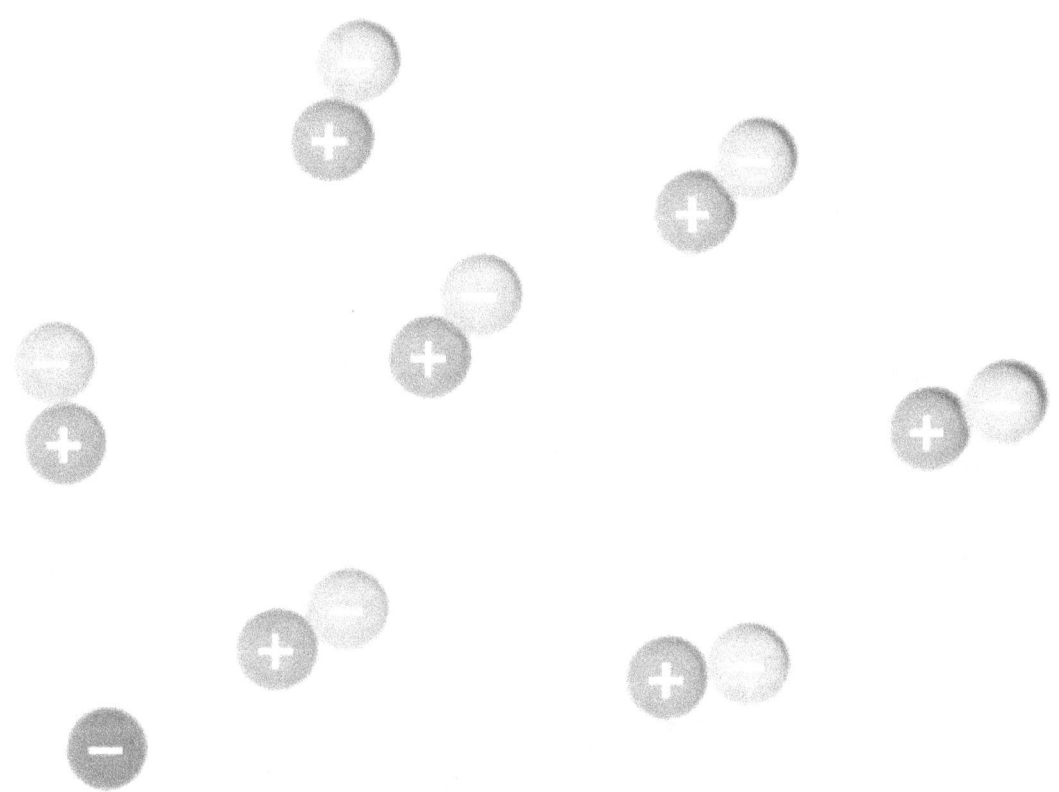

A schematic depiction of virtual electron–positron pairs appearing at random near an electron (at lower left)

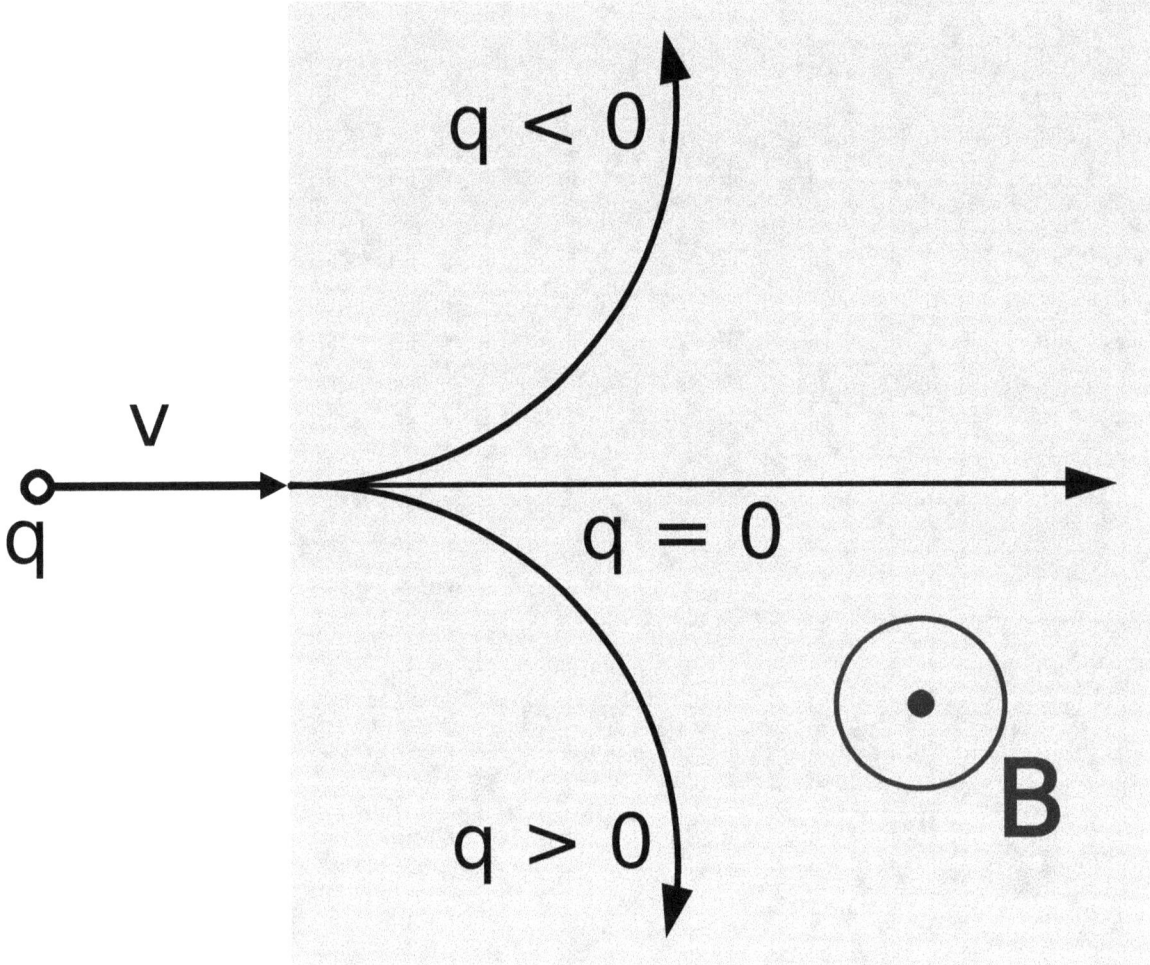

A particle with charge q *(at left) is moving with velocity* v *through a magnetic field* B *that is oriented toward the viewer. For an electron,* q *is negative so it follows a curved trajectory toward the top.*

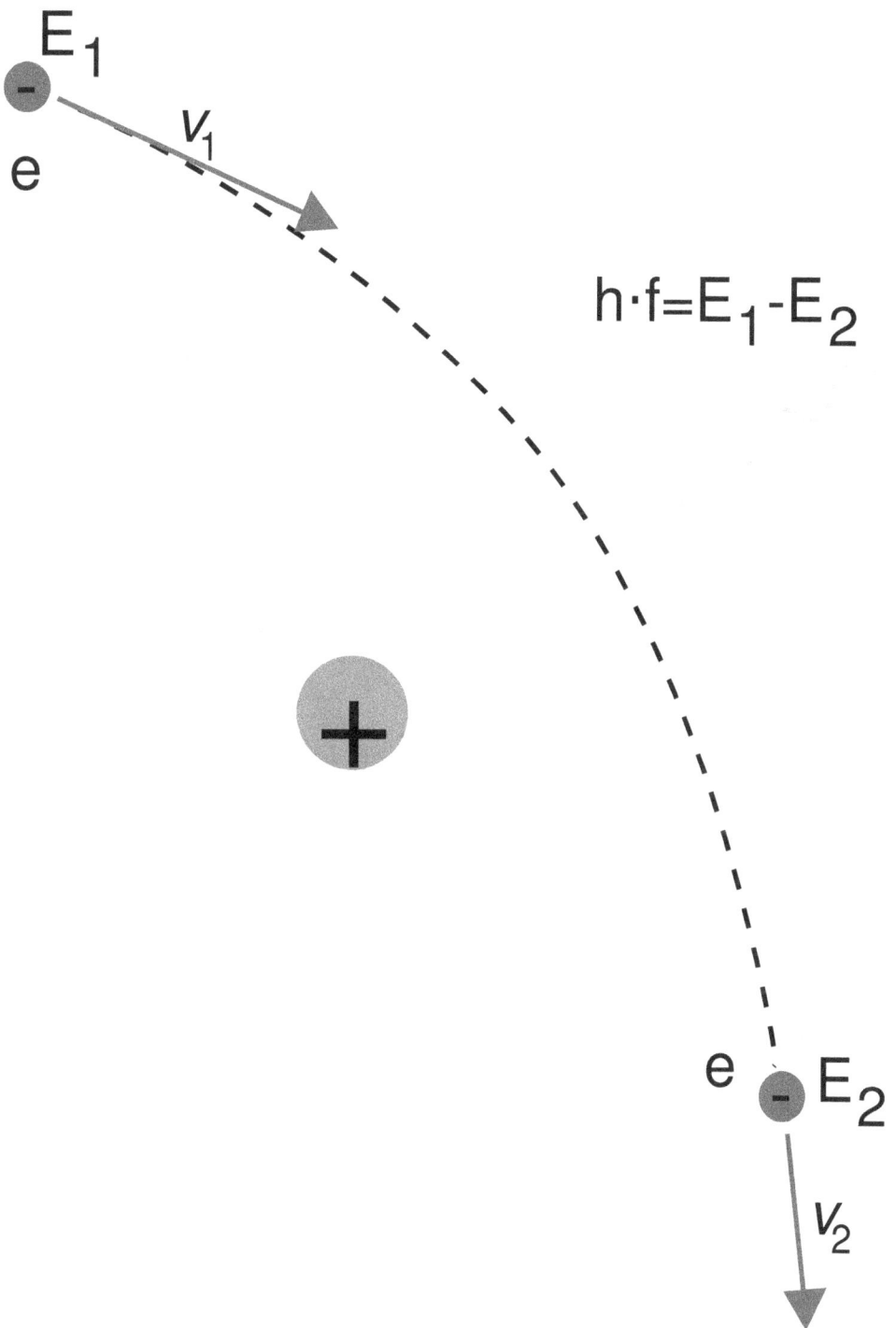

$$h \cdot f = E_1 - E_2$$

Here, Bremsstrahlung is produced by an electron e *deflected by the electric field of an atomic nucleus. The energy change* $E_2 - E_1$ *determines the frequency* f *of the emitted photon.*

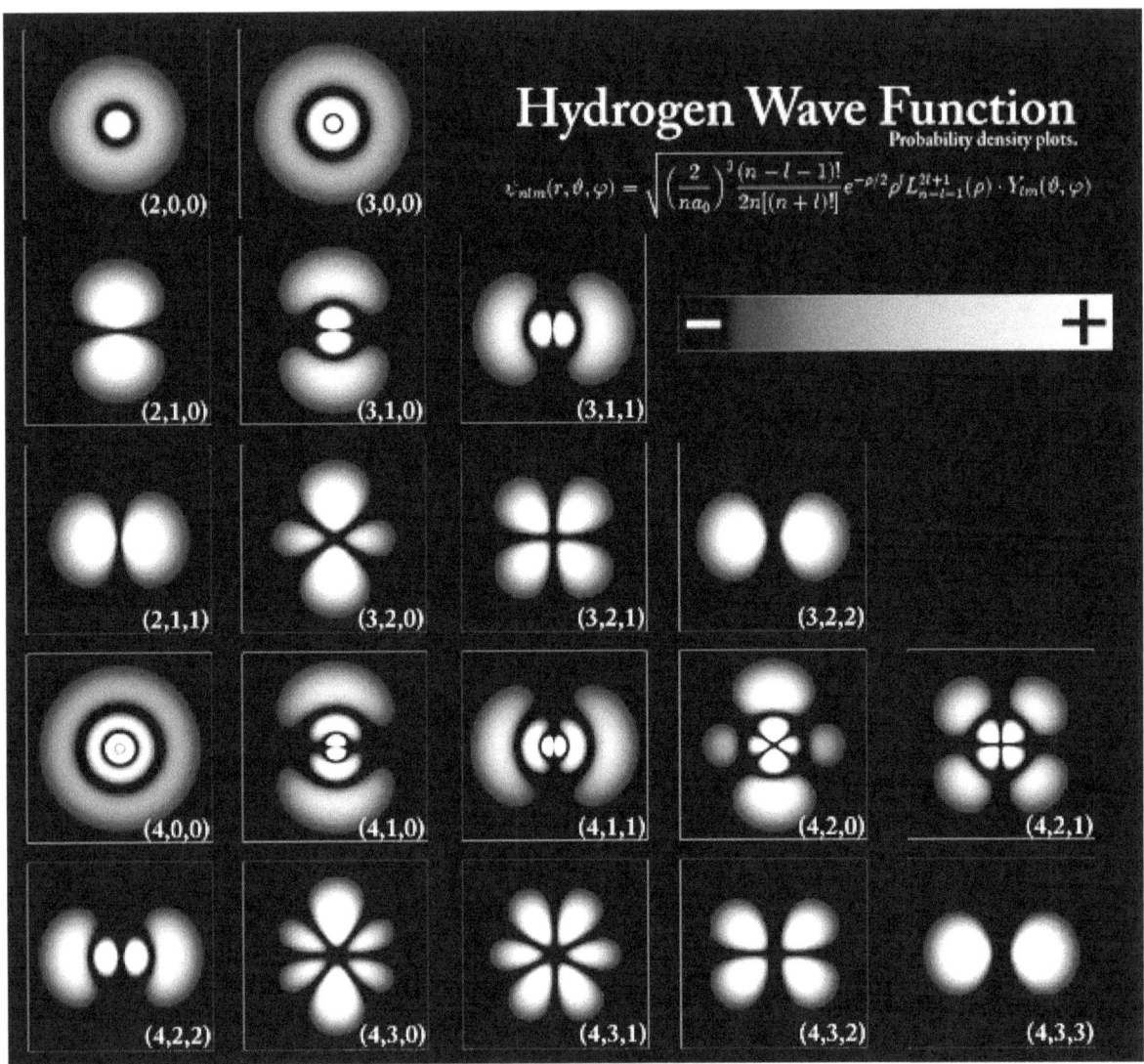

Probability densities for the first few hydrogen atom orbitals, seen in cross-section. The energy level of a bound electron determines the orbital it occupies, and the color reflects the probability of finding the electron at a given position.

A lightning discharge consists primarily of a flow of electrons.[112] *The electric potential needed for lightning may be generated by a triboelectric effect.*[113][114]

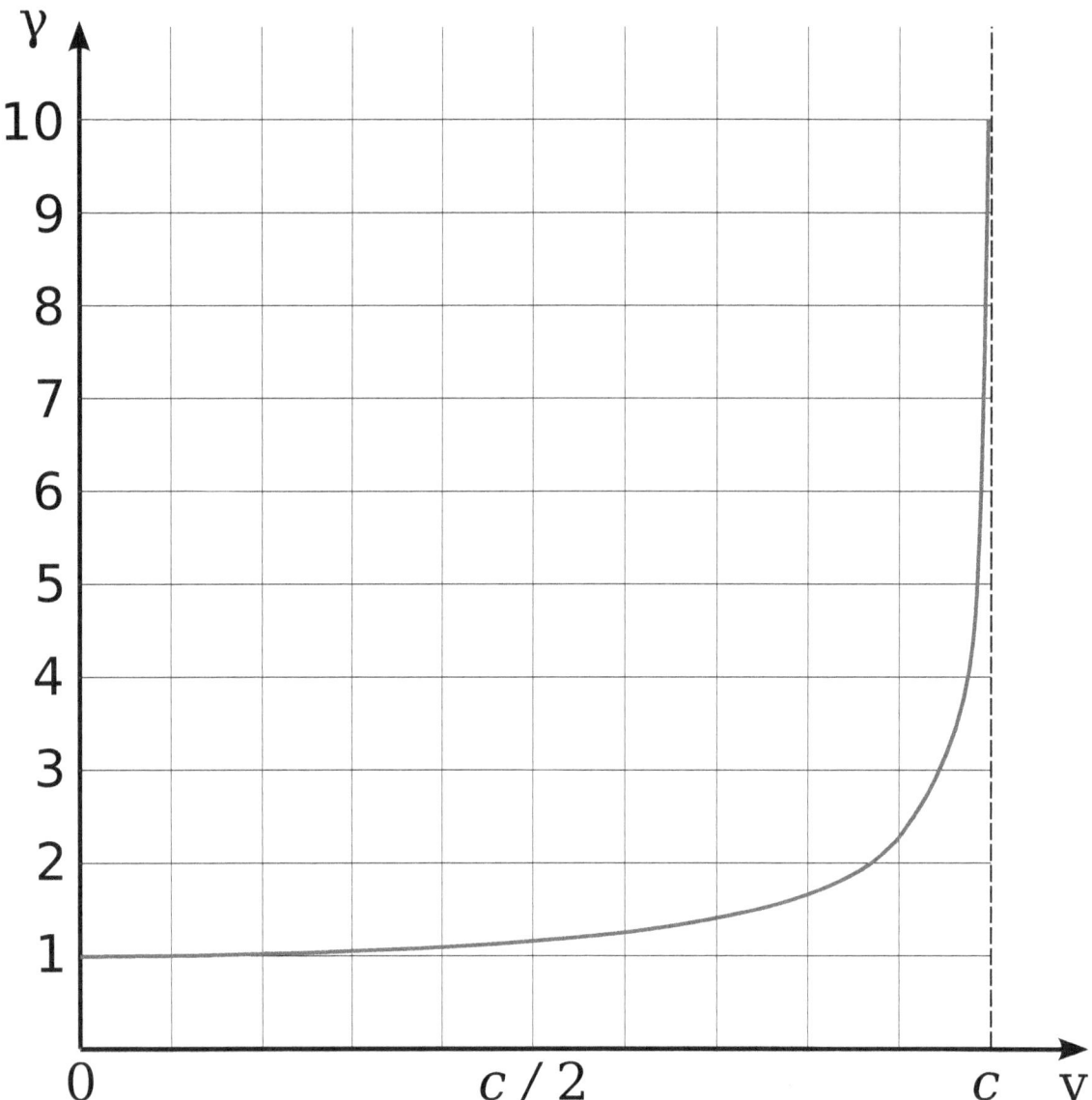

Lorentz factor as a function of velocity. It starts at value 1 and goes to infinity as v approaches c.

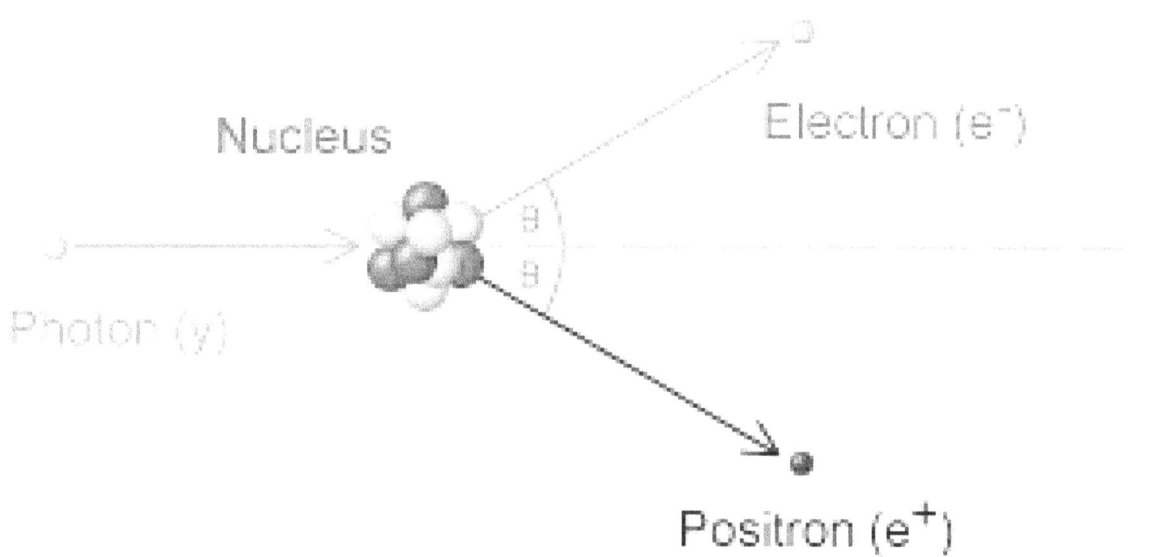

Pair production caused by the collision of a photon with an atomic nucleus

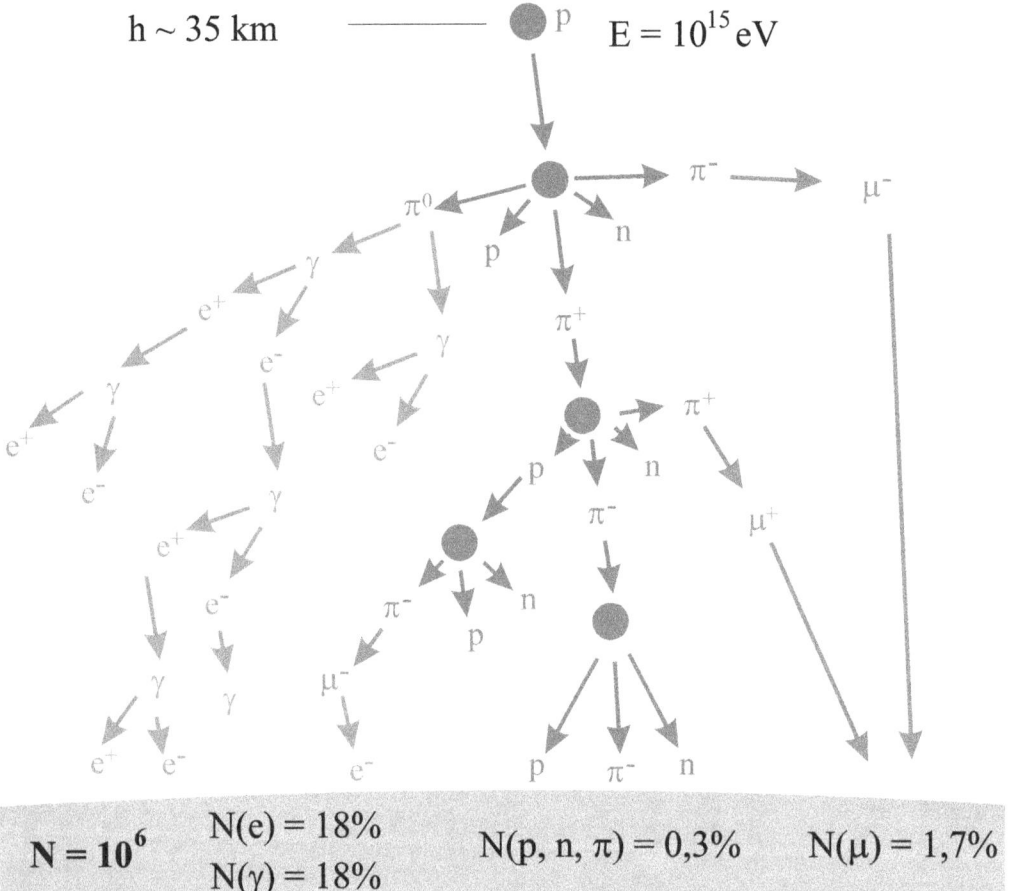

An extended air shower generated by an energetic cosmic ray striking the Earth's atmosphere

Aurorae are mostly caused by energetic electrons precipitating into the atmosphere.[146]

During a NASA wind tunnel test, a model of the Space Shuttle is targeted by a beam of electrons, simulating the effect of ionizing gases during re-entry.[155]

Chapter 8

Electron configuration

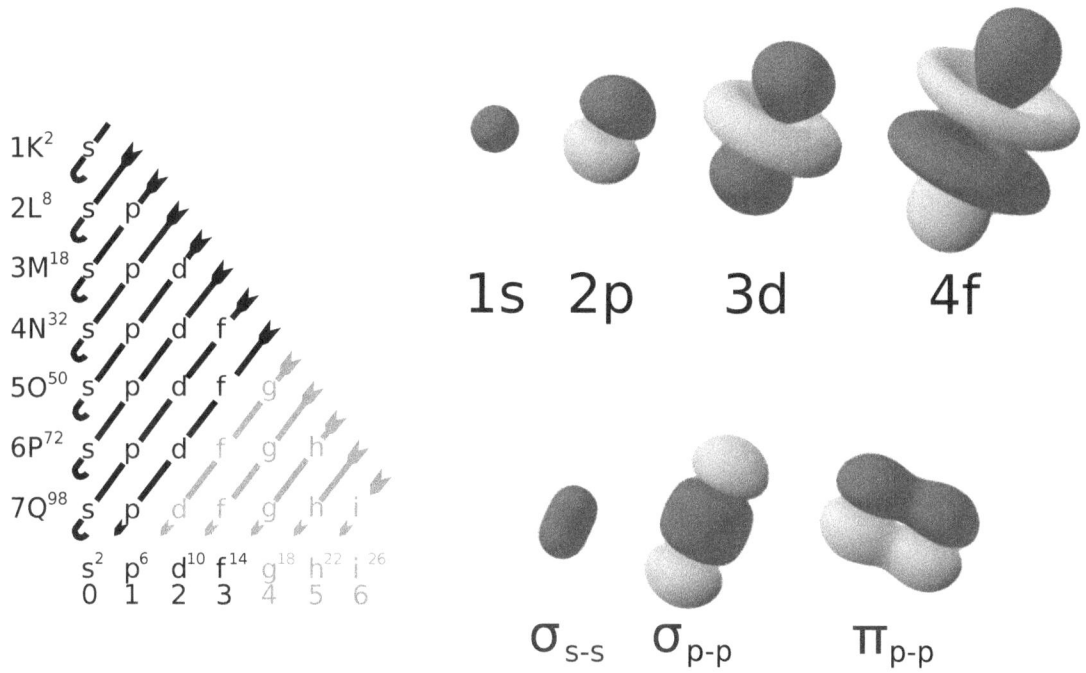

Electron atomic and molecular orbitals

In atomic physics and quantum chemistry, the **electron configuration** is the distribution of electrons of an atom or molecule (or other physical structure) in atomic or molecular orbitals.[1] For example, the electron configuration of the neon atom is $1s^2\,2s^2\,2p^6$.

Electronic configurations describe electrons as each moving independently in an orbital, in an average field created by all other orbitals. Mathematically, configurations are described by Slater determinants or configuration state functions.

According to the laws of quantum mechanics, for systems with only one electron, an energy is associated with each electron configuration and, upon certain conditions, electrons are able to move from one configuration to another by the emission or absorption of a quantum of energy, in the form of a photon.

Knowledge of the electron configuration of different atoms is useful in understanding the structure of the periodic table

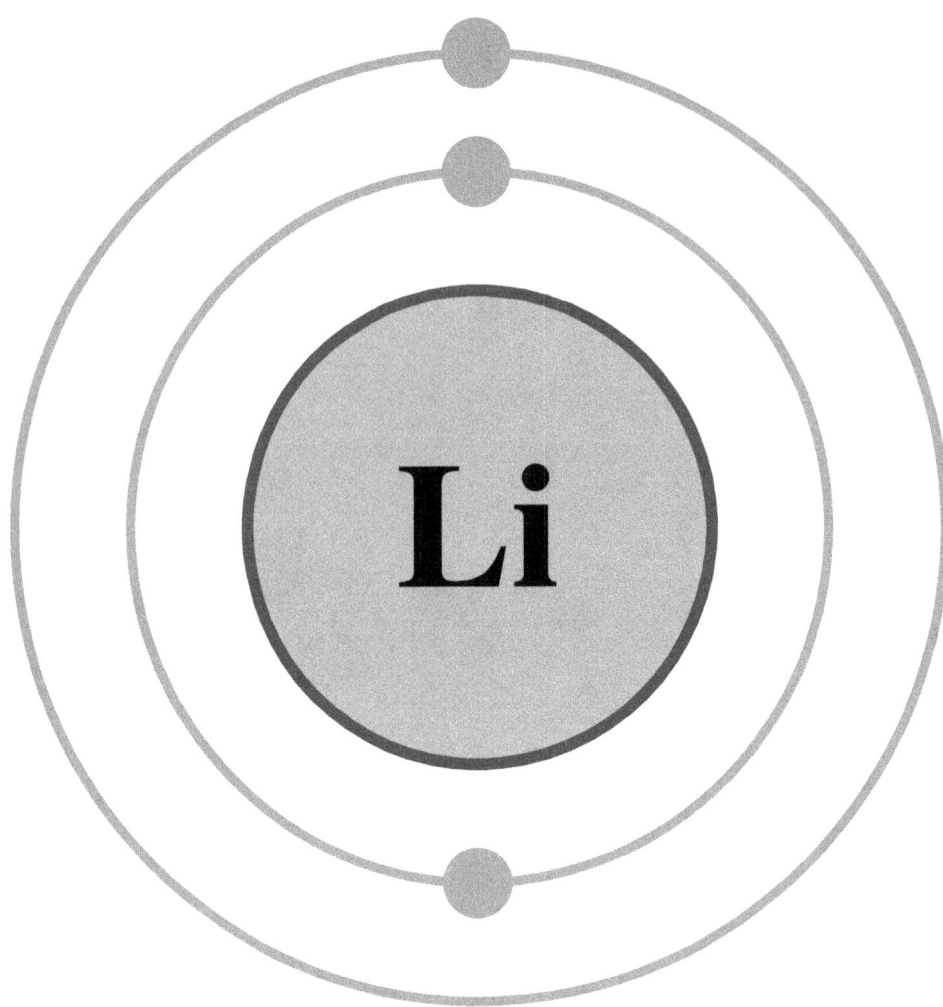

A Bohr diagram of lithium

of elements. The concept is also useful for describing the chemical bonds that hold atoms together. In bulk materials, this same idea helps explain the peculiar properties of lasers and semiconductors.

8.1   Shells and subshells

See also: Electron shell

Electron configuration was first conceived of under the Bohr model of the atom, and it is still common to speak of shells and subshells despite the advances in understanding of the quantum-mechanical nature of electrons.

An electron shell is the set of allowed states that share the same principal quantum number, n (the number before the letter in the orbital label), that electrons may occupy. An atom's nth electron shell can accommodate $2n^2$ electrons, *e.g.*

the first shell can accommodate 2 electrons, the second shell 8 electrons, and the third shell 18 electrons. The factor of two arises because the allowed states are doubled due to electron spin—each atomic orbital admits up to two otherwise identical electrons with opposite spin, one with a spin +1/2 (usually denoted by an up-arrow) and one with a spin −1/2 (with a down-arrow).

A subshell is the set of states defined by a common azimuthal quantum number, ℓ, within a shell. The values $\ell = 0$, 1, 2, 3 correspond to the s, p, d, and f labels, respectively. The maximum number of electrons that can be placed in a subshell is given by $2(2\ell + 1)$. This gives two electrons in an s subshell, six electrons in a p subshell, ten electrons in a d subshell and fourteen electrons in an f subshell.

The numbers of electrons that can occupy each shell and each subshell arise from the equations of quantum mechanics,[2] in particular the Pauli exclusion principle, which states that no two electrons in the same atom can have the same values of the four quantum numbers.[3]

8.2   Notation

See also: Atomic orbital

Physicists and chemists use a standard notation to indicate the electron configurations of atoms and molecules. For atoms, the notation consists of a sequence of atomic orbital labels (e.g. for phosphorus the sequence 1s, 2s, 2p, 3s, 3p) with the number of electrons assigned to each orbital (or set of orbitals sharing the same label) placed as a superscript. For example, hydrogen has one electron in the s-orbital of the first shell, so its configuration is written $1s^1$. Lithium has two electrons in the 1s-subshell and one in the (higher-energy) 2s-subshell, so its configuration is written $1s^2\, 2s^1$ (pronounced "one-s-two, two-s-one"). Phosphorus (atomic number 15) is as follows: $1s^2\, 2s^2\, 2p^6\, 3s^2\, 3p^3$.

For atoms with many electrons, this notation can become lengthy and so an abbreviated notation is used, since all but the last few subshells are identical to those of one or another of the noble gases. Phosphorus, for instance, differs from neon ($1s^2\, 2s^2\, 2p^6$) only by the presence of a third shell. Thus, the electron configuration of neon is pulled out, and phosphorus is written as follows: $[\text{Ne}]\, 3s^2\, 3p^3$. This convention is useful as it is the electrons in the outermost shell that most determine the chemistry of the element.

For a given configuration, the order of writing the orbitals is not completely fixed since only the orbital occupancies have physical significance. For example, the electron configuration of the titanium ground state can be written as either $[\text{Ar}]\, 4s^2\, 3d^2$ or $[\text{Ar}]\, 3d^2\, 4s^2$. The first notation follows the order based on the Madelung rule for the configurations of neutral atoms; 4s is filled before 3d in the sequence Ar, K, Ca, Sc, Ti. The second notation groups all orbitals with the same value of n together, corresponding to the "spectroscopic" order of orbital energies that is the reverse of the order in which electrons are removed from a given atom to form positive ions; 3d is filled before 4s in the sequence Ti^{4+}, Ti^{3+}, Ti^{2+}, Ti^+, Ti.

The superscript 1 for a singly occupied orbital is not compulsory. It is quite common to see the letters of the orbital labels (s, p, d, f) written in an italic or slanting typeface, although the International Union of Pure and Applied Chemistry (IUPAC) recommends a normal typeface (as used here). The choice of letters originates from a now-obsolete system of categorizing spectral lines as "sharp", "principal", "diffuse" and "fundamental" (or "fine"), based on their observed fine structure: their modern usage indicates orbitals with an azimuthal quantum number, l, of 0, 1, 2 or 3 respectively. After "f", the sequence continues alphabetically "g", "h", "i"... ($l = 4$, 5, 6...), skipping "j", although orbitals of these types are rarely required.[4][5]

The electron configurations of molecules are written in a similar way, except that molecular orbital labels are used instead of atomic orbital labels (see below).

8.3   Energy — ground state and excited states

The energy associated to an electron is that of its orbital. The energy of a configuration is often approximated as the sum of the energy of each electron, neglecting the electron-electron interactions. The configuration that corresponds to the lowest electronic energy is called the ground state. Any other configuration is an excited state.

As an example, the ground state configuration of the sodium atom is $1s^2 2s^2 2p^6 3s$, as deduced from the Aufbau principle (see below). The first excited state is obtained by promoting a 3s electron to the 3p orbital, to obtain the $1s^2 2s^2 2p^6 3p$ configuration, abbreviated as the 3p level. Atoms can move from one configuration to another by absorbing or emitting energy. In a sodium-vapor lamp for example, sodium atoms are excited to the 3p level by an electrical discharge, and return to the ground state by emitting yellow light of wavelength 589 nm.

Usually, the excitation of valence electrons (such as 3s for sodium) involves energies corresponding to photons of visible or ultraviolet light. The excitation of core electrons is possible, but requires much higher energies, generally corresponding to x-ray photons. This would be the case for example to excite a 2p electron to the 3s level and form the excited $1s^2 2s^2 2p^5 3s^2$ configuration.

The remainder of this article deals only with the ground-state configuration, often referred to as "the" configuration of an atom or molecule.

8.4 History

Niels Bohr (1923) was the first to propose that the periodicity in the properties of the elements might be explained by the electronic structure of the atom.[6] His proposals were based on the then current Bohr model of the atom, in which the electron shells were orbits at a fixed distance from the nucleus. Bohr's original configurations would seem strange to a present-day chemist: sulfur was given as 2.4.4.6 instead of $1s^2 2s^2 2p^6 3s^2 3p^4$ (2.8.6).

The following year, E. C. Stoner incorporated Sommerfeld's third quantum number into the description of electron shells, and correctly predicted the shell structure of sulfur to be 2.8.6.[7] However neither Bohr's system nor Stoner's could correctly describe the changes in atomic spectra in a magnetic field (the Zeeman effect).

Bohr was well aware of this shortcoming (and others), and had written to his friend Wolfgang Pauli to ask for his help in saving quantum theory (the system now known as "old quantum theory"). Pauli realized that the Zeeman effect must be due only to the outermost electrons of the atom, and was able to reproduce Stoner's shell structure, but with the correct structure of subshells, by his inclusion of a fourth quantum number and his exclusion principle (1925):[8]

> *It should be forbidden for more than one electron with the same value of the main quantum number n to have the same value for the other three quantum numbers k [l], j [m$_l$] and m [m$_s$].*

The Schrödinger equation, published in 1926, gave three of the four quantum numbers as a direct consequence of its solution for the hydrogen atom:[2] this solution yields the atomic orbitals that are shown today in textbooks of chemistry (and above). The examination of atomic spectra allowed the electron configurations of atoms to be determined experimentally, and led to an empirical rule (known as Madelung's rule (1936),[9] see below) for the order in which atomic orbitals are filled with electrons.

8.5 Atoms: Aufbau principle and Madelung rule

The Aufbau principle (from the German *Aufbau*, "building up, construction") was an important part of Bohr's original concept of electron configuration. It may be stated as:[10]

> *a maximum of two electrons are put into orbitals in the order of increasing orbital energy: the lowest-energy orbitals are filled before electrons are placed in higher-energy orbitals.*

The principle works very well (for the ground states of the atoms) for the first 18 elements, then decreasingly well for the following 100 elements. The modern form of the Aufbau principle describes an order of orbital energies given by Madelung's rule (or Klechkowski's rule). This rule was first stated by Charles Janet in 1929, rediscovered by Erwin Madelung in 1936,[9] and later given a theoretical justification by V.M. Klechkowski[11]

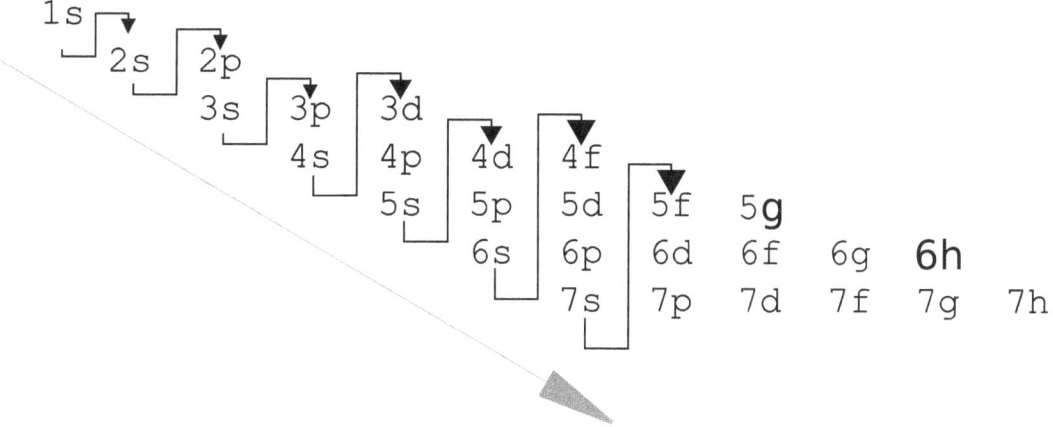

The approximate order of filling of atomic orbitals, following the arrows from 1s to 7p. (After 7p the order includes orbitals outside the range of the diagram, starting with 8s.)

1. Orbitals are filled in the order of increasing $n+l$;

2. Where two orbitals have the same value of $n+l$, they are filled in order of increasing n.

This gives the following order for filling the orbitals:

1s, 2s, 2p, 3s, 3p, 4s, 3d, 4p, 5s, 4d, 5p, 6s, 4f, 5d, 6p, 7s, 5f, 6d, 7p, (8s, 5g, 6f, 7d, 8p, and 9s)

In this list the orbitals in parentheses are not occupied in the ground state of the heaviest atom now known (Uuo, Z = 118).

The Aufbau principle can be applied, in a modified form, to the protons and neutrons in the atomic nucleus, as in the shell model of nuclear physics and nuclear chemistry.

8.5.1 Periodic table

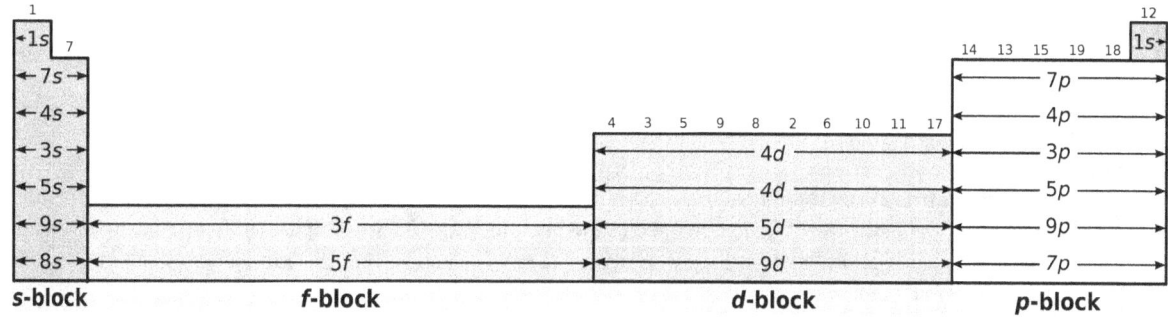

Electron configuration table

The form of the periodic table is closely related to the electron configuration of the atoms of the elements. For example, all the elements of group 2 have an electron configuration of [E] ns^2 (where [E] is an inert gas configuration), and have notable similarities in their chemical properties. In general, the periodicity of the periodic table in terms of periodic table blocks is clearly due to the number of electrons (2, 6, 10, 14...) needed to fill s, p, d, and f subshells.

The outermost electron shell is often referred to as the "valence shell" and (to a first approximation) determines the chemical properties. It should be remembered that the similarities in the chemical properties were remarked on more than a century before the idea of electron configuration.[12] It is not clear how far Madelung's rule *explains* (rather than simply describes) the periodic table,[13] although some properties (such as the common +2 oxidation state in the first row of the transition metals) would obviously be different with a different order of orbital filling.

8.5.2 Shortcomings of the Aufbau principle

The Aufbau principle rests on a fundamental postulate that the order of orbital energies is fixed, both for a given element and between different elements; in both cases this is only approximately true. It considers atomic orbitals as "boxes" of fixed energy into which can be placed two electrons and no more. However, the energy of an electron "in" an atomic orbital depends on the energies of all the other electrons of the atom (or ion, or molecule, etc.). There are no "one-electron solutions" for systems of more than one electron, only a set of many-electron solutions that cannot be calculated exactly[14] (although there are mathematical approximations available, such as the Hartree–Fock method).

The fact that the Aufbau principle is based on an approximation can be seen from the fact that there is an almost-fixed filling order at all, that, within a given shell, the s-orbital is always filled before the p-orbitals. In a hydrogen-like atom, which only has one electron, the s-orbital and the p-orbitals of the same shell have exactly the same energy, to a very good approximation in the absence of external electromagnetic fields. (However, in a real hydrogen atom, the energy levels are slightly split by the magnetic field of the nucleus, and by the quantum electrodynamic effects of the Lamb shift.)

8.5.3 Ionization of the transition metals

The naïve application of the Aufbau principle leads to a well-known paradox (or apparent paradox) in the basic chemistry of the transition metals. Potassium and calcium appear in the periodic table before the transition metals, and have electron configurations [Ar] $4s^1$ and [Ar] $4s^2$ respectively, i.e. the 4s-orbital is filled before the 3d-orbital. This is in line with Madelung's rule, as the 4s-orbital has $n+l = 4$ ($n = 4$, $l = 0$) while the 3d-orbital has $n+l = 5$ ($n = 3$, $l = 2$). After calcium, most neutral atoms in the first series of transition metals (Sc-Zn) have configurations with two 4s electrons, but there are two exceptions. Chromium and copper have electron configurations [Ar] $3d^5$ $4s^1$ and [Ar] $3d^{10}$ $4s^1$ respectively, i.e. one electron has passed from the 4s-orbital to a 3d-orbital to generate a half-filled or filled subshell. In this case, the usual explanation is that "half-filled or completely filled subshells are particularly stable arrangements of electrons".

The apparent paradox arises when electrons are *removed* from the transition metal atoms to form ions. The first electrons to be ionized come not from the 3d-orbital, as one would expect if it were "higher in energy", but from the 4s-orbital. This interchange of electrons between 4s and 3d is found for all atoms of the first series of transition metals.[15] The configurations of the neutral atoms (K, Ca, Sc, Ti, V, Cr, ...) usually follow the order 1s, 2s, 2p, 3s, 3p, 4s, 3d, ...; however the successive stages of ionization of a given atom (such as Fe^{4+}, Fe^{3+}, Fe^{2+}, Fe^+, Fe) usually follow the order 1s, 2s, 2p, 3s, 3p, 3d, 4s, ...

This phenomenon is only paradoxical if it is assumed that the energy order of atomic orbitals is fixed and unaffected by the nuclear charge or by the presence of electrons in other orbitals. If that were the case, the 3d-orbital would have the same energy as the 3p-orbital, as it does in hydrogen, yet it clearly doesn't. There is no special reason why the Fe^{2+} ion should have the same electron configuration as the chromium atom, given that iron has two more protons in its nucleus than chromium, and that the chemistry of the two species is very different. Melrose and Eric Scerri have analyzed the changes of orbital energy with orbital occupations in terms of the two-electron repulsion integrals of the Hartree-Fock method of atomic structure calculation.[16]

Similar ion-like $3d^x4s^0$ configurations occur in transition metal complexes as described by the simple crystal field theory, even if the metal has oxidation state 0. For example, chromium hexacarbonyl can be described as a chromium atom (not ion) surrounded by six carbon monoxide ligands. The electron configuration of the central chromium atom is described as $3d^6$ with the six electrons filling the three lower-energy d orbitals between the ligands. The other two d orbitals are at higher energy due to the crystal field of the ligands. This picture is consistent with the experimental fact that the complex is diamagnetic, meaning that it has no unpaired electrons. However, in a more accurate description using molecular orbital theory, the d-like orbitals occupied by the six electrons are no longer identical with the d orbitals of the free atom.

8.5.4   Other exceptions to Madelung's rule

There are several more exceptions to Madelung's rule among the heavier elements, and it is more and more difficult to resort to simple explanations, such as the stability of half-filled subshells. It is possible to predict most of the exceptions by Hartree–Fock calculations,[17] which are an approximate method for taking account of the effect of the other electrons on orbital energies. For the heavier elements, it is also necessary to take account of the effects of Special Relativity on the energies of the atomic orbitals, as the inner-shell electrons are moving at speeds approaching the speed of light. In general, these relativistic effects[18] tend to decrease the energy of the s-orbitals in relation to the other atomic orbitals.[19] The table below shows the ground state configuration in terms of orbital occupancy, but it does not show the ground state in terms of the sequence of orbital energies as determined spectroscopically. For example, in the transition metals, the 4s orbital is of a higher energy than the 3d orbitals; and in the lanthanides, the 6s is higher than the 4f and 5d. The ground states can be seen in the Electron configurations of the elements (data page).

The electron-shell configuration of elements beyond rutherfordium has not yet been empirically verified, but they are expected to follow Madelung's rule without exceptions until element 120.[22]

8.6   Electron configuration in molecules

In molecules, the situation becomes more complex, as each molecule has a different orbital structure. The molecular orbitals are labelled according to their symmetry,[23] rather than the atomic orbital labels used for atoms and monatomic ions: hence, the electron configuration of the dioxygen molecule, O_2, is written $1\sigma_g^2 1\sigma_u^2 2\sigma_g^2 2\sigma_u^2 3\sigma_g^2 1\pi_u^4 1\pi_g^2$,[24][25] or equivalently $1\sigma_g^2 1\sigma_u^2 2\sigma_g^2 2\sigma_u^2 1\pi_u^4 3\sigma_g^2 1\pi_g^2$.[1] The term $1\pi_g^2$ represents the two electrons in the two degenerate π*-orbitals (antibonding). From Hund's rules, these electrons have parallel spins in the ground state, and so dioxygen has a net magnetic moment (it is paramagnetic). The explanation of the paramagnetism of dioxygen was a major success for molecular orbital theory.

The electronic configuration of polyatomic molecules can change without absorption or emission of a photon through vibronic couplings.

8.6.1   Electron configuration in solids

In a solid, the electron states become very numerous. They cease to be discrete, and effectively blend into continuous ranges of possible states (an electron band). The notion of electron configuration ceases to be relevant, and yields to band theory.

8.7   Applications

The most widespread application of electron configurations is in the rationalization of chemical properties, in both inorganic and organic chemistry. In effect, electron configurations, along with some simplified form of molecular orbital theory, have become the modern equivalent of the valence concept, describing the number and type of chemical bonds that an atom can be expected to form.

This approach is taken further in computational chemistry, which typically attempts to make quantitative estimates of chemical properties. For many years, most such calculations relied upon the "linear combination of atomic orbitals" (LCAO) approximation, using an ever larger and more complex basis set of atomic orbitals as the starting point. The last step in such a calculation is the assignment of electrons among the molecular orbitals according to the Aufbau principle. Not all methods in calculational chemistry rely on electron configuration: density functional theory (DFT) is an important example of a method that discards the model.

For atoms or molecules with more than one electron, the motion of electrons are correlated and such a picture is no longer exact. A very large number of electronic configurations are needed to exactly describe any multi-electron system, and no energy can be associated with one single configuration. However, the electronic wave function is usually dominated by a very small number of configurations and therefore the notion of electronic configuration remains essential for multi-electron systems.

A fundamental application of electron configurations is in the interpretation of atomic spectra. In this case, it is necessary to supplement the electron configuration with one or more term symbols, which describe the different energy levels available to an atom. Term symbols can be calculated for any electron configuration, not just the ground-state configuration listed in tables, although not all the energy levels are observed in practice. It is through the analysis of atomic spectra that the ground-state electron configurations of the elements were experimentally determined.

8.8   See also

- Born–Oppenheimer approximation

- Electron configurations of the elements (data page)

- Periodic table (electron configurations)

- Atomic orbital

- Energy level

- Term symbol

- Molecular term symbol

- HOMO/LUMO

- Periodic Table Group

- d electron count

- Extension of the periodic table beyond the seventh period Discusses the limits of the periodic table

8.9   Notes

[1] IUPAC, *Compendium of Chemical Terminology*, 2nd ed. (the "Gold Book") (1997). Online corrected version: (2006–) "configuration (electronic)".

[2] In formal terms, the quantum numbers n, ℓ and $m\ell$ arise from the fact that the solutions to the time-independent Schrödinger equation for hydrogen-like atoms are based on spherical harmonics.

[3] IUPAC, *Compendium of Chemical Terminology*, 2nd ed. (the "Gold Book") (1997). Online corrected version: (2006–) "Pauli exclusion principle".

[4] Weisstein, Eric W. (2007). "Electron Orbital". *wolfram*.

[5] Ebbing, Darrell D.; Gammon, Steven D. (2007-01-12). *General Chemistry*. p. 284. ISBN 978-0-618-73879-3.

[6] Bohr, Niels (1923). "Über die Anwendung der Quantumtheorie auf den Atombau. I". *Zeitschrift für Physik* **13**: 117. Bibcode: doi:10.1007/BF01328209.

[7] Stoner, E.C. (1924). "The distribution of electrons among atomic levels". *Philosophical Magazine (6th Ser.)* **48** (286): 719–36. doi:10.1080/14786442408634535.

[8] Pauli, Wolfgang (1925). "Über den Einfluss der Geschwindigkeitsabhändigkeit der elektronmasse auf den Zeemaneffekt". *Zeitschrift für Physik* **31**: 373. Bibcode:1925ZPhy...31..373P. doi:10.1007/BF02980592. English translation from Scerri, Eric R. (1991). "The Electron Configuration Model, Quantum Mechanics and Reduction" (PDF). *Br. J. Phil. Sci.* **42** (3): 309–25. doi:10.1093/bjps/42.3.309.

[9] Madelung, Erwin (1936). *Mathematische Hilfsmittel des Physikers*. Berlin: Springer.

[10] IUPAC, *Compendium of Chemical Terminology*, 2nd ed. (the "Gold Book") (1997). Online corrected version: (2006–) "aufbau principle".

[11] Wong, D. Pan (1979). "Theoretical justification of Madelung's rule". *Journal of Chemical Education* **56** (11): 714–18. Bibcode:1979JChEd..56..714W. doi:10.1021/ed056p714.

[12] The similarities in chemical properties and the numerical relationship between the atomic weights of calcium, strontium and barium was first noted by Johann Wolfgang Döbereiner in 1817.

[13] Scerri, Eric R. (1998). "How Good Is the Quantum Mechanical Explanation of the Periodic System?" (PDF). *Journal of Chemical Education* **75** (11): 1384–85. Bibcode:1998JChEd..75.1384S. doi:10.1021/ed075p1384. Ostrovsky, V.N. (2005). "On Recent Discussion Concerning Quantum Justification of the Periodic Table of the Elements". *Foundations of Chemistry* **7** (3): 235–39. doi:10.1007/s10698-005-2141-y.

[14] Electrons are identical particles, a fact that is sometimes referred to as "indistinguishability of electrons". A one-electron solution to a many-electron system would imply that the electrons could be distinguished from one another, and there is strong experimental evidence that they can't be. The exact solution of a many-electron system is a n-body problem with $n \geq 3$ (the nucleus counts as one of the "bodies"): such problems have evaded analytical solution since at least the time of Euler.

[15] There are some cases in the second and third series where the electron remains in an s-orbital.

[16] Melrose, Melvyn P.; Scerri, Eric R. (1996). "Why the 4s Orbital is Occupied before the 3d". *Journal of Chemical Education* **73** (6): 498–503. Bibcode:1996JChEd..73..498M. doi:10.1021/ed073p498.

[17] Meek, Terry L.; Allen, Leland C. (2002). "Configuration irregularities: deviations from the Madelung rule and inversion of orbital energy levels". *Chem. Phys. Lett.* **362** (5–6): 362–64. Bibcode:2002CPL...362..362M. doi:10.1016/S0009-2614(02)00919-3.

[18] IUPAC, *Compendium of Chemical Terminology*, 2nd ed. (the "Gold Book") (1997). Online corrected version: (2006–) "relativistic effects".

[19] Pyykkö, Pekka (1988). "Relativistic effects in structural chemistry". *Chem. Rev.* **88** (3): 563–94. doi:10.1021/cr00085a006.

[20] Miessler, G. L.; Tarr, D. A. (1999). *Inorganic Chemistry* (2nd ed.). Prentice-Hall. p. 38.

[21] Scerri, Eric R. (2007). *The periodic table: its story and its significance*. Oxford University Press. pp. 239–240. ISBN 0-19-530573-6.

[22] Haire, Richard G. (2006). "Transactinides and the future elements". In Morss; Edelstein, Norman M.; Fuger, Jean. *The Chemistry of the Actinide and Transactinide Elements* (3rd ed.). Dordrecht, The Netherlands: Springer Science+Business Media. ISBN 1-4020-3555-1.

[23] The labels are written in lowercase to indicate that they correspond to one-electron functions. They are numbered consecutively for each symmetry type (irreducible representation in the character table of the point group for the molecule), starting from the orbital of lowest energy for that type.

[24] Levine I.N. *Quantum Chemistry* (4th ed., Prentice Hall 1991) p.376 ISBN 0-205-12770-3

[25] Miessler G.L. and Tarr D.A. *Inorganic Chemistry* (2nd ed., Prentice Hall 1999) p.118 ISBN 0-13-841891-8

8.10   References

- Jolly, William L. (1991). *Modern Inorganic Chemistry* (2nd ed.). New York: McGraw-Hill. pp. 1–23. ISBN 0-07-112651-1.

- Scerri, Eric (2007). *The Periodic System, Its Story and Its Significance*. New York: Oxford University Press. ISBN 0-19-530573-6.

8.11 External links

- What does an atom look like? Configuration in 3D

- The trouble with the aufbau principle Eric Scerri, *Education in Chemistry*, 7 November 2013

Chapter 9

Electric charge

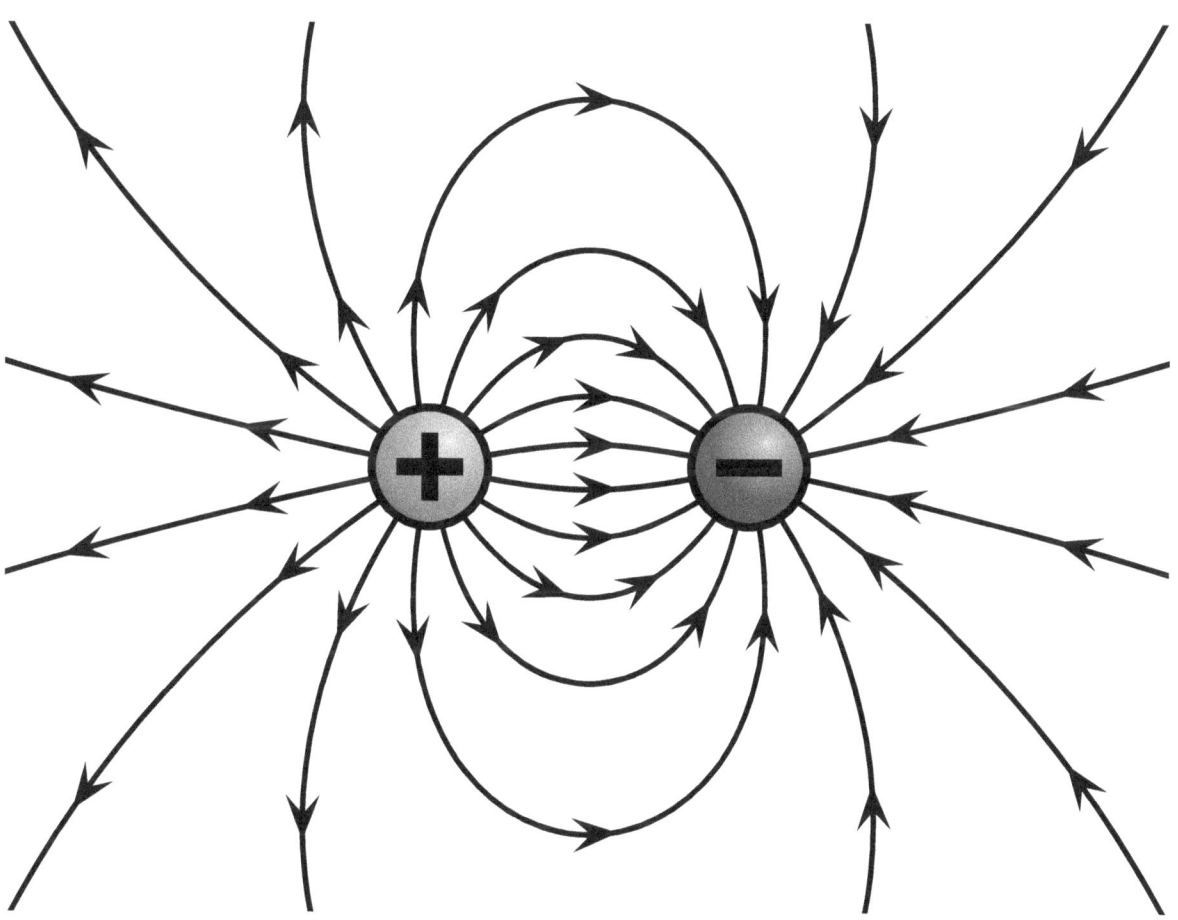

Electric field of a positive and a negative point charge.

Electric charge is the physical property of matter that causes it to experience a force when placed in an electromagnetic field. There are two types of electric charges: positive and negative. Positively charged substances are repelled from other positively charged substances, but attracted to negatively charged substances; negatively charged substances are repelled from negative and attracted to positive. An object is negatively charged if it has an excess of electrons, and is otherwise positively charged or uncharged. The SI derived unit of electric charge is the coulomb (C), although in electrical engineering it is also common to use the ampere-hour (Ah), and in chemistry it is common to use the elementary charge (e) as a unit. The symbol Q is often used to denote charge. The early knowledge of how charged substances interact is now called classical electrodynamics, and is still very accurate if quantum effects do not need to be considered. The *electric charge* is a fundamental conserved property of some subatomic particles, which determines their electromagnetic interaction. Electrically charged matter is influenced by, and produces, electromagnetic fields. The interaction between a moving charge and an electromagnetic field is the source of the electromagnetic force, which is one of the four fundamental forces (See also:magnetic field).

Twentieth-century experiments demonstrated that electric charge is *quantized*; that is, it comes in integer multiples of individual small units called the elementary charge, e, approximately equal to 1.602×10^{-19} coulombs (except for particles called quarks, which have charges that are integer multiples of $e/3$). The proton has a charge of $+e$, and the electron has a charge of $-e$. The study of charged particles, and how their interactions are mediated by photons, is called quantum electrodynamics.

9.1 Overview

Charge is the fundamental property of forms of matter that exhibit electrostatic attraction or repulsion in the presence of other matter. Electric charge is a characteristic property of many subatomic particles. The charges of free-standing particles are integer multiples of the elementary charge e; we say that electric charge is *quantized*. Michael Faraday, in his electrolysis experiments, was the first to note the discrete nature of electric charge. Robert Millikan's oil-drop experiment demonstrated this fact directly, and measured the elementary charge.

By convention, the charge of an electron is -1, while that of a proton is $+1$. Charged particles whose charges have the same sign repel one another, and particles whose charges have different signs attract. Coulomb's law quantifies the electrostatic force between two particles by asserting that the force is proportional to the product of their charges, and inversely proportional to the square of the distance between them.

The charge of an antiparticle equals that of the corresponding particle, but with opposite sign. Quarks have fractional charges of either $-\frac{1}{3}$ or $+\frac{2}{3}$, but free-standing quarks have never been observed (the theoretical reason for this fact is asymptotic freedom).

The electric charge of a macroscopic object is the sum of the electric charges of the particles that make it up. This charge is often small, because matter is made of atoms, and atoms typically have equal numbers of protons and electrons, in which case their charges cancel out, yielding a net charge of zero, thus making the atom neutral.

An *ion* is an atom (or group of atoms) that has lost one or more electrons, giving it a net positive charge (cation), or that has gained one or more electrons, giving it a net negative charge (anion). *Monatomic ions* are formed from single atoms, while *polyatomic ions* are formed from two or more atoms that have been bonded together, in each case yielding an ion with a positive or negative net charge.

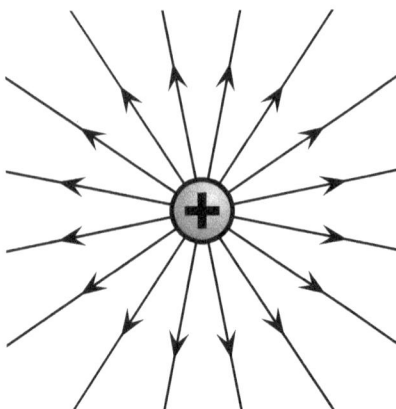

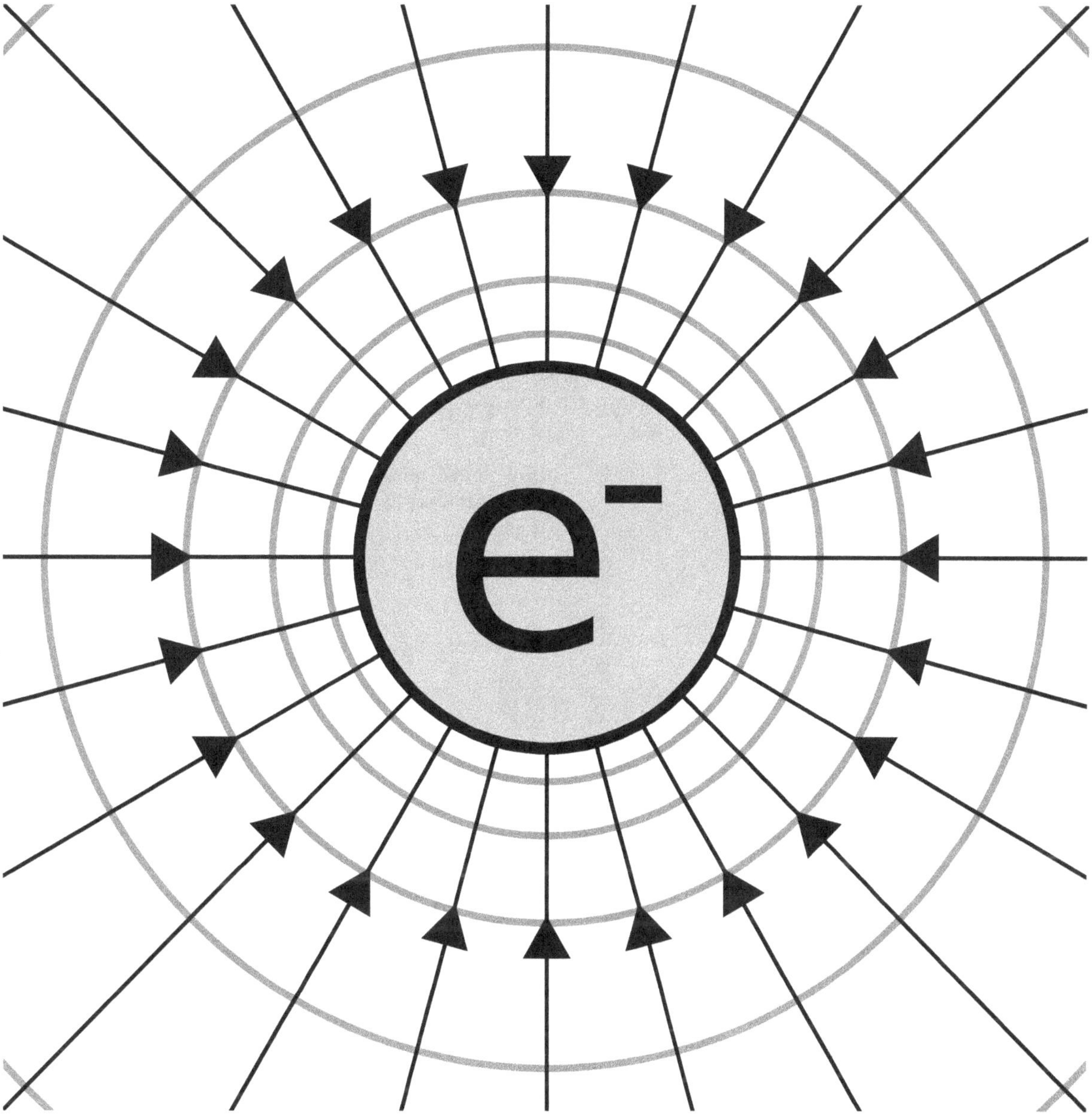

Diagram showing field lines and equipotentials around an electron, a negatively charged particle. In an electrically neutral atom, the number of electrons is equal to the number of protons (which are positively charged), resulting in a net zero overall charge

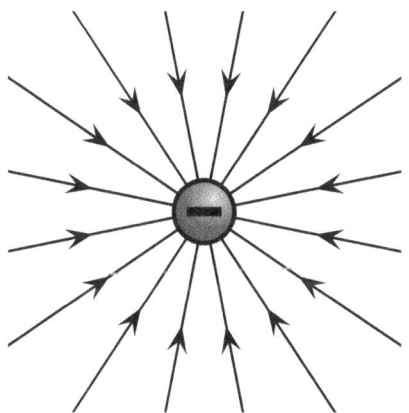

Electric field induced by a positive electric charge (left) and a field induced by a negative electric charge (right).

During formation of macroscopic objects, constituent atoms and ions usually combine to form structures composed of neutral *ionic compounds* electrically bound to neutral atoms. Thus macroscopic objects tend toward being neutral overall, but macroscopic objects are rarely perfectly net neutral.

Sometimes macroscopic objects contain ions distributed throughout the material, rigidly bound in place, giving an overall net positive or negative charge to the object. Also, macroscopic objects made of conductive elements, can more or less easily (depending on the element) take on or give off electrons, and then maintain a net negative or positive charge indefinitely. When the net electric charge of an object is non-zero and motionless, the phenomenon is known as static electricity. This can easily be produced by rubbing two dissimilar materials together, such as rubbing amber with fur or glass with silk. In this way non-conductive materials can be charged to a significant degree, either positively or negatively. Charge taken from one material is moved to the other material, leaving an opposite charge of the same magnitude behind. The law of *conservation of charge* always applies, giving the object from which a negative charge has been taken a positive charge of the same magnitude, and vice versa.

Even when an object's net charge is zero, charge can be distributed non-uniformly in the object (e.g., due to an external electromagnetic field, or bound polar molecules). In such cases the object is said to be polarized. The charge due to polarization is known as bound charge, while charge on an object produced by electrons gained or lost from outside the object is called *free charge*. The motion of electrons in conductive metals in a specific direction is known as electric current.

9.2   Units

The SI unit of quantity of electric charge is the coulomb, which is equivalent to about 6.242×10^{18} e (e is the charge of a proton). Hence, the charge of an electron is approximately -1.602×10^{-19} C. The coulomb is defined as the quantity of charge that has passed through the cross section of an electrical conductor carrying one ampere within one second. The symbol Q is often used to denote a quantity of electricity or charge. The quantity of electric charge can be directly measured with an electrometer, or indirectly measured with a ballistic galvanometer.

After finding the quantized character of charge, in 1891 George Stoney proposed the unit 'electron' for this fundamental unit of electrical charge. This was before the discovery of the particle by J.J. Thomson in 1897. The unit is today treated as nameless, referred to as "elementary charge", "fundamental unit of charge", or simply as "e". A measure of charge should be a multiple of the elementary charge e, even if at large scales charge seems to behave as a real quantity. In some contexts it is meaningful to speak of fractions of a charge; for example in the charging of a capacitor, or in the fractional quantum Hall effect.

In systems of units other than SI such as cgs, electric charge is expressed as combination of only three fundamental quantities such as length, mass and time and not four as in SI where electric charge is a combination of length, mass, time and electric current.

9.3   History

As reported by the ancient Greek mathematician Thales of Miletus around 600 BC, charge (or *electricity*) could be accumulated by rubbing fur on various substances, such as amber. The Greeks noted that the charged amber buttons could attract light objects such as hair. They also noted that if they rubbed the amber for long enough, they could even get an electric spark to jump. This property derives from the triboelectric effect.

In 1600, the English scientist William Gilbert returned to the subject in *De Magnete*, and coined the New Latin word *electricus* from ηλεκτρον (*elektron*), the Greek word for *amber*, which soon gave rise to the English words "electric" and "electricity." He was followed in 1660 by Otto von Guericke, who invented what was probably the first electrostatic generator. Other European pioneers were Robert Boyle, who in 1675 stated that electric attraction and repulsion can act across a vacuum; Stephen Gray, who in 1729 classified materials as conductors and insulators; and C. F. du Fay, who

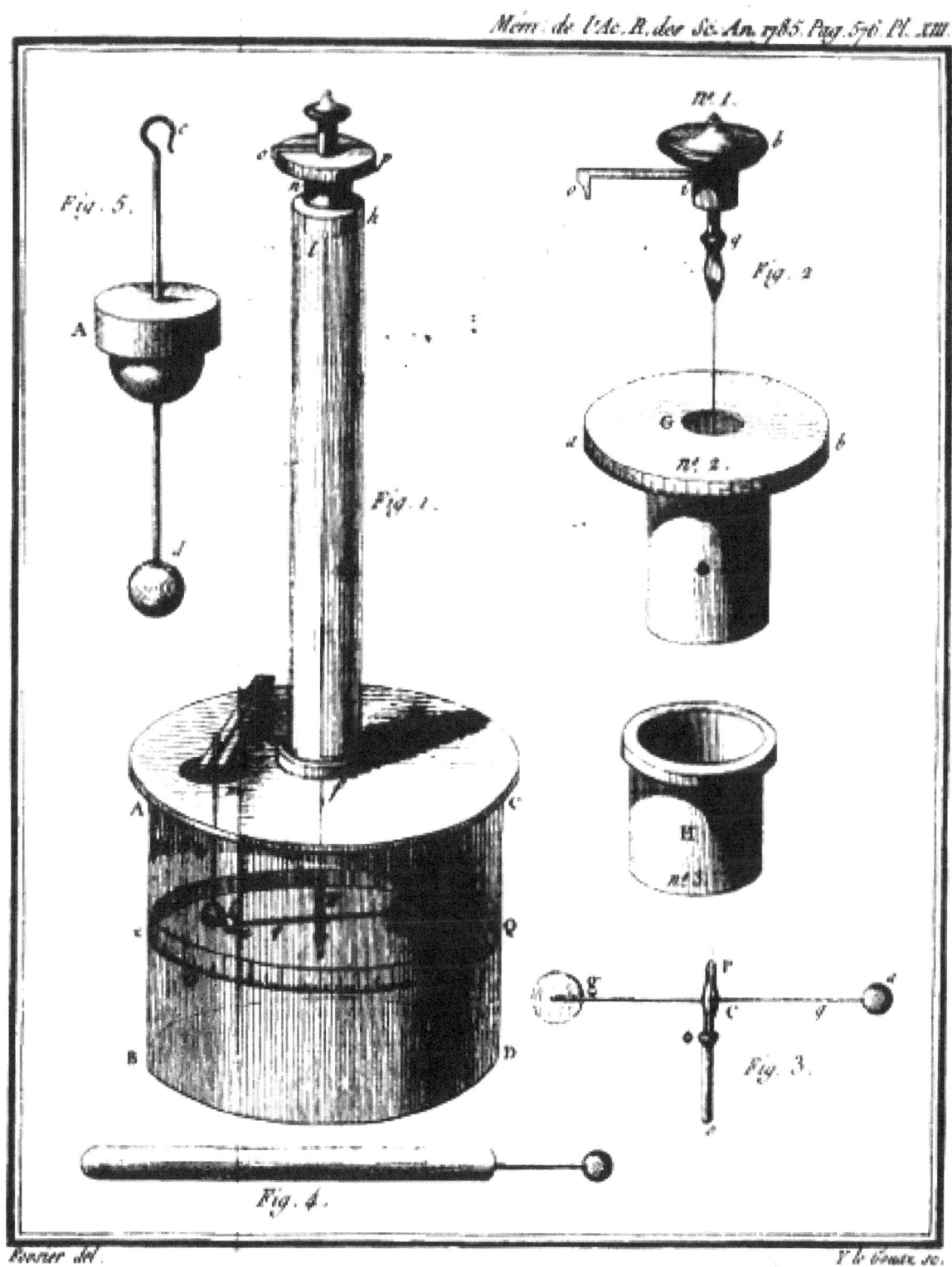

Coulomb's torsion balance

proposed in 1733[1] that electricity comes in two varieties that cancel each other, and expressed this in terms of a two-fluid theory. When glass was rubbed with silk, du Fay said that the glass was charged with *vitreous electricity*, and, when amber was rubbed with fur, the amber was said to be charged with *resinous electricity*. In 1839, Michael Faraday showed that the apparent division between static electricity, current electricity, and bioelectricity was incorrect, and all were a consequence of the behavior of a single kind of electricity appearing in opposite polarities. It is arbitrary which polarity is called positive and which is called negative. Positive charge can be defined as the charge left on a glass rod after being rubbed with silk.[2]

One of the foremost experts on electricity in the 18th century was Benjamin Franklin, who argued in favour of a one-fluid theory of electricity. Franklin imagined electricity as being a type of invisible fluid present in all matter; for example, he believed that it was the glass in a Leyden jar that held the accumulated charge. He posited that rubbing insulating surfaces together caused this fluid to change location, and that a flow of this fluid constitutes an electric current. He also posited that when matter contained too little of the fluid it was "negatively" charged, and when it had an excess it was "positively" charged. For a reason that was not recorded, he identified the term "positive" with vitreous electricity and "negative" with resinous electricity. William Watson arrived at the same explanation at about the same time.

9.4 Static electricity and electric current

Static electricity and electric current are two separate phenomena. They both involve electric charge, and may occur simultaneously in the same object. Static electricity refers to the electric charge of an object and the related electrostatic discharge when two objects are brought together that are not at equilibrium. An electrostatic discharge creates a change in the charge of each of the two objects. In contrast, electric current is the flow of electric charge through an object, which produces no net loss or gain of electric charge.

9.4.1 Electrification by friction

Further information: triboelectric effect

When a piece of glass and a piece of resin—neither of which exhibit any electrical properties—are rubbed together and left with the rubbed surfaces in contact, they still exhibit no electrical properties. When separated, they attract each other.

A second piece of glass rubbed with a second piece of resin, then separated and suspended near the former pieces of glass and resin causes these phenomena:

- The two pieces of glass repel each other.
- Each piece of glass attracts each piece of resin.
- The two pieces of resin repel each other.

This attraction and repulsion is an *electrical phenomena,* and the bodies that exhibit them are said to be *electrified,* or *electrically charged.* Bodies may be electrified in many other ways, as well as by friction. The electrical properties of the two pieces of glass are similar to each other but opposite to those of the two pieces of resin: The glass attracts what the resin repels and repels what the resin attracts.

If a body electrified in any manner whatsoever behaves as the glass does, that is, if it repels the glass and attracts the resin, the body is said to be 'vitreously' electrified, and if it attracts the glass and repels the resin it is said to be 'resinously' electrified. All electrified bodies are found to be either vitreously or resinously electrified.

It is the established convention of the scientific community to define the vitreous electrification as positive, and the resinous electrification as negative. The exactly opposite properties of the two kinds of electrification justify our indicating them by opposite signs, but the application of the positive sign to one rather than to the other kind must be considered as a matter of arbitrary convention, just as it is a matter of convention in mathematical diagram to reckon positive distances towards the right hand.

No force, either of attraction or of repulsion, can be observed between an electrified body and a body not electrified.[3]

Actually, all bodies are electrified, but may appear not to be so by the relative similar charge of neighboring objects in the environment. An object further electrified + or − creates an equivalent or opposite charge by default in neighboring objects, until those charges can equalize. The effects of attraction can be observed in high-voltage experiments, while lower voltage effects are merely weaker and therefore less obvious. The attraction and repulsion forces are codified by Coulomb's Law (attraction falls off at the square of the distance, which has a corollary for acceleration in a gravitational field, suggesting that gravitation may be merely electrostatic phenomenon between relatively weak charges in terms of scale). See also the Casimir effect.

It is now known that the Franklin/Watson model was fundamentally correct. There is only one kind of electrical charge, and only one variable is required to keep track of the amount of charge.[4] On the other hand, just knowing the charge is not a complete description of the situation. Matter is composed of several kinds of electrically charged particles, and these particles have many properties, not just charge.

The most common charge carriers are the positively charged proton and the negatively charged electron. The movement of any of these charged particles constitutes an electric current. In many situations, it suffices to speak of the *conventional current* without regard to whether it is carried by positive charges moving in the direction of the conventional current or by negative charges moving in the opposite direction. This macroscopic viewpoint is an approximation that simplifies electromagnetic concepts and calculations.

At the opposite extreme, if one looks at the microscopic situation, one sees there are many ways of carrying an electric current, including: a flow of electrons; a flow of electron "holes" that act like positive particles; and both negative and positive particles (ions or other charged particles) flowing in opposite directions in an electrolytic solution or a plasma.

Beware that, in the common and important case of metallic wires, the direction of the conventional current is opposite to the drift velocity of the actual charge carriers, i.e., the electrons. This is a source of confusion for beginners.

9.5   Properties

Aside from the properties described in articles about electromagnetism, charge is a relativistic invariant. This means that any particle that has charge Q, no matter how fast it goes, always has charge Q. This property has been experimentally verified by showing that the charge of *one* helium nucleus (two protons and two neutrons bound together in a nucleus and moving around at high speeds) is the same as *two* deuterium nuclei (one proton and one neutron bound together, but moving much more slowly than they would if they were in a helium nucleus).

9.6   Conservation of electric charge

Main article: Charge conservation

The total electric charge of an isolated system remains constant regardless of changes within the system itself. This law is inherent to all processes known to physics and can be derived in a local form from gauge invariance of the wave function. The conservation of charge results in the charge-current continuity equation. More generally, the net change in charge density ϱ within a volume of integration V is equal to the area integral over the current density $\mathbf{J}$ through the closed surface $S = \partial V$, which is in turn equal to the net current I:

$$-\tfrac{d}{dt}\int_V \rho\,\mathrm{d}V = \oiint_{\partial V} \mathbf{J}\cdot\mathbf{dS} = \int J dS \cos\theta = I.$$

Thus, the conservation of electric charge, as expressed by the continuity equation, gives the result:

$$I = \frac{dQ}{dt}.$$

The charge transferred between times t_i and t_f is obtained by integrating both sides:

$$Q = \int_{t_i}^{t_f} I \, dt$$

where I is the net outward current through a closed surface and Q is the electric charge contained within the volume defined by the surface.

9.7   See also

- Quantity of electricity

- SI electromagnetism units

9.8   References

[1] Two Kinds of Electrical Fluid: Vitreous and Resinous – 1733

[2] Electromagnetic Fields (2nd Edition), Roald K. Wangsness, Wiley, 1986. ISBN 0-471-81186-6 (intermediate level textbook)

[3] James Clerk Maxwell *A Treatise on Electricity and Magnetism*, pp. 32-33, Dover Publications Inc., 1954 ASIN: B000HFDK0K, 3rd ed. of 1891

[4] One Kind of Charge

9.9   External links

- How fast does a charge decay?

- Science Aid: Electrostatic charge Easy-to-understand page on electrostatic charge.

- History of the electrical units.

Chapter 10

Mass number

The **mass number (A)**, also called **atomic mass number** or **nucleon number**, is the total number of protons and neutrons (together known as nucleons) in an atomic nucleus. It determines the atomic mass of atoms. Because protons and neutrons both are baryons, the mass number A is identical with the baryon number B as of the nucleus as of the whole atom or ion. The mass number is different for each different isotope of a chemical element. This is not the same as the atomic number (**Z**) which denotes the number of protons in a nucleus, and thus uniquely identifies an element. Hence, the difference between the mass number and the atomic number gives the number of neutrons (N) in a given nucleus: $N=A-Z$.[1]

The mass number is written either after the element name or as a superscript to the left of an element's symbol. For example, the most common isotope of carbon is carbon-12, or 12C, which has 6 protons and 6 neutrons. The full isotope symbol would also have the atomic number (**Z**) as a subscript to the left of the element symbol directly below the mass number: 12
6C.[2] This is technically redundant, as each element is defined by its atomic number, so it is often omitted.

10.1   Mass number changes in radioactive decay

Different types of radioactive decay are characterized by their changes in mass number as well as atomic number, according to the radioactive displacement law of Fajans and Soddy. For example, uranium-238 usually decays by alpha decay, where the nucleus loses two neutrons and two protons in the form of an alpha particle. Thus the atomic number and the number of neutrons each decrease by 2 (Z: 92→90, N: 146→144), so that the mass number decreases by 4 (A = 238→234); the result is an atom of thorium-234 and an alpha particle (4
2He2+):[3]

On the other hand, carbon-14 naturally decays by radioactive beta decay, whereby one neutron is transmuted into a proton with the emission of an electron and an anti-neutrino. Thus the atomic number increases by 1 (Z: 6→7) and the mass number remains the same (A = 14), while the number of neutrons decreases by 1 (N: 8→7).[4] The resulting atom is nitrogen-14, with seven protons and seven neutrons:

Another type of radioactive decay without change in mass number is emission of a gamma ray from a nuclear isomer or metastable excited state of an atomic nucleus. Since all the protons and neutrons remain in the nucleus unchanged in this process, the mass number is also unchanged.

10.2 Mass number and isotopic mass

The mass number gives an estimate of the isotopic mass measured in atomic mass units (u). For ^{12}C the isotopic mass is exactly 12, since the atomic mass unit is defined as 1/12 of the mass of ^{12}C. For other isotopes, the isotopic mass is usually within 0.1 u of the mass number. For example, ^{35}Cl has a mass number of 35 and an isotopic mass of 34.96885.

There are two reasons for the difference between mass number and isotopic mass, known as the mass defect:

1. The neutron is slightly heavier than the proton. This increases the mass of nuclei with more neutrons than protons relative to the atomic mass unit scale based on ^{12}C with equal numbers of protons and neutrons.

2. The nuclear binding energy varies between nuclei. A nucleus with greater binding energy has a lower total energy, and therefore a lower mass according to Einstein's mass-energy equivalence relation $E = mc^2$. For ^{35}Cl the isotopic mass is less than 35 so this must be the dominant factor.

10.3 Relative Atomic Mass of an element

The mass number should also not be confused with the relative atomic mass (also called atomic weight) of an element, which is the ratio of the average atomic mass of the different isotopes of that element (weighted by abundance) to the unified atomic mass unit.[5] This weighted average can be quite different from the near-integer values for individual isotopic masses.

For instance, there are two main isotopes of chlorine: chlorine-35 and chlorine-37. In any given sample of chlorine that has not been subjected to mass separation there will be roughly 75% of chlorine atoms which are chlorine-35 and only 25% of chlorine atoms which are chlorine-37. This gives chlorine a relative atomic mass of 35.5 (actually 35.4527 g/mol).

Moreover, the weighted average mass can be near-integer, but at the same time not corresponding to the mass of any natural isotope. For example, bromine has only two stable isotopes, ^{79}Br and ^{81}Br, naturally present in approximately equal fractions, which leads to the standard atomic mass of bromine close to 80 (79.904 g/mol),[6] even though the isotope ^{80}Br with such mass is unstable.

10.4 References

[1] "How many protons, electrons and neutrons are in an atom of krypton, carbon, oxygen, neon,platnum, gold, etc...?". Thomas Jefferson National Accelerator Facility. Retrieved 2008-08-27.

[2] "Elemental Notation and Isotopes". Science Help Online. Retrieved 2008-08-27.

[3] Suchocki, John. *Conceptual Chemistry*, 2007. Page 119.

[4] Curran, Greg (2004). *Homework Helpers*. Career Press. pp. 78–79. ISBN 1-56414-721-5.

[5] "IUPAC Definition of Relative Atomic Mass". International Union of Pure and Applied Chemistry. Retrieved 2008-08-27.

[6] "Atomic Weights and Isotopic Compositions for All Elements". NIST.

10.5 Further reading

- Bishop, Mark. "The Structure of Matter and Chemical Elements (ch. 3)". *An Introduction to Chemistry*. Chiral Publishing. p. 93. ISBN 978-0-9778105-4-3. Retrieved 2008-07-08.

Chapter 11

Atomic number

See also: List of elements by atomic number
In chemistry and physics, the **atomic number** of a chemical element (also known as its **proton number**) is the number of

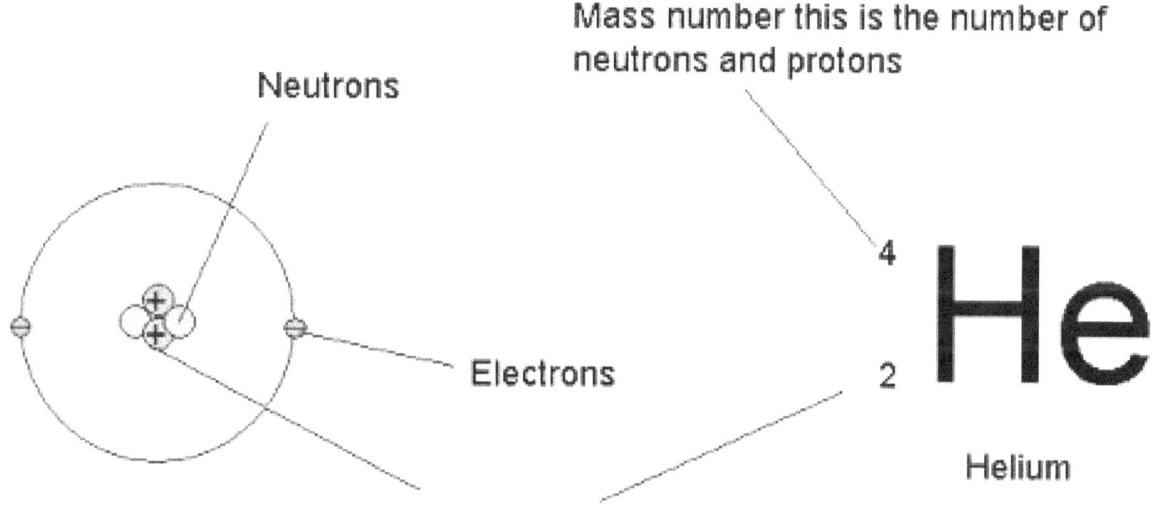

Mass number this is the number of neutrons and protons

Neutrons

Electrons

$^{4}_{2}$He

Helium

Protons:
This number lets us know how many protons there are. In a neutral atom this is also the same as the number of electrons.

An explanation of the superscripts and subscripts seen in atomic number notation. Atomic number is the number of protons, and therefore also the total positive charge, in the atomic nucleus.

protons found in the nucleus of an atom of that element, and therefore identical to the charge number of the nucleus. It is conventionally represented by the symbol Z. The atomic number uniquely identifies a chemical element. In an uncharged atom, the atomic number is also equal to the number of electrons.

The atomic number, Z, should not be confused with the mass number, A, which is the number of nucleons, the total number of protons and neutrons in the nucleus of an atom. The number of neutrons, N, is known as the neutron number of the atom; thus, $A = Z + N$ (these quantities are always whole numbers). Since protons and neutrons have approximately the same mass (and the mass of the electrons is negligible for many purposes) and the mass defect of nucleon binding is always small compared to the nucleon mass, the atomic mass of any atom, when expressed in unified atomic mass units

146

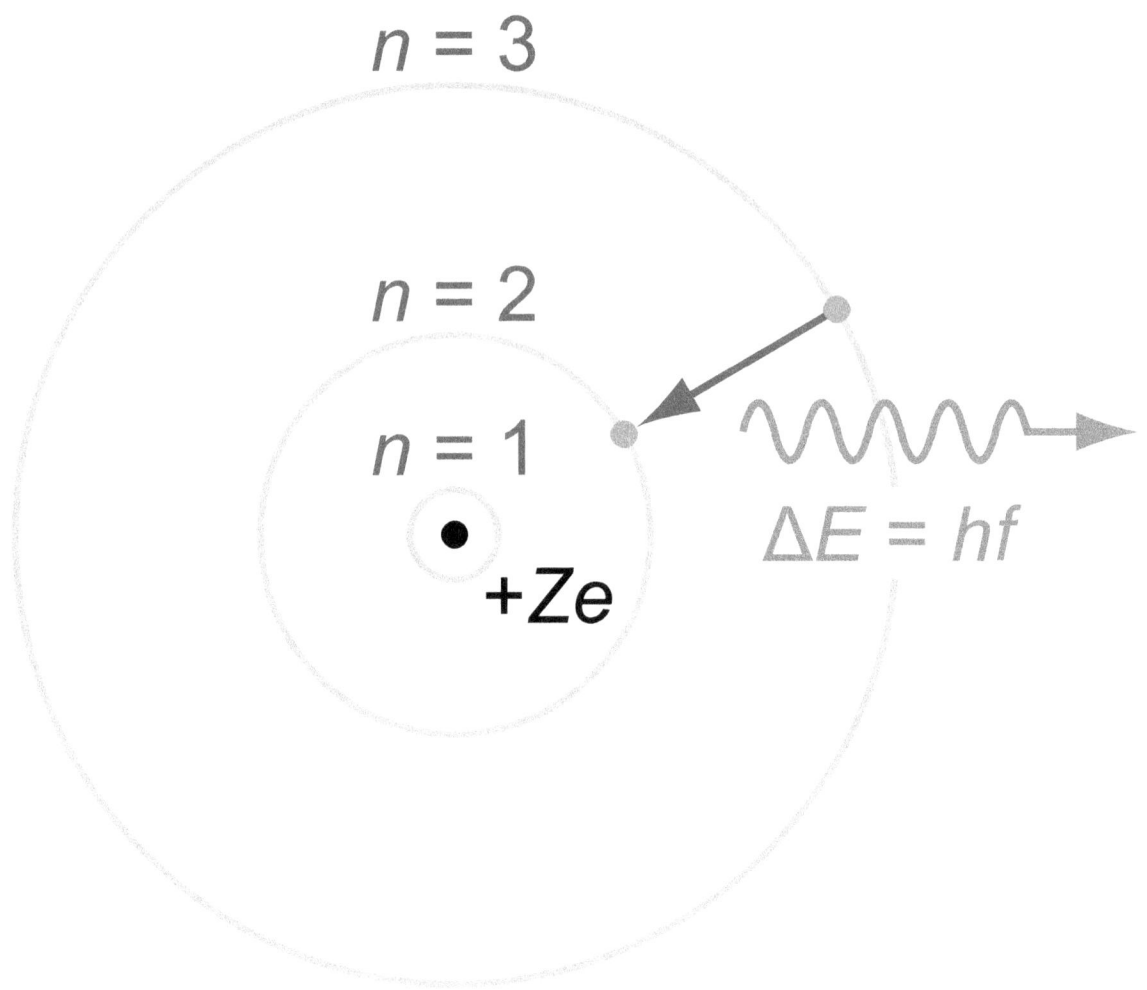

$$n = 3$$

$$n = 2$$

$$n = 1$$

$$+Ze$$

$$\Delta E = hf$$

*The **Rutherford–Bohr model** of the hydrogen atom (Z = 1) or a hydrogen-like ion (Z > 1). In this model it is an essential feature that the photon energy (or frequency) of the electromagnetic radiation emitted (shown) when an electron jumps from one orbital to another, be proportional to the mathematical square of atomic charge (Z^2). Experimental measurement by Henry Moseley of this radiation for many elements (from Z = 13 to 92) showed the results as predicted by Bohr. Both the concept of atomic number and the Bohr model were thereby given scientific credence.*

(making a quantity called the "relative isotopic mass"), is roughly (to within 1%) equal to the whole number A.

Atoms with the same atomic number Z but different neutron numbers N, and hence different atomic masses, are known as isotopes. A little more than three-quarters of naturally occurring elements exist as a mixture of isotopes (see monoisotopic elements), and the average isotopic mass of an isotopic mixture for an element (called the relative atomic mass) in a defined environment on Earth, determines the element's standard atomic weight. Historically, it was these atomic weights of elements (in comparison to hydrogen) that were the quantities measurable by chemists in the 19th century.

The conventional symbol Z comes from the German word **Zahl** meaning number/numeral/figure, which, prior to the modern synthesis of ideas from chemistry and physics, merely denoted an element's numerical place in the periodic table, whose order is approximately, but not completely, consistent with the order of the elements by atomic weights. Only after 1915, with the suggestion and evidence that this Z number was also the nuclear charge and a physical characteristic of atoms, did the word *Atomzahl* (and its English equivalent *atomic number*) come into common use in this context.

11.1   History

11.1.1   The periodic table and a natural number for each element

Loosely speaking, the existence or construction of a periodic table of elements creates an ordering of the elements, and so they can be numbered in order.

Dmitri Mendeleev claimed that he arranged his first periodic tables in order of atomic weight ("Atomgewicht").[1] However, in consideration of the elements' observed chemical properties, he changed the order slightly and placed tellurium (atomic weight 127.6) ahead of iodine (atomic weight 126.9).[1][2] This placement is consistent with the modern practice of ordering the elements by proton number, Z, but that number was not known or suspected at the time.

A simple numbering based on periodic table position was never entirely satisfactory, however. Besides the case of iodine and tellurium, later several other pairs of elements (such as argon and potassium, cobalt and nickel) were known to have nearly identical or reversed atomic weights, thus requiring their placement in the periodic table to be determined by their chemical properties. However the gradual identification of more and more chemically similar lanthanide elements, whose atomic number was not obvious, led to inconsistency and uncertainty in the periodic numbering of elements at least from lutetium (element 71) onwards (hafnium was not known at this time).

11.1.2   The Rutherford-Bohr model and van den Broek

In 1911, Ernest Rutherford gave a model of the atom in which a central core held most of the atom's mass and a positive charge which, in units of the electron's charge, was to be approximately equal to half of the atom's atomic weight, expressed in numbers of hydrogen atoms. This central charge would thus be approximately half the atomic weight (though it was almost 25% different from the atomic number of gold ($Z = 79$, $A = 197$), the single element from which Rutherford made his guess). Nevertheless, in spite of Rutherford's estimation that gold had a central charge of about 100 (but was element $Z = 79$ on the periodic table), a month after Rutherford's paper appeared, Antonius van den Broek first formally suggested that the central charge and number of electrons in an atom was *exactly* equal to its place in the periodic table (also known as element number, atomic number, and symbolized Z). This proved eventually to be the case.

11.1.3   Moseley's 1913 experiment

The experimental position improved dramatically after research by Henry Moseley in 1913.[3] Moseley, after discussions with Bohr who was at the same lab (and who had used Van den Broek's hypothesis in his Bohr model of the atom), decided to test Van den Broek's and Bohr's hypothesis directly, by seeing if spectral lines emitted from excited atoms fitted the Bohr theory's postulation that the frequency of the spectral lines be proportional to the square of Z.

To do this, Moseley measured the wavelengths of the innermost photon transitions (K and L lines) produced by the elements from aluminum ($Z = 13$) to gold ($Z = 79$) used as a series of movable anodic targets inside an x-ray tube.[4] The square root of the frequency of these photons (x-rays) increased from one target to the next in an arithmetic progression. This led to the conclusion (Moseley's law) that the atomic number does closely correspond (with an offset of one unit for K-lines, in Moseley's work) to the calculated electric charge of the nucleus, i.e. the element number Z. Among other things, Moseley demonstrated that the lanthanide series (from lanthanum to lutetium inclusive) must have 15 members— no fewer and no more—which was far from obvious from the chemistry at that time.

11.1.4   The proton and the idea of nuclear electrons

In 1915 the reason for nuclear charge being quantized in units of Z, which were now recognized to be the same as the element number, was not understood. An old idea called Prout's hypothesis had postulated that the elements were all made of residues (or "protyles") of the lightest element hydrogen, which in the Bohr-Rutherford model had a single electron and a nuclear charge of one. However, as early as 1907 Rutherford and Thomas Royds had shown that alpha particles, which had a charge of +2, were the nuclei of helium atoms, which had a mass four times that of hydrogen, not two times.

If Prout's hypothesis were true, something had to be neutralizing some of the charge of the hydrogen nuclei present in the nuclei of heavier atoms.

In 1917 Rutherford succeeded in generating hydrogen nuclei from a nuclear reaction between alpha particles and nitrogen gas, and believed he had proven Prout's law. He called the new heavy nuclear particles protons in 1920 (alternate names being proutons and protyles). It had been immediately apparent from the work of Moseley that the nuclei of heavy atoms have more than twice as much mass as would be expected from their being made of hydrogen nuclei, and thus there was required a hypothesis for the neutralization of the extra protons presumed present in all heavy nuclei. A helium nucleus was presumed to be composed of four protons plus two "nuclear electrons" (electrons bound inside the nucleus) to cancel two of the charges. At the other end of the periodic table, a nucleus of gold with a mass 197 times that of hydrogen, was thought to contain 118 nuclear electrons in the nucleus to give it a residual charge of $+79$, consistent with its atomic number.

11.1.5 The discovery of the neutron makes Z the proton number

All consideration of nuclear electrons ended with James Chadwick's discovery of the neutron in 1932. An atom of gold now was seen as containing 118 neutrons rather than 118 nuclear electrons, and its positive charge now was realized to come entirely from a content of 79 protons. After 1932, therefore, an element's atomic number Z was also realized to be identical to the proton number of its nuclei.

11.2 The symbol of Z

The conventional symbol Z possibly comes from the German word *Atomzahl* (atomic number).[5] However, prior to 1915, the word *Zahl* (simply *number*) was used for an element's assigned number in the periodic table.

11.3 Chemical properties

Each element has a specific set of chemical properties as a consequence of the number of electrons present in the neutral atom, which is Z (the atomic number). The configuration of these electrons follows from the principles of quantum mechanics. The number of electrons in each element's electron shells, particularly the outermost valence shell, is the primary factor in determining its chemical bonding behavior. Hence, it is the atomic number alone that determines the chemical properties of an element; and it is for this reason that an element can be defined as consisting of *any* mixture of atoms with a given atomic number.

11.4 New elements

The quest for new elements is usually described using atomic numbers. As of 2010, elements with atomic numbers 1 to 118 have been observed. Synthesis of new elements is accomplished by bombarding target atoms of heavy elements with ions, such that the sum of the atomic numbers of the target and ion elements equals the atomic number of the element being created. In general, the half-life becomes shorter as atomic number increases, though an "island of stability" may exist for undiscovered isotopes with certain numbers of protons and neutrons.

11.5 See also

- History of the periodic table
- Effective atomic number
- Atomic theory

- Prout's hypothesis

11.6   References

[1] The Periodic Table of Elements, American Institute of Physics

[2] The Development of the Periodic Table, Chemsoc

[3] Ordering the Elements in the Periodic Table, Royal Chemical Society

[4] Moseley's paper with illustrations

[5] Origin of symbol Z

Russian chemist Dmitri Mendeleev created a periodic table of the elements that ordered them numerically by atomic weight, yet occasionally used chemical properties in contradiction to weight.

Niels Bohr's 1913 Bohr model of the atom required van den Broek's atomic number of nuclear charges, and Bohr believed that Moseley's work contributed greatly to the acceptance of the model.

Henry Moseley helped develop the concept of atomic number by showing experimentally (1913) that Van den Broek's 1911 hypothesis combined with the Bohr model nearly correctly predicted atomic X-ray emissions.

Chapter 12

Atomic mass

This article is about the mass of an atomic particle. For the band that became Def Leppard, see Atomic Mass (band). Not to be confused with relative atomic mass.

 The **atomic mass** (m_a) is the mass of an atomic particle, sub-atomic particle, or molecule. It is commonly expressed in unified atomic mass units (u) where by international agreement, 1 unified atomic mass unit is defined as 1/12 of the mass of a single carbon-12 atom (at rest).[1] For atoms, the protons and neutrons of the nucleus account for almost all of the mass, and the atomic mass measured in u has nearly the same value as the mass number.

When divided by unified atomic mass units or daltons to form a pure number ratio, the atomic mass of an atom becomes a dimensionless number called the **relative isotopic mass** (see section below). Thus, the atomic mass of a carbon-12 atom is 12 **u** or 12 daltons (Da), but the relative isotopic mass of a carbon-12 atom is simply 12.

The atomic mass or relative isotopic mass refers to the mass of a single particle, and is fundamentally different from the quantities **elemental atomic weight** (also called "relative *atomic* mass") and standard atomic weight, both of which refer to averages (mathematical means) of naturally-occurring atomic mass values for samples of elements. Most elements have more than one stable nuclide; for those elements, such an average depends on the mix of nuclides present, which may vary to some limited extent depending on the source of the sample, as each nuclide has a different mass. (However, a typical value can be established, which is called the standard atomic weight.) By contrast, atomic mass figures refer to an individual particle species: as atoms of the same species are identical, atomic mass values are expected to have no intrinsic variance at all. Atomic mass figures are thus commonly reported to many more significant figures than atomic weights. Standard atomic weight is related to atomic mass by the abundance ranking of isotopes for each element. It is usually *about* the same value as the atomic mass of the most abundant isotope, other than what looks like (but is not actually) a rounding difference.

The atomic mass of atoms, ions, or atomic nuclei is slightly less than the sum of the masses of their constituent protons, neutrons, and electrons, due to binding energy mass loss (as per $E=mc^2$).[2]

12.1 Relative isotopic mass: the same quantity as atomic mass, but with different units

Relative *isotopic* mass (a property of a single atom) is not to be confused with the averaged quantity "relative *atomic* mass," which is the same as atomic weight (see above), and is an average of values for many atoms in a given sample of a chemical element.

Relative isotopic mass is similar to atomic mass and has exactly the same numerical value as atomic mass, whenever atomic mass is expressed in unified atomic mass units. The only difference in that case, is that relative isotopic mass is a pure number with no units. This loss of units results from the use of a scaling ratio with respect to a carbon-12 standard, and the word "relative" in the term "relative isotopic mass" refers to this scaling *relative* to carbon-12.

The relative isotopic mass, then, is the mass of a given isotope (specifically, any single nuclide), when this value is scaled

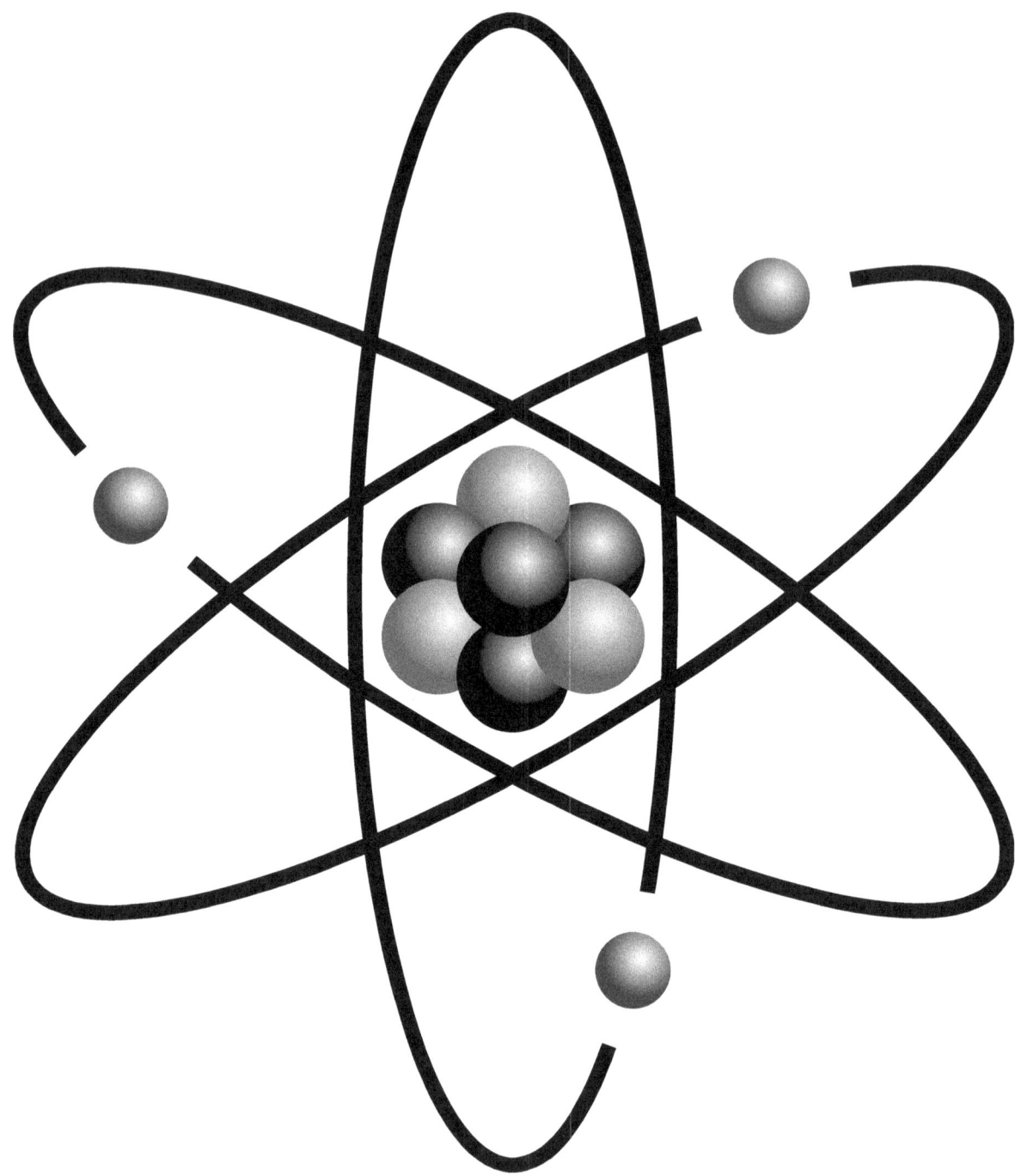

*Stylized lithium−7 atom: 3 protons, 4 neutrons, & 3 electrons (total electrons are ~1/4300th of the mass of the nucleus). It has a mass of 7.016 **u**. Rare lithium-6 (mass of 6.015 **u**) has only 3 neutrons, reducing the atomic weight (average) of lithium to 6.941.*

by the mass of carbon-12, when the latter is set equal to 12. Equivalently, the relative isotopic mass of an isotope or nuclide is the mass of the isotope relative to 1/12 of the mass of a carbon-12 atom.

For example, the **relative isotopic mass** of a carbon-12 atom is exactly 12. For comparison, the **atomic mass** of a carbon-12 atom is exactly 12 daltons or 12 unified atomic mass units. Alternately, the atomic mass of a carbon-12 atom may be expressed in any other mass units: for example, the atomic mass of a carbon-12 atom is about $1.998467052 \times 10^{-26}$ kilogram.

As in the case of atomic mass, no nuclides other than carbon-12 have exactly whole-number values of relative isotopic mass. As is the case for the related *atomic mass* when expressed in unified atomic mass units or daltons, the relative isotopic mass numbers of nuclides other than carbon-12 are not whole numbers, but are always close to whole numbers. This is discussed more fully below.

12.2   Similar terms for different quantities

The atomic mass and relative isotopic mass are sometimes confused, or incorrectly used, as synonyms of relative atomic mass (also known as atomic weight) and the standard atomic weight (a particular variety of atomic weight). However, as noted in the introduction, atomic weight and standard atomic weight represent terms for (abundance-weighted) averages of atomic masses in elemental samples, not for single nuclides. As such, atomic weight and standard atomic weight often differ numerically from relative isotopic mass and atomic mass, and they can also have different units than atomic mass when this quantity is not expressed in unified atomic mass units (see the linked article for atomic weight).

The atomic mass (relative isotopic mass) is defined as the mass of a single atom, which can only be one isotope (nuclide) at a time, and is not an abundance-weighted average, as in the case of relative atomic mass/atomic weight. The atomic mass or relative isotopic mass of each isotope and nuclide of a chemical element is therefore a number that can in principle be measured to a very great precision, since every specimen of such a nuclide is expected to be exactly identical to every other specimen, as all atoms of a given type in the same energy state, and every specimen of a particular nuclide, are expected to be exactly identical in mass to every other specimen of that nuclide. For example, every atom of oxygen-16 is expected to have exactly the same atomic mass (relative isotopic mass) as every other atom of oxygen-16.

In the case of many elements that have one naturally occurring isotope (mononuclidic elements) or one dominant isotope, the actual numerical similarity/difference between the atomic mass of the most common isotope, and the relative atomic mass or (standard) atomic weight can be small or even nil, and does affect most bulk calculations. However, such an error can exist and even be important when considering individual atoms for elements that are not mononuclidic.

For non-mononuclidic elements that have more than one common isotope, the numerical difference in relative atomic mass (atomic weight) from even the most common relative isotopic mass, can be half a mass unit or more (e.g. see the case of chlorine where atomic weight and standard atomic weight are about 35.45). The atomic mass (relative isotopic mass) of an uncommon isotope can differ from the relative atomic mass, atomic weight, or standard atomic weight, by several mass units.

Atomic masses expressed in unified atomic mass units (i.e. relative isotopic masses) are always close to whole-number values, but never (except in the case of carbon-12) exactly a whole number, for two reasons:

- protons and neutrons have different masses, and different nuclides have different ratios of protons and neutrons.

- atomic masses are reduced, to different extents, by their binding energies.

The ratio of atomic mass to mass number varies from about 0.99884 for ^{56}Fe to 1.00782505 for ^{1}H.

Any mass defect due to nuclear binding energy is experimentally a small fraction (less than 1%) of the mass of equal number of free nucleons. When compared to the average mass per nucleon in carbon-12, which is moderately strongly-bound compared with other atoms, the mass defect of binding for most atoms is an even smaller fraction of a dalton (unified atomic mass unit, based on carbon-12). Since free protons and neutrons differ from each other in mass by a small fraction of a dalton (about 0.0014 **u**), rounding the relative isotopic mass, or the atomic mass of any given nuclide given in daltons to the nearest whole number always gives the nucleon count, or mass number. The neutron count (neutron number) may then be derived by subtracting the number of protons (atomic number) from the mass number.

12.3   Mass defects in atomic masses

The amount that the ratio of atomic masses to mass number deviates from 1 is as follows: the deviation starts positive at hydrogen$-$1, then decreases until it reaches a local minimum at helium-4. Isotopes of lithium, beryllium, and boron are

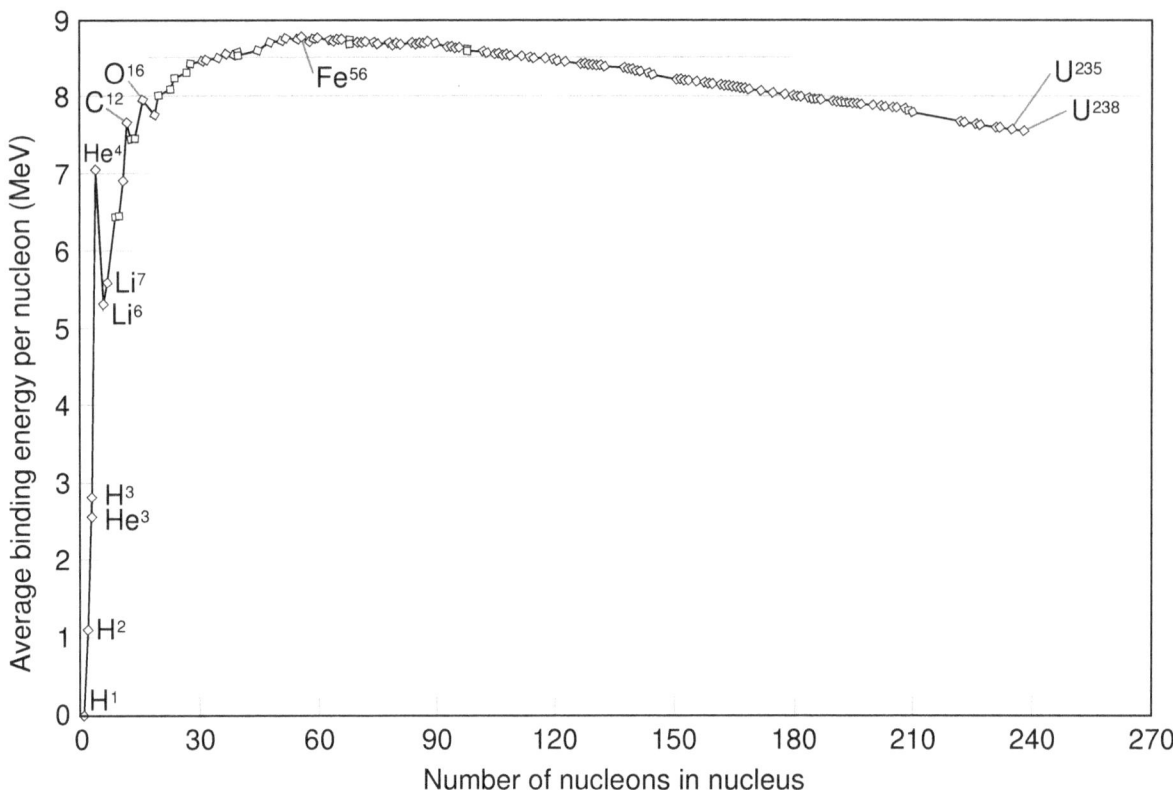

Binding energy per nucleon of common isotopes. A graph of the ratio of mass number to atomic mass would be similar.

less strongly bound than helium, as shown by their increasing mass-to-mass number ratios.

At carbon, the ratio of mass (in daltons) to mass number is defined as 1, and after carbon it becomes less than one until a minimum is reached at iron−56 (with only slightly higher values for iron-58 and nickel−62), then increases to positive values in the heavy isotopes, with increasing atomic number. This corresponds to the fact that nuclear fission in an element heavier than zirconium produces energy, and fission in any element lighter than niobium requires energy. On the other hand, nuclear fusion of two atoms of an element lighter than scandium (except for helium) produces energy, whereas fusion in elements heavier than calcium requires energy. The fusion of two atoms of He-4 to give beryllium−8 would require energy, and the beryllium would quickly fall apart again. He-4 can fuse with tritium (H-3) or with He-3, and these processes occurred during Big Bang nucleosynthesis. The formation of elements with more than seven nucleons requires the fusion of three atoms of He-4 in the so-called triple alpha process, skipping over lithium, beryllium, and boron to produce carbon.

Here are some values of the ratio of atomic mass to mass number:

12.4   Measurement of atomic masses

Direct comparison and measurement of the masses of atoms is achieved with mass spectrometry.

12.5   Conversion factor between atomic mass units and grams

The standard scientific unit used to quantify the amount of a substance in macroscopic quantities is the mole (symbol: mol), which is defined arbitrarily as the amount of a substance which has as many atoms or molecules as there are atoms in 12 grams of the carbon isotope C-12. The number of atoms in a mole is called Avogadro's number, the value of which is approximately 6.022×10^{23}.

One mole of a substance always contains almost exactly the *relative atomic mass* or *molar mass* of that substance; however, this may or may not be true for the *atomic mass*, depending on whether or not the element exists naturally in more than one isotope. For example, the relative atomic mass of iron is 55.847 g/mol, and therefore one mole of iron as commonly found on earth has a mass of 55.847 grams. The *atomic mass* of the ^{56}Fe isotope is 55.935 u and one mole of ^{56}Fe atoms would then in theory have a mass of 55.935 g, but such amounts of pure ^{56}Fe have never been found (or separated out) on Earth. However there are 22 mononuclidic elements of which essentially only a single isotope is found in nature (common examples are fluorine, sodium, aluminum and phosphorus) and for these elements the relative atomic mass and atomic mass are the same. Samples of these elements therefore may serve as reference standards for certain atomic mass values.

The formula for conversion between atomic mass units and SI mass in grams for a single atom is:

$$1 \text{ u} = \frac{M_\text{u}}{N_\text{A}} = \frac{1 \text{ g/mol}}{N_\text{A}}$$

where M_u is the Molar mass constant and N_A is the Avogadro constant.

12.6   Relationship between atomic and molecular masses

Similar definitions apply to molecules. One can compute the molecular mass of a compound by adding the atomic masses of its constituent atoms (nuclides). One can compute the molar mass of a compound by adding the relative atomic masses of the elements given in the chemical formula. In both cases the multiplicity of the atoms (the number of times it occurs) must be taken into account, usually by multiplication of each unique mass by its multiplicity.

12.7   History

Main articles: History of chemistry and Unified atomic mass unit

The first scientists to determine relative atomic masses were John Dalton and Thomas Thomson between 1803 and 1805 and Jöns Jakob Berzelius between 1808 and 1826. Relative atomic mass (*Atomic weight*) was originally defined relative to that of the lightest element, hydrogen, which was taken as 1.00, and in the 1820s Prout's hypothesis stated that atomic masses of all elements would prove to be exact multiples of that of hydrogen. Berzelius, however, soon proved that this was not even approximately true, and for some elements, such as chlorine, relative atomic mass, at about 35.5, falls almost exactly halfway between two integral multiples of that of hydrogen. Still later, this was shown to be largely due to a mix of isotopes, and that the atomic masses of pure isotopes, or nuclides, are multiples of the hydrogen mass, to within about 1%.

In the 1860s Stanislao Cannizzaro refined relative atomic masses by applying Avogadro's law (notably at the Karlsruhe Congress of 1860). He formulated a law to determine relative atomic masses of elements: *the different quantities of the same element contained in different molecules are all whole multiples of the atomic weight* and determined relative atomic masses and molecular masses by comparing the vapor density of a collection of gases with molecules containing one or more of the chemical element in question.[3]

In the 20th century, until the 1960s chemists and physicists used two different atomic-mass scales. The chemists used a "atomic mass unit" (amu) scale such that the natural mixture of oxygen isotopes had an atomic mass 16, while the physicists assigned the same number 16 to only the atomic mass of the most common oxygen isotope (O-16, containing eight protons and eight neutrons). However, because oxygen-17 and oxygen-18 are also present in natural oxygen this led to two different tables of atomic mass. The unified scale based on carbon-12, ^{12}C, met the physicists' need to base the scale on a pure isotope, while being numerically close to the chemists' scale.

The term *atomic weight* is being phased out slowly and being replaced by *relative atomic mass*, in most current usage. This shift in nomenclature reaches back to the 1960s and has been the source of much debate in the scientific community, which was triggered by the adoption of the unified atomic mass unit and the realization that weight was in some ways an inappropriate term. The argument for keeping the term "atomic weight" was primarily that it was a well understood term to those in the field, that the term "atomic mass" was already in use (as it is currently defined) and that the term "relative atomic mass" might be easily confused with *relative isotopic mass* (the mass of a single atom of a given nuclide, expressed dimensionlessly relative to 1/12 of the mass of carbon-12; see section above).

In 1979, as a compromise, the term "relative atomic mass" was introduced as a secondary synonym for atomic weight. Twenty years later the primacy of these synonyms was reversed, and the term "relative atomic mass" is now the preferred term.

However, the term "*standard* atomic weights" (referring to the standardized expectation atomic weights of differing samples) have maintained the same name.[4] In the case of this latter term, simple replacement of the "atomic weight" term with "relative atomic mass" would have resulted in the term "standard relative atomic mass."

12.8   See also

- Atomic number

- Atomic mass unit

- Isotope

- Isotope geochemistry

- Molecular mass

- Jean Stas

12.9   References

[1] IUPAC, *Compendium of Chemical Terminology*, 2nd ed. (the "Gold Book") (1997). Online corrected version: (2006–) "atomic mass".

[2] Atomic mass, Encyclopædia Britannica on-line

[3] Williams, Andrew (2007). "Origin of the Formulas of Dihydrogen and Other Simple Molecules". *J. Chem. Ed.* **84** (11): 1779. Bibcode:2007JChEd..84.1779W. doi:10.1021/ed084p1779.

[4] De Bievre, P.; Peiser, H. S. (1992). "'Atomic weight': The name, its history, definition, and units" (PDF). *Pure&App. Chem.* **64** (10): 1535. doi:10.1351/pac199264101535.

12.10   External links

- NIST relative atomic masses of all isotopes and the standard atomic weights of the elements

- AME2003 Atomic Mass Evaluation from the National Nuclear Data Center

Chapter 13

Atomic radius

The **atomic radius** of a chemical element is a measure of the size of its atoms, usually the mean or typical distance from the center of the nucleus to the boundary of the surrounding cloud of electrons. Since the boundary is not a well-defined physical entity, there are various non-equivalent definitions of atomic radius. Three widely used definitions of atomic radius are Van der Waals radius, ionic radius, and covalent radius.

Depending on the definition, the term may apply only to isolated atoms, or also to atoms in condensed matter, covalently bound in molecules, or in ionized and excited states; and its value may be obtained through experimental measurements, or computed from theoretical models. Under some definitions, the value of the radius may depend on the atom's state and context.[1]

Electrons do not have definite orbits, or sharply defined ranges. Rather, their positions must be described as probability distributions that taper off gradually as one moves away from the nucleus, without a sharp cutoff. Moreover, in condensed matter and molecules, the electron clouds of the atoms usually overlap to some extent, and some of the electrons may roam over a large region encompassing two or more atoms.

Under most definitions the radii of isolated neutral atoms range between 30 and 300 pm (trillionths of a meter), or between 0.3 and 3 angstroms. Therefore, the radius of an atom is more than 10,000 times the radius of its nucleus (1–10 fm),[2] and less than 1/1000 of the wavelength of visible light (400–700 nm).

For many purposes, atoms can be modeled as spheres. This is only a crude approximation, but it can provide quantitative explanations and predictions for many phenomena, such as the density of liquids and solids, the diffusion of fluids through molecular sieves, the arrangement of atoms and ions in crystals, and the size and shape of molecules.

Atomic radii vary in a predictable and explicable manner across the periodic table. For instance, the radii generally decrease along each period (row) of the table, from the alkali metals to the noble gases; and increase down each group (column). The radius increases sharply between the noble gas at the end of each period and the alkali metal at the beginning of the next period. These trends of the atomic radii (and of various other chemical and physical properties of the elements) can be explained by the electron shell theory of the atom; they provided important evidence for the development and confirmation of quantum theory. The atomic radii decrease across the Periodic Table because as the atomic number increases, the number of protons increases across the period, but the extra electrons are only added to the same quantum shell. Therefore, the effective nuclear charge towards the outermost electrons increases, drawing the outermost electrons closer. As a result, the electron cloud contracts and the atomic radius decreases.

13.1 History

In 1920, shortly after it had become possible to determine the sizes of atoms using X-ray crystallography, it was suggested that all atoms of the same element have the same radii.[3] However, in 1923, when more crystal data had become available, it was found that the approximation of an atom as a sphere does not necessarily hold when comparing the same atom in different crystal structures.[4]

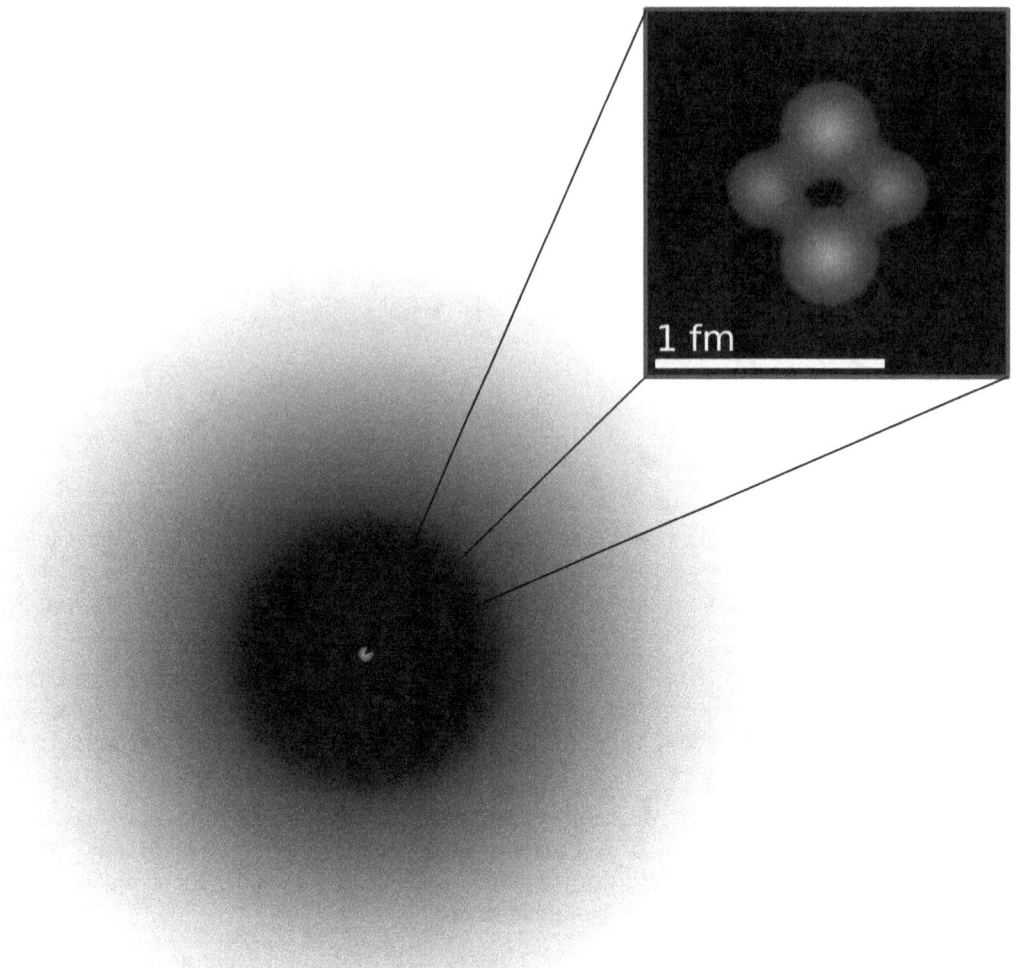

$$1\ \text{Å} = 100{,}000\ \text{fm}$$

Diagram of a helium atom, showing the electron probability density as shades of gray.

13.2 Definitions

Widely used definitions of atomic radius include:

- Van der Waals radius: in principle, half the minimum distance between the nuclei of two atoms of the element that are not bound to the same molecule.[5]

- Ionic radius: the nominal radius of the ions of an element in a specific ionization state, deduced from the spacing of atomic nuclei in crystalline salts that include that ion. In principle, the spacing between two adjacent oppositely charged ions (the length of the ionic bond between them) should equal the sum of their ionic radii.[5]

- Covalent radius: the nominal radius of the atoms of an element when covalently bound to other atoms, as deduced from the separation between the atomic nuclei in molecules. In principle, the distance between two atoms that are bound to each other in a molecule (the length of that covalent bond) should equal the sum of their covalent radii.[5]

The approximate shape of a molecule of ethanol, CH₃CH₂OH. Each atom is modeled by a sphere with the element's Van der Waals radius.

- Metallic radius: the nominal radius of atoms of an element when joined to other atoms by metallic bonds.

- Bohr radius: the radius of the lowest-energy electron orbit predicted by Bohr model of the atom (1913).[6][7] It is only applicable to atoms and ions with a single electron, such as hydrogen, singly ionized helium, and positronium. Although the model itself is now obsolete, the Bohr radius for the hydrogen atom is still regarded as an important physical constant.

13.3   Empirically measured atomic radii

The following table shows empirically measured **covalent** radii for the elements, as published by J. C. Slater in 1964.[8] The values are in picometers (pm or 1×10^{-12} m,), with an accuracy of about 5 pm. The shade of the box ranges from red to yellow as the radius increases; gray indicates lack of data.

13.4   Explanation of the general trends

The way the atomic radius varies with increasing atomic number can be explained by the arrangement of electrons in shells of fixed capacity. The shells are generally filled in order of increasing radius, since the negatively charged electrons are

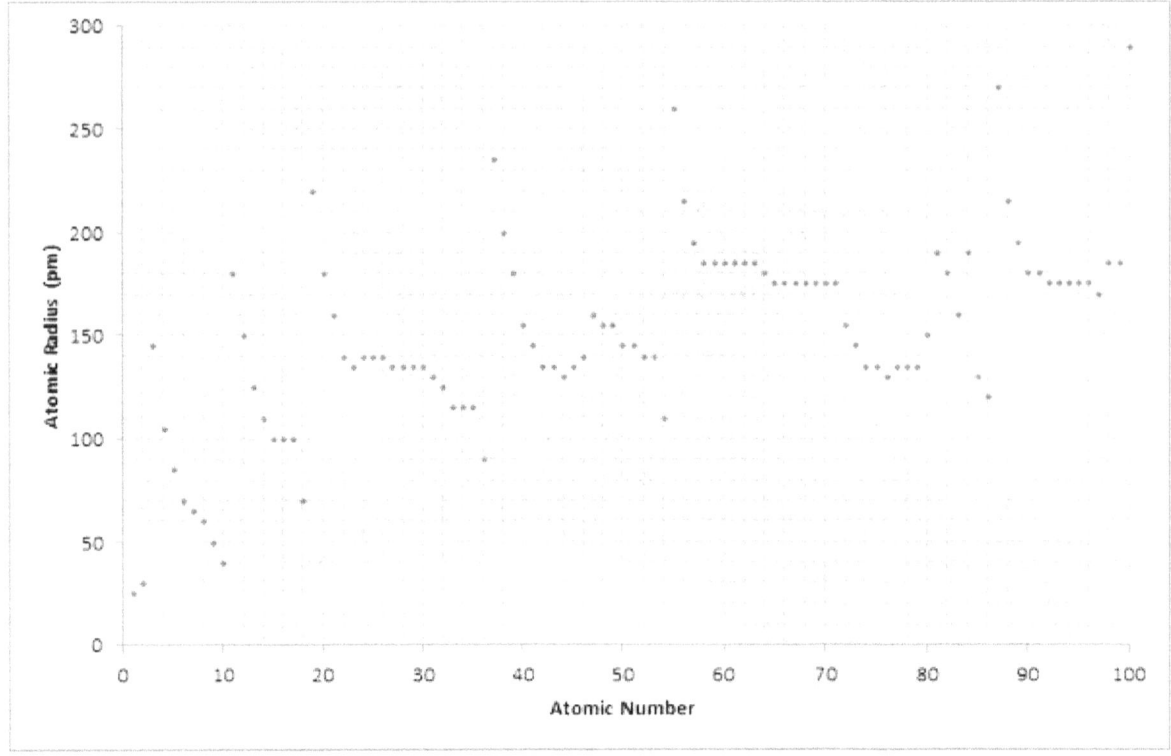

A graph comparing the calculated atomic radius of elements with atomic numbers 1–100. Accuracy of ±5 pm.

attracted by the positively charged protons in the nucleus. As the atomic number increases along each row of the periodic table, the additional electrons go into the same outermost shell; whose radius gradually contracts, due to the increasing nuclear charge. In a noble gas, the outermost shell is completely filled; therefore, the additional electron of next alkali metal will go into the next outer shell, accounting for the sudden increase in the atomic radius.

The increasing nuclear charge is partly counterbalanced by the increasing number of electrons, a phenomenon that is known as shielding; which explains why the size of atoms usually increases down each column. However, there is one notable exception, known as the lanthanide contraction: the 5d block of elements are much smaller than one would expect, due to the shielding caused by the 4f electrons.

The following table summarizes the main phenomena that influence the atomic radius of an element:

13.4.1 Lanthanide contraction

Main article: Lanthanide contraction

The electrons in the 4f-subshell, which is progressively filled from cerium ($Z = 58$) to lutetium ($Z = 71$), are not particularly effective at shielding the increasing nuclear charge from the sub-shells further out. The elements immediately following the lanthanides have atomic radii which are smaller than would be expected and which are almost identical to the atomic radii of the elements immediately above them.[9] Hence hafnium has virtually the same atomic radius (and chemistry) as zirconium, and tantalum has an atomic radius similar to niobium, and so forth. The effect of the lanthanide contraction is noticeable up to platinum ($Z = 78$), after which it is masked by a relativistic effect known as the inert pair effect.

Due to lanthanide contraction, the 5 following observations can be drawn:

1. The size of Ln^{3+} ions regularly decreases with atomic number. According to Fajans' rules, decrease in size of Ln^{3+} ions increases the covalent character and decreases the basic character between Ln^{3+} and OH^- ions in $Ln(OH)_3$.

Hence the order of size of Ln^{3+} is given:
$La^{3+} > Ce^{3+} > ..., ... > Lu^{3+}$.

2. There is a regular decrease in their ionic radii.

3. There is a regular decrease in their tendency to act as a reducing agent, with increase in atomic number.

4. The second and third rows of d-block transition elements are quite close in properties.

5. Consequently, these elements occur together in natural minerals and are difficult to separate.

13.4.2　d-Block contraction

Main article: d-block contraction

The d-block contraction is less pronounced than the lanthanide contraction but arises from a similar cause. In this case, it is the poor shielding capacity of the 3d-electrons which affects the atomic radii and chemistries of the elements immediately following the first row of the transition metals, from gallium ($Z = 31$) to bromine ($Z = 35$).[9]

13.5　Calculated atomic radii

The following table shows atomic radii computed from theoretical models, as published by Enrico Clementi and others in 1967.[10] The values are in picometres (pm).

13.6　See also

- Atomic radii of the elements (data page)

- Chemical bond

- Covalent radius

- Bond length

- Steric hindrance

13.7　References

[1] Cotton, F. A.; Wilkinson, G. (1988). *Advanced Inorganic Chemistry* (5th ed.). Wiley. p. 1385. ISBN 978-0-471-84997-1.

[2] Basdevant, J.-L.; Rich, J.; Spiro, M. (2005). *Fundamentals in Nuclear Physics.* Springer. p. 13, fig 1.1. ISBN 978-0-387-01672-6.

[3] Bragg, W. L. (1920). "The arrangement of atoms in crystals".*Philosophical Magazine.* **640**(236): 169–189.doi:10.1080/146111.

[4] Wyckoff, R. W. G. (1923). "On the Hypothesis of Constant Atomic Radii". *Proceedings of the National Academy of Sciences of the United States of America* **9** (2): 33–38. Bibcode:1923PNAS....9...33W. doi:10.1073/pnas.9.2.33. PMC 1085234. PMID 16576657.

[5] Pauling, L. (1945). *The Nature of the Chemical Bond* (2nd ed.). Cornell University Press. LCCN 42034474.

[6] Bohr, N. (1913). "On the Constitution of Atoms and Molecules, Part I. – Binding of Electrons by Positive Nuclei" (PDF). *Philosophical Magazine.* **6 26** (151): 1–24. doi:10.1080/14786441308634955. Retrieved 8 June 2011.

[7] Bohr, N. (1913). "On the Constitution of Atoms and Molecules, Part II. – Systems containing only a Single Nucleus" (PDF). *Philosophical Magazine.* 6 **26** (153): 476–502. doi:10.1080/14786441308634993. Retrieved 8 June 2011.

[8] Slater, J. C. (1964). "Atomic Radii in Crystals". *Journal of Chemical Physics* **41**(10): 3199–3205. Bibcode:1964JChPh..41.319S. doi:10.1063/1.1725697.

[9] Jolly, W. L. (1991). *Modern Inorganic Chemistry* (2nd ed.). McGraw-Hill. p. 22. ISBN 978-0-07-112651-9.

[10] Clementi, E.; Raimond, D. L.; Reinhardt, W. P. (1967). "Atomic Screening Constants from SCF Functions. II. Atoms with 37 to 86 Electrons". *Journal of Chemical Physics* **47** (4): 1300–1307. Bibcode:1967JChPh..47.1300C. doi:10.1063/1.1712084.

Chapter 14

Atomic orbital

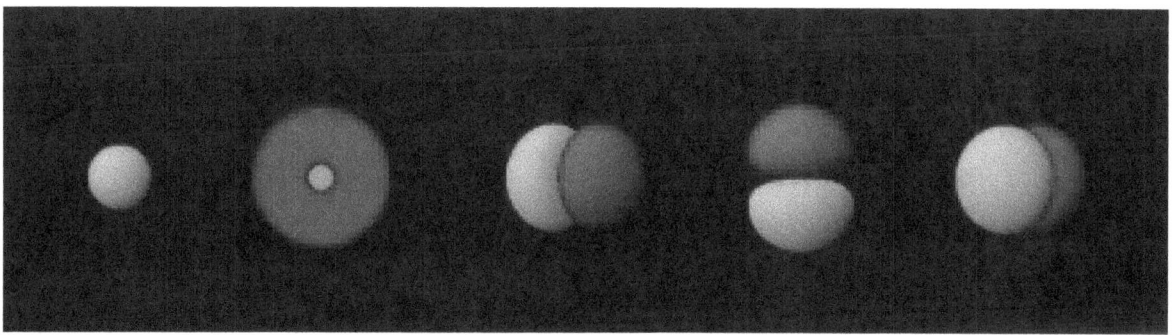

The shapes of the first five atomic orbitals: 1s, 2s, 2p$_x$, 2p$_y$, and 2p$_z$. The colors show the wave function phase. These are graphs of ψ(x, y, z) functions which depend on the coordinates of one electron. To see the elongated shape of ψ(x, y, z)^2 functions that show probability density more directly, see the graphs of d-orbitals below.

An **atomic orbital** is a mathematical function that describes the wave-like behavior of either one electron or a pair of electrons in an atom.[1] This function can be used to calculate the probability of finding any electron of an atom in any specific region around the atom's nucleus. The term may also refer to the physical region or space where the electron can be calculated to be present, as defined by the particular mathematical form of the orbital.[2]

Each orbital in an atom is characterized by a unique set of values of the three quantum numbers n, ℓ, and m, which respectively correspond to the electron's energy, angular momentum, and an angular momentum vector component (the magnetic quantum number). Any orbital can be occupied by a maximum of two electrons, each with its own spin quantum number. The simple names **s orbital**, **p orbital**, **d orbital** and **f orbital** refer to orbitals with angular momentum quantum number ℓ = 0, 1, 2 and 3 respectively. These names, together with the value of n, are used to describe the electron configurations of atoms. They are derived from the description by early spectroscopists of certain series of alkali metal spectroscopic lines as **s**harp, **p**rincipal, **d**iffuse, and **f**undamental. Orbitals for ℓ > 3 continue alphabetically, omitting j (g, h, i, k, …).[3][4][5]

Atomic orbitals are the basic building blocks of the **atomic orbital model** (alternatively known as the electron cloud or wave mechanics model), a modern framework for visualizing the submicroscopic behavior of electrons in matter. In this model the electron cloud of a multi-electron atom may be seen as being built up (in approximation) in an electron configuration that is a product of simpler hydrogen-like atomic orbitals. The repeating *periodicity* of the blocks of 2, 6, 10, and 14 elements within sections of the periodic table arises naturally from the total number of electrons that occupy a complete set of **s**, **p**, **d** and **f** atomic orbitals, respectively.

14.1 Electron properties

With the development of quantum mechanics and experimental findings (such as the two slits diffraction of electrons), it was found that the orbiting electrons around a nucleus could not be fully described as particles, but needed to be explained by the wave-particle duality. In this sense, the electrons have the following properties:

Wave-like properties:

1. The electrons do not orbit the nucleus in the sense of a planet orbiting the sun, but instead exist as standing waves. The lowest possible energy an electron can take is therefore analogous to the fundamental frequency of a wave on a string. Higher energy states are then similar to harmonics of the fundamental frequency.

2. The electrons are never in a single point location, although the probability of interacting with the electron at a single point can be found from the wave function of the electron.

Particle-like properties:

1. There is always an integer number of electrons orbiting the nucleus.

2. Electrons jump between orbitals in a particle-like fashion. For example, if a single photon strikes the electrons, only a single electron changes states in response to the photon.

3. The electrons retain particle like-properties such as: each wave state has the same electrical charge as the electron particle. Each wave state has a single discrete spin (spin up or spin down).

Thus, despite the obvious analogy to planets revolving around the Sun, electrons cannot be described simply as solid particles. In addition, atomic orbitals do not closely resemble a planet's elliptical path in ordinary atoms. A more accurate analogy might be that of a large and often oddly shaped "atmosphere" (the electron), distributed around a relatively tiny planet (the atomic nucleus). Atomic orbitals exactly describe the shape of this "atmosphere" only when a single electron is present in an atom. When more electrons are added to a single atom, the additional electrons tend to more evenly fill in a volume of space around the nucleus so that the resulting collection (sometimes termed the atom's "electron cloud"[6]) tends toward a generally spherical zone of probability describing where the atom's electrons will be found.

14.1.1 Formal quantum mechanical definition

Atomic orbitals may be defined more precisely in formal quantum mechanical language. Specifically, in quantum mechanics, the state of an atom, i.e., an eigenstate of the atomic Hamiltonian, is approximated by an expansion (see configuration interaction expansion and basis set) into linear combinations of anti-symmetrized products (Slater determinants) of one-electron functions. The spatial components of these one-electron functions are called atomic orbitals. (When one considers also their spin component, one speaks of **atomic spin orbitals**.) A state is actually a function of the coordinates of all the electrons, so that their motion is correlated, but this is often approximated by this independent-particle model of products of single electron wave functions.[7] (The London dispersion force, for example, depends on the correlations of the motion of the electrons.)

In atomic physics, the atomic spectral lines correspond to transitions (quantum leaps) between quantum states of an atom. These states are labeled by a set of quantum numbers summarized in the term symbol and usually associated with particular electron configurations, i.e., by occupation schemes of atomic orbitals (for example, $1s^2\ 2s^2\ 2p^6$ for the ground state of neon—term symbol: $^1S_0$).

This notation means that the corresponding Slater determinants have a clear higher weight in the configuration interaction expansion. The atomic orbital concept is therefore a key concept for visualizing the excitation process associated with a given transition. For example, one can say for a given transition that it corresponds to the excitation of an electron from an occupied orbital to a given unoccupied orbital. Nevertheless, one has to keep in mind that electrons are fermions ruled by the Pauli exclusion principle and cannot be distinguished from the other electrons in the atom. Moreover, it sometimes happens that the configuration interaction expansion converges very slowly and that one cannot speak about simple one-determinant wave function at all. This is the case when electron correlation is large.

Fundamentally, an atomic orbital is a one-electron wave function, even though most electrons do not exist in one-electron atoms, and so the one-electron view is an approximation. When thinking about orbitals, we are often given an orbital vision which (even if it is not spelled out) is heavily influenced by this Hartree–Fock approximation, which is one way to reduce the complexities of molecular orbital theory.

14.1.2   Types of orbitals

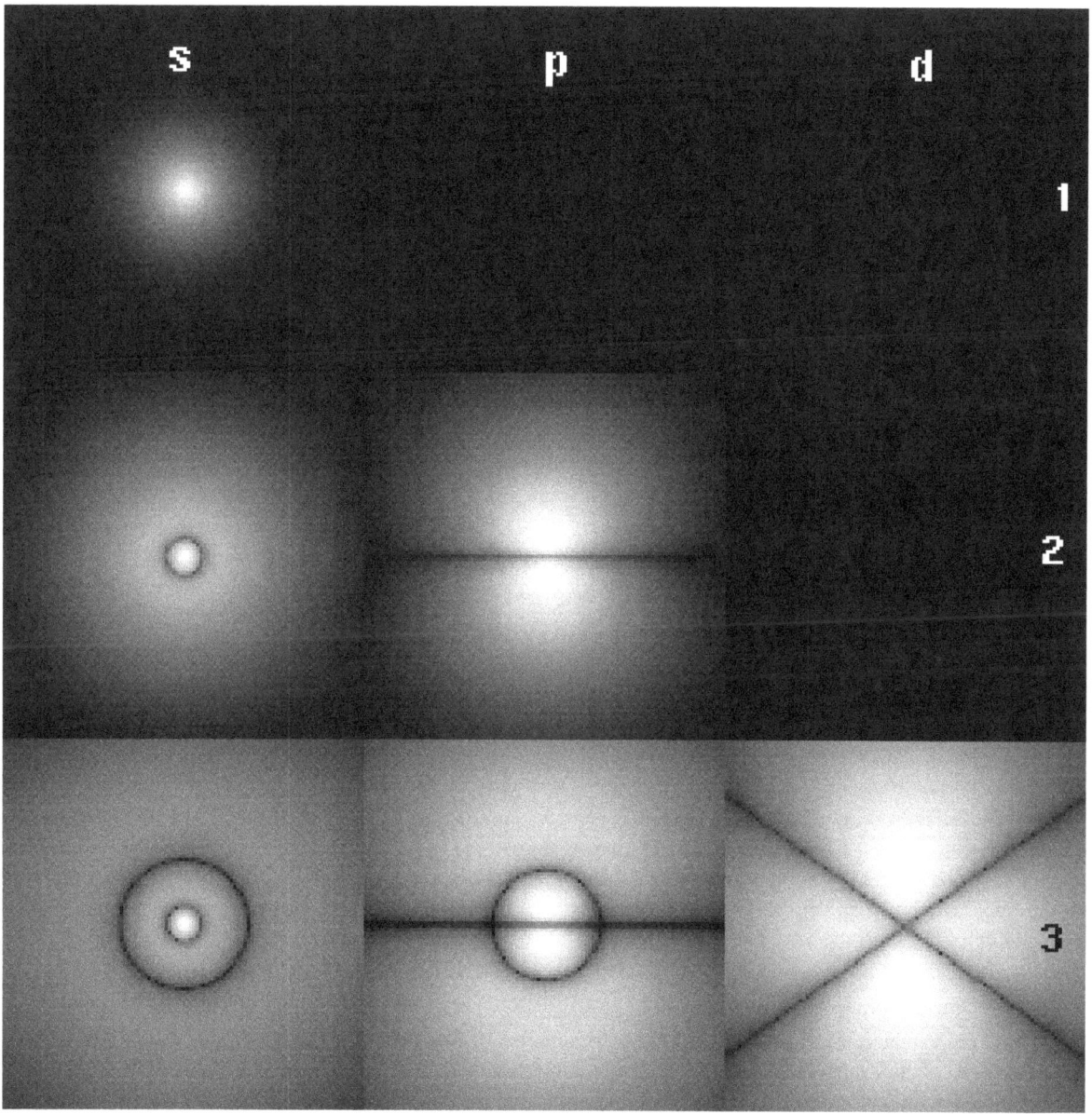

False-color density images of some hydrogen-like atomic orbitals (f orbitals and higher are not shown)

Atomic orbitals can be the hydrogen-like "orbitals" which are exact solutions to the Schrödinger equation for a hydrogen-like "atom" (i.e., an atom with one electron). Alternatively, atomic orbitals refer to functions that depend on the coordinates of one electron (i.e., orbitals) but are used as starting points for approximating wave functions that depend on the simultaneous coordinates of all the electrons in an atom or molecule. The coordinate systems chosen for atomic orbitals are usually spherical coordinates (r, θ, φ) in atoms and cartesians (x, y, z) in polyatomic molecules. The advantage of spherical coordinates (for atoms) is that an orbital wave function is a product of three factors each dependent on a single coordinate: $\psi(r, \theta, \varphi) = R(r)\,\Theta(\theta)\,\Phi(\varphi)$.

The angular factors of atomic orbitals $\Theta(\theta)\,\Phi(\varphi)$ generate s, p, d, etc. functions as real combinations of spherical harmonics $Y\ell m(\theta, \varphi)$ (where ℓ and m are quantum numbers). There are typically three mathematical forms for the radial functions $R(r)$ which can be chosen as a starting point for the calculation of the properties of atoms and molecules with many electrons.

1. the *hydrogen-like atomic orbitals* are derived from the exact solution of the Schrödinger Equation for one electron and a nucleus, for a hydrogen-like atom. The part of the function that depends on the distance from the nucleus has nodes (radial nodes) and decays as $e^{-(\text{constant} \times \text{distance})}$.

2. The Slater-type orbital (STO) is a form without radial nodes but decays from the nucleus as does the hydrogen-like orbital.

3. The form of the Gaussian type orbital (Gaussians) has no radial nodes and decays as $e^{(-\text{distance squared})}$.

Although hydrogen-like orbitals are still used as pedagogical tools, the advent of computers has made STOs preferable for atoms and diatomic molecules since combinations of STOs can replace the nodes in hydrogen-like atomic orbital. Gaussians are typically used in molecules with three or more atoms. Although not as accurate by themselves as STOs, combinations of many Gaussians can attain the accuracy of hydrogen-like orbitals.

14.2 History

Main article: Atomic theory

The term "orbital" was coined by Robert Mulliken in 1932 as an abbreviation for *one-electron orbital wave function*.[8] However, the idea that electrons might revolve around a compact nucleus with definite angular momentum was convincingly argued at least 19 years earlier by Niels Bohr,[9] and the Japanese physicist Hantaro Nagaoka published an orbit-based hypothesis for electronic behavior as early as 1904.[10] Explaining the behavior of these electron "orbits" was one of the driving forces behind the development of quantum mechanics.[11]

14.2.1 Early models

With J.J. Thomson's discovery of the electron in 1897,[12] it became clear that atoms were not the smallest building blocks of nature, but were rather composite particles. The newly discovered structure within atoms tempted many to imagine how the atom's constituent parts might interact with each other. Thomson theorized that multiple electrons revolved in orbit-like rings within a positively charged jelly-like substance,[13] and between the electron's discovery and 1909, this "plum pudding model" was the most widely accepted explanation of atomic structure.

Shortly after Thomson's discovery, Hantaro Nagaoka, a Japanese physicist, predicted a different model for electronic structure.[10] Unlike the plum pudding model, the positive charge in Nagaoka's "Saturnian Model" was concentrated into a central core, pulling the electrons into circular orbits reminiscent of Saturn's rings. Few people took notice of Nagaoka's work at the time,[14] and Nagaoka himself recognized a fundamental defect in the theory even at its conception, namely that a classical charged object cannot sustain orbital motion because it is accelerating and therefore loses energy due to electromagnetic radiation.[15] Nevertheless, the Saturnian model turned out to have more in common with modern theory than any of its contemporaries.

14.2.2 Bohr atom

In 1909, Ernest Rutherford discovered that bulk of the atomic mass was tightly condensed into a nucleus, which was also found to be positively charged. It became clear from his analysis in 1911 that the plum pudding model could not explain atomic structure. Shortly after, in 1913, Rutherford's post-doctoral student Niels Bohr proposed a new model of the atom, wherein electrons orbited the nucleus with classical periods, but were only permitted to have discrete values of angular momentum, quantized in units $h/2\pi$.[9] This constraint automatically permitted only certain values of electron energies. The Bohr model of the atom fixed the problem of energy loss from radiation from a ground state (by declaring that there was no state below this), and more importantly explained the origin of spectral lines.

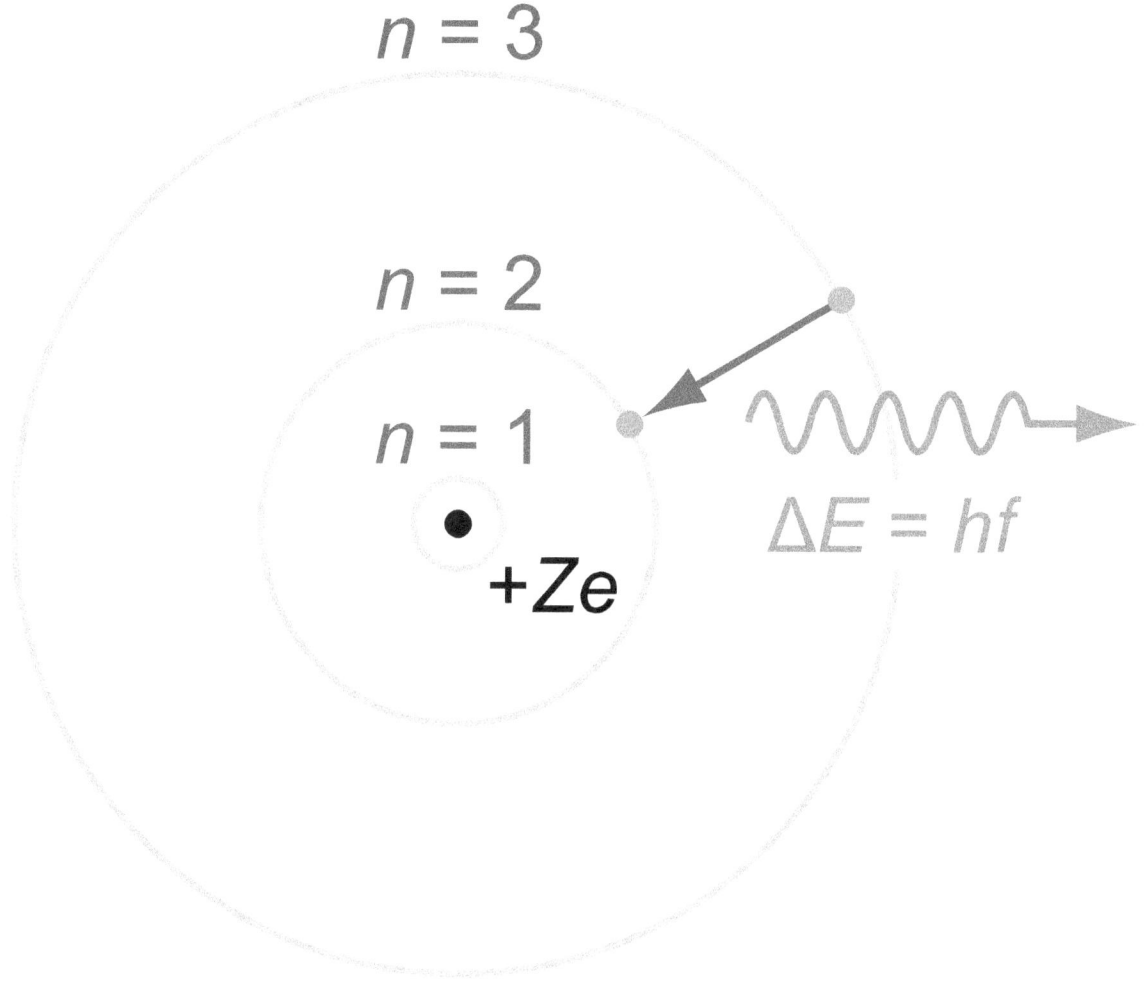

The Rutherford–Bohr model of the hydrogen atom.

After Bohr's use of Einstein's explanation of the photoelectric effect to relate energy levels in atoms with the wavelength of emitted light, the connection between the structure of electrons in atoms and the emission and absorption spectra of atoms became an increasingly useful tool in the understanding of electrons in atoms. The most prominent feature of emission and absorption spectra (known experimentally since the middle of the 19th century), was that these atomic spectra contained discrete lines. The significance of the Bohr model was that it related the lines in emission and absorption spectra to the energy differences between the orbits that electrons could take around an atom. This was, however, *not* achieved by Bohr through giving the electrons some kind of wave-like properties, since the idea that electrons could behave as matter waves was not suggested until eleven years later. Still, the Bohr model's use of quantized angular momenta and therefore quantized energy levels was a significant step towards the understanding of electrons in atoms, and also a significant step towards the development of quantum mechanics in suggesting that quantized restraints must account for all discontinuous energy levels and spectra in atoms.

With de Broglie's suggestion of the existence of electron matter waves in 1924, and for a short time before the full 1926 Schrödinger equation treatment of hydrogen-like atom, a Bohr electron "wavelength" could be seen to be a function of its momentum, and thus a Bohr orbiting electron was seen to orbit in a circle at a multiple of its half-wavelength (this physically incorrect Bohr model is still often taught to beginning students). The Bohr model for a short time could be seen as a classical model with an additional constraint provided by the 'wavelength' argument. However, this period was immediately superseded by the full three-dimensional wave mechanics of 1926. In our current understanding of physics, the Bohr model is called a semi-classical model because of its quantization of angular momentum, not primarily because of its relationship with electron wavelength, which appeared in hindsight a dozen years after the Bohr model was proposed.

The Bohr model was able to explain the emission and absorption spectra of hydrogen. The energies of electrons in the n = 1, 2, 3, etc. states in the Bohr model match those of current physics. However, this did not explain similarities between different atoms, as expressed by the periodic table, such as the fact that helium (two electrons), neon (10 electrons), and argon (18 electrons) exhibit similar chemical inertness. Modern quantum mechanics explains this in terms of electron shells and subshells which can each hold a number of electrons determined by the Pauli exclusion principle. Thus the n = 1 state can hold one or two electrons, while the n = 2 state can hold up to eight electrons in 2s and 2p subshells. In helium, all n = 1 states are fully occupied; the same for n = 1 and n = 2 in neon. In argon the 3s and 3p subshells are similarly fully occupied by eight electrons; quantum mechanics also allows a 3d subshell but this is at higher energy than the 3s and 3p in argon (contrary to the situation in the hydrogen atom) and remains empty.

14.2.3 Modern conceptions and connections to the Heisenberg Uncertainty Principle

Immediately after Heisenberg discovered his uncertainty relation,[16] it was noted by Bohr that the existence of any sort of wave packet implies uncertainty in the wave frequency and wavelength, since a spread of frequencies is needed to create the packet itself.[17] In quantum mechanics, where all particle momenta are associated with waves, it is the formation of such a wave packet which localizes the wave, and thus the particle, in space. In states where a quantum mechanical particle is bound, it must be localized as a wave packet, and the existence of the packet and its minimum size implies a spread and minimal value in particle wavelength, and thus also momentum and energy. In quantum mechanics, as a particle is localized to a smaller region in space, the associated compressed wave packet requires a larger and larger range of momenta, and thus larger kinetic energy. Thus, the binding energy to contain or trap a particle in a smaller region of space, increases without bound, as the region of space grows smaller. Particles cannot be restricted to a geometric point in space, since this would require an infinite particle momentum.

In chemistry, Schrödinger, Pauling, Mulliken and others noted that the consequence of Heisenberg's relation was that the electron, as a wave packet, could not be considered to have an exact location in its orbital. Max Born suggested that the electron's position needed to be described by a probability distribution which was connected with finding the electron at some point in the wave-function which described its associated wave packet. The new quantum mechanics did not give exact results, but only the probabilities for the occurrence of a variety of possible such results. Heisenberg held that the path of a moving particle has no meaning if we cannot observe it, as we cannot with electrons in an atom.

In the quantum picture of Heisenberg, Schrödinger and others, the Bohr atom number n for each orbital became known as an *n-sphere* in a three dimensional atom and was pictured as the mean energy of the probability cloud of the electron's wave packet which surrounded the atom.

14.3 Orbital names

Orbitals are given names in the form:

X type^y

where X is the energy level corresponding to the principal quantum number n, **type** is a lower-case letter denoting the shape or subshell of the orbital and it corresponds to the angular quantum number ℓ, and y is the number of electrons in that orbital.

For example, the orbital $1s^2$ (pronounced "one ess two") has two electrons and is the lowest energy level (n = 1) and has an angular quantum number of ℓ = 0. In X-ray notation, the principal quantum number is given a letter associated with it. For n = 1, 2, 3, 4, 5, ..., the letters associated with those numbers are K, L, M, N, O, ... respectively.

14.4   Hydrogen-like orbitals

Main article: Hydrogen-like atom

The simplest atomic orbitals are those that are calculated for systems with a single electron, such as the hydrogen atom. An atom of any other element ionized down to a single electron is very similar to hydrogen, and the orbitals take the same form. In the Schrödinger equation for this system of one negative and one positive particle, the atomic orbitals are the eigenstates of the Hamiltonian operator for the energy. They can be obtained analytically, meaning that the resulting orbitals are products of a polynomial series, and exponential and trigonometric functions. (see hydrogen atom).

For atoms with two or more electrons, the governing equations can only be solved with the use of methods of iterative approximation. Orbitals of multi-electron atoms are *qualitatively* similar to those of hydrogen, and in the simplest models, they are taken to have the same form. For more rigorous and precise analysis, the numerical approximations must be used.

A given (hydrogen-like) atomic orbital is identified by unique values of three quantum numbers: n, ℓ, and mℓ. The rules restricting the values of the quantum numbers, and their energies (see below), explain the electron configuration of the atoms and the periodic table.

The stationary states (quantum states) of the hydrogen-like atoms are its atomic orbitals. However, in general, an electron's behavior is not fully described by a single orbital. Electron states are best represented by time-depending "mixtures" (linear combinations) of multiple orbitals. See Linear combination of atomic orbitals molecular orbital method.

The quantum number n first appeared in the Bohr model where it determines the radius of each circular electron orbit. In modern quantum mechanics however, n determines the mean distance of the electron from the nucleus; all electrons with the same value of *n* lie at the same average distance. For this reason, orbitals with the same value of *n* are said to comprise a "shell". Orbitals with the same value of *n* and also the same value of ℓ are even more closely related, and are said to comprise a "subshell".

14.5   Quantum numbers

Main article: Quantum number

Because of the quantum mechanical nature of the electrons around a nucleus, atomic orbitals can be uniquely defined by a set of integers known as quantum numbers. These quantum numbers only occur in certain combinations of values, and their physical interpretation changes depending on whether real or complex versions of the atomic orbitals are employed.

14.5.1   Complex orbitals

In physics, the most common orbital descriptions are based on the solutions to the hydrogen atom, where orbitals are given by the product between a radial function and a pure spherical harmonic. The quantum numbers, together with the rules governing their possible values, are as follows:

The principal quantum number n describes the energy of the electron and is always a positive integer. In fact, it can be any positive integer, but for reasons discussed below, large numbers are seldom encountered. Each atom has, in general, many orbitals associated with each value of *n*; these orbitals together are sometimes called *electron shells*.

The azimuthal quantum number ℓ describes the orbital angular momentum of each electron and is a non-negative integer. Within a shell where n is some integer n_0, ℓ ranges across all (integer) values satisfying the relation $0 \leq \ell \leq n_0 - 1$. For instance, the $n = 1$ shell has only orbitals with $\ell = 0$, and the $n = 2$ shell has only orbitals with $\ell = 0$, and $\ell = 1$. The set of orbitals associated with a particular value of ℓ are sometimes collectively called a *subshell*.

The magnetic quantum number, m_ℓ, describes the magnetic moment of an electron in an arbitrary direction, and is also always an integer. Within a subshell where ℓ is some integer ℓ_0, m_ℓ ranges thus: $-\ell_0 \leq m_\ell \leq \ell_0$.

The above results may be summarized in the following table. Each cell represents a subshell, and lists the values of m_ℓ

available in that subshell. Empty cells represent subshells that do not exist.

Subshells are usually identified by their n - and ℓ -values. n is represented by its numerical value, but ℓ is represented by a letter as follows: 0 is represented by 's', 1 by 'p', 2 by 'd', 3 by 'f', and 4 by 'g'. For instance, one may speak of the subshell with $n = 2$ and $\ell = 0$ as a '2s subshell'.

Each electron also has a spin quantum number, **s**, which describes the spin of each electron (spin up or spin down). The number **s** can be $+\frac{1}{2}$ or $-\frac{1}{2}$.

The Pauli exclusion principle states that no two electrons can occupy the same quantum state: every electron in an atom must have a unique combination of quantum numbers.

The above conventions imply a preferred axis (for example, the z direction in Cartesian coordinates), and they also imply a preferred direction along this preferred axis. Otherwise there would be no sense in distinguishing $m = +1$ from $m = -1$. As such, the model is most useful when applied to physical systems that share these symmetries. The Stern–Gerlach experiment — where an atom is exposed to a magnetic field — provides one such example.[18]

14.5.2   Real orbitals

An atom that is embedded in a crystalline solid feels multiple preferred axes, but no preferred direction. Instead of building atomic orbitals out of the product of radial functions and a single spherical harmonic, linear combinations of spherical harmonics are typically used, designed so that the imaginary part of the spherical harmonics cancel out. These **real orbitals** are the building blocks most commonly shown in orbital visualizations.

In the real hydrogen-like orbitals, for example, n and ℓ have the same interpretation and significance as their complex counterparts, but m is no longer a good quantum number (though its absolute value is). The orbitals are given new names based on their shape with respect to a standardized Cartesian basis. The real hydrogen-like p orbitals are given by the following[19][20]

$$p_z = p_0$$

$$p_x = \frac{1}{\sqrt{2}}\left(p_1 + p_{-1}\right)$$

$$p_y = \frac{1}{i\sqrt{2}}\left(p_1 - p_{-1}\right)$$

where $p_0 = Rn_1\,Y_{1\,0}$, $p_1 = Rn_1\,Y_{1\,1}$, and $p_{-1} = Rn_1\,Y_{1\,-1}$, are the complex orbitals corresponding to $\ell = 1$.

14.6   Shapes of orbitals

Simple pictures showing orbital shapes are intended to describe the angular forms of regions in space where the electrons occupying the orbital are likely to be found. The diagrams cannot, however, show the entire region where an electron can be found, since according to quantum mechanics there is a non-zero probability of finding the electron (almost) anywhere in space. Instead the diagrams are approximate representations of boundary or contour surfaces where the probability density $|\psi(r, \theta, \varphi)|^2$ has a constant value, chosen so that there is a certain probability (for example 90%) of finding the electron within the contour. Although $|\psi|^2$ as the square of an absolute value is everywhere non-negative, the sign of the wave function $\psi(r, \theta, \varphi)$ is often indicated in each subregion of the orbital picture.

Sometimes the ψ function will be graphed to show its phases, rather than the $|\psi(r, \theta, \varphi)|^2$ which shows probability density but has no phases (which have been lost in the process of taking the absolute value, since $\psi(r, \theta, \varphi)$ is a complex number). $|\psi(r, \theta, \varphi)|^2$ orbital graphs tend to have less spherical, thinner lobes than $\psi(r, \theta, \varphi)$ graphs, but have the same number of lobes in the same places, and otherwise are recognizable. This article, in order to show wave function phases, shows mostly $\psi(r, \theta, \varphi)$ graphs.

Cross-section of computed hydrogen atom orbital ($\psi(r, \theta, \varphi)^2$) for the 6s (n = 6, ℓ = 0, m = 0) orbital. Note that s orbitals, though spherically symmetrical, have radially placed wave-nodes for n > 1. However, only s orbitals invariably have a center anti-node; the other types never do.

The lobes can be viewed as interference patterns between the two counter rotating "m" and "−*m*" modes, with the projection of the orbital onto the xy plane having a resonant "m" wavelengths around the circumference. For each m there are two of these $\langle m \rangle + \langle -m \rangle$ and $\langle m \rangle - \langle -m \rangle$. For the case where $m = 0$ the orbital is vertical, counter rotating information is unknown, and the orbital is z-axis symmetric. For the case where $\ell = 0$ there are no counter rotating modes. There are only radial modes and the shape is spherically symmetric. For any given n, the smaller ℓ is, the more radial nodes there are. Loosely speaking n is energy, ℓ is analogous to eccentricity, and m is orientation.

Generally speaking, the number n determines the size and energy of the orbital for a given nucleus: as n increases, the size of the orbital increases. However, in comparing different elements, the higher nuclear charge Z of heavier elements causes their orbitals to contract by comparison to lighter ones, so that the overall size of the whole atom remains very roughly constant, even as the number of electrons in heavier elements (higher Z) increases.

Also in general terms, ℓ determines an orbital's shape, and mℓ its orientation. However, since some orbitals are described by equations in complex numbers, the shape sometimes depends on mℓ also. Together, the whole set of orbitals for a given ℓ and n fill space as symmetrically as possible, though with increasingly complex sets of lobes and nodes.

The single s-orbitals ($\ell = 0$) are shaped like spheres. For $n = 1$ it is roughly a solid ball (it is most dense at the center and fades exponentially outwardly), but for $n = 2$ or more, each single s-orbital is composed of spherically symmetric surfaces which are nested shells (i.e., the "wave-structure" is radial, following a sinusoidal radial component as well). See illustration of a cross-section of these nested shells, at right. The s-orbitals for all n numbers are the only orbitals with an anti-node (a region of high wave function density) at the center of the nucleus. All other orbitals (p, d, f, etc.) have angular momentum, and thus avoid the nucleus (having a wave node *at* the nucleus).

The three p-orbitals for $n = 2$ have the form of two ellipsoids with a point of tangency at the nucleus (the two-lobed shape is sometimes referred to as a "dumbbell"—there are two lobes pointing in opposite directions from each other). The three p-orbitals in each shell are oriented at right angles to each other, as determined by their respective linear combination of values of mℓ. The overall result is a lobe pointing along each direction of the primary axes.

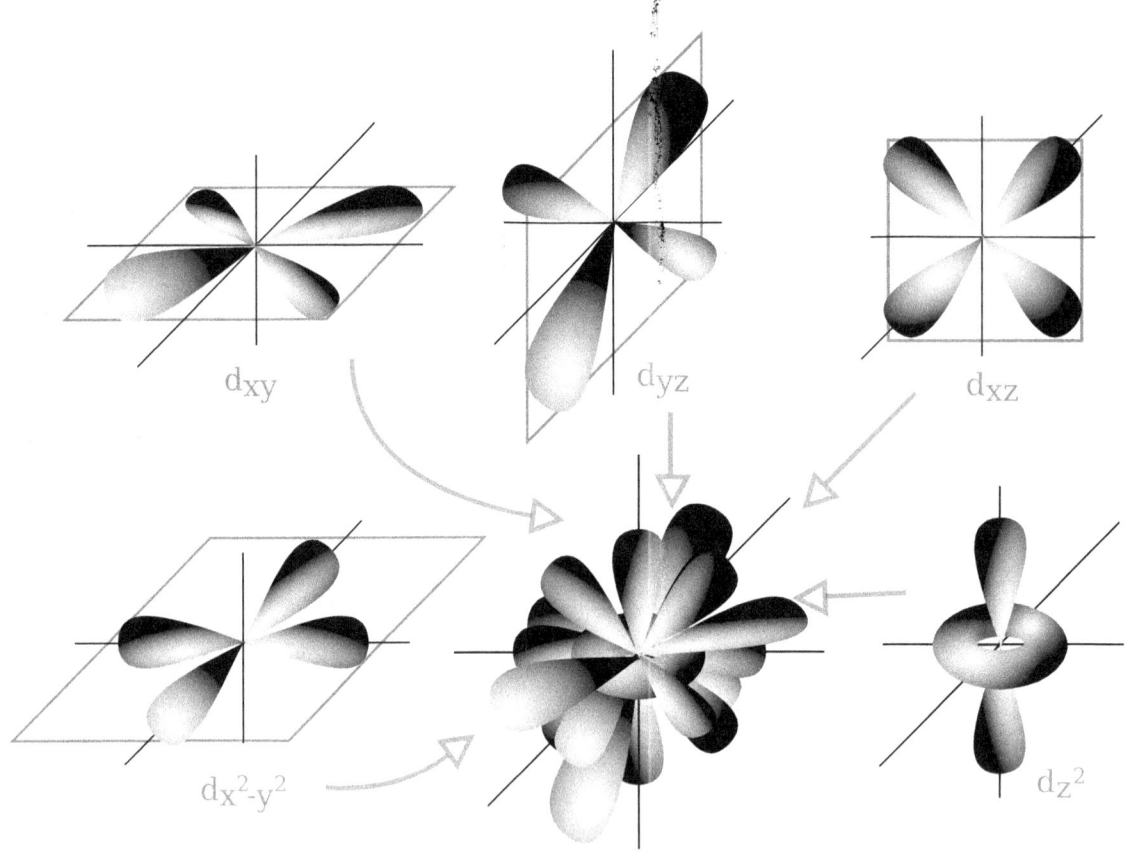

The five d orbitals in $\psi(x, y, z)^2$ form, with a combination diagram showing how they fit together to fill space around an atomic nucleus.

Four of the five d-orbitals for $n = 3$ look similar, each with four pear-shaped lobes, each lobe tangent at right angles to two others, and the centers of all four lying in one plane. Three of these planes are the xy-, xz-, and yz-planes—the lobes are between the pairs of primary axes—and the fourth has the centres along the x and y axes themselves. The fifth and final d-orbital consists of three regions of high probability density: a torus with two pear-shaped regions placed symmetrically on its z axis. The overall total of 18 directional lobes point in every primary axis direction and between every pair.

There are seven f-orbitals, each with shapes more complex than those of the d-orbitals.

Additionally, as is the case with the s orbitals, individual p, d, f and g orbitals with n values higher than the lowest possible value, exhibit an additional radial node structure which is reminiscent of harmonic waves of the same type, as compared with the lowest (or fundamental) mode of the wave. As with s orbitals, this phenomenon provides p, d, f, and g orbitals at the next higher possible value of n (for example, 3p orbitals vs. the fundamental 2p), an additional node in each lobe.

Still higher values of n further increase the number of radial nodes, for each type of orbital.

The shapes of atomic orbitals in one-electron atom are related to 3-dimensional spherical harmonics. These shapes are not unique, and any linear combination is valid, like a transformation to cubic harmonics, in fact it is possible to generate sets where all the d's are the same shape, just like the *px*, *py*, and *pz* are the same shape.[21][22]

14.6.1 Orbitals table

This table shows all orbital configurations for the real hydrogen-like wave functions up to 7s, and therefore covers the simple electronic configuration for all elements in the periodic table up to radium. "ψ" graphs are shown with − and + wave function phases shown in two different colors (arbitrarily red and blue). The *pz* orbital is the same as the *p0* orbital, but the *px* and *py* are formed by taking linear combinations of the *p+1* and *p−1* orbitals (which is why they are listed under the $m = \pm 1$ label). Also, the *p+1* and *p−1* are not the same shape as the *p0*, since they are pure spherical harmonics.

14.6.2 Qualitative understanding of shapes

The shapes of atomic orbitals can be understood qualitatively by considering the analogous case of standing waves on a circular drum.[23] To see the analogy, the mean vibrational displacement of each bit of drum membrane from the equilibrium point over many cycles (a measure of average drum membrane velocity and momentum at that point) must be considered relative to that point's distance from the center of the drum head. If this displacement is taken as being analogous to the probability of finding an electron at a given distance from the nucleus, then it will be seen that the many modes of the vibrating disk form patterns that trace the various shapes of atomic orbitals. The basic reason for this correspondence lies in the fact that the distribution of kinetic energy and momentum in a matter-wave is predictive of where the particle associated with the wave will be. That is, the probability of finding an electron at a given place is also a function of the electron's average momentum at that point, since high electron momentum at a given position tends to "localize" the electron in that position, via the properties of electron wave-packets (see the Heisenberg uncertainty principle for details of the mechanism).

This relationship means that certain key features can be observed in both drum membrane modes and atomic orbitals. For example, in all of the modes analogous to **s** orbitals (the top row in the animated illustration below), it can be seen that the very center of the drum membrane vibrates most strongly, corresponding to the antinode in all **s** orbitals in an atom. This antinode means the electron is most likely to be at the physical position of the nucleus (which it passes straight through without scattering or striking it), since it is moving (on average) most rapidly at that point, giving it maximal momentum.

A mental "planetary orbit" picture closest to the behavior of electrons in **s** orbitals, all of which have no angular momentum, might perhaps be that of a Keplerian orbit with the orbital eccentricity of 1 but a finite major axis, not physically possible (because particles were to collide), but can be imagined as a limit of orbits with equal major axes but increasing eccentricity.

Below, a number of drum membrane vibration modes are shown. The analogous wave functions of the hydrogen atom are indicated. A correspondence can be considered where the wave functions of a vibrating drum head are for a two-coordinate system $\psi(r, \theta)$ and the wave functions for a vibrating sphere are three-coordinate $\psi(r, \theta, \varphi)$.

s-type modes

- Mode (1s orbital)
- Mode (2s orbital)
- Mode (3s orbital)

None of the other sets of modes in a drum membrane have a central antinode, and in all of them the center of the drum does not move. These correspond to a node at the nucleus for all non-**s** orbitals in an atom. These orbitals all have some angular momentum, and in the planetary model, they correspond to particles in orbit with eccentricity less than 1.0, so that they do not pass straight through the center of the primary body, but keep somewhat away from it.

In addition, the drum modes analogous to **p** and **d** modes in an atom show spatial irregularity along the different radial directions from the center of the drum, whereas all of the modes analogous to **s** modes are perfectly symmetrical in radial direction. The non radial-symmetry properties of non-**s** orbitals are necessary to localize a particle with angular momentum and a wave nature in an orbital where it must tend to stay away from the central attraction force, since any particle localized at the point of central attraction could have no angular momentum. For these modes, waves in the drum head tend to avoid the central point. Such features again emphasize that the shapes of atomic orbitals are a direct consequence of the wave nature of electrons.

p-type modes

- Mode (2p orbital)

- Mode (3p orbital)

- Mode (4p orbital)

d-type modes

- Mode (3d orbital)

- Mode (4d orbital)

- Mode (5d orbital)

14.7 Orbital energy

Main article: Electron shell

In atoms with a single electron (hydrogen-like atoms), the energy of an orbital (and, consequently, of any electrons in the orbital) is determined exclusively by n. The $n = 1$ orbital has the lowest possible energy in the atom. Each successively higher value of n has a higher level of energy, but the difference decreases as n increases. For high n, the level of energy becomes so high that the electron can easily escape from the atom. In single electron atoms, all levels with different ℓ within a given n are (to a good approximation) degenerate, and have the same energy. This approximation is broken to a slight extent by the effect of the magnetic field of the nucleus, and by quantum electrodynamics effects. The latter induce tiny binding energy differences especially for **s** electrons that go nearer the nucleus, since these feel a very slightly different nuclear charge, even in one-electron atoms; see Lamb shift.

In atoms with multiple electrons, the energy of an electron depends not only on the intrinsic properties of its orbital, but also on its interactions with the other electrons. These interactions depend on the detail of its spatial probability distribution, and so the energy levels of orbitals depend not only on n but also on ℓ. Higher values of ℓ are associated with higher values of energy; for instance, the 2p state is higher than the 2s state. When $\ell = 2$, the increase in energy of the orbital becomes so large as to push the energy of orbital above the energy of the s-orbital in the next higher shell; when $\ell = 3$ the energy is pushed into the shell two steps higher. The filling of the 3d orbitals does not occur until the 4s orbitals have been filled.

The increase in energy for subshells of increasing angular momentum in larger atoms is due to electron–electron interaction effects, and it is specifically related to the ability of low angular momentum electrons to penetrate more effectively toward the nucleus, where they are subject to less screening from the charge of intervening electrons. Thus, in atoms of higher atomic number, the ℓ of electrons becomes more and more of a determining factor in their energy, and the principal quantum numbers n of electrons becomes less and less important in their energy placement.

The energy sequence of the first 35 subshells (e.g., 1s, 2p, 3d, etc.) is given in the following table. Each cell represents a subshell with n and ℓ given by its row and column indices, respectively. The number in the cell is the subshell's position in the sequence. For a linear listing of the subshells in terms of increasing energies in multielectron atoms, see the section below.

Note: empty cells indicate non-existent sublevels, while numbers in italics indicate sublevels that could (potentially) exist, but which do not hold electrons in any element currently known.

14.8   Electron placement and the periodic table

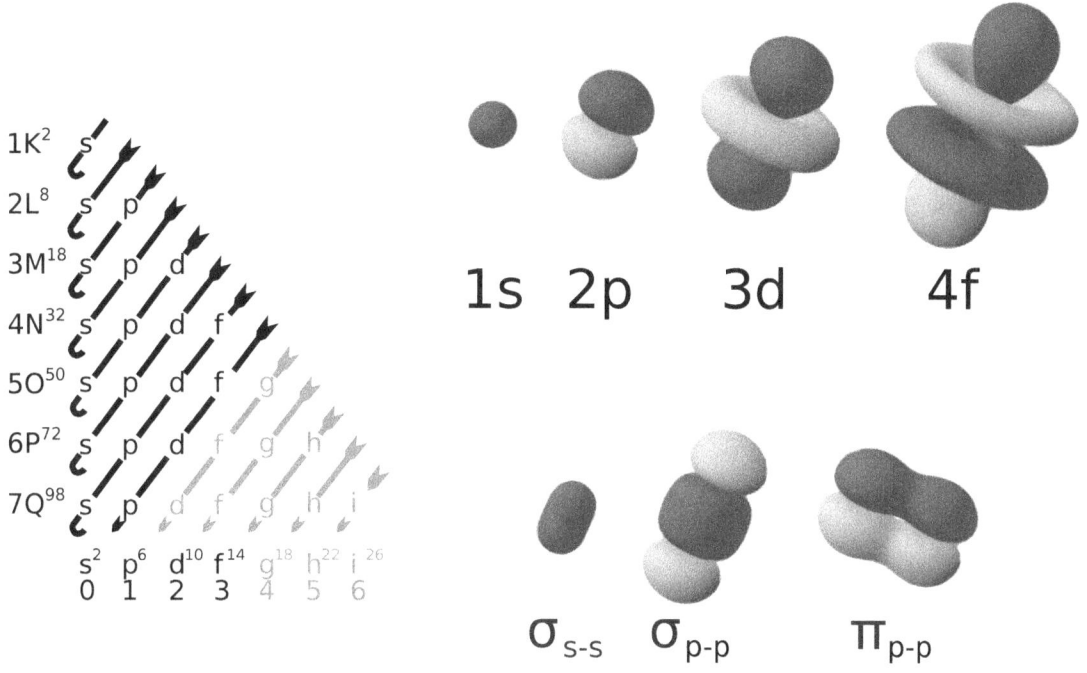

$$1s_2^2\,2s_4^2\,2p_{10}^6\,3s_{12}^2\,3p_{18}^6\,4s_{20}^2\,3d_{30}^{10}\,4p_{36}^6\,5s_{38}^2\,4d_{48}^{10}\,5p_{54}^6\,6s_{56}^2\,4f_{70}^{14}\,5d_{80}^{10}\,6p_{86}^6\,7s_{88}^2\,5f_{102}^{14}\,6d_{112}^{10}\,7p_{118}^6$$

*Electron atomic and molecular orbitals. The chart of orbitals (**left**) is arranged by increasing energy (see Madelung rule).* Note that atomic orbits are functions of three variables (two angles, and the distance r from the nucleus). These images are faithful to the angular component of the orbital, but not entirely representative of the orbital as a whole.

Main articles: electron configuration and electron shell

Several rules govern the placement of electrons in orbitals (*electron configuration*). The first dictates that no two electrons in an atom may have the same set of values of quantum numbers (this is the Pauli exclusion principle). These quantum numbers include the three that define orbitals, as well as s, or spin quantum number. Thus, two electrons may occupy a single orbital, so long as they have different values of s. However, *only* two electrons, because of their spin, can be associated with each orbital.

Additionally, an electron always tends to fall to the lowest possible energy state. It is possible for it to occupy any orbital so long as it does not violate the Pauli exclusion principle, but if lower-energy orbitals are available, this condition is unstable. The electron will eventually lose energy (by releasing a photon) and drop into the lower orbital. Thus, electrons fill orbitals in the order specified by the energy sequence given above.

This behavior is responsible for the structure of the periodic table. The table may be divided into several rows (called 'periods'), numbered starting with 1 at the top. The presently known elements occupy seven periods. If a certain period has number i, it consists of elements whose outermost electrons fall in the ith shell. Niels Bohr was the first to propose (1923) that the periodicity in the properties of the elements might be explained by the periodic filling of the electron energy levels, resulting in the electronic structure of the atom.[24]

The periodic table may also be divided into several numbered rectangular 'blocks'. The elements belonging to a given

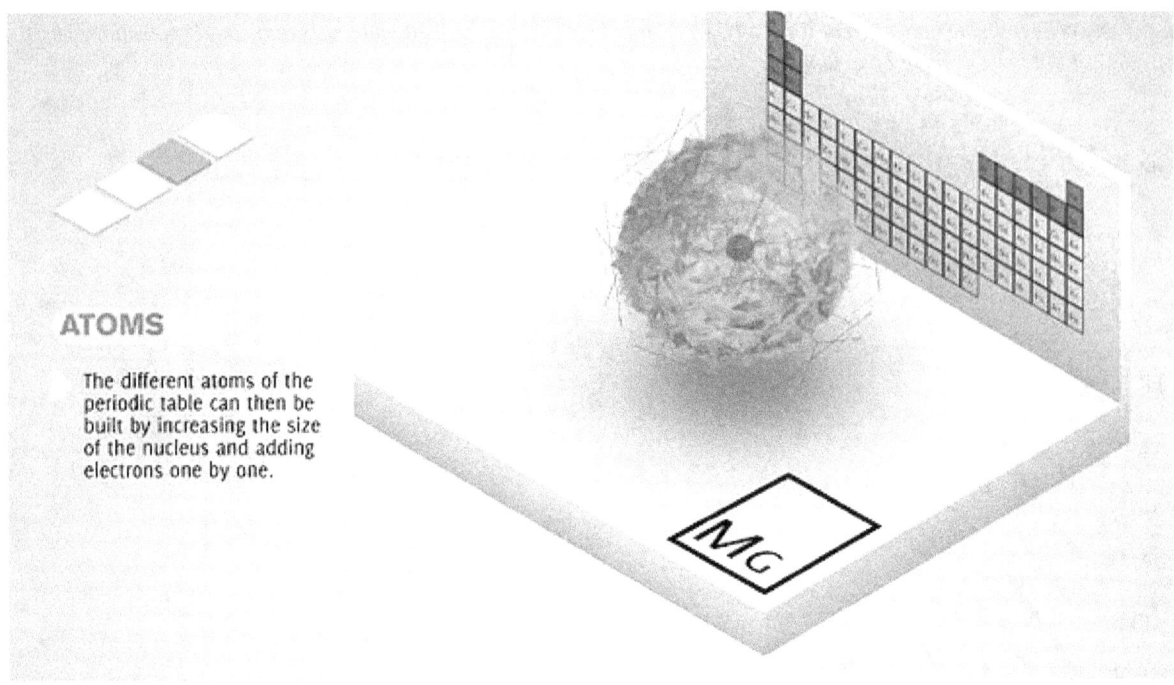

ATOMS

The different atoms of the periodic table can then be built by increasing the size of the nucleus and adding electrons one by one.

Atomic orbitals and periodic table construction

block have this common feature: their highest-energy electrons all belong to the same ℓ-state (but the n associated with that ℓ-state depends upon the period). For instance, the leftmost two columns constitute the 's-block'. The outermost electrons of Li and Be respectively belong to the 2s subshell, and those of Na and Mg to the 3s subshell.

The following is the order for filling the "subshell" orbitals, which also gives the order of the "blocks" in the periodic table:

1s, 2s, 2p, 3s, 3p, 4s, 3d, 4p, 5s, 4d, 5p, 6s, 4f, 5d, 6p, 7s, 5f, 6d, 7p

The "periodic" nature of the filling of orbitals, as well as emergence of the **s**, **p**, **d** and **f** "blocks", is more obvious if this order of filling is given in matrix form, with increasing principal quantum numbers starting the new rows ("periods") in the matrix. Then, each subshell (composed of the first two quantum numbers) is repeated as many times as required for each pair of electrons it may contain. The result is a compressed periodic table, with each entry representing two successive elements:

Although this is the general order of orbital filling according to the Madelung rule, there are exceptions, and the actual electronic energies of each element are also dependent upon additional details of the atoms (see Electron configuration#Atoms: Aufbau principle and Madelung rule).

The number of electrons in an electrically neutral atom increases with the atomic number. The electrons in the outermost shell, or *valence electrons*, tend to be responsible for an element's chemical behavior. Elements that contain the same number of valence electrons can be grouped together and display similar chemical properties.

14.8.1 Relativistic effects

Main article: Relativistic quantum chemistry

For elements with high atomic number Z, the effects of relativity become more pronounced, and especially so for s electrons, which move at relativistic velocities as they penetrate the screening electrons near the core of high-Z atoms. This relativistic increase in momentum for high speed electrons causes a corresponding decrease in wavelength and contraction of 6s orbitals relative to 5d orbitals (by comparison to corresponding s and d electrons in lighter elements in the same column of the periodic table); this results in 6s valence electrons becoming lowered in energy.

Examples of significant physical outcomes of this effect include the lowered melting temperature of mercury (which results from 6s electrons not being available for metal bonding) and the golden color of gold and caesium (which results from narrowing of 6s to 5d transition energy to the point that visible light begins to be absorbed).[25]

In the Bohr Model, an $n = 1$ electron has a velocity given by $v = Z\alpha c$, where Z is the atomic number, α is the fine-structure constant, and c is the speed of light. In non-relativistic quantum mechanics, therefore, any atom with an atomic number greater than 137 would require its 1s electrons to be traveling faster than the speed of light. Even in the Dirac equation, which accounts for relativistic effects, the wavefunction of the electron for atoms with **Z > 137** is oscillatory and unbounded. The significance of element 137, also known as untriseptium, was first pointed out by the physicist Richard Feynman. Element 137 is sometimes informally called feynmanium (symbol Fy) . However, Feynman's approximation fails to predict the exact critical value of Z due to the non-point-charge nature of the nucleus and very small orbital radius of inner electrons, resulting in a potential seen by inner electrons which is effectively less than Z. The critical Z value which makes the atom unstable with regard to high-field breakdown of the vacuum and production of electron-positron pairs, does not occur until Z is about 173. These conditions are not seen except transiently in collisions of very heavy nuclei such as lead or uranium in accelerators, where such electron-positron production from these effects has been claimed to be observed. See Extension of the periodic table beyond the seventh period.

There are no nodes in relativistic orbital densities, although individual components of the wavefunction will have nodes.[26]

14.9   Transitions between orbitals

Main article: Atomic electron transition

Bound quantum states have discrete energy levels. When applied to atomic orbitals, this means that the energy differences between states are also discrete. A transition between these states (i.e., an electron absorbing or emitting a photon) can thus only happen if the photon has an energy corresponding with the exact energy difference between said states.

Consider two states of the hydrogen atom:

State 1) $n = 1$, $\ell = 0$, $m\ell = 0$ and $s = +\frac{1}{2}$

State 2) $n = 2$, $\ell = 0$, $m\ell = 0$ and $s = +\frac{1}{2}$

By quantum theory, state 1 has a fixed energy of E_1, and state 2 has a fixed energy of E_2. Now, what would happen if an electron in state 1 were to move to state 2? For this to happen, the electron would need to gain an energy of exactly $E_2 - E_1$. If the electron receives energy that is less than or greater than this value, it cannot jump from state 1 to state 2. Now, suppose we irradiate the atom with a broad-spectrum of light. Photons that reach the atom that have an energy of exactly $E_2 - E_1$ will be absorbed by the electron in state 1, and that electron will jump to state 2. However, photons that are greater or lower in energy cannot be absorbed by the electron, because the electron can only jump to one of the orbitals, it cannot jump to a state between orbitals. The result is that only photons of a specific frequency will be absorbed by the atom. This creates a line in the spectrum, known as an absorption line, which corresponds to the energy difference between states 1 and 2.

The atomic orbital model thus predicts line spectra, which are observed experimentally. This is one of the main validations of the atomic orbital model.

The atomic orbital model is nevertheless an approximation to the full quantum theory, which only recognizes many electron states. The predictions of line spectra are qualitatively useful but are not quantitatively accurate for atoms and ions other than those containing only one electron.

14.10 See also

- 3D hydrogen orbitals on Wikimedia Commons

- Atomic electron configuration table

- Condensed matter physics

- Electron configuration

- Energy level

- Hund's rules

- Molecular orbital

- Quantum chemistry

- Quantum chemistry computer programs

- Solid state physics

- Orbital resonance

- Wave function collapse

14.11 References

[1] Orchin, Milton; Macomber, Roger S.; Pinhas, Allan; Wilson, R. Marshall (2005). *Atomic Orbital Theory* (PDF).

[2] Daintith, J. (2004). *Oxford Dictionary of Chemistry*. New York: Oxford University Press. ISBN 0-19-860918-3.

[3] Griffiths, David (1995). *Introduction to Quantum Mechanics*. Prentice Hall. pp. 190–191. ISBN 0-13-124405-1.

[4] Levine, Ira (2000). *Quantum Chemistry* (5 ed.). Prentice Hall. pp. 144–145. ISBN 0-13-685512-1.

[5] Laidler, Keith J.; Meiser, John H. (1982). *Physical Chemistry*. Benjamin/Cummings. p. 488. ISBN 0-8053-5682-7.

[6] Feynman, Richard; Leighton, Robert B.; Sands, Matthew (2006). *The Feynman Lectures on Physics -The Definitive Edition, Vol 1 lect 6*. Pearson PLC, Addison Wesley. p. 11. ISBN 0-8053-9046-4.

[7] Roger Penrose, *The Road to Reality*

[8] Mulliken, Robert S. (July 1932). "Electronic Structures of Polyatomic Molecules and Valence. II. General Considerations". *Physical Review* **41** (1): 49–71. Bibcode:1932PhRv...41...49M. doi:10.1103/PhysRev.41.49.

[9] Bohr, Niels (1913). "On the Constitution of Atoms and Molecules". *Philosophical Magazine* **26** (1): 476. Bibcode:1914Natur..93. doi:10.1038/093268a0.

[10] Nagaoka, Hantaro (May 1904). "Kinetics of a System of Particles illustrating the Line and the Band Spectrum and the Phenomena of Radioactivity". *Philosophical Magazine* **7** (41): 445–455. doi:10.1080/14786440409463141.

[11] Bryson, Bill (2003). *A Short History of Nearly Everything*. Broadway Books. pp. 141–143. ISBN 0-7679-0818-X.

[12] Thomson, J. J. (1897). "Cathode rays". *Philosophical Magazine* **44** (269): 293. doi:10.1080/14786449708621070.

[13] Thomson, J. J. (1904). "On the Structure of the Atom: an Investigation of the Stability and Periods of Oscillation of a numbe rof Corpuscles arranged at equal intervals around the Circumference of a Circle; with Application of the Results to the Theoryo f Atomic Structure" (extract of paper). *Philosophical Magazine Series* **67** (39): 237. doi:10.1080/14786440409463107.

[14] Rhodes, Richard (1995). *The Making of the Atomic Bomb*. Simon & Schuster. pp. 50–51. ISBN 978-0-684-81378-3.

[15] Nagaoka, Hantaro (May 1904). "Kinetics of a System of Particles illustrating the Line and the Band Spectrum and the Phenomena of Radioactivity". *Philosophical Magazine* **7**: 446. doi:10.1080/14786440409463141.

[16] Heisenberg, W. (March 1927). "Über den anschaulichen Inhalt der quantentheoretischen Kinematik und Mechanik". *Zeitschrift für Physik A* **43** (3–4): 172–198. Bibcode:1927ZPhy...43..172H. doi:10.1007/BF01397280.

[17] Bohr, Niels (April 1928). "The Quantum Postulate and the Recent Development of Atomic Theory". *Nature* **121** (3050): 580–590. Bibcode:1928Natur.121..580B. doi:10.1038/121580a0.

[18] Gerlach, W.; Stern, O. (1922). "Das magnetische Moment des Silberatoms". *Zeitschrift für Physik* **9**: 353–355. Bibcode: doi:10.1007/BF01326984.

[19] Levine, Ira (2014). *Quantum Chemistry* (7th ed.). Pearson Education. pp. 141–2. ISBN 0-321-80345-0.

[20] Blanco, Miguel A.; Flórez, M.; Bermejo, M. (December 1997). "Evaluation of the rotation matrices in the basis of real spherical harmonics". *Journal of Molecular Structure: THEOCHEM* **419** (1-3): 19–27. doi:10.1016/S0166-1280(97)00185-1.

[21] Powell, Richard E. (1968). "The five equivalent d orbitals". *Journal of Chemical Education* **45**(1): 45. Bibcode:1968JChEd..45.5P. doi:10.1021/ed045p45.

[22] Kimball, George E. (1940). "Directed Valence". *The Journal of Chemical Physics* **8** (2): 188. Bibcode:1940JChPh...8..188K. doi:10.1063/1.1750628.

[23] Cazenave, Lions, T., P.; Lions, P. L. (1982). "Orbital stability of standing waves for some nonlinear Schrödinger equations". *Communications in Mathematical Physics* **85** (4): 549–561. Bibcode:1982CMaPh..85..549C. doi:10.1007/BF01403504.

[24] Bohr, Niels (1923). "Über die Anwendung der Quantumtheorie auf den Atombau. I". *Zeitschrift für Physik* **13**: 117. Bibcode: doi:10.1007/BF01328209.

[25] Lower, Stephen. "Primer on Quantum Theory of the Atom".

[26] Szabo, Attila (1969). "Contour diagrams for relativistic orbitals". *Journal of Chemical Education* **46**(10): 678. Bibcode:196678S. doi:10.1021/ed046p678.

14.12   Further reading

- Tipler, Paul; Llewellyn, Ralph (2003). *Modern Physics* (4 ed.). New York: W. H. Freeman and Company. ISBN 0-7167-4345-0.

- Scerri, Eric (2007). *The Periodic Table, Its Story and Its Significance*. New York: Oxford University Press. ISBN 978-0-19-530573-9.

- Levine, Ira (2014). *Quantum Chemistry* (7th ed.). Pearson Education. ISBN 0-321-80345-0.

- Griffiths, David (2000). *Introduction to Quantum Mechanics* (2 ed.). Benjamin Cummings. ISBN 978-0-13-111892-8.

- Cohen, Irwin; Bustard, Thomas (1966). "Atomic Orbitals: Limitations and Variations". *J. Chem. Educ.* **43** (4): 187. Bibcode:1966JChEd..43..187C. doi:10.1021/ed043p187.

14.13   External links

- Guide to atomic orbitals

- Covalent Bonds and Molecular Structure

- Animation of the time evolution of an hydrogenic orbital

- The Orbitron, a visualization of all common and uncommon atomic orbitals, from 1s to 7g

- Grand table Still images of many orbitals

- David Manthey's Orbital Viewer renders orbitals with $n \leq 30$

- Java orbital viewer applet

- What does an atom look like? Orbitals in 3D

- Atom Orbitals v.1.5 visualization software

- Crystal-Style Orbital Viewer

Chapter 15

Atomism

Atomism (from Greek ἄτομον, *atomon*, i.e. "uncuttable", "indivisible"[1][2][3]) is a natural philosophy that developed in several ancient traditions. The atomists theorized that nature consists of two fundamental principles: *atom* and *void*. Unlike their modern scientific namesake in atomic theory, philosophical atoms come in an infinite variety of shapes and sizes, each indestructible, immutable and surrounded by a void where they collide with the others or hook together forming a cluster. Clusters of different shapes, arrangements, and positions give rise to the various macroscopic substances in the world.[4][5]

References to the concept of atomism and its atoms are found in ancient India and ancient Greece. In India the Jain,[6][7] Ajivika and Carvaka schools of atomism may date back to the 4th century BCE.[8] The Nyaya and Vaisheshika schools later developed theories on how atoms combined into more complex objects.[9] In the West, atomism emerged in the 5th century BCE with Leucippus and Democritus.[10] Whether Indian culture influenced Greek or vice versa or whether both evolved independently is a matter of dispute.[11]

The particles of chemical matter for which chemists and other natural philosophers of the early 19th century found experimental evidence were thought to be indivisible, and therefore were given the name "atom", long used by the atomist philosophy.

However, in the 20th century, the "atoms" of the chemists were found to be composed of even smaller entities: electrons, neutrons, and protons, and further experiments showed that protons and neutrons are made of quarks. Although the connection to historical atomism is at best tenuous, elementary particles have thus become a modern analog of philosophical atoms, despite the misnomer in chemistry.

15.1   Reductionism

Philosophical atomism is a reductive argument: not only that everything is composed of atoms and void, but that nothing they compose really exists: the only things that really exist are atoms ricocheting off each other mechanistically in an otherwise empty void. Atomism stands in contrast to a substance theory wherein a prime material continuum remains qualitatively invariant under division (for example, the ratio of the four classical elements would be the same in any portion of a homogeneous material).

Indian Buddhists, such as Dharmakirti and others, also developed distinctive theories of atomism, for example, involving momentary (instantaneous) atoms, that flash in and out of existence (Kalapas).

15.2   Greek atomism

In the 5th century BC, Leucippus and his pupil Democritus proposed that all matter was composed of small indivisible particles called atoms, in order to reconcile two conflicting schools of thought on the nature of reality. On one side was

Heraclitus, who believed that the nature of all existence is change. On the other side was Parmenides, who believed instead that all change is illusion.

Parmenides denied the existence of motion, change and void. He believed all existence to be a single, all-encompassing and unchanging mass (a concept known as monism), and that change and motion were mere illusions. This conclusion, as well as the reasoning that led to it, may indeed seem baffling to the modern empirical mind, but Parmenides explicitly rejected sensory experience as the path to an understanding of the universe, and instead used purely abstract reasoning. Firstly, he believed there is no such thing as void, equating it with non-being (i.e. "if the void *is*, then it is not nothing; therefore it is not the void"). This in turn meant that motion is impossible, because there is no void to move into.[12] [13] He also wrote all that *is* must be an indivisible unity, for if it were manifold, then there would have to be a void that could divide it (and he did not believe the void exists). Finally, he stated that the all encompassing Unity is unchanging, for the Unity already encompasses all that is and can be.[14]

Democritus accepted most of Parmenides' arguments, except for the idea that change is an illusion. He believed change was real, and if it was not then at least the illusion had to be explained. He thus supported the concept of void, and stated that the universe is made up of many Parmenidean entities that move around in the void.[12] The void is infinite and provides the space in which the atoms can pack or scatter differently. The different possible packings and scatterings within the void make up the shifting outlines and bulk of the objects that organisms feel, see, eat, hear, smell, and taste. While organisms may feel hot or cold, hot and cold actually have no real existence. They are simply sensations produced in organisms by the different packings and scatterings of the atoms in the void that compose the object that organisms sense as being "hot" or "cold".

The work of Democritus only survives in secondhand reports, some of which are unreliable or conflicting. Much of the best evidence of Democritus' theory of atomism is reported by Aristotle in his discussions of Democritus' and Plato's contrasting views on the types of indivisibles composing the natural world.[15]

15.2.1 Geometry and atoms

Plato (c. 427 — c. 347 BC), were he familiar with the atomism of Democritus, would have objected to its mechanistic materialism. He argued that atoms just crashing into other atoms could never produce the beauty and form of the world. In Plato's *Timaeus*, (28B – 29A) the character of Timeaus insisted that the cosmos was not eternal but was created, although its creator framed it after an eternal, unchanging model.

One part of that creation were the four simple bodies of fire, air, water, and earth. But Plato did not consider these corpuscles to be the most basic level of reality, for in his view they were made up of an unchanging level of reality, which was mathematical. These simple bodies were geometric solids, the faces of which were, in turn, made up of triangles. The square faces of the cube were each made up of four isosceles right-angled triangles and the triangular faces of the tetrahedron, octahedron, and icosahedron were each made up of six right-angled triangles.

He postulated the geometric structure of the simple bodies of the four elements as summarized in the table to the right. The cube, with its flat base and stability, was assigned to earth; the tetrahedron was assigned to fire because its penetrating points and sharp edges made it mobile. The points and edges of the octahedron and icosahedron were blunter and so these less mobile bodies were assigned to air and water. Since the simple bodies could be decomposed into triangles, and the triangles reassembled into atoms of different elements, Plato's model offered a plausible account of changes among the primary substances.[16][17]

15.2.2 The rejection of atoms

Sometime before 330 BC Aristotle asserted that the elements of fire, air, earth, and water were not made of atoms, but were continuous. Aristotle considered the existence of a void, which was required by atomic theories, to violate physical principles. Change took place not by the rearrangement of atoms to make new structures, but by transformation of matter from what it was in potential to a new actuality. A piece of wet clay, when acted upon by a potter, takes on its potential to be an actual drinking mug. Aristotle has often been criticized for rejecting atomism, but in ancient Greece the atomic theories of Democritus remained "pure speculations, incapable of being put to any experimental test. Granted that atomism was, in the long run, to prove far more fruitful than any qualitative theory of matter, in the short run the theory that Aristotle proposed must have seemed in some respects more promising".[18][19]

15.2.3   Later ancient atomism

Epicurus (341–270) studied atomism with Nausiphanes who had been a student of Democritus. Although Epicurus was certain of the existence of atoms and the void, he was less sure we could adequately explain specific natural phenomena such as earthquakes, lightning, comets, or the phases of the Moon (Lloyd 1973, 25–6). Few of Epicurus's writings survive and those that do reflect his interest in applying Democritus's theories to assist people in taking responsibility for themselves and for their own happiness—since he held there are no gods around that can help them. He understood gods' role as moral ideals.

His ideas are also represented in the works of his follower Lucretius, who wrote *On the Nature of Things*. This scientific work in poetic form illustrates several segments of Epicurean theory on how the universe came into its current stage and it shows that the phenomena we perceive are actually composite forms. The atoms and the void are eternal and in constant motion. Atomic collisions create objects, which are still composed of the same eternal atoms whose motion for a while is incorporated into the created entity. Human sensations and meteorological phenomena are also explained by Lucretius in terms of atomic motion.

15.2.4   Atomism and ethics

Some later philosophers attributed the idea that man created gods; the gods did not create man to Democritus. For example, Sextus Empiricus noted:

> Some people think that we arrived at the idea of gods from the remarkable things that happen in the world. Democritus ... says that the people of ancient times were frightened by happenings in the heavens such as thunder, lightning, ..., and thought that they were caused by gods.[20]

Three hundred years after Epicurus, Lucretius in his epic poem *On the Nature of Things* would depict him as the hero who crushed the monster Religion through educating the people in what was possible in the atoms and what was *not* possible in the atoms. However, Epicurus expressed a non-aggressive attitude characterized by his statement: "The man who best knows how to meet external threats makes into one family all the creatures he can; and those he can not, he at any rate does not treat as aliens; and where he finds even this impossible, he avoids all dealings, and, so far as is advantageous, excludes them from his life."

15.3   Indian atomism

The Indian atomistic position, like many movements in Indian Philosophy and Mathematics, starts with an argument from Linguistics. The Vedic etymologist and grammarian Yaska (c. 7th century BC) in his Nirukta, in dealing with models for how linguistic structures get to have their meanings, takes the atomistic position that words are the "primary" carrier of meaning – i.e. words have a preferred ontological status in defining meaning. This position was to be the subject of a fierce debate in the Indian tradition from the early Christian era till the 18th century, involving different philosophers from the Nyaya, Mimamsa and Buddhist schools.

In the pratishakhya text (c. 2nd century BCE), the gist of the controversy was stated cryptically in the sutra form as "saMhitA pada-prakr^tiH".[21] According to the atomist view, the words (*pada*) would be the primary elements (*prakrti*) out of which the sentence is constructed, while the holistic view considers the sentence as the primary entity, originally "given" in its context of utterance, and the words are arrived at only through analysis and abstraction.[22]

These two positions came to be called *a-kShaNDa-pakSha* (indivisibility or sentence-holism), a position developed later by Bhartrihari (c. 500 AD), vs. *kShaNDa-pakSha* (atomism), a position adopted by the Mimamsa and Nyaya schools (Note: *kShanDa* = fragmented; "a-kShanDa" = whole).

Between the 5th and 3rd centuries BC, the atom (anu or aṇor) is mentioned in the Bhagavad Gita (Chapter 8, Verse 9):

*kaviṁ purāṇam anuśāsitāram **aṇor** aṇīyāṁsam anusmared yaḥ sarvasya dhātāram acintya-rūpam āditya-varṇaṁ tamasaḥ parastāt*

One meditates on the omniscient, primordial, the controller, smaller than the atom, yet the maintainer of everything; whose form is inconceivable, resplendent like the sun and totally transcendental to material nature

The ancient *shāshvata-vāda* doctrine of eternalism, which held that elements are eternal, is also suggestive of a possible starting point for atomism (Gangopadhyaya 1981).

The atomist position had transcended language into epistemology by the time that Nyaya–Vaisesika, Buddhist and Jaina theology were developing mature philosophical positions.

Will Durant wrote in *Our Oriental Heritage*:

> "Two systems of Indian thought propound physical theories suggestively similar to those of Greece. Kanada, founder of the Vaisheshika philosophy, held that the world was composed of atoms as many in kind as the various elements. The Jains more nearly approximated to Democritus by teaching that all atoms were of the same kind, producing different effects by diverse modes of combinations. Kanada believed light and heat to be varieties of the same substance; Udayana taught that all heat comes from the sun; and Vachaspati, like Newton, interpreted light as composed of minute particles emitted by substances and striking the eye."

Indian atomism in the Middle Ages was still mostly philosophical and/or religious in intent, though it was also scientific. Because the "infallible Vedas", the oldest Hindu texts, do not mention atoms (though they do mention elements), atomism was not orthodox in many schools of Hindu philosophy, although accommodationist interpretations or assumptions of lost text justified the use of atomism for non-orthodox schools of Hindu thought. The Buddhist and Jaina schools, however, were more willing to accept the ideas of atomism.

15.3.1 Nyaya–Vaisesika school

Main articles: Nyaya and Vaisesika

The Nyaya–Vaisesika school developed one of the earliest forms of atomism; scholars date the Nyaya and Vaisesika texts from the 6th to 1st centuries BC. Like the Buddhist atomists, the Vaisesika had a pseudo-Aristotelian theory of atomism. They posited the four elemental atom types, but in Vaisesika physics atoms had 24 different possible qualities, divided between general extensive properties and specific (intensive) properties. Like the Jaina school, the Nyaya–Vaisesika atomists had elaborate theories of how atoms combine. In both Jaina and Vaisesika atomism, atoms first combine in pairs (dyads), and then group into trios of pairs (triads), which are the smallest visible units of matter.[23]

15.3.2 Buddhist school

Main article: Buddhist atomism

The Buddhist atomists had very qualitative, Aristotelian-style atomic theory. According to ancient Buddhist atomism, which probably began developing before the 4th century BC, there are four kinds of atoms, corresponding to the standard elements. Each of these elements has a specific property, such as solidity or motion, and performs a specific function in mixtures, such as providing support or causing growth. Like the Hindu Jains, the Buddhists were able to integrate a theory of atomism with their theological presuppositions. Later Indian Buddhist philosophers, such as Dharmakirti and Dignāga, considered atoms to be point-sized, durationless, and made of energy.

15.3.3 Jaina school

Further information: Jain cosmology, Dravya (Jainism) and Karma in Jainism

The most elaborate and well-preserved Indian theory of atomism comes from the philosophy of the Jaina school, dating back to at least the 6th century BC. Some of the Jain texts that refer to matter and atoms are Pancastikayasara, Kalpasutra, Tattvarthasutra and Pannavana Suttam. The Jains envisioned the world as consisting wholly of atoms, except for souls. Paramāṇus or atoms were considered as the basic building blocks of all matter. Their concept of atoms was very similar to classical atomism, differing primarily in the specific properties of atoms. Each atom, according to Jain philosophy, has one kind of taste, one smell, one color, and two kinds of touch, though it is unclear what was meant by "kind of touch". Atoms can exist in one of two states: subtle, in which case they can fit in infinitesimally small spaces, and gross, in which case they have extension and occupy a finite space. Certain characteristics of Paramāṇu correspond with that sub-atomic particles. For example Paramāṇu is characterized by continuous motion either in a straight line or in case of attractions from other Paramāṇus, it follows a curved path. This corresponds with the description of orbit of electrons across the Nucleus. Ultimate particles are also described as particles with positive (Snigdha i.e. smooth charge) and negative (Rūksa – rough) charges that provide them the binding force. Although atoms are made of the same basic substance, they can combine based on their eternal properties to produce any of six "aggregates", which seem to correspond with the Greek concept of "elements": earth, water, shadow, sense objects, karmic matter, and unfit matter. To the Jains, karma was real, but was a naturalistic, mechanistic phenomenon caused by buildups of subtle karmic matter within the soul. They also had detailed theories of how atoms could combine, react, vibrate, move, and perform other actions, all of which were thoroughly deterministic.

15.3.4 Time as explained in *Srimad-Bhagavatam*

"The material manifestation's ultimate particle, which is indivisible and not formed into a body, is called the atom (*parama-aṇuh*). It exists always as an invisible identity, even after the dissolution of all forms. The material body is but a combination of such atoms, but it is misunderstood by the common man.

Atoms are the ultimate state of the manifest universe. When they stay in their own forms without forming different bodies, they are called the unlimited oneness. There are certainly different bodies in physical forms, but the atoms themselves form the complete manifestation.

One can estimate time by measuring the movement of the atomic combination of bodies. Time is the potency of the almighty Personality of Godhead, Hari, who controls all physical movement although He is not visible in the physical world.

The division of gross time is calculated as follows: two atoms make one double atom, and three double atoms make one hexatom. This hexatom is visible in the sunshine which enters through the holes of a window screen. One can clearly see that the hexatom goes up towards the sky.

The atom is described as an invisible particle, but when six such atoms combine together, they are called a *trasareṇu*, and this is visible in the sunshine pouring through the holes of a window screen."

— Excerpted from *Srimad-Bhagavatam* (*SB* 3.11.1–5) by A. C. Bhaktivedanta Swami Prabhupada, courtesy of the Bhaktivedanta Book Trust International (prabhupadabooks.com)

15.4 Islamic atomism

See also: Early Islamic philosophy: Atomism and Alchemy and chemistry in medieval Islam

Atomistic philosophies are found very early in Islamic philosophy and was influenced by earlier Greek and to some extent Indian philosophy.[24][25] Like both the Greek and Indian versions, Islamic atomism was a charged topic that had the potential for conflict with the prevalent religious orthodoxy, but it was instead more often favoured by orthodox Islamic theologians. It was such a fertile and flexible idea that, as in Greece and India, it flourished in some leading schools of Islamic thought.

15.4.1 Asharite atomism

See also: Ash'ari

The most successful form of Islamic atomism was in the Asharite school of Islamic theology, most notably in the work of the theologian al-Ghazali (1058–1111). In Asharite atomism, atoms are the only perpetual, material things in existence, and all else in the world is "accidental" meaning something that lasts for only an instant. Nothing accidental can be the cause of anything else, except perception, as it exists for a moment. Contingent events are not subject to natural physical causes, but are the direct result of God's constant intervention, without which nothing could happen. Thus nature is completely dependent on God, which meshes with other Asharite Islamic ideas on causation, or the lack thereof (Gardet 2001). Al-Ghazali also used the theory to support his theory of occasionalism. In a sense, the Asharite theory of atomism has far more in common with Indian atomism than it does with Greek atomism.[26]

15.4.2 Averroism

See also: Averroism

Other traditions in Islam rejected the atomism of the Asharites and expounded on many Greek texts, especially those of Aristotle. An active school of philosophers in Spain, including the noted commentator Averroes (AD 1126–1198) explicitly rejected the thought of al-Ghazali and turned to an extensive evaluation of the thought of Aristotle. Averroes commented in detail on most of the works of Aristotle and his commentaries did much to guide the interpretation of Aristotle in later Jewish and Christian scholastic thought.

15.5 Medieval European speculations

While Aristotelian philosophy eclipsed the importance of the atomists in late Roman and medieval Europe, their work was still preserved and exposited through commentaries on the works of Aristotle. In the 2nd century, Galen (AD 129–216) presented extensive discussions of the Greek atomists, especially Epicurus, in his Aristotle commentaries. According to historian of atomism Joshua Gregory, there was no serious work done with atomism from the time of Galen until Gassendi and Descartes resurrected it in the 17th century; "the gap between these two 'modern naturalists' and the ancient Atomists marked "the exile of the atom" and "it is universally admitted that the Middle Ages had abandoned Atomism, and virtually lost it."

However, although the ancient atomists' works were unavailable, Scholastic thinkers still had Aristotle's critiques of atomism. In the medieval universities there were expressions of atomism. For example, in the 14th century Nicholas of Autrecourt considered that matter, space, and time were all made up of indivisible atoms, points, and instants and that all generation and corruption took place by the rearrangement of material atoms. The similarities of his ideas with those of al-Ghazali suggest that Nicholas may have been familiar with Ghazali's work, perhaps through Averroes' refutation of it (Marmara, 1973–74).

15.5.1 Scholastic *minima naturalia*

Main article: Minima naturalia

Although the atomism of Epicurus had fallen out of favor in the centuries of Scholasticism, a related Aristotelian concept, that of *minima naturalia* (natural minima) received extensive consideration. *Minima naturalia* were theorized by Aristotle as the smallest parts into which a homogeneous natural substance (e.g., flesh, bone, or wood) could be divided and still retain its essential character. Speculation on *minima naturalia* provided philosophical background for the mechanistic philosophy of early modern thinkers like Descartes, and for the alchemical works of Geber and Daniel Sennert, who in turn influenced the corpuscularian alchemist Robert Boyle, one of the founders of modern chemistry.[27][28]

Unlike the atomism of Leucippus, Democritus, and Epicurus, and also unlike the later atomic theory of John Dalton, the Aristotelian natural minimum was not conceptualized as physically indivisible--"atomic" in the contemporary sense. Instead, the concept was rooted in Aristotle's hylomorphic worldview, which held that every physical thing is a compound of matter (Greek *hyle*) and an immaterial substantial form (Greek *morphe*) that imparts its essential nature and structure. For instance, a rubber ball for a hylomorphist like Aristotle would be rubber (matter) structured by spherical shape (form).

Aristotle's intuition was that there is some smallest size beyond which matter could no longer be structured as flesh, or bone, or wood, or some other such organic substance that for Aristotle, living before the microscope, could be considered homogeneous. For instance, if flesh were divided beyond its natural minimum, what would be left might be a large amount of the element water, and smaller amounts of the other elements. But whatever water or other elements were left, they would no longer have the "nature" of flesh: in hylomorphic terms, they would no longer be matter structured by the form of flesh; instead the remaining water, e.g., would be matter structured by the form of water, not the form of flesh. This is suggestive of modern chemistry, in which, e.g., a bar of gold can be continually divided until one has a single atom of gold, but further division yields only subatomic particles (electrons, quarks, etc.) which are no longer "gold." However, the parallel is not exact: *minima naturalia* are not a direct anticipation of modern chemical and physical concepts.

A chief theme in late Roman and Scholastic commentary on this concept is reconciling *minima naturalia* with the general Aristotelian principle of infinite divisibility. Commentators like John Philoponus and Thomas Aquinas reconciled these aspects of Aristotle's thought by distinguishing between mathematical and "natural" divisibility. With few exceptions, much of the curriculum in the universities of Europe was based on such Aristotelianism for most of the Middle Ages (Kargon 1966). Scholasticism was standard science in the time of Isaac Newton, but in the 17th century, a renewed interest in Epicurean atomism and corpuscularianism as a hybrid or an alternative to Aristotelian physics had begun to mount outside the classroom.

15.6   Atomic renaissance

The main figures in the rebirth of atomism were René Descartes, Pierre Gassendi, and Robert Boyle, as well as other notable figures.

One of the first groups of atomists in England was a cadre of amateur scientists known as the Northumberland circle, led by Henry Percy (1585–1632), the 9th Earl of Northumberland. Although they published little of account, they helped to disseminate atomistic ideas among the burgeoning scientific culture of England, and may have been particularly influential to Francis Bacon, who became an atomist around 1605, though he later rejected some of the claims of atomism. Though they revived the classical form of atomism, this group was among the scientific avant-garde: the Northumberland circle contained nearly half of the confirmed Copernicans prior to 1610 (the year of Galileo's The Starry Messenger). Other influential atomists of late 16th and early 17th centuries include Giordano Bruno, Thomas Hobbes (who also changed his stance on atomism late in his career), and Thomas Hariot. A number of different atomistic theories were blossoming in France at this time, as well (Clericuzio 2000).

Galileo Galilei (1564–1642) was an advocate of atomism in his 1612, *Discourse on Floating Bodies* (Redondi 1969). In The Assayer, Galileo offered a more complete physical system based on a corpuscular theory of matter, in which all phenomena—with the exception of sound—are produced by "matter in motion". Galileo identified some basic problems with Aristotelian physics through his experiments. He utilized a theory of atomism as a partial replacement, but he was never unequivocally committed to it. For example, his experiments with falling bodies and inclined planes led him to the concepts of circular inertial motion and accelerating free-fall. The current Aristotelian theories of impetus and terrestrial motion were inadequate to explain these. While atomism did not explain the law of fall either, it was a more promising framework in which to develop an explanation because motion was conserved in ancient atomism (unlike Aristotelian physics).

René Descartes' (1596–1650) "mechanical" philosophy of corpuscularism had much in common with atomism, and is considered, in some senses, to be a different version of it. Descartes thought everything physical in the universe to be made of tiny *vortices* of matter. Like the ancient atomists, Descartes claimed that sensations, such as taste or temperature, are caused by the shape and size of tiny pieces of matter. The main difference between atomism and Descartes' concept was the existence of the void. For him, there could be no vacuum, and all matter was constantly swirling to prevent a void as corpuscles moved through other matter. Another key distinction between Descartes' view and classical atomism is the mind/body duality of Descartes, which allowed for an independent realm of existence for thought, soul, and most importantly, God. Gassendi's concept was closer to classical atomism, but with no atheistic overtone.

Pierre Gassendi (1592–1655) was a Catholic priest from France who was also an avid natural philosopher. He was particularly intrigued by the Greek atomists, so he set out to "purify" atomism from its heretical and atheistic philosophical conclusions (Dijksterhius 1969). Gassendi formulated his atomistic conception of mechanical philosophy partly in response to Descartes; he particularly opposed Descartes' reductionist view that only purely mechanical explanations of physics are valid, as well as the application of geometry to the whole of physics (Clericuzio 2000).

15.6.1 Corpuscularianism

Main article: Corpuscularianism

Corpuscularianism is similar to atomism, except that where atoms were supposed to be indivisible, corpuscles could in principle be divided. In this manner, for example, it was theorized that mercury could penetrate into metals and modify their inner structure, a step on the way towards transmutative production of gold. Corpuscularianism was associated by its leading proponents with the idea that some of the properties that objects appear to have are artifacts of the perceiving mind: 'secondary' qualities as distinguished from 'primary' qualities.[29] Not all corpuscularianism made use of the primary-secondary quality distinction, however. An influential tradition in medieval and early modern alchemy argued that chemical analysis revealed the existence of robust corpuscles that retained their identity in chemical compounds (to use the modern term). William R. Newman has dubbed this approach to matter theory "chymical atomism," and has argued for its significance to both the mechanical philosophy and to the chemical atomism that emerged in the early 19th century.[30] Corpuscularianism stayed a dominant theory over the next several hundred years and retained its links with alchemy in the work of scientists such as Robert Boyle and Isaac Newton in the 17th century.[31][32] It was used by Newton, for instance, in his development of the corpuscular theory of light. The form that came to be accepted by most English scientists after Robert Boyle (1627–1692) was an amalgam of the systems of Descartes and Gassendi. In The Sceptical Chymist (1661), Boyle demonstrates problems that arise from chemistry, and offers up atomism as a possible explanation. The unifying principle that would eventually lead to the acceptance of a hybrid corpuscular–atomism was mechanical philosophy, which became widely accepted by physical sciences.

15.7 Atomic theory

Main article: Atomic theory

By the late 18th century, the useful practices of engineering and technology began to influence philosophical explanations for the composition of matter. Those who speculated on the ultimate nature of matter began to verify their "thought experiments" with some repeatable demonstrations, when they could.

Roger Boscovich provided the first general mathematical theory of atomism, based on the ideas of Newton and Leibniz but transforming them so as to provide a programme for atomic physics.[33]

In 1808, John Dalton assimilated the known experimental work of many people to summarize the empirical evidence on the composition of matter. He noticed that distilled water everywhere analyzed to the same elements, hydrogen and oxygen. Similarly, other purified substances decomposed to the same elements in the same proportions by weight.

> Therefore we may conclude that the ultimate particles of all homogeneous bodies are perfectly alike in weight, figure, etc. In other words, every particle of water is like every other particle of water; every particle of hydrogen is like every other particle of hydrogen, etc.

Furthermore, he concluded that there was a unique atom for each element, using Lavoisier's definition of an element as a substance that could not be analyzed into something simpler. Thus, Dalton concluded the following.

> Chemical analysis and synthesis go no farther than to the separation of particles one from another, and to their reunion. No new creation or destruction of matter is within the reach of chemical agency. We might as well attempt to introduce a new planet into the solar system, or to annihilate one already in existence, as to create or destroy a particle of hydrogen. All the changes we can produce, consist in separating particles that are in a state of cohesion or combination, and joining those that were previously at a distance.

And then he proceeded to give a list of relative weights in the compositions of several common compounds, summarizing:

> 1st. That water is a binary compound of hydrogen and oxygen, and the relative weights of the two elementary atoms are as 1:7, nearly;

> 2nd. That ammonia is a binary compound of hydrogen and azote nitrogen, and the relative weights of the two atoms are as 1:5, nearly...

Dalton concluded that the fixed proportions of elements by weight suggested that the atoms of one element combined with only a limited number of atoms of the other elements to form the substances that he listed.

15.7.1 Atomic theory controversy

Dalton's atomic theory remained controversial throughout the 19th century.[34] Whilst the Law of definite proportion were accepted, the hypothesis that this was due to atoms was not so widely accepted. For example in 1826 when Sir Humphry Davy presented Dalton the Royal Medal from the Royal Society, Davy said that the theory only became useful when the atomic conjecture was ignored.[35] Sir Benjamin Collins Brodie in 1866 published the first part of his Calculus of Chemical Operations[36] as a non atomic alternative to the Atomic Theory. He described atomic theory as a 'Thoroughly materialistic bit of joiners work'.[37] Alexander Williamson used his Presidential Address to the London Chemical Society in 1869[38] to defend the Atomic Theory against its critics and doubters. This in turn led to further meetings at which the positivists again attacked the supposition that there were atoms. The matter was finally resolved in Dalton's favour in the early 20th century with the rise of atomic physics.

15.8 See also

- Becoming (philosophy)
- History of chemistry
- Infinite divisibility
- Ontological pluralism
- Physical ontology

15.9 Notes

[1] ἄτομον. Liddell, Henry George; Scott, Robert; *A Greek–English Lexicon* at the Perseus Project

[2] "atom". Online Etymology Dictionary.

[3] The term 'atomism' is recorded in English since 1670–80 (*Random House Webster's Unabridged Dictionary*, 2001, "atomism").

[4] Aristotle, *Metaphysics* I, 4, 985^b 10–15.

[5] Berryman, Sylvia, "Ancient Atomism", *The Stanford Encyclopedia of Philosophy* (Fall 2008 Edition), Edward N. Zalta (ed.), http://plato.stanford.edu/archives/fall2008/entries/atomism-ancient/

[6] Gangopadhyaya, Mrinalkanti (1981). *Indian Atomism: History and Sources*. Atlantic Highlands, New Jersey: Humanities Press. ISBN 0-391-02177-X. OCLC 10916778.

[7] Iannone, A. Pablo (2001). *Dictionary of World Philosophy*. Routledge. pp. 83,356. ISBN 0-415-17995-5. OCLC 44541769.

[8] Thomas McEvilley, The Shape of Ancient Thought: Comparative Studies in Greek and Indian Philosophies ISBN 1-58115-203-5, Allwarth Press, 2002, p. 317-321.

[9] Richard King, Indian philosophy: an introduction to Hindu and Buddhist thought, , Edinburgh University Press, 1999, ISBN 0-7486-0954-7, pp. 105-107.

[10] The atomists, Leucippus and Democritus: fragments, a text and translation with a commentary by C.C.W. Taylor, University of Toronto Press Incorporated 1999, ISBN 0-8020-4390-9, pp. 157-158.

[11] Teresi, Dick (2003). *Lost Discoveries: The Ancient Roots of Modern Science*. Simon & Schuster. pp. 213–214. ISBN 0-7432-4379-X.

[12] Andrew G. van Melsen (1952). *From Atomos to Atom*. Mineola, N.Y.: Dover Publications. ISBN 0486495841.

[13] Bertrand Russel (1946). *History of Western Philosophy*. London: Routledge. p. 75. ISBN 0415325056.

[14] Andrew G. van Melsen. (1952). *From Atomos to Atom: The History and Concept of the Atom*. Dover Phoenix Editions. ISBN 0-486-49584-1

[15] Berryman, Sylvia, "Democritus", *The Stanford Encyclopedia of Philosophy* (Fall 2008 Edition), Edward N. Zalta (ed.), http://plato.stanford.edu/archives/fall2008/entries/democritus

[16] Lloyd, Geoffrey (1970). *Early Greek Science: Thales to Aristotle*. London; New York: Chatto and Windus; W. W. Norton & Company. pp. 74–77. ISBN 0-393-00583-6.

[17] Cornford, Francis Macdonald (1957). *Plato's Cosmology: The* Timaeus *of Plato*. New York: Liberal Arts Press. pp. 210–239. ISBN 0-87220-386-7.

[18] Lloyd, Geoffrey (1968). *Aristotle: The Growth and Structure of his Thought*. Cambridge: Cambridge University Press. p. 165. ISBN 0-521-09456-9.

[19] Lloyd, Geoffrey (1970). *Early Greek Science: Thales to Aristotle*. London; New York: Chatto and Windus; W. W. Norton & Company. pp. 108–109. ISBN 0-393-00583-6.

[20] Taylor, C. C. W. (1999). *The Atomists, Leucippus and Democritus: a text and translation with commentary by C. C. W. Taylor*. Toronto; Buffalo: University of Toronto Press. ISBN 0-8020-4390-9.

[21] Bimal Krishna Matilal (1990). *The word and the world: India's contribution to the study of language*. Oxford. Yaska is dealt with in Chapter 3. ISBN 0-19-562515-3.

[22] McEvilley (2002), 317–320

[23] Teresi, Dick (2003). *Lost Discoveries: The Ancient Roots of Modern Science*. Simon & Schuster. pp. 213–214. ISBN 0-7432-4379-X.

[24] Saeed, Abdullah (2006). *Islamic Thought: An Introduction*. Routledge. p. 95. ISBN 978-0415364096.

[25] Michael Marmura (1976). "God and his creation:Two medieval Islamic views". In R. M. Savory. *Introduction to Islamic Civilization*. Cambridge University Press. p. 49.

[26] Shlomo Pines (1986). *Studies in Arabic versions of Greek texts and in mediaeval science* **2**. Brill Publishers. pp. 355–6. ISBN 965-223-626-8.

[27] John Emery Murdoch; Christoph Herbert Lüthy; William Royall Newman (1 January 2001). "The Medieval and Renaissance Tradition of Minima Naturalia". *Late Medieval and Early Modern Corpuscular Matter Theories*. BRILL. pp. 91–133. ISBN 90-04-11516-1.

[28] Alan Chalmers (4 June 2009). *The Scientist's Atom and the Philosopher's Stone: How Science Succeeded and Philosophy Failed to Gain Knowledge of Atoms.* Springer. pp. 75–96. ISBN 978-90-481-2362-9.

[29] The Mechanical Philosophy - Early modern 'atomism' ("corpuscularianism" as it was known)

[30] William R. Newman, "The Significance of 'Chymical Atomism'," in Edith Sylla and W. R. Newman, eds., *Evidence and Interpretation: Studies on Early Science and Medicine in Honor of John E. Murdoch* (Leiden: Brill, 2009), pp. 248-264 and Newman, *Atoms and Alchemy: Chymistry and the Experimental Origins of the Scientific Revolution* (Chicago: University of Chicago Press, 2006)

[31] Levere, Trevor, H. (2001). *Transforming Matter – A History of Chemistry for Alchemy to the Buckyball.* The Johns Hopkins University Press. ISBN 0-8018-6610-3.

[32] Corpuscularianism - Philosophical Dictionary

[33] Lancelot Law Whyte Essay on Atomism, 1961, p 54.

[34] Brock(ed), W.H. (1967). *The Atomic Debates.* Leicester University Press. p. 1.

[35] Davy(ed), J. *Collected Works of Sir Humphrey Davy.* Bart. p. 93 vol 8.

[36] Brodie, Sir Benjamin Collins (1866). *Philosophical Transactions of the Royal Society.* pp. 781–859 vol I56.

[37] Brock(ed), W.H. (1967). *The Atomic Debates.* Leicester University Press. p. 12.

[38] Brock(ed), W.H. (1967). *The Atomic Debates.* Leicester University Press. p. 15.

15.10   References

- Clericuzio, Antonio. *Elements, Principles, and Corpuscles; a study of atomism and chemistry in the seventeenth century.* Dordrecht; Boston: Kluwer Academic Publishers, 2000.

- Cornford, Francis MacDonald. *Plato's Cosmology: The* Timaeus *of Plato.* New York: Liberal Arts Press, 1957.

- Dijksterhuis, E. *The Mechanization of the World Picture.* Trans. by C. Dikshoorn. New York: Oxford University Press, 1969. ISBN 0-691-02396-4

- Firth, Raymond. *Religion: A Humanist Interpretation.* Routledge, 1996. ISBN 0-415-12897-8.

- Gangopadhyaya, Mrinalkanti. *Indian Atomism: history and sources.* Atlantic Highlands, New Jersey: Humanities Press, 1981. ISBN 0-391-02177-X

- Gardet, L. "djuz'" in *Encyclopaedia of Islam CD-ROM Edition, v. 1.1.* Leiden: Brill, 2001.

- Gregory, Joshua C. *A Short History of Atomism.* London: A. and C. Black, Ltd, 1981.

- Kargon, Robert Hugh. *Atomism in England from Hariot to Newton.* Oxford: Clarendon Press, 1966.

- Lloyd, G. E. R. *Aristotle: The Growth and Structure of his Thought.* Cambridge: Cambridge University Press, 1968. ISBN 0-521-09456-9

- Lloyd, G. E. R. *Greek Science After Aristotle.* New York: W. W. Norton, 1973. ISBN 0-393-00780-4

- Marmara, Michael E. "Causation in Islamic Thought." *Dictionary of the History of Ideas.* New York: Charles Scribner's Sons, 1973-74. online at the of Virginia Electronic Text Center.

- McEvilley, Thomas (2002). *The Shape of Ancient Thought: Comparative Studies in Greek and Indian Philosophies.* New York: Allworth Communications Inc. ISBN 1-58115-203-5.

- Redondi, Pietro. *Galileo Heretic.* Translated by Raymond Rosenthal. Princeton, NJ: Princeton University Press, 1987. ISBN 0-691-02426-X

15.11 External links

- The dictionary definition of atomism at Wiktionary

- *Dictionary of the History of Ideas*: Atomism: Antiquity to the Seventeenth Century

- *Dictionary of the History of Ideas*: Atomism in the Seventeenth Century

- Jonathan Schaffer, "Is There a Fundamental Level?" *Nous* 37 (2003): 498–517. Article by a philosopher who opposes atomism

- Article on traditional Greek atomism

- Atomism from the 17th to the 20th Century at Stanford Encyclopedia of Philosophy

Chapter 16

Bohr model

'Rutherford–Bohr model' and 'Bohr–Rutherford diagram' redirect to this page. 'Bohr model' is not to be confused with Bohr equation.

In atomic physics, the Rutherford–Bohr model or **Bohr model**, introduced by Niels Bohr in 1913, depicts the atom as a small, positively charged nucleus surrounded by electrons that travel in circular orbits around the nucleus—similar in structure to the solar system, but with attraction provided by electrostatic forces rather than gravity. After the cubic model (1902), the plum-pudding model (1904), the Saturnian model (1904), and the Rutherford model (1911) came the **Rutherford–Bohr model** or just *Bohr model* for short (1913). The improvement to the Rutherford model is mostly a quantum physical interpretation of it. The Bohr model has been superseded, but the quantum theory remains sound.

The model's key success lay in explaining the Rydberg formula for the spectral emission lines of atomic hydrogen. While the Rydberg formula had been known experimentally, it did not gain a theoretical underpinning until the Bohr model was introduced. Not only did the Bohr model explain the reason for the structure of the Rydberg formula, it also provided a justification for its empirical results in terms of fundamental physical constants.

The Bohr model is a relatively primitive model of the hydrogen atom, compared to the valence shell atom. As a theory, it can be derived as a first-order approximation of the hydrogen atom using the broader and much more accurate quantum mechanics and thus may be considered to be an obsolete scientific theory. However, because of its simplicity, and its correct results for selected systems (see below for application), the Bohr model is still commonly taught to introduce students to quantum mechanics or energy level diagrams before moving on to the more accurate, but more complex, valence shell atom. A related model was originally proposed by Arthur Erich Haas in 1910, but was rejected. The quantum theory of the period between Planck's discovery of the quantum (1900) and the advent of a full-blown quantum mechanics (1925) is often referred to as the old quantum theory.

16.1 Origin

In the early 20th century, experiments by Ernest Rutherford established that atoms consisted of a diffuse cloud of negatively charged electrons surrounding a small, dense, positively charged nucleus.[2] Given this experimental data, Rutherford naturally considered a planetary-model atom, the Rutherford model of 1911 – electrons orbiting a solar nucleus – however, said planetary-model atom has a technical difficulty. The laws of classical mechanics (i.e. the Larmor formula), predict that the electron will release electromagnetic radiation while orbiting a nucleus. Because the electron would lose energy, it would rapidly spiral inwards, collapsing into the nucleus on a timescale of around 16 picoseconds.[3] This atom model is disastrous, because it predicts that all atoms are unstable.[4]

Also, as the electron spirals inward, the emission would rapidly increase in frequency as the orbit got smaller and faster. This would produce a continuous smear, in frequency, of electromagnetic radiation. However, late 19th century experiments with electric discharges have shown that atoms will only emit light (that is, electromagnetic radiation) at certain discrete frequencies.

To overcome this difficulty, Niels Bohr proposed, in 1913, what is now called the *Bohr model of the atom*. He suggested

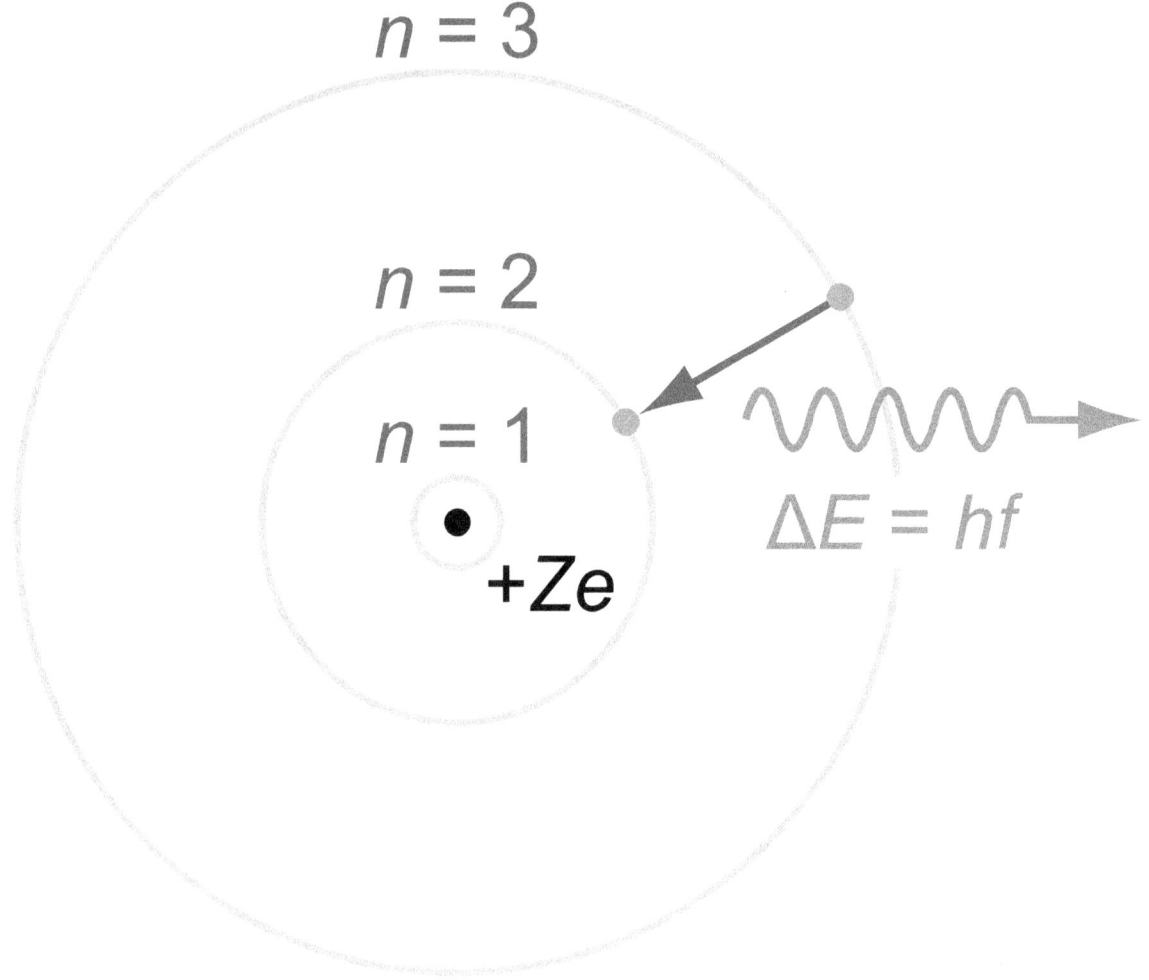

*The **Rutherford–Bohr model** of the hydrogen atom (Z = 1) or a hydrogen-like ion (Z > 1), where the negatively charged electron confined to an atomic shell encircles a small, positively charged atomic nucleus and where an electron jump between orbits is accompanied by an emitted or absorbed amount of electromagnetic energy (hv).[1] The orbits in which the electron may travel are shown as grey circles; their radius increases as n^2, where n is the principal quantum number. The 3 → 2 transition depicted here produces the first line of the Balmer series, and for hydrogen (Z = 1) it results in a photon of wavelength 656 nm (red light).*

that electrons could only have certain *classical* motions:

1. Electrons in atoms orbit the nucleus.

2. The electrons can only orbit stably, without radiating, in certain orbits (called by Bohr the "stationary orbits"[5]) at a certain discrete set of distances from the nucleus. These orbits are associated with definite energies and are also called energy shells or energy levels. In these orbits, the electron's acceleration does not result in radiation and energy loss as required by classical electromagnetics. The Bohr model of an atom was based upon Planck's quantum theory of radiation.

3. Electrons can only gain and lose energy by jumping from one allowed orbit to another, absorbing or emitting electromagnetic radiation with a frequency ν determined by the energy difference of the levels according to the Planck relation:

$$\Delta E = E_2 - E_1 = h\nu \,,$$

where h is Planck's constant. The frequency of the radiation emitted at an orbit of period T is as it would be in classical mechanics; it is the reciprocal of the classical orbit period:

$$\nu = \tfrac{1}{T}.$$

The significance of the Bohr model is that the laws of classical mechanics apply to the motion of the electron about the nucleus *only when restricted by a quantum rule*. Although Rule 3 is not completely well defined for small orbits, because the emission process involves two orbits with two different periods, Bohr could determine the energy spacing between levels using Rule 3 and come to an exactly correct quantum rule: the angular momentum L is restricted to be an integer multiple of a fixed unit:

$$L = n\frac{h}{2\pi} = n\hbar$$

where $n = 1, 2, 3, ...$ is called the principal quantum number, and $\hbar = h/2\pi$. The lowest value of n is 1; this gives a smallest possible orbital radius of 0.0529 nm known as the Bohr radius. Once an electron is in this lowest orbit, it can get no closer to the proton. Starting from the angular momentum quantum rule, Bohr[2] was able to calculate the energies of the allowed orbits of the hydrogen atom and other hydrogen-like atoms and ions.

Other points are:

1. Like Einstein's theory of the Photoelectric effect, Bohr's formula assumes that during a quantum jump a *discrete* amount of energy is radiated. However, unlike Einstein, Bohr stuck to the *classical* Maxwell theory of the electromagnetic field. Quantization of the electromagnetic field was explained by the discreteness of the atomic energy levels; Bohr did not believe in the existence of photons.

2. According to the Maxwell theory the frequency ν of classical radiation is equal to the rotation frequency ν_{rot} of the electron in its orbit, with harmonics at integer multiples of this frequency. This result is obtained from the Bohr model for jumps between energy levels En and En_k when k is much smaller than n. These jumps reproduce the frequency of the k-th harmonic of orbit n. For sufficiently large values of n (so-called Rydberg states), the two orbits involved in the emission process have nearly the same rotation frequency, so that the classical orbital frequency is not ambiguous. But for small n (or large k), the radiation frequency has no unambiguous classical interpretation. This marks the birth of the correspondence principle, requiring quantum theory to agree with the classical theory only in the limit of large quantum numbers.

3. The Bohr-Kramers-Slater theory (BKS theory) is a failed attempt to extend the Bohr model, which violates the conservation of energy and momentum in quantum jumps, with the conservation laws only holding on average.

Bohr's condition, that the angular momentum is an integer multiple of $\hbar$ was later reinterpreted in 1924 by de Broglie as a standing wave condition: the electron is described by a wave and a whole number of wavelengths must fit along the circumference of the electron's orbit:

$$n\lambda = 2\pi r.$$

Substituting de Broglie's wavelength of $\lambda = h/p$ reproduces Bohr's rule. In 1913, however, Bohr justified his rule by appealing to the correspondence principle, without providing any sort of wave interpretation. In 1913, the wave behavior of matter particles such as the electron (i.e., matter waves) was not suspected.

In 1925 a new kind of mechanics was proposed, quantum mechanics, in which Bohr's model of electrons traveling in quantized orbits was extended into a more accurate model of electron motion. The new theory was proposed by Werner Heisenberg. Another form of the same theory, wave mechanics, was discovered by the Austrian physicist Erwin Schrödinger independently, and by different reasoning. Schrödinger employed de Broglie's matter waves, but sought wave solutions of a three-dimensional wave equation describing electrons that were constrained to move about the nucleus of a hydrogen-like atom, by being trapped by the potential of the positive nuclear charge.

16.2 Electron energy levels

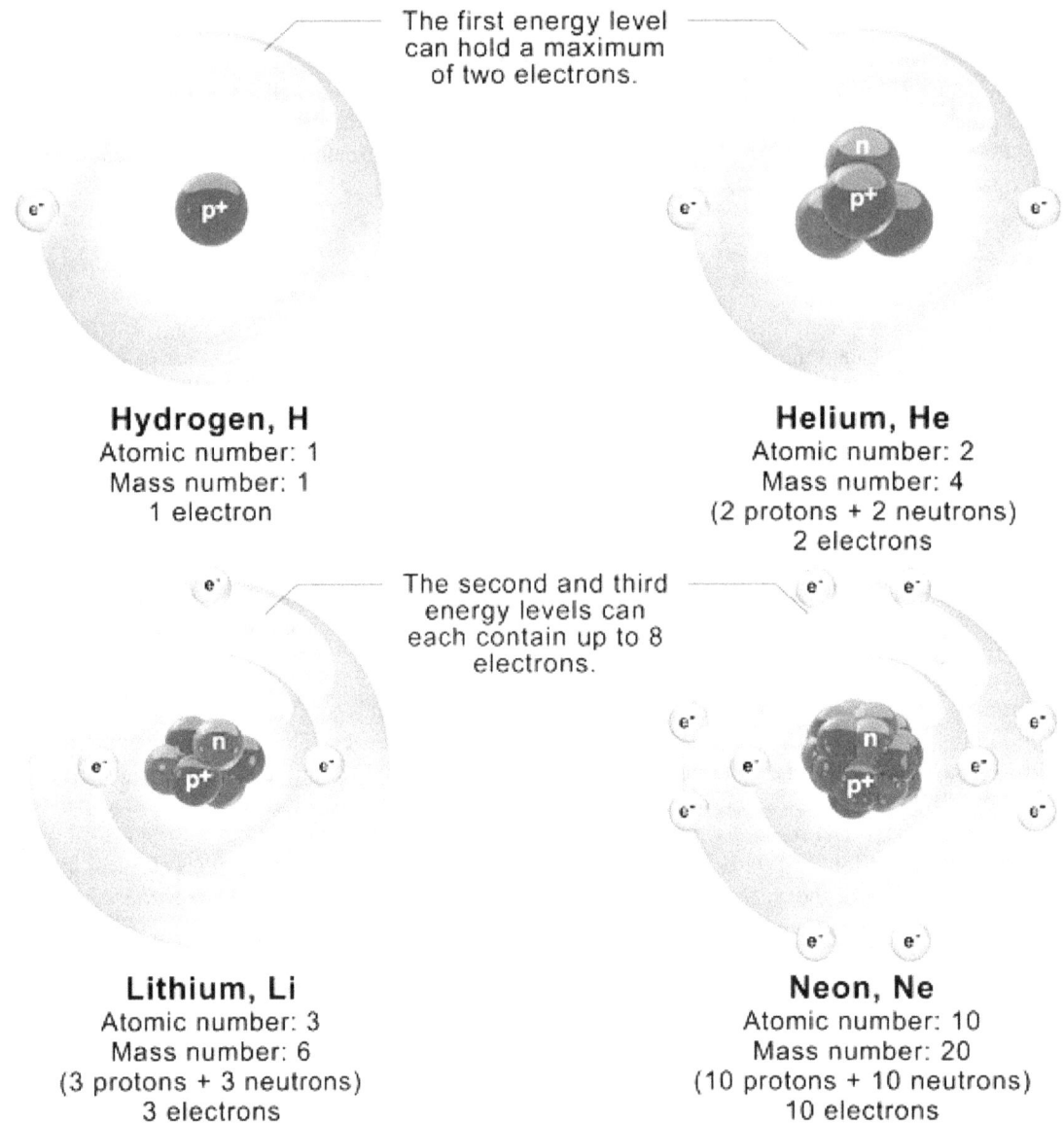

Models depicting electron energy levels in hydrogen, helium, lithium, and neon

The Bohr model gives almost exact results only for a system where two charged points orbit each other at speeds much less than that of light. This not only includes one-electron systems such as the hydrogen atom, singly ionized helium, doubly ionized lithium, but it includes positronium and Rydberg states of any atom where one electron is far away from everything else. It can be used for K-line X-ray transition calculations if other assumptions are added (see Moseley's law below). In high energy physics, it can be used to calculate the masses of heavy quark mesons.

Calculation of the orbits requires two assumptions.

- **Classical mechanics**

  The electron is held in a circular orbit by electrostatic attraction. The centripetal force is equal to the Coulomb force.

$$\frac{m_e v^2}{r} = \frac{Zk_e e^2}{r^2}$$

where m_e is the electron's mass, e is the charge of the electron, k_e is Coulomb's constant and Z is the atom's atomic number. It is assumed here that the mass of the nucleus is much larger than the electron mass (which is a good assumption). This equation determines the electron's speed at any radius:

$$v = \sqrt{\frac{Zk_e e^2}{m_e r}}.$$

It also determines the electron's total energy at any radius:

$$E = \frac{1}{2} m_e v^2 - \frac{Zk_e e^2}{r} = -\frac{Zk_e e^2}{2r}.$$

The total energy is negative and inversely proportional to r. This means that it takes energy to pull the orbiting electron away from the proton. For infinite values of r, the energy is zero, corresponding to a motionless electron infinitely far from the proton. The total energy is half the potential energy, which is also true for noncircular orbits by the virial theorem.

- **A quantum rule**

The angular momentum $L = m_e vr$ is an integer multiple of $\hbar$:

$$m_e vr = n\hbar$$

Substituting the expression for the velocity gives an equation for r in terms of n:

$$m_e \sqrt{\frac{k_e Z e^2}{m_e r}} r = n\hbar$$

so that the allowed orbit radius at any n is:

$$r_n = \frac{n^2 \hbar^2}{Zk_e e^2 m_e}$$

The smallest possible value of r in the hydrogen atom (Z=1) is called the Bohr radius and is equal to:

$$r_1 = \frac{\hbar^2}{k_e e^2 m_e} \approx 5.29 \times 10^{-11} \text{m}$$

The energy of the n-th level for any atom is determined by the radius and quantum number:

$$E = -\frac{Zk_e e^2}{2r_n} = -\frac{Z^2 (k_e e^2)^2 m_e}{2\hbar^2 n^2} \approx \frac{-13.6Z^2}{n^2} \text{eV}$$

An electron in the lowest energy level of hydrogen ($n = 1$) therefore has about 13.6 eV less energy than a motionless electron infinitely far from the nucleus. The next energy level ($n = 2$) is −3.4 eV. The third ($n = 3$) is −1.51 eV, and so on. For larger values of n, these are also the binding energies of a highly excited atom with one electron in a large circular orbit around the rest of the atom.

The combination of natural constants in the energy formula is called the Rydberg energy (RE):

$$R_E = \frac{(k_e e^2)^2 m_e}{2\hbar^2}$$

This expression is clarified by interpreting it in combinations that form more natural units:

$m_e c^2$ is the rest mass energy of the electron (511 keV)

$\frac{k_e e^2}{\hbar c} = \alpha \approx \frac{1}{137}$ is the fine structure constant

$R_E = \frac{1}{2}(m_e c^2)\alpha^2$

Since this derivation is with the assumption that the nucleus is orbited by one electron, we can generalize this result by letting the nucleus have a charge $q = Ze$ where Z is the atomic number. This will now give us energy levels for hydrogenic atoms, which can serve as a rough order-of-magnitude approximation of the actual energy levels. So for nuclei with Z protons, the energy levels are (to a rough approximation):

$$E_n = -\frac{Z^2 R_E}{n^2}$$

The actual energy levels cannot be solved analytically for more than one electron (see *n*-body problem) because the electrons are not only affected by the nucleus but also interact with each other via the Coulomb Force.

When $Z = 1/\alpha$ ($Z \approx 137$), the motion becomes highly relativistic, and Z^2 cancels the α^2 in R; the orbit energy begins to be comparable to rest energy. Sufficiently large nuclei, if they were stable, would reduce their charge by creating a bound electron from the vacuum, ejecting the positron to infinity. This is the theoretical phenomenon of electromagnetic charge screening which predicts a maximum nuclear charge. Emission of such positrons has been observed in the collisions of heavy ions to create temporary super-heavy nuclei.

The Bohr formula properly uses the reduced mass of electron and proton in all situations, instead of the mass of the electron: $m_{red} = \frac{m_e m_p}{m_e + m_p} = m_e \frac{1}{1 + m_e/m_p}$. However, these numbers are very nearly the same, due to the much larger mass of the proton, about 1836.1 times the mass of the electron, so that the reduced mass in the system is the mass of the electron multiplied by the constant 1836.1/(1+1836.1) = 0.99946. This fact was historically important in convincing Rutherford of the importance of Bohr's model, for it explained the fact that the frequencies of lines in the spectra for singly ionized helium do not differ from those of hydrogen by a factor of exactly 4, but rather by 4 times the ratio of the reduced mass for the hydrogen vs. the helium systems, which was much closer to the experimental ratio than exactly 4.

For positronium, the formula uses the reduced mass also, but in this case, it is exactly the electron mass divided by 2. For any value of the radius, the electron and the positron are each moving at half the speed around their common center of mass, and each has only one fourth the kinetic energy. The total kinetic energy is half what it would be for a single electron moving around a heavy nucleus.

$$E_n = \frac{R_E}{2n^2}$$

16.3 Rydberg formula

The Rydberg formula, which was known empirically before Bohr's formula, is seen in Bohr's theory as describing the energies of transitions or quantum jumps between one orbital energy levels. Bohr's formula gives the numerical value of the already-known and measured Rydberg's constant, but in terms of more fundamental constants of nature, including the electron's charge and Planck's constant.

When the electron gets moved from its original energy level to a higher one, it then jumps back each level till it comes to the original position, which results in a photon being emitted. Using the derived formula for the different energy levels of hydrogen one may determine the wavelengths of light that a hydrogen atom can emit.

The energy of a photon emitted by a hydrogen atom is given by the difference of two hydrogen energy levels:

$$E = E_i - E_f = R_E \left(\frac{1}{n_f^2} - \frac{1}{n_i^2} \right)$$

where nf is the final energy level, and ni is the initial energy level.

Since the energy of a photon is

$$E = \frac{hc}{\lambda},$$

the wavelength of the photon given off is given by

$$\frac{1}{\lambda} = R \left(\frac{1}{n_f^2} - \frac{1}{n_i^2} \right).$$

This is known as the Rydberg formula, and the Rydberg constant R is R_E/hc, or $R_E/2\pi$ in natural units. This formula was known in the nineteenth century to scientists studying spectroscopy, but there was no theoretical explanation for this form or a theoretical prediction for the value of R, until Bohr. In fact, Bohr's derivation of the Rydberg constant, as well as the concomitant agreement of Bohr's formula with experimentally observed spectral lines of the Lyman ($n_f = 1$), Balmer ($n_f = 2$), and Paschen ($n_f = 3$) series, and successful theoretical prediction of other lines not yet observed, was one reason that his model was immediately accepted.

To apply to atoms with more than one electron, the Rydberg formula can be modified by replacing "Z" with "Z – b" or "n" with "n – b" where b is constant representing a screening effect due to the inner-shell and other electrons (see Electron shell and the later discussion of the "Shell Model of the Atom" below). This was established empirically before Bohr presented his model.

16.4   Shell model of heavier atoms

Bohr extended the model of hydrogen to give an approximate model for heavier atoms. This gave a physical picture that reproduced many known atomic properties for the first time.

Heavier atoms have more protons in the nucleus, and more electrons to cancel the charge. Bohr's idea was that each discrete orbit could only hold a certain number of electrons. After that orbit is full, the next level would have to be used. This gives the atom a shell structure, in which each shell corresponds to a Bohr orbit.

This model is even more approximate than the model of hydrogen, because it treats the electrons in each shell as non-interacting. But the repulsions of electrons are taken into account somewhat by the phenomenon of screening. The electrons in outer orbits do not only orbit the nucleus, but they also move around the inner electrons, so the effective charge Z that they feel is reduced by the number of the electrons in the inner orbit.

For example, the lithium atom has two electrons in the lowest 1s orbit, and these orbit at Z=2. Each one sees the nuclear charge of Z=3 minus the screening effect of the other, which crudely reduces the nuclear charge by 1 unit. This means that the innermost electrons orbit at approximately 1/4 the Bohr radius. The outermost electron in lithium orbits at roughly Z=1, since the two inner electrons reduce the nuclear charge by 2. This outer electron should be at nearly one Bohr radius from the nucleus. Because the electrons strongly repel each other, the effective charge description is very approximate; the effective charge Z doesn't usually come out to be an integer. But Moseley's law experimentally probes the innermost pair of electrons, and shows that they do see a nuclear charge of approximately Z–1, while the outermost electron in an atom or ion with only one electron in the outermost shell orbits a core with effective charge Z–k where k is the total number of electrons in the inner shells.

The shell model was able to qualitatively explain many of the mysterious properties of atoms which became codified in the late 19th century in the periodic table of the elements. One property was the size of atoms, which could be determined approximately by measuring the viscosity of gases and density of pure crystalline solids. Atoms tend to get smaller toward the right in the periodic table, and become much larger at the next line of the table. Atoms to the right of the table tend to gain electrons, while atoms to the left tend to lose them. Every element on the last column of the table is chemically inert (noble gas).

In the shell model, this phenomenon is explained by shell-filling. Successive atoms become smaller because they are filling orbits of the same size, until the orbit is full, at which point the next atom in the table has a loosely bound outer electron, causing it to expand. The first Bohr orbit is filled when it has two electrons, which explains why helium is inert. The second orbit allows eight electrons, and when it is full the atom is neon, again inert. The third orbital contains eight again, except that in the more correct Sommerfeld treatment (reproduced in modern quantum mechanics) there are extra "d" electrons. The third orbit may hold an extra 10 d electrons, but these positions are not filled until a few more orbitals from the next level are filled (filling the n=3 d orbitals produces the 10 transition elements). The irregular filling pattern is an effect of interactions between electrons, which are not taken into account in either the Bohr or Sommerfeld models and which are difficult to calculate even in the modern treatment.

16.5   Moseley's law and calculation of K-alpha X-ray emission lines

Niels Bohr said in 1962, "You see actually the Rutherford work [the nuclear atom] was not taken seriously. We cannot understand today, but it was not taken seriously at all. There was no mention of it any place. The great change came from Moseley."

In 1913 Henry Moseley found an empirical relationship between the strongest X-ray line emitted by atoms under electron bombardment (then known as the K-alpha line), and their atomic number Z. Moseley's empiric formula was found to be derivable from Rydberg and Bohr's formula (Moseley actually mentions only Ernest Rutherford and Antonius Van den Broek in terms of models). The two additional assumptions that [1] this X-ray line came from a transition between energy levels with quantum numbers 1 and 2, and [2], that the atomic number Z when used in the formula for atoms heavier than hydrogen, should be diminished by 1, to $(Z-1)^2$.

Moseley wrote to Bohr, puzzled about his results, but Bohr was not able to help. At that time, he thought that the postulated innermost "K" shell of electrons should have at least four electrons, not the two which would have neatly explained the result. So Moseley published his results without a theoretical explanation.

Later, people realized that the effect was caused by charge screening, with an inner shell containing only 2 electrons. In the experiment, one of the innermost electrons in the atom is knocked out, leaving a vacancy in the lowest Bohr orbit, which contains a single remaining electron. This vacancy is then filled by an electron from the next orbit, which has n=2. But the n=2 electrons see an effective charge of Z−1, which is the value appropriate for the charge of the nucleus, when a single electron remains in the lowest Bohr orbit to screen the nuclear charge +Z, and lower it by −1 (due to the electron's negative charge screening the nuclear positive charge). The energy gained by an electron dropping from the second shell to the first gives Moseley's law for K-alpha lines:

$$E = h\nu = E_i - E_f = R_E(Z-1)^2 \left(\frac{1}{1^2} - \frac{1}{2^2} \right)$$

or

$$f = \nu = R_v \left(\frac{3}{4} \right) (Z-1)^2 = (2.46 \times 10^{15}\,\text{Hz})(Z-1)^2.$$

Here, $\boldsymbol{R_v} = \boldsymbol{RE/h}$ is the Rydberg constant, in terms of frequency equal to 3.28 x 10^{15} Hz. For values of Z between 11 and 31 this latter relationship had been empirically derived by Moseley, in a simple (linear) plot of the square root of X-ray frequency against atomic number (however, for silver, Z = 47, the experimentally obtained screening term should be replaced by 0.4). Notwithstanding its restricted validity,[6] Moseley's law not only established the objective meaning of atomic number (see Henry Moseley for detail) but, as Bohr noted, it also did more than the Rydberg derivation to establish the validity of the Rutherford/Van den Broek/Bohr nuclear model of the atom, with atomic number (place on the periodic table) standing for whole units of nuclear charge.

The K-alpha line of Moseley's time is now known to be a pair of close lines, written as ($\mathbf{K\alpha_1}$ and $\mathbf{K\alpha_2}$) in Siegbahn notation.

16.6 Shortcomings

The Bohr model gives an incorrect value $\mathrm{L}{=}\hbar$ for the ground state orbital angular momentum. The angular momentum in the true ground state is known to be zero from experiment.[7] Although mental pictures fail somewhat at these levels of scale, an electron in the lowest modern "orbital" with no orbital momentum, may be thought of as not to rotate "around" the nucleus at all, but merely to go tightly around it in an ellipse with zero area (this may be pictured as "back and forth", without striking or interacting with the nucleus). This is only reproduced in a more sophisticated semiclassical treatment like Sommerfeld's. Still, even the most sophisticated semiclassical model fails to explain the fact that the lowest energy state is spherically symmetric - it doesn't point in any particular direction. Nevertheless, in the modern *fully quantum treatment in phase space*, the proper deformation (full extension) of the semi-classical result adjusts the angular momentum value to the correct effective one. As a consequence, the physical ground state expression is obtained through a shift of the vanishing quantum angular momentum expression, which corresponds to spherical symmetry.

In modern quantum mechanics, the electron in hydrogen is a spherical cloud of probability that grows denser near the nucleus. The rate-constant of probability-decay in hydrogen is equal to the inverse of the Bohr radius, but since Bohr worked with circular orbits, not zero area ellipses, the fact that these two numbers exactly agree is considered a "coincidence". (However, many such coincidental agreements are found between the semiclassical vs. full quantum mechanical treatment of the atom; these include identical energy levels in the hydrogen atom and the derivation of a fine structure constant, which arises from the relativistic Bohr–Sommerfeld model (see below) and which happens to be equal to an entirely different concept, in full modern quantum mechanics).

The Bohr model also has difficulty with, or else fails to explain:

- Much of the spectra of larger atoms. At best, it can make predictions about the K-alpha and some L-alpha X-ray emission spectra for larger atoms, if *two* additional ad hoc assumptions are made (see Moseley's law above). Emission spectra for atoms with a single outer-shell electron (atoms in the lithium group) can also be approximately predicted. Also, if the empiric electron–nuclear screening factors for many atoms are known, many other spectral lines can be deduced from the information, in similar atoms of differing elements, via the Ritz–Rydberg combination principles (see Rydberg formula). All these techniques essentially make use of Bohr's Newtonian energy-potential picture of the atom.

- the relative intensities of spectral lines; although in some simple cases, Bohr's formula or modifications of it, was able to provide reasonable estimates (for example, calculations by Kramers for the Stark effect).

- The existence of fine structure and hyperfine structure in spectral lines, which are known to be due to a variety of relativistic and subtle effects, as well as complications from electron spin.

- The Zeeman effect – changes in spectral lines due to external magnetic fields; these are also due to more complicated quantum principles interacting with electron spin and orbital magnetic fields.

- The model also violates the uncertainty principle in that it considers electrons to have known orbits and locations, two things which can not be measured simultaneously.

- Doublets and Triplets: Appear in the spectra of some atoms: Very close pairs of lines. Bohr's model cannot say why some energy levels should be very close together.

- Multi-electron Atoms: don't have energy levels predicted by the model. It doesn't work for (neutral) helium.

- A rotating charge, such as the electron classically orbiting around the nucleus, would constantly lose energy in form of electromagnetic radiation (via various mechanisms: dipole radiation, Bremsstrahlung,...). But such radiation is not observed.

16.7 Refinements

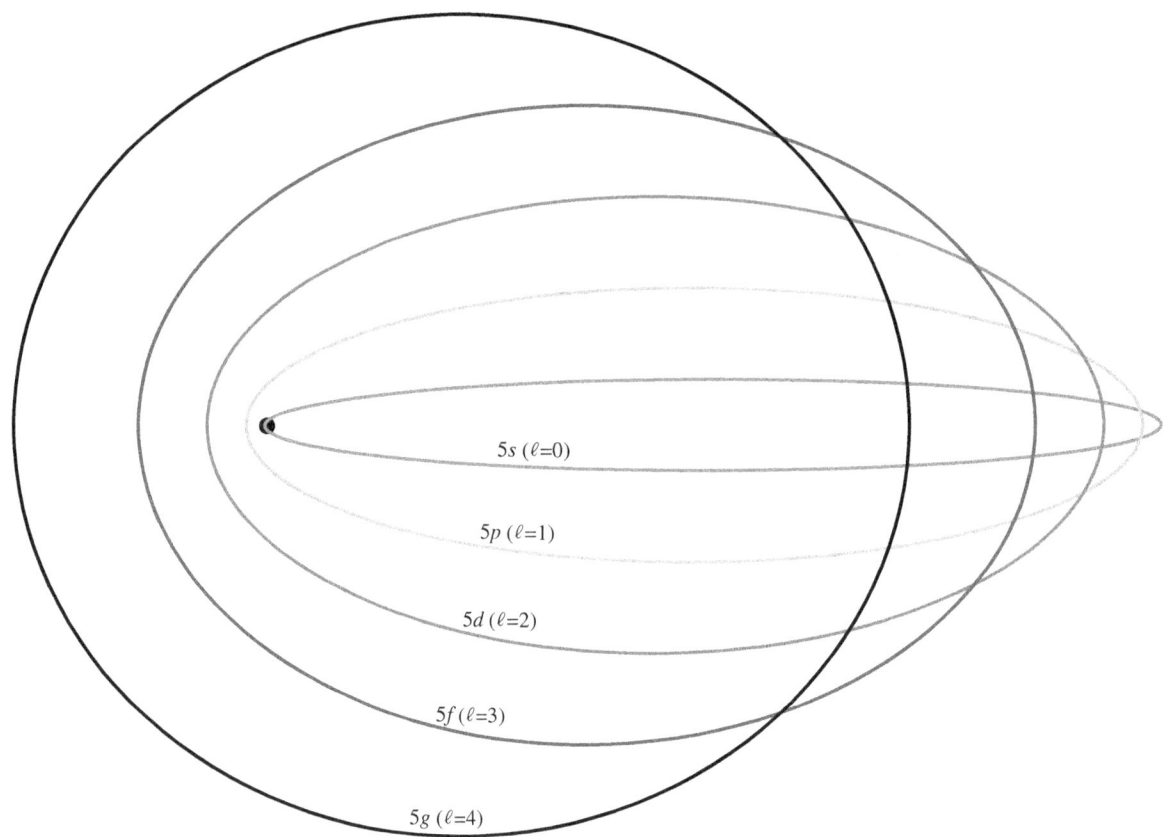

5s (ℓ=0)

5p (ℓ=1)

5d (ℓ=2)

5f (ℓ=3)

5g (ℓ=4)

Elliptical orbits with the same energy and quantized angular momentum

Several enhancements to the Bohr model were proposed, most notably the **Sommerfeld model** or **Bohr–Sommerfeld model**, which suggested that electrons travel in elliptical orbits around a nucleus instead of the Bohr model's circular orbits.[1] This model supplemented the quantized angular momentum condition of the Bohr model with an additional radial quantization condition, the **Sommerfeld–Wilson quantization condition**[8][9]

$$\int_0^T p_r \, dq_r = nh$$

where pr is the radial momentum canonically conjugate to the coordinate q which is the radial position and T is one full orbital period. The integral is the action of action-angle coordinates. This condition, suggested by the correspondence principle, is the only one possible, since the quantum numbers are adiabatic invariants.

The Bohr–Sommerfeld model was fundamentally inconsistent and led to many paradoxes. The magnetic quantum number measured the tilt of the orbital plane relative to the xy-plane, and it could only take a few discrete values. This contradicted the obvious fact that an atom could be turned this way and that relative to the coordinates without restriction. The Sommerfeld quantization can be performed in different canonical coordinates and sometimes gives different answers. The incorporation of radiation corrections was difficult, because it required finding action-angle coordinates for a combined radiation/atom system, which is difficult when the radiation is allowed to escape. The whole theory did not extend to non-integrable motions, which meant that many systems could not be treated even in principle. In the end, the model was replaced by the modern quantum mechanical treatment of the hydrogen atom, which was first given by Wolfgang Pauli in 1925, using Heisenberg's matrix mechanics. The current picture of the hydrogen atom is based on the atomic orbitals of wave mechanics which Erwin Schrödinger developed in 1926.

However, this is not to say that the Bohr model was without its successes. Calculations based on the Bohr–Sommerfeld model were able to accurately explain a number of more complex atomic spectral effects. For example, up to first-order perturbations, the Bohr model and quantum mechanics make the same predictions for the spectral line splitting in the Stark effect. At higher-order perturbations, however, the Bohr model and quantum mechanics differ, and measurements of the Stark effect under high field strengths helped confirm the correctness of quantum mechanics over the Bohr model. The prevailing theory behind this difference lies in the shapes of the orbitals of the electrons, which vary according to the energy state of the electron.

The Bohr–Sommerfeld quantization conditions lead to questions in modern mathematics. Consistent semiclassical quantization condition requires a certain type of structure on the phase space, which places topological limitations on the types of symplectic manifolds which can be quantized. In particular, the symplectic form should be the curvature form of a connection of a Hermitian line bundle, which is called a prequantization.

16.8   See also

16.9   References

16.9.1   Footnotes

[1] Akhlesh Lakhtakia (Ed.); Salpeter, Edwin E. (1996). "Models and Modelers of Hydrogen". *American Journal of Physics* (World Scientific) **65** (9): 933. Bibcode:1997AmJPh..65..933L. doi:10.1119/1.18691. ISBN 981-02-2302-1.

[2] Niels Bohr (1913). "On the Constitution of Atoms and Molecules, Part I" (PDF). *Philosophical Magazine* **26** (151): 1–24. doi:10.1080/14786441308634955.

[3] Olsen and McDonald 2005

[4] "CK12 – Chemistry Flexbook Second Edition – The Bohr Model of the Atom". Retrieved 30 September 2014.

[5] Niels Bohr (1913). "On the Constitution of Atoms and Molecules, Part II Systems Containing Only a Single Nucleus" (PDF). *Philosophical Magazine* **26** (153): 476–502. doi:10.1080/14786441308634993.

[6] M.A.B. Whitaker (1999). "The Bohr–Moseley synthesis and a simple model for atomic x-ray energies". *European Journal of Physics* **20** (3): 213–220. Bibcode:1999EJPh...20..213W. doi:10.1088/0143-0807/20/3/312.

[7] Smith, Brian. "Quantum Ideas: Week 2" Lecture Notes, p.17. University of Oxford. Retrieved Jan. 23, 2015.

[8] A. Sommerfeld (1916). "Zur Quantentheorie der Spektrallinien". *Annalen der Physik* **51** (17): 1. Bibcode:1916AnP...356....1S. doi:10.1002/andp.19163561702.

[9] W. Wilson (1915). "The quantum theory of radiation and line spectra".*Philosophical Magazine***29**(174): 795–802.doi8635362.

16.9.2   Primary sources

- Niels Bohr (1913). "On the Constitution of Atoms and Molecules, Part I" (PDF). *Philosophical Magazine* **26** (151): 1–24. doi:10.1080/14786441308634955.

- Niels Bohr (1913). "On the Constitution of Atoms and Molecules, Part II Systems Containing Only a Single Nucleus" (PDF). *Philosophical Magazine* **26** (153): 476–502. doi:10.1080/14786441308634993.

- Niels Bohr (1913). "On the Constitution of Atoms and Molecules, Part III Systems containing several nuclei". *Philosophical Magazine* **26**: 857–875. doi:10.1080/14786441308635031.

- Niels Bohr (1914). "The spectra of helium and hydrogen".*Nature***92**(2295): 231–232.Bibcode:1913Natur..92..23B. doi:10.1038/092231d0.

- Niels Bohr (1921)."Atomic Structure".*Nature***107**(2682): 104–107.Bibcode:1921Natur.107..104B.doi:10.1104a0.

- A. Einstein (1917). "Zum Quantensatz von Sommerfeld und Epstein". *Verhandlungen der Deutschen Physikalischen Gesellschaft* **19**: 82–92. Reprinted in *The Collected Papers of Albert Einstein*, A. Engel translator, (1997) Princeton University Press, Princeton. **6** p. 434. (provides an elegant reformulation of the Bohr–Sommerfeld quantization conditions, as well as an important insight into the quantization of non-integrable (chaotic) dynamical systems.)

16.10   Further reading

- Linus Carl Pauling (1970). "Chapter 5-1". *General Chemistry* (3rd ed.). San Francisco: W.H. Freeman & Co.

  - Reprint: Linus Pauling (1988). *General Chemistry*. New York: Dover Publications. ISBN 0-486-65622-5.

- George Gamow (1985). "Chapter 2". *Thirty Years That Shook Physics*. Dover Publications.

- Walter J. Lehmann (1972). "Chapter 18". *Atomic and Molecular Structure: the development of our concepts*. John Wiley and Sons.

- Paul Tipler and Ralph Llewellyn (2002). *Modern Physics* (4th ed.). W. H. Freeman. ISBN 0-7167-4345-0.

- Klaus Hentschel: Elektronenbahnen, Quantensprünge und Spektren, in: Charlotte Bigg & Jochen Hennig (eds.) Atombilder. Ikonografien des Atoms in Wissenschaft und Öffentlichkeit des 20. Jahrhunderts, Göttingen: Wallstein-Verlag 2009, pp. 51–61

- Steven and Susan Zumdahl (2010). "Chapter 7.4". *Chemistry* (8th ed.). Brooks/Cole. ISBN 978-0-495-82992-8.

- Helge Kragh (2011). "Conceptual objections to the Bohr atomic theory — do electrons have a "free will" ?". *European Physical Journal H* **36** (3): 327. Bibcode:2011EPJH...36..327K. doi:10.1140/epjh/e2011-20031-x.

16.11   External links

- Standing waves in Bohr's atomic model An interactive simulation to intuitively explain the quantization condition of standing waves in Bohr's atomic model

Chapter 17

Matter

This article is about the concept in the physical sciences. For other uses, see Matter (disambiguation).

Before the 20th century, the term **matter** included **ordinary matter** composed of atoms and excluded other energy phenomena such as light or sound. This concept of matter may be generalized from atoms to include any objects having mass even when at rest, but this is ill-defined because an object's mass can arise from its (possibly massless) constituents' motion and interaction energies. Thus, matter does not have a universal definition, nor is it a fundamental concept in physics today. Matter is also used loosely as a general term for the substance that makes up all observable physical objects.[1][2]

All the objects from everyday life that we can bump into, touch or squeeze are composed of atoms. This atomic matter is in turn made up of interacting subatomic particles—usually a nucleus of protons and neutrons, and a cloud of orbiting electrons.[3][4] Typically, science considers these composite particles matter because they have both rest mass and volume. By contrast, massless particles, such as photons, are not considered matter, because they have neither rest mass nor volume. However, not all particles with rest mass have a classical volume, since fundamental particles such as quarks and leptons (sometimes equated with matter) are considered "point particles" with no effective size or volume. Nevertheless, quarks and leptons together make up "ordinary matter", and their interactions contribute to the effective volume of the composite particles that make up ordinary matter.

Matter commonly exists in four *states* (or *phases*): solid, liquid and gas, and plasma. However, advances in experimental techniques have revealed other previously theoretical phases, such as Bose–Einstein condensates and fermionic condensates. A focus on an elementary-particle view of matter also leads to new phases of matter, such as the quark–gluon plasma.[5] For much of the history of the natural sciences people have contemplated the exact nature of matter. The idea that matter was built of discrete building blocks, the so-called *particulate theory of matter*, was first put forward by the Greek philosophers Leucippus (~490 BC) and Democritus (~470–380 BC).[6]

Matter should not be confused with mass, as the two are not quite the same in modern physics.[7] For example, mass is a conserved quantity, which means that its value is unchanging through time, within closed systems. However, matter is *not* conserved in such systems, although this is not obvious in ordinary conditions on Earth, where matter is approximately conserved. Still, special relativity shows that matter may disappear by conversion into energy, even inside closed systems, and it can also be created from energy, within such systems. However, because *mass* (like energy) can neither be created nor destroyed, the quantity of mass and the quantity of energy remain the same during a transformation of matter (which represents a certain amount of energy) into non-material (i.e., non-matter) energy. This is also true in the reverse transformation of energy into matter.

Different fields of science use the term matter in different, and sometimes incompatible, ways. Some of these ways are based on loose historical meanings, from a time when there was no reason to distinguish mass and matter. As such, there is no single universally agreed scientific meaning of the word "matter". Scientifically, the term "mass" is well-defined, but "matter" is not. Sometimes in the field of physics "matter" is simply equated with particles that exhibit rest mass (i.e., that cannot travel at the speed of light), such as quarks and leptons. However, in both physics and chemistry, matter exhibits both wave-like and particle-like properties, the so-called wave–particle duality.[8][9][10]

17.1 Definition

17.1.1 Common definition

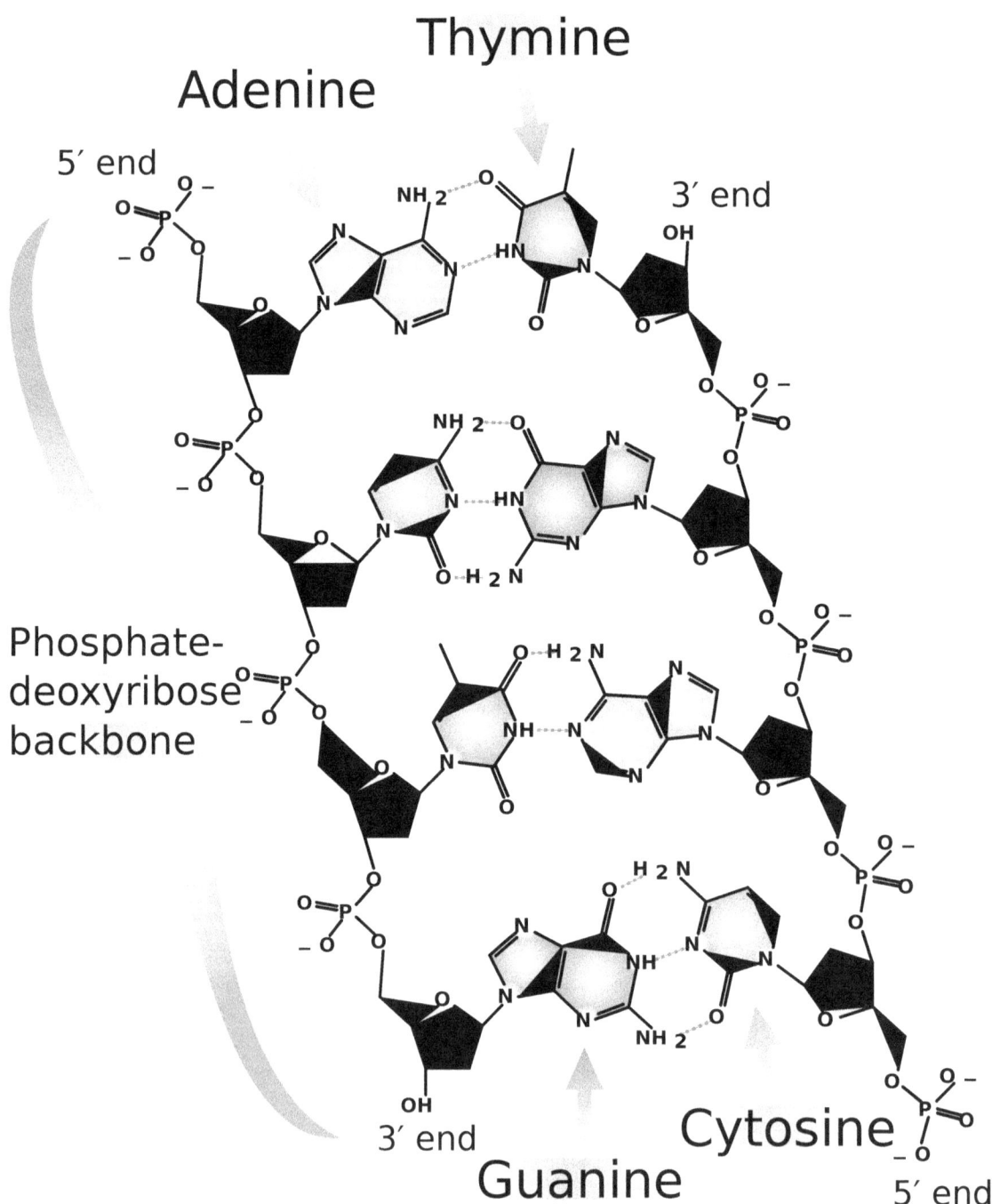

The DNA molecule is an example of matter under the "atoms and molecules" definition.

The common definition of matter is *anything that has mass and volume (occupies space)*.[11][12] For example, a car would be said to be made of matter, as it occupies space, and has mass.

The observation that matter occupies space goes back to antiquity. However, an explanation for why matter occupies space is recent, and is argued to be a result of the phenomenon described in the Pauli exclusion principle.[13][14] Two particular examples where the exclusion principle clearly relates matter to the occupation of space are white dwarf stars and neutron stars, discussed further below.

17.1.2 Relativity

Main article: Mass–energy equivalence

In the context of relativity, mass is not an additive quantity, in the sense that one can add the rest masses of particles in a system to get the total rest mass of the system.[1] Thus, in relativity usually a more general view is that it is not the sum of rest masses, but the energy–momentum tensor that quantifies the amount of matter. This tensor gives the rest mass for the entire system. "Matter" therefore is sometimes considered as anything that contributes to the energy–momentum of a system, that is, anything that is not purely gravity.[15][16] This view is commonly held in fields that deal with general relativity such as cosmology. In this view, light and other massless particles and fields are part of matter.

The reason for this is that in this definition, electromagnetic radiation (such as light) as well as the energy of electromagnetic fields contributes to the mass of systems, and therefore appears to add matter to them. For example, light radiation (or thermal radiation) trapped inside a box would contribute to the mass of the box, as would any kind of energy inside the box, including the kinetic energy of particles held by the box. Nevertheless, isolated individual particles of light (photons) and the isolated kinetic energy of massive particles, are normally not considered to be *matter*.

A difference between matter and mass therefore may seem to arise when single particles are examined. In such cases, the mass of single photons is zero. For particles with rest mass, such as leptons and quarks, isolation of the particle in a frame where it is not moving, removes its kinetic energy.

A source of definition difficulty in relativity arises from two definitions of mass in common use, one of which is formally equivalent to total energy (and is thus observer dependent), and the other of which is referred to as rest mass or invariant mass and is independent of the observer. Only "rest mass" is loosely equated with matter (since it can be weighed). Invariant mass is usually applied in physics to unbound systems of particles. However, energies which contribute to the "invariant mass" may be weighed also in special circumstances, such as when a system that has invariant mass is confined and has no net momentum (as in the box example above). Thus, a photon with no mass may (confusingly) still add mass to a system in which it is trapped. The same is true of the kinetic energy of particles, which by definition is not part of their rest mass, but which does add rest mass to systems in which these particles reside (an example is the mass added by the motion of gas molecules of a bottle of gas, or by the thermal energy of any hot object).

Since such mass (kinetic energies of particles, the energy of trapped electromagnetic radiation and stored potential energy of repulsive fields) is measured as part of the mass of ordinary *matter* in complex systems, the "matter" status of "massless particles" and fields of force becomes unclear in such systems. These problems contribute to the lack of a rigorous definition of matter in science, although mass is easier to define as the total stress–energy above (this is also what is weighed on a scale, and what is the source of gravity).

17.1.3 Atoms definition

A definition of "matter" based on its physical and chemical structure is: *matter is made up of atoms*.[17] As an example, deoxyribonucleic acid molecules (DNA) are matter under this definition because they are made of atoms. This definition can extend to include charged atoms and molecules, so as to include plasmas (gases of ions) and electrolytes (ionic solutions), which are not obviously included in the atoms definition. Alternatively, one can adopt the *protons, neutrons, and electrons* definition.

17.1.4 Protons, neutrons and electrons definition

A definition of "matter" more fine-scale than the atoms and molecules definition is: *matter is made up of what atoms and molecules are made of*, meaning anything made of positively charged protons, neutral neutrons, and negatively charged electrons.[18] This definition goes beyond atoms and molecules, however, to include substances made from these building blocks that are *not* simply atoms or molecules, for example white dwarf matter—typically, carbon and oxygen nuclei in a sea of degenerate electrons. At a microscopic level, the constituent "particles" of matter such as protons, neutrons, and electrons obey the laws of quantum mechanics and exhibit wave–particle duality. At an even deeper level, protons and neutrons are made up of quarks and the force fields (gluons) that bind them together (see Quarks and leptons definition below).

17.1.5 Quarks and leptons definition

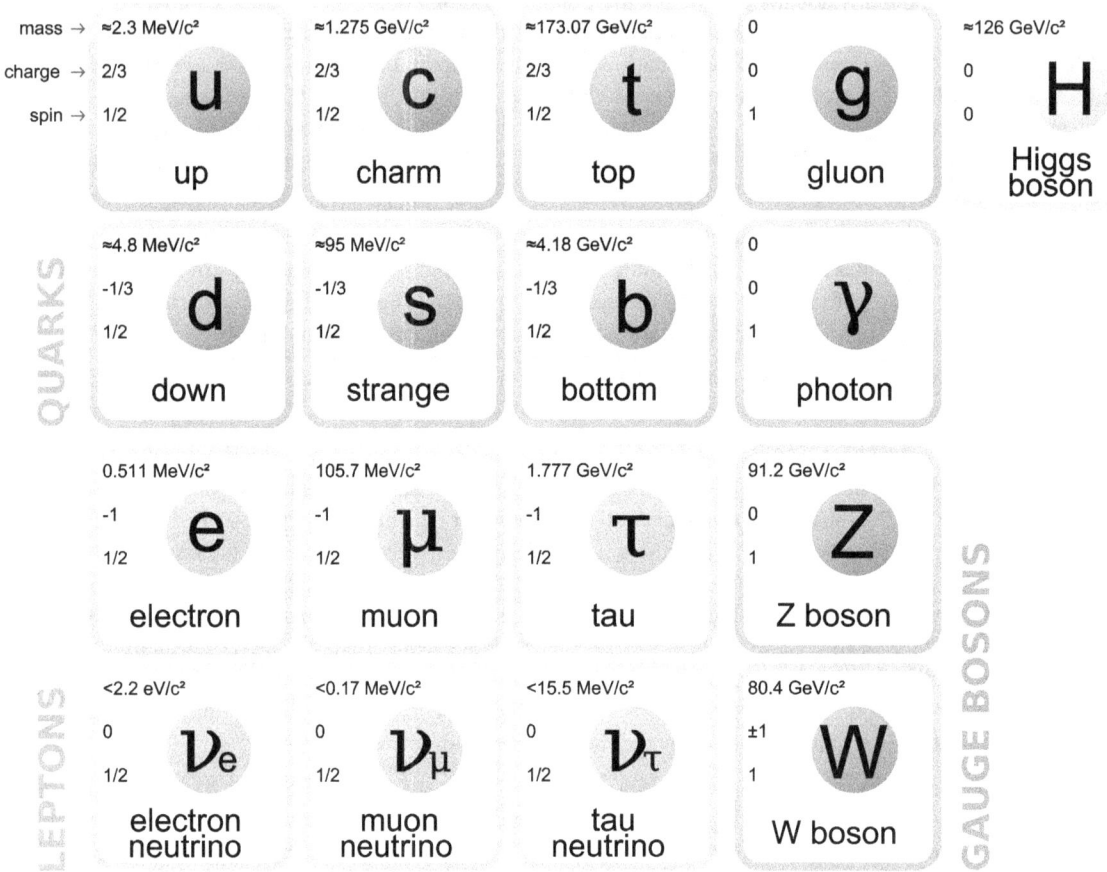

Under the "quarks and leptons" definition, the elementary and composite particles made of the quarks (in purple) and leptons (in green) would be matter—while the gauge bosons (in red) would not be matter. However, interaction energy inherent to composite particles (for example, gluons involved in neutrons and protons) contribute to the mass of ordinary matter.

As seen in the above discussion, many early definitions of what can be called *ordinary matter* were based upon its structure or *building blocks*. On the scale of elementary particles, a definition that follows this tradition can be stated as: *ordinary matter is everything that is composed of elementary fermions, namely quarks and leptons.*[19][20] The connection between these formulations follows.

Leptons (the most famous being the electron), and quarks (of which baryons, such as protons and neutrons, are made) combine to form atoms, which in turn form molecules. Because atoms and molecules are said to be matter, it is natural to phrase the definition as: *ordinary matter is anything that is made of the same things that atoms and molecules are made of*. (However, notice that one also can make from these building blocks matter that is *not* atoms or molecules.) Then, because electrons are leptons, and protons, and neutrons are made of quarks, this definition in turn leads to the definition of matter as being *quarks and leptons*, which are the two types of elementary fermions. Carithers and Grannis state: *Ordinary matter is composed entirely of first-generation particles, namely the [up] and [down] quarks, plus the electron and its neutrino.*[20] (Higher generations particles quickly decay into first-generation particles, and thus are not commonly encountered.[21])

This definition of ordinary matter is more subtle than it first appears. All the particles that make up ordinary matter (leptons and quarks) are elementary fermions, while all the force carriers are elementary bosons.[22] The W and Z bosons that mediate the weak force are not made of quarks or leptons, and so are not ordinary matter, even if they have mass.[23] In other words, mass is not something that is exclusive to ordinary matter.

The quark–lepton definition of ordinary matter, however, identifies not only the elementary building blocks of matter, but also includes composites made from the constituents (atoms and molecules, for example). Such composites contain an interaction energy that holds the constituents together, and may constitute the bulk of the mass of the composite. As an example, to a great extent, the mass of an atom is simply the sum of the masses of its constituent protons, neutrons and electrons. However, digging deeper, the protons and neutrons are made up of quarks bound together by gluon fields (see dynamics of quantum chromodynamics) and these gluons fields contribute significantly to the mass of hadrons.[24] In other words, most of what composes the "mass" of ordinary matter is due to the binding energy of quarks within protons and neutrons.[25] For example, the sum of the mass of the three quarks in a nucleon is approximately 12.5 MeV/c^2, which is low compared to the mass of a nucleon (approximately 938 MeV/c^2).[21][26] The bottom line is that most of the mass of everyday objects comes from the interaction energy of its elementary components.

17.1.6   Smaller building blocks issue

The Standard Model groups matter particles into three generations, where each generation consists of two quarks and two leptons. The first generation is the *up* and *down* quarks, the *electron* and the *electron neutrino*; the second includes the *charm* and *strange* quarks, the *muon* and the *muon neutrino*; the third generation consists of the *top* and *bottom* quarks and the *tau* and *tau neutrino*.[27] The most natural explanation for this would be that quarks and leptons of higher generations are excited states of the first generations. If this turns out to be the case, it would imply that quarks and leptons are composite particles, rather than elementary particles.[28]

17.2   Structure

In particle physics, fermions are particles that obey Fermi–Dirac statistics. Fermions can be elementary, like the electron—or composite, like the proton and neutron. In the Standard Model, there are two types of elementary fermions: quarks and leptons, which are discussed next.

17.2.1   Quarks

Main article: Quark

Quarks are particles of spin-$\frac{1}{2}$, implying that they are fermions. They carry an electric charge of $-\frac{1}{3}$ e (down-type quarks) or $+\frac{2}{3}$ e (up-type quarks). For comparison, an electron has a charge of -1 e. They also carry colour charge, which is the equivalent of the electric charge for the strong interaction. Quarks also undergo radioactive decay, meaning that they are subject to the weak interaction. Quarks are massive particles, and therefore are also subject to gravity.

Baryonic matter

Main article: Baryon

Baryons are strongly interacting fermions, and so are subject to Fermi–Dirac statistics. Amongst the baryons are the protons and neutrons, which occur in atomic nuclei, but many other unstable baryons exist as well. The term baryon usually refers to triquarks—particles made of three quarks. "Exotic" baryons made of four quarks and one antiquark are known as the pentaquarks, but their existence is not generally accepted.

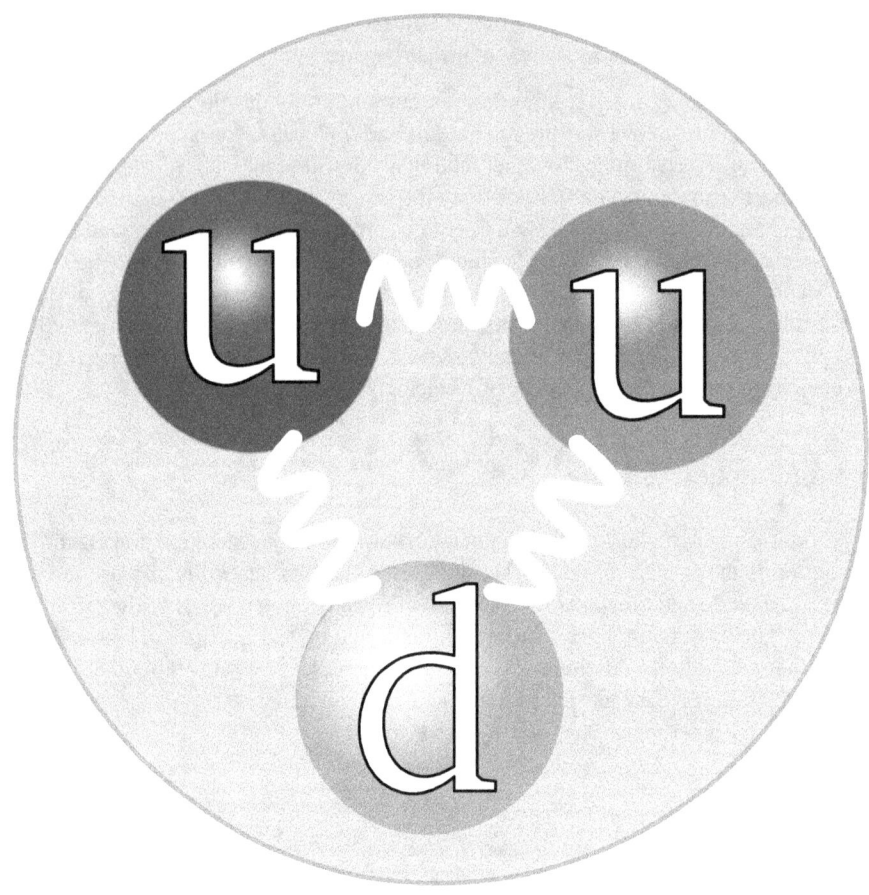

Quark structure of a proton: 2 up quarks and 1 down quark.

Baryonic matter is the part of the universe that is made of baryons (including all atoms). This part of the universe does not include dark energy, dark matter, black holes or various forms of degenerate matter, such as compose white dwarf stars and neutron stars. Microwave light seen by Wilkinson Microwave Anisotropy Probe (WMAP), suggests that only about 4.6% of that part of the universe within range of the best telescopes (that is, matter that may be visible because light could reach us from it), is made of baryonic matter. About 23% is dark matter, and about 72% is dark energy.[30]

Degenerate matter

Main article: Degenerate matter

In physics, **degenerate matter** refers to the ground state of a gas of fermions at a temperature near absolute zero.[31] The Pauli exclusion principle requires that only two fermions can occupy a quantum state, one spin-up and the other spin-down.

A comparison between the white dwarf IK Pegasi B (center), its A-class companion IK Pegasi A (left) and the Sun (right). This white dwarf has a surface temperature of 35,500 K.

Hence, at zero temperature, the fermions fill up sufficient levels to accommodate all the available fermions—and in the case of many fermions, the maximum kinetic energy (called the *Fermi energy*) and the pressure of the gas becomes very large, and depends on the number of fermions rather than the temperature, unlike normal states of matter.

Degenerate matter is thought to occur during the evolution of heavy stars.[32] The demonstration by Subrahmanyan Chandrasekhar that white dwarf stars have a maximum allowed mass because of the exclusion principle caused a revolution in the theory of star evolution.[33]

Degenerate matter includes the part of the universe that is made up of neutron stars and white dwarfs.

Strange matter

Main article: Strange matter

Strange matter is a particular form of quark matter, usually thought of as a *liquid* of up, down, and strange quarks. It is contrasted with nuclear matter, which is a liquid of neutrons and protons (which themselves are built out of up and down quarks), and with non-strange quark matter, which is a quark liquid that contains only up and down quarks. At high enough density, strange matter is expected to be color superconducting. Strange matter is hypothesized to occur in the core of neutron stars, or, more speculatively, as isolated droplets that may vary in size from femtometers (strangelets) to kilometers (quark stars).

Two meanings of the term "strange matter"   In particle physics and astrophysics, the term is used in two ways, one broader and the other more specific.

1. The broader meaning is just quark matter that contains three flavors of quarks: up, down, and strange. In this definition, there is a critical pressure and an associated critical density, and when nuclear matter (made of protons and neutrons) is compressed beyond this density, the protons and neutrons dissociate into quarks, yielding quark matter (probably strange matter).

2. The narrower meaning is quark matter that is *more stable than nuclear matter*. The idea that this could happen is the "strange matter hypothesis" of Bodmer[34] and Witten.[35] In this definition, the critical pressure is zero: the true ground state of matter is *always* quark matter. The nuclei that we see in the matter around us, which are droplets of nuclear matter, are actually metastable, and given enough time (or the right external stimulus) would decay into droplets of strange matter, i.e. strangelets.

17.2.2 Leptons

Main article: Lepton

Leptons are particles of spin-$\frac{1}{2}$, meaning that they are fermions. They carry an electric charge of -1 e (charged leptons) or 0 e (neutrinos). Unlike quarks, leptons do not carry colour charge, meaning that they do not experience the strong interaction. Leptons also undergo radioactive decay, meaning that they are subject to the weak interaction. Leptons are massive particles, therefore are subject to gravity.

17.3 Phases

Main article: Phase (matter)
See also: Phase diagram and State of matter
In bulk, matter can exist in several different forms, or states of aggregation, known as *phases*,[39] depending on ambient pressure, temperature and volume.[40] A phase is a form of matter that has a relatively uniform chemical composition and physical properties (such as density, specific heat, refractive index, and so forth). These phases include the three familiar ones (solids, liquids, and gases), as well as more exotic states of matter (such as plasmas, superfluids, supersolids, Bose–Einstein condensates, ...). A *fluid* may be a liquid, gas or plasma. There are also paramagnetic and ferromagnetic phases of magnetic materials. As conditions change, matter may change from one phase into another. These phenomena are called phase transitions, and are studied in the field of thermodynamics. In nanomaterials, the vastly increased ratio of surface area to volume results in matter that can exhibit properties entirely different from those of bulk material, and not well described by any bulk phase (see nanomaterials for more details).

Phases are sometimes called *states of matter*, but this term can lead to confusion with thermodynamic states. For example, two gases maintained at different pressures are in different *thermodynamic states* (different pressures), but in the same *phase* (both are gases).

17.4 Antimatter

Main article: Antimatter

In particle physics and quantum chemistry, **antimatter** is matter that is composed of the antiparticles of those that constitute ordinary matter. If a particle and its antiparticle come into contact with each other, the two annihilate; that is, they may both be converted into other particles with equal energy in accordance with Einstein's equation $E = mc^2$. These new particles may be high-energy photons (gamma rays) or other particle–antiparticle pairs. The resulting particles are endowed with an amount of kinetic energy equal to the difference between the rest mass of the products of the annihilation and the rest mass of the original particle–antiparticle pair, which is often quite large.

Antimatter is not found naturally on Earth, except very briefly and in vanishingly small quantities (as the result of radioactive decay, lightning or cosmic rays). This is because antimatter that came to exist on Earth outside the confines

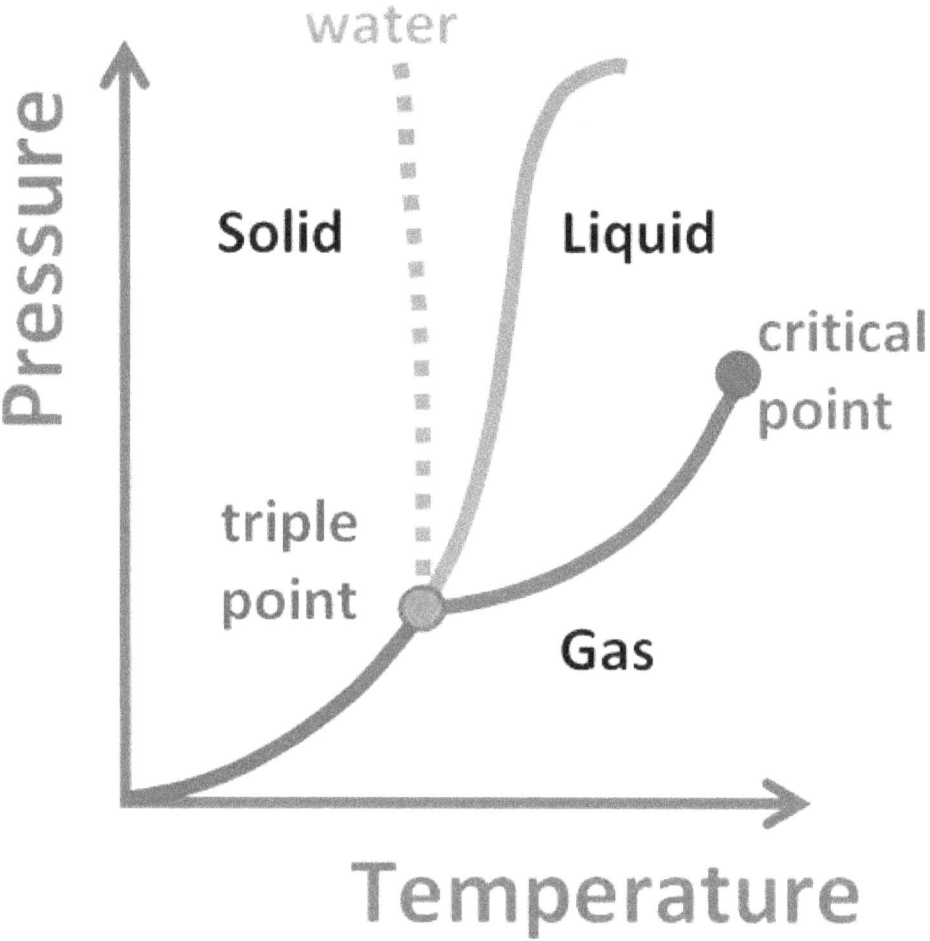

Phase diagram for a typical substance at a fixed volume. Vertical axis is Pressure, horizontal axis is Temperature. The green line marks the freezing point (above the green line is solid, *below it is* liquid*) and the blue line the boiling point (above it is* liquid *and below it is* gas*). So, for example, at higher* T, *a higher* P *is necessary to maintain the substance in liquid phase. At the triple point the three phases; liquid, gas and solid; can coexist. Above the critical point there is no detectable difference between the phases. The dotted line shows the anomalous behavior of water: ice melts at constant temperature with increasing pressure.*[38]

of a suitable physics laboratory would almost instantly meet the ordinary matter that Earth is made of, and be annihilated. Antiparticles and some stable antimatter (such as antihydrogen) can be made in tiny amounts, but not in enough quantity to do more than test a few of its theoretical properties.

There is considerable speculation both in science and science fiction as to why the observable universe is apparently almost entirely matter, and whether other places are almost entirely antimatter instead. In the early universe, it is thought that matter and antimatter were equally represented, and the disappearance of antimatter requires an asymmetry in physical laws called the charge parity (or CP symmetry) violation. CP symmetry violation can be obtained from the Standard Model,[41] but at this time the apparent asymmetry of matter and antimatter in the visible universe is one of the great unsolved problems in physics. Possible processes by which it came about are explored in more detail under baryogenesis.

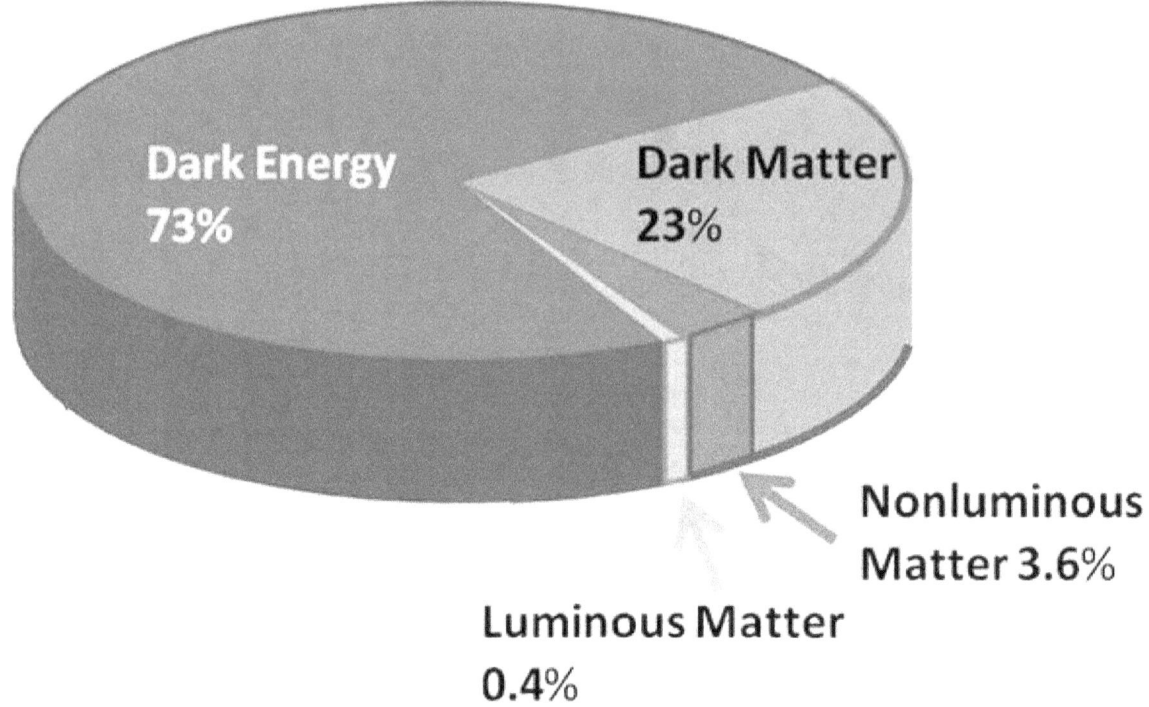

Pie chart showing the fractions of energy in the universe contributed by different sources. Ordinary matter is divided into luminous matter (the stars and luminous gases and 0.005% radiation) and nonluminous matter (intergalactic gas and about 0.1% neutrinos and 0.04% supermassive black holes). Ordinary matter is uncommon. Modeled after Ostriker and Steinhardt.[42] For more information, see NASA.

17.5 Other types

Ordinary matter, in the quarks and leptons definition, constitutes about 4% of the energy of the observable universe. The remaining energy is theorized to be due to exotic forms, of which 23% is dark matter[43][44] and 73% is dark energy.[45][46]

17.5.1 Dark matter

Main articles: Dark matter, Lambda-CDM model and WIMPs
See also: Galaxy formation and evolution and Dark matter halo

In astrophysics and cosmology, **dark matter** is matter of unknown composition that does not emit or reflect enough electromagnetic radiation to be observed directly, but whose presence can be inferred from gravitational effects on visible matter.[50][51] Observational evidence of the early universe and the big bang theory require that this matter have energy and mass, but is not composed of either elementary fermions (as above) OR gauge bosons. The commonly accepted view is that most of the dark matter is non-baryonic in nature.[50] As such, it is composed of particles as yet unobserved in the laboratory. Perhaps they are supersymmetric particles,[52] which are not Standard Model particles, but relics formed at very high energies in the early phase of the universe and still floating about.[50]

17.5.2 Dark energy

Main article: Dark energy
See also: Big bang § Dark energy

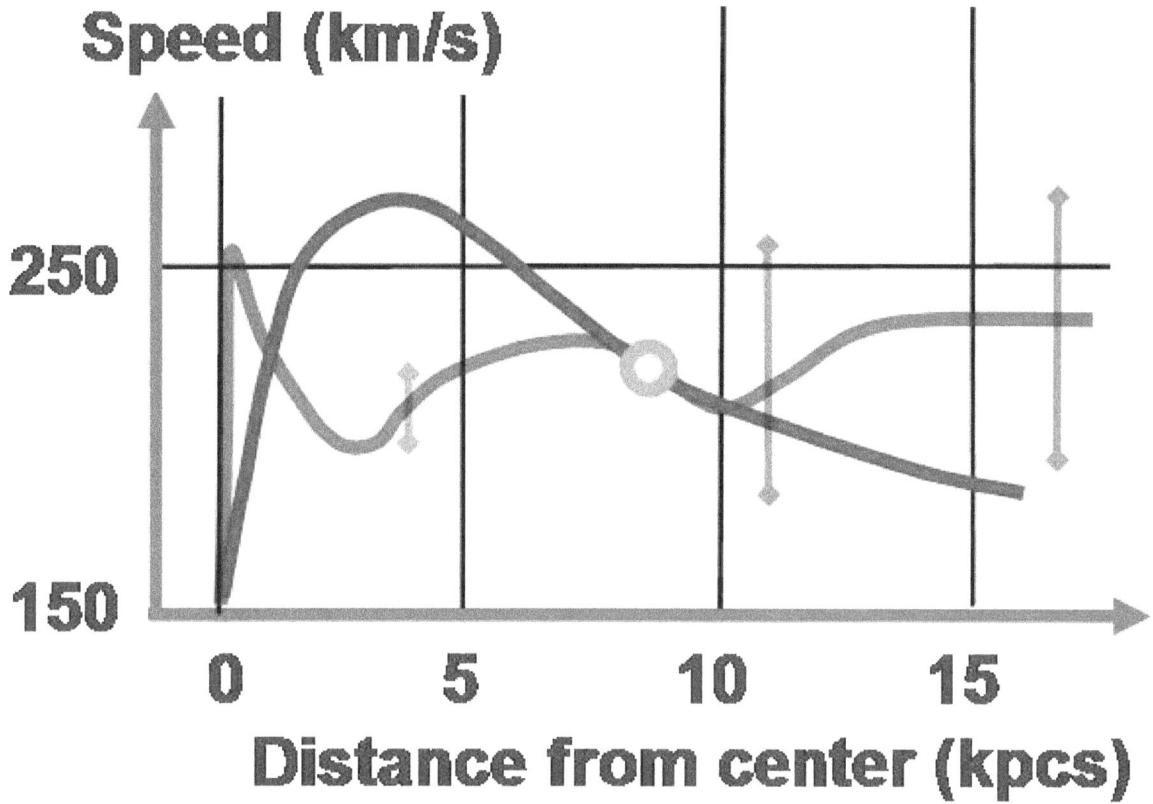

Galaxy rotation curve for the Milky Way. Vertical axis is speed of rotation about the galactic center. Horizontal axis is distance from the galactic center. The sun is marked with a yellow ball. The observed curve of speed of rotation is blue. The predicted curve based upon stellar mass and gas in the Milky Way is red. The difference is due to dark matter or perhaps a modification of the law of gravity.[47][48][49] Scatter in observations is indicated roughly by gray bars.

In cosmology, **dark energy** is the name given to the antigravitating influence that is accelerating the rate of expansion of the universe. It is known not to be composed of known particles like protons, neutrons or electrons, nor of the particles of dark matter, because these all gravitate.[53][54]

> Fully 70% of the matter density in the universe appears to be in the form of dark energy. Twenty-six percent is dark matter. Only 4% is ordinary matter. So less than 1 part in 20 is made out of matter we have observed experimentally or described in the standard model of particle physics. Of the other 96%, apart from the properties just mentioned, we know absolutely nothing.
> — Lee Smolin: *The Trouble with Physics*, p. 16

17.5.3   Exotic matter

Main article: Exotic matter

Exotic matter is a hypothetical concept of particle physics. It covers any material that violates one or more classical conditions or is not made of known baryonic particles. Such materials would possess qualities like negative mass or being repelled rather than attracted by gravity.

17.6 Historical development

17.6.1 Origins

The pre-Socratics were among the first recorded speculators about the underlying nature of the visible world. Thales (c. 624 BC–c. 546 BC) regarded water as the fundamental material of the world. Anaximander (c. 610 BC–c. 546 BC) posited that the basic material was wholly characterless or limitless: the Infinite (*apeiron*). Anaximenes (flourished 585 BC, d. 528 BC) posited that the basic stuff was *pneuma* or air. Heraclitus (c. 535–c. 475 BC) seems to say the basic element is fire, though perhaps he means that all is change. Empedocles (c. 490–430 BC) spoke of four elements of which everything was made: earth, water, air, and fire.[55] Meanwhile, Parmenides argued that change does not exist, and Democritus argued that everything is composed of minuscule, inert bodies of all shapes called atoms, a philosophy called atomism. All of these notions had deep philosophical problems.[56]

Aristotle (384 BC – 322 BC) was the first to put the conception on a sound philosophical basis, which he did in his natural philosophy, especially in *Physics* book I.[57] He adopted as reasonable suppositions the four Empedoclean elements, but added a fifth, aether. Nevertheless, these elements are not basic in Aristotle's mind. Rather they, like everything else in the visible world, are composed of the basic *principles* matter and form.

The word Aristotle uses for matter, ὕλη (*hyle* or *hule*), can be literally translated as wood or timber, that is, "raw material" for building.[58] Indeed, Aristotle's conception of matter is intrinsically linked to something being made or composed. In other words, in contrast to the early modern conception of matter as simply occupying space, matter for Aristotle is definitionally linked to process or change: matter is what underlies a change of substance.

For example, a horse eats grass: the horse changes the grass into itself; the grass as such does not persist in the horse, but some aspect of it—its matter—does. The matter is not specifically described (e.g., as atoms), but consists of whatever persists in the change of substance from grass to horse. Matter in this understanding does not exist independently (i.e., as a substance), but exists interdependently (i.e., as a "principle") with form and only insofar as it underlies change. It can be helpful to conceive of the relationship of matter and form as very similar to that between parts and whole. For Aristotle, matter as such can only *receive* actuality from form; it has no activity or actuality in itself, similar to the way that parts as such only have their existence *in* a whole (otherwise they would be independent wholes).

17.6.2 Early modernity

René Descartes (1596–1650) originated the modern conception of matter. He was primarily a geometer. Instead of, like Aristotle, deducing the existence of matter from the physical reality of change, Descartes arbitrarily postulated matter to be an abstract, mathematical substance that occupies space:

> So, extension in length, breadth, and depth, constitutes the nature of bodily substance; and thought constitutes the nature of thinking substance. And everything else attributable to body presupposes extension, and is only a mode of extended
> — René Descartes, *Principles of Philosophy*[59]

For Descartes, matter has only the property of extension, so its only activity aside from locomotion is to exclude other bodies:[60] this is the mechanical philosophy. Descartes makes an absolute distinction between mind, which he defines as unextended, thinking substance, and matter, which he defines as unthinking, extended substance.[61] They are independent things. In contrast, Aristotle defines matter and the formal/forming principle as complementary *principles* that together compose one independent thing (substance). In short, Aristotle defines matter (roughly speaking) as what things are actually made of (with a *potential* independent existence), but Descartes elevates matter to an actual independent thing in itself.

The continuity and difference between Descartes' and Aristotle's conceptions is noteworthy. In both conceptions, matter is passive or inert. In the respective conceptions matter has different relationships to intelligence. For Aristotle, matter and intelligence (form) exist together in an interdependent relationship, whereas for Descartes, matter and intelligence (mind) are definitionally opposed, independent substances.[62]

Descartes' justification for restricting the inherent qualities of matter to extension is its permanence, but his real criterion is not permanence (which equally applied to color and resistance), but his desire to use geometry to explain all material properties.[63] Like Descartes, Hobbes, Boyle, and Locke argued that the inherent properties of bodies were limited to extension, and that so-called secondary qualities, like color, were only products of human perception.[64]

Isaac Newton (1643–1727) inherited Descartes' mechanical conception of matter. In the third of his "Rules of Reasoning in Philosophy", Newton lists the universal qualities of matter as "extension, hardness, impenetrability, mobility, and inertia".[65] Similarly in *Optics* he conjectures that God created matter as "solid, massy, hard, impenetrable, movable particles", which were "...even so very hard as never to wear or break in pieces".[66] The "primary" properties of matter were amenable to mathematical description, unlike "secondary" qualities such as color or taste. Like Descartes, Newton rejected the essential nature of secondary qualities.[67]

Newton developed Descartes' notion of matter by restoring to matter intrinsic properties in addition to extension (at least on a limited basis), such as mass. Newton's use of gravitational force, which worked "at a distance", effectively repudiated Descartes' mechanics, in which interactions happened exclusively by contact.[68]

Though Newton's gravity would seem to be a *power* of bodies, Newton himself did not admit it to be an *essential* property of matter. Carrying the logic forward more consistently, Joseph Priestley argued that corporeal properties transcend contact mechanics: chemical properties require the *capacity* for attraction.[68] He argued matter has other inherent powers besides the so-called primary qualities of Descartes, et al.[69]

Since Priestley's time, there has been a massive expansion in knowledge of the constituents of the material world (viz., molecules, atoms, subatomic particles), but there has been no further development in the *definition* of matter. Rather the question has been set aside. Noam Chomsky summarizes the situation that has prevailed since that time:

> What is the concept of body that finally emerged?[...] The answer is that there is no clear and definite conception of body.[...] Rather, the material world is whatever we discover it to be, with whatever properties it must be assumed to have for the purposes of explanatory theory. Any intelligible theory that offers genuine explanations and that can be assimilated to the core notions of physics becomes part of the theory of the material world, part of our account of body. If we have such a theory in some domain, we seek to assimilate it to the core notions of physics, perhaps modifying these notions as we carry out this enterprise.
> — Noam Chomsky, '**Language and problems of knowledge: the Managua lectures,** *p. 144*[68]

So matter is whatever physics studies and the object of study of physics is matter: there is no independent general definition of matter, apart from its fitting into the methodology of measurement and controlled experimentation. In sum, the boundaries between what constitutes matter and everything else remains as vague as the demarcation problem of delimiting science from everything else.[70]

17.6.3   Late nineteenth and early twentieth centuries

In the 19th century, following the development of the periodic table, and of atomic theory, atoms were seen as being the fundamental constituents of matter; atoms formed molecules and compounds.[71]

The common definition in terms of occupying space and having mass is in contrast with most physical and chemical definitions of matter, which rely instead upon its structure and upon attributes not necessarily related to volume and mass. At the turn of the nineteenth century, the knowledge of matter began a rapid evolution.

Aspects of the Newtonian view still held sway. James Clerk Maxwell discussed matter in his work *Matter and Motion*.[72] He carefully separates "matter" from space and time, and defines it in terms of the object referred to in Newton's first law of motion.

However, the Newtonian picture was not the whole story. In the 19th century, the term "matter" was actively discussed by a host of scientists and philosophers, and a brief outline can be found in Levere.[73] A textbook discussion from 1870 suggests matter is what is made up of atoms:[74]

> Three divisions of matter are recognized in science: masses, molecules and atoms.
> A Mass of matter is any portion of matter appreciable by the senses.

A Molecule is the smallest particle of matter into which a body can be divided without losing its identity.
An Atom is a still smaller particle produced by division of a molecule.

Rather than simply having the attributes of mass and occupying space, matter was held to have chemical and electrical properties. The famous physicist J. J. Thomson wrote about the "constitution of matter" and was concerned with the possible connection between matter and electrical charge.[75]

17.6.4   Later developments

There is an entire literature concerning the "structure of matter", ranging from the "electrical structure" in the early 20th century,[76] to the more recent "quark structure of matter", introduced today with the remark: *Understanding the quark structure of matter has been one of the most important advances in contemporary physics*.[77] In this connection, physicists speak of *matter fields*, and speak of particles as "quantum excitations of a mode of the matter field".[8][9] And here is a quote from de Sabbata and Gasperini: "With the word "matter" we denote, in this context, the sources of the interactions, that is spinor fields (like quarks and leptons), which are believed to be the fundamental components of matter, or scalar fields, like the Higgs particles, which are used to introduced mass in a gauge theory (and that, however, could be composed of more fundamental fermion fields)."[78]

The modern conception of matter has been refined many times in history, in light of the improvement in knowledge of just *what* the basic building blocks are, and in how they interact.

In the late 19th century with the discovery of the electron, and in the early 20th century, with the discovery of the atomic nucleus, and the birth of particle physics, matter was seen as made up of electrons, protons and neutrons interacting to form atoms. Today, we know that even protons and neutrons are not indivisible, they can be divided into quarks, while electrons are part of a particle family called leptons. Both quarks and leptons are elementary particles, and are currently seen as being the fundamental constituents of matter.[79]

These quarks and leptons interact through four fundamental forces: gravity, electromagnetism, weak interactions, and strong interactions. The Standard Model of particle physics is currently the best explanation for all of physics, but despite decades of efforts, gravity cannot yet be accounted for at the quantum level; it is only described by classical physics (see quantum gravity and graviton).[80] Interactions between quarks and leptons are the result of an exchange of force-carrying particles (such as photons) between quarks and leptons.[81] The force-carrying particles are not themselves building blocks. As one consequence, mass and energy (which cannot be created or destroyed) cannot always be related to matter (which can be created out of non-matter particles such as photons, or even out of pure energy, such as kinetic energy). Force carriers are usually not considered matter: the carriers of the electric force (photons) possess energy (see Planck relation) and the carriers of the weak force (W and Z bosons) are massive, but neither are considered matter either.[82] However, while these particles are not considered matter, they do contribute to the total mass of atoms, subatomic particles, and all systems that contain them.[83][84]

17.6.5   Summary

The term "matter" is used throughout physics in a bewildering variety of contexts: for example, one refers to "condensed matter physics",[85] "elementary matter",[86] "partonic" matter, "dark" matter, "anti"-matter, "strange" matter, and "nuclear" matter. In discussions of matter and antimatter, normal matter has been referred to by Alfvén as *koinomatter* (Gk. *common matter*).[87] It is fair to say that in physics, there is no broad consensus as to a general definition of matter, and the term "matter" usually is used in conjunction with a specifying modifier.

17.7   See also

17.8   References

[1]   R. Penrose (1991). "The mass of the classical vacuum". In S. Saunders, H.R. Brown. *The Philosophy of Vacuum*. Oxford University Press. p. 21. ISBN 0-19-824449-5.

[2]   "Matter (physics)". *McGraw-Hill's Access Science: Encyclopedia of Science and Technology Online*. Retrieved 2009-05-24.

[3]   P. Davies (1992). *The New Physics: A Synthesis*. Cambridge University Press. p. 1. ISBN 0-521-43831-4.

[4]   G. 't Hooft (1997). *In search of the ultimate building blocks*. Cambridge University Press. p. 6. ISBN 0-521-57883-3.

[5]   "RHIC Scientists Serve Up "Perfect" Liquid" (Press release). Brookhaven National Laboratory. 18 April 2005. Retrieved 2009-09-15.

[6]   J. Olmsted; G.M. Williams (1996). *Chemistry: The Molecular Science* (2nd ed.). Jones & Bartlett. p. 40. ISBN 0-8151-8450-6.

[7]   J. Mongillo (2007). *Nanotechnology 101*. Greenwood Publishing. p. 30. ISBN 0-313-33880-9.

[8]   P.C.W. Davies (1979). *The Forces of Nature*. Cambridge University Press. p. 116. ISBN 0-521-22523-X.

[9]   S. Weinberg (1998). *The Quantum Theory of Fields*. Cambridge University Press. p. 2. ISBN 0-521-55002-5.

[10]   M. Masujima (2008). *Path Integral Quantization and Stochastic Quantization*. Springer. p. 103. ISBN 3-540-87850-5.

[11]   S.M. Walker; A. King (2005). *What is Matter?*. Lerner Publications. p. 7. ISBN 0-8225-5131-4.

[12]   J.Kenkel; P.B. Kelter; D.S. Hage (2000). *Chemistry: An Industry-based Introduction with CD-ROM*. CRC Press. p. 2. ISBN 1-56670-303-4. All basic science textbooks define *matter* as simply the collective aggregate of all material substances that occupy space and have mass or weight.

[13]   K.A. Peacock (2008). *The Quantum Revolution: A Historical Perspective*. Greenwood Publishing Group. p. 47. ISBN 0-313-33448-X.

[14]   M.H. Krieger (1998). *Constitutions of Matter: Mathematically Modeling the Most Everyday of Physical Phenomena*. University of Chicago Press. p. 22. ISBN 0-226-45305-7.

[15]   S.M. Caroll (2004). *Spacetime and Geometry*. Addison Wesley. pp. 163–164. ISBN 0-8053-8732-3.

[16]   P. Davies (1992). *The New Physics: A Synthesis*. Cambridge University Press. p. 499. ISBN 0-521-43831-4. **Matter fields**: the fields whose quanta describe the elementary particles that make up the material content of the Universe (as opposed to the gravitons and their supersymmetric partners).

[17]   G. F. Barker (1870). "Divisions of matter". *A text-book of elementary chemistry: theoretical and inorganic*. John F Morton & Co. p. 2. ISBN 978-1-4460-2206-1.

[18]   M. de Podesta (2002). *Understanding the Properties of Matter* (2nd ed.). CRC Press. p. 8. ISBN 0-415-25788-3.

[19]   B. Povh; K. Rith; C. Scholz; F. Zetsche; M. Lavelle (2004). "Part I: Analysis: The building blocks of matter". *Particles and Nuclei: An Introduction to the Physical Concepts* (4th ed.). Springer. ISBN 3-540-20168-8.

[20]   B. Carithers, P. Grannis (1995). "Discovery of the Top Quark" (PDF). *Beam Line* (SLAC National Accelerator Laboratory) **25** (3): 4–16.

[21]   D. Green (2005). *High PT physics at hadron colliders*. Cambridge University Press. p. 23. ISBN 0-521-83509-7.

[22]   L. Smolin (2007). *The Trouble with Physics: The Rise of String Theory, the Fall of a Science, and What Comes Next*. Mariner Books. p. 67. ISBN 0-618-91868-X.

[23]   The W boson mass is 80.398 GeV; see Figure 1 in C. Amsler *et al.* (Particle Data Group) (2008). "Review of Particle Physics: The Mass and Width of the W Boson" (PDF). *Physics Letters B* **667**: 1. Bibcode:2008PhLB..667....1P. doi:10.1016/j.phys8.

[24]   I.J.R. Aitchison; A.J.G. Hey (2004). *Gauge Theories in Particle Physics*. CRC Press. p. 48. ISBN 0-7503-0864-8.

[25] B. Povh; K. Rith; C. Scholz; F. Zetsche; M. Lavelle (2004). *Particles and Nuclei: An Introduction to the Physical Concepts.* Springer. p. 103. ISBN 3-540-20168-8.

[26] T. Hatsuda (2008). "Quark–gluon plasma and QCD". In H. Akai. *Condensed matter theories* **21**. Nova Publishers. p. 296. ISBN 1-60021-501-7.

[27] K.W Staley (2004). "Origins of the Third Generation of Matter". *The Evidence for the Top Quark.* Cambridge University Press. p. 8. ISBN 0-521-82710-8.

[28] Y. Ne'eman; Y. Kirsh (1996). *The Particle Hunters* (2nd ed.). Cambridge University Press. p. 276. ISBN 0-521-47686-0. [T]he most natural explanation to the existence of higher generations of quarks and leptons is that they correspond to excited states of the first generation, and experience suggests that excited systems must be composite

[29] C. Amsler *et al.* (Particle Data Group) (2008). "Reviews of Particle Physics: Quarks" (PDF). *Physics Letters B* **667**: 1. Bibcode:2008PhLB..667....1P. doi:10.1016/j.physletb.2008.07.018.

[30] "Five Year Results on the Oldest Light in the Universe". NASA. 2008. Retrieved 2008-05-02.

[31] H.S. Goldberg; M.D. Scadron (1987). *Physics of Stellar Evolution and Cosmology.* Taylor & Francis. p. 202. ISBN 0-677-05540-4.

[32] H.S. Goldberg; M.D. Scadron (1987). *Physics of Stellar Evolution and Cosmology.* Taylor & Francis. p. 233. ISBN 0-677-05540-4.

[33] J.-P. Luminet; A. Bullough; A. King (1992). *Black Holes.* Cambridge University Press. p. 75. ISBN 0-521-40906-3.

[34] A. Bodmer (1971). "Collapsed Nuclei". *Physical Review D* **4** (6): 1601. Bibcode:1971PhRvD...4.1601B. doi:10.1103/Ph01.

[35] E. Witten (1984). "Cosmic Separation of Phases". *Physical Review D* **30** (2): 272. Bibcode:1984PhRvD..30..272W. doi:

[36] C. Amsler *et al.* (Particle Data Group) (2008). "Review of Particle Physics: Leptons" (PDF). *Physics Letters B* **667**: 1. Bibcode:2008PhLB..667....1P. doi:10.1016/j.physletb.2008.07.018.

[37] C. Amsler *et al.* (Particle Data Group) (2008). "Review of Particle Physics: Neutrinos Properties" (PDF). *Physics Letters B* **667**: 1. Bibcode:2008PhLB..667....1P. doi:10.1016/j.physletb.2008.07.018.

[38] S. R. Logan (1998). *Physical Chemistry for the Biomedical Sciences.* CRC Press. pp. 110–111. ISBN 0-7484-0710-3.

[39] P.J. Collings (2002). "Chapter 1: States of Matter". *Liquid Crystals: Nature's Delicate Phase of Matter.* Princeton University Press. ISBN 0-691-08672-9.

[40] D.H. Trevena (1975). "Chapter 1.2: Changes of phase". *The Liquid Phase.* Taylor & Francis. ISBN 978-0-85109-031-3.

[41] National Research Council (US) (2006). *Revealing the hidden nature of space and time.* National Academies Press. p. 46. ISBN 0-309-10194-8.

[42] J.P. Ostriker; P.J. Steinhardt (2003). "New Light on Dark Matter". *Science* **300** (5627): 1909–13. arXiv:astro-ph/0306402. Bibcode:2003Sci...300.1909O. doi:10.1126/science.1085976. PMID 12817140.

[43] K. Pretzl (2004). "Dark Matter, Massive Neutrinos and Susy Particles". *Structure and Dynamics of Elementary Matter.* Walter Greiner. p. 289. ISBN 1-4020-2446-0.

[44] K. Freeman; G. McNamara (2006). "What can the matter be?". *In Search of Dark Matter.* Birkhäuser Verlag. p. 105. ISBN 0-387-27616-5.

[45] J.C. Wheeler (2007). *Cosmic Catastrophes: Exploding Stars, Black Holes, and Mapping the Universe.* Cambridge University Press. p. 282. ISBN 0-521-85714-7.

[46] J. Gribbin (2007). *The Origins of the Future: Ten Questions for the Next Ten Years.* Yale University Press. p. 151. ISBN 0-300-12596-8.

[47] P. Schneider (2006). *Extragalactic Astronomy and Cosmology.* Springer. p. 4, Fig. 1.4. ISBN 3-540-33174-3.

[48] T. Koupelis; K.F. Kuhn (2007). *In Quest of the Universe.* Jones & Bartlett Publishers. p. 492; Fig. 16.13. ISBN 0-7637-4387-9.

[49] M. H. Jones; R. J. Lambourne; D. J. Adams (2004). *An Introduction to Galaxies and Cosmology*. Cambridge University Press. p. 21; Fig. 1.13. ISBN 0-521-54623-0.

[50] D. Majumdar (2007). "Dark matter — possible candidates and direct detection". arXiv:hep-ph/0703310 [hep-ph].

[51] K.A. Olive (2003). "Theoretical Advanced Study Institute lectures on dark matter". arXiv:astro-ph/0301505 [astro-ph].

[52] K.A. Olive (2009). "Colliders and Cosmology". *European Physical Journal C* **59**(2): 269–295.arXiv:0806.1208.Bibc69O. doi:10.1140/epjc/s10052-008-0738-8.

[53] J.C. Wheeler (2007). *Cosmic Catastrophes*. Cambridge University Press. p. 282. ISBN 0-521-85714-7.

[54] L. Smolin (2007). *The Trouble with Physics*. Mariner Books. p. 16. ISBN 0-618-91868-X.

[55] S. Toulmin; J. Goodfield (1962). *The Architecture of Matter*. University of Chicago Press. pp. 48–54.

[56] Discussed by Aristotle in *Physics*, esp. book I, but also later; as well as *Metaphysics* I–II.

[57] For a good explanation and elaboration, see R.J. Connell (1966). *Matter and Becoming*. Priory Press.

[58] H. G. Liddell; R. Scott; J. M. Whiton (1891). *A lexicon abridged from Liddell & Scott's Greek–English lexicon*. Harper and Brothers. p. 725.

[59] R. Descartes (1644). "The Principles of Human Knowledge". *Principles of Philosophy I*. p. 53.

[60] though even this property seems to be non-essential (René Descartes, *Principles of Philosophy* II [1644], "On the Principles of Material Things", no. 4.)

[61] R. Descartes (1644). "The Principles of Human Knowledge". *Principles of Philosophy I*. pp. 8, 54, 63.

[62] D.L. Schindler (1986). "The Problem of Mechanism". In D.L. Schindler. *Beyond Mechanism*. University Press of America.

[63] E.A. Burtt, *Metaphysical Foundations of Modern Science* (Garden City, New York: Doubleday and Company, 1954), 117–118.

[64] J.E. McGuire and P.M. Heimann, "The Rejection of Newton's Concept of Matter in the Eighteenth Century", *The Concept of Matter in Modern Philosophy* ed. Ernan McMullin (Notre Dame: University of Notre Dame Press, 1978), 104–118 (105).

[65] Isaac Newton, *Mathematical Principles of* Natural Philosophy, *trans. A. Motte, revised by F. Cajori (Berkeley: University of California Press, 1934), pp. 398–400. Further analyzed by Maurice A. Finocchiaro, "Newton's Third Rule of Philosophizing: A Role for Logic in Historiography", Isis 65:1 (Mar. 1974), pp. 66–73.*

[66] Isaac Newton, *Optics*, Book III, pt. 1, query 31.

[67] McGuire and Heimann, 104.

[68] N. Chomsky (1988). *Language and problems of knowledge: the Managua lectures* (2nd ed.). MIT Press. p. 144. ISBN 0-262-53070-8.

[69] McGuire and Heimann, 113.

[70] Nevertheless, it remains true that the mathematization regarded as requisite for a modern physical theory carries its own implicit notion of matter, which is very like Descartes', despite the demonstrated vacuity of the latter's notions.

[71] M. Wenham (2005). *Understanding Primary Science: Ideas, Concepts and Explanations* (2nd ed.). Paul Chapman Educational Publishing. p. 115. ISBN 1-4129-0163-4.

[72] J.C. Maxwell (1876). *Matter and Motion*. Society for Promoting Christian Knowledge. p. 18. ISBN 0-486-66895-9.

[73] T.H. Levere (1993). "Introduction". *Affinity and Matter: Elements of Chemical Philosophy, 1800–1865*. Taylor & Francis. ISBN 2-88124-583-8.

[74] G.F. Barker (1870). "Introduction". *A Text Book of Elementary Chemistry: Theoretical and Inorganic*. John P. Morton and Company. p. 2.

[75] J. J. Thomson (1909). "Preface". *Electricity and Matter*. A. Constable.

[76] O.W. Richardson (1914). "Chapter 1". *The Electron Theory of Matter*. The University Press.

[77] M. Jacob (1992). *The Quark Structure of Matter*. World Scientific. ISBN 981-02-3687-5.

[78] V. de Sabbata; M. Gasperini (1985). *Introduction to Gravitation*. World Scientific. p. 293. ISBN 9971-5-0049-3.

[79] The history of the concept of matter is a history of the fundamental *length scales* used to define matter. Different building blocks apply depending upon whether one defines matter on an atomic or elementary particle level. One may use a definition that matter is atoms, or that matter is hadrons, or that matter is leptons and quarks depending upon the scale at which one wishes to define matter. B. Povh; K. Rith; C. Scholz; F. Zetsche; M. Lavelle (2004). "Fundamental constituents of matter". *Particles and Nuclei: An Introduction to the Physical Concepts* (4th ed.). Springer. ISBN 3-540-20168-8.

[80] J. Allday (2001). *Quarks, Leptons and the Big Bang*. CRC Press. p. 12. ISBN 0-7503-0806-0.

[81] B.A. Schumm (2004). *Deep Down Things: The Breathtaking Beauty of Particle Physics*. Johns Hopkins University Press. p. 57. ISBN 0-8018-7971-X.

[82] See for example, M. Jibu; K. Yasue (1995). *Quantum Brain Dynamics and Consciousness*. John Benjamins Publishing Company. p. 62. ISBN 1-55619-183-9., B. Martin (2009). *Nuclear and Particle Physics* (2nd ed.). John Wiley & Sons. p. 125. ISBN 0-470-74275-5. and K. W. Plaxco; M. Gross (2006). *Astrobiology: A Brief Introduction*. Johns Hopkins University Press. p. 23. ISBN 0-8018-8367-9.

[83] P. A. Tipler; R. A. Llewellyn (2002). *Modern Physics*. Macmillan. pp. 89–91, 94–95. ISBN 0-7167-4345-0.

[84] P. Schmüser; H. Spitzer (2002). "Particles". In L. Bergmann et al. *Constituents of Matter: Atoms, Molecules, Nuclei*. CRC Press. pp. 773 *ff*. ISBN 0-8493-1202-7.

[85] P. M. Chaikin; T. C. Lubensky (2000). *Principles of Condensed Matter Physics*. Cambridge University Press. p. xvii. ISBN 0-521-79450-1.

[86] W. Greiner (2003). W. Greiner, M.G. Itkis, G. Reinhardt, M.C. Güçlü, ed. *Structure and Dynamics of Elementary Matter*. Springer. p. xii. ISBN 1-4020-2445-2.

[87] P. Sukys (1999). *Lifting the Scientific Veil: Science Appreciation for the Nonscientist*. Rowman & Littlefield. p. 87. ISBN 0-8476-9600-6.

17.9   Further reading

- Lillian Hoddeson; Michael Riordan, eds. (1997). *The Rise of the Standard Model*. Cambridge University Press. ISBN 0-521-57816-7.

- Timothy Paul Smith (2004). "The search for quarks in ordinary matter". *Hidden Worlds*. Princeton University Press. p. 1. ISBN 0-691-05773-7.

- Harald Fritzsch (2005). *Elementary Particles: Building blocks of matter*. World Scientific. p. 1. ISBN 981-256-141-2.

- Bertrand Russell (1992). "The philosophy of matter". *A Critical Exposition of the Philosophy of Leibniz* (Reprint of 1937 2nd ed.). Routledge. p. 88. ISBN 0-415-08296-X.

- Stephen Toulmin and June Goodfield, *The Architecture of Matter* (Chicago: University of Chicago Press, 1962).

- Richard J. Connell, *Matter and Becoming* (Chicago: The Priory Press, 1966).

- Ernan McMullin, *The Concept of Matter in Greek and Medieval Philosophy* (Notre Dame, Indiana: Univ. of Notre Dame Press, 1965).

- Ernan McMullin, *The Concept of Matter in Modern Philosophy* (Notre Dame, Indiana: University of Notre Dame Press, 1978).

17.10   External links

- Visionlearning Module on Matter

- Matter in the universe How much Matter is in the Universe?

- NASA on superfluid core of neutron star

- Matter and Energy: A False Dichotomy – Conversations About Science with Theoretical Physicist Matt Strassler

Chapter 18

State of matter

Not to be confused with Phase (matter).

In physics, a **state of matter** is one of the distinct forms that matter takes on. Four states of matter are observable in everyday life: solid, liquid, gas, and plasma. Many other states are known, such as Bose–Einstein condensates and neutron-degenerate matter, but these only occur in extreme situations such as ultra cold or ultra dense matter. Other states, such as quark–gluon plasmas, are believed to be possible but remain theoretical for now. For a complete list of all exotic states of matter, see the list of states of matter.

Historically, the distinction is made based on qualitative differences in properties. Matter in the solid state maintains a fixed volume and shape, with component particles (atoms, molecules or ions) close together and fixed into place. Matter in the liquid state maintains a fixed volume, but has a variable shape that adapts to fit its container. Its particles are still close together but move freely. Matter in the gaseous state has both variable volume and shape, adapting both to fit its container. Its particles are neither close together nor fixed in place. Matter in the plasma state has variable volume and shape, but as well as neutral atoms, it contains a significant number of ions and electrons, both of which can move around freely. Plasma is the most common form of visible matter in the universe.[1]

The term phase is sometimes used as a synonym for state of matter, but a system can contain several immiscible phases of the same state of matter (see Phase (matter) for more discussion of the difference between the two terms).

18.1   The four fundamental states

18.1.1   Solid

Main article: Solid

In a solid the particles (ions, atoms or molecules) are closely packed together. The forces between particles are strong so that the particles cannot move freely but can only vibrate. As a result, a solid has a stable, definite shape, and a definite volume. Solids can only change their shape by force, as when broken or cut.

In crystalline solids, the particles (atoms, molecules, or ions) are packed in a regularly ordered, repeating pattern. There are various different crystal structures, and the same substance can have more than one structure (or solid phase). For example, iron has a body-centred cubic structure at temperatures below 912 °C, and a face-centred cubic structure between 912 and 1394 °C. Ice has fifteen known crystal structures, or fifteen solid phases, which exist at various temperatures and pressures.[2]

Glasses and other non-crystalline, amorphous solids without long-range order are not thermal equilibrium ground states; therefore they are described below as nonclassical states of matter.

The four fundamental states of matter. Clockwise from top left, they are solid, liquid, plasma, and gas, represented by an ice sculpture, a drop of water, electrical arcing from a tesla coil, and the air around clouds, respectively.

A crystalline solid: atomic resolution image of strontium titanate. Brighter atoms are Sr and darker ones are Ti.

Solids can be transformed into liquids by melting, and liquids can be transformed into solids by freezing. Solids can also change directly into gases through the process of sublimation, and gases can likewise change directly into solids through deposition.

18.1.2 Liquid

Structure of a classical monatomic liquid. Atoms have many nearest neighbors in contact, yet no long-range order is present.

Main article: Liquid

A liquid is a nearly incompressible fluid that conforms to the shape of its container but retains a (nearly) constant volume independent of pressure. The volume is definite if the temperature and pressure are constant. When a solid is heated above its melting point, it becomes liquid, given that the pressure is higher than the triple point of the substance. Intermolecular (or interatomic or interionic) forces are still important, but the molecules have enough energy to move relative to each other and the structure is mobile. This means that the shape of a liquid is not definite but is determined by its container. The volume is usually greater than that of the corresponding solid, the best known exception being water, H_2O. The highest temperature at which a given liquid can exist is its critical temperature.[3]

18.1.3 Gas

Main article: Gas

A gas is a compressible fluid. Not only will a gas conform to the shape of its container but it will also expand to fill the container.

In a gas, the molecules have enough kinetic energy so that the effect of intermolecular forces is small (or zero for an ideal gas), and the typical distance between neighboring molecules is much greater than the molecular size. A gas has no definite shape or volume, but occupies the entire container in which it is confined. A liquid may be converted to a gas by heating at constant pressure to the boiling point, or else by reducing the pressure at constant temperature.

At temperatures below its critical temperature, a gas is also called a vapor, and can be liquefied by compression alone

The spaces between gas molecules are very big. Gas molecules have very weak or no bonds at all. The molecules in "gas" can move freely and fast.

without cooling. A vapor can exist in equilibrium with a liquid (or solid), in which case the gas pressure equals the vapor pressure of the liquid (or solid).

A supercritical fluid (SCF) is a gas whose temperature and pressure are above the critical temperature and critical pressure respectively. In this state, the distinction between liquid and gas disappears. A supercritical fluid has the physical properties of a gas, but its high density confers solvent properties in some cases, which leads to useful applications. For example, supercritical carbon dioxide is used to extract caffeine in the manufacture of decaffeinated coffee.[4]

18.1.4 Plasma

Main article: Plasma (physics)

Like a gas, plasma does not have definite shape or volume. Unlike gases, plasmas are electrically conductive, produce magnetic fields and electric currents, and respond strongly to electromagnetic forces. Positively charged nuclei swim in a "sea" of freely-moving disassociated electrons, similar to the way such charges exist in conductive metal. In fact it is this electron "sea" that allows matter in the plasma state to conduct electricity.

The plasma state is often misunderstood, but it is actually quite common on Earth, and the majority of people observe it on a regular basis without even realizing it. Lightning, electric sparks, fluorescent lights, neon lights, plasma televisions, some types of flame and the stars are all examples of illuminated matter in the plasma state.

A gas is usually converted to a plasma in one of two ways, either from a huge voltage difference between two points, or by exposing it to extremely high temperatures.

Heating matter to high temperatures causes electrons to leave the atoms, resulting in the presence of free electrons. At very high temperatures, such as those present in stars, it is assumed that essentially all electrons are "free", and that a very high-energy plasma is essentially bare nuclei swimming in a sea of electrons.

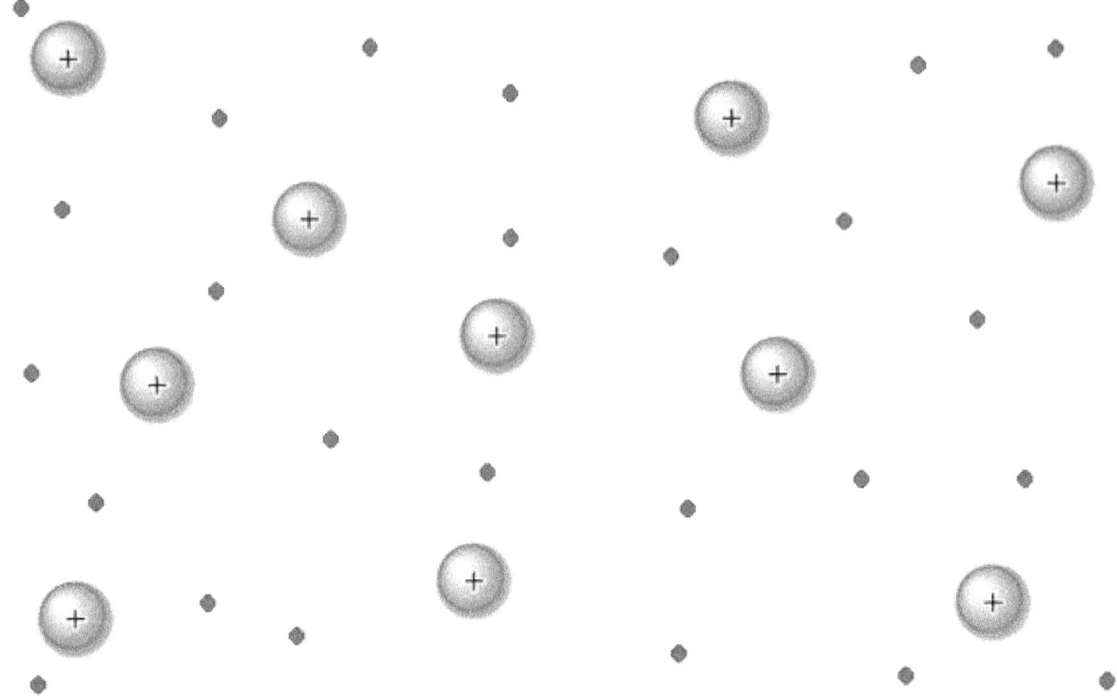

In a plasma, electrons are ripped away from their nuclei, forming an electron "sea". This gives it the ability to conduct electricity.

18.2 Phase transitions

Main article: Phase transitions

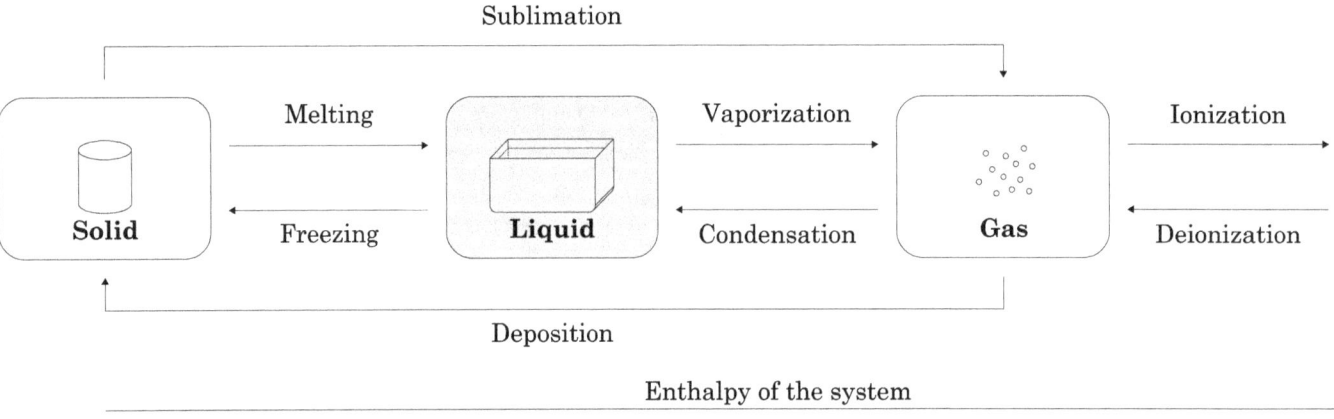

This diagram illustrates transitions between the four fundamental states of matter.

A state of matter is also characterized by phase transitions. A phase transition indicates a change in structure and can be recognized by an abrupt change in properties. A distinct state of matter can be defined as any set of states distinguished from any other set of states by a phase transition. Water can be said to have several distinct solid states.[5] The appearance of superconductivity is associated with a phase transition, so there are superconductive states. Likewise, ferromagnetic states are demarcated by phase transitions and have distinctive properties. When the change of state occurs in stages the intermediate steps are called mesophases. Such phases have been exploited by the introduction of liquid crystal technology. [6][7]

The state or *phase* of a given set of matter can change depending on pressure and temperature conditions, transitioning to other phases as these conditions change to favor their existence; for example, solid transitions to liquid with an increase in temperature. Near absolute zero, a substance exists as a solid. As heat is added to this substance it melts into a liquid at its melting point, boils into a gas at its boiling point, and if heated high enough would enter a plasma state in which the electrons are so energized that they leave their parent atoms.

Forms of matter that are not composed of molecules and are organized by different forces can also be considered different states of matter. Superfluids (like Fermionic condensate) and the quark–gluon plasma are examples.

In a chemical equation, the state of matter of the chemicals may be shown as (s) for solid, (l) for liquid, and (g) for gas. An aqueous solution is denoted (aq). Matter in the plasma state is seldom used (if at all) in chemical equations, so there is no standard symbol to denote it. In the rare equations that plasma is used in plasma is symbolized as (p).

18.3   Non-classical states

18.3.1   Glass

Main article: Glass

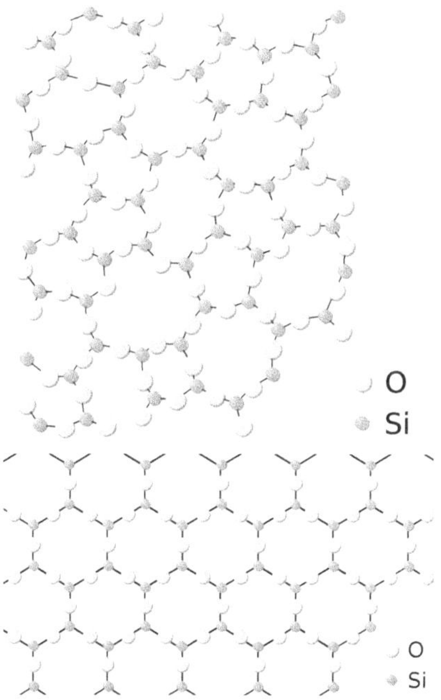

Schematic representation of a random-network glassy form (left) and ordered crystalline lattice (right) of identical chemical composition.

Glass is a non-crystalline or amorphous solid material that exhibits a glass transition when heated towards the liquid state. Glasses can be made of quite different classes of materials: inorganic networks (such as window glass, made of silicate plus additives), metallic alloys, ionic melts, aqueous solutions, molecular liquids, and polymers. Thermodynamically, a glass is in a metastable state with respect to its crystalline counterpart. The conversion rate, however, is practically zero.

18.3.2   Crystals with some degree of disorder

A plastic crystal is a molecular solid with long-range positional order but with constituent molecules retaining rotational freedom; in an orientational glass this degree of freedom is frozen in a quenched disordered state.

Similarly, in a spin glass magnetic disorder is frozen.

18.3.3 Liquid crystal states

Main article: Liquid crystal

Liquid crystal states have properties intermediate between mobile liquids and ordered solids. Generally, they are able to flow like a liquid, but exhibiting long-range order. For example, the nematic phase consists of long rod-like molecules such as para-azoxyanisole, which is nematic in the temperature range 118–136 °C.[8] In this state the molecules flow as in a liquid, but they all point in the same direction (within each domain) and cannot rotate freely.

Other types of liquid crystals are described in the main article on these states. Several types have technological importance, for example, in liquid crystal displays.

18.3.4 Magnetically ordered

Transition metal atoms often have magnetic moments due to the net spin of electrons that remain unpaired and do not form chemical bonds. In some solids the magnetic moments on different atoms are ordered and can form a ferromagnet, an antiferromagnet or a ferrimagnet.

In a ferromagnet—for instance, solid iron—the magnetic moment on each atom is aligned in the same direction (within a magnetic domain). If the domains are also aligned, the solid is a permanent magnet, which is magnetic even in the absence of an external magnetic field. The magnetization disappears when the magnet is heated to the Curie point, which for iron is 768 °C.

An antiferromagnet has two networks of equal and opposite magnetic moments, which cancel each other out so that the net magnetization is zero. For example, in nickel(II) oxide (NiO), half the nickel atoms have moments aligned in one direction and half in the opposite direction.

In a ferrimagnet, the two networks of magnetic moments are opposite but unequal, so that cancellation is incomplete and there is a non-zero net magnetization. An example is magnetite (Fe_3O_4), which contains Fe^{2+} and Fe^{3+} ions with different magnetic moments.

18.3.5 Microphase-separated

Main article: Copolymer

Copolymers can undergo microphase separation to form a diverse array of periodic nanostructures, as shown in the example of the styrene-butadiene-styrene block copolymer shown at right. Microphase separation can be understood by analogy to the phase separation between oil and water. Due to chemical incompatibility between the blocks, block copolymers undergo a similar phase separation. However, because the blocks are covalently bonded to each other, they cannot demix macroscopically as water and oil can, and so instead the blocks form nanometer-sized structures. Depending on the relative lengths of each block and the overall block topology of the polymer, many morphologies can be obtained, each its own phase of matter.

18.3.6 Quantum spin liquid

Main article: Quantum spin liquid

A disordered state in a system of interacting quantum spins which preserves its disorder to very low temperatures, unlike other disordered states.

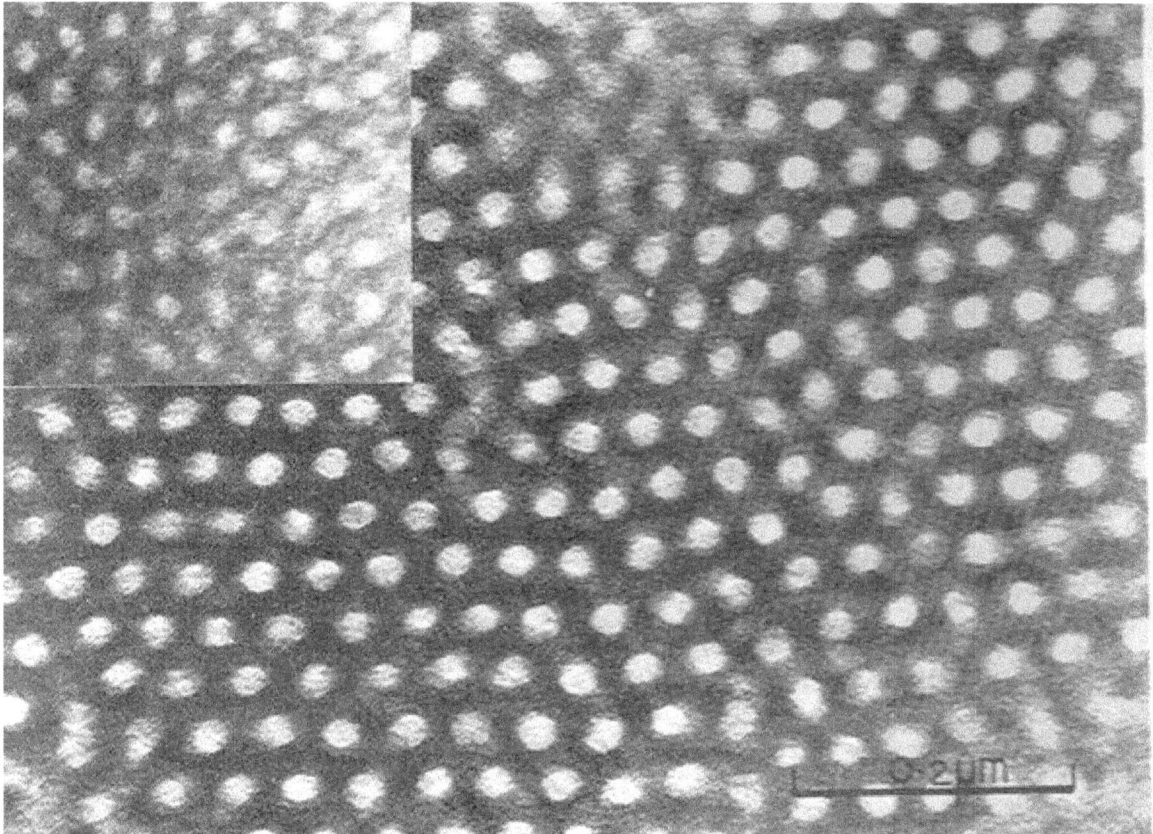

SBS block copolymer in TEM

18.4   Low-temperature states

18.4.1   Superfluid

Main article: Superfluid

Close to absolute zero, some liquids form a second liquid state described as **superfluid** because it has zero viscosity (or infinite fluidity; i.e., flowing without friction). This was discovered in 1937 for helium, which forms a superfluid below the lambda temperature of 2.17 K. In this state it will attempt to "climb" out of its container.[9] It also has infinite thermal conductivity so that no temperature gradient can form in a superfluid. Placing a superfluid in a spinning container will result in quantized vortices.

These properties are explained by the theory that the common isotope helium-4 forms a Bose–Einstein condensate (see next section) in the superfluid state. More recently, Fermionic condensate superfluids have been formed at even lower temperatures by the rare isotope helium-3 and by lithium-6.[10]

18.4.2   Bose–Einstein condensate

Main article: Bose–Einstein condensate

In 1924, Albert Einstein and Satyendra Nath Bose predicted the "Bose–Einstein condensate" (BEC), sometimes referred to as the fifth state of matter. In a BEC, matter stops behaving as independent particles, and collapses into a single quantum state that can be described with a single, uniform wavefunction.

Liquid helium in a superfluid phase creeps up on the walls of the cup in a Rollin film, eventually dripping out from the cup.

In the gas phase, the Bose–Einstein condensate remained an unverified theoretical prediction for many years. In 1995, the research groups of Eric Cornell and Carl Wieman, of JILA at the University of Colorado at Boulder, produced the first such condensate experimentally. A Bose–Einstein condensate is "colder" than a solid. It may occur when atoms have very similar (or the same) quantum levels, at temperatures very close to absolute zero (-273.15 °C).

18.4.3 Fermionic condensate

Main article: Fermionic condensate

A *fermionic condensate* is similar to the Bose–Einstein condensate but composed of fermions. The Pauli exclusion principle prevents fermions from entering the same quantum state, but a pair of fermions can behave as a boson, and multiple such pairs can then enter the same quantum state without restriction.

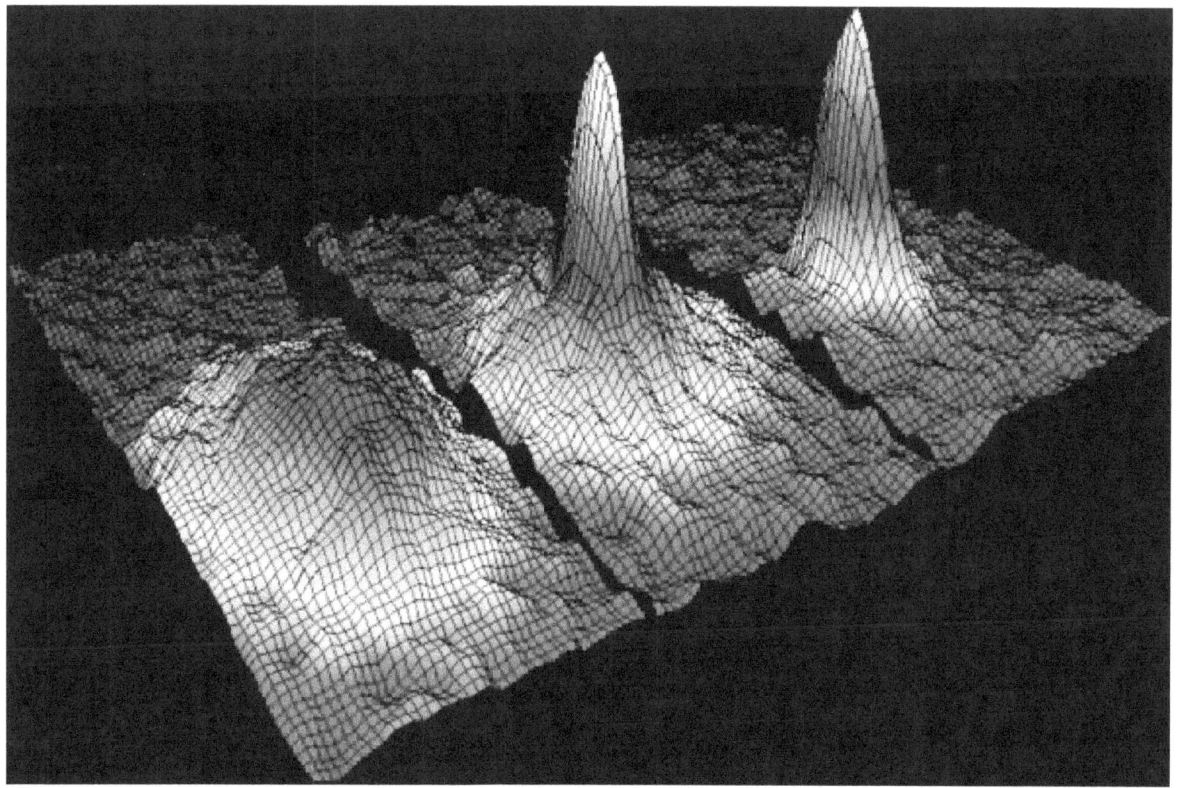

Velocity in a gas of rubidium as it is cooled: the starting material is on the left, and Bose–Einstein condensate is on the right.

18.4.4   Rydberg molecule

One of the metastable states of strongly non-ideal plasma is Rydberg matter, which forms upon condensation of excited atoms. These atoms can also turn into ions and electrons if they reach a certain temperature. In April 2009, *Nature* reported the creation of Rydberg molecules from a Rydberg atom and a ground state atom,[11] confirming that such a state of matter could exist.[12] The experiment was performed using ultracold rubidium atoms.

18.4.5   Quantum Hall state

Main article: Quantum Hall effect

A *quantum Hall state* gives rise to quantized Hall voltage measured in the direction perpendicular to the current flow. A *quantum spin Hall state* is a theoretical phase that may pave the way for the development of electronic devices that dissipate less energy and generate less heat. This is a derivation of the Quantum Hall state of matter.

18.4.6   Strange matter

Main article: Strange matter

Strange matter is a type of quark matter that may exist inside some neutron stars close to the Tolman–Oppenheimer–Volkoff limit (approximately 2–3 solar masses). It may be stable at lower energy states once formed.

18.4.7 Photonic matter

Main article: Photonic matter

In photonic matter, photons behave as if they had mass, and can interact with each other, even forming photonic "molecules". This is in contrast to the usual properties of photons, which have no rest mass, and cannot interact.

18.4.8 Dropleton

Main article: Dropleton

A "quantum fog" of electrons and holes that flow around each other and even ripple like a liquid, rather than existing as discrete pairs.[13]

18.5 High-energy states

18.5.1 Degenerate matter

Main article: Degenerate matter

Under extremely high pressure, ordinary matter undergoes a transition to a series of exotic states of matter collectively known as degenerate matter. In these conditions, the structure of matter is supported by the Pauli exclusion principle. These are of great interest to astrophysicists, because these high-pressure conditions are believed to exist inside stars that have used up their nuclear fusion "fuel", such as the white dwarfs and neutron stars.

Electron-degenerate matter is found inside white dwarf stars. Electrons remain bound to atoms but are able to transfer to adjacent atoms. Neutron-degenerate matter is found in neutron stars. Vast gravitational pressure compresses atoms so strongly that the electrons are forced to combine with protons via inverse beta-decay, resulting in a superdense conglomeration of neutrons. (Normally free neutrons outside an atomic nucleus will decay with a half life of just under 15 minutes, but in a neutron star, as in the nucleus of an atom, other effects stabilize the neutrons.)

18.5.2 Quark–gluon plasma

Main article: Quark–gluon plasma

Quark–gluon plasma is a phase in which quarks become free and able to move independently (rather than being perpetually bound into particles) in a sea of gluons (subatomic particles that transmit the strong force that binds quarks together); this is similar to splitting molecules into atoms. This state may be briefly attainable in particle accelerators, and allows scientists to observe the properties of individual quarks, and not just theorize. See also Strangeness production.

Quark–gluon plasma was discovered at CERN in 2000.

18.5.3 Color-glass condensate

Main article: Color-glass condensate

Color-glass condensate is a type of matter theorized to exist in atomic nuclei traveling near the speed of light. According to Einstein's theory of relativity, a high-energy nucleus appears length contracted, or compressed, along its direction of motion. As a result, the gluons inside the nucleus appear to a stationary observer as a "gluonic wall" traveling near the speed of light. At very high energies, the density of the gluons in this wall is seen to increase greatly. Unlike the quark–gluon plasma produced in the collision of such walls, the color-glass condensate describes the walls themselves, and is an intrinsic property of the particles that can only be observed under high-energy conditions such as those at RHIC and possibly at the Large Hadron Collider as well.

18.6   Very high energy states

The **gravitational singularity** predicted by general relativity to exist at the center of a black hole is *not* a phase of matter; it is not a material object at all (although the mass-energy of matter contributed to its creation) but rather a property of spacetime at a location. It could be argued, of course, that all particles are properties of spacetime at a location,[14] leaving a half-note of controversy on the subject.

18.7   Other proposed states

18.7.1   Supersolid

Main article: Supersolid

A supersolid is a spatially ordered material (that is, a solid or crystal) with superfluid properties. Similar to a superfluid, a supersolid is able to move without friction but retains a rigid shape. Although a supersolid is a solid, it exhibits so many characteristic properties different from other solids that many argue it is another state of matter.[15]

18.7.2   String-net liquid

Main article: String-net liquid

In a string-net liquid, atoms have apparently unstable arrangement, like a liquid, but are still consistent in overall pattern, like a solid. When in a normal solid state, the atoms of matter align themselves in a grid pattern, so that the spin of any electron is the opposite of the spin of all electrons touching it. But in a string-net liquid, atoms are arranged in some pattern that requires some electrons to have neighbors with the same spin. This gives rise to curious properties, as well as supporting some unusual proposals about the fundamental conditions of the universe itself.

18.7.3   Superglass

Main article: Superglass

A superglass is a phase of matter characterized, at the same time, by superfluidity and a frozen amorphous structure.

18.7.4   Dark matter

Main article: Dark matter

While dark matter is estimated to comprise 83% of the mass of matter in the universe, most of its properties remain a mystery due to the fact that it neither absorbs nor emits electromagnetic radiation, and there are many competing theories regarding what dark matter is actually made of. Thus, while it is hypothesized to exist and comprise the vast majority of matter in the universe, almost all of its properties are unknown and a matter of speculation, because it has only been observed through its gravitational effects.[16][17]

18.7.5   Equilibrium gel

Main article: Equilibrium gel

Equilibrium gel is made from a synthetic clay called Laponite. Unlike other gels, it maintains the same consistency throughout its structure and is stable, which means it does not separate into sections of solid mass and those of more liquid mass. Equilibrium gel filtration liquid chromatography is a technique used for the quantitation of ligand binding.[18]

18.8   See also

- Hidden states of matter
- Classical element
- Condensed matter physics
- Cooling curve
- Phase (matter)
- Supercooling
- Superheating

18.9   Notes and references

[1] It is often stated that more than 99% of the material in the visible universe is plasma. See, for instance, D. A. Gurnett; A. Bhattacharjee (2005). *Introduction to Plasma Physics: With Space and Laboratory Applications.* Cambridge, UK: Cambridge University Press. p. 2. ISBN 0-521-36483-3. and K Scherer; H Fichtner; B Heber (2005). *Space Weather: The Physics Behind a Slogan.* Berlin: Springer. p. 138. ISBN 3-540-22907-8.. Essentially, all of the visible light from space comes from stars, which are plasmas with a temperature such that they radiate strongly at visible wavelengths. Most of the ordinary (or baryonic) matter in the universe, however, is found in the intergalactic medium, which is also a plasma, but much hotter, so that it radiates primarily as X-rays. The current scientific consensus is that about 96% of the total energy density in the universe is not plasma or any other form of ordinary matter, but a combination of cold dark matter and dark energy.

[2] M.A. Wahab (2005). *Solid State Physics: Structure and Properties of Materials.* Alpha Science. pp. 1–3. ISBN 1-84265-218-4.

[3] F. White (2003). *Fluid Mechanics.* McGraw-Hill. p. 4. ISBN 0-07-240217-2.

[4] G. Turrell (1997). *Gas Dynamics: Theory and Applications.* John Wiley & Sons. pp. 3–5. ISBN 0-471-97573-7.

[5] M. Chaplin (20 August 2009). "Water phase Diagram". *Water Structure and Science.* Retrieved 23 February 2010.

[6] D.L. Goodstein (1985). *States of Matter.* Dover Phoenix. ISBN 978-0-486-49506-4.

[7] A.P. Sutton (1993). *Electronic Structure of Materials.* Oxford Science Publications. pp. 10–12. ISBN 978-0-19-851754-2.

[8] Shao, Y.; Zerda, T. W. (1998). "Phase Transitions of Liquid Crystal PAA in Confined Geometries". *Journal of Physical Chemistry B* **102** (18): 3387–3394. doi:10.1021/jp9734437.

[9] J.R. Minkel (20 February 2009). "Strange but True: Superfluid Helium Can Climb Walls". *Scientific American.* Retrieved 23 February 2010.

[10] L. Valigra (22 June 2005). "MIT physicists create new form of matter". MIT News. Retrieved 23 February 2010.

[11]V. Bendkowsky et al. (2009). "Observation of Ultralong-Range Rydberg Molecules".*Nature***458**(7241): 1005–.1005B. doi:10.1038/nature07945. PMID 19396141.

[12] V. Gill (23 April 2009). "World First for Strange Molecule". BBC News. Retrieved 23 February 2010.

[13] http://www.iflscience.com/physics/new-state-matter-discovered#3Oe9x65kkHViXABt.99

[14] David Chalmers; David Manley; Ryan Wasserman (2009). *Metametaphysics: New Essays on the Foundations of Ontology.* Oxford University Press. pp. 378–. ISBN 978-0-19-954604-6.

[15] G. Murthy et al. (1997). "Superfluids and Supersolids on Frustrated Two-Dimensional Lattices". *Physical Review B* **55** (5): 3104. arXiv:cond-mat/9607217. Bibcode:1997PhRvB..55.3104M. doi:10.1103/PhysRevB.55.3104.

[16] Trimble, Virginia (1987). "Existence and nature of dark matter in the universe". *Annual Review of Astronomy and Astrophysics* **25**: 425–472. Bibcode:1987ARA&A..25..425T. doi:10.1146/annurev.aa.25.090187.002233.

[17] Hinshaw, Gary F. (29 January 2010). "What is the universe made of?". *Universe 101.* NASA website. Retrieved 17 March 2010.

[18] Cartlidge, Edwin (12 January 2012). "New State of Matter Seen in Clay". *Technology.* Science Now website. Retrieved 10 September 2013.

18.10   External links

- 2005-06-22, MIT News: MIT physicists create new form of matter Citat: "... They have become the first to create a new type of matter, a gas of atoms that shows high-temperature superfluidity."

- 2003-10-10, Science Daily: Metallic Phase For Bosons Implies New State Of Matter

- 2004-01-15, ScienceDaily: Probable Discovery Of A New, Supersolid, Phase Of Matter Citat: "...We apparently have observed, for the first time, a solid material with the characteristics of a superfluid...but because all its particles are in the identical quantum state, it remains a solid even though its component particles are continually flowing..."

- 2004-01-29, ScienceDaily: NIST/University Of Colorado Scientists Create New Form Of Matter: A Fermionic Condensate

- Short videos demonstrating of States of Matter, solids, liquids and gases by Prof. J M Murrell, University of Sussex

Chapter 19

Phase (matter)

Not to be confused with State of matter.

In the physical sciences, a **phase** is a region of space (a thermodynamic system), throughout which all physical properties of a material are essentially uniform.[1][2]:86[3]:3 Examples of physical properties include density, index of refraction, magnetization and chemical composition. A simple description is that a phase is a region of material that is chemically uniform, physically distinct, and (often) mechanically separable. In a system consisting of ice and water in a glass jar, the ice cubes are one phase, the water is a second phase, and the humid air over the water is a third phase. The glass of the jar is another separate phase. (See State of Matter#Glass)

The term *phase* is sometimes used as a synonym for state of matter, but there can be several immiscible phases of the same state of matter. Also, the term *phase* is sometimes used to refer to a set of equilibrium states demarcated in terms of state variables such as pressure and temperature by a phase boundary on a phase diagram. Because phase boundaries relate to changes in the organization of matter, such as a change from liquid to solid or a more subtle change from one crystal structure to another, this latter usage is similar to the use of "phase" as a synonym for state of matter. However, the state of matter and phase diagram usages are not commensurate with the formal definition given above and the intended meaning must be determined in part from the context in which the term is used.

19.1 Types of phases

Distinct phases may be described as different states of matter such as gas, liquid, solid, plasma or Bose–Einstein condensate. Useful mesophases between solid and liquid form other states of matter.

Distinct phases may also exist within a given state of matter. As shown in the diagram for iron alloys, several phases exist for both the solid and liquid states. Phases may also be differentiated based on solubility as in polar (hydrophilic) or non-polar (hydrophobic). A mixture of water (a polar liquid) and oil (a non-polar liquid) will spontaneously separate into two phases. Water has a very low solubility (is insoluble) in oil, and oil has a low solubility in water. Solubility is the maximum amount of a solute that can dissolve in a solvent before the solute ceases to dissolve and remains in a separate phase. A mixture can separate into more than two liquid phases and the concept of phase separation extends to solids, i.e., solids can form solid solutions or crystallize into distinct crystal phases. Metal pairs that are mutually soluble can form alloys, whereas metal pairs that are mutually insoluble cannot.

As many as eight immiscible liquid phases have been observed.[4] Mutually immiscible liquid phases are formed from water (aqueous phase), hydrophobic organic solvents, perfluorocarbons (fluorous phase), silicones, several different metals, and also from molten phosphorus. Not all organic solvents are completely miscible, e.g. a mixture of ethylene glycol and toluene may separate into two distinct organic phases.[5]

Phases do not need to macroscopically separate spontaneously. Emulsions and colloids are examples of immiscible phase pair combinations that do not physically separate.

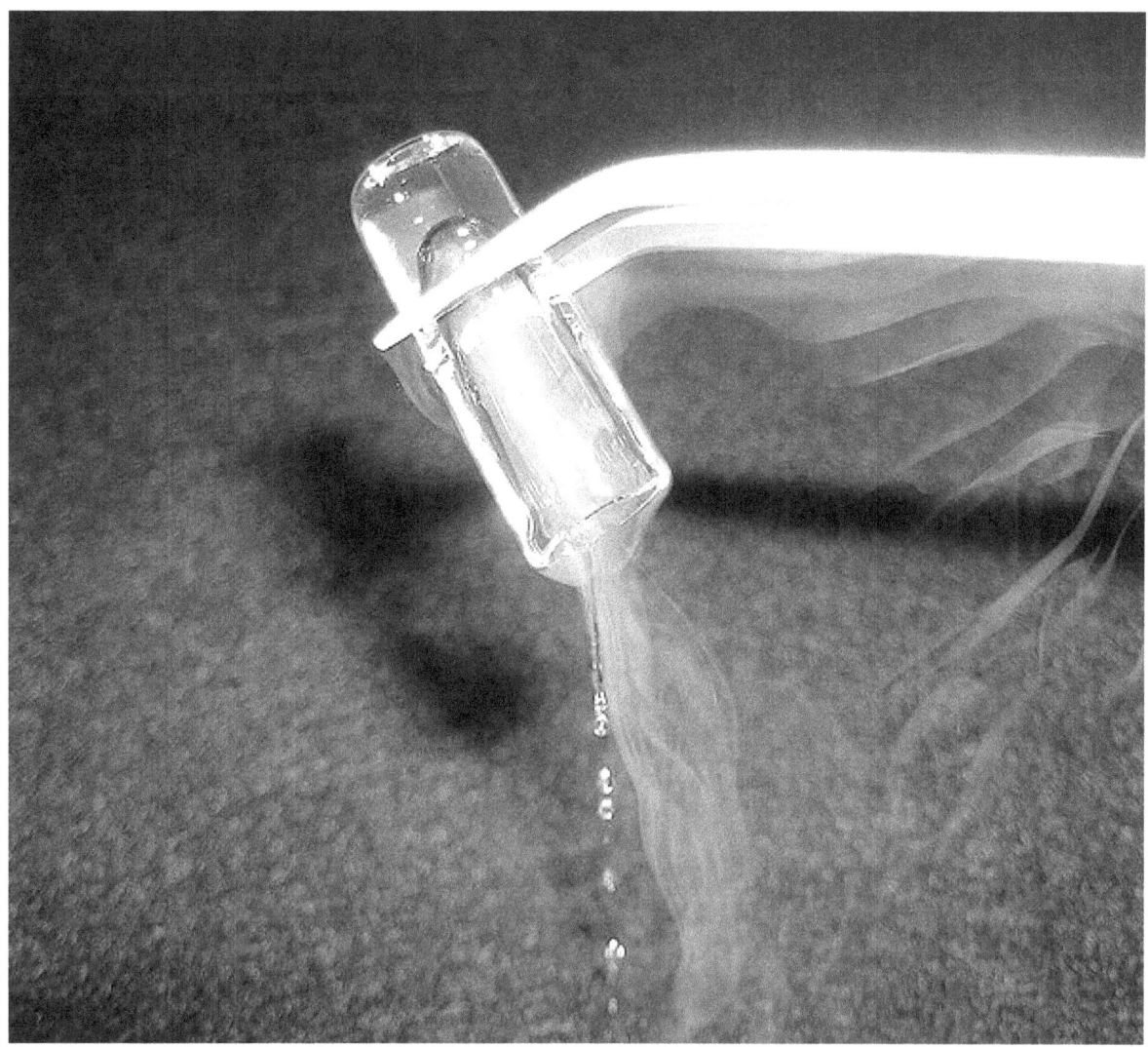

A small piece of rapidly melting argon ice shows the transition from solid to liquid.

19.2   Phase equilibrium

Left to equilibration, many compositions will form a uniform single phase, but depending on the temperature and pressure even a single substance may separate into two or more distinct phases. Within each phase, the properties are uniform but between the two phases properties differ.

Water in a closed jar with an air space over it forms a two phase system. Most of the water is in the liquid phase, where it is held by the mutual attraction of water molecules. Even at equilibrium molecules are constantly in motion and, once in a while, a molecule in the liquid phase gains enough kinetic energy to break away from the liquid phase and enter the gas phase. Likewise, every once in a while a vapor molecule collides with the liquid surface and condenses into the liquid. At equilibrium, evaporation and condensation processes exactly balance and there is no net change in the volume of either phase.

At room temperature and pressure, the water jar reaches equilibrium when the air over the water has a humidity of about 3%. This percentage increases as the temperature goes up. At 100 °C and atmospheric pressure, equilibrium is not reached until the air is 100% water. If the liquid is heated a little over 100 °C, the transition from liquid to gas will occur not only at the surface, but throughout the liquid volume: the water boils.

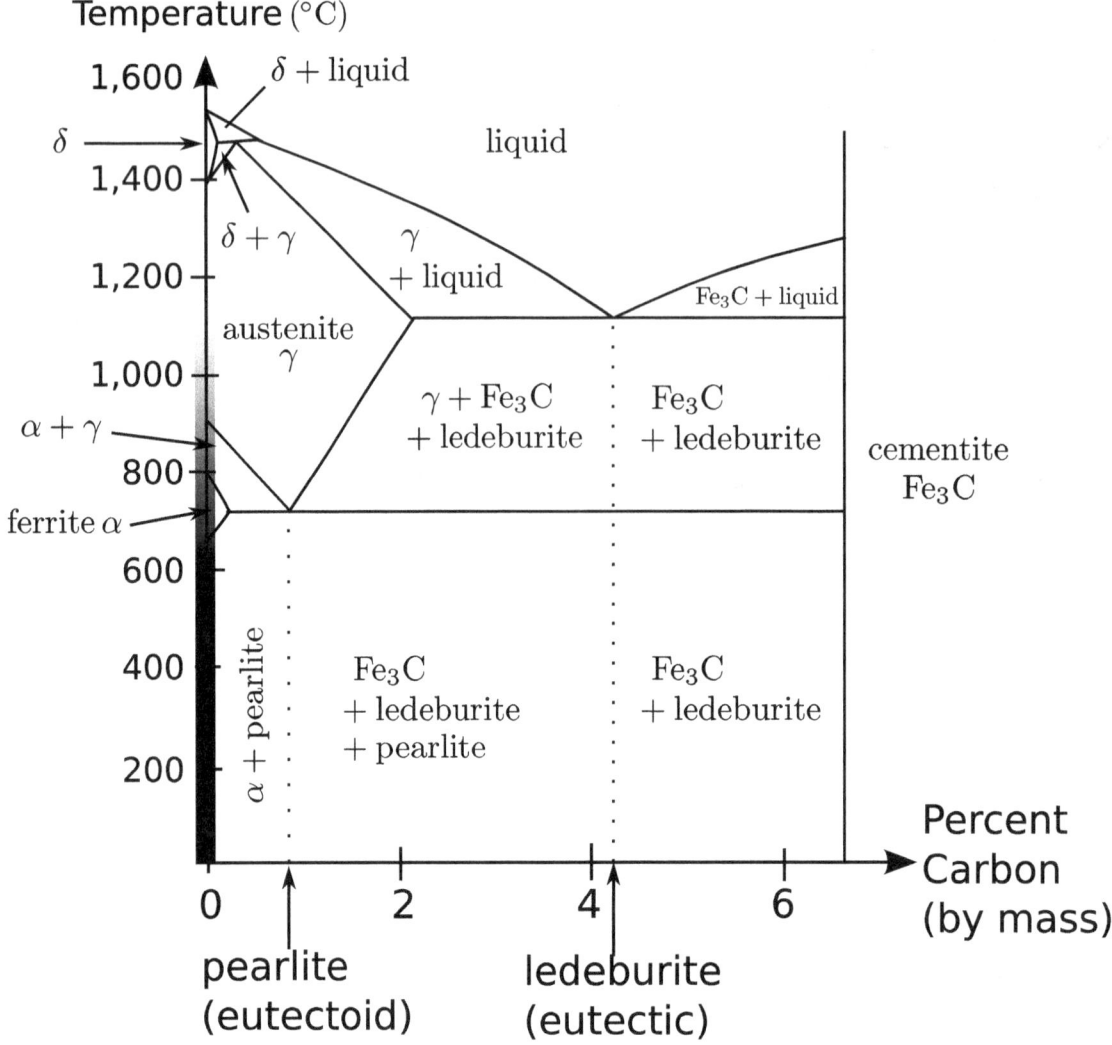

Iron-carbon phase diagram, showing the conditions necessary to form different phases

19.3 Number of phases

See also: Multiphasic liquid

For a given composition, only certain phases are possible at a given temperature and pressure. The number and type of phases that will form is hard to predict and is usually determined by experiment. The results of such experiments can be plotted in phase diagrams.

The phase diagram shown here is for a single component system. In this simple system, which phases that are possible depends only on pressure and temperature. The markings show points where two or more phases can co-exist in equilibrium. At temperatures and pressures away from the markings, there will be only one phase at equilibrium.

In the diagram, the blue line marking the boundary between liquid and gas does not continue indefinitely, but terminates at a point called the critical point. As the temperature and pressure approach the critical point, the properties of the liquid and gas become progressively more similar. At the critical point, the liquid and gas become indistinguishable. Above the critical point, there are no longer separate liquid and gas phases: there is only a generic fluid phase referred to as a supercritical fluid. In water, the critical point occurs at around 647 K (374 °C or 705 °F) and 22.064 MPa.

An unusual feature of the water phase diagram is that the solid–liquid phase line (illustrated by the dotted green line) has

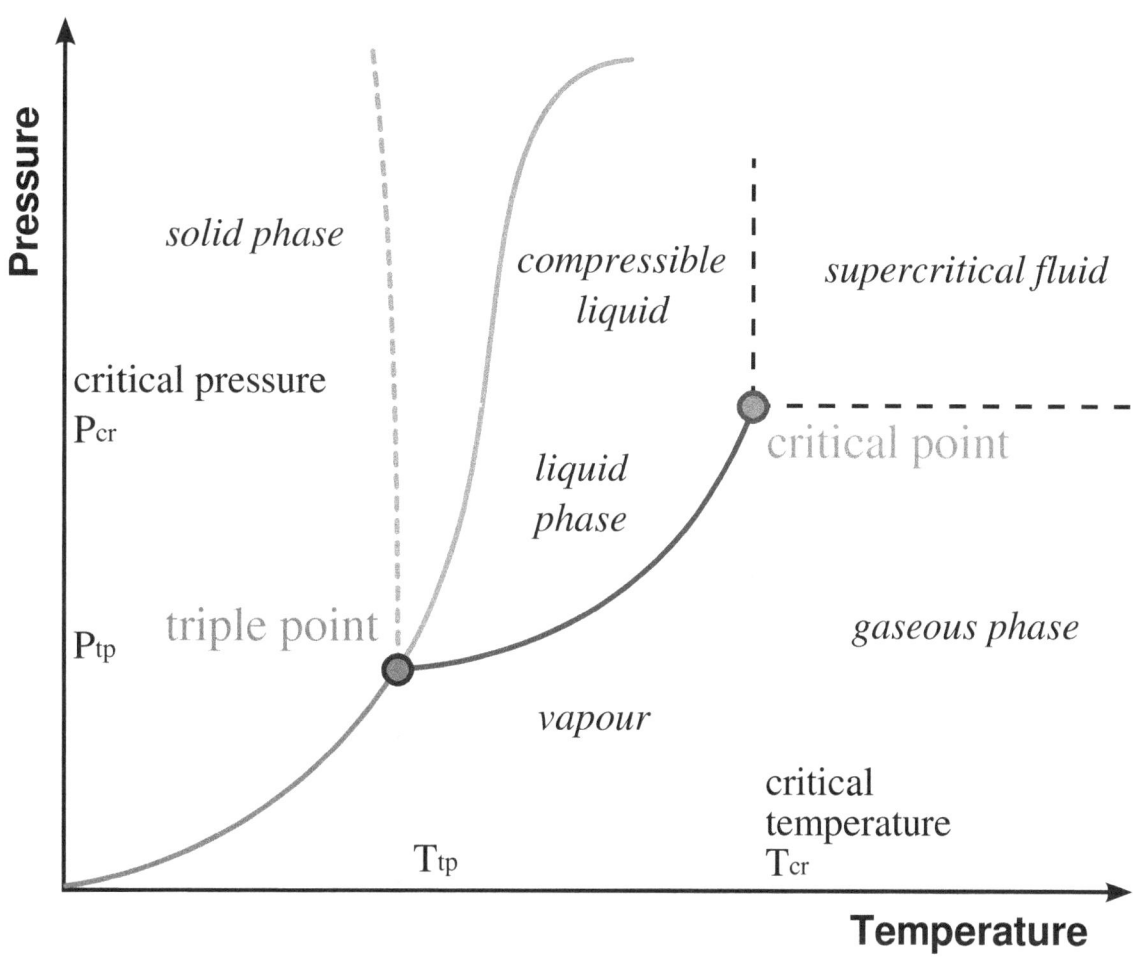

A typical phase diagram for a single-component material, exhibiting solid, liquid and gaseous phases. The solid green line shows the usual shape of the liquid–solid phase line. The dotted green line shows the anomalous behavior of water when the pressure increases. The triple point and the critical point are shown as red dots.

a negative slope. For most substances, the slope is positive as exemplified by the dark green line. This unusual feature of water is related to ice having a lower density than liquid water. Increasing the pressure drives the water into the higher density phase, which causes melting.

Another interesting though not unusual feature of the phase diagram is the point where the solid–liquid phase line meets the liquid–gas phase line. The intersection is referred to as the triple point. At the triple point, all three phases can coexist.

Experimentally, the phase lines are relatively easy to map due to the interdependence of temperature and pressure that develops when multiple phases forms. See Gibbs' phase rule. Consider a test apparatus consisting of a closed and well insulated cylinder equipped with a piston. By charging the right amount of water and applying heat, the system can be brought to any point in the gas region of the phase diagram. If the piston is slowly lowered, the system will trace a curve of increasing temperature and pressure within the gas region of the phase diagram. At the point where gas begins to condense to liquid, the direction of the temperature and pressure curve will abruptly change to trace along the phase line until all of the water has condensed.

19.4 Interfacial phenomena

Between two phases in equilibrium there is a narrow region where the properties are not that of either phase. Although this region may be very thin, it can have significant and easily observable effects, such as causing a liquid to exhibit surface tension. In mixtures, some components may preferentially move toward the interface. In terms of modeling, describing, or understanding the behavior of a particular system, it may be efficacious to treat the interfacial region as a separate phase.

19.5 Crystal phases

A single material may have several distinct solid states capable of forming separate phases. Water is a well-known example of such a material. For example, water ice is ordinarily found in the hexagonal form ice Ih, but can also exist as the cubic ice Ic, the rhombohedral ice II, and many other forms. Polymorphism is the ability of a solid to exist in more than one crystal form. For pure chemical elements, polymorphism is known as allotropy. For example, diamond, graphite, and fullerenes are different allotropes of carbon.

19.6 Phase transitions

Main article: Phase transition

When a substance undergoes a phase transition (changes from one state of matter to another) it usually either takes up or releases energy. For example, when water evaporates, the kinetic energy expended as the evaporating molecules escape the attractive forces of the liquid is reflected in a decrease in temperature. The amount of energy required to induce the transition is more than the amount required to heat the water from room temperature to just short of boiling temperature, which is why evaporation is useful for cooling. See Enthalpy of vaporization. The reverse process, condensation, releases heat. The heat energy, or enthalpy, associated with a solid to liquid transition is the enthalpy of fusion and that associated with a solid to gas transition is the enthalpy of sublimation.

19.7 See also

- Phase boundary

- Literature of phase boundaries

19.8 Notes and references

[1] Modell, Michael; Robert C. Reid (1974). *Thermodynamics and Its Applications*. Englewood Cliffs, NJ: Prentice-Hall. ISBN 0-13-914861-2.

[2] Enrico Fermi (25 April 2012). *Thermodynamics*. Courier Corporation. ISBN 978-0-486-13485-7.

[3] Clement John Adkins (14 July 1983). *Equilibrium Thermodynamics*. Cambridge University Press. ISBN 978-0-521-27456-2.

[4] One such system is, from the top: mineral oil, silicone oil, water, aniline, perfluoro(dimethylcyclohexane), white phosphorus, gallium, and mercury. The system remains indefinitely separated at 45 °C, where gallium and phosphorus are in the molten state. From Reichardt, C. (2006). *Solvents and Solvent Effects in Organic Chemistry*. Wiley-VCH. pp. 9–10. ISBN 3-527-60567-3.

[5] This phenomenon can be used to help with catalyst recycling in Heck vinylation. See Bhanage, B.M. et al. (1998). "Comparison of activity and selectivity of various metal-TPPTS complex catalysts in ethylene glycol — toluene biphasic Heck vinylation reactions of iodobenzene". *Tetrahedron Letters* **39** (51): 9509–9512. doi:10.1016/S0040-4039(98)02225-4.

19.9   External links

- French physicists find a solution that reversibly solidifies with a *rise* in temperature – α-cyclodextrin, water, and 4-methylpyridine

Chapter 20

Isotope

This article is about the atomic variants of chemical elements. For other uses, see Isotope (disambiguation).
  Isotopes are variants of a particular chemical element which differ in neutron number, although all isotopes of a given

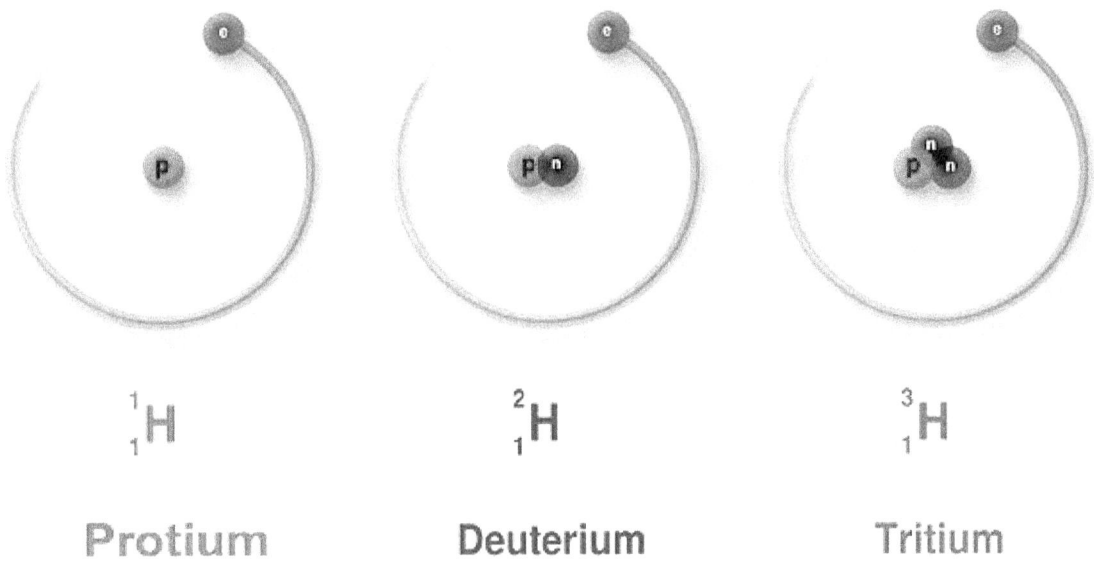

$$^1_1H \qquad ^2_1H \qquad ^3_1H$$

Protium   Deuterium   Tritium

The three naturally-occurring isotopes of hydrogen. The fact that each isotope has one proton makes them all variants of hydrogen: the identity of the isotope is given by the number of neutrons. From left to right, the isotopes are protium (^1H) with zero neutrons, deuterium (^2H) with one neutron, and tritium (^3H) with two neutrons.

element have the same number of protons in each atom. The term isotope is formed from the Greek roots isos (ἴσος "equal") and topos (τόπος "place"), meaning "the same place". Thus, different isotopes of a single element occupy the same position on the periodic table. The number of protons within the atom's nucleus is called atomic number and is equal to the number of electrons in the neutral (non-ionized) atom. Each atomic number identifies a specific element, but not the isotope; an atom of a given element may have a wide range in its number of neutrons. The number of nucleons (both protons and neutrons) in the nucleus is the atom's mass number, and each isotope of a given element has a different mass number.

For example, carbon-12, carbon-13 and carbon-14 are three isotopes of the element carbon with mass numbers 12, 13 and 14 respectively. The atomic number of carbon is 6, which means that every carbon atom has 6 protons, so that the neutron numbers of these isotopes are 6, 7 and 8 respectively.

20.1   Isotope vs. nuclide

Nuclide refers to a nucleus rather than to an atom. Identical nuclei belong to one nuclide, for example each nucleus of the carbon-13 nuclide is composed of 6 protons and 7 neutrons. The *nuclide* concept (referring to individual nuclear species) emphasizes nuclear properties over chemical properties, while the *isotope* concept (grouping all atoms of each element) emphasizes chemical over nuclear. The neutron number has large effects on nuclear properties, but its effect on chemical properties is negligible for most elements. Even in the case of the very lightest elements where the ratio of neutron number to atomic number varies the most between isotopes it usually has only a small effect, although it does matter in some circumstances (for hydrogen, the lightest element, the isotope effect is large enough to strongly affect biology). Since *isotope* is the older term, it is better known than *nuclide*, and is still sometimes used in contexts where *nuclide* might be more appropriate, such as nuclear technology and nuclear medicine.

20.2   Notation

An isotope and/or nuclide is specified by the name of the particular element (this indicates the atomic number) followed by a hyphen and the mass number (e.g. helium-3, helium-4, carbon-12, carbon-14, uranium-235 and uranium-239).[1] When a chemical symbol is used, e.g., "C" for carbon, standard notation (now known as "AZE notation" because A is the mass number, Z the atomic number, and E for element) is to indicate the mass number (number of nucleons) with a superscript at the upper left of the chemical symbol and to indicate the atomic number with a subscript at the lower left (e.g. $^{3}_{2}$He, $^{4}_{2}$He, $^{12}_{6}$C, $^{14}_{6}$C, $^{235}_{92}$U, and $^{239}_{92}$U).[2] Since the atomic number is given by the element symbol, it is common to state only the mass number in the superscript and leave out the atomic number subscript (e.g. 3He, 4He, 12C, 14C, 235U, and 239U). The letter m is sometimes appended after the mass number to indicate a nuclear isomer, a metastable or energetically-excited nuclear state (as opposed to the lowest-energy ground state), for example $^{180m}_{73}$Ta (tantalum-180m).

20.3   Radioactive, primordial, and stable isotopes

Some isotopes are radioactive, and are therefore described as radioisotopes or radionuclides, while others have never been observed to undergo radioactive decay and are described as stable isotopes or stable nuclides. For example, 14C is a radioactive form of carbon while 12C and 13C are stable isotopes. There are about 339 naturally occurring nuclides on Earth,[3] of which 288 are primordial nuclides, meaning that they have existed since the solar system's formation.

Primordial nuclides include 34 nuclides with very long half-lives (over 80 million years) and 254 that are formally considered as "stable nuclides",[3] since they have not been observed to decay. In most cases, for obvious reasons, if an element has stable isotopes, those isotopes predominate in the elemental abundance found on Earth and in the solar system. However, in the cases of three elements (tellurium, indium, and rhenium) the most abundant isotope found in nature is actually one (or two) extremely long lived radioisotope(s) of the element, despite these elements having one or more stable isotopes.

Theory predicts that many apparently "stable" isotopes/nuclides are radioactive, with extremely long half-lives (discounting the possibility of proton decay, which would make all nuclides ultimately unstable). Of the 254 nuclides never observed to decay, only 90 of these (all from the first 40 elements) are theoretically stable to all known forms of decay. Element 41 (niobium) is theoretically unstable via spontaneous fission, but this has never been detected. Many other stable nuclides are in theory energetically susceptible to other known forms of decay, such as alpha decay or double beta decay, but no decay products have yet been observed, and so these isotopes are described as "observationally stable". The predicted half-lives for these nuclides often greatly exceed the estimated age of the universe, and in fact there are also 27 known radionuclides (see primordial nuclide) with half-lives longer than the age of the universe.

Adding in the radioactive nuclides that have been created artificially, there are more than 3100 currently known nuclides.[4] These include 905 nuclides that are either stable or have half-lives longer than 60 minutes. See list of nuclides for details.

20.4 History

20.4.1 Radioactive isotopes

The existence of isotopes was first suggested in 1913 by the radiochemist Frederick Soddy, based on studies of radioactive decay chains that indicated about 40 different species described as *radioelements* (i.e. radioactive elements) between uranium and lead, although the periodic table only allowed for 11 elements from uranium to lead.[5][6]

Several attempts to separate these new radioelements chemically had failed.[7] For example, Soddy had shown in 1910 that mesothorium (later shown to be ^{228}Ra), radium (^{226}Ra, the longest-lived isotope), and thorium X (^{224}Ra) are impossible to separate.[8] Attempts to place the radioelements in the periodic table led Soddy and Kazimierz Fajans independently to propose their radioactive displacement law in 1913, to the effect that alpha decay produced an element two places to the left in the periodic table, while beta decay emission produced an element one place to the right.[9] Soddy recognized that emission of an alpha particle followed by two beta particles led to the formation of an element chemically identical to the initial element but with a mass four units lighter and with different radioactive properties.

Soddy proposed that several types of atoms (differing in radioactive properties) could occupy the same place in the table. For example, the alpha-decay of uranium-235 forms thorium-231, while the beta decay of actinium-230 forms thorium-230.[7] The term "isotope", Greek for "at the same place", was suggested to Soddy by Margaret Todd, a Scottish physician and family friend, during a conversation in which he explained his ideas to her.[8][10][11][12][13][14]

In 1914 T. W. Richards found variations between the atomic weight of lead from different mineral sources, attributable to variations in isotopic composition due to different radioactive origins.[7][15]

20.4.2 Stable isotopes

The first evidence for multiple isotopes of a stable (non-radioactive) element was found by J. J. Thomson in 1913 as part of his exploration into the composition of canal rays (positive ions).[16][17] Thomson channeled streams of neon ions through a magnetic and an electric field and measured their deflection by placing a photographic plate in their path. Each stream created a glowing patch on the plate at the point it struck. Thomson observed two separate patches of light on the photographic plate (see image), which suggested two different parabolas of deflection. Thomson eventually concluded that some of the atoms in the neon gas were of higher mass than the rest.

F. W. Aston subsequently discovered multiple stable isotopes for numerous elements using a mass spectrograph. In 1919 Aston studied neon with sufficient resolution to show that the two isotopic masses are very close to the integers 20 and 22, and that neither is equal to the known molar mass (20.2) of neon gas. This is an example of Aston's whole number rule for isotopic masses, which states that large deviations of elemental molar masses from integers are primarily due to the fact that the element is a mixture of isotopes. Aston similarly showed that the molar mass of chlorine (35.45) is a weighted average of the almost integral masses for the two isotopes ^{35}Cl and ^{37}Cl.[18]

20.5 Variation in properties between isotopes

See also: Neutron-proton ratio

20.5.1 Chemical and molecular properties

A neutral atom has the same number of electrons as protons. Thus different isotopes of a given element all have the same number of electrons and share a similar electronic structure. Because the chemical behavior of an atom is largely determined by its electronic structure, different isotopes exhibit nearly identical chemical behavior. The main exception to this is the kinetic isotope effect: due to their larger masses, heavier isotopes tend to react somewhat more slowly than lighter isotopes of the same element. This is most pronounced by far for protium (1H), deuterium (2H), and tritium (3H), because deuterium has twice the mass of protium and tritium has three times the mass of protium. These mass differences also affect the behavior of their respective chemical bonds, by changing the center of gravity (reduced mass) of the atomic systems. However, for heavier elements the relative mass difference between isotopes is much less, so that the mass-difference effects on chemistry are usually negligible. (Heavy elements also have relatively more neutrons than lighter elements, so the ratio of the nuclear mass to the collective electronic mass is slightly greater.)

Similarly, two molecules that differ only in the isotopes of their atoms (isotopologues) have identical electronic structure, and therefore almost indistinguishable physical and chemical properties (again with deuterium and tritium being the primary exceptions). The **vibrational modes** of a molecule are determined by its shape and by the masses of its constituent atoms; so different isotopologues have different sets of vibrational modes. Since vibrational modes allow a molecule to absorb photons of corresponding energies, isotopologues have different optical properties in the infrared range.

20.5.2 Nuclear properties and stability

See also: Stable isotope, List of nuclides and List of elements by stability of isotopes

Atomic nuclei consist of protons and neutrons bound together by the residual strong force. Because protons are positively charged, they repel each other. Neutrons, which are electrically neutral, stabilize the nucleus in two ways. Their copresence pushes protons slightly apart, reducing the electrostatic repulsion between the protons, and they exert the attractive nuclear force on each other and on protons. For this reason, one or more neutrons are necessary for two or more protons to bind into a nucleus. As the number of protons increases, so does the ratio of neutrons to protons necessary to ensure a stable nucleus (see graph at right). For example, although the neutron:proton ratio of 3
2He is 1:2, the neutron:proton ratio of 238
92U is greater than 3:2. A number of lighter elements have stable nuclides with the ratio 1:1 ($Z = N$). The nuclide 40
20Ca (calcium-40) is observationally the heaviest stable nuclide with the same number of neutrons and protons; (theoretically, the heaviest stable one is sulfur-32). All stable nuclides heavier than calcium-40 contain more neutrons than protons.

20.5.3 Numbers of isotopes per element

Of the 81 elements with a stable isotope, the largest number of stable isotopes observed for any element is ten (for the element tin). No element has nine stable isotopes. Xenon is the only element with eight stable isotopes. Four elements have seven stable isotopes, eight have six stable isotopes, ten have five stable isotopes, nine have four stable isotopes, five have three stable isotopes, 16 have two stable isotopes (counting 180m
73Ta as stable), and 26 elements have only a single stable isotope (of these, 19 are so-called mononuclidic elements, having a single primordial stable isotope that dominates and fixes the atomic weight of the natural element to high precision; 3 radioactive **mononuclidic** elements occur as well).[19] In total, there are 254 nuclides that have not been observed to decay. For the 80 elements that have one or more stable isotopes, the average number of stable isotopes is 254/80 = 3.2 isotopes per element.

20.5.4 Even and odd nucleon numbers

Main article: Even and odd atomic nuclei

The proton:neutron ratio is not the only factor affecting nuclear stability. It depends also on evenness or oddness of its atomic number Z, neutron number N and, consequently, of their sum, the mass number A. Oddness of both Z and N tends to lower the nuclear binding energy, making odd nuclei, generally, less stable. This remarkable difference of nuclear binding energy between neighbouring nuclei, especially of odd-A isobars, has important consequences: unstable isotopes with a nonoptimal number of neutrons or protons decay by beta decay (including positron decay), electron capture or other exotic means, such as spontaneous fission and cluster decay.

The majority of stable nuclides are even-proton-even-neutron, where all numbers Z, N, and A are even. The odd-A stable nuclides are divided (roughly evenly) into odd-proton-even-neutron, and even-proton-odd-neutron nuclides. Odd-proton-odd-neutron nuclei are the least common.

Even atomic number

The 148 even-proton, even-neutron (EE) nuclides comprise ~ 58% of all stable nuclides and all have spin 0 because of pairing. There are also 22 primordial long-lived even-even nuclides. As a result, each of the 41 even-numbered elements from 2 to 82 has at least one stable isotope, and most of these elements have *several* primordial isotopes. Half of these even-numbered elements have six or more stable isotopes. The extreme stability of helium-4 due to a double pairing of 2 protons and 2 neutrons prevents *any* nuclides containing five or eight nucleons from existing for long enough to serve as platforms for the buildup of heavier elements via nuclear fusion in stars (see triple alpha process).

These 53 stable nuclides have an even number of protons and an odd number of neutrons. They are a minority in comparison to the even-even isotopes, which are about 3 times as numerous. Among the 41 even-Z elements that have a stable nuclide, only three elements (argon, cerium, and lead) have no even-odd stable nuclides. One element (tin) has three. There are 24 elements that have one even-odd nuclide and 13 that have two odd-even nuclides. Of 35 primordial radionuclides there exist four even-odd nuclides (see table at right), including the fissile 235
92U. Because of their odd neutron numbers, the even-odd nuclides tend to have large neutron capture cross sections, due to the energy that results from neutron-pairing effects. These stable even-proton odd-neutron nuclides tend to be uncommon by abundance in nature, generally because, to form and enter into primordial abundance, they must have escaped capturing neutrons to form yet other stable even-even isotopes, during both the s-process and r-process of neutron capture, during nucleosynthesis in stars. For this reason, only 195
78Pt and 9
4Be are the most naturally abundant isotopes of their element.

Odd atomic number

48 stable odd-proton-even-neutron nuclides, stabilized by their even numbers of paired neutrons, form most of the stable isotopes of the odd-numbered elements; the very few odd-odd nuclides comprise the others. There are 41 odd-numbered elements with $Z = 1$ through 81, of which 39 have stable isotopes (the elements technetium (
43Tc) and promethium (
61Pm) have no stable isotopes). Of these 39 odd Z elements, 30 elements (including hydrogen-1 where 0 neutrons is even) have one stable odd-even isotope, and nine elements: chlorine (
17Cl), potassium (
19K), copper (
29Cu), gallium (
31Ga), bromine (
35Br), silver (
47Ag), antimony (
51Sb), iridium (
77Ir), and thallium (
81Tl), have two odd-even stable isotopes each. This makes a total $30 + 2(9) = 48$ stable odd-even isotopes.

There are also five primordial long-lived radioactive odd-even isotopes, 87
37Rb, 115
49In, 187
75Re, 151

63Eu, and 209

83Bi. The last two were only recently found to decay, with half-lives greater than 10^{18} years.

Only five stable nuclides contain both an odd number of protons *and* an odd number of neutrons. The first four "odd-odd" nuclides occur in low mass nuclides, for which changing a proton to a neutron or vice versa would lead to a very lopsided proton-neutron ratio (2

1H, 6

3Li, 10

5B, and 14

7N; spins 1, 1, 3, 1). The only other entirely "stable" odd-odd nuclide is 180m

73Ta (spin 9), the only primordial nuclear isomer, which has not yet been observed to decay despite experimental attempts.[20] Hence, all observationally stable odd-odd nuclides have nonzero integer spin. This is because the single unpaired neutron and unpaired proton have a larger nuclear force attraction to each other if their spins are aligned (producing a total spin of at least 1 unit), instead of anti-aligned. See deuterium for the simplest case of this nuclear behavior.

Many odd-odd radionuclides (like tantalum-180) with comparatively short half lives are known. Usually, they beta-decay to their nearby even-even isobars that have paired protons and paired neutrons. Of the nine primordial odd-odd nuclides (five stable and four radioactive with long half lives), only 14

7N is the most common isotope of a common element. This is the case because it is a part of the CNO cycle. The nuclides 6

3Li and 10

5B are minority isotopes of elements that are themselves rare compared to other light elements, while the other six isotopes make up only a tiny percentage of the natural abundance of their elements. For example, 180m

73Ta is thought to be the rarest of the 254 stable isotopes.

Odd neutron number

Actinides with odd neutron number are generally fissile (with thermal neutrons), while those with even neutron number are generally not, though they are fissionable with fast neutrons. Only 195

78Pt, 9

4Be and 14

7N have odd neutron number and are the most naturally abundant isotope of their element.

20.6   Occurrence in nature

See also: Abundance of the chemical elements

Elements are composed of one or more naturally occurring isotopes. The unstable (radioactive) isotopes are either primordial or postprimordial. Primordial isotopes were a product of stellar nucleosynthesis or another type of nucleosynthesis such as cosmic ray spallation, and have persisted down to the present because their rate of decay is so slow (e.g., uranium-238 and potassium-40). Postprimordial isotopes were created by cosmic ray bombardment as cosmogenic nuclides (e.g., tritium, carbon-14), or by the decay of a radioactive primordial isotope to a radioactive radiogenic nuclide daughter (e.g., uranium to radium). A few isotopes are naturally synthesized as nucleogenic nuclides, by some other natural nuclear reaction, such as when neutrons from natural nuclear fission are absorbed by another atom.

As discussed above, only 80 elements have any stable isotopes, and 26 of these have only one stable isotope. Thus, about two-thirds of stable elements occur naturally on Earth in multiple stable isotopes, with the largest number of stable isotopes for an element being ten, for tin (

50Sn). There are about 94 elements found naturally on Earth (up to plutonium inclusive), though some are detected only in very tiny amounts, such as plutonium-244. Scientists estimate that the elements that occur naturally on Earth (some only as radioisotopes) occur as 339 isotopes (nuclides) in total.[21] Only 254 of these naturally occurring isotopes are stable in the sense of never having been observed to decay as of the present time. An additional 35 primordial nuclides (to a total of 289 primordial nuclides), are radioactive with known half-lives, but have half-lives longer than 80 million years, allowing them to exist from the beginning of the solar system. See list of nuclides for details.

All the known stable isotopes occur naturally on Earth; the other naturally occurring-isotopes are radioactive but occur on Earth due to their relatively long half-lives, or else due to other means of ongoing natural production. These include the afore-mentioned cosmogenic nuclides, the nucleogenic nuclides, and any radiogenic radioisotopes formed by ongoing decay of a primordial radioactive isotope, such as radon and radium from uranium.

An additional ~3000 radioactive isotopes not found in nature have been created in nuclear reactors and in particle accelerators. Many short-lived isotopes not found naturally on Earth have also been observed by spectroscopic analysis, being naturally created in stars or supernovae. An example is aluminium-26, which is not naturally found on Earth, but is found in abundance on an astronomical scale.

The tabulated atomic masses of elements are averages that account for the presence of multiple isotopes with different masses. Before the discovery of isotopes, empirically determined noninteger values of atomic mass confounded scientists. For example, a sample of chlorine contains 75.8% chlorine-35 and 24.2% chlorine-37, giving an average atomic mass of 35.5 atomic mass units.

According to generally accepted cosmology theory, only isotopes of hydrogen and helium, traces of some isotopes of lithium and beryllium, and perhaps some boron, were created at the Big Bang, while all other isotopes were synthesized later, in stars and supernovae, and in interactions between energetic particles such as cosmic rays, and previously produced isotopes. (See nucleosynthesis for details of the various processes thought responsible for isotope production.) The respective abundances of isotopes on Earth result from the quantities formed by these processes, their spread through the galaxy, and the rates of decay for isotopes that are unstable. After the initial coalescence of the solar system, isotopes were redistributed according to mass, and the isotopic composition of elements varies slightly from planet to planet. This sometimes makes it possible to trace the origin of meteorites.

20.7   Atomic mass of isotopes

The atomic mass (m_r) of an isotope is determined mainly by its mass number (i.e. number of nucleons in its nucleus). Small corrections are due to the binding energy of the nucleus (see mass defect), the slight difference in mass between proton and neutron, and the mass of the electrons associated with the atom, the latter because the electron:nucleon ratio differs among isotopes.

The mass number is a dimensionless quantity. The atomic mass, on the other hand, is measured using the atomic mass unit based on the mass of the carbon-12 atom. It is denoted with symbols "u" (for unified atomic mass unit) or "Da" (for dalton).

The atomic masses of naturally occurring isotopes of an element determine the atomic mass of the element. When the element contains N isotopes, the expression below is applied for the average atomic mass $\overline{m}_a$:

$$\overline{m}_a = m_1 x_1 + m_2 x_2 + ... + m_N x_N$$

where m_1, m_2, ..., mN are the atomic masses of each individual isotope, and x_1, ..., xN are the relative abundances of these isotopes.

20.8   Applications of isotopes

20.8.1   Purification of isotopes

Main article: isotope separation

Several applications exist that capitalize on properties of the various isotopes of a given element. Isotope separation is a significant technological challenge, particularly with heavy elements such as uranium or plutonium. Lighter elements such as lithium, carbon, nitrogen, and oxygen are commonly separated by gas diffusion of their compounds such as CO and NO. The separation of hydrogen and deuterium is unusual since it is based on chemical rather than physical properties, for example in the Girdler sulfide process. Uranium isotopes have been separated in bulk by gas diffusion, gas centrifugation, laser ionization separation, and (in the Manhattan Project) by a type of production mass spectrometry.

20.8.2   Use of chemical and biological properties

Main articles: isotope geochemistry, cosmochemistry and paleoclimatology

- Isotope analysis is the determination of isotopic signature, the relative abundances of isotopes of a given element in a particular sample. For biogenic substances in particular, significant variations of isotopes of C, N and O can occur. Analysis of such variations has a wide range of applications, such as the detection of adulteration in food products[22] or the geographic origins of products using isoscapes. The identification of certain meteorites as having originated on Mars is based in part upon the isotopic signature of trace gases contained in them.[23]

- Isotopic substitution can be used to determine the mechanism of a chemical reaction via the kinetic isotope effect.

- Another common application is isotopic labeling, the use of unusual isotopes as tracers or markers in chemical reactions. Normally, atoms of a given element are indistinguishable from each other. However, by using isotopes of different masses, even different nonradioactive stable isotopes can be distinguished by mass spectrometry or infrared spectroscopy. For example, in 'stable isotope labeling with amino acids in cell culture (SILAC)' stable isotopes are used to quantify proteins. If radioactive isotopes are used, they can be detected by the radiation they emit (this is called *radioisotopic labeling*).

- Isotopes are commonly used to determine the concentration of various elements or substances using the isotope dilution method, whereby known amounts of isotopically-substituted compounds are mixed with the samples and the isotopic signatures of the resulting mixtures are determined with mass spectrometry.

20.8.3   Use of nuclear properties

- A technique similar to radioisotopic labeling is radiometric dating: using the known half-life of an unstable element, one can calculate the amount of time that has elapsed since a known level of isotope existed. The most widely known example is radiocarbon dating used to determine the age of carbonaceous materials.

- Several forms of spectroscopy rely on the unique nuclear properties of specific isotopes, both radioactive and stable. For example, nuclear magnetic resonance (NMR) spectroscopy can be used only for isotopes with a nonzero nuclear spin. The most common isotopes used with NMR spectroscopy are ^{1}H, ^{2}D, ^{15}N, ^{13}C, and ^{31}P.

- Mössbauer spectroscopy also relies on the nuclear transitions of specific isotopes, such as ^{57}Fe.

- Radionuclides also have important uses. Nuclear power and nuclear weapons development require relatively large quantities of specific isotopes. Nuclear medicine and radiation oncology utilize radioisotopes respectively for medical diagnosis and treatment.

20.9   See also

- Abundance of the chemical elements

- Atom

- Table of nuclides

- Table of nuclides (complete)

- List of isotopes

- List of isotopes by half-life

- List of elements by stability of isotopes

- Isotones

- Isobars

- Radionuclide (or radioisotope)

- Nuclear medicine (includes medical isotopes)

- Isotopomer

- List of particles

- Geotraces

- Isotope dilution

20.10   Notes

- Isotopes are nuclides having the same number of protons; compare:

  - Isotones are nuclides having the same number of neutrons. $N = A - Z$

  - Isobars are nuclides having the same mass number, i.e. sum of protons plus neutrons. A

  - Nuclear isomers are different excited states of the same type of nucleus. A transition from one isomer to another is accompanied by emission or absorption of a gamma ray, or the process of internal conversion. Isomers are by definition both isotopic and isobaric. (Not to be confused with chemical isomers.)

  - Isodiaphers are nuclides having the same neutron excess, i.e. number of neutrons minus number of protons. $D = N - Z$

- Bainbridge mass spectrometer

20.11   References

[1] IUPAC (Connelly, N. G.; Damhus, T.; Hartshorn, R. M.; and Hutton, A. T.), *Nomenclature of Inorganic Chemistry – IUPAC Recommendations 2005*, The Royal Society of Chemistry, 2005; IUPAC (McCleverty, J. A.; and Connelly, N. G.), *Nomenclature of Inorganic Chemistry II. Recommendations 2000*, The Royal Society of Chemistry, 2001; IUPAC (Leigh, G. J.), *Nomenclature of Inorganic Chemistry (recommendations 1990)*, Blackwell Science, 1990; IUPAC, *Nomenclature of Inorganic Chemistry, Second Edition*, 1970; probably in the 1958 first edition as well

[2] This notation seems to have been introduced in the second half of the 1930s. Before that, various notations were used, such as Ne(22) for neon-22 (1934), Ne^{22} for neon-22 (1935), or even Pb_{210} for lead-210 (1933).

[3] "Radioactives Missing From The Earth".

[4] "NuDat 2 Description".

[5] Choppin, G.; Liljenzin, J. O. and Rydberg, J. (1995) *Radiochemistry and Nuclear Chemistry* (2nd ed.) Butterworth-Heinemann, pp. 3–5

[6] Others had also suggested the possibility of isotopes; e.g.,

  - Strömholm, Daniel and Svedberg, Theodor (1909) "Untersuchungen über die Chemie der radioactiven Grundstoffe II." (Investigations into the chemistry of the radioactive elements, part 2), *Zeitschrift für anorganischen Chemie*, **63**: 197–206; see especially page 206.

- Alexander Thomas Cameron, *Radiochemistry* (London, England: J. M. Dent & Sons, 1910), p. 141. (Cameron also anticipated the displacement law.)

[7]   Scerri, Eric R. (2007) *The Periodic Table* Oxford University Press, pp. 176–179 ISBN 0195305736

[8]   Nagel, Miriam C. (1982). "Frederick Soddy: From Alchemy to Isotopes". *Journal of Chemical Education* **59** (9): 739–740. Bibcode:1982JChEd..59..739N. doi:10.1021/ed059p739.

[9]   See:

  - Kasimir Fajans (1913) "Über eine Beziehung zwischen der Art einer radioaktiven Umwandlung und dem elektrochemischen Verhalten der betreffenden Radioelemente" (On a relation between the type of radioactive transformation and the electrochemical behavior of the relevant radioactive elements), *Physikalische Zeitschrift*, **14**: 131–136.

  - Soddy announced his "displacement law" in: Soddy, Frederick (1913). "The Radio-Elements and the Periodic Law". *Nature* **91** (2264): 57. Bibcode:1913Natur..91...57S. doi:10.1038/091057a0..

  - Soddy elaborated his displacement law in: Soddy, Frederick (1913) "Radioactivity," *Chemical Society Annual Report*, **10**: 262–288.

  - Alexander Smith Russell (1888–1972) also published a displacement law: Russell, Alexander S. (1913) "The periodic system and the radio-elements," *Chemical News and Journal of Industrial Science*, **107**: 49–52.

[10]Soddy first used the word "isotope" in: Soddy, Frederick (1913)."Intra-atomic charge".*Nature***92**(2301): 399–400.Bibcod. doi:10.1038/092399c0.

[11]Fleck, Alexander (1957). "Frederick Soddy".*Biographical Memoirs of Fellows of the Royal Society***3**: 203–216.doi:10.104. p. 208: Up to 1913 we used the phrase 'radio elements chemically non-separable' and at that time the word isotope was suggested in a drawing-room discussion with Dr. Margaret Todd in the home of Soddy's father-in-law, Sir George Beilby.

[12] Budzikiewicz H and Grigsby RD (2006). "Mass spectrometry and isotopes: a century of research and discussion". *Mass spectrometry reviews* **25** (1): 146–57. doi:10.1002/mas.20061. PMID 16134128.

[13] Scerri, Eric R. (2007) *The Periodic Table*, Oxford University Press, ISBN 0195305736, Ch. 6, note 44 (p. 312) citing Alexander Fleck, described as a former student of Soddy's.

[14] In his 1893 book, William T. Preyer also used the word "isotope" to denote similarities among elements. From p. 9 of William T. Preyer, *Das genetische System der chemischen Elemente* [The genetic system of the chemical elements] (Berlin, Germany: R. Friedländer & Sohn, 1893): "Die ersteren habe ich der Kürze wegen isotope Elemente genannt, weil sie in jedem der sieben Stämmme der gleichen Ort, nämlich dieselbe Stuffe, einnehmen." (For the sake of brevity, I have named the former "isotopic" elements, because they occupy the same place in each of the seven families [i.e., columns of the periodic table], namely the same step [i.e., row of the periodic table].)

[15] The origins of the conceptions of isotopes Frederick Soddy, Nobel prize lecture

[16]Thomson, J. J. (1912). "XIX. Further experiments on positive rays".*Philosophical Magazine Series 6***24**(140): 209.doi:8637325.

[17]Thomson, J. J. (1910)."LXXXIII. Rays of positive electricity".*Philosophical Magazine Series 6***20**(118): 752.doi:10.106962.

[18] Mass spectra and isotopes Francis W. Aston, Nobel prize lecture 1922

[19] **Sonzogni**, Alejandro (2008). "Interactive Chart of Nuclides". National Nuclear Data Center: Brookhaven National Laboratory. Retrieved 2013-05-03.

[20] Hult, Mikael; Wieslander, J. S.; Marissens, Gerd; Gasparro, Joël; Wätjen, Uwe; Misiaszek, Marcin (2009). "Search for the radioactivity of 180mTa using an underground HPGe sandwich spectrometer". *Applied Radiation and Isotopes* **67** (5): 918–21. doi:10.1016/j.apradiso.2009.01.057. PMID 19246206.

[21] "Radioactives Missing From The Earth". Don-lindsay-archive.org. Retrieved 2012-06-16.

[22] E. Jamin; Guérin, Régis; Rétif, Mélinda; Lees, Michèle; Martin, Gérard J. et al. (2003). "Improved Detection of Added Water in Orange Juice by Simultaneous Determination of the Oxygen-18/Oxygen-16 Isotope Ratios of Water and Ethanol Derived from Sugars". *J. Agric. Food Chem.* **51** (18): 5202. doi:10.1021/jf030167m.

[23] A. H. Treiman, J. D. Gleason and D. D. Bogard (2000). "The SNC meteorites are from Mars". *Planet. Space Sci.* **48** (12–14): 1213. Bibcode:2000P&SS...48.1213T. doi:10.1016/S0032-0633(00)00105-7.

20.12 External links

- The Nuclear Science web portal Nucleonica

- The Karlsruhe Nuclide Chart

- National Nuclear Data Center Portal to large repository of free data and analysis programs from NNDC

- National Isotope Development Center Coordination and management of the production, availability, and distribution of isotopes, and reference information for the isotope community

- Isotope Development & Production for Research and Applications (IDPRA) U.S. Department of Energy program for isotope production and production research and development

- International Atomic Energy Agency Homepage of International Atomic Energy Agency (IAEA), an Agency of the United Nations (UN)

- Atomic Weights and Isotopic Compositions for All Elements Static table, from NIST (National Institute of Standards and Technology)

- Atomgewichte, Zerfallsenergien und Halbwertszeiten aller Isotope

- Exploring the Table of the Isotopes at the LBNL

- Current isotope research and information isotope.info

- Emergency Preparedness and Response: Radioactive Isotopes by the CDC (Centers for Disease Control and Prevention)

- Chart of Nuclides Interactive Chart of Nuclides (National Nuclear Data Center)

- Interactive Chart of the nuclides, isotopes and Periodic Table

- The LIVEChart of Nuclides – IAEA with isotope data.

- Annotated bibliography for isotopes from the Alsos Digital Library for Nuclear Issues

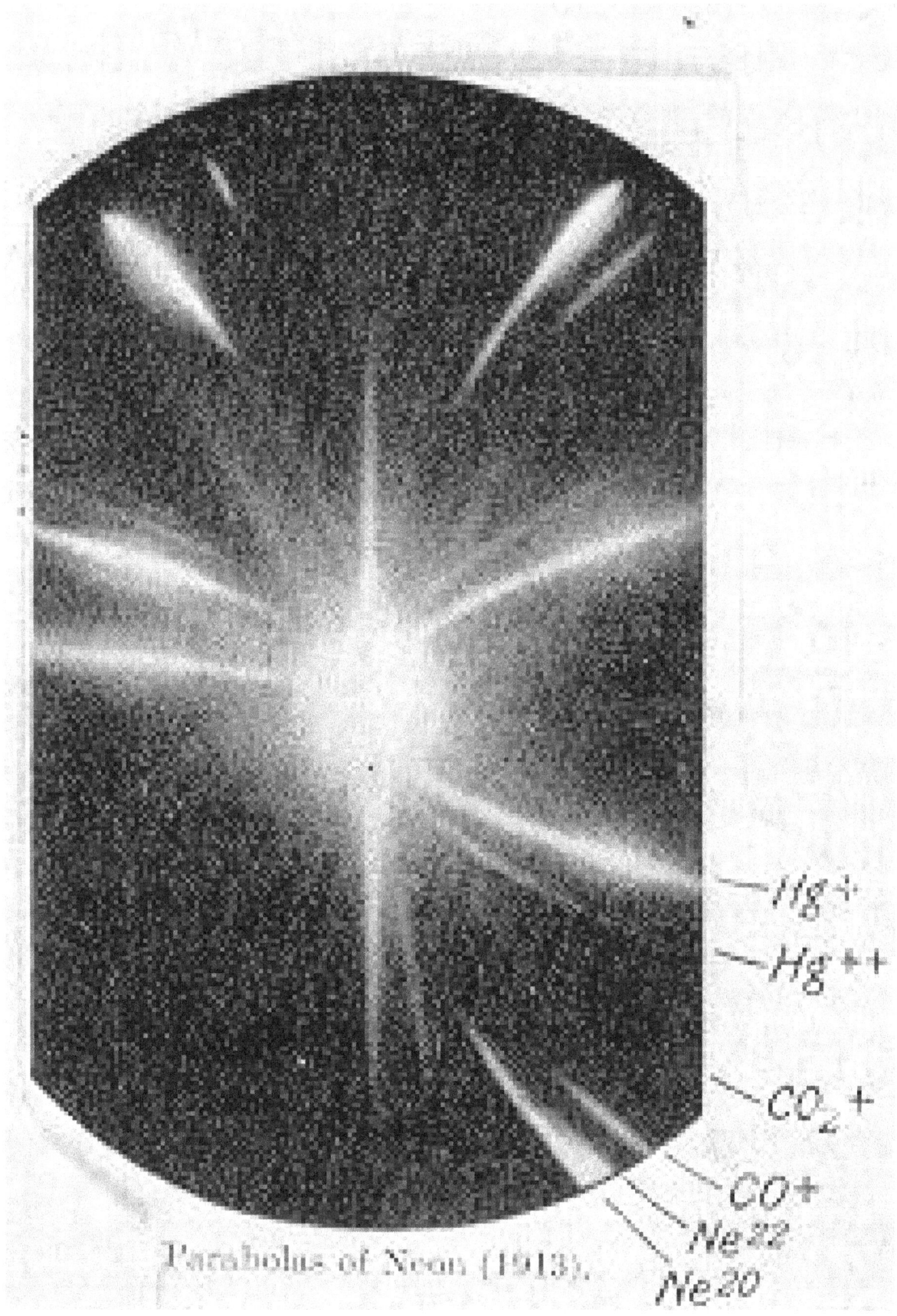

In the bottom right corner of J. J. Thomson's photographic plate are the separate impact marks for the two isotopes of neon: neon-20 and neon-22.

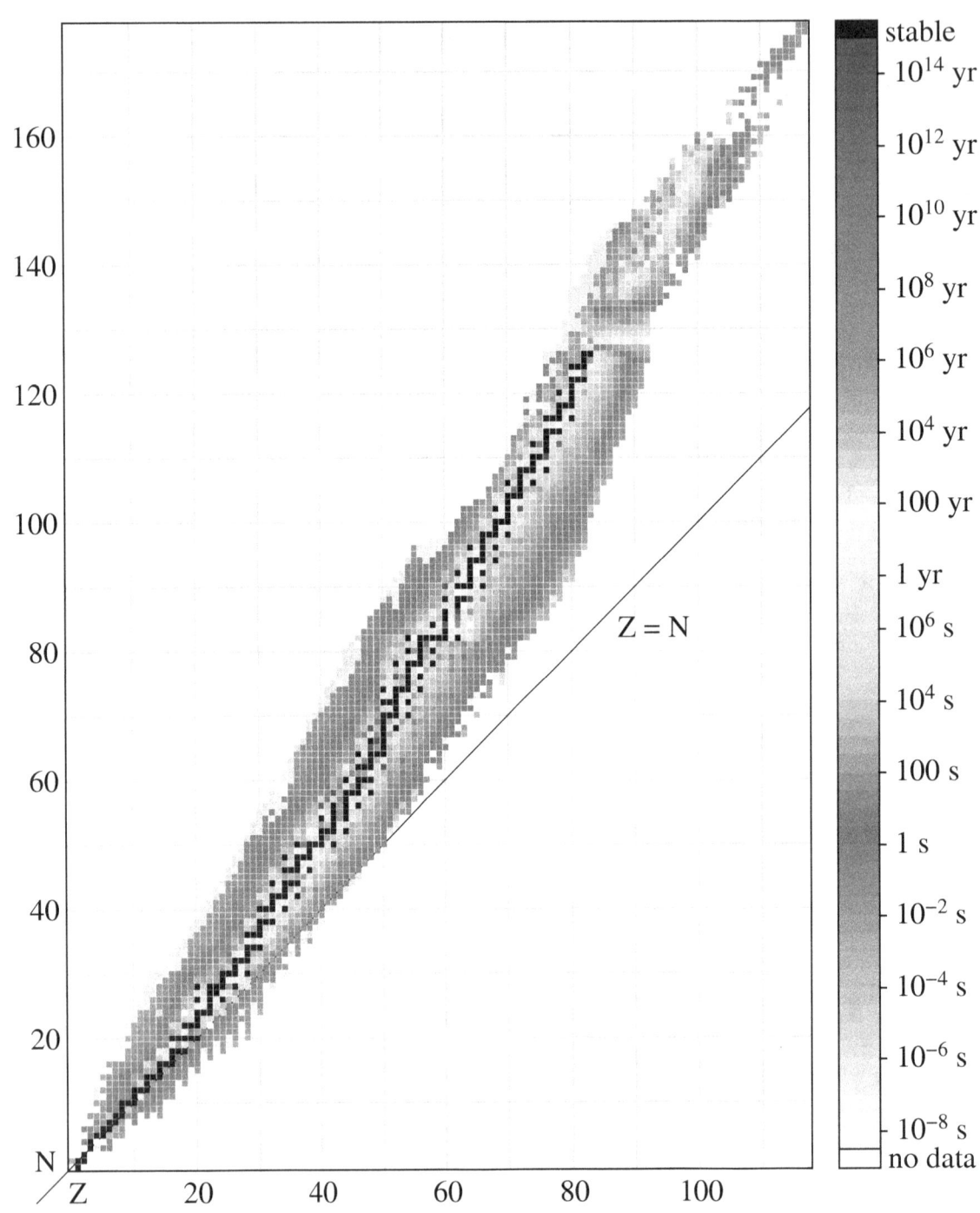

Isotope half-lives. The plot for stable isotopes diverges from the line Z = N *as the element number* Z *becomes larger*

Chapter 21

Chemical bond

A **chemical bond** is an attraction between atoms that allows the formation of chemical substances that contain two or more atoms. The bond is caused by the electrostatic force of attraction between opposite charges, either between electrons and nuclei, or as the result of a dipole attraction. The strength of chemical bonds varies considerably; there are "strong bonds" such as covalent or ionic bonds and "weak bonds" such as dipole–dipole interactions, the London dispersion force and hydrogen bonding.

Since opposite charges attract via a simple electromagnetic force, the negatively charged electrons that are orbiting the nucleus and the positively charged protons in the nucleus attract each other. An electron positioned between two nuclei will be attracted to both of them, and the nuclei will be attracted toward electrons in this position. This attraction constitutes the chemical bond. Due to the matter wave nature of electrons and their smaller mass, they must occupy a much larger amount of volume compared with the nuclei, and this volume occupied by the electrons keeps the atomic nuclei relatively far apart, as compared with the size of the nuclei themselves. This phenomenon limits the distance between nuclei and atoms in a bond.

In general, strong chemical bonding is associated with the sharing or transfer of electrons between the participating atoms. The atoms in molecules, crystals, metals and diatomic gases—indeed most of the physical environment around us—are held together by chemical bonds, which dictate the structure and the bulk properties of matter.

All bonds can be explained by quantum theory, but, in practice, simplification rules allow chemists to predict the strength, directionality, and polarity of bonds. The octet rule and VSEPR theory are two examples. More sophisticated theories are valence bond theory which includes orbital hybridization and resonance, and the linear combination of atomic orbitals molecular orbital method which includes ligand field theory. Electrostatics are used to describe bond polarities and the effects they have on chemical substances.

21.1   Overview of main types of chemical bonds

A chemical bond is an attraction between atoms. This attraction may be seen as the result of different behaviors of the outermost or valence electrons of atoms. Although all of these behaviors merge into each other seamlessly in various bonding situations so that there is no clear line to be drawn between them, the behaviors of atoms become so **qualitatively** different as the character of the bond changes **quantitatively**, that it remains useful and customary to differentiate between the bonds that cause these different properties of condensed matter.

In the simplest view of a so-called 'covalent' bond, one or more electrons (often a pair of electrons) are drawn into the space between the two atomic nuclei. Here the negatively charged electrons are attracted to the positive charges of *both* nuclei, instead of just their own. This overcomes the repulsion between the two positively charged nuclei of the two atoms, and so this overwhelming attraction holds the two nuclei in a fixed configuration of equilibrium, even though they will still vibrate at equilibrium position. Thus, covalent bonding involves sharing of electrons in which the positively charged nuclei of two or more atoms simultaneously attract the negatively charged electrons that are being shared between them. These bonds exist between two particular identifiable atoms and have a direction in space, allowing them to be shown as single

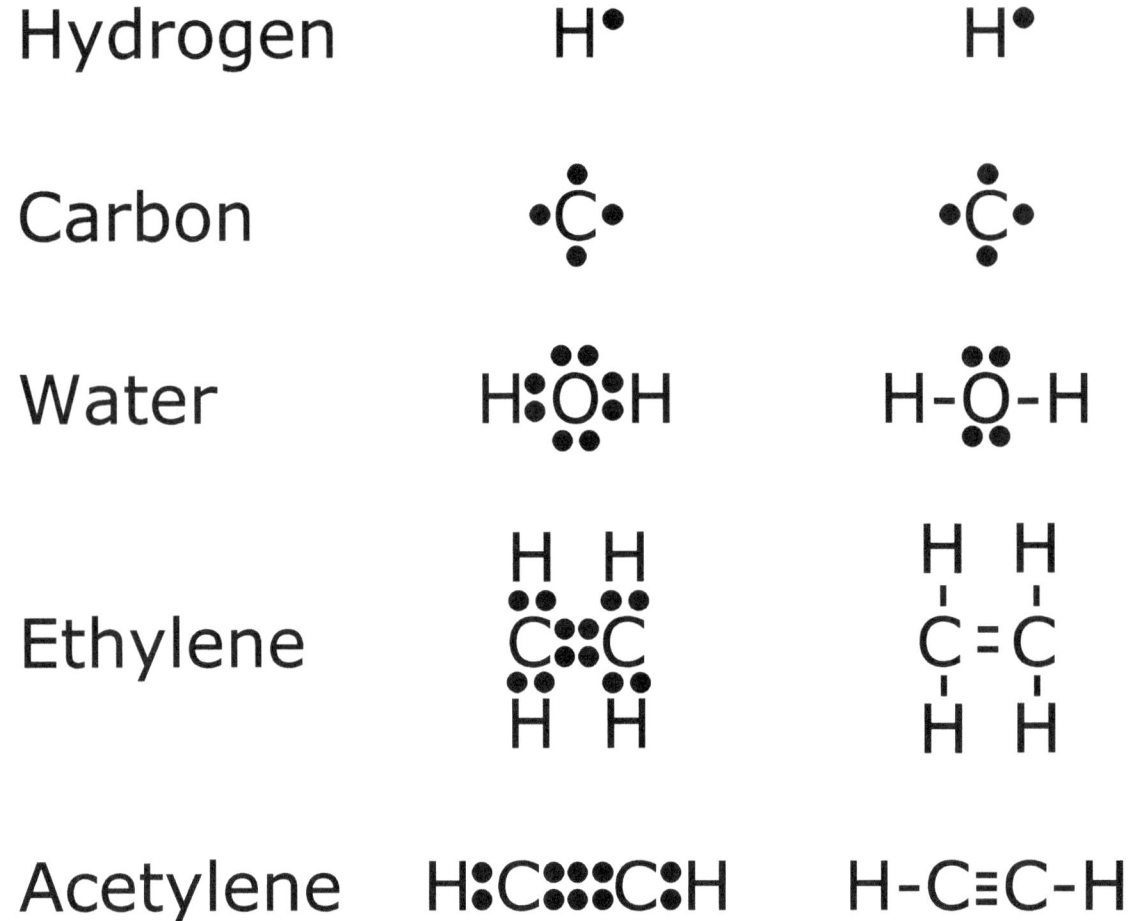

Examples of Lewis dot-style representations of chemical bonds between carbon (C), hydrogen (H), and oxygen (O). Lewis dot diagrams were an early attempt to describe chemical bonding and are still widely used today.

connecting lines between atoms in drawings, or modeled as sticks between spheres in models. In a polar covalent bond, one or more electrons are unequally shared between two nuclei. Covalent bonds often result in the formation of small collections of better-connected atoms called molecules, which in solids and liquids are bound to other molecules by forces that are often much weaker than the covalent bonds that hold the molecules internally together. Such weak intermolecular bonds give organic molecular substances, such as waxes and oils, their soft bulk character, and their low melting points (in liquids, molecules must cease most structured or oriented contact with each other). When covalent bonds link long chains of atoms in large molecules, however (as in polymers such as nylon), or when covalent bonds extend in networks through solids that are not composed of discrete molecules (such as diamond or quartz or the silicate minerals in many types of rock) then the structures that result may be both strong and tough, at least in the direction oriented correctly with networks of covalent bonds. Also, the melting points of such covalent polymers and networks increase greatly.

In a simplified view of an *ionic* bond, the bonding electron is not shared at all, but transferred. In this type of bond, the outer atomic orbital of one atom has a vacancy which allows the addition of one or more electrons. These newly added electrons potentially occupy a lower energy-state (effectively closer to more nuclear charge) than they experience in a different atom. Thus, one nucleus offers a more tightly bound position to an electron than does another nucleus, with the result that one atom may transfer an electron to the other. This transfer causes one atom to assume a net positive charge, and the other to assume a net negative charge. The *bond* then results from electrostatic attraction between atoms and the atoms become positive or negatively charged ions. Ionic bonds may be seen as extreme examples of polarization in covalent bonds. Often, such bonds have no particular orientation in space, since they result from equal electrostatic attraction of each ion to all ions around them. Ionic bonds are strong (and thus ionic substances require high temperatures to melt) but also brittle, since the forces between ions are short-range and do not easily bridge cracks and fractures. This type of bond gives rise to the physical characteristics of crystals of classic mineral salts, such as table salt.

A less often mentioned type of bonding is *metallic* bonding. In this type of bonding, each atom in a metal donates one or more electrons to a "sea" of electrons that reside between many metal atoms. In this sea, each electron is free (by virtue of its wave nature) to be associated with great many atoms at once. The bond results because the metal atoms become somewhat positively charged due to loss of their electrons while the electrons remain attracted to many atoms, without being part of any given atom. Metallic bonding may be seen as an extreme example of delocalization of electrons over a large system of covalent bonds, in which every atom participates. This type of bonding is often very strong (resulting in the tensile strength of metals). However, metallic bonding is more collective in nature than other types, and so they allow metal crystals to more easily deform, because they are composed of atoms attracted to each other, but not in any particularly-oriented ways. This results in the malleability of metals. The sea of electrons in metallic bonding causes the characteristically good electrical and thermal conductivity of metals, and also their "shiny" reflection of most frequencies of white light.

21.2 History

Main articles: History of chemistry and History of the molecule

Early speculations into the nature of the **chemical bond**, from as early as the 12th century, supposed that certain types of chemical species were joined by a type of chemical affinity. In 1704, Sir Isaac Newton famously outlined his atomic bonding theory, in "Query 31" of his Opticks, whereby atoms attach to each other by some "force". Specifically, after acknowledging the various popular theories in vogue at the time, of how atoms were reasoned to attach to each other, i.e. "hooked atoms", "glued together by rest", or "stuck together by conspiring motions", Newton states that he would rather infer from their cohesion, that "particles attract one another by some force, which in immediate contact is exceedingly strong, at small distances performs the chemical operations, and reaches not far from the particles with any sensible effect."

In 1819, on the heels of the invention of the voltaic pile, Jöns Jakob Berzelius developed a theory of chemical combination stressing the electronegative and electropositive character of the combining atoms. By the mid 19th century, Edward Frankland, F.A. Kekulé, A.S. Couper, Alexander Butlerov, and Hermann Kolbe, building on the theory of radicals, developed the theory of valency, originally called "combining power", in which compounds were joined owing to an attraction of positive and negative poles. In 1916, chemist Gilbert N. Lewis developed the concept of the electron-pair bond, in which two atoms may share one to six electrons, thus forming the single electron bond, a single bond, a double bond, or a triple bond; in Lewis's own words, "An electron may form a part of the shell of two different atoms and cannot be said to belong to either one exclusively."[1]

That same year, Walther Kossel put forward a theory similar to Lewis' only his model assumed complete transfers of electrons between atoms, and was thus a model of ionic bonding. Both Lewis and Kossel structured their bonding models on that of Abegg's rule (1904).

In 1927, the first mathematically complete quantum description of a simple chemical bond, i.e. that produced by one electron in the hydrogen molecular ion, H_2^+, was derived by the Danish physicist Oyvind Burrau.[2] This work showed that the quantum approach to chemical bonds could be fundamentally and quantitatively correct, but the mathematical methods used could not be extended to molecules containing more than one electron. A more practical, albeit less quantitative, approach was put forward in the same year by Walter Heitler and Fritz London. The Heitler-London method forms the basis of what is now called valence bond theory. In 1929, the linear combination of atomic orbitals molecular orbital method (LCAO) approximation was introduced by Sir John Lennard-Jones, who also suggested methods to derive electronic structures of molecules of F_2 (fluorine) and O_2 (oxygen) molecules, from basic quantum principles. This molecular orbital theory represented a covalent bond as an orbital formed by combining the quantum mechanical Schrödinger atomic orbitals which had been hypothesized for electrons in single atoms. The equations for bonding electrons in multi-electron atoms could not be solved to mathematical perfection (i.e., *analytically*), but approximations for them still gave many good qualitative predictions and results. Most quantitative calculations in modern quantum chemistry use either valence bond or molecular orbital theory as a starting point, although a third approach, density functional theory, has become increasingly popular in recent years.

In 1933, H. H. James and A. S. Coolidge carried out a calculation on the dihydrogen molecule that, unlike all previous calculation which used functions only of the distance of the electron from the atomic nucleus, used functions which also explicitly added the distance between the two electrons.[3] With up to 13 adjustable parameters they obtained a result very close to the experimental result for the dissociation energy. Later extensions have used up to 54 parameters and gave excellent agreement with experiments. This calculation convinced the scientific community that quantum theory could give agreement with experiment. However this approach has none of the physical pictures of the valence bond and molecular orbital theories and is difficult to extend to larger molecules.

21.3 Bonds in chemical formulas

The fact that atoms and molecules are three-dimensional makes it difficult to use a single technique for indicating orbitals and bonds. In **molecular formulas** the chemical bonds (binding orbitals) between atoms are indicated by various methods according to the type of discussion. Sometimes, they are completely neglected. For example, in organic chemistry chemists are sometimes concerned only with the functional groups of the molecule. Thus, the molecular formula of ethanol may be written in a paper in conformational form, three-dimensional form, full two-dimensional form (indicating every bond with no three-dimensional directions), compressed two-dimensional form ($CH_3–CH_2–OH$), by separating the functional group from another part of the molecule (C_2H_5OH), or by its atomic constituents (C_2H_6O), according to what is discussed. Sometimes, even the non-bonding valence shell electrons (with the two-dimensional approximate directions) are marked, i.e. for elemental carbon $.\dot{C}.$ Some chemists may also mark the respective orbitals, i.e. the hypothetical ethene^{-4} anion ($\sqrt{}C=C\wedge^{-4}$) indicating the possibility of bond formation.

21.4 Strong chemical bonds

Strong chemical bonds are the *intramolecular* forces which hold atoms together in molecules. A strong chemical bond is formed from the transfer or sharing of electrons between atomic centers and relies on the electrostatic attraction between the protons in nuclei and the electrons in the orbitals.

The types of strong bond differ due to the difference in electronegativity of the constituent elements. A large difference in electronegativity leads to more polar (ionic) character in the bond.

21.4.1 Ionic bond

Main article: Ionic bonding

Ionic bonding is a type of electrostatic interaction between atoms which have a large electronegativity difference. There is no precise value that distinguishes ionic from covalent bonding, but a difference of electronegativity of over 1.7 is likely to be ionic, and a difference of less than 1.7 is likely to be covalent.[5] Ionic bonding leads to separate positive and negative ions. Ionic charges are commonly between −3e to +3e. Ionic bonding commonly occurs in metal salts such as sodium chloride (table salt). A typical feature of ionic bonds is that the species form into ionic crystals, in which no ion is specifically paired with any single other ion, in a specific directional bond. Rather, each species of ion is surrounded by ions of the opposite charge, and the spacing between it and each of the oppositely charged ions near it, is the same for all surrounding atoms of the same type. It is thus no longer possible to associate an ion with any specific other single ionized atom near it. This is a situation unlike that in covalent crystals, where covalent bonds between specific atoms are still discernible from the shorter distances between them, as measured via such techniques as X-ray diffraction.

Ionic crystals may contain a mixture of covalent and ionic species, as for example salts of complex acids, such as sodium cyanide, NaCN. Many minerals are of this type. X-ray diffraction shows that in NaCN, for example, the bonds between sodium cations (Na^+) and the cyanide anions (CN^-) are *ionic*, with no sodium ion associated with any particular cyanide. However, the bonds between C and N atoms in cyanide are of the *covalent* type, making each of the carbon and nitrogen associated with *just one* of its opposite type, to which it is physically much closer than it is to other carbons or nitrogens in a sodium cyanide crystal.

When such crystals are melted into liquids, the ionic bonds are broken first because they are non-directional and allow the charged species to move freely. Similarly, when such salts dissolve into water, the ionic bonds are typically broken by the interaction with water, but the covalent bonds continue to hold. For example, in solution, the cyanide ions, still bound together as single CN^- ions, move independently through the solution, as do sodium ions, as Na^+. In water, charged ions move apart because each of them are more strongly attracted to a number of water molecules, than to each other. The attraction between ions and water molecules in such solutions is due to a type of weak dipole-dipole type chemical bond. In melted ionic compounds, the ions continue to be attracted to each other, but not in any ordered or crystalline way.

21.4.2   Covalent bond

Main article: Covalent bond

Covalent bonding is a common type of bonding, in which the electronegativity difference between the bonded atoms is small or nonexistent. Bonds within most organic compounds are described as covalent. See sigma bonds and pi bonds for LCAO-description of such bonding.

A polar covalent bond is a covalent bond with a significant ionic character. This means that the electrons are closer to one of the atoms than the other, creating an imbalance of charge. They occur as a bond between two atoms with moderately different electronegativities and give rise to dipole-dipole interactions. The electronegativity of these bonds is 0.3 to 1.7.

A coordinate covalent bond is one where both bonding electrons are from one of the atoms involved in the bond. These bonds give rise to Lewis acids and bases. The electrons are shared roughly equally between the atoms in contrast to ionic bonding. Such bonding occurs in molecules such as the ammonium ion (NH_4^+) and are shown by an arrow pointing to the Lewis acid. Also known as non-polar covalent bond, the electronegativity of these bonds range from 0 to 0.3.

Molecules which are formed primarily from non-polar covalent bonds are often immiscible in water or other polar solvents, but much more soluble in non-polar solvents such as hexane.

Single and multiple bonds

A single bond between two atoms corresponds to the sharing of one pair of electrons. The electron density of these two bonding electrons is concentrated in the region between the two atoms, which is the defining quality of a sigma bond.

A double bond between two atoms is formed by the sharing of two pairs of electrons, one in a sigma bond and one in a pi bond, with electron density concentrated on two opposite sides of the internuclear axis. A triple bond consists of three shared electron pairs, forming one sigma and two pi bonds.

Quadruple and higher bonds are very rare and occur only between certain transition metal atoms.

21.4.3   Metallic bonding

Main article: Metallic bonding

In metallic bonding, bonding electrons are delocalized over a lattice of atoms. By contrast, in ionic compounds, the locations of the binding electrons and their charges are static. The freely-moving or delocalization of bonding electrons leads to classical metallic properties such as luster (surface light reflectivity), electrical and thermal conductivity, ductility, and high tensile strength.

21.5   Intermolecular bonding

Main article: Intermolecular Force

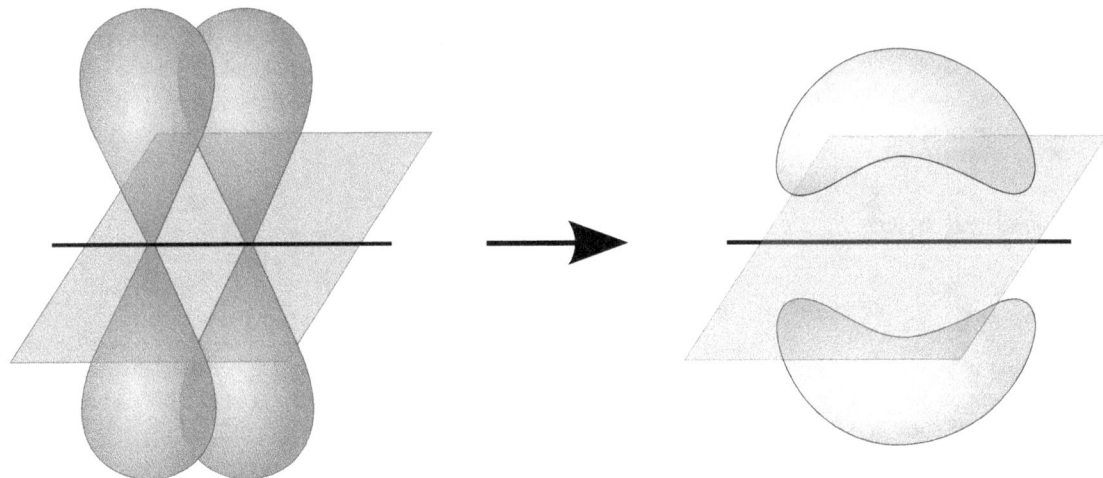

Two p-orbitals forming a pi-bond.

There are four basic types of bonds that can be formed between two or more (otherwise non-associated) molecules, ions or atoms. Intermolecular forces cause molecules to be attracted or repulsed by each other. Often, these define some of the physical characteristics (such as the melting point) of a substance.

- A large difference in electronegativity between two bonded atoms will cause a permanent charge separation, or dipole, in a molecule or ion. Two or more molecules or ions with permanent dipoles can interact within **dipole-dipole interactions**. The bonding electrons in a molecule or ion will, on average, be closer to the more electronegative atom more frequently than the less electronegative one, giving rise to partial charges on each atom, and causing electrostatic forces between molecules or ions.

- A **hydrogen bond** is effectively a strong example of an interaction between two permanent dipoles. The large difference in electronegativities between hydrogen and any of fluorine, nitrogen and oxygen, coupled with their lone pairs of electrons cause strong electrostatic forces between molecules. Hydrogen bonds are responsible for the high boiling points of water and ammonia with respect to their heavier analogues.

- The **London dispersion force** arises due to instantaneous dipoles in neighbouring atoms. As the negative charge of the electron is not uniform around the whole atom, there is always a charge imbalance. This small charge will induce a corresponding dipole in a nearby molecule; causing an attraction between the two. The electron then moves to another part of the electron cloud and the attraction is broken.

- A **cation–pi interaction** occurs between a pi bond and a cation.

21.6 Theories of chemical bonding

In the (unrealistic) limit of "pure" ionic bonding, electrons are perfectly localized on one of the two atoms in the bond. Such bonds can be understood by classical physics. The forces between the atoms are characterized by isotropic continuum electrostatic potentials. Their magnitude is in simple proportion to the charge difference.

Covalent bonds are better understood by valence bond theory or molecular orbital theory. The properties of the atoms involved can be understood using concepts such as oxidation number. The electron density within a bond is not assigned to individual atoms, but is instead delocalized between atoms. In valence bond theory, the two electrons on the two atoms are coupled together with the bond strength depending on the overlap between them. In molecular orbital theory, the linear combination of atomic orbitals (LCAO) helps describe the delocalized molecular orbital structures and energies based on the atomic orbitals of the atoms they came from. Unlike pure ionic bonds, covalent bonds may have directed anisotropic properties. These may have their own names, such as sigma bond and pi bond.

In the general case, atoms form bonds that are intermediate between ionic and covalent, depending on the relative electronegativity of the atoms involved. This type of bond is sometimes called polar covalent.

21.7   References

[1] Lewis, Gilbert N. (1916). "The Atom and the Molecule". *Journal of the American Chemical Society* **38** (4): 772. doi:1a002. a copy

[2] Laidler, K. J. (1993). *The World of Physical Chemistry*. Oxford University Press. p. 346. ISBN 0-19-855919-4.

[3] James, H. H.; Coolidge, A. S. (1933). "The Ground State of the Hydrogen Molecule". *Journal of Chemical Physics* (American Institute of Physics) **1** (12): 825–835. doi:10.1063/1.1749252.

[4] "Bond Lengths and Energies". Science.uwaterloo.ca. Retrieved 2013-10-15.

[5] Atkins, Peter; Loretta Jones (1997). *Chemistry: Molecules, Matter and Change*. New York: W. H. Freeman & Co. pp. 294–295. ISBN 0-7167-3107-X.

21.8   External links

- W. Locke (1997). Introduction to Molecular Orbital Theory. Retrieved May 18, 2005.

- Carl R. Nave (2005). HyperPhysics. Retrieved May 18, 2005.

- Linus Pauling and the Nature of the Chemical Bond: A Documentary History. Retrieved February 29, 2008.

Chapter 22

Valence (chemistry)

For other uses, see Valence (disambiguation).
"Multivalent" redirects here. For other uses, see Polyvalence (disambiguation).

In chemistry, the **valence** (or **valency**) of an element is a measure of its combining power with other atoms when it forms chemical compounds or molecules. The concept of valence was developed in the second half of the 19th century and was successful in explaining the molecular structure of inorganic and organic compounds. [1] The quest for the underlying causes of valence led to the modern theories of chemical bonding, including Lewis structures (1916), valence bond theory (1927), molecular orbitals (1928), valence shell electron pair repulsion theory (1958), and all of the advanced methods of quantum chemistry.

22.1 Description

The combining power or affinity of an atom of an element was determined by the number of hydrogen atoms that it combined with. In methane, carbon has a valence of 4; in ammonia, nitrogen has a valence of 3; in water, oxygen has a valence of two; and in hydrogen chloride, chlorine has a valence of 1. Chlorine, as it has a valence of one, can be substituted for hydrogen, so phosphorus has a valence of 5 in phosphorus pentachloride, PCl_5. Valence diagrams of a compound represent the connectivity of the elements, with lines drawn between two elements, sometimes called bonds, representing a saturated valency for each element.[1] Examples are

Valence only describes connectivity; it does not describe the geometry of molecular compounds, or what are now known to be ionic compounds or giant covalent structures. A line between atoms does not represent a pair of electrons as it does in Lewis diagrams.

22.2 Modern definitions

Valence is defined by the IUPAC as:-

> *The maximum number of univalent atoms (originally hydrogen or chlorine atoms) that may combine with an atom of the element under consideration, or with a fragment, or for which an atom of this element can be substituted.* .[2]

An alternative modern description is:-[3]

> *The number of hydrogen atoms that can combine with an element in a binary hydride or twice the number of oxygen atoms combining with an element in its oxide or oxides.* This definition differs from the IUPAC definition as an element can be said to have more than one valence.

22.3   Historical development

The etymology of the word "valence" traces back to 1425, meaning "extract, preparation," from Latin *valentia* "strength, capacity," and the chemical meaning referring to the "combining power of an element" is recorded from 1884, from German *Valenz*.[4]

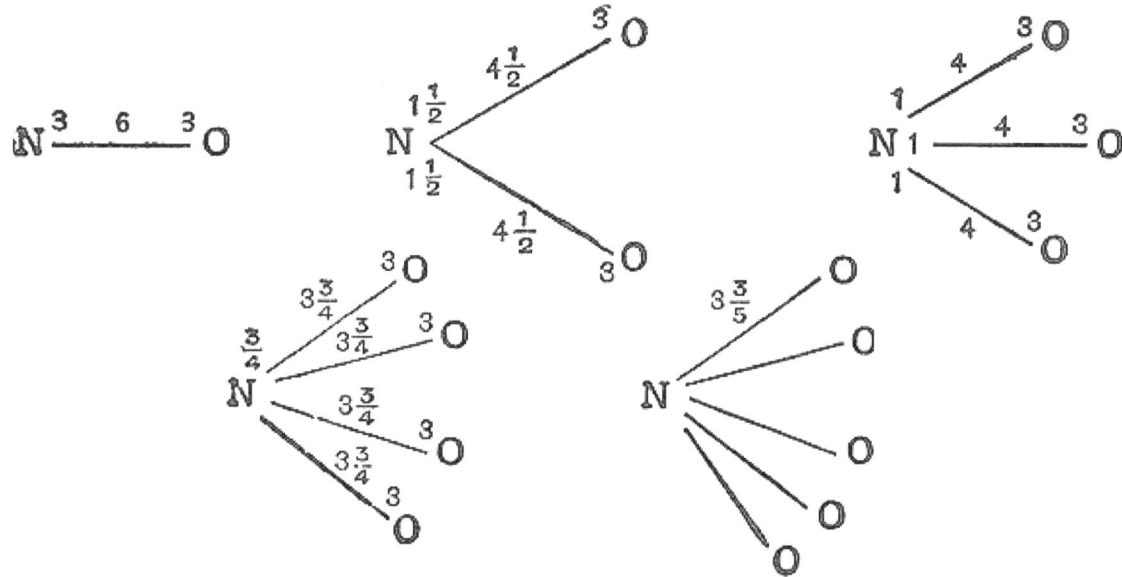

William Higgins' combinations of ultimate particles (1789)

In 1789, William Higgins published views on what he called combinations of "ultimate" particles, which foreshadowed the concept of valency bonds.[5] If, for example, according to Higgins, the force between the ultimate particle of oxygen and the ultimate particle of nitrogen were 6, then the strength of the force would be divided accordingly, and likewise for the other combinations of ultimate particles (see illustration).

The exact inception, however, of the theory of chemical valencies can be traced to an 1852 paper by Edward Frankland, in which he combined the older theories of free radicals and "type theory" with thoughts on chemical affinity to show that certain elements have the tendency to combine with other elements to form compounds containing 3, i.e., in the three atom groups (e.g., NO_3, NH_3, NI_3, etc.) or 5, i.e., in the five atom groups (e.g., NO_5, NH_4O, PO_5, etc.), equivalents of the attached elements. It is in this manner, according to Frankland, that their affinities are best satisfied. Following these examples and postulates, Frankland declares how obvious it is that[6]

This "combining power" was afterwards called **quantivalence** or valency (and valence by American chemists).[5] In 1857 August Kekulé proposed fixed valences for many elements, such as four for carbon, and used them to propose structural formulas for many organic molecules which are still accepted today.

Most 19th-century chemists defined the valence of an element as the number of its bonds without distinguishing different types of valence or of bond. However, in 1893 Alfred Werner described transition metal coordination complexes such as $[Co(NH_3)_6]Cl_3$ in which he distinguished *primary* and *secondary* valences, corresponding to the modern concepts of oxidation state and coordination number respectively.

For main-group elements, in 1904 Richard Abegg considered *positive* and *negative* valences (maximum and minimum oxidation states), and proposed Abegg's rule to the effect that their difference is often eight.

22.3.1 Electrons and valence

The Rutherford model of the nuclear atom (1911) showed that the exterior of an atom is occupied by electrons, which suggests that electrons are responsible for the interaction of atoms and the formation of chemical bonds. In 1916 Gilbert N. Lewis explained valence and chemical bonding in terms of a tendency of (main-group) atoms to achieve a stable octet of eight valence-shell electrons. According to Lewis, covalent bonding leads to octets by sharing of electrons, and ionic bonding leads to octets by transfer of electrons from one atom to the other. The term covalence is attributed to Irving Langmuir, who stated in 1919 that "the number of pairs of electrons which any given atom shares with the adjacent atoms is called the *covalence* of that atom."[7] The prefix *co-* means "together", so that a co-valent bond means that the atoms share valence. Subsequent to this, it is now more common to speak of covalent bonds rather than "valence", which has fallen out of use in higher level work with the advances in the theory of chemical bonding, but is still widely used in elementary studies where it provides a heuristic introduction to the subject.

In the 1930s Linus Pauling proposed that there are also polar covalent bonds which are intermediate between covalent and ionic, and that the degree of ionic character depends on the difference of electronegativity of the two bonded atoms.

Pauling also considered hypervalent molecules in which main-group elements have apparent valences greater than the maximum of four allowed by the octet rule. For example, in the sulfur hexafluoride molecule (SF_6), Pauling considered that the sulfur forms six true two-electron bonds using so-called sp^3d^2 hybrid atomic orbitals which combine one s, three p and two d orbitals. However more recently, quantum-mechanical calculations on this and similar molecules have shown that the role of d orbitals in the bonding is minimal, and that the SF_6 molecule should be described as having six polar covalent (partly ionic) bonds made from only four orbitals on sulfur (one s and three p) in accordance with the octet rule, together with six orbitals on the fluorines.[8] Similar calculations on transition metal molecules show that the role of p orbitals is minimal, so that at most one s and five d orbitals on the metal participate in bonding.

22.4 Common valences

For elements in the main groups of the periodic table, the valence can vary between one and seven.

Many elements have a common valence related to their position in the periodic table, and nowadays this is rationalised by the octet rule. The Latin/Greek prefixes uni/mono, bi/di, ter/tri, quadri/tetra, quinque/penta are used to describe ions in the one, two, three, four or five charge states. *Polyvalence* or *multivalence* refers to species that are not restricted to a specific number of valence bonds. Species with a single charge are univalent (monovalent)). For example, the Cs^+ cation is a univalent or monovalent cation, whereas the Ca^{2+} cation is a divalent cation, and the Fe^{3+} cation is a trivalent cation. Unlike Cs and Ca, Fe can also exist in other charge states, notably 2+ and 4+, and is thus known as a multivalent (polyvalent) ion.

22.5 Valence versus oxidation state

Because of the ambiguity of the term valence,[9] nowadays other notations are used in practice. Beside the system of oxidation numbers as used in *Stock nomenclature* for coordination compounds,[10] and the lambda notation, as used in the IUPAC nomenclature of inorganic chemistry,[11] "oxidation state" is a more clear indication of the electronic state of atoms in a molecule.

The "oxidation state" of an atom in a molecule gives the number of valence electrons it has gained or lost.[12] In contrast to the *valency number*, the *oxidation state* can be positive (for an electropositive atom) or negative (for an electronegative atom).

Elements in a high oxidation state can have a valence higher than four. For example, in perchlorates, chlorine has seven valence bonds and ruthenium, in the +8 oxidation state in ruthenium tetroxide, has eight valence bonds.

22.5.1 Examples

(valencies according to the *number of valence bonds* definition and conform oxidation states)

* The univalent perchlorate ion (ClO_4^-) has valence 1.
** Iron oxide appears in a crystal structure, so no typical molecule can be identified.
In ferrous oxide, Fe has oxidation number II, in ferric oxide, oxidation number III.

Examples where valences and oxidation states differ due to bonds between identical atoms:

Valences may also be different from absolute values of oxidation states due to different polarity of bonds. For example, in dichloromethane, CH_2Cl_2, carbon has valence 4 but oxidation state 0.

22.6 "Maximum number of bonds" definition

Frankland took the view that the valence (he used the term "atomicity") of an element was a single value that corresponded to the maximum value observed. The number of unused valencies on atoms of what are now called the p-block elements is generally even, and Frankland suggested that the unused valencies saturated one another. For example, nitrogen has a maximum valence of 5, in forming ammonia two valencies are left unattached; sulfur has a maximum valence of 6, in forming hydrogen sulphide four valencies are left unattached.[13][14]

The International Union of Pure and Applied Chemistry (IUPAC) has made several attempts to arrive at an unambiguous definition of valence. The current version, adopted in 1994:[15]

> *The maximum number of univalent atoms (originally hydrogen or chlorine atoms) that may combine with an atom of the element under consideration, or with a fragment, or for which an atom of this element can be substituted.*[2]

Hydrogen and chlorine were originally used as examples of univalent atoms, because of their nature to form only one single bond. Hydrogen has only one *valence electron* and can form only one bond with an atom that has an incomplete outer shell. Chlorine has seven *valence electrons* and can form only one bond with an atom that donates a *valence electron* to complete chlorine's outer shell. However, chlorine can also have oxidation states from +1 to +7 and can form more than one bond by donating *valence electrons*.

Although hydrogen has only one valence electron, it can form bonds with more than one atom. In the bifluoride ion ([HF 2]−
), for example, it forms a three-center four-electron bond with two fluoride atoms:

[F–H F⁻ ↔ F⁻ H–F]

Another example is the Three-center two-electron bond in diborane (B_2H_6).

22.7 Maximum valences of the elements

Maximum valences for the elements are based on the data from list of oxidation states of the elements.

22.8 See also

- Valence electron
- Abegg's rule
- Oxidation state

22.9 References

[1] Partington, James Riddick (1921). *A text-book of inorganic chemistry for university students* (1st ed.). Retrieved April 13, 2014.

[2] IUPAC Gold Book definition: valence

[3] Greenwood, Norman N.; Earnshaw, Alan (1997). *Chemistry of the Elements* (2nd ed.). Butterworth-Heinemann. ISBN 0080379419.

[4] Harper, Douglas. "valence". *Online Etymology Dictionary*.

[5] Partington, J.R. (1989). *A Short History of Chemistry*. Dover Publications, Inc. ISBN 0-486-65977-1.

[6] Frankland, E. (1852). Phil. Trans., vol. cxlii, 417.

[7] Langmuir, Irving (1919). "The Arrangement of Electrons in Atoms and Molecules". *Journal of the American Chemical Society* **41** (6): 868–934. doi:10.1021/ja02227a002.

[8] E. Magnusson. Hypercoordinate molecules of second-row elements: d functions or d orbitals? *J. Am. Chem. Soc.* **1990**, *112*, 7940–7951. doi:10.1021/ja00178a014

[9] The Free Dictionary: *valence*

[10] IUPAC, Gold Book definition: *oxidation number*

[11] IUPAC, Gold Book definition: *lambda*

[12] IUPAC Gold Book definition: oxidation state

[13] Frankland, E. (1870). *Lecture notes for chemical students(Google eBook)* (2d ed.). J. Van Voorst. p. 21.

[14] Frankland, E.; Japp, F.R (1885). *Inorganic chemistry* (1st ed.). pp. 75–85. Retrieved April 8, 2014.

[15] Muller, P. (1994). "Glossary of terms used in physical organic chemistry (IUPAC Recommendations 1994)". *Pure and Applied Chemistry* **66** (5). doi:10.1351/pac199466051077.

Chapter 23

Ion

For other uses, see Ion (disambiguation).

An **ion** (/ˈaɪən, -ɒn/)[1] is an atom or molecule in which the total number of electrons is not equal to the total number of protons, giving the atom or molecule a net positive or negative electrical charge.

Ions can be created, by either chemical or physical means, via ionization. In chemical terms, if a neutral atom loses one or more electrons, it has a net positive charge and is known as a cation. If an atom gains electrons, it has a net negative charge and is known as an anion. An ion consisting of a single atom is an atomic or monatomic ion; if it consists of two or more atoms, it is a molecular or polyatomic ion. Because of their electric charges, cations and anions attract each other and readily form ionic compounds, such as salts.

In the case of physical ionization of a medium, such as a gas, what are known as "ion pairs" are created by ion impact, and each pair consists of a free electron and a positive ion.[2]

23.1   History of discovery

The word *ion* is the Greek ἰόν, *ion*, "going", the present participle of ἰέναι, *ienai*, "to go". This term was introduced by English physicist and chemist Michael Faraday in 1834 for the then-unknown species that *goes* from one electrode to the other through an aqueous medium.[3][4] Faraday did not know the nature of these species, but he knew that since metals dissolved into and entered a solution at one electrode, and new metal came forth from a solution at the other electrode, that some kind of substance moved through the solution in a current, conveying matter from one place to the other.

Faraday also introduced the words *anion* for a negatively charged ion, and *cation* for a positively charged one. In Faraday's nomenclature, cations were named because they were attracted to the cathode in a galvanic device and anions were named due to their attraction to the anode.

23.2   Characteristics

Ions in their gas-like state are highly reactive, and do not occur in large amounts on Earth, except in flames, lightning, electrical sparks, and other plasmas. These gas-like ions rapidly interact with ions of opposite charge to give neutral molecules or ionic salts. Ions are also produced in the liquid or solid state when salts interact with solvents (for example, water) to produce "solvated ions," which are more stable, for reasons involving a combination of energy and entropy changes as the ions move away from each other to interact with the liquid. These stabilized species are more commonly found in the environment at low temperatures. A common example is the ions present in seawater, which are derived from the dissolved salts.

All ions are charged, which means that like all charged objects they are:

- attracted to opposite electric charges (positive to negative, and vice versa),

- repelled by like charges

- when moving, travel in trajectories that are deflected by a magnetic field.

Electrons, due to their smaller mass and thus larger space-filling properties as matter waves, determine the size of atoms and molecules that possess any electrons at all. Thus, anions (negatively charged ions) are larger than the parent molecule or atom, as the excess electron(s) repel each other, and add to the physical size of the ion, because its size is determined by its electron cloud. As such, in general, cations are smaller than the corresponding parent atom or molecule due to the smaller size of its electron cloud. One particular cation (that of hydrogen) contains no electrons, and thus consists of a single proton - *very much smaller* than the parent hydrogen atom.

23.2.1 Anions and cations

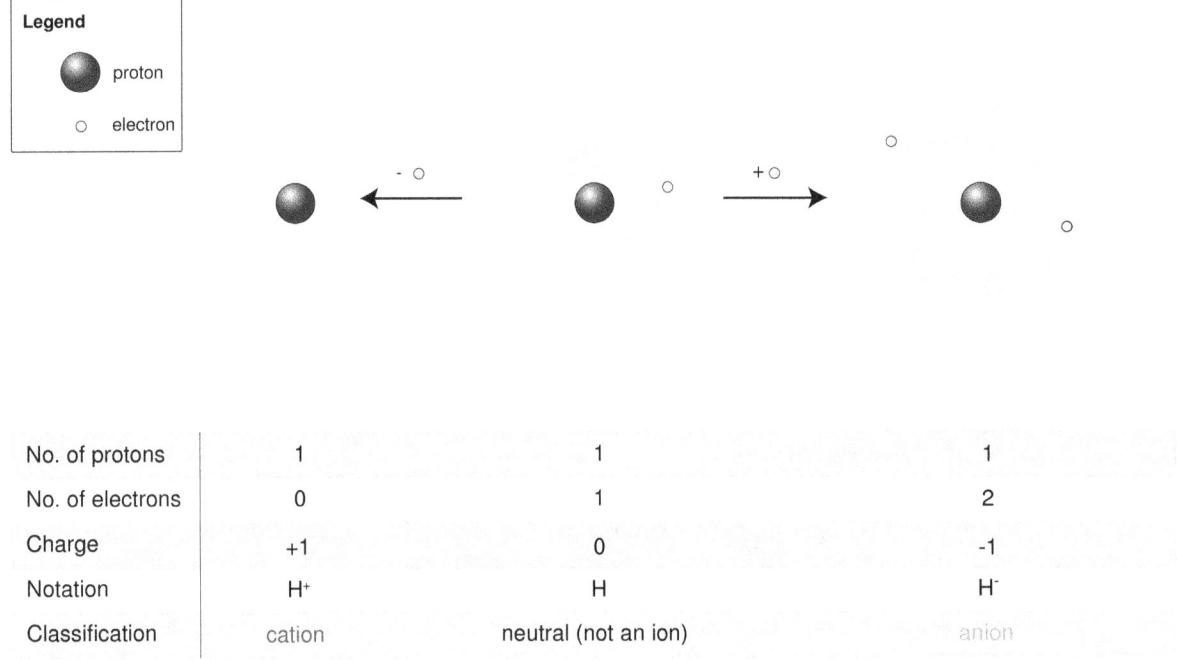

No. of protons	1	1	1
No. of electrons	0	1	2
Charge	+1	0	-1
Notation	H^+	H	H^-
Classification	cation	neutral (not an ion)	anion

Hydrogen atom (center) contains a single proton and a single electron. Removal of the electron gives a cation (left), whereas addition of an electron gives an anion (right). The hydrogen anion, with its loosely held two-electron cloud, has a larger radius than the neutral atom, which in turn is much larger than the bare proton of the cation. Hydrogen forms the only cation that has no electrons, but even cations that (unlike hydrogen) still retain one or more electrons are still smaller than the neutral atoms or molecules from which they are derived.

"Cation" and "Anion" redirect here. For the particle physics/quantum computing concept, see Anyon. For other uses, see Ion (disambiguation).

Since the electric charge on a proton is equal in magnitude to the charge on an electron, the net electric charge on an ion is equal to the number of protons in the ion minus the number of electrons.

An *anion* (−) (/ˈæn.aɪ.ən/ *AN-eye-ən*), from the Greek word ἄνω (*ánō*), meaning "up",[5] is an ion with more electrons than protons, giving it a net negative charge (since electrons are negatively charged and protons are positively charged).[6]

A *cation* (+) (/ˈkæt.aɪ.ən/ *KAT-eye-ən*), from the Greek word κατά (*katá*), meaning "down",[7] is an ion with fewer electrons than protons, giving it a positive charge.[8]

There are additional names used for ions with multiple charges. For example, an ion with a −2 charge is known as a dianion and an ion with a +2 charge is known as a dication. A zwitterion is a neutral molecule with positive and negative charges at different locations within that molecule.[9]

23.2.2   Natural occurrences

Ions are ubiquitous in nature and are responsible for diverse phenomena from the luminescence of the Sun to the existence of the Earth's ionosphere. Atoms in their ionic state may have a different color from neutral atoms, and thus light absorption by metal ions gives the color of gemstones. In both inorganic and organic chemistry (including biochemistry), the interaction of water and ions is extremely important; an example is the energy that drives breakdown of adenosine triphosphate (ATP). The following sections describe contexts in which ions feature prominently; these are arranged in decreasing physical length-scale, from the astronomical to the microscopic.

Astronomical

A collection of non-aqueous gas-like ions, or even a gas containing a proportion of charged particles, is called a plasma. Greater than 99.9% of visible matter in the Universe may be in the form of plasmas.[10] These include our Sun and other stars and the space between planets, as well as the space in between stars. Plasmas are often called the *fourth state of matter* because their properties are substantially different from those of solids, liquids, and gases. Astrophysical plasmas predominantly contain a mixture of electrons and protons (ionized hydrogen).

23.3   Related technology

Ions can be non-chemically prepared using various ion sources, usually involving high voltage or temperature. These are used in a multitude of devices such as mass spectrometers, optical emission spectrometers, particle accelerators, ion implanters, and ion engines.

As reactive charged particles, they are also used in air purification by disrupting microbes, and in household items such as smoke detectors.

As signaling and metabolism in organisms are controlled by a precise ionic gradient across membranes, the disruption of this gradient contributes to cell death. This is a common mechanism exploited by natural and artificial biocides, including the ion channels gramicidin and amphotericin (a fungicide).

Inorganic dissolved ions are a component of total dissolved solids, an indicator of water quality in the world.

23.3.1   Detection of ionising radiation

The ionising effect of radiation on a gas is extensively used for the detection of radiation such as alpha, beta, gamma and X-rays. The original ionisation event in these instruments results in the formation of an "ion pair"; a positive ion and a free electron, by ion impact by the radiation on the gas molecules. The ionization chamber is the simplest of these detectors, and collects all the charges created by *direct ionisation* within the gas through the application of an electric field.[2]

The Geiger–Müller tube and the proportional counter both use a phenomenon known as a Townsend avalanche to multiply the effect of the original ionising event by means of a cascade effect whereby the free electrons are given sufficient energy by the electric field to release further electrons by ion impact.

23.4   Chemistry

23.4.1   Notation

Visualisation of ion chamber operation

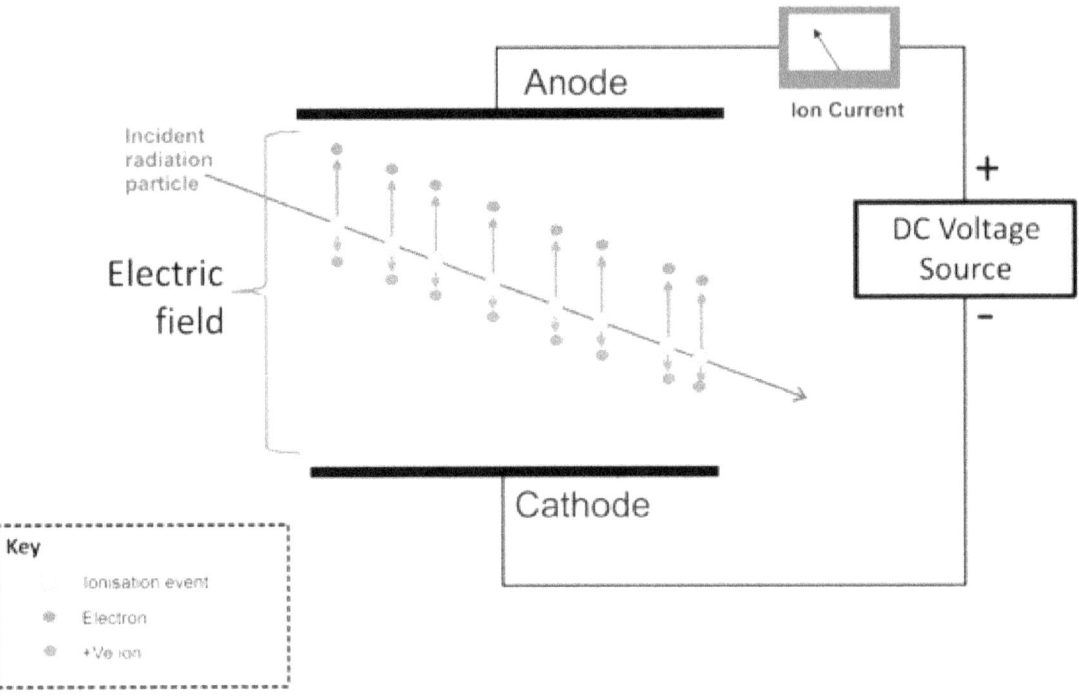

Schematic of an ion chamber, showing drift of ions. Electrons drift faster than positive ions due to their much smaller mass.[2]

Denoting the charged state

When writing the chemical formula for an ion, its net charge is written in superscript immediately after the chemical structure for the molecule/atom. The net charge is written with the magnitude *before* the sign; that is, a doubly charged cation is indicated as **2+** instead of **+2**. However, the magnitude of the charge is omitted for singly charged molecules/atoms; for example, the sodium cation is indicated as Na^+ and *not* Na^{1+}.

An alternative (and acceptable) way of showing a molecule/atom with multiple charges is by drawing out the signs multiple times; this is often seen with transition metals. Chemists sometimes circle the sign; this is merely ornamental and does not alter the chemical meaning. All three representations of Fe2+ shown in the figure are, thus, equivalent.

Monatomic ions are sometimes also denoted with Roman numerals; for example, the Fe2+ example seen above is occasionally referred to as Fe(II) or Fe^{II}. The Roman numeral designates the *formal oxidation state* of an element, whereas the superscripted numerals denotes the net charge. The two notations are, therefore, exchangeable for monatomic ions, but the Roman numerals *cannot* be applied to polyatomic ions. However, it is possible to mix the notations for the individual metal center with a polyatomic complex, as shown by the uranyl ion example.

Sub-classes

If an ion contains unpaired electrons, it is called a *radical* ion. Just like uncharged radicals, radical ions are very reactive. Polyatomic ions containing oxygen, such as carbonate and sulfate, are called *oxyanions*. Molecular ions that contain at least one carbon to hydrogen bond are called *organic ions*. If the charge in an organic ion is formally centered on a carbon, it is termed a *carbocation* (if positively charged) or *carbanion* (if negatively charged).

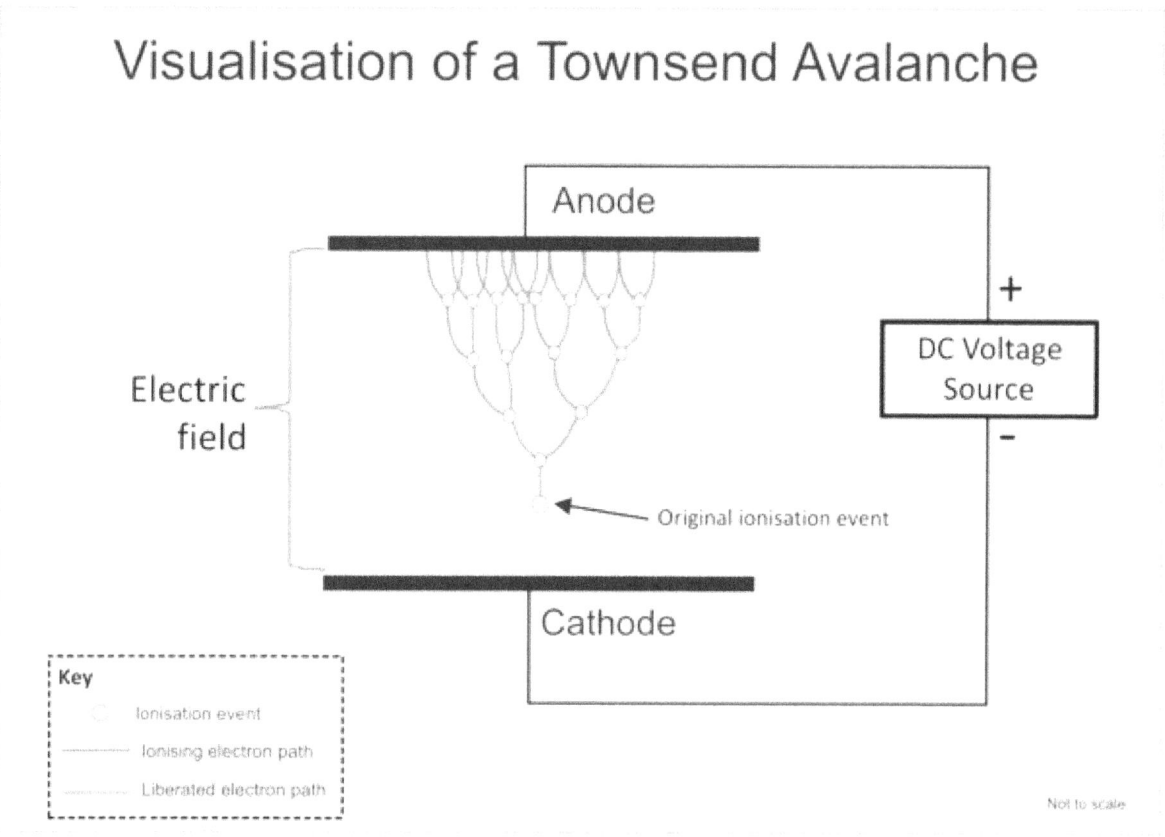

Avalanche effect between two electrodes. The original ionisation event liberates one electron, and each subsequent collision liberates a further electron, so two electrons emerge from each collision: the ionising electron and the liberated electron.

$$\mathrm{Fe}^{2+} \qquad \mathrm{Fe}^{++} \qquad \mathrm{Fe}^{\oplus\oplus}$$

Equivalent notations for an iron atom (Fe) that lost two electrons, referred to as ferrous.

23.4.2   Formation

Formation of monatomic ions

Monatomic ions are formed by the gain or loss of electrons to the valence shell (the outer-most electron shell) in an atom. The inner shells of an atom are filled with electrons that are tightly bound to the positively charged atomic nucleus, and so do not participate in this kind of chemical interaction. The process of gaining or losing electrons from a neutral atom or molecule is called *ionization*.

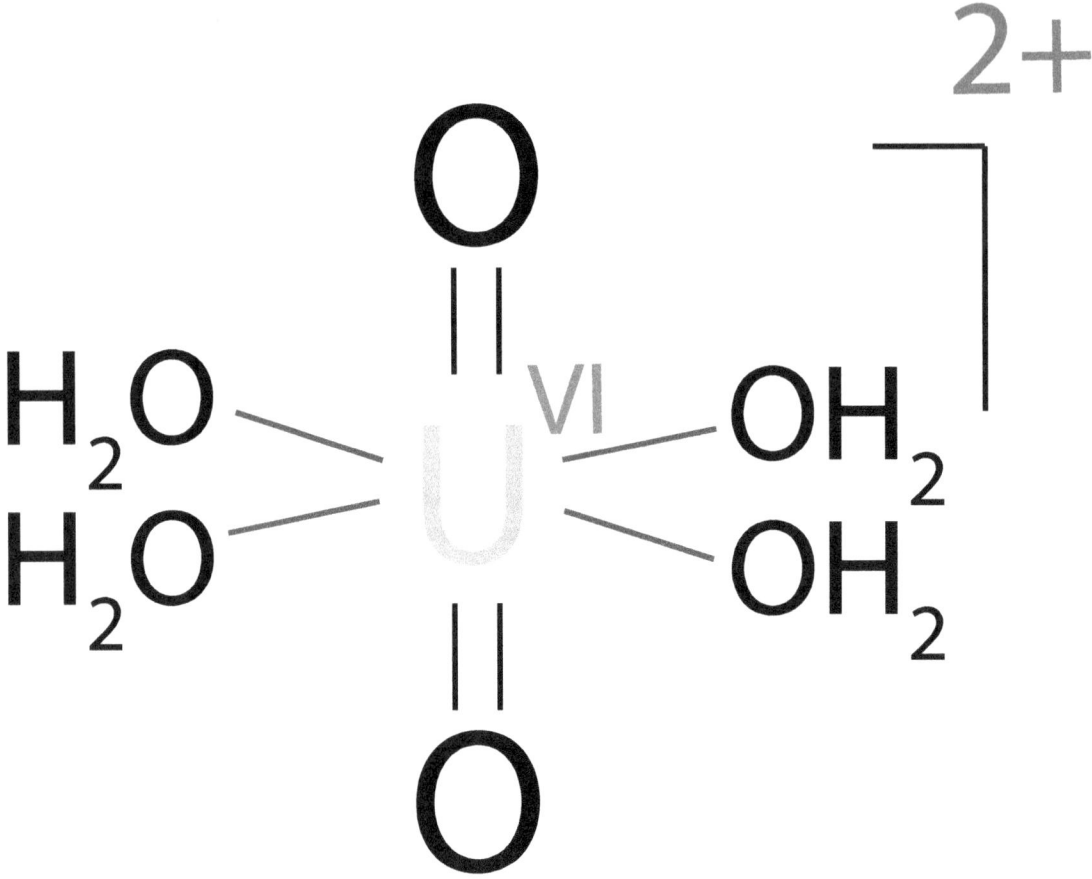

Mixed Roman numerals and charge notations for the uranyl ion. The oxidation state of the metal is shown as superscripted Roman numerals, whereas the charge of the entire complex is shown by the angle symbol together with the magnitude and sign of the net charge.

Atoms can be ionized by bombardment with radiation, but the more usual process of ionization encountered in chemistry is the transfer of electrons between atoms or molecules. This transfer is usually driven by the attaining of stable ("closed shell") electronic configurations. Atoms will gain or lose electrons depending on which action takes the least energy.

For example, a sodium atom, Na, has a single electron in its valence shell, surrounding 2 stable, filled inner shells of 2 and 8 electrons. Since these filled shells are very stable, a sodium atom tends to lose its extra electron and attain this stable configuration, becoming a sodium cation in the process

Na → Na+
+ e−

On the other hand, a chlorine atom, Cl, has 7 electrons in its valence shell, which is one short of the stable, filled shell with 8 electrons. Thus, a chlorine atom tends to *gain* an extra electron and attain a stable 8-electron configuration, becoming a chloride anion in the process:

Cl + e− → Cl−

This driving force is what causes sodium and chlorine to undergo a chemical reaction, wherein the "extra" electron is transferred from sodium to chlorine, forming sodium cations and chloride anions. Being oppositely charged, these cations and anions form ionic bonds and combine to form sodium chloride, NaCl, more commonly known as table salt.

Na+
+ Cl–
→ NaCl

Formation of polyatomic and molecular ions

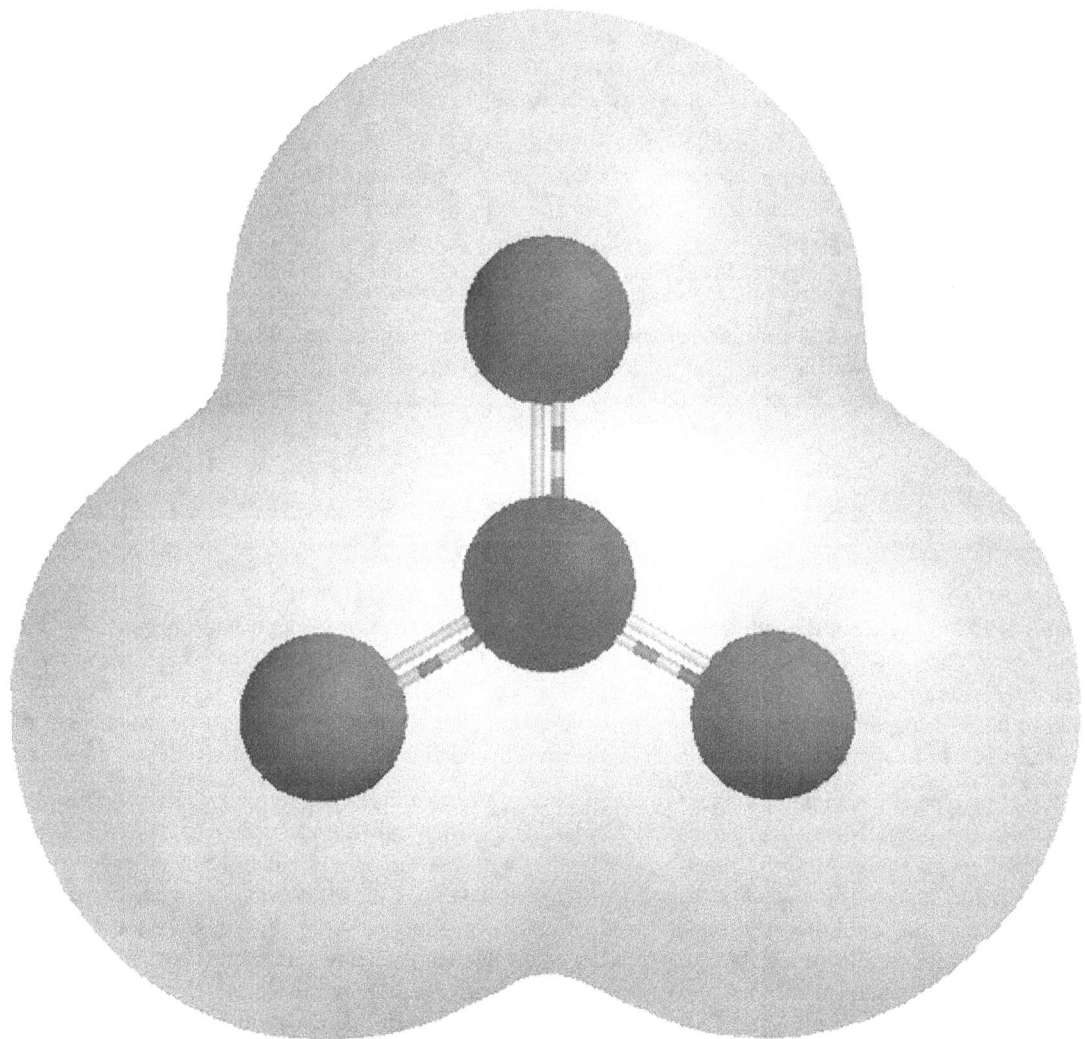

An electrostatic potential map of the nitrate ion (NO_3^-). The 3-dimensional shell represents a single arbitrary isopotential.

Polyatomic and molecular ions are often formed by the gaining or losing of elemental ions such as a proton, H^+, in neutral molecules. For example, when ammonia, NH_3, accepts a proton, H^+—a process called protonation—it forms the ammonium ion, NH_4^+. Ammonia and ammonium have the same number of electrons in essentially the same electronic configuration, but ammonium has an extra proton that gives it a net positive charge.

Ammonia can also lose an electron to gain a positive charge, forming the ion $\cdot NH+$

3. However, this ion is unstable, because it has an incomplete valence shell around the nitrogen atom, making it a very reactive radical ion.

Due to the instability of radical ions, polyatomic and molecular ions are usually formed by gaining or losing elemental ions such as H+
, rather than gaining or losing electrons. This allows the molecule to preserve its stable electronic configuration while acquiring an electrical charge.

Ionization potential

Main article: Ionization potential

The energy required to detach an electron in its lowest energy state from an atom or molecule of a gas with less net electric charge is called the *ionization potential*, or *ionization energy*. The nth ionization energy of an atom is the energy required to detach its nth electron after the first $n - 1$ electrons have already been detached.

Each successive ionization energy is markedly greater than the last. Particularly great increases occur after any given block of atomic orbitals is exhausted of electrons. For this reason, ions tend to form in ways that leave them with full orbital blocks. For example, sodium has one *valence electron* in its outermost shell, so in ionized form it is commonly found with one lost electron, as Na+
. On the other side of the periodic table, chlorine has seven valence electrons, so in ionized form it is commonly found with one gained electron, as Cl−
. Caesium has the lowest measured ionization energy of all the elements and helium has the greatest.[11] In general, the ionization energy of metals is much lower than the ionization energy of nonmetals, which is why, in general, metals will lose electrons to form positively charged ions and nonmetals will gain electrons to form negatively charged ions.

23.4.3 Ionic bonding

Main article: Ionic bond

Ionic bonding is a kind of chemical bonding that arises from the mutual attraction of oppositely charged ions. Ions of like charge repel each other, and ions of opposite charge attract each other. Therefore ions do not usually exist on their own, but will bind with ions of opposite charge to form a crystal lattice. The resulting compound is called an *ionic compound*, and is said to be held together by *ionic bonding*. In ionic compounds there arise characteristic distances between ion neighbors from which the spatial extension and the ionic radius of individual ions may be derived.

The most common type of ionic bonding is seen in compounds of metals and nonmetals (except noble gases, which rarely form chemical compounds). Metals are characterized by having a small number of electrons in excess of a stable, closed-shell electronic configuration. As such, they have the tendency to lose these extra electrons in order to attain a stable configuration. This property is known as *electropositivity*. Non-metals, on the other hand, are characterized by having an electron configuration just a few electrons short of a stable configuration. As such, they have the tendency to gain more electrons in order to achieve a stable configuration. This tendency is known as *electronegativity*. When a highly electropositive metal is combined with a highly electronegative nonmetal, the extra electrons from the metal atoms are transferred to the electron-deficient nonmetal atoms. This reaction produces metal cations and nonmetal anions, which are attracted to each other to form a *salt*.

23.4.4 Common ions

23.5 See also

- Air ionizer
- Auroras

- Gaseous ionisation detectors

- Ion beam

- Ion exchange

- Ionising radiation

- List of plasma (physics) articles

- Stopping power of radiation particles

23.6   References

[1] "Ion" entry in *Collins English Dictionary*, HarperCollins Publishers, 1998.

[2] Knoll, Glenn F (1999). *Radiation detection and measurement* (3rd ed.). New York: Wiley. ISBN 0-471-07338-5.

[3] *Michael Faraday (1791-1867)*. UK: BBC.

[4] "Online etymology dictionary". Retrieved 2011-01-07.

[5]*Oxford University Press*(2013). "Oxford Reference: OVERVIEW anion". oxfordreference.com.doi:10.1093/oi/autho4 (inactive 2015-01-01).

[6] *University of Colorado Boulder* (November 21, 2013). "Atoms and Elements, Isotopes and Ions". colorado.edu.

[7]*Oxford University Press*(2013). "Oxford Reference: OVERVIEW cation". oxfordreference.com.doi:10.1093/oi/aut (inactive 2015-01-01).

[8] Douglas W. Haywick, Ph.D.; University of South Alabama (2007–2008). "Elemental Chemistry" (PDF). usouthal.edu.

[9] *Purdue University* (November 21, 2013). "Amino Acids". purdue.edu.

[10] Plasma, Plasma, Everywhere Science@NASA Headline news, Space Science n° 158, September 7, 1999.

[11] Chemical elements listed by ionization energy. Lenntech.com

Chapter 24

Chemical element

A **chemical element** (or **element**) is a chemical substance consisting of atoms having the same number of protons in their atomic nuclei (i.e. the same atomic number, Z).[1] There are 118 elements that have been identified, of which the first 98 occur naturally on Earth with the remaining 20 being Synthetic elements. There are 80 elements that have at least one stable isotope and 38 that have only radioactive isotopes, which decay over time into other elements. Iron is the most abundant element (by mass) making up the Earth, while oxygen is the most common element in the crust of the earth.[2]

Chemical elements constitute approximately 15% of the matter in the universe: the remainder is dark matter, the composition of it is unknown, but it is not composed of chemical elements.[3] The two lightest elements, hydrogen and helium were mostly formed in the Big Bang and are the most common elements in the universe. The next three elements (lithium, beryllium and boron) were formed mostly by cosmic ray spallation, and are thus more rare than those that follow. Formation of elements with from six to twenty six protons occurred and continues to occur in main sequence stars via stellar nucleosynthesis. The high abundance of oxygen, silicon, and iron on Earth reflects their common production in such stars. Elements with greater than twenty six protons are formed by supernova nucleosynthesis in supernovae, which, when they explode, blast these elements far into space as planetary nebulae, where they may become incorporated into planets when they are formed.[4]

When different elements are chemically combined, with the atoms held together by chemical bonds, they form chemical compounds. Only a minority of elements are found uncombined as relatively pure minerals. Among the more common of such "native elements" are copper, silver, gold, carbon (as coal, graphite, or diamonds), and sulfur. All but a few of the most inert elements, such as noble gases and noble metals, are usually found on Earth in chemically combined form, as chemical compounds. While about 32 of the chemical elements occur on Earth in native uncombined forms, most of these occur as mixtures. For example, atmospheric air is primarily a mixture of nitrogen, oxygen, and argon, and native solid elements occur in alloys, such as that of iron and nickel.

The history of the discovery and use of the elements began with primitive human societies that found native elements like carbon, sulfur, copper and gold. Later civilizations extracted elemental copper, tin, lead and iron from their ores by smelting, using charcoal. Alchemists and chemists subsequently identified many more, with almost all of the naturally-occurring elements becoming known by 1900.

The properties of the chemical elements are summarized on the periodic table, which organizes the elements by increasing atomic number into rows ("periods") in which the columns ("groups") share recurring ("periodic") physical and chemical properties. Save for unstable radioactive elements with short half-lives, all of the elements are available industrially, most of them in high degrees of purity.

24.1 Description

The lightest chemical elements are hydrogen and helium, both created by Big Bang nucleosynthesis during the first 20 minutes of the universe[5] in a ratio of around 3:1 by mass (or 12:1 by number of atoms).[6][7] Almost all other elements found in nature were made by various natural methods of nucleosynthesis.[8] On Earth, small amounts of new atoms are naturally produced in nucleogenic reactions, or in cosmogenic processes, such as cosmic ray spallation. New atoms are also naturally produced on Earth as radiogenic daughter isotopes of ongoing radioactive decay processes such as alpha decay, beta decay, spontaneous fission, cluster decay, and other rarer modes of decay.

Of the 98 naturally occurring elements, those with atomic numbers 1 through 82 each have at least one stable isotope, (except for technetium, element 43 and promethium, element 61, which have no stable isotopes). Isotopes considered stable are those for which no radioactive decay has yet been observed. Elements with atomic numbers 83 through 98 are unstable to the point that radioactive decay of all isotopes can be detected. Some of these elements, notably bismuth (atomic number 83), thorium (atomic number 90), uranium (atomic number 92) and plutonium (atomic number 94), have one or more isotopes with half-lives long enough to survive as remnants of the explosive stellar nucleosynthesis that produced the heavy elements before the formation of our solar system. For example, at over 1.9×10^{19} years, over a billion times longer than the current estimated age of the universe, bismuth-209 (atomic number 83) has the longest known alpha decay half-life of any naturally occurring element.[9][10] The very heaviest elements (those beyond californium, atomic number 98) undergo radioactive decay with half-lives so short that they do not occur in nature and must be synthesized.

As of 2010, there are 118 known elements (in this context, "known" means observed well enough, even from just a few decay products, to have been differentiated from other elements).[11][12] Of these 118 elements, 98 occur naturally on Earth.[13] Ten of these occur in extreme trace quantities: technetium, atomic number 43; promethium, number 61; astatine, number 85; francium, number 87; neptunium, number 93; plutonium, number 94; americium, number 95; curium, number 96; berkelium, number 97; and californium, number 98. These 98 elements have been detected in the universe at large, in the spectra of stars and also supernovae, where short-lived radioactive elements are newly being made. The first 98 elements have been detected directly on Earth as primordial nuclides present from the formation of the solar system, or as naturally-occurring fission or transmutation products of uranium and thorium.

The remaining 20 heavier elements, not found today either on Earth or in astronomical spectra, have been produced artificially: these are all radioactive, with very short half-lives; if any atoms of these elements were present at the formation of Earth, they are extremely likely, to the point of certainty, to have already decayed, and if present in novae, have been in quantities too small to have been noted. Technetium was the first purportedly non-naturally occurring element synthesized, in 1937, although trace amounts of technetium have since been found in nature (and also the element may have been discovered naturally in 1925).[14] This pattern of artificial production and later natural discovery has been repeated with several other radioactive naturally-occurring rare elements.[15]

Lists of the elements are available by name, by symbol, by atomic number, by density, by melting point, and by boiling point as well as ionization energies of the elements. The nuclides of stable and radioactive elements are also available as a list of nuclides, sorted by length of half-life for those that are unstable. One of the most convenient, and certainly the most traditional presentation of the elements, is in the form of the periodic table, which groups together elements with similar chemical properties (and usually also similar electronic structures).

24.1.1 Atomic number

Main article: atomic number

The atomic number of an element is equal to the number of protons in each atom, and defines the element.[16] For example, all carbon atoms contain 6 protons in their atomic nucleus; so the atomic number of carbon is 6.[17] Carbon atoms may have different numbers of neutrons; atoms of the same element having different numbers of neutrons are known as isotopes of the element.[18]

The number of protons in the atomic nucleus also determines its electric charge, which in turn determines the number of electrons of the atom in its non-ionized state. The electrons are placed into atomic orbitals that determine the atom's various chemical properties. The number of neutrons in a nucleus usually has very little effect on an element's chemical properties (except in the case of hydrogen and deuterium). Thus, all carbon isotopes have nearly identical chemical properties because they all have six protons and six electrons, even though carbon atoms may, for example, have 6 or 8 neutrons. That is why the atomic number, rather than mass number or atomic weight, is considered the identifying characteristic of a chemical element.

The symbol for atomic number is Z.

24.1.2 Isotopes

Main articles: Isotope, Stable isotope and List of nuclides

Isotopes are atoms of the same element (that is, with the same number of protons in their atomic nucleus), but having *different* numbers of neutrons. Most (66 of 94) naturally occurring elements have more than one stable isotope. Thus, for example, there are three main isotopes of carbon. All carbon atoms have 6 protons in the nucleus, but they can have either 6, 7, or 8 neutrons. Since the mass numbers of these are 12, 13 and 14 respectively, the three isotopes of carbon are known as carbon-12, carbon-13, and carbon-14, often abbreviated to ^{12}C, ^{13}C, and ^{14}C. Carbon in everyday life and in chemistry is a mixture of ^{12}C (about 98.9%), ^{13}C (about 1.1%) and about 1 atom per trillion of ^{14}C.

Except in the case of the isotopes of hydrogen (which differ greatly from each other in relative mass—enough to cause chemical effects), the isotopes of a given element are chemically nearly indistinguishable.

All of the elements have some isotopes that are radioactive (radioisotopes), although not all of these radioisotopes occur naturally. The radioisotopes typically decay into other elements upon radiating an alpha or beta particle. If an element has isotopes that are not radioactive, these are termed "stable" isotopes. All of the known stable isotopes occur naturally (see primordial isotope). The many radioisotopes that are not found in nature have been characterized after being artificially made. Certain elements have no stable isotopes and are composed *only* of radioactive isotopes: specifically the elements without any stable isotopes are technetium (atomic number 43), promethium (atomic number 61), and all observed elements with atomic numbers greater than 82.

Of the 80 elements with at least one stable isotope, 26 have only one single stable isotope. The mean number of stable isotopes for the 80 stable elements is 3.1 stable isotopes per element. The largest number of stable isotopes that occur for a single element is 10 (for tin, element 50).

24.1.3 Isotopic mass and atomic mass

Main articles: atomic mass and relative atomic mass

The mass number of an element, A, is the number of nucleons (protons and neutrons) in the atomic nucleus. Different isotopes of a given element are distinguished by their mass numbers, which are conventionally written as a superscript on the left hand side of the atomic symbol (e.g., ^{238}U). The mass number is always a simple whole number and has units of "nucleons." An example of a referral to a mass number is "magnesium-24," which is an atom with 24 nucleons (12 protons and 12 neutrons).

Whereas the mass number simply counts the total number of neutrons and protons and is thus a natural (or whole) number, the atomic mass of a single atom is a real number for the mass of a particular isotope of the element, the unit being **u**. In general, when expressed in **u** it differs in value slightly from the mass number for a given nuclide (or isotope) since the mass of the protons and neutrons is not exactly 1 **u**, since the electrons contribute a lesser share to the atomic mass as neutron number exceeds proton number, and (finally) because of the nuclear binding energy. For example, the atomic mass of chlorine-35 to five significant digits is 34.969 **u** and that of chlorine-37 is 36.966 **u**. However, the atomic mass in **u** of each isotope is quite close to its simple mass number (always within 1%). The only isotope whose atomic mass is exactly a natural number is ^{12}C, which by definition has a mass of exactly 12, because **u** is defined as 1/12 of the mass of a free neutral carbon-12 atom in the ground state.

The relative atomic mass (historically and commonly also called "atomic weight") of an element is the *average* of the atomic masses of all the chemical element's isotopes as found in a particular environment, weighted by isotopic abundance, relative to the atomic mass unit (**u**). This number may be a fraction that is *not* close to a whole number, due to the averaging process. For example, the relative atomic mass of chlorine is 35.453 **u**, which differs greatly from a whole number due to being made of an average of 76% chlorine-35 and 24% chlorine-37. Whenever a relative atomic mass value differs by more than 1% from a whole number, it is due to this averaging effect resulting from significant amounts of more than one isotope being naturally present in the sample of the element in question.

24.1.4   Chemically pure and isotopically pure

Chemists and nuclear scientists have different definitions of a *pure element*. In chemistry, a pure element means a substance whose atoms all (or in practice almost all) have the same atomic number, or number of protons. Nuclear scientists, however, define a pure element as one that consists of only one stable isotope.[19]

For example, a copper wire is 99.99% chemically pure if 99.99% of its atoms are copper, with 29 protons each. However it is not isotopically pure since ordinary copper consists of two stable isotopes, 69% ^{63}Cu and 31% ^{65}Cu, with different numbers of neutrons.

24.1.5   Allotropes

Main article: Allotropy

Atoms of chemically pure elements may bond to each other chemically in more than one way, allowing the pure element to exist in multiple structures (spatial arrangements of atoms), known as allotropes, which differ in their properties. For example, carbon can be found as diamond, which has a tetrahedral structure around each carbon atom; graphite, which has layers of carbon atoms with a hexagonal structure stacked on top of each other; graphene, which is a single layer of graphite that is very strong; fullerenes, which have nearly spherical shapes; and carbon nanotubes, which are tubes with a hexagonal structure (even these may differ from each other in electrical properties). The ability of an element to exist in one of many structural forms is known as 'allotropy'.

The standard state, also known as reference state, of an element is defined as its thermodynamically most stable state at 1 bar at a given temperature (typically at 298.15 K). In thermochemistry, an element is defined to have an enthalpy of formation of zero in its standard state. For example, the reference state for carbon is graphite, because the structure of graphite is more stable than that of the other allotropes.

24.1.6   Properties

Several kinds of descriptive categorizations can be applied broadly to the elements, including consideration of their general physical and chemical properties, their states of matter under familiar conditions, their melting and boiling points, their densities, their crystal structures as solids, and their origins.

General properties

Several terms are commonly used to characterize the general physical and chemical properties of the chemical elements. A first distinction is between metals, which readily conduct electricity, nonmetals, which do not, and a small group, (the *metalloids*), having intermediate properties and often behaving as semiconductors.

A more refined classification is often shown in colored presentations of the periodic table. This system restricts the terms "metal" and "nonmetal" to only certain of the more broadly defined metals and nonmetals, adding additional terms for certain sets of the more broadly viewed metals and nonmetals. The version of this classification used in the periodic tables presented here includes: actinides, alkali metals, alkaline earth metals, halogens, lanthanides, transition metals, post-transition metals; metalloids, noble gases, polyatomic nonmetals, diatomic nonmetals, and transition metals. In this system, the alkali metals, alkaline earth metals, and transition metals, as well as the lanthanides and the actinides, are special groups of the metals viewed in a broader sense. Similarly, the polyatomic nonmetals, diatomic nonmetals and the noble gases are nonmetals viewed in the broader sense. In some presentations, the halogens are not distinguished, with astatine identified as a metalloid and the others identified as nonmetals.

States of matter

Another commonly used basic distinction among the elements is their state of matter (phase), whether solid, liquid, or gas, at a selected standard temperature and pressure (STP). Most of the elements are solids at conventional temperatures and atmospheric pressure, while several are gases. Only bromine and mercury are liquids at 0 degrees Celsius (32 degrees Fahrenheit) and normal atmospheric pressure; caesium and gallium are solids at that temperature, but melt at 28.4 °C (83.2 °F) and 29.8 °C (85.6 °F), respectively.

Melting and boiling points

Melting and boiling points, typically expressed in degrees Celsius at a pressure of one atmosphere, are commonly used in characterizing the various elements. While known for most elements, either or both of these measurements is still undetermined for some of the radioactive elements available in only tiny quantities. Since helium remains a liquid even at absolute zero at atmospheric pressure, it has only a boiling point, and not a melting point, in conventional presentations.

Densities

Main article: Densities of the elements (data page)

The density at a selected standard temperature and pressure (STP) is frequently used in characterizing the elements. Density is often expressed in grams per cubic centimeter (g/cm^3). Since several elements are gases at commonly encountered temperatures, their densities are usually stated for their gaseous forms; when liquefied or solidified, the gaseous elements have densities similar to those of the other elements.

When an element has allotropes with different densities, one representative allotrope is typically selected in summary presentations, while densities for each allotrope can be stated where more detail is provided. For example, the three familiar allotropes of carbon (amorphous carbon, graphite, and diamond) have densities of 1.8–2.1, 2.267, and 3.515 g/cm^3, respectively.

Crystal structures

The elements studied to date as solid samples have eight kinds of crystal structures: cubic, body-centered cubic, face-centered cubic, hexagonal, monoclinic, orthorhombic, rhombohedral, and tetragonal. For some of the synthetically produced transuranic elements, available samples have been too small to determine crystal structures.

Occurrence and origin on Earth

Chemical elements may also be categorized by their origin on Earth, with the first 98 considered naturally occurring, while those with atomic numbers beyond 98 have only been produced artificially as the synthetic products of man-made nuclear reactions.

Of the 98 naturally occurring elements, 84 are considered primordial and either stable or weakly radioactive. The remaining 14 naturally occurring elements possess half lives too short for them to have been present at the beginning of the Solar System, and are therefore considered transient elements. Of these 14 transient elements, 7 (polonium, astatine, radon, francium, radium, actinium, and protactinium) are relatively common decay products of thorium, uranium, and plutonium. The remaining 7 transient elements (technetium, promethium, neptunium, americium, curium, berkelium, and californium) occur only rarely, as products of rare nuclear reaction processes involving uranium or other heavy elements.

Elements with atomic numbers 1 through 40 are all stable, while those with atomic numbers 41 through 82 (except technetium and promethium) are metastable. The half-lives of these metastable "theoretical radionuclides" are so long (at least 100 million times longer than the estimated age of the universe) that their radioactive decay has yet to be detected by experiment. Elements with atomic numbers 83 through 98 are unstable to the point that their radioactive decay can be detected. Some of these elements, notably thorium (atomic number 90) and uranium (atomic number 92), have one or more isotopes with half-lives long enough to survive as remnants of the explosive stellar nucleosynthesis that produced the heavy elements before the formation of our solar system. For example, at over 1.9×10^{19} years, over a billion times longer than the current estimated age of the universe, bismuth-209 (atomic number 83) has the longest known alpha decay half-life of any naturally occurring element.[9][10] The very heaviest elements (those beyond californium, atomic number 98) undergo radioactive decay with short half-lives and do not occur in nature.

24.1.7 The periodic table

Main article: Periodic table

The properties of the chemical elements are often summarized using the periodic table, which powerfully and elegantly organizes the elements by increasing atomic number into rows ("periods") in which the columns ("groups") share recurring ("periodic") physical and chemical properties. The current standard table contains 118 confirmed elements as of 10 April 2010.

Although earlier precursors to this presentation exist, its invention is generally credited to the Russian chemist Dmitri Mendeleev in 1869, who intended the table to illustrate recurring trends in the properties of the elements. The layout of the table has been refined and extended over time as new elements have been discovered and new theoretical models have been developed to explain chemical behavior.

Use of the periodic table is now ubiquitous within the academic discipline of chemistry, providing an extremely useful framework to classify, systematize and compare all the many different forms of chemical behavior. The table has also found wide application in physics, geology, biology, materials science, engineering, agriculture, medicine, nutrition, environmental health, and astronomy. Its principles are especially important in chemical engineering.

24.2 Nomenclature and symbols

The various chemical elements are formally identified by their unique atomic numbers, by their accepted names, and by their symbols.

24.2.1 Atomic numbers

The known elements have atomic numbers from 1 through 118, conventionally presented as Arabic numerals. Since the elements can be uniquely sequenced by atomic number, conventionally from lowest to highest (as in a periodic table), sets of elements are sometimes specified by such notation as "through", "beyond", or "from ... through", as in "through iron", "beyond uranium", or "from lanthanum through lutetium". The terms "light" and "heavy" are sometimes also used informally to indicate relative atomic numbers (not densities!), as in "lighter than carbon" or "heavier than lead", although technically the weight or mass of atoms of an element (their atomic weights or atomic masses) do not always increase monotonically with their atomic numbers.

24.2.2 Element names

Main article: List of chemical element name etymologies

The naming of various substances now known as elements precedes the atomic theory of matter, as names were given locally by various cultures to various minerals, metals, compounds, alloys, mixtures, and other materials, although at the time it was not known which chemicals were elements and which compounds. As they were identified as elements, the existing names for anciently-known elements (e.g., gold, mercury, iron) were kept in most countries. National differences emerged over the names of elements either for convenience, linguistic niceties, or nationalism. For a few illustrative examples: German speakers use "Wasserstoff" (water substance) for "hydrogen", "Sauerstoff" (acid substance) for "oxygen" and "Stickstoff" (smothering substance) for "nitrogen", while English and some romance languages use "sodium" for "natrium" and "potassium" for "kalium", and the French, Italians, Greeks, Portuguese and Poles prefer "azote/azot/azoto" (from roots meaning "no life") for "nitrogen".

For purposes of international communication and trade, the official names of the chemical elements both ancient and more recently recognized are decided by the International Union of Pure and Applied Chemistry (IUPAC), which has decided on a sort of international English language, drawing on traditional English names even when an element's chemical symbol is based on a Latin or other traditional word, for example adopting "gold" rather than "aurum" as the name for the 79th element (Au). IUPAC prefers the British spellings "aluminium" and "caesium" over the U.S. spellings "aluminum" and "cesium", and the U.S. "sulfur" over the British "sulphur". However, elements that are practical to sell in bulk in many countries often still have locally used national names, and countries whose national language does not use the Latin alphabet are likely to use the IUPAC element names.

According to IUPAC, chemical elements are not proper nouns in English; consequently, the full name of an element is not routinely capitalized in English, even if derived from a proper noun, as in californium and einsteinium. Isotope names of chemical elements are also uncapitalized if written out, *e.g.,* carbon-12 or uranium-235. Chemical element *symbols* (such as Cf for californium and Es for einsteinium), are always capitalized (see below).

In the second half of the twentieth century, physics laboratories became able to produce nuclei of chemical elements with half-lives too short for an appreciable amount of them to exist at any time. These are also named by IUPAC, which generally adopts the name chosen by the discoverer. This practice can lead to the controversial question of which research group actually discovered an element, a question that has delayed naming of elements with atomic number of 104 and higher for a considerable time. (See element naming controversy).

Precursors of such controversies involved the nationalistic namings of elements in the late 19th century. For example, *lutetium* was named in reference to Paris, France. The Germans were reluctant to relinquish naming rights to the French, often calling it *cassiopeium*. Similarly, the British discoverer of *niobium* originally named it *columbium,* in reference to the New World. It was used extensively as such by American publications prior to international standardization.

24.2.3 Chemical symbols

For listings of current chemical symbols, symbols not currently used, and other symbols that may look like chemical symbols, see Symbol (chemistry).

Specific chemical elements

Before chemistry became a science, alchemists had designed arcane symbols for both metals and common compounds. These were however used as abbreviations in diagrams or procedures; there was no concept of atoms combining to form molecules. With his advances in the atomic theory of matter, John Dalton devised his own simpler symbols, based on circles, to depict molecules.

The current system of chemical notation was invented by Berzelius. In this typographical system, chemical symbols are not mere abbreviations—though each consists of letters of the Latin alphabet. They are intended as universal symbols for people of all languages and alphabets.

The first of these symbols were intended to be fully universal. Since Latin was the common language of science at that time, they were abbreviations based on the Latin names of metals. Cu comes from Cuprum, Fe comes from Ferrum, Ag from Argentum. The symbols were not followed by a period (full stop) as with abbreviations. Later chemical elements were also assigned unique chemical symbols, based on the name of the element, but not necessarily in English. For example, sodium has the chemical symbol 'Na' after the Latin *natrium*. The same applies to "W" (wolfram) for tungsten, "Fe" (ferrum) for iron, "Hg" (hydrargyrum) for mercury, "Sn" (stannum) for tin, "K" (kalium) for potassium, "Au" (aurum) for gold, "Ag" (argentum) for silver, "Pb" (plumbum) for lead, "Cu" (cuprum) for copper, and "Sb" (stibium) for antimony.

Chemical symbols are understood internationally when element names might require translation. There have sometimes been differences in the past. For example, Germans in the past have used "J" (for the alternate name Jod) for iodine, but now use "I" and "Iod."

The first letter of a chemical symbol is always capitalized, as in the preceding examples, and the subsequent letters, if any, are always lower case (small letters). Thus, the symbols for californium or einsteinium are Cf and Es.

General chemical symbols

There are also symbols in chemical equations for groups of chemical elements, for example in comparative formulas. These are often a single capital letter, and the letters are reserved and not used for names of specific elements. For example, an "**X**" indicates a variable group (usually a halogen) in a class of compounds, while "**R**" is a radical, meaning a compound structure such as a hydrocarbon chain. The letter "**Q**" is reserved for "heat" in a chemical reaction. "**Y**" is also often used as a general chemical symbol, although it is also the symbol of yttrium. "**Z**" is also frequently used as a general variable group. "**E**" is used in organic chemistry to denote an electron-withdrawing group. "**L**" is used to represent a general ligand in inorganic and organometallic chemistry. "**M**" is also often used in place of a general metal.

At least two additional, two-letter generic chemical symbols are also in informal usage, "**Ln**" for any lanthanide element and "**An**" for any actinide element. "**Rg**" was formerly used for any rare gas element, but the group of rare gases has now been renamed noble gases and the symbol "**Rg**" has now been assigned to the element roentgenium.

Isotope symbols

Isotopes are distinguished by the atomic mass number (total protons and neutrons) for a particular isotope of an element, with this number combined with the pertinent element's symbol. IUPAC prefers that isotope symbols be written in superscript notation when practical, for example ^{12}C and ^{235}U. However, other notations, such as carbon-12 and uranium-235, or C-12 and U-235, are also used.

As a special case, the three naturally occurring isotopes of the element hydrogen are often specified as **H** for ^{1}H (protium), **D** for ^{2}H (deuterium), and **T** for ^{3}H (tritium). This convention is easier to use in chemical equations, replacing the need to write out the mass number for each atom. For example, the formula for heavy water may be written D_2O instead of $^{2}H_2O$.

24.3   Origin of the elements

Only about 4% of the total mass of the universe is made of atoms or ions, and thus represented by chemical elements. This fraction is about 15% of the total matter, with the remainder of the matter (85%) being dark matter. The nature of dark matter is unknown, but it is not composed of atoms of chemical elements because it contains no protons, neutrons, or electrons. (The remaining non-matter part of the mass of the universe is composed of the even more mysterious dark energy).

The universe's 98 naturally occurring chemical elements are thought to have been produced by at least four cosmic processes. Most of the hydrogen and helium in the universe was produced primordially in the first few minutes of the Big Bang. Three recurrently occurring later processes are thought to have produced the remaining elements. Stellar nucleosynthesis, an ongoing process, produces all elements from carbon through iron in atomic number, but little lithium, beryllium, or boron. Elements heavier in atomic number than iron, as heavy as uranium and plutonium, are produced by explosive nucleosynthesis in supernovas and other cataclysmic cosmic events. Cosmic ray spallation (fragmentation) of carbon, nitrogen, and oxygen is important to the production of lithium, beryllium and boron.

During the early phases of the Big Bang, nucleosynthesis of hydrogen nuclei resulted in the production of hydrogen-1 (protium, ^{1}H) and helium-4 (^{4}He), as well as a smaller amount of deuterium (^{2}H) and very minuscule amounts (on the order of 10^{-10}) of lithium and beryllium. Even smaller amounts of boron may have been produced in the Big Bang, since it has been observed in some very old stars, while carbon has not.[20] It is generally agreed that no heavier elements than boron were produced in the Big Bang. As a result, the primordial abundance of atoms (or ions) consisted of roughly 75% ^{1}H, 25% ^{4}He, and 0.01% deuterium, with only tiny traces of lithium, beryllium, and perhaps boron.[21] Subsequent enrichment of galactic halos occurred due to stellar nucleosynthesis and supernova nucleosynthesis.[22] However, the element abundance in intergalactic space can still closely resemble primordial conditions, unless it has been enriched by some means.

On Earth (and elsewhere), trace amounts of various elements continue to be produced from other elements as products of natural transmutation processes. These include some produced by cosmic rays or other nuclear reactions (see cosmogenic and nucleogenic nuclides), and others produced as decay products of long-lived primordial nuclides.[23] For example, trace (but detectable) amounts of carbon-14 (^{14}C) are continually produced in the atmosphere by cosmic rays impacting nitrogen atoms, and argon-40 (^{40}Ar) is continually produced by the decay of primordially occurring but unstable potassium-40 (^{40}K). Also, three primordially occurring but radioactive actinides, thorium, uranium, and plutonium, decay through a series of recurrently produced but unstable radioactive elements such as radium and radon, which are transiently present in any sample of these metals or their ores or compounds. Seven other radioactive elements, technetium, promethium, neptunium, americium, curium, berkelium, and californium, occur only incidentally in natural materials, produced as individual atoms by natural fission of the nuclei of various heavy elements or in other rare nuclear processes.

Human technology has produced various additional elements beyond these first 98, with those through atomic number 118 now known.

24.4   Abundance

Main article: Abundance of the chemical elements

The following graph (note log scale) shows the abundance of elements in our solar system. The table shows the twelve most common elements in our galaxy (estimated spectroscopically), as measured in parts per million, by mass.[24] Nearby galaxies that have evolved along similar lines have a corresponding enrichment of elements heavier than hydrogen and helium. The more distant galaxies are being viewed as they appeared in the past, so their abundances of elements appear closer to the primordial mixture. As physical laws and processes appear common throughout the visible universe, however, scientist expect that these galaxies evolved elements in similar abundance.

The abundance of elements in the Solar System is in keeping with their origin from nucleosynthesis in the Big Bang and a number of progenitor supernova stars. Very abundant hydrogen and helium are products of the Big Bang, but the next three elements are rare since they had little time to form in the Big Bang and are not made in stars (they are, however, produced in small quantities by the breakup of heavier elements in interstellar dust, as a result of impact by cosmic rays). Beginning with carbon, elements are produced in stars by buildup from alpha particles (helium nuclei), resulting in an alternatingly larger abundance of elements with even atomic numbers (these are also more stable). In general, such elements up to iron are made in large stars in the process of becoming supernovas. Iron-56 is particularly common, since it is the most stable element that can easily be made from alpha particles (being a product of decay of radioactive nickel-56, ultimately made from 14 helium nuclei). Elements heavier than iron are made in energy-absorbing processes in large stars, and their abundance in the universe (and on Earth) generally decreases with their atomic number.

The abundance of the chemical elements on **Earth** varies from air to crust to ocean, and in various types of life. The abundance of elements in Earth's crust differs from that in the Solar system (as seen in the Sun and heavy planets like Jupiter) mainly in selective loss of the very lightest elements (hydrogen and helium) and also volatile neon, carbon (as hydrocarbons), nitrogen and sulfur, as a result of solar heating in the early formation of the solar system. Oxygen, the most abundant Earth element by mass, is retained on Earth by combination with silicon. Aluminum at 8% by mass is more common in the Earth's crust than in the universe and solar system, but the composition of the far more bulky mantle, which has magnesium and iron in place of aluminum (which occurs there only at 2% of mass) more closely mirrors the elemental composition of the solar system, save for the noted loss of volatile elements to space, and loss of iron which has migrated to the Earth's core.

The composition of the human body, by contrast, more closely follows the composition of seawater—save that the human body has additional stores of carbon and nitrogen necessary to form the proteins and nucleic acids, together with phosphorus in the nucleic acids and energy transfer molecule adenosine triphosphate (ATP) that occurs in the cells of all living organisms. Certain kinds of organisms require particular additional elements, for example the magnesium in chlorophyll in green plants, the calcium in mollusc shells, or the iron in the hemoglobin in vertebrate animals' red blood cells.

24.5   History

24.5.1   Evolving definitions

The concept of an "element" as an undivisible substance has developed through three major historical phases: Classical definitions (such as those of the ancient Greeks), chemical definitions, and atomic definitions.

Classical definitions

Ancient philosophy posited a set of classical elements to explain observed patterns in nature. These *elements* originally referred to *earth*, *water*, *air* and *fire* rather than the chemical elements of modern science.

The term 'elements' (*stoicheia*) was first used by the Greek philosopher Plato in about 360 BCE in his dialogue Timaeus, which includes a discussion of the composition of inorganic and organic bodies and is a speculative treatise on chemistry. Plato believed the elements introduced a century earlier by Empedocles were composed of small polyhedral forms: tetrahedron (fire), octahedron (air), icosahedron (water), and cube (earth).[25][26]

Aristotle, c. 350 BCE, also used the term *stoicheia* and added a fifth element called aether, which formed the heavens. Aristotle defined an element as:

> Element – one of those bodies into which other bodies can decompose, and that itself is not capable of being divided into other.[27]

Chemical definitions

In 1661, Robert Boyle proposed his theory of corpuscularism which favoured the analysis of matter as constituted by irreducible units of matter (atoms) and, choosing to side with neither Aristotle's view of the four elements nor Paracelsus' view of three fundamental elements, left open the question of the number of elements.[28] The first modern list of chemical elements was given in Antoine Lavoisier's 1789 *Elements of Chemistry*, which contained thirty-three elements, including light and caloric.[29] By 1818, Jöns Jakob Berzelius had determined atomic weights for forty-five of the forty-nine then-accepted elements. Dmitri Mendeleev had sixty-six elements in his periodic table of 1869.

From Boyle until the early 20th century, an element was defined as a pure substance that could not be decomposed into any simpler substance.[28] Put another way, a chemical element cannot be transformed into other chemical elements by chemical processes. Elements during this time were generally distinguished by their atomic weights, a property measurable with fair accuracy by available analytical techniques.

Atomic definitions

The 1913 discovery by English physicist Henry Moseley that the nuclear charge is the physical basis for an atom's atomic number, further refined when the nature of protons and neutrons became appreciated, eventually led to the current definition of an element based on atomic number (number of protons per atomic nucleus). The use of atomic numbers, rather than atomic weights, to distinguish elements has greater predictive value (since these numbers are integers), and also resolves some ambiguities in the chemistry-based view due to varying properties of isotopes and allotropes within the same element. Currently, IUPAC defines an element to exist if it has isotopes with a lifetime longer than the 10^{-14} seconds it takes the nucleus to form an electronic cloud.[30]

By 1914, seventy-two elements were known, all naturally occurring.[31] The remaining naturally occurring elements were discovered or isolated in subsequent decades, and various additional elements have also been produced synthetically, with much of that work pioneered by Glenn T. Seaborg. In 1955, element 101 was discovered and named mendelevium in honor of D.I. Mendeleev, the first to arrange the elements in a periodic manner. Most recently, the synthesis of element 118 was reported in October 2006, and the synthesis of element 117 was reported in April 2010.[32]

24.5.2 Discovery and recognition of various elements

See also: Timeline of chemical elements discoveries

Ten materials familiar to various prehistoric cultures are now known to be chemical elements: Carbon, copper, gold, iron, lead, mercury, silver, sulfur, tin, and zinc. Three additional materials now accepted as elements, arsenic, antimony, and bismuth, were recognized as distinct substances prior to 1500 AD. Phosphorus, cobalt, and platinum were isolated before 1750.

Most of the remaining naturally occurring chemical elements were identified and characterized by 1900, including:

- Such now-familiar industrial materials as aluminium, silicon, nickel, chromium, magnesium, and tungsten

- Reactive metals such as lithium, sodium, potassium, and calcium

- The halogens fluorine, chlorine, bromine, and iodine

- Gases such as hydrogen, oxygen, nitrogen, helium, argon, and neon

- Most of the rare-earth elements, including cerium, lanthanum, gadolinium, and neodymium.

- The more common radioactive elements, including uranium, thorium, radium, and radon

Elements isolated or produced since 1900 include:

- The three remaining undiscovered regularly occurring stable natural elements: hafnium, lutetium, and rhenium

- Plutonium, which was first produced synthetically in 1940 by Glenn T. Seaborg, but is now also known from a few long-persisting natural occurrences

- The seven incidentally occurring natural elements (americium, berkelium, californium, curium, neptunium, promet ,and technetium), which were all first produced synthetically but later discovered in trace amounts in certain geological samples

- Three scarce decay products of uranium or thorium, (astatine, francium, and protactinium), and

- Various synthetic transuranic elements, beginning with einsteinium, fermium, mendelevium, nobelium, and lawrencium

24.5.3 Recently discovered elements

The first transuranium element (element with atomic number greater than 92) discovered was neptunium in 1940. Since 1999 claims for the discovery of new elements have been considered by the IUPAC/IUPAP Joint Working Party. As of May 2012, only the elements up to 112, copernicium, as well as element 114 Flerovium and element 116 Livermorium have been confirmed as discovered by IUPAC, while claims have been made for synthesis of elements 113, 115, 117[33] and 118. The discovery of element 112 was acknowledged in 2009, and the name 'copernicium' and the atomic symbol 'Cn' were suggested for it.[34] The name and symbol were officially endorsed by IUPAC on 19 February 2010.[35] The heaviest element that is believed to have been synthesized to date is element 118, ununoctium, on 9 October 2006, by the Flerov Laboratory of Nuclear Reactions in Dubna, Russia.[12][36] Element 117 was the latest element claimed to be discovered, in 2009.[33] IUPAC officially recognized flerovium and livermorium, elements 114 and 116, in June 2011 and approved their names in May 2012.[37]

24.6 List of the 118 known chemical elements

The following sortable table includes the 118 known chemical elements, with the names linking to the *Wikipedia* articles on each.

- **Atomic number**, **name**, and **symbol** all serve independently as unique identifiers.

- **Names** are those accepted by IUPAC; provisional names for recently produced elements not yet formally named are in parentheses.

- **Group, period,** and **block** refer to an element's position in the periodic table. Group numbers here show the currently accepted numbering; for older alternate numberings, see Group (periodic table).

- **State of matter** *(solid, liquid,* or *gas)* applies at standard temperature and pressure conditions (STP).

- **Occurrence** distinguishes naturally occurring elements, categorized as either *primordial* or *transient* (from decay), and additional *synthetic* elements that have been produced technologically, but are not known to occur naturally.

- **Description** summarizes an element's properties using the broad categories commonly presented in periodic tables: Actinide, alkali metal, alkaline earth metal, halogen, lanthanide, metal, metalloid, noble gas, non-metal, and transition metal.

24.7 See also

- Discovery of the chemical elements

- Element collecting

- Fictional element

- Goldschmidt classification

- Island of stability

- List of elements by name

- List of the elements' densities

- List of nuclides

- Periodic Systems of Small Molecules

- Prices of elements and their compounds

- Symbol (chemical element)#Symbols not currently used

- Systematic element name

- Table of nuclides

24.8 References

[1] IUPAC (ed.). "chemical element". *http://iupac.org". doi:10.1351/goldbook.C01022.*

[2] Los Alamos National Laboratory (2011). "Periodic Table of Elements: Oxygen". Los Alamos, New Mexico: Los Alamos National Security, LLC. Retrieved 7 May 2011.

[3] Oerter, Robert (2006). *The Theory of Almost Everything: The Standard Model, the Unsung Triumph of Modern Physics.* Penguin. p. 223. ISBN 978-0-452-28786-0.

[4] E. M. Burbidge, G. R. Burbidge, W. A. Fowler, F. Hoyle (1957). "Synthesis of the Elements in Stars". *Reviews of Modern Physics* **29** (4): 547–650. Bibcode:1957RvMP...29..547B. doi:10.1103/RevModPhys.29.547.

[5] See the timeline on p.10 in Oganessian, Yu. Ts.; Utyonkov, V.; Lobanov, Yu.; Abdullin, F.; Polyakov, A.; Sagaidak, R.; Shirokovsky, I.; Tsyganov, Yu. et al. (2006). "Evidence for Dark Matter" (PDF). *Physical Review C* **74** (4): 044602. Bibcode:2006PhRvC..74d4602O. doi:10.1103/PhysRevC.74.044602.

[6] lbl.gov (2005). "The Universe Adventure Hydrogen and Helium". *Lawrence Berkeley National Laboratory U.S. Department of Energy.*

[7] astro.soton.ac.uk (January 3, 2001). "Formation of the light elements". *University of Southampton.*

[8] foothill.edu (October 18, 2006). "How Stars Make Energy and New Elements" (PDF). *Foothill College.*

[9] Dumé, B (23 April 2003). "Bismuth breaks half-life record for alpha decay". *Physicsworld.com* (Bristol, England: Institute of Physics). Retrieved 14 July 2015.

[10] de Marcillac, P; Coron, N; Dambier, G; Leblanc, J; Moalic, J-P (2003). "Experimental detection of alpha-particles from the radioactive decay of natural bismuth". *Nature* **422** (6934): 876–8. Bibcode:2003Natur.422..876D. doi:10.1038/nature01541. PMID 12712201.

[11] Sanderson, K (17 October 2006). "Heaviest element made – again". Nature News. doi:10.1038/news061016-4.

[12] Schewe, P; Stein, B (17 October 200). "Elements 116 and 118 Are Discovered". *Physics News Update*. American Institute of Physics. Retrieved 19 October 2006. Check date values in: |date= (help)

[13] scienceline.ucsb.edu/. "How many elements are there in the known universe?". *University of California, Santa Cruz.*

[14] *United States Environmental Protection Agency.* "Technetium-99". epa.gov. Retrieved 26 February 2013.

[15] *Harvard–Smithsonian Center for Astrophysics.* "ORIGIN OF HEAVY ELEMENTS". cfa.harvard.edu. Retrieved 26 February 2013.

[16] "ATOMIC NUMBER AND MASS NUMBERS". ndt-ed.org. Retrieved 17 February 2013.

[17] periodic.lanl.gov. "PERIODIC TABLE OF ELEMENTS: LANL Carbon". *Los Alamos National Laboratory.*

[18] Katsuya Yamada. "Atomic mass, isotopes, and mass number." (PDF). *Los Angeles Pierce College.*

[19] "Pure element". European Nuclear Society.

[20] Wilford, JN (14 January 1992). "Hubble Observations Bring Some Surprises". *New York Times.*

[21] Wright, EL (12 September 2004). "Big Bang Nucleosynthesis". UCLA, Division of Astronomy. Retrieved 22 February 2007.

[22] Wallerstein, George; Iben, Icko; Parker, Peter; Boesgaard, Ann; Hale, Gerald; Champagne, Arthur; Barnes, Charles; Käppeler, Franz et al. (1999). "Synthesis of the elements in stars: forty years of progress" (PDF). *Reviews of Modern Physics* **69** (4): 995–1084. Bibcode:1997RvMP...69..995W. doi:10.1103/RevModPhys.69.995.

[23] Earnshaw, A; Greenwood, N (1997). *Chemistry of the Elements* (2nd ed.). Butterworth-Heinemann.

[24] Croswell, K (1996). *Alchemy of the Heavens.* Anchor. ISBN 0-385-47214-5.

[25] Plato (2008) [c. 360 BC]. *Timaeus.* Forgotten Books. p. 45. ISBN 978-1-60620-018-6.

[26] Hillar, M (2004). "The Problem of the Soul in Aristotle's De anima". NASA/WMAP. Retrieved 10 August 2006.

[27] Partington, JR (1937). *A Short History of Chemistry.* New York: Dover Publications. ISBN 0-486-65977-1.

[28] Boyle, R (1661). *The Sceptical Chymist.* London. ISBN 0-922802-90-4.

[29] Lavoisier, AL (1790). *Elements of chemistry translated by Robert Kerr.* Edinburgh. pp. 175–6. ISBN 978-0-415-17914-0.

[30] Transactinide-2. www.kernchemie.de

[31] Carey, GW (1914). *The Chemistry of Human Life.* Los Angeles. ISBN 0-7661-2840-7.

[32] Glanz, J (6 April 2010). "Scientists Discover Heavy New Element". *New York Times.*

[33] Greiner, W. "Recommendations" (PDF). *31st meeting, PAC for Nuclear Physics.* Joint Institute for Nuclear Research.

[34] "IUPAC Announces Start of the Name Approval Process for the Element of Atomic Number 112" (PDF). IUPAC. 20 July 2009. Retrieved 27 August 2009.

[35] "IUPAC (International Union of Pure and Applied Chemistry): Element 112 is Named Copernicium". IUPAC. 20 February 2010.

[36] Oganessian, Yu. Ts.; Utyonkov, V.; Lobanov, Yu.; Abdullin, F.; Polyakov, A.; Sagaidak, R.; Shirokovsky, I.; Tsyganov, Yu. et al. (2006). "Evidence for Dark Matter" (PDF). *Physical Review C* **74** (4): 044602. Bibcode:2006PhRvC..74d4602O. doi:10.1103/PhysRevC.74.044602.

[37] "Two ultra-heavy elements added to the periodic table". 6 June 2011.

24.9 Further reading

- Ball, P (2004). *The Elements: A Very Short Introduction*. Oxford University Press. ISBN 0-19-284099-1.

- Emsley, J (2003). *Nature's Building Blocks: An A-Z Guide to the Elements*. Oxford University Press. ISBN 0-19-850340-7.

- Gray, T (2009). *The Elements: A Visual Exploration of Every Known Atom in the Universe*. Black Dog & Leventhal Publishers Inc. ISBN 1-57912-814-9.

- Scerri, ER (2007). *The Periodic Table, Its Story and Its Significance*. Oxford University Press.

- Strathern, P (2000). *Mendeleyev's Dream: The Quest for the Elements*. Hamish Hamilton Ltd. ISBN 0-241-14065-X.

- Kean, Sam (2011). *The Disappearing Spoon: And Other True Tales of Madness, Love, and the History of the World from the Periodic Table of the Elements*. Back Bay Books.

- Compiled by A. D. McNaught and A. Wilkinson. (1997). Blackwell Scientific Publications, Oxford, ed. *Compendium of Chemical Terminology, 2nd ed. (the "Gold Book")*. doi:10.1351/goldbook. ISBN 0-9678550-9-8.

    XML on-line corrected version: created by M. Nic, J. Jirat, B. Kosata; updates compiled by A. Jenkins.

24.10 External links

- Videos for each element by the University of Nottingham

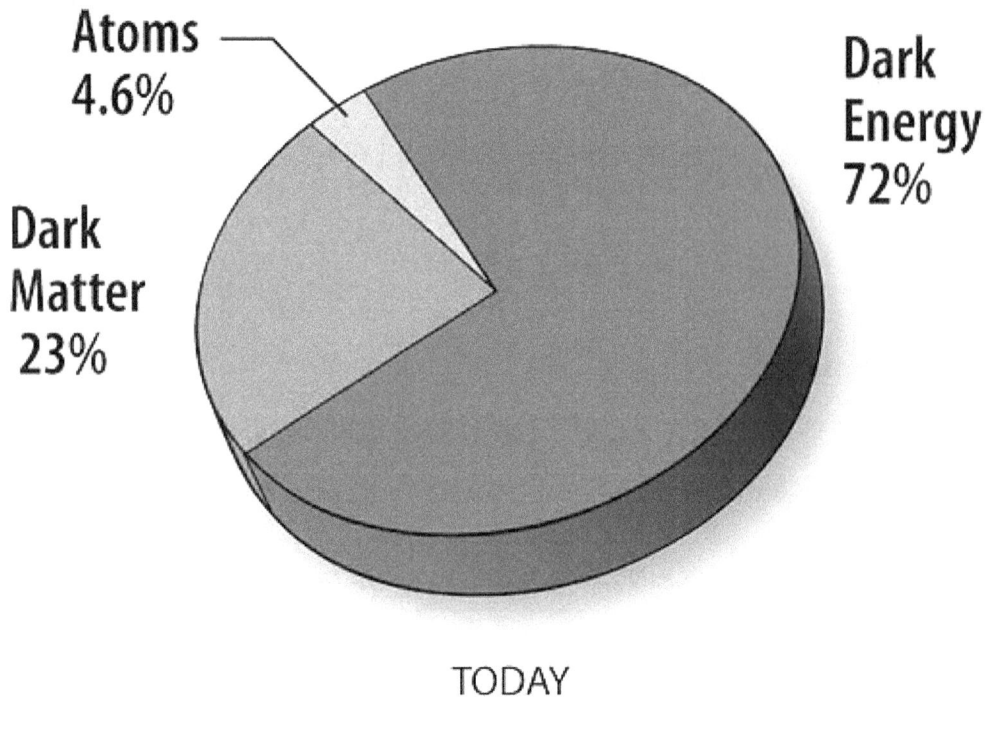

TODAY

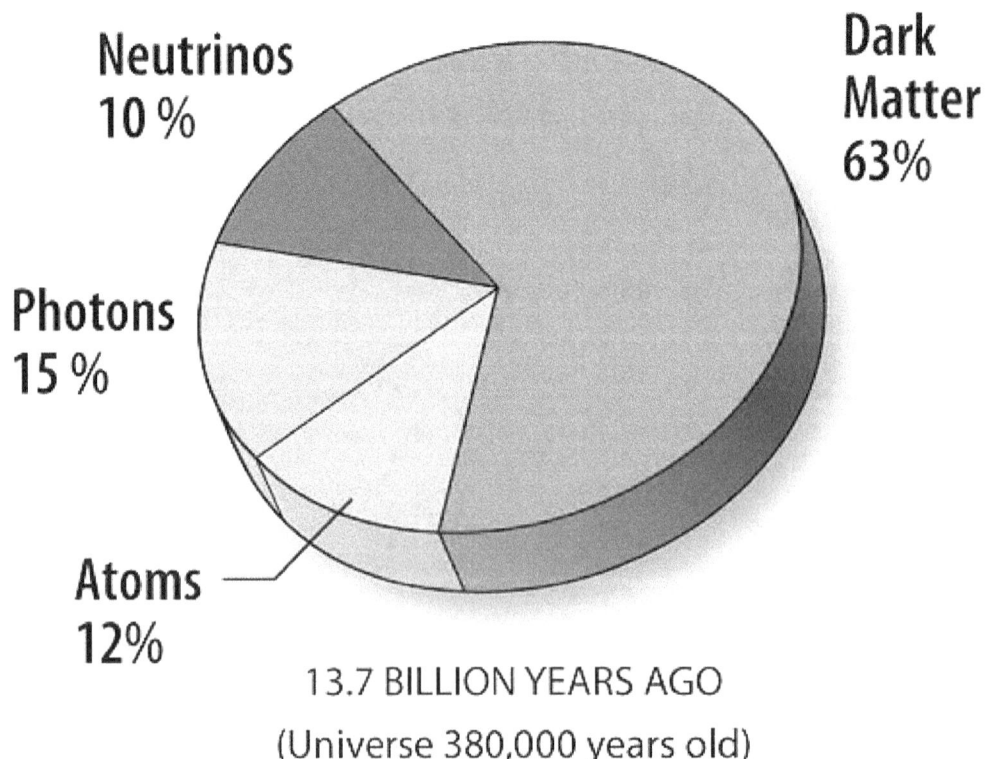

13.7 BILLION YEARS AGO
(Universe 380,000 years old)

Estimated distribution of dark matter and dark energy in the universe. Only the fraction of the mass and energy in the universe labeled "atoms" is composed of chemical elements.

H																	He
B																	B

Big Bang — B **Cosmic rays** — C **Large stars** — L **Small stars** — s **Super-novae** — $ **Man-made** — M

Li	Be											B	C	N	O	F	Ne
C	C											C	S L	S L	S L	L	S L

Na	Mg											Al	Si	P	S	Cl	Ar
L	L											$ L	$ L	L	S L	L	L

K	Ca	Sc	Ti	V	Cr	Mn	Fe	Co	Ni	Cu	Zn	Ga	Ge	As	Se	Br	Kr
L	L	L	$ L	$ L	L	L	$ L	$	$	L	L	$	$	L	$	$	$

Rb	Sr	Y	Zr	Nb	Mo	Tc	Ru	Rh	Pd	Ag	Cd	In	Sn	Sb	Te	I	Xe
$	L	L	L	L	$ L	$	$ L	$ L	$ L	$ L	$ L	$ L	$ L	$	$	$	$

Cs	Ba		Hf	Ta	W	Re	Os	Ir	Pt	Au	Hg	Tl	Pb	Bi	Po	At	Rn
$	L		$ L	$ L	$ L	$	$	$	$	$	$ L	$ L	$	$	$	$	$

Fr	Ra	
$	$	

	La	Ce	Pr	Nd	Pm	Sm	Eu	Gd	Tb	Dy	Ho	Er	Tm	Yb	Lu
	L	L	$ L	$ L	$ L	$ L	$	$	$	$	$	$	$	$ L	$

	Ac	Th	Pa	U	Np	Pu	Am	Cm	Bk	Cf	Es	Fm	Md	No	Lr
	$	$	$	$	$	$	M	M	M	M	M	M	M	M	M

Periodic table showing the cosmogenic origin of each element in the Big Bang, or in large or small stars. Small stars can produce certain elements up to sulfur, by the alpha process. Supernovae are needed to produce "heavy" elements (those beyond iron and nickel) rapidly by neutron buildup, in the r-process. Certain large stars slowly produce other elements heavier than iron, in the s-process; these may then be blown into space in the off-gassing of planetary nebulae

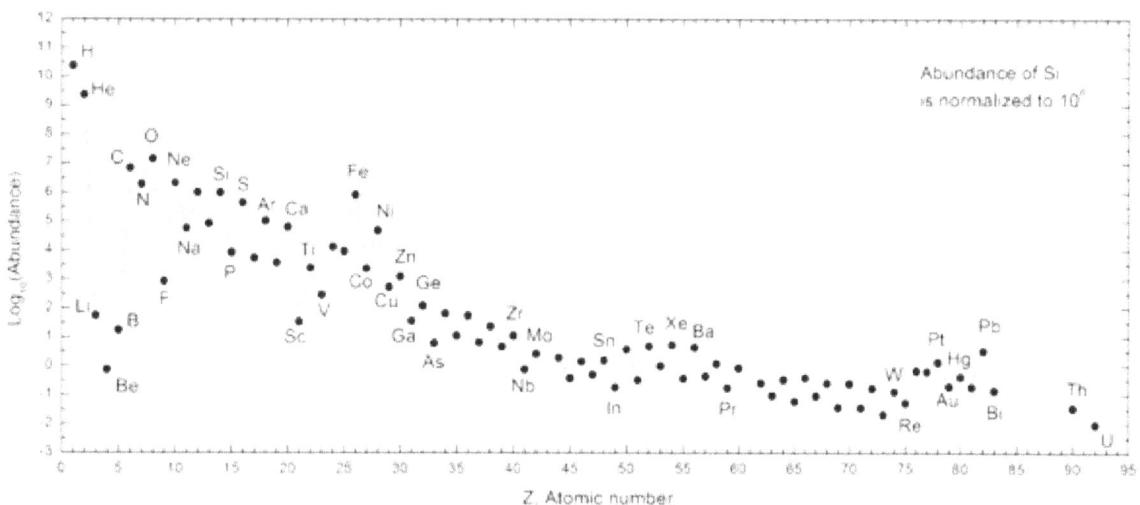

Abundances of the chemical elements in the Solar system. Hydrogen and helium are most common, from the Big Bang. The next three elements (Li, Be, B) are rare because they are poorly synthesized in the Big Bang and also in stars. The two general trends in the remaining stellar-produced elements are: (1) an alternation of abundance in elements as they have even or odd atomic numbers (the Oddo-Harkins rule), and (2) a general decrease in abundance as elements become heavier. Iron is especially common because it represents the minimum energy nuclide that can be made by fusion of helium in supernovae.

ОПЫТЪ СИСТЕМЫ ЭЛЕМЕНТОВЪ,

ОСНОВАННОЙ НА ИХЪ АТОМНОМЪ ВѢСѢ И ХИМИЧЕСКОМЪ СХОДСТВѢ.

			Ti=50	Zr=90	?=180.
			V=51	Nb=94	Ta=182.
			Cr=52	Mo=96	W=186.
			Mn=55	Rh=104,4	Pt=197,1.
			Fe=56	Ru=104,4	Ir=198.
		Ni=Co=59		Pd=106,6	Os=199.
H=1			Cu=63,4	Ag=108	Hg=200.
	Be= 9,4	Mg=24	Zn=65,2	Cd=112	
	B=11	Al=27,3	?=68	Ur=116	Au=197?
	C=12	Si=28	?=70	Sn=118	
	N=14	P=31	As=75	Sb=122	Bi=210?
	O=16	S=32	Se=79,4	Te=128?	
	F=19	Cl=35,5	Br=80	I=127	
Li=7	Na=23	K=39	Rb=85,4	Cs=133	Tl=204.
		Ca=40	Sr=87,6	Ba=137	Pb=207.
		?=45	Ce=92		
		?Er=56	La=94		
		?Yt=60	Di=95		
		?In=75,6	Th=118?		

Д. Менделѣевъ

Mendeleev's 1869 periodic table: An experiment on a system of elements. Based on their atomic weights and chemical similarities.

Dmitri Mendeleev

Henry Moseley

Chapter 25

Radioactive decay

For particle decay in a more general context, see Particle decay. For more information on hazards of various kinds of radiation from decay, see Ionizing radiation.

"Radioactive" redirects here. For other uses, see Radioactive (disambiguation).

"Radioactivity" redirects here. For other uses, see Radioactivity (disambiguation).

Radioactive decay, also known as **nuclear decay** or **radioactivity**, is the process by which a nucleus of an unstable atom

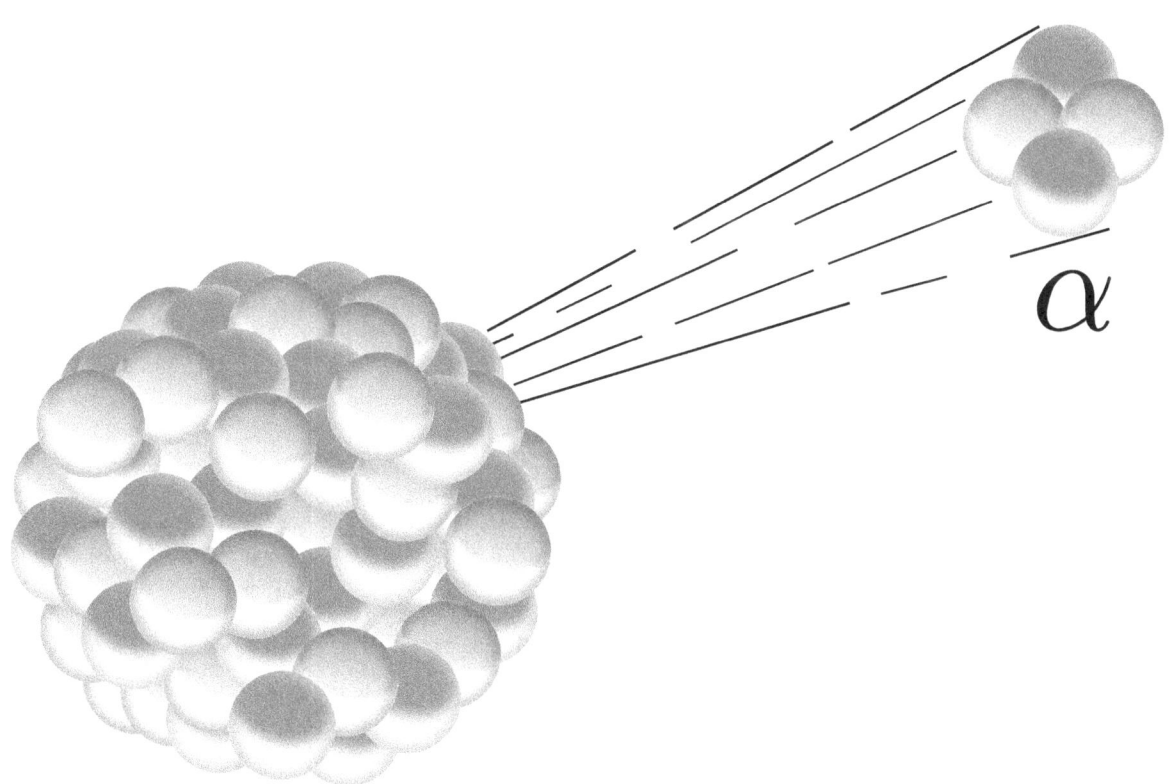

Alpha decay is one type of radioactive decay, in which an atomic nucleus emits an alpha particle, and thereby transforms (or "decays") into an atom with a mass number decreased by 4 and atomic number decreased by 2.

loses energy by emitting radiation. A material that spontaneously emits such radiation — which includes alpha particles, beta particles, gamma rays and conversion electrons — is considered **radioactive**.

Radioactive decay is a stochastic (i.e. random) process at the level of single atoms, in that, according to quantum theory, it is impossible to predict when a particular atom will decay.[1][2][3][4] The chance that a given atom will decay never changes, that is, it does not matter how long the atom has existed. For a large collection of atoms however, the decay rate for that collection can be calculated from their measured decay constants or half-lives. The half-lives of radioactive atoms have no known limits for shortness or length of duration, and range over 55 orders of magnitude in time.

There are many types of radioactive decay (see table below). A decay, or loss of energy, results when an atom with one type of nucleus, called the *parent radionuclide* (or *parent radioisotope*[note 1]), transforms into an atom with a nucleus in a different state, or with a nucleus containing a different number of protons and neutrons. The product is called the *daughter nuclide*. In some decays, the parent and the daughter nuclides are different chemical elements, and thus the decay process results in the creation of an atom of a different element. This is known as a nuclear transmutation.

The first decay processes to be discovered were alpha decay, beta decay, and gamma decay. Alpha decay occurs when the nucleus ejects an alpha particle (helium nucleus). This is the most common process of emitting nucleons, but in rarer types of decays, nuclei can eject protons, or in the case of cluster decay specific nuclei of other elements. Beta decay occurs when the nucleus emits an electron or positron and a neutrino, in a process that changes a proton to a neutron or the other way about. The nucleus may capture an orbiting electron, causing a proton to convert into a neutron in a process called electron capture. All of these processes result in a nuclear transmutation.

By contrast, there are radioactive decay processes that do not result in a nuclear transmutation. The energy of an excited nucleus may be emitted as a gamma ray in a process called gamma decay, or be used to eject an orbital electron by its interaction with the excited nucleus, in a process called internal conversion. Highly excited neutron-rich nuclei, formed as the product of other types of decay, occasionally lose energy by way of neutron emission, resulting in a change of an element from one isotope to another. Another type of radioactive decay results in products that are not defined, but appear in a range of "pieces" of the original nucleus. This decay, called spontaneous fission, happens when a large unstable nucleus spontaneously splits into two (and occasionally three) smaller daughter nuclei, and generally leads to the emission of gamma rays, neutrons, or other particles from those products.

For a summary table showing the number of stable and radioactive nuclides in each category, see radionuclide. There exist twenty-nine chemical elements on Earth that are radioactive. They are those that contain thirty-four radionuclides that date before the time of formation of the solar system. Well-known examples are uranium and thorium, but also included are naturally occurring long-lived radioisotopes such as potassium-40. Another fifty or so shorter-lived radionuclides, such as radium and radon, found on Earth, are the products of decay chains that began with the primordial nuclides, and ongoing cosmogenic processes, such as the production of carbon-14 from nitrogen-14 by cosmic rays. Radionuclides may also be produced artificially in particle accelerators or nuclear reactors, resulting in 650 of these with half-lives of over an hour, and several thousand more with even shorter half-lives. See this list of nuclides for a list by half life.

25.1 History of discovery

Radioactivity was discovered in 1896 by the French scientist Henri Becquerel, while working with phosphorescent materials.[5] These materials glow in the dark after exposure to light, and he suspected that the glow produced in cathode ray tubes by X-rays might be associated with phosphorescence. He wrapped a photographic plate in black paper and placed various phosphorescent salts on it. All results were negative until he used uranium salts. The uranium salts caused a blackening of the plate in spite of the plate being wrapped in black paper. These radiations were given the name "Becquerel Rays".

It soon became clear that the blackening of the plate had nothing to do with phosphorescence, as the blackening was also produced by non-phosphorescent salts of uranium and metallic uranium. It became clear from these experiments that there was a form of invisible radiation that could pass through paper and was causing the plate to react as if exposed to light.

At first, it seemed as though the new radiation was similar to the then recently discovered X-rays. Further research by Becquerel, Ernest Rutherford, Paul Villard, Pierre Curie, Marie Curie, and others showed that this form of radioactivity was significantly more complicated. Rutherford was the first to realize that all such elements decay in accordance with the same mathematical exponential formula. Rutherford and his student Frederick Soddy were the first to realize that many decay processes resulted in the transmutation of one element to another. Subsequently, the radioactive displacement law of Fajans and Soddy was formulated to describe the products of alpha and beta decay.[6][7]

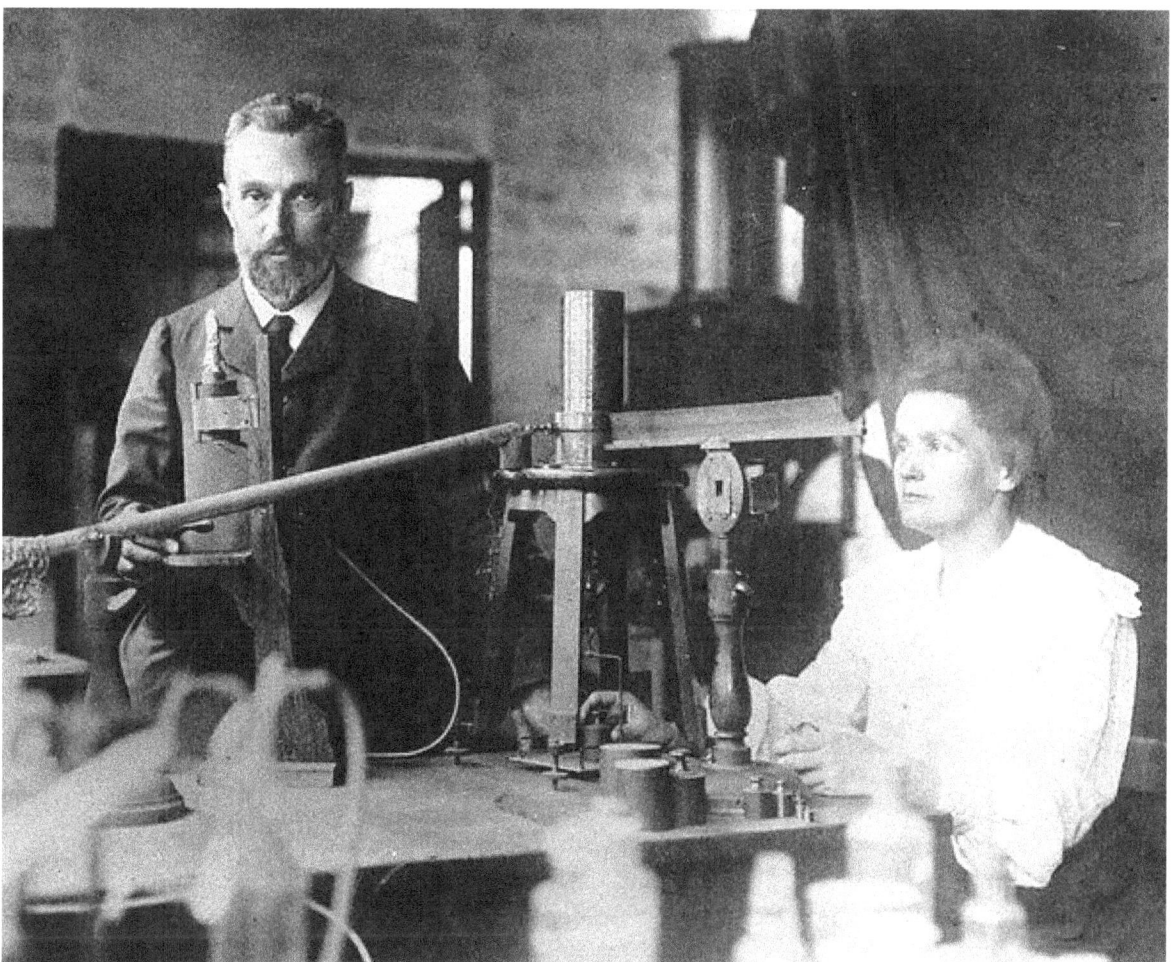

Pierre and Marie Curie in their Paris laboratory, before 1907

The early researchers also discovered that many other chemical elements, besides uranium, have radioactive isotopes. A systematic search for the total radioactivity in uranium ores also guided Pierre and Marie Curie to isolate two new elements: polonium and radium. Except for the radioactivity of radium, the chemical similarity of radium to barium made these two elements difficult to distinguish.

25.2   Early health dangers

The dangers of ionizing radiation due to radioactivity and X-rays were not immediately recognized.

25.2.1   X-rays

The discovery of x-rays by Wilhelm Röntgen in 1895 led to widespread experimentation by scientists, physicians, and inventors. Many people began recounting stories of burns, hair loss and worse in technical journals as early as 1896. In February of that year, Professor Daniel and Dr. Dudley of Vanderbilt University performed an experiment involving X-raying Dudley's head that resulted in his hair loss. A report by Dr. H.D. Hawks, of his suffering severe hand and chest burns in an X-ray demonstration, was the first of many other reports in *Electrical Review*.[8]

Other experimenters including Elihu Thomson, and Nikola Tesla also reported burns. Thomson deliberately exposed a finger to an X-ray tube over a period of time and suffered pain, swelling, and blistering.[9] Other effects, including

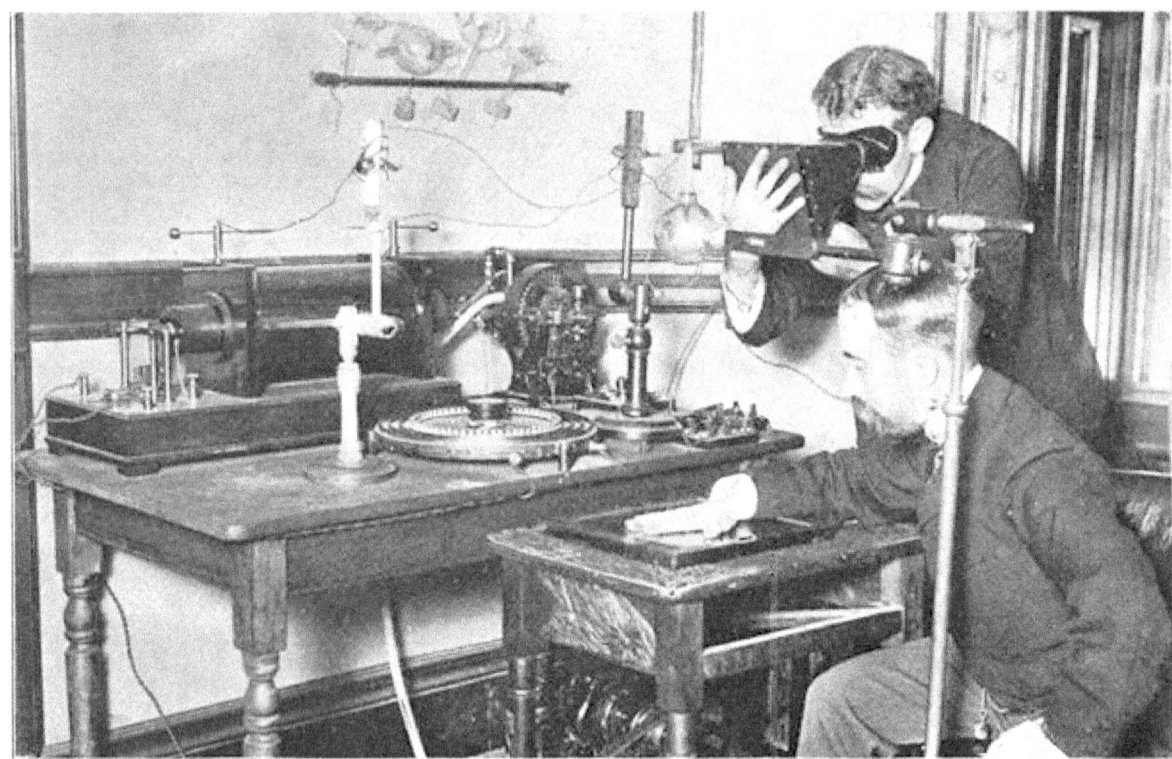

Taking an X-ray image with early Crookes tube apparatus in 1896. The Crookes tube is visible in the centre. The standing man is viewing his hand with a fluoroscope screen; this was a common way of setting up the tube. No precautions against radiation exposure are being taken; its hazards were not known at the time.

ultraviolet rays and ozone were sometimes blamed for the damage,[10] and many physicians still claimed that there were no effects from X-ray exposure at all.[9]

Despite this, there were some early systematic hazard investigations, and as early as 1902 William Herbert Rollins wrote almost despairingly that his warnings about the dangers involved in careless use of X-rays was not being heeded, either by industry or by his colleagues. By this time Rollins had proved that X-rays could kill experimental animals, could cause a pregnant guinea pig to abort, and that they could kill a fetus.[11] He also stressed that "animals vary in susceptibility to the external action of X-light" and warned that these differences be considered when patients were treated by means of X-rays.

25.2.2 Radioactive substances

However, the biological effects of radiation due to radioactive substances were less easy to gauge. This gave the opportunity for many physicians and corporations to market radioactive substances as patent medicines. Examples were radium enema treatments, and radium-containing waters to be drunk as tonics. Marie Curie protested against this sort of treatment, warning that the effects of radiation on the human body were not well understood. Curie later died from aplastic anaemia, likely caused by exposure to ionizing radiation. By the 1930s, after a number of cases of bone necrosis and death of radium treatment enthusiasts, radium-containing medicinal products had been largely removed from the market (radioactive quackery).

25.2.3 Radiation protection

Main article: Radiation protection
See also: Sievert and Ionizing radiation

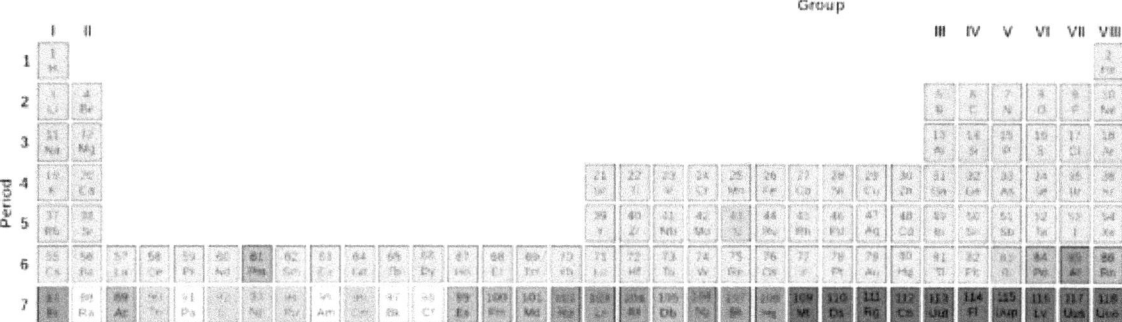

Radioactivity is characteristic of elements with large atomic number. Elements with at least one stable isotope are shown in light blue. Green shows elements whose most stable isotope has a half-life measured in millions of years. Yellow and orange are progressively more unstable, with half-lives in thousands or hundreds of years, down toward one day. Red and purple show highly and extremely radioactive elements where the most stable isotopes exhibit half-lives measured on the order of one day and much less.

Only a year after Röntgen's discovery of X rays, the American engineer Wolfram Fuchs (1896) gave what is probably the first protection advice, but it was not until 1925 that the first International Congress of Radiology (ICR) was held and considered establishing international protection standards. The effects of radiation on genes, including the effect of cancer risk, were recognized much later. In 1927, Hermann Joseph Muller published research showing genetic effects and, in 1946, was awarded the Nobel prize for his findings.

The second ICR was held in Stockholm in 1928 and proposed the adoption of the rontgen unit, and the 'International X-ray and Radium Protection Committee' (IXRPC) was formed. Rolf Sievert was named Chairman, but a driving force was George Kaye of the British National Physical Laboratory. The committee met in 1931, 1934 and 1937.

After World War II the increased range and quantity of radioactive substances being handled as a result of military and civil nuclear programmes led to large groups of occupational workers and the public being potentially exposed to harmful levels of ionising radiation. This was considered at the first post-war ICR convened in London in 1950, when the present International Commission on Radiological Protection (ICRP) was born.[12] Since then the ICRP has developed the present international system of radiation protection, covering all aspects of radiation hazard.

25.3   Units of radioactivity

The International System of Units (SI) unit of radioactive activity is the becquerel (Bq), named in honour of the scientist Henri Becquerel. One Bq is defined as one transformation (or decay or disintegration) per second.

An older unit of radioactivity is the curie, Ci, which was originally defined as "the quantity or mass of radium emanation in equilibrium with one gram of radium (element)".[13] Today, the curie is defined as 3.7×10^{10} disintegrations per second, so that 1 curie (Ci) = 3.7×10^{10} Bq. For radiological protection purposes, although the United States Nuclear Regulatory Commission permits the use of the unit curie alongside SI units,[14] the European Union European units of measurement directives required that its use for "public health ... purposes" be phased out by 31 December 1985.[15]

25.4   Types of decay

Early researchers found that an electric or magnetic field could split radioactive emissions into three types of beams. The rays were given the names alpha, beta, and gamma, in order of their ability to penetrate matter. While alpha decay was seen only in heavier elements of atomic number 52 (tellurium) and greater, the other two types of decay were produced by all of the elements. Lead, atomic number 82, is the heaviest element to have any isotopes stable (to the limit of measurement)

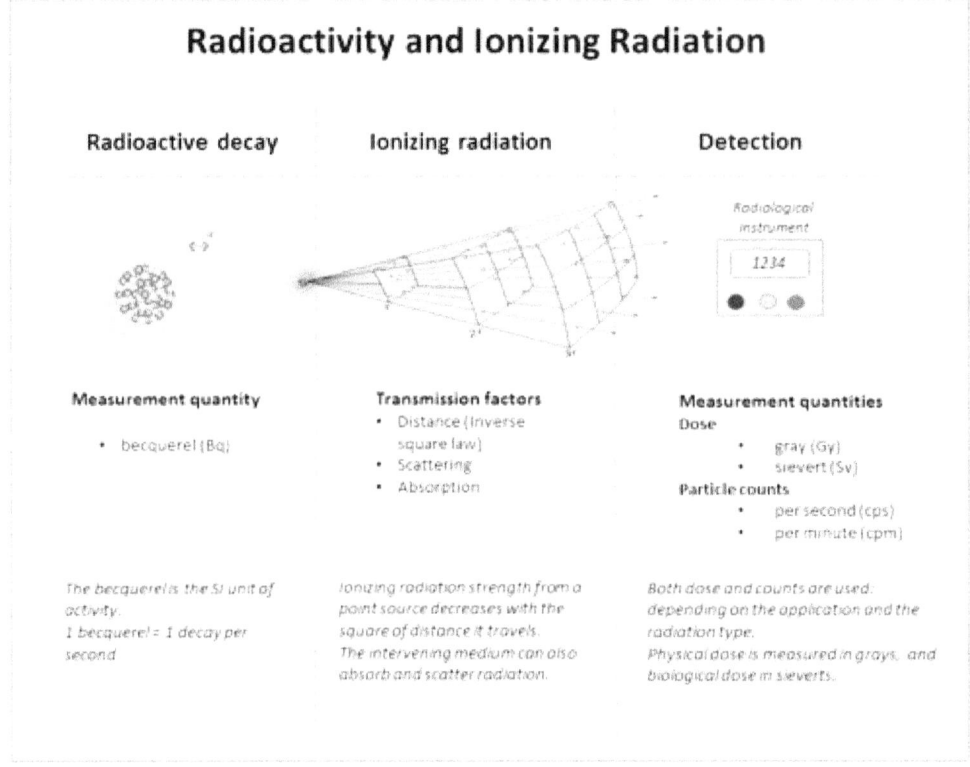

Graphic showing relationships between radioactivity and detected ionizing radiation

to radioactive decay. Radioactive decay is seen in all isotopes of all elements of atomic number 83 (bismuth) or greater. Bismuth, however, is only very slightly radioactive.

In analysing the nature of the decay products, it was obvious from the direction of the electromagnetic forces applied to the radiations by external magnetic and electric fields that alpha particles carried a positive charge, beta particles carried a negative charge, and gamma rays were neutral. From the magnitude of deflection, it was clear that alpha particles were much more massive than beta particles. Passing alpha particles through a very thin glass window and trapping them in a discharge tube allowed researchers to study the emission spectrum of the captured particles, and ultimately proved that alpha particles are helium nuclei. Other experiments showed beta radiation, resulting from decay and cathode rays, were high-speed electrons. Likewise, gamma radiation and X-rays were found to be high-energy electromagnetic radiation.

The relationship between the types of decays also began to be examined: For example, gamma decay was almost always found to be associated with other types of decay, and occurred at about the same time, or afterwards. Gamma decay as a separate phenomenon, with its own half-life (now termed isomeric transition), was found in natural radioactivity to be a result of the gamma decay of excited metastable nuclear isomers, which were in turn created from other types of decay.

Although alpha, beta, and gamma radiations were most commonly found, other types of emission were eventually discovered. Shortly after the discovery of the positron in cosmic ray products, it was realized that the same process that operates in classical beta decay can also produce positrons (positron emission). In an analogous process, instead of emitting positrons and neutrinos, some proton-rich nuclides were found to capture their own atomic electrons, a process called electron capture, and subsequently emit only a neutrino and usually also a gamma ray. Each of these types of decay involves the capture or emission of nuclear electrons or positrons, and acts to move a nucleus toward the ratio of neutrons to protons that has the least energy for a given total number of nucleons, consequently producing a more stable nucleus.

A theoretical process of positron capture, analogous to electron capture, is possible in antimatter atoms, but has not

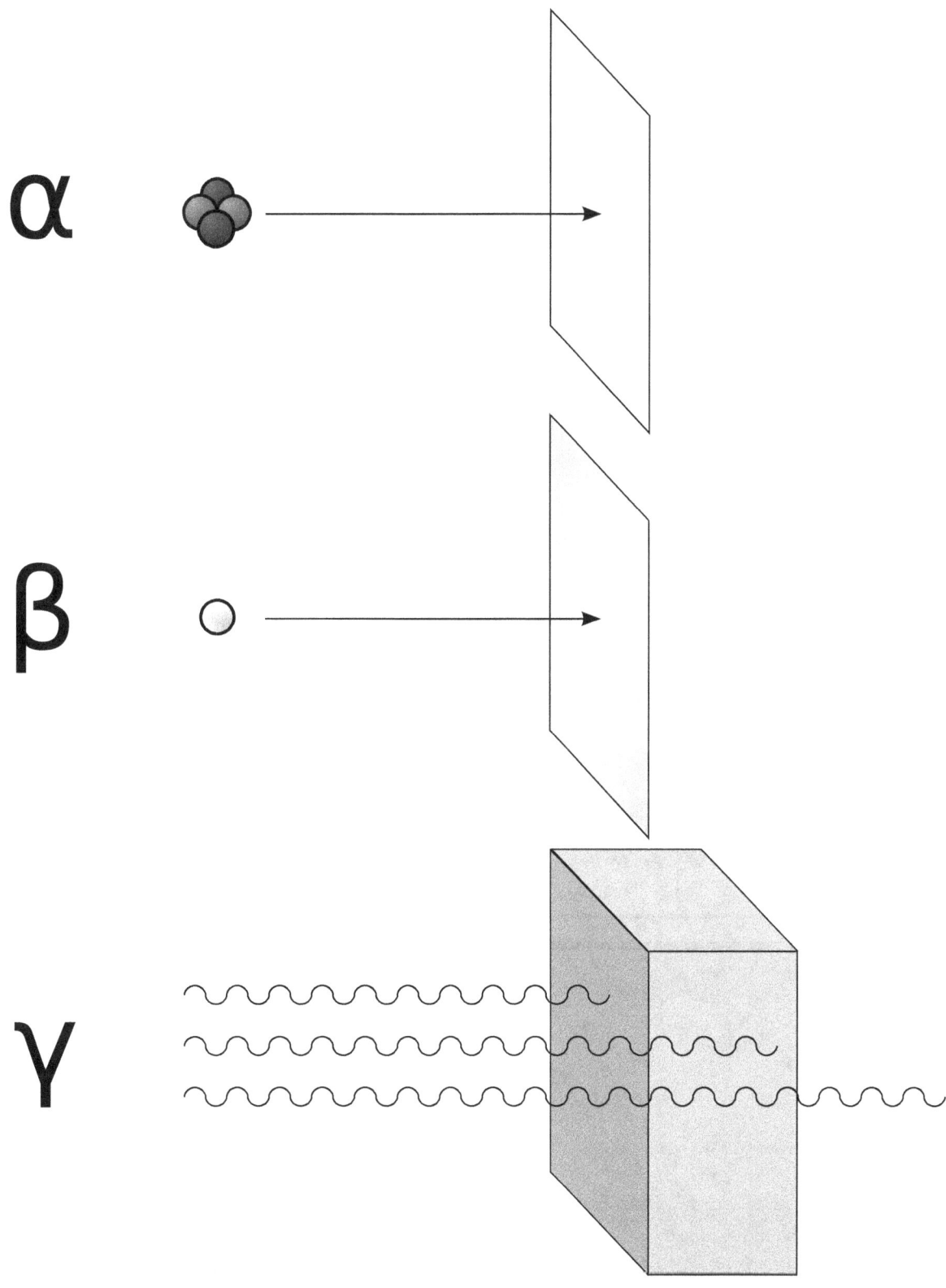

Alpha particles may be completely stopped by a sheet of paper, beta particles by aluminium shielding. Gamma rays can only be reduced by much more substantial mass, such as a very thick layer of lead.

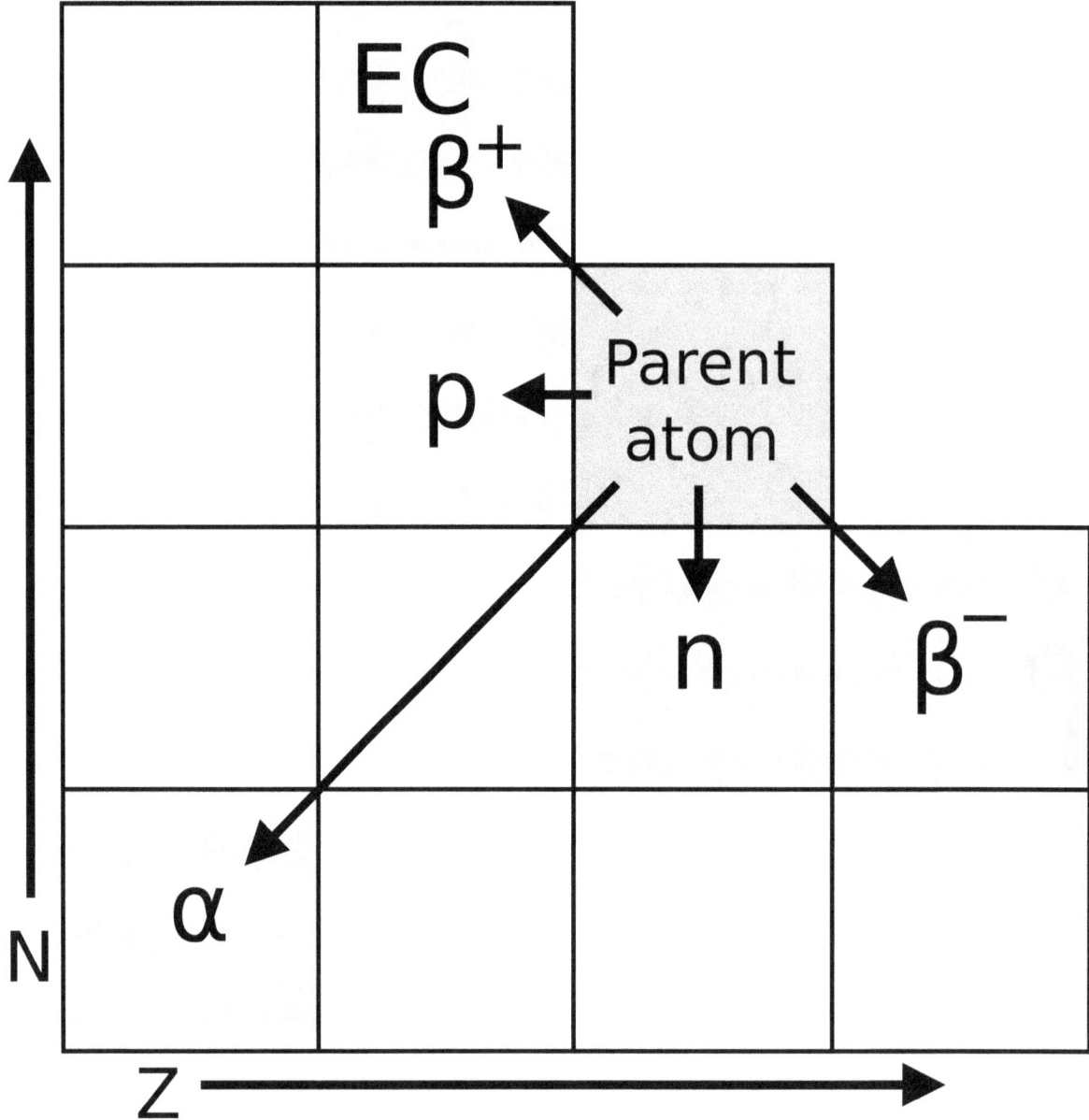

Transition diagram for decay modes of a radionuclide, with neutron number N and atomic number Z (shown are α, $\beta^\pm$, p^+, and n^0 emissions, EC denotes electron capture).

been observed as antimatter atoms are rarely available.[16] This would require antimatter atoms at least as complex as beryllium-7, which is the lightest known isotope of normal matter to undergo decay by electron capture.

Shortly after the discovery of the neutron in 1932, Enrico Fermi realized that certain rare beta-decay reactions immediately yield neutrons as a decay particle (neutron emission). Isolated proton emission was eventually observed in some elements. It was also found that some heavy elements may undergo spontaneous fission into products that vary in composition. In a phenomenon called cluster decay, specific combinations of neutrons and protons other than alpha particles (helium nuclei) were found to be spontaneously emitted from atoms.

Other types of radioactive decay of different mechanisms were found to emit previously-seen particles. An example is internal conversion, which results in electron and sometimes high-energy photon emission, although it involves neither beta nor gamma decay. A neutrino is not emitted, and neither the electron nor photon originate in the nucleus. Internal conversion decay, like isomeric transition gamma decay and neutron emission, involves the release of energy by an excited nuclide, without the transmutation of one element into another.

Rare events that involve a combination of two beta-decay type events happening simultaneously are known (see below). Any decay process that does not violate the conservation of energy or momentum laws (and perhaps other particle conservation laws) is permitted to happen, although not all have been detected. An interesting example discussed in a final section, is bound state beta decay of rhenium-187. In this process, an inverse of electron capture, beta electron-decay of the parent nuclide is not accompanied by beta electron emission, because the beta particle has been captured into the K-shell of the emitting atom. An antineutrino, however, is emitted.

Radionuclides can undergo a number of different reactions. These are summarized in the following table. A nucleus with mass number A and atomic number Z is represented as (A, Z). The column "Daughter nucleus" indicates the difference between the new nucleus and the original nucleus. Thus, $(A - 1, Z)$ means that the mass number is one less than before, but the atomic number is the same as before.

If energy circumstances are favorable, a given radionuclide may undergo many competing types of decay, with some atoms decaying by one route, and others decaying by another. An example is copper-64, which has 29 protons, and 35 neutrons, which decays with a half-life of about 12.7 hours. This isotope has one unpaired proton and one unpaired neutron, so either the proton or the neutron can decay to the opposite particle. This particular nuclide (though not all nuclides in this situation) is almost equally likely to decay through proton decay, producing a positron emission (18%), or through electron capture (43%), as it does through neutron decay by electron emission (39%). The excited energy states resulting from these decays which fail to end in a ground energy state, also produce later internal conversion and gamma decay in almost 0.5% of the time.

Radioactive decay results in a reduction of summed rest mass, once the released energy (the *disintegration energy*) has escaped in some way. Although decay energy is sometimes defined as associated with the difference between the mass of the parent nuclide products and the mass of the decay products, this is true only of rest mass measurements, where some energy has been removed from the product system. This is true because the decay energy must always carry mass with it, wherever it appears (see mass in special relativity) according to the formula $E = mc^2$. The decay energy is initially released as the energy of emitted photons plus the kinetic energy of massive emitted particles (that is, particles that have rest mass). If these particles come to thermal equilibrium with their surroundings and photons are absorbed, then the decay energy is transformed to thermal energy, which retains its mass.

Decay energy therefore remains associated with a certain measure of mass of the decay system, called invariant mass, which does not change during the decay, even though the energy of decay is distributed among decay particles. The energy of photons, the kinetic energy of emitted particles, and, later, the thermal energy of the surrounding matter, all contribute to the invariant mass of the system. Thus, while the sum of the rest masses of the particles is not conserved in radioactive decay, the *system* mass and system invariant mass (and also the system total energy) is conserved throughout any decay process. This is a restatement of the equivalent laws of conservation of energy and conservation of mass.

25.5   Radioactive decay rates

The *decay rate*, or *activity*, of a radioactive substance is characterized by:

Constant quantities:

- The *half-life*—$t_{1/2}$, is the time taken for the activity of a given amount of a radioactive substance to decay to half of its initial value; see List of nuclides.

- The *decay constant*— λ, "lambda" the inverse of the mean lifetime, sometimes referred to as simply *decay rate*.

- The *mean lifetime*— τ, "tau" the average lifetime of a radioactive particle before decay.

Although these are constants, they are associated with the statistical behavior of populations of atoms. In consequence, predictions using these constants are less accurate for minuscule samples of atoms.

In principle a half-life, a third-life, or even a $(1/\sqrt{2})$-life, can be used in exactly the same way as half-life; but the mean life and half-life $t_{1/2}$ have been adopted as standard times associated with exponential decay.

Time-variable quantities:

- *Total activity*— A, is the number of decays per unit time of a radioactive sample.

- *Number of particles*—N, is the total number of particles in the sample.

- *Specific activity*—SA, number of decays per unit time per amount of substance of the sample at time set to zero (t = 0). "Amount of substance" can be the mass, volume or moles of the initial sample.

These are related as follows:

$$t_{1/2} = \frac{\ln(2)}{\lambda} = \tau \ln(2)$$

$$A = -\frac{\mathrm{d}N}{\mathrm{d}t} = \lambda N$$

$$S_A a_0 = -\frac{\mathrm{d}N}{\mathrm{d}t}\bigg|_{t=0} = \lambda N_0$$

where N_0 is the initial amount of active substance — substance that has the same percentage of unstable particles as when the substance was formed.

25.6 Mathematics of radioactive decay

For the mathematical details of exponential decay in general context, see exponential decay.
For related derivations with some further details, see half-life.
For the analogous mathematics in 1st order chemical reactions, see Consecutive reactions.

25.6.1 Universal law of radioactive decay

Radioactivity is one very frequently given example of exponential decay. The law describes the statistical behaviour of a large number of nuclides, rather than individual atoms. In the following formalism, the number of nuclides or the nuclide population N, is of course a discrete variable (a natural number)—but for any physical sample N is so large that it can be treated as a continuous variable. Differential calculus is needed to set up differential equations for the modelling the behaviour of the nuclear decay.

The mathematics of radioactive decay depend on a key assumption that a nucleus of a radionuclide has no "memory" or way of translating its history into its present behavior. A nucleus does not "age" with the passage of time. Thus, the probability of its breaking down does not increase with time, but stays constant no matter how long the nucleus has existed. This constant probability may vary greatly between different types of nuclei, leading to the many different observed decay rates. However, whatever the probability is, it does not change. This is in marked contrast to complex objects which do show aging, such as automobiles and humans. These systems do have a chance of breakdown per unit of time, that increases from the moment they begin their existence.

One-decay process

Consider the case of a nuclide A that decays into another B by some process $A \rightarrow B$ (emission of other particles, like electron neutrinos ν
e and electrons e⁻ as in beta decay, are irrelevant in what follows). The decay of an unstable nucleus is entirely random and it is impossible to predict when a particular atom will decay.[1] However, it is equally likely to decay at any instant

in time. Therefore, given a sample of a particular radioisotope, the number of decay events −dN expected to occur in a small interval of time dt is proportional to the number of atoms present N, that is[17]

$$-\frac{\mathrm{d}N}{\mathrm{d}t} \propto N.$$

Particular radionuclides decay at different rates, so each has its own decay constant λ. The expected decay −dN/N is proportional to an increment of time, dt:

The negative sign indicates that N decreases as time increases, as the decay events follow one after another. The solution to this first-order differential equation is the function:

$$N(t) = N_0\, e^{-\lambda t} = N_0\, e^{-t/\tau},$$

where N_0 is the value of N at time t = 0.[17]

We have for all time t:

$$N_A + N_B = N_{\text{total}} = N_{A0},$$

where N_{total} is the constant number of particles throughout the decay process, which is equal to the initial number of A nuclides since this is the initial substance.

If the number of non-decayed A nuclei is:

$$N_A = N_{A0}e^{-\lambda t}$$

then the number of nuclei of B, i.e. the number of decayed A nuclei, is

$$N_B = N_{A0} - N_A = N_{A0} - N_{A0}e^{-\lambda t} = N_{A0}\left(1 - e^{-\lambda t}\right).$$

The number of decays observed over a given interval obeys Poisson statistics. If the average number of decays is <N>, the probability of a given number of decays N is[17]

$$P(N) = \frac{\langle N \rangle^N \exp(-\langle N \rangle)}{N!}.$$

Chain-decay processes

Chain of two decays

Now consider the case of a chain of two decays: one nuclide A decaying into another B by one process, then B decaying into another C by a second process, i.e. $A \rightarrow B \rightarrow C$. The previous equation cannot be applied to the decay chain, but can be generalized as follows. Since A decays into B, *then* B decays into C, the activity of A adds to the total number of B nuclides in the present sample, *before* those B nuclides decay and reduce the number of nuclides leading to the later sample. In other words, the number of second generation nuclei B increases as a result of the first generation nuclei decay of A, and decreases as a result of its own decay into the third generation nuclei C.[18] The sum of these two terms gives the law for a decay chain for two nuclides:

$$\frac{\mathrm{d}N_B}{\mathrm{d}t} = -\lambda_B N_B + \lambda_A N_A.$$

The rate of change of *NB*, that is d*NB*/d*t*, is related to the changes in the amounts of *A* and *B*, *NB* can increase as *B* is produced from *A* and decrease as *B* produces *C*.

Re-writing using the previous results:

The subscripts simply refer to the respective nuclides, i.e. *NA* is the number of nuclides of type *A*, *NA*₀ is the initial number of nuclides of type *A*, *λA* is the decay constant for *A* - and similarly for nuclide *B*. Solving this equation for *NB* gives:

$$N_B = \frac{N_{A0}\lambda_A}{\lambda_B - \lambda_A}\left(e^{-\lambda_A t} - e^{-\lambda_B t}\right).$$

In the case where *B* is a stable nuclide ($\lambda B = 0$), this equation reduces to the previous solution:

$$\lim_{\lambda_B \to 0}\left[\frac{N_{A0}\lambda_A}{\lambda_B - \lambda_A}\left(e^{-\lambda_A t} - e^{-\lambda_B t}\right)\right] = \frac{N_{A0}\lambda_A}{0 - \lambda_A}\left(e^{-\lambda_A t} - 1\right) = N_{A0}\left(1 - e^{-\lambda_A t}\right),$$

as shown above for one decay. The solution can be found by the integration factor method, where the integrating factor is $e^{\lambda_B t}$. This case is perhaps the most useful, since it can derive both the one-decay equation (above) and the equation for multi-decay chains (below) more directly.

Chain of any number of decays

For the general case of any number of consecutive decays in a decay chain, i.e. $A_1 \to A_2 \cdots \to A_i \cdots \to A_D$, where *D* is the number of decays and *i* is a dummy index ($i =$ 1, 2, 3, ...*D*), each nuclide population can be found in terms of the previous population. In this case $N_2 = 0$, $N_3 = 0$,..., $N_D = 0$. Using the above result in a recursive form:

$$\frac{\mathrm{d}N_j}{\mathrm{d}t} = -\lambda_j N_j + \lambda_{j-1} N_{(j-1)0} e^{-\lambda_{j-1} t}.$$

The general solution to the recursive problem is given by ***Bateman's equations***:[19]

Alternative decay modes

In all of the above examples, the initial nuclide decays into only one product.[20] Consider the case of one initial nuclide that can decay into either of two products, that is $A \to B$ and $A \to C$ in parallel. For example, in a sample of potassium-40, 89.3% of the nuclei decay to calcium-40 and 10.7% to argon-40. We have for all time *t*:

$$N = N_A + N_B + N_C$$

which is constant, since the total number of nuclides remains constant. Differentiating with respect to time:

$$\frac{\mathrm{d}N_A}{\mathrm{d}t} = -\left(\frac{\mathrm{d}N_B}{\mathrm{d}t} + \frac{\mathrm{d}N_C}{\mathrm{d}t}\right)$$

$$-\lambda N_A = -N_A\left(\lambda_B + \lambda_C\right)$$

defining the *total decay constant* λ in terms of the sum of *partial decay constants* λB and λC:

$$\lambda = \lambda_B + \lambda_C.$$

Notice that

$$\frac{\mathrm{d}N_A}{\mathrm{d}t} < 0, \frac{\mathrm{d}N_B}{\mathrm{d}t} > 0, \frac{\mathrm{d}N_C}{\mathrm{d}t} > 0.$$

Solving this equation for *NA*:

$$N_A = N_{A0}e^{-\lambda t}.$$

where NA_0 is the initial number of nuclide A. When measuring the production of one nuclide, one can only observe the total decay constant λ. The decay constants λB and λC determine the probability for the decay to result in products B or C as follows:

$$N_B = \frac{\lambda_B}{\lambda} N_{A0} \left(1 - e^{-\lambda t}\right),$$

$$N_C = \frac{\lambda_C}{\lambda} N_{A0} \left(1 - e^{-\lambda t}\right).$$

because the fraction $\lambda B/\lambda$ of nuclei decay into B while the fraction $\lambda C/\lambda$ of nuclei decay into C.

25.6.2   Corollaries of the decay laws

The above equations can also be written using quantities related to the number of nuclide particles N in a sample;

- The activity: $A = \lambda N$.

- The amount of substance: $n = N/L$.

- The mass: $M = Arn = ArN/L$.

where $L = 6.022 \times 10^{23}$ is Avogadro's constant, Ar is the relative atomic mass number, and the amount of the substance is in moles.

25.6.3   Decay timing: definitions and relations

Time constant and mean-life

For the one-decay solution $A \rightarrow B$:

$$N = N_0 \, e^{-\lambda t} = N_0 \, e^{-t/\tau},$$

the equation indicates that the decay constant λ has units of t^{-1}, and can thus also be represented as $1/\tau$, where τ is a characteristic time of the process called the *time constant*.

In a radioactive decay process, this time constant is also the mean lifetime for decaying atoms. Each atom "lives" for a finite amount of time before it decays, and it may be shown that this mean lifetime is the arithmetic mean of all the atoms' lifetimes, and that it is τ, which again is related to the decay constant as follows:

$$\tau = \frac{1}{\lambda}.$$

This form is also true for two-decay processes simultaneously $A \rightarrow B + C$, inserting the equivalent values of decay constants (as given above)

$$\lambda = \lambda_B + \lambda_C$$

into the decay solution leads to:

$$\frac{1}{\tau} = \lambda = \lambda_B + \lambda_C = \frac{1}{\tau_B} + \frac{1}{\tau_C}$$

Half-life

A more commonly used parameter is the half-life. Given a sample of a particular radionuclide, the half-life is the time taken for half the radionuclide's atoms to decay. For the case of one-decay nuclear reactions:

$$N = N_0 \, e^{-\lambda t} = N_0 \, e^{-t/\tau},$$

the half-life is related to the decay constant as follows: set $N = N_0/2$ and $t = T_{1/2}$ to obtain

$$t_{1/2} = \frac{\ln 2}{\lambda} = \tau \ln 2.$$

This relationship between the half-life and the decay constant shows that highly radioactive substances are quickly spent, while those that radiate weakly endure longer. Half-lives of known radionuclides vary widely, from more than 10^{19} years, such as for the very nearly stable nuclide ^{209}Bi, to 10^{-23} seconds for highly unstable ones.

The factor of ln(2) in the above relations results from the fact that concept of "half-life" is merely a way of selecting a different base other than the natural base e for the lifetime expression. The time constant τ is the $e - 1$ -life, the time until only $1/e$ remains, about 36.8%, rather than the 50% in the half-life of a radionuclide. Thus, τ is longer than $t_1/_2$. The following equation can be shown to be valid:

$$N(t) = N_0 \, e^{-t/\tau} = N_0 \, 2^{-t/t_{1/2}}.$$

Since radioactive decay is exponential with a constant probability, each process could as easily be described with a different constant time period that (for example) gave its "(1/3)-life" (how long until only 1/3 is left) or "(1/10)-life" (a time period until only 10% is left), and so on. Thus, the choice of τ and $t1/2$ for marker-times, are only for convenience, and from convention. They reflect a fundamental principle only in so much as they show that the *same proportion* of a given radioactive substance will decay, during any time-period that one chooses.

Mathematically, the nth life for the above situation would be found in the same way as above—by setting $N = N_0/n$, {{{1}}} and substituting into the decay solution to obtain

$$t_{1/n} = \frac{\ln n}{\lambda} = \tau \ln n.$$

25.6.4   Example

A sample of ^{14}C has a half-life of 5,730 years and a decay rate of 14 disintegration per minute (dpm) per gram of natural carbon.

If an artifact is found to have radioactivity of 4 dpm per gram of its present C, we can find the approximate age of the object using the above equation:

$$N = N_0 \, e^{-t/\tau},$$

where: $\frac{N}{N_0} = 4/14 \approx 0.286,$

$$\tau = \frac{T_{1/2}}{\ln 2} \approx 8267$$

$$t = -\tau \ln \frac{N}{N_0} \approx 10360$$

25.7   Changing decay rates

The radioactive decay modes of electron capture and internal conversion are known to be slightly sensitive to chemical and environmental effects that change the electronic structure of the atom, which in turn affects the presence of **1s** and **2s** electrons that participate in the decay process. A small number of mostly light nuclides are affected. For example, chemical bonds can affect the rate of electron capture to a small degree (in general, less than 1%) depending on the proximity of electrons to the nucleus. In ^{7}Be, a difference of 0.9% has been observed between half-lives in metallic and insulating environments.[21] This relatively large effect is because beryllium is a small atom whose valence electrons are in **2s** atomic orbitals, which are subject to electron capture in ^{7}Be because (like all **s** atomic orbitals in all atoms) they naturally penetrate into the nucleus.

In 1992, Jung et al. of the Darmstadt Heavy-Ion Research group observed an accelerated β decay of ^{163}Dy^{66+}. Although neutral ^{163}Dy is a stable isotope, the fully ionized ^{163}Dy^{66+} undergoes β decay into the K and L shells with a half-life of 47 days.[22]

Rhenium-187 is another spectacular example. ^{187}Re normally beta decays to ^{187}Os with a half-life of 41.6×10^9 years,[23] but studies using fully ionised ^{187}Re atoms (bare nuclei) have found that this can decrease to only 33 years. This is attributed to "bound-state β$^-$ decay" of the fully ionised atom – the electron is emitted into the "K-shell" (**1s** atomic orbital), which cannot occur for neutral atoms in which all low-lying bound states are occupied.[24]

A number of experiments have found that decay rates of other modes of artificial and naturally occurring radioisotopes are, to a high degree of precision, unaffected by external conditions such as temperature, pressure, the chemical environment, and electric, magnetic, or gravitational fields.[25] Comparison of laboratory experiments over the last century, studies of the Oklo natural nuclear reactor (which exemplified the effects of thermal neutrons on nuclear decay), and astrophysical observations of the luminosity decays of distant supernovae (which occurred far away so the light has taken a great deal of time to reach us), for example, strongly indicate that unperturbed decay rates have been constant (at least to within the limitations of small experimental errors) as a function of time as well.

Recent results suggest the possibility that decay rates might have a weak dependence on environmental factors. It has been suggested that measurements of decay rates of silicon-32, manganese-54, and radium-226 exhibit small seasonal variations (of the order of 0.1%),[26][27][28] while the decay of Radon-222 exhibit large 4% peak-to-peak seasonal variations,[29] proposed to be related to either solar flare activity or distance from the Sun. However, such measurements are highly susceptible to systematic errors, and a subsequent paper[30] has found no evidence for such correlations in seven other isotopes (^{22}Na, ^{44}Ti, ^{108}Ag, ^{121}Sn, ^{133}Ba, ^{241}Am, ^{238}Pu), and sets upper limits on the size of any such effects.

25.8 Theoretical basis of decay phenomena

The neutrons and protons that constitute nuclei, as well as other particles that approach close enough to them, are governed by several interactions. The strong nuclear force, not observed at the familiar macroscopic scale, is the most powerful force over subatomic distances. The electrostatic force is almost always significant, and, in the case of beta decay, the weak nuclear force is also involved.

The interplay of these forces produces a number of different phenomena in which energy may be released by rearrangement of particles in the nucleus, or else the change of one type of particle into others. These rearrangements and transformations may be hindered energetically, so that they do not occur immediately. In certain cases, random quantum vacuum fluctuations are theorized to promote relaxation to a lower energy state (the "decay") in a phenomenon known as quantum tunneling. Radioactive decay half-life of nuclides has been measured over timescales of 55 orders of magnitude, from 2.3×10^{-23} seconds (for hydrogen-7) to 6.9×10^{31} seconds (for tellurium-128).[31] The limits of these timescales are set by the sensitivity of instrumentation only, and there are no known natural limits to how brief or long a decay half life for radioactive decay of a radionuclide may be.

The decay process, like all hindered energy transformations, may be analogized by a snowfield on a mountain. While friction between the ice crystals may be supporting the snow's weight, the system is inherently unstable with regard to a state of lower potential energy. A disturbance would thus facilitate the path to a state of greater entropy: The system will move towards the ground state, producing heat, and the total energy will be distributable over a larger number of quantum states. Thus, an avalanche results. The **total** energy does not change in this process, but, because of the second law of thermodynamics, avalanches have only been observed in one direction and that is toward the "ground state" — the state with the largest number of ways in which the available energy could be distributed.

Such a collapse (a *decay event*) requires a specific activation energy. For a snow avalanche, this energy comes as a disturbance from outside the system, although such disturbances can be arbitrarily small. In the case of an excited atomic nucleus, the arbitrarily small disturbance comes from quantum vacuum fluctuations. A radioactive nucleus (or any excited system in quantum mechanics) is unstable, and can, thus, *spontaneously* stabilize to a less-excited system. The resulting transformation alters the structure of the nucleus and results in the emission of either a photon or a high-velocity particle that has mass (such as an electron, alpha particle, or other type).

25.9 Occurrence and applications

According to the Big Bang theory, stable isotopes of the lightest five elements (H, He, and traces of Li, Be, and B) were produced very shortly after the emergence of the universe, in a process called Big Bang nucleosynthesis. These lightest stable nuclides (including deuterium) survive to today, but any radioactive isotopes of the light elements produced in the Big Bang (such as tritium) have long since decayed. Isotopes of elements heavier than boron were not produced at all in the Big Bang, and these first five elements do not have any long-lived radioisotopes. Thus, all radioactive nuclei are, therefore, relatively young with respect to the birth of the universe, having formed later in various other types of nucleosynthesis in stars (in particular, supernovae), and also during ongoing interactions between stable isotopes and energetic particles. For example, carbon-14, a radioactive nuclide with a half-life of only 5,730 years, is constantly produced in Earth's upper atmosphere due to interactions between cosmic rays and nitrogen.

Nuclides that are produced by radioactive decay are called radiogenic nuclides, whether they themselves are stable or not. There exist stable radiogenic nuclides that were formed from short-lived extinct radionuclides in the early solar system.[32][33] The extra presence of these stable radiogenic nuclides (such as Xe-129 from primordial I-129) against the background of primordial stable nuclides can be inferred by various means.

Radioactive decay has been put to use in the technique of radioisotopic labeling, which is used to track the passage of a chemical substance through a complex system (such as a living organism). A sample of the substance is synthesized with a high concentration of unstable atoms. The presence of the substance in one or another part of the system is determined by detecting the locations of decay events.

On the premise that radioactive decay is truly random (rather than merely chaotic), it has been used in hardware random-number generators. Because the process is not thought to vary significantly in mechanism over time, it is also a valuable tool in estimating the absolute ages of certain materials. For geological materials, the radioisotopes and some of their decay products become trapped when a rock solidifies, and can then later be used (subject to many well-known qualifications) to estimate the date of the solidification. These include checking the results of several simultaneous processes and their products against each other, within the same sample. In a similar fashion, and also subject to qualification, the rate of formation of carbon-14 in various eras, the date of formation of organic matter within a certain period related to the isotope's half-life may be estimated, because the carbon-14 becomes trapped when the organic matter grows and incorporates the new carbon-14 from the air. Thereafter, the amount of carbon-14 in organic matter decreases according to decay processes that may also be independently cross-checked by other means (such as checking the carbon-14 in individual tree rings, for example).

25.10 Origins of radioactive nuclides

Main article: nucleosynthesis

Radioactive primordial nuclides found in the Earth are residues from ancient supernova explosions which occurred before the formation of the solar system. They are the long-lived fraction of radionuclides surviving in the primordial solar nebula through planet accretion until the present. The naturally occurring short-lived radiogenic radionuclides found in rocks are the daughters of these radioactive primordial nuclides. Another minor source of naturally occurring radioactive nuclides are cosmogenic nuclides, formed by cosmic ray bombardment of material in the Earth's atmosphere or crust. The radioactive decay of these radionuclides in rocks within Earth's mantle and crust contribute significantly to Earth's internal heat budget.

25.11 Decay chains and multiple modes

The daughter nuclide of a decay event may also be unstable (radioactive). In this case, it will also decay, producing radiation. The resulting second daughter nuclide may also be radioactive. This can lead to a sequence of several decay events. Eventually, a stable nuclide is produced. This is called a *decay chain* (see this article for specific details of important natural decay chains).

An example is the natural decay chain of ^{238}U, which is as follows:

- decays, through alpha-emission, with a half-life of 4.5 billion years to thorium-234

- which decays, through beta-emission, with a half-life of 24 days to protactinium-234

- which decays, through beta-emission, with a half-life of 1.2 minutes to uranium-234

- which decays, through alpha-emission, with a half-life of 240 thousand years to thorium-230

- which decays, through alpha-emission, with a half-life of 77 thousand years to radium-226

- which decays, through alpha-emission, with a half-life of 1.6 thousand years to radon-222

- which decays, through alpha-emission, with a half-life of 3.8 days to polonium-218

- which decays, through alpha-emission, with a half-life of 3.1 minutes to lead-214

- which decays, through beta-emission, with a half-life of 27 minutes to bismuth-214

- which decays, through beta-emission, with a half-life of 20 minutes to polonium 214

- which decays, through alpha-emission, with a half-life of 160 microseconds to lead-210

- which decays, through beta-emission, with a half-life of 22 years to bismuth-210

- which decays, through beta-emission, with a half-life of 5 days to polonium-210

- which decays, through alpha-emission, with a half-life of 140 days to lead-206, which is a stable nuclide.

Some radionuclides may have several different paths of decay. For example, approximately 36% of bismuth-212 decays, through alpha-emission, to thallium-208 while approximately 64% of bismuth-212 decays, through beta-emission, to polonium-212. Both thallium-208 and polonium-212 are radioactive daughter products of bismuth-212, and both decay directly to stable lead-208.

25.12 Associated hazard warning signs

- The trefoil symbol used to indicate ionising radiation.

- 2007 ISO radioactivity danger symbol intended for IAEA Category 1, 2 and 3 sources defined as dangerous sources capable of death or serious injury.[1]

- The dangerous goods transport classification sign for radioactive materials

1. ^ IAEA news release Feb 2007

25.13 See also

- Actinides in the environment

- Background radiation

- Chernobyl disaster

- Crimes involving radioactive substances

- Decay chain

- Fallout shelter

- Half-life

- Lists of nuclear disasters and radioactive incidents

- National Council on Radiation Protection and Measurements

- Nuclear engineering

- Nuclear medicine

- Nuclear pharmacy

- Nuclear physics

- Nuclear power

- Particle decay

- Poisson process

- Radiation

- Radiation therapy

- Radioactive contamination

- Radioactivity in biology

- Radiometric dating

- Radionuclide a.k.a. "radio-isotope"

- Secular equilibrium

- Transient equilibrium

25.14   Notes

[1] Radionuclide is the more correct term, but radioisotope is also used. The difference between isotope and nuclide is explained at Isotope#Isotope vs. nuclide.

25.15   References

25.15.1   Inline

[1] "Decay and Half Life". Retrieved 2009-12-14.

[2] Stabin, Michael G. (2007). "3".*Radiation Protection and Dosimetry: An Introduction to Health Physics*.Springer.doi:10.1007/9-0-387-49983-3. ISBN 978-0387499826.

[3] Best, Lara; Rodrigues, George; Velker, Vikram (2013). "1.3". *Radiation Oncology Primer and Review*. Demos Medical Publishing. ISBN 978-1620700044.

[4] Loveland, W.; Morrissey, D.; Seaborg, G.T. (2006). *Modern Nuclear Chemistry*. Wiley-Interscience. p. 57. ISBN 0-471-11532-0.

[5] Mould, Richard F. (1995). *A century of X-rays and radioactivity in medicine : with emphasis on photographic records of the early years* (Reprint. with minor corr ed.). Bristol: Inst. of Physics Publ. p. 12. ISBN 9780750302241.

[6] Kasimir Fajans, "Radioactive transformations and the periodic system of the elements". Berichte der Deutschen Chemischen Gesellschaft, Nr. 46, 1913, p. 422–439

[7] Frederick Soddy, "The Radio Elements and the Periodic Law", Chem. News, Nr. 107, 1913, p.97–99

[8] Sansare, K.; Khanna, V.; Karjodkar, F. (2011). "Early victims of X-rays: a tribute and current perception". *Dentomaxillofacial Radiology* **40** (2): 123–125. doi:10.1259/dmfr/73488299. ISSN 0250-832X. PMC 3520298. PMID 21239576.

[9] Ronald L. Kathern and Paul L. Ziemer, he First Fifty Years of Radiation Protection, physics.isu.edu

[10] Hrabak, M.; Padovan, R. S.; Kralik, M.; Ozretic, D.; Potocki, K. (July 2008). "Nikola Tesla and the Discovery of X-rays". *RadioGraphics* **28** (4): 1189–92. doi:10.1148/rg.284075206. PMID 18635636.

[11] Geoff Meggitt (2008), *Taming the Rays - A history of Radiation and Protection.*, Lulu.com, ISBN 978-1-4092-4667-1

[12] Clarke, R.H.; J. Valentin (2009). "The History of ICRP and the Evolution of its Policies" (PDF). *Annals of the ICRP*. ICRP Publication 109 **39** (1): pp. 75–110. doi:10.1016/j.icrp.2009.07.009. Retrieved 12 May 2012.

[13] Rutherford, Ernest (6 October 1910). "Radium Standards and Nomenclature". *Nature* **84** (2136): 430–431.

[14] *10 CFR 20.1005*. US Nuclear Regulatory Commission. 2009.

[15] The Council of the European Communities (1979-12-21). "Council Directive 80/181/EEC of 20 December 1979 on the approximation of the laws of the Member States relating to Unit of measurement and on the repeal of Directive 71/354/EEC". Retrieved 19 May 2012.

[16] Radioactive Decay

[17] Patel, S.B. (2000). *Nuclear physics : an introduction*. New Delhi: New Age International. pp. 62–72. ISBN 9788122401257.

[18] Introductory Nuclear Physics, K.S. Krane, 1988, John Wiley & Sons Inc, ISBN 978-0-471-80553-3

[19] Cetnar, Jerzy (May 2006). "General solution of Bateman equations for nuclear transmutations". *Annals of Nuclear Energy* **33** (7): 640–645. doi:10.1016/j.anucene.2006.02.004.

[20] K.S. Krane (1988). *Introductory Nuclear Physics*. John Wiley & Sons Inc. p. 164. ISBN 978-0-471-80553-3.

[21] Wang, B.; Yan, S.; Limata, B. et al. (2006). "Change of the 7Be electron capture half-life in metallic environments". *The European Physical Journal A* **28** (3): 375–377. Bibcode:2006EPJA...28..375W. doi:10.1140/epja/i2006-10068-x. ISSN 1434-6001.

[22] Jung, M.; Bosch, F.; Beckert, K. et al. (1992). "First observation of bound-state β^- decay". *Physical Review Letters* **69** (15): 2164–2167. Bibcode:1992PhRvL..69.2164J. doi:10.1103/PhysRevLett.69.2164. ISSN 0031-9007. PMID 10046415.

[23] Smoliar, M.I.; Walker, R.J.; Morgan, J.W. (1996). "Re-Os ages of group IIA, IIIA, IVA, and IVB iron meteorites". *Science* **271** (5252): 1099–1102. Bibcode:1996Sci...271.1099S. doi:10.1126/science.271.5252.1099.

[24] Bosch, F.; Faestermann, T.; Friese, J.; Heine, F.; Kienle, P.; Wefers, E.; Zeitelhack, K.; Beckert, K.; Franzke, B.; Klepper, O.; Kozhuharov, C.; Menzel, G.; Moshammer, R.; Nolden, F.; Reich, H.; Schlitt, B.; Steck, M.; Stöhlker, T.; Winkler, T.; Takahashi, K. (1996). "Observation of bound-state β– decay of fully ionized ^{187}Re:^{187}Re-^{187}Os Cosmochronometry". *Physical Review Letters* **77** (26): 5190–5193. Bibcode:1996PhRvL..77.5190B. doi:10.1103/PhysRevLett.77.5190. PMID 10062738.

[25] Emery, G.T. (1972). "Perturbation of Nuclear Decay Rates" (PDF). *Annual Review of Nuclear Science* (ACS Publications) **22**: 165–202. Bibcode:1972ARNPS..22..165E. doi:10.1146/annurev.ns.22.120172.001121. Retrieved 6 August 2012.

[26] "The mystery of varying nuclear decay". *Physics World*. 2 October 2008.

[27] Jenkins, Jere H.; Fischbach, Ephraim (2009). "Perturbation of Nuclear Decay Rates During the Solar Flare of 13 December 2006". *Astroparticle Physics* **31** (6): 407–411. arXiv:0808.3156. Bibcode:2009APh....31..407J. doi:10.1016/j.astropartphys.

[28] Jenkins, J. H.; Buncher, John B.; Gruenwald, John T.; Krause, Dennis E.; Mattes, Joshua J. et al. (2009). "Evidence of correlations between nuclear decay rates and Earth–Sun distance". *Astroparticle Physics* **32** (1): 42–46. arXiv:0808.3283. Bibcode:2009APh....32...42J. doi:10.1016/j.astropartphys.2009.05.004.

[29] Peter A. Sturrock, Gideon Steinitz, Ephraim Fischbach, Daniel Javorsek, II, Jere H. Jenkins, Analysis of Gamma Radiation from a Radon Source: Indications of a Solar Influence, Accessed on line September 2, 2012.

[30] Norman, E. B.; Shugart, Howard A.; Joshi, Tenzing H.; Firestone, Richard B. et al. (2009). "Evidence against correlations between nuclear decay rates and Earth–Sun distance" (PDF). *Astroparticle Physics* **31** (2): 135–137. arXiv:0810.3265. Bibcode:2009APh....31..135N. doi:10.1016/j.astropartphys.2008.12.004.

[31] NUBASE evaluation of nuclear and decay properties

[32] Clayton, Donald D. (1983). *Principles of Stellar Evolution and Nucleosynthesis* (2nd ed.). University of Chicago Press. p. 75. ISBN 0-226-10953-4.

[33] Bolt, B. A.; Packard, R. E.; Price, P. B. (2007). "John H. Reynolds, Physics: Berkeley". The University of California, Berkeley. Retrieved 2007-10-01.

25.15.2 General

- "Radioactivity", Encyclopædia Britannica. 2006. Encyclopædia Britannica Online. December 18, 2006

- Radio-activity by Ernest Rutherford Phd, Encyclopædia Britannica Eleventh Edition

25.16   External links

- The Lund/LBNL Nuclear Data Search – Contains tabulated information on radioactive decay types and energies.

- Nomenclature of nuclear chemistry

- Specific activity and related topics.

- The Live Chart of Nuclides – IAEA

- Health Physics Society Public Education Website

- Beach, Chandler B., ed. (1914). "Becquerel Rays". *The New Student's Reference Work*. Chicago: F. E. Compton and Co.

- Annotated bibliography for radioactivity from the Alsos Digital Library for Nuclear Issues

- Stochastic Java applet on the decay of radioactive atoms by Wolfgang Bauer

- Stochastic Flash simulation on the decay of radioactive atoms by David M. Harrison

- "Henri Becquerel: The Discovery of Radioactivity", Becquerel's 1896 articles online and analyzed on *BibNum* [click 'à télécharger' for English version].

- "Radioactive change", Rutherford & Soddy article (1903), online and analyzed on *Bibnum* [click 'à télécharger' for English version].

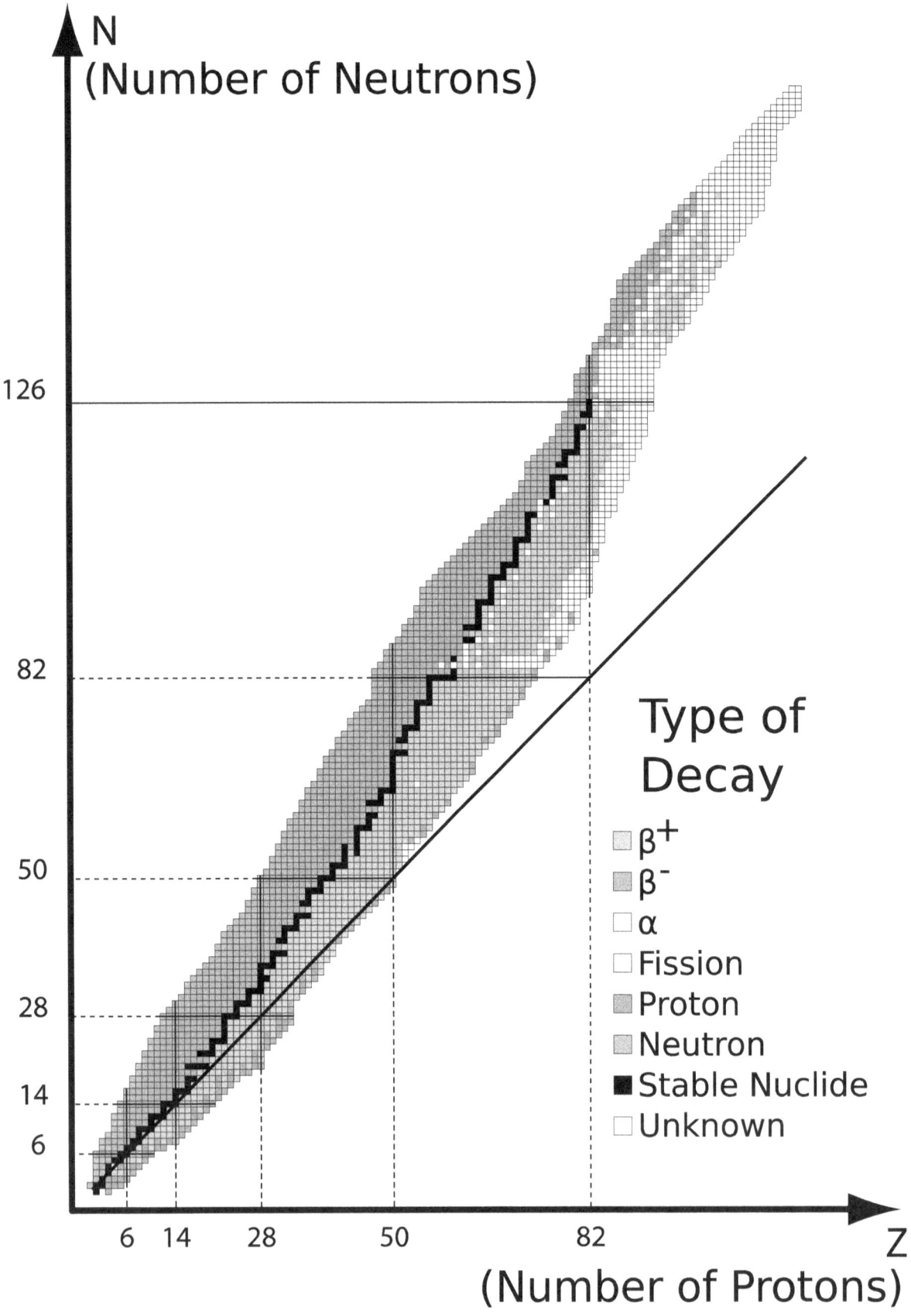

Types of radioactive decay related to N and Z numbers

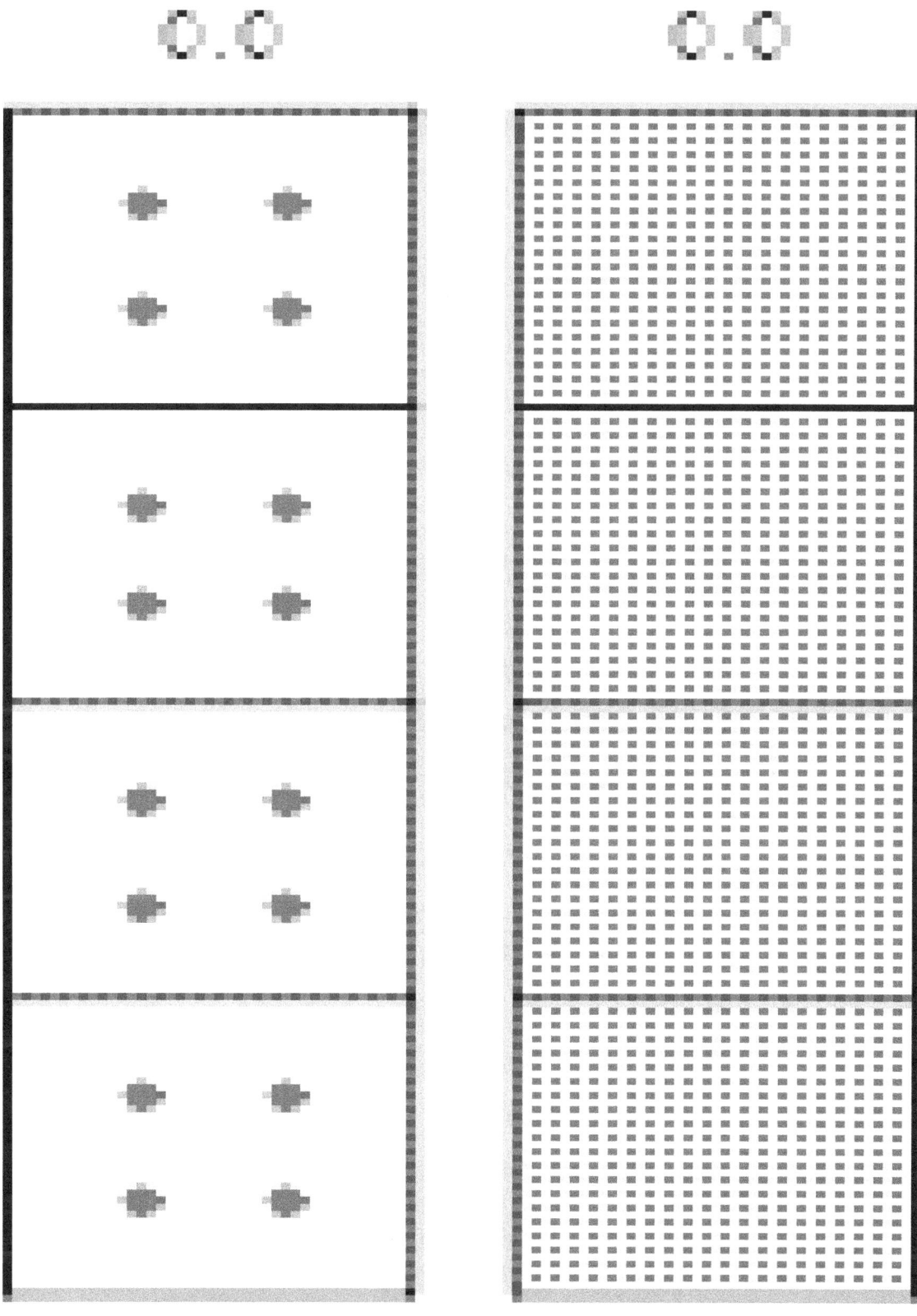

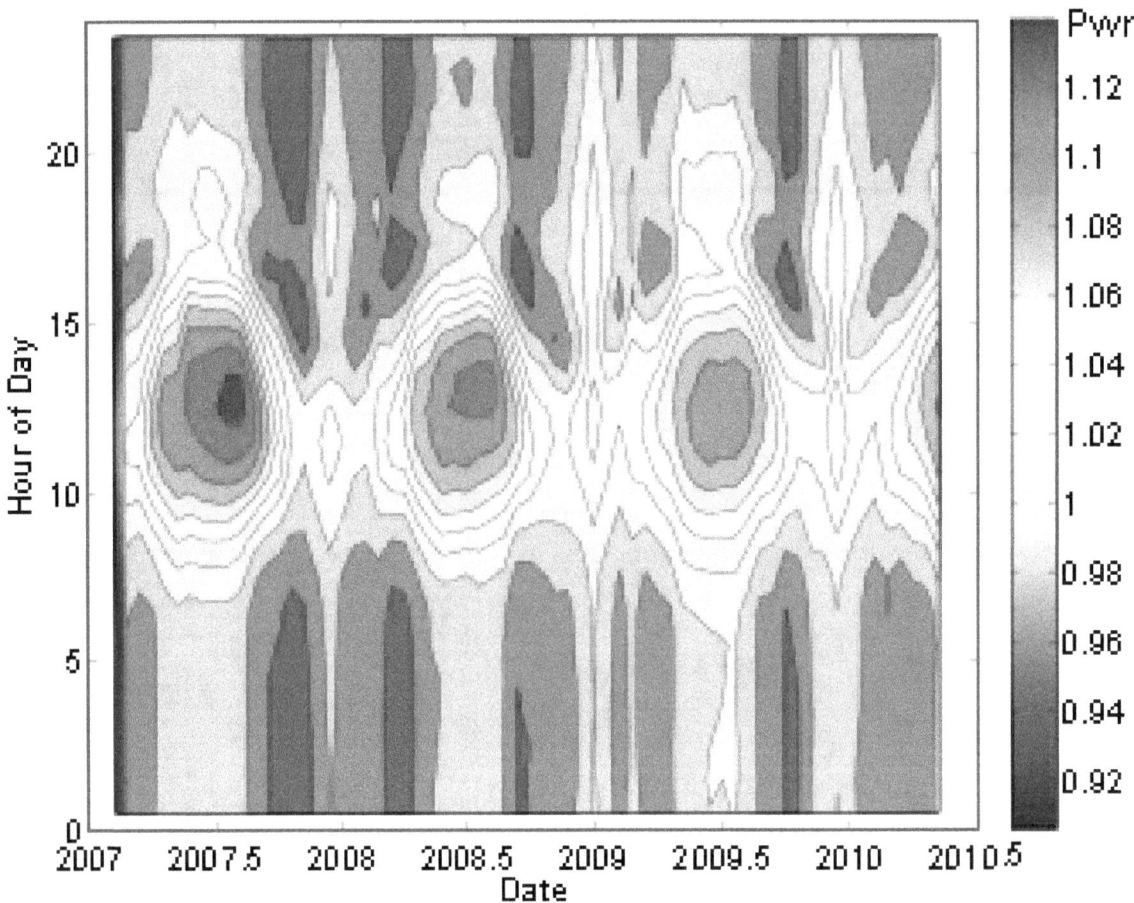

Decay Rate of Radon-222 as a function of date and time of day. The color-bar gives the power of the observed signal and represents ~4% seasonal decay rate variation.

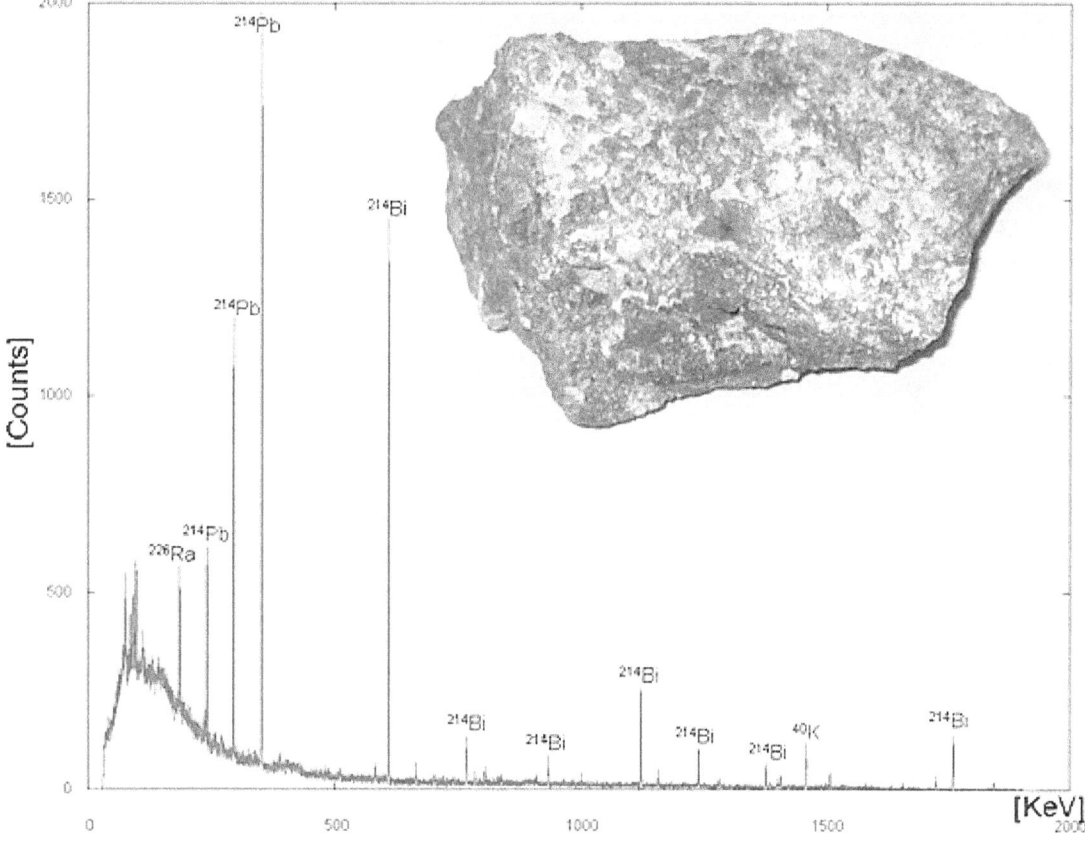

Gamma-ray energy spectrum of uranium ore (inset). Gamma-rays are emitted by decaying nuclides, and the gamma-ray energy can be used to characterize the decay (which nuclide is decaying to which). Here, using the gamma-ray spectrum, several nuclides that are typical of the decay chain of ^{238}U have been identified: ^{226}Ra, ^{214}Pb, ^{214}Bi.

Chapter 26

Stable isotope ratio

This article is about stable isotope ratios. For the general category of stable nuclides, see Stable nuclide.

The term **stable isotope** has a similar meaning to stable nuclide, but is preferably used when speaking of nuclides of a specific element. Hence, the plural form **stable isotopes** usually refers to isotopes of the same element. The relative abundance of such stable isotopes can be measured experimentally (isotope analysis), yielding an isotope ratio that can be used as a research tool. Theoretically, such stable isotopes could include the radiogenic daughter products of radioactive decay, used in radiometric dating. However, the expression **stable isotope ratio** is preferably used to refer to isotopes whose relative abundances are affected by isotope fractionation in nature. This field is termed stable isotope geochemistry.

26.1 Stable isotope ratios

See also: Isotope fractionation

Measurement of the ratios of naturally occurring stable isotopes (isotope analysis) plays an important role in isotope geochemistry, but stable isotopes (mostly carbon, nitrogen and oxygen) are also finding uses in ecological and biological studies. Other workers have used oxygen isotope ratios to reconstruct historical atmospheric temperatures, making them important tools for paleoclimatology.

Commonly analysed stable isotope systems include those of oxygen, carbon, nitrogen, hydrogen and sulfur. These isotope systems for lighter elements that exhibit more than one primordial isotope for each element, have been under investigation for many years in order to study processes of isotope fractionation in natural systems. The long history of study of these elements is in part because the proportions of stable isotopes in these light and volatile elements is relatively easy to measure. However, recent advances in isotope ratio mass spectrometry (i.e. multiple-collector inductively coupled plasma mass spectrometry) now enable the measurement of isotope ratios in heavier stable elements, such as iron, copper, zinc, molybdenum, etc.

26.2 Applications

The variations in oxygen and hydrogen isotope ratios have applications in hydrology since most samples will lie between two extremes, ocean water and Arctic/Antarctic snow.[1] Given a sample of water from an aquifer, and a sufficiently sensitive tool to measure the variation in the isotopic ratio of hydrogen in the sample, it is possible to infer the source, be it ocean water seeping into the aquifer or precipitation seeping into the aquifer, and even to estimate the proportions from each source.[2]

Another application is in paleotemperature measurement for Paleoclimatology. For example one technique is based on the variation in isotopic fractionation of oxygen by biological systems with temperature.[3] Species of foraminifera incorporate oxygen as calcium carbonate in their shells. The ratio of the oxygen isotopes oxygen-16 and oxygen-18 incorporated into the calcium carbonate varies with temperature and the oxygen isotopic composition of the water. This oxygen remains "fixed" in the calcium carbonate when the forminifera dies, falls to the sea bed, and its shell becomes part of the sediment. It is possible to select standard species of forminifera from sections through the sediment column, and by mapping the variation in oxygen isotopic ratio, deduce the temperature that the forminifera encountered during life if changes in the oxygen isotopic composition of the water can be constrained.[4]

In forensic science, research suggests that the variation in certain isotope ratios in drugs derived from plant sources (cannabis, cocaine) can be used to determine the drug's continent of origin.[5]

It also has applications in "doping control," to distinguish between endogenous and exogenous (synthetic) sources of hormones.[6][7]

Chondrite meteorites are classified using the oxygen isotope ratios. In addition, an unusual signature of carbon-13 confirms the non-terrestrial origin for organic compounds found in carbonaceous chondrites, as in the Murchison meteorite.

26.3 See also

- Radiocarbon dating

26.4 Bibliography

- Allègre C.J., 2008. *Isotope Geology* (Cambridge University Press).

- Faure G., Mensing T.M. (2004), *Isotopes: Principles and Applications* (John Wiley & Sons).

- Hoefs J., 2004. *Stable Isotope Geochemistry* (Springer Verlag).

- Sharp Z., 2006. *Principles of Stable Isotope Geochemistry* (Prentice Hall).

26.5 References

[1] Han LF, Gröning M, Aggarwal P, Helliker BR (2006). "Reliable determination of oxygen and hydrogen isotope ratios in atmospheric water vapour adsorbed on 3A molecular sieve". *Rapid Commun. Mass Spectrom.* **20** (23): 3612–8. doi:10.1002/rcm.2772. PMID 17091470.

[2] Weldeab S, Lea DW, Schneider RR, Andersen N (2007). "155,000 years of West African monsoon and ocean thermal evolution". *Science* **316** (5829): 1303–7. Bibcode:2007Sci...316.1303W. doi:10.1126/science.1140461. PMID 17540896.

[3] Tolosa I, Lopez JF, Bentaleb I, Fontugne M, Grimalt JO (1999). "Carbon isotope ratio monitoring-gas chromatography mass spectrometric measurements in the marine environment: biomarker sources and paleoclimate applications". *Sci. Total Environ.* 237-238: 473–81. doi:10.1016/S0048-9697(99)00159-X. PMID 10568296.

[4] Shen JJ, You CF (2003). "A 10-fold improvement in the precision of boron isotopic analysis by negative thermal ionization mass spectrometry". *Anal. Chem.* **75** (9): 1972–7. doi:10.1021/ac020589f. PMID 12720329.

[5] Casale J, Casale E, Collins M, Morello D, Cathapermal S, Panicker S (2006). "Stable isotope analyses of heroin seized from the merchant vessel Pong Su". *J. Forensic Sci.* **51** (3): 603–6. doi:10.1111/j.1556-4029.2006.00123.x. PMID 16696708.

[6] Author, A. "Stable isotope ratio analysis in sports anti-doping". *Drug Testing and Analysis* **4** (12): 893–896. doi:10.1002/dta.1399.

[7] Cawley, Adam T.; Kazlauskas, Rymantas; Trout, Graham J.; Rogerson, Jill H.; George, Adrian V. "Isotopic Fractionation of Endogenous Anabolic Androgenic Steroids and Its Relationship to Doping Control in Sports" (PDF). *Journal of Chromatographic Science* **43**: 32–38. doi:10.1093/chromsci/43.1.32.

Chapter 27

Transuranium element

The **transuranium elements** (also known as **transuranic elements**) are the chemical elements with atomic numbers greater than 92 (the atomic number of uranium). All of these elements are unstable and decay radioactively into other elements.

27.1 Overview

Of the elements with atomic numbers 1 to 92, all can be found in nature, having stable (such as hydrogen), or very long half-life (such as uranium) isotopes, or are created as common products of the decay of uranium and thorium (such as radon) — only technetium, of the elements below uranium, was man-made for its discovery in 1936.

All of the elements with higher atomic numbers, however, have been first discovered in the laboratory, with neptunium, plutonium, americium, curium, berkelium and californium later also discovered in nature. They are all radioactive, with a half-life much shorter than the age of the Earth, so any atoms of these elements, if they ever were present at the Earth's formation, have long since decayed. Trace amounts of these six elements form in some uranium-rich rock, and small amounts are produced during atmospheric tests of atomic weapons. The Np, Pu, Am, Cm, Bk, and Cf are generated from neutron capture in uranium ore with subsequent beta decays (e.g. $^{238}U + n \rightarrow {}^{239}U \rightarrow {}^{239}Np \rightarrow {}^{239}Pu$).

Transuranic elements can be artificially generated synthetic elements, via nuclear reactors or particle accelerators. The half lives of these elements show a general trend of decreasing as atomic numbers increase. There are exceptions, however, including dubnium and several isotopes of curium. Further anomalous elements in this series have been predicted by Glenn T. Seaborg, and are categorised as the "island of stability."[1]

Heavy transuranic elements are difficult and expensive to produce, and their prices increase rapidly with atomic number. As of 2008, weapons-grade plutonium cost around \$4,000/gram,[2] and californium cost \$60,000,000/gram.[3] Due to production difficulties, none of the elements beyond californium have industrial applications, and of them, only einsteinium has ever been produced in macroscopic quantities.[4]

Transuranic elements that have not been discovered, or have been discovered but are not yet officially named, use IUPAC's systematic element names. The naming of transuranic elements may be a source of controversy.

27.2 Discovery and naming of transuranium elements

So far, essentially all the transuranium elements have been produced at three laboratories:

- The Radiation Laboratory (now Lawrence Berkeley National Laboratory) at the University of California, Berkeley, led principally by Edwin McMillan, Glenn Seaborg, and Albert Ghiorso, during 1945-1974:

Group

	I	II											III	IV	V	VI	VII	VIII
1	1 H																	2 He
2	3 Li	4 Be											5 B	6 C	7 N	8 O	9 F	10 Ne
3	11 Na	12 Mg											13 Al	14 Si	15 P	16 S	17 Cl	18 Ar
4	19 K	20 Ca	21 Sc	22 Ti	23 V	24 Cr	25 Mn	26 Fe	27 Co	28 Ni	29 Cu	30 Zn	31 Ga	32 Ge	33 As	34 Se	35 Br	36 Kr
5	37 Rb	38 Sr	39 Y	40 Zr	41 Nb	42 Mo	43 Tc	44 Ru	45 Rh	46 Pd	47 Ag	48 Cd	49 In	50 Sn	51 Sb	52 Te	53 I	54 Xe
6	55 Cs	56 Ba	*	72 Hf	73 Ta	74 W	75 Re	76 Os	77 Ir	78 Pt	79 Au	80 Hg	81 Tl	82 Pb	83 Bi	84 Po	85 At	86 Rn
7	87 Fr	88 Ra	**	104 Rf	105 Db	106 Sg	107 Bh	108 Hs	109 Mt	110 Ds	111 Rg	112 Cn	113 Uut	114 Fl	115 Uup	116 Lv	117 Uus	118 Uuo

Period

* Lanthanides	57 La	58 Ce	59 Pr	60 Nd	61 Pm	62 Sm	63 Eu	64 Gd	65 Tb	66 Dy	67 Ho	68 Er	69 Tm	70 Yb	71 Lu
** Actinides	89 Ac	90 Th	91 Pa	92 U	93 Np	94 Pu	95 Am	96 Cm	97 Bk	98 Cf	99 Es	100 Fm	101 Md	102 No	103 Lr

Periodic table with elements colored according to the half-life of their most stable isotope.
Elements which contain at least one stable isotope.
Slightly radioactive elements: the most stable isotope is very long-lived, with a half-life of over four million years.
Significantly radioactive elements: the most stable isotope has half-life between 800 and 34,000 years.
Radioactive elements: the most stable isotope has half-life between one day and 103 years.
Highly radioactive elements: the most stable isotope has half-life between several minutes and one day.
Extremely radioactive elements: the most stable isotope has half-life less than several minutes.

- 93. neptunium, Np, named after the planet Neptune, as it follows uranium and Neptune follows Uranus in the planetary sequence (1940).

- 94. plutonium, Pu, named after the dwarf planet Pluto, following the same naming rule as it follows neptunium and Pluto follows Neptune in the pre-2006 planetary sequence (1940).

- 95. americium, Am, named because it is an analog to europium, and so was named after the continent where it was first produced (1944).

- 96. curium, Cm, named after Pierre and Marie Curie, famous scientists who separated out the first radioactive elements (1944).

- 97. berkelium, Bk, named after the city of Berkeley, where the University of California, Berkeley is located (1949).

- 98. californium, Cf, named after the state of California, where the university is located (1950).

- 99. einsteinium, Es, named after the theoretical physicist Albert Einstein (1952).

- 100. fermium, Fm, named after Enrico Fermi, the physicist who produced the first controlled chain reaction (1952).

- 101. mendelevium, Md, named after the Russian chemist Dmitri Mendeleev, credited for being the primary creator of the periodic table of the chemical elements (1955).

- 102. nobelium, No, named after Alfred Nobel (1956).

- 103. lawrencium, Lr, named after Ernest O. Lawrence, a physicist best known for development of the cyclotron, and the person for whom the Lawrence Livermore National Laboratory and the Lawrence Berkeley National Laboratory (which hosted the creation of these transuranium elements) are named (1961).

- 104. rutherfordium, Rf, named after Ernest Rutherford, who was responsible for the concept of the atomic nucleus (1968). This discovery was also claimed by the Joint Institute for Nuclear Research (JINR) in Dubna, Russia (then the Soviet Union), led principally by G. N. Flerov.

- 105. dubnium, Db, an element that is named after the city of Dubna, where the JINR is located. Originally named "hahnium" in honor of Otto Hahn (1970) but renamed by the International Union of Pure and Applied Chemistry. This discovery was also claimed by the JINR.

- 106. seaborgium, Sg, named after Glenn T. Seaborg. This name caused controversy because Seaborg was still alive, but eventually became accepted by international chemists (1974). This discovery was also claimed by the JINR.

- The Gesellschaft für Schwerionenforschung (Society for Heavy Ion Research) in Darmstadt, Hessen, Germany, led principally by Peter Armbruster and Sigurd Hofmann, during 1980-2000:

  - 107. bohrium, Bh, named after the Danish physicist Niels Bohr, important in the elucidation of the structure of the atom (1981). This discovery was also claimed by the JINR.

  - 108. hassium, Hs, named after the Latin form of the name of Hessen, the German *Bundesland* where this work was performed (1984).

  - 109. meitnerium, Mt, named after Lise Meitner, an Austrian physicist who was one of the earliest scientists to become involved in the study of nuclear fission (1982).

  - 110. darmstadtium, Ds, named after Darmstadt, Germany, the city in which this work was performed (1994).

  - 111. roentgenium, Rg, named after Wilhelm Conrad Röntgen, discoverer of X-rays (1994).

  - 112. copernicium, Cn, named after astronomer Nicolaus Copernicus (1996).

- The Joint Institute for Nuclear Research (JINR) in Dubna, Russia, led principally by Y. Oganessian, in collaboration with several other laboratories including the Lawrence Livermore National Laboratory (LLNL), since 2000:

  - 113. ununtrium, Uut, temporary name, (2003).

  - 114. flerovium, Fl, named after Soviet physicist Georgy Flyorov, founder of the JINR (1999).

  - 115. ununpentium, Uup, temporary name, (2003).

  - 116. livermorium, Lv, named after the Lawrence Livermore National Laboratory, a collaborator with JINR in the discovery, (2000).

  - 117. ununseptium, Uus, temporary name, (2010).

  - 118. ununoctium, Uuo, temporary name, (2002).

The temporary names listed above are generic names assigned according to a convention (the systematic element names). They will be replaced by permanent names as the elements are confirmed by independent work.

27.3 List of the transuranic elements by chemical series

*The existence of these elements has been claimed and generally accepted, but not yet acknowledged by the IUPAC.

The names and symbols of elements 113, 115, 117, and 118 are provisional until permanent names for the elements are decided on, usually within a year after the discovery acknowledgement by IUPAC.

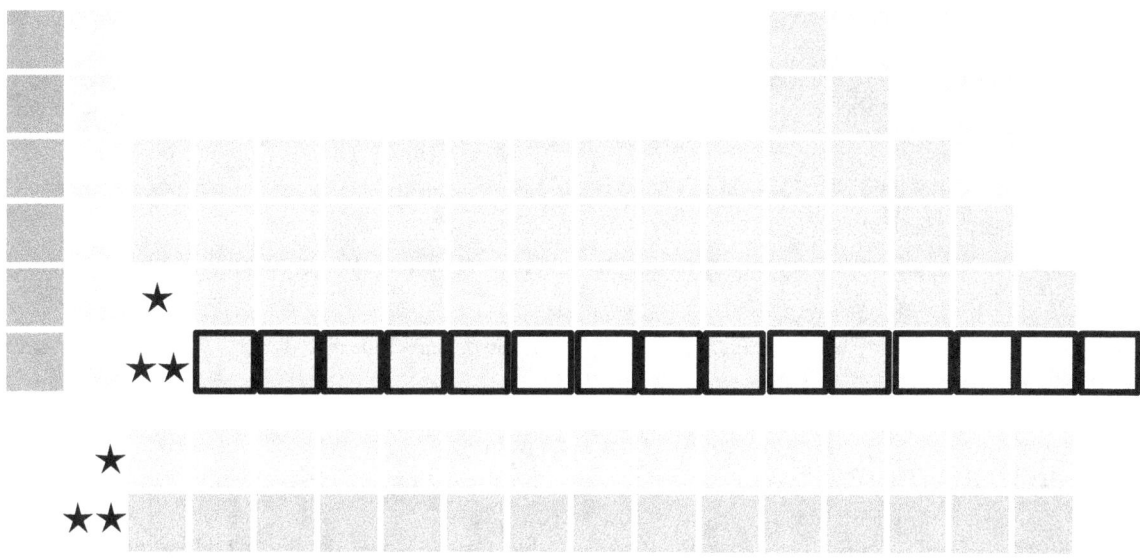

Position of the transactinide elements in the periodic table.

27.4   Super-heavy elements

Super-heavy elements, (also known as *super heavy atoms*, commonly abbreviated **SHE**) may refer to elements beyond atomic number 100, but also may refer to all transuranium elements. The transactinide elements begin with rutherfordium (atomic number 104).[5] They have only been made artificially, and currently serve no practical purpose because their short half-lives cause them to decay after a very short time, ranging from a few minutes to just a few milliseconds (except for dubnium, which has a half life of over a day), which also makes them extremely hard to study.[6][7]

Super-heavy atoms have all been created during the latter half of the 20th century and are continually being created during the 21st century as technology advances. They are created through the bombardment of elements in a particle accelerator. For example, the nuclear fusion of californium−249 and carbon−12 creates rutherfordium. These elements are created in quantities on the atomic scale and no method of mass creation has been found.[6]

27.5   See also

- Bose–Einstein condensate (also known as *Superatom*)

- Island of stability

- Minor actinides

- Waste Isolation Pilot Plant, repository for transuranic waste

- Extension of the periodic table beyond the seventh period

27.6   References

[1] Considine, Glenn, ed. (2002). *Van Nostrand's Scientific Encyclopedia* (9th ed.). New York: Wiley Interscience. p. 738. ISBN 0-471-33230-5.

[2] "Price of Plutonium". The Physics Factbook.

[3] Rodger C. Martin and Steven E. Kos. "Applications and Availability of Californium-252 Neutron Sources for Waste Characterization" (pdf).

[4] Haire, Richard G. (2006). "Fermium, Mendelevium, Nobelium and Lawrencium". In Morss; Edelstein, Norman M.; Fuger, Jean. *The Chemistry of the Actinide and Transactinide Elements* (3rd ed.). Dordrecht, The Netherlands: Springer Science+Business Media. ISBN 1-4020-3555-1.

[5] Uzi Kaldor; Stephen Wilson (2003). *Theoretical chemistry and physics of heavy and superheavy elements.* Springer. pp. 5, 8. ISBN 978-1-4020-1371-3. Retrieved 10 June 2011.

[6] Heenen, P. H.; Nazarewicz, W. (2002). "Quest for superheavy nuclei". *Europhysics News* **33**: 5. Bibcode:2002ENews..33....5H. doi:10.1051/epn:2002102.

[7] Greenwood, N. N. (1997). "Recent developments concerning the discovery of elements 100–111". *Pure and Applied Chemistry* **69**: 179. doi:10.1351/pac199769010179.

27.7 Further reading

- Eric Scerri, A Very Short Introduction to the Periodic Table, Oxford University Press, Oxford, 2011.

- The Superheavy Elements

- Annotated bibliography for the transuranic elements from the Alsos Digital Library for Nuclear Issues.

- Transuranium elements

- Super Heavy Elements network official website (network of the European integrated infrastructure initiative EURONS)

- Darmstadium and beyond

- Christian Schnier, Joachim Feuerborn, Bong-Jun Lee: Traces of transuranium elements in terrestrial minerals? (Online, PDF-Datei, 493 kB)

- Christian Schnier, Joachim Feuerborn, Bong-Jun Lee: The search for super heavy elements (SHE) in terrestrial minerals using XRF with high energy synchrotron radiation. (Online, PDF-Datei, 446 kB)

Chapter 28

Timeline of atomic and subatomic physics

A timeline of atomic and subatomic physics.

28.1 Early beginnings

- 600 BCE[1] Kanada theorizes the existence of four kinds of atoms, which could combine to produce diatomic and triatomic molecules.

- 430 BCE[1] Democritus speculates about fundamental indivisible particles—calls them "atoms"

- 200 BCE[1] Jainism calls atom Paramanu which can neither be created nor destroyed. It is eternal, i.e., it existed in the past, exists in the present and will continue to exist in the future. It is the permanent basis of the physical existence. The entire physical existence is composed of these ultimate atoms.

28.2 The beginning of chemistry

- 1766 Henry Cavendish discovers and studies hydrogen

- 1778 Carl Scheele and Antoine Lavoisier discover that air is composed mostly of nitrogen and oxygen

- 1781 Joseph Priestley creates water by igniting hydrogen and oxygen

- 1800 William Nicholson and Anthony Carlisle use electrolysis to separate water into hydrogen and oxygen

- 1803 John Dalton introduces atomic ideas into chemistry and states that matter is composed of atoms of different weights

- 1805 (approximate time) Thomas Young conducts the double-slit experiment with light

- 1811 Amedeo Avogadro claims that equal volumes of gases should contain equal numbers of molecules

- 1832 Michael Faraday states his laws of electrolysis

- 1871 Dmitri Mendeleyev systematically examines the periodic table and predicts the existence of gallium, scandium, and germanium

- 1873 Johannes van der Waals introduces the idea of weak attractive forces between molecules

- 1885 Johann Balmer finds a mathematical expression for observed hydrogen line wavelengths

- 1887 Heinrich Hertz discovers the photoelectric effect

- 1894 Lord Rayleigh and William Ramsay discover argon by spectroscopically analyzing the gas left over after nitrogen and oxygen are removed from air

- 1895 William Ramsay discovers terrestrial helium by spectroscopically analyzing gas produced by decaying uranium

- 1896 Antoine Becquerel discovers the radioactivity of uranium

- 1896 Pieter Zeeman studies the splitting of sodium D lines when sodium is held in a flame between strong magnetic poles

- 1897 J.J. Thomson discovers the electron

- 1898 William Ramsay and Morris Travers discover neon, and negatively charged beta particles

28.3 Timeline of classical mechanics

Main article: Timeline of classical mechanics

28.4 The age of quantum mechanics

- 1887 Heinrich Rudolf Hertz discovers the photoelectric effect that will play a very important role in the development of the quantum theory with Einstein's explanation of this effect in terms of *quanta* of light

- 1896 Wilhelm Conrad Röntgen discovers the X-rays while studying electrons in plasma; scattering X-rays—that were considered as 'waves' of high-energy electromagnetic radiation—Arthur Compton will be able to demonstrate in 1922 the 'particle' aspect of electromagnetic radiation.

- 1900 Paul Villard discovers gamma-rays while studying uranium decay

- 1900 Johannes Rydberg refines the expression for observed hydrogen line wavelengths

- 1900 Max Planck states his quantum hypothesis and blackbody radiation law

- 1902 Philipp Lenard observes that maximum photoelectron energies are independent of illuminating intensity but depend on frequency

- 1902 Theodor Svedberg suggests that fluctuations in molecular bombardment cause the Brownian motion

- 1905 Albert Einstein explains the photoelectric effect

- 1906 Charles Barkla discovers that each element has a characteristic X-ray and that the degree of penetration of these X-rays is related to the atomic weight of the element

- 1909 Hans Geiger and Ernest Marsden discover large angle deflections of alpha particles by thin metal foils

- 1909 Ernest Rutherford and Thomas Royds demonstrate that alpha particles are doubly ionized helium atoms

- 1911 Ernest Rutherford explains the Geiger–Marsden experiment by invoking a nuclear atom model and derives the Rutherford cross section

- 1911 Jean Perrin proves the existence of atoms and molecules

- 1911 Ştefan Procopiu measures the magnetic dipole moment of the electron

- 1912 Max von Laue suggests using crystal lattices to diffract X-rays

- 1912 Walter Friedrich and Paul Knipping diffract X-rays in zinc blende

- 1913 William Henry Bragg and William Lawrence Bragg work out the Bragg condition for strong X-ray reflection

- 1913 Henry Moseley shows that nuclear charge is the real basis for numbering the elements

- 1913 Niels Bohr presents his quantum model of the atom[2]

- 1913 Robert Millikan measures the fundamental unit of electric charge

- 1913 Johannes Stark demonstrates that strong electric fields will split the Balmer spectral line series of hydrogen

- 1914 James Franck and Gustav Hertz observe atomic excitation

- 1914 Ernest Rutherford suggests that the positively charged atomic nucleus contains protons

- 1915 Arnold Sommerfeld develops a modified Bohr atomic model with elliptic orbits to explain relativistic fine structure

- 1916 Gilbert N. Lewis and Irving Langmuir formulate an electron shell model of chemical bonding

- 1917 Albert Einstein introduces the idea of stimulated radiation emission

- 1918 Ernest Rutherford notices that, when alpha particles were shot into nitrogen gas, his scintillation detectors showed the signatures of hydrogen nuclei.

- 1921 Alfred Landé introduces the Landé g-factor

- 1922 Arthur Compton studies X-ray photon scattering by electrons demonstrating the 'particle' aspect of electromagnetic radiation.

- 1922 Otto Stern and Walther Gerlach show "spin quantization"

- 1923 Lise Meitner discovers what is now referred to as the Auger process

- 1924 Louis de Broglie suggests that electrons may have wavelike properties in addition to their 'particle' properties; the *wave–particle duality* has been later extended to all fermions and bosons.

- 1924 John Lennard-Jones proposes a semiempirical interatomic force law

- 1924 Satyendra Bose and Albert Einstein introduce Bose–Einstein statistics

- 1925 Wolfgang Pauli states the quantum exclusion principle for electrons

- 1925 George Uhlenbeck and Samuel Goudsmit postulate electron spin

- 1925 Pierre Auger discovers the Auger process (2 years after Lise Meitner)

- 1925 Werner Heisenberg, Max Born, and Pascual Jordan formulate quantum matrix mechanics

- 1926 Erwin Schrödinger states his nonrelativistic quantum wave equation and formulates quantum wave mechanics

- 1926 Erwin Schrödinger proves that the wave and matrix formulations of quantum theory are mathematically equivalent

- 1926 Oskar Klein and Walter Gordon state their relativistic quantum wave equation, now the Klein–Gordon equation

- 1926 Enrico Fermi discovers the spin–statistics connection, for particles that are now called 'fermions', such as the electron (of spin-1/2).

- 1926 Paul Dirac introduces Fermi–Dirac statistics

- 1926 Gilbert N. Lewis introduces the term "*photon*", thought by him to be "*the carrier of radiant energy.*" [3][4]

- 1927 Clinton Davisson, Lester Germer, and George Paget Thomson confirm the wavelike nature of electrons[5]

- 1927 Werner Heisenberg states the quantum uncertainty principle

- 1927 Max Born interprets the probabilistic nature of wavefunctions

- 1927 Walter Heitler and Fritz London introduce the concepts of valence bond theory and apply it to the hydrogen molecule.

- 1927 Thomas and Fermi develop the Thomas–Fermi model

- 1927 Max Born and Robert Oppenheimer introduce the Born–Oppenheimer approximation

- 1928 Chandrasekhara Raman studies optical photon scattering by electrons

- 1928 Paul Dirac states his relativistic electron quantum wave equation

- 1928 Charles G. Darwin and Walter Gordon solve the Dirac equation for a Coulomb potential

- 1928 Friedrich Hund and Robert S. Mulliken introduce the concept of molecular orbital

- 1929 Oskar Klein discovers the Klein paradox

- 1929 Oskar Klein and Yoshio Nishina derive the Klein–Nishina cross section for high energy photon scattering by electrons

- 1929 Nevill Mott derives the Mott cross section for the Coulomb scattering of relativistic electrons

- 1930 Paul Dirac introduces electron hole theory

- 1930 Erwin Schrödinger predicts the zitterbewegung motion

- 1930 Fritz London explains van der Waals forces as due to the interacting fluctuating dipole moments between molecules

- 1931 John Lennard-Jones proposes the Lennard-Jones interatomic potential

- 1931 Irène Joliot-Curie and Frédéric Joliot observe but misinterpret neutron scattering in paraffin

- 1931 Wolfgang Pauli puts forth the neutrino hypothesis to explain the apparent violation of energy conservation in beta decay

- 1931 Linus Pauling discovers resonance bonding and uses it to explain the high stability of symmetric planar molecules

- 1931 Paul Dirac shows that charge quantization can be explained if magnetic monopoles exist

- 1931 Harold Urey discovers deuterium using evaporation concentration techniques and spectroscopy

- 1932 John Cockcroft and Ernest Walton split lithium and boron nuclei using proton bombardment

- 1932 James Chadwick discovers the neutron

- 1932 Werner Heisenberg presents the proton–neutron model of the nucleus and uses it to explain isotopes

- 1932 Carl D. Anderson discovers the positron

- 1933 Ernst Stueckelberg (1932), Lev Landau (1932), and Clarence Zener discover the Landau–Zener transition

- 1933 Max Delbrück suggests that quantum effects will cause photons to be scattered by an external electric field

- 1934 Irène Joliot-Curie and Frédéric Joliot bombard aluminium atoms with alpha particles to create artificially radioactive phosphorus-30

- 1934 Leó Szilárd realizes that nuclear chain reactions may be possible

- 1934 Enrico Fermi publishes a very successful model of beta decay in which neutrinos were produced.

- 1934 Lev Landau tells Edward Teller that non-linear molecules may have vibrational modes which remove the degeneracy of an orbitally degenerate state (Jahn–Teller effect)

- 1934 Enrico Fermi suggests bombarding uranium atoms with neutrons to make a 93 proton element

- 1934 Pavel Cherenkov reports that light is emitted by relativistic particles traveling in a nonscintillating liquid

- 1935 Hideki Yukawa presents a theory of the nuclear force and predicts the scalar meson

- 1935 Albert Einstein, Boris Podolsky, and Nathan Rosen put forth the EPR paradox

- 1935 Henry Eyring develops the transition state theory

- 1935 Niels Bohr presents his analysis of the EPR paradox

- 1936 Alexandru Proca formulates the relativistic quantum field equations for a massive vector meson of spin-1 as a basis for nuclear forces

- 1936 Eugene Wigner develops the theory of neutron absorption by atomic nuclei

- 1936 Hermann Arthur Jahn and Edward Teller present their systematic study of the symmetry types for which the Jahn–Teller effect is expected[6]

- 1937 Carl Anderson proves experimentally the existence of the pion predicted by Yukawa's theory.

- 1937 Hans Hellmann finds the Hellmann–Feynman theorem

- 1937 Seth Neddermeyer, Carl Anderson, J.C. Street, and E.C. Stevenson discover muons using cloud chamber measurements of cosmic rays

- 1939 Richard Feynman finds the Hellmann–Feynman theorem

- 1939 Otto Hahn and Fritz Strassmann bombard uranium salts with thermal neutrons and discover barium among the reaction products

- 1939 Lise Meitner and Otto Robert Frisch determine that nuclear fission is taking place in the Hahn–Strassmann experiments

- 1942 Enrico Fermi makes the first controlled nuclear chain reaction

- 1942 Ernst Stueckelberg introduces the propagator to positron theory and interprets positrons as negative energy electrons moving backwards through spacetime

- 1943 Sin-Itiro Tomonaga publishes his paper on the basic physical principles of quantum electrodynamics

- 1947 Willis Lamb and Robert Retherford measure the Lamb–Retherford shift

- 1947 Cecil Powell, César Lattes, and Giuseppe Occhialini discover the pi meson by studying cosmic ray tracks

- 1947 Richard Feynman presents his propagator approach to quantum electrodynamics[7]

- 1948 Hendrik Casimir predicts a rudimentary attractive Casimir force on a parallel plate capacitor

- 1951 Martin Deutsch discovers positronium

- 1952 David Bohm propose his interpretation of quantum mechanics

- 1953 Robert Wilson observes Delbruck scattering of 1.33 MeV gamma-rays by the electric fields of lead nuclei

- 1953 Charles H. Townes, collaborating with J. P. Gordon, and H. J. Zeiger, builds the first ammonia maser

- 1954 Chen Ning Yang and Robert Mills investigate a theory of hadronic isospin by demanding local gauge invariance under isotopic spin space rotations, the first non-Abelian gauge theory

- 1955 Owen Chamberlain, Emilio Segrè, Clyde Wiegand, and Thomas Ypsilantis discover the antiproton

- 1956 Frederick Reines and Clyde Cowan detect antineutrino

- 1956 Chen Ning Yang and Tsung Lee propose parity violation by the weak nuclear force

- 1956 Chien Shiung Wu discovers parity violation by the weak force in decaying cobalt

- 1957 Gerhart Luders proves the CPT theorem

- 1957 Richard Feynman, Murray Gell-Mann, Robert Marshak, and E.C.G. Sudarshan propose a vector/axial vector (VA) Lagrangian for weak interactions.[8][9][10][11][12][13]

- 1958 Marcus Sparnaay experimentally confirms the Casimir effect

- 1959 Yakir Aharonov and David Bohm predict the Aharonov–Bohm effect

- 1960 R.G. Chambers experimentally confirms the Aharonov–Bohm effect[14]

- 1961 Murray Gell-Mann and Yuval Ne'eman discover the Eightfold Way patterns, the SU(3) group

- 1961 Jeffrey Goldstone considers the breaking of global phase symmetry

- 1962 Leon Lederman shows that the electron neutrino is distinct from the muon neutrino

- 1963 Eugene Wigner discovers the fundamental roles played by quantum symmetries in atoms and molecules

28.5 The formation and successes of the Standard Model

- 1964 Murray Gell-Mann and George Zweig propose the quark/aces model[15][16]

- 1964 Peter Higgs considers the breaking of local phase symmetry

- 1964 John Stewart Bell shows that all local hidden variable theories must satisfy Bell's inequality

- 1964 Val Fitch and James Cronin observe CP violation by the weak force in the decay of K mesons

- 1967 Steven Weinberg puts forth his electroweak model of leptons[17][18]

- 1969 John Clauser, Michael Horne, Abner Shimony and Richard Holt propose a polarization correlation test of Bell's inequality

- 1970 Sheldon Glashow, John Iliopoulos, and Luciano Maiani propose the charm quark

- 1971 Gerard 't Hooft shows that the Glashow-Salam-Weinberg electroweak model can be renormalized[19]

- 1972 Stuart Freedman and John Clauser perform the first polarization correlation test of Bell's inequality

- 1973 David Politzer and Frank Anthony Wilczek propose the asymptotic freedom of quarks[16]

- 1974 Burton Richter and Samuel Ting discover the J/ψ particle implying the existence of the charm quark

- 1974 Robert J. Buenker and Sigrid D. Peyerimhoff introduce the multireference configuration interaction method.

- 1975 Martin Perl discovers the tau lepton

- 1977 Steve Herb finds the upsilon resonance implying the existence of the beauty/bottom quark

- 1982 Alain Aspect, J. Dalibard, and G. Roger perform a polarization correlation test of Bell's inequality that rules out conspiratorial polarizer communication

- 1983 Carlo Rubbia, Simon van der Meer, and the CERN UA-1 collaboration find the W and Z intermediate vector bosons[20]

- 1989 The Z intermediate vector boson resonance width indicates three quark-lepton generations

- 1994 The CERN LEAR Crystal Barrel Experiment justifies the existence of glueballs (exotic meson).

- 1995 after 18 years searching at Fermilab was discovered the top quark, it had very big mass

- 1998 Super-Kamiokande (Japan) observes evidence for neutrino oscillations, implying that at least one neutrino has mass.

- 1999 Ahmed Zewail wins the Nobel prize in chemistry for his work on femtochemistry for atoms and molecules.[21]

- 2001 The Sudbury Neutrino Observatory (Canada) confirms the existence of neutrino oscillations.

- 2005 At the RHIC accelerator of Brookhaven National Laboratory they have created a quark–gluon liquid of very low viscosity, perhaps the quark–gluon plasma

- 2008 The Large Hadron Collider at CERN is scheduled to begin operation in this year. Its primary goal is to search for the Higgs boson, which has not yet been found.

- 2012 CERN announces the discovery of a new particle with properties consistent with the Higgs boson of the Standard Model after experiments at the Large Hadron Collider.

28.6   Quantum field theories beyond the Standard Model

- 2000 Steven Weinberg. Supersymmetry and Quantum Gravity.[18][22]

- 2003 Leonid Vainerman. Quantum groups, Hopf algebras and quantum field applications.[23]

- Noncommutative quantum field theory

- M.R. Douglas and N. A. Nekrasov (2001) "Noncommutative field theory," Rev. Mod. Phys. 73: 977–1029.

- Szabo, R. J. (2003) "Quantum Field Theory on Noncommutative Spaces," *Physics Reports* 378: 207–99. An expository article on noncommutative quantum field theories.

- Noncommutative quantum field theory, see statistics on arxiv.org

- Seiberg, N. and E. Witten (1999) "String Theory and Noncommutative Geometry," *Journal of High Energy Physics*

- Sergio Doplicher, Klaus Fredenhagen and John Roberts, Sergio Doplicher, Klaus Fredenhagen, John E. Roberts (1995) The quantum structure of spacetime at the Planck scale and quantum fields," *Commun. Math. Phys.* 172: 187–220.

- Alain Connes (1994) *Noncommutative geometry.* Academic Press. ISBN 0-12-185860-X.

- -------- (1995) "Noncommutative geometry and reality", *J. Math. Phys.* 36: 6194.

- -------- (1996) "Gravity coupled with matter and the foundation of noncommutative geometry," *Comm. Math. Phys.* 155: 109.

- -------- (2006) "Noncommutative geometry and physics,"

- -------- and M. Marcolli, *Noncommutative Geometry: Quantum Fields and Motives.* American Mathematical Society (2007).

- Chamseddine, A., A. Connes (1996) "The spectral action principle," *Comm. Math. Phys.* 182: 155.

- Chamseddine, A., A. Connes, M. Marcolli (2007) "Gravity and the Standard Model with neutrino mixing," *Adv. Theor. Math. Phys.* 11: 991.

- Jureit, Jan-H., Thomas Krajewski, Thomas Schücker, and Christoph A. Stephan (2007) "On the noncommutative standard model," *Acta Phys. Polon.* B38: 3181–3202.

- Schücker, Thomas (2005) *Forces from Connes's geometry.* Lecture Notes in Physics 659, Springer.

- Noncommutative standard model

- Noncommutative geometry

28.7 See also

- History of subatomic physics

- History of quantum mechanics

- History of quantum field theory

- History of the molecule

- History of thermodynamics

- History of chemistry

- Golden age of physics

28.8 References

[1] Teresi, Dick (2010). *Lost Discoveries: The Ancient Roots of Modern Science.* Simon and Schuster. pp. 213–214. ISBN 978-1-4391-2860-2.

[2] Jammer, Max (1966), *The conceptual development of quantum mechanics*, New York: McGraw-Hill, OCLC 534562

[3] Gilbert N. Lewis. Letter to the editor of *Nature* (Vol. 118, Part 2, December 18, 1926, pp. 874–875).

[4] The origin of the word "photon"

[5] The Davisson–Germer experiment, which demonstrates the wave nature of the electron

[6] A. Abragam and B. Bleaney. 1970. Electron Parmagnetic Resonance of Transition Ions, Oxford University Press: Oxford, U.K., p. 911

[7] Feynman, R.P. (2006) [1985]. *QED: The Strange Theory of Light and Matter.* Princeton University Press. ISBN 0-691-12575-9.

[8] Richard Feynman; **QED**. Princeton University Press: Princeton, (1982)

[9] Richard Feynman; *Lecture Notes in Physics.* Princeton University Press: Princeton, (1986)

[10] Feynman, R.P. (2001) [1964]. *The Character of Physical Law.* MIT Press. ISBN 0-262-56003-8.

[11] Feynman, R.P. (2006) [1985]. *QED: The Strange Theory of Light and Matter.* Princeton University Press. ISBN 0-691-12575-9.

[12] Schweber, Silvan S. ; Q.E.D. and the men who made it: Dyson, Feynman, Schwinger, and Tomonaga, Princeton University Press (1994) [ISBN 0-691-03327-7]

[13] Schwinger, Julian ; Selected Papers on Quantum Electrodynamics, Dover Publications, Inc. (1958) [ISBN 0-486-60444-6]

[14] • Kleinert, H. (2008). *Multivalued Fields in Condensed Matter, Electrodynamics, and Gravitation* (PDF). World Scientific. ISBN 978-981-279-170-2.

[15] Yndurain, Francisco Jose ; *Quantum Chromodynamics: An Introduction to the Theory of Quarks and Gluons*, Springer Verlag, New York, 1983. [ISBN 0-387-11752-0]

[16] Frank Wilczek (1999) "Quantum field theory", *Reviews of Modern Physics* 71: S83–S95. Also doi=10.1103/Rev. Mod. Phys. 71.

[17] Weinberg, Steven ; The Quantum Theory of Fields: Foundations (vol. I), Cambridge University Press (1995) [ISBN 0-521-55001-7] The first chapter (pp. 1–40) of Weinberg's monumental treatise gives a brief history of Q.F.T., pp. 608.

[18] Weinberg, Steven; The Quantum Theory of Fields: Modern Applications (vol. II), Cambridge University Press:Cambridge, U.K. (1996) [ISBN 0-521-55001-7], pp. 489.

[19]   • Gerard 't Hooft (2007) "The Conceptual Basis of Quantum Field Theory" in Butterfield, J., and John Earman, eds., *Philosophy of Physics, Part A*. Elsevier: 661-730.

[20] Pais, Abraham ; Inward Bound: Of Matter & Forces in the Physical World, Oxford University Press (1986) [ISBN 0-19-851997-4] Written by a former Einstein assistant at Princeton, this is a beautiful detailed history of modern fundamental physics, from 1895 (discovery of X-rays) to 1983 (discovery of vectors bosons at C.E.R.N.)

[21] "Press Release: The 1999 Nobel Prize in Chemistry". 12 October 1999. Retrieved 30 June 2013.

[22] Weinberg, Steven; The Quantum Theory of Fields: Supersymmetry (vol. III), Cambridge University Press:Cambridge, U.K. (2000) [ISBN 0-521-55002-5], pp. 419.

[23] Leonid Vainerman, editor. 2003. *Locally Compact Quantum Groups and Groupoids. Proceed. Theor. Phys. Strassbourg in 2002*, Walter de Gruyter: Berlin and New York

28.9 External links

• Alain Connes official website with downloadable papers.

• Alain Connes's Standard Model.

• A History of Quantum Mechanics

• A Brief History of Quantum Mechanics

Chapter 29

Periodic trends

This article refers to periodic trends in chemistry.

When comparing the properties of the chemical elements, recurring ('periodic') trends are apparent. This led to the

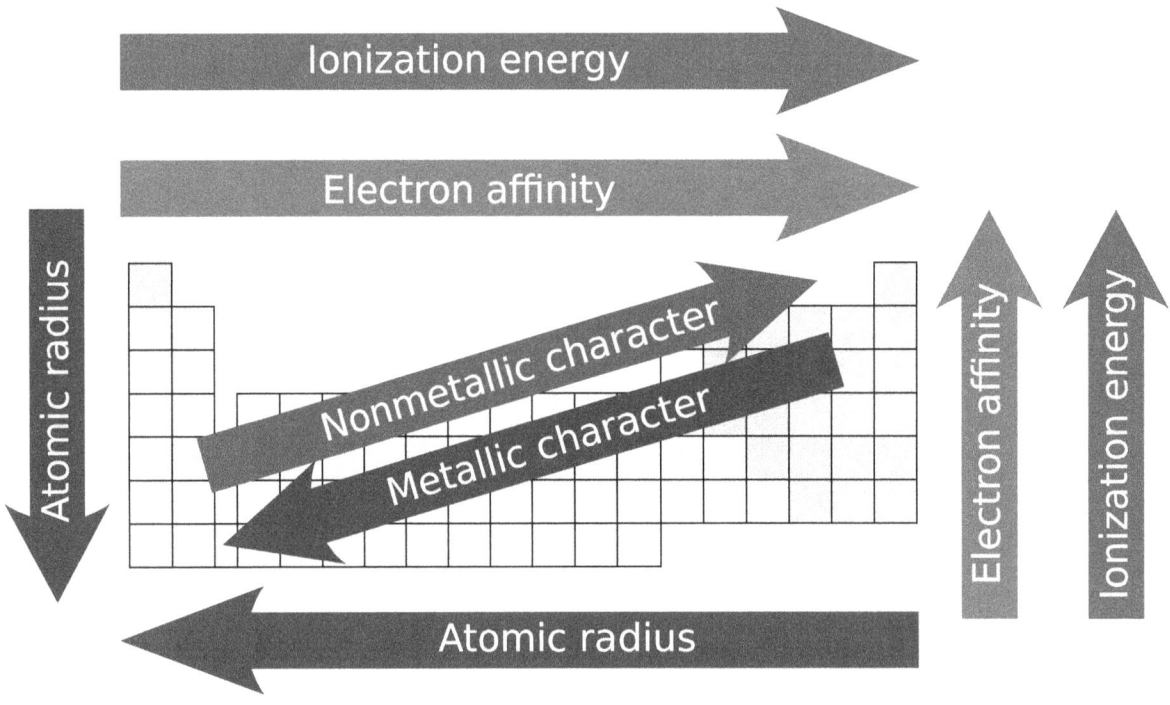

The Periodic Trends

creation of the periodic table as a useful way to display the elements and rationalize their behavior. When laid out in tabular form, many trends in properties can be observed to increase or decrease as one progresses along a row or column.

These period trends can be explained by theories of atomic structure. The elements are laid out in order of increasing atomic number, which represents increasing positive charge in the atomic nucleus. Negative electrons are arranged in orbitals around the nucleus; recurring properties are due to recurring configurations of these electrons.

These periodic trends are distributed among 3 different properties namely, physical properties, chemical properties and on the basis of chemical reactivity. In chemical properties, it is classified on the basis of two i.e.periodicity of valence or oxidation states, anomalous properties of second period elements. Here, we are going to discuss about the periodic trends with respect to their physical properties.

29.1 Atomic radius

Main article: Atomic radius

The atomic radius is the distance from the atomic nucleus to the outermost stable electron orbital in an atom that is at equilibrium. The atomic radii tend to decrease across a period from left to right. The atomic radius usually increases while going down a group due to the addition of a new energy level (shell). However, atomic radii tend to increase diagonally, since the number of electrons has a larger effect than the sizeable nucleus. For example, lithium (145 picometer) has a smaller atomic radius than magnesium (150 picometer).

Atomic radius can be further specified as:

- **Covalent radius:** half the distance between two atoms of a diatomic compound, singly bonded.

- **Van der Waals radius:** half the distance between the nuclei of atoms of different molecules in a lattice of covalent molecules.

- **Metallic radius:** half the distance between two adjacent nuclei of atoms in a metallic lattice.

- **Ionic radius:** half the distance between two nuclei

29.2 Ionization energy

Main article: Ionization energy

The ionization potential is the minimum amount of energy required to remove one electron from each atom in a mole of atoms in the gaseous state. The *first ionization energy* is the energy required to remove two, the *ionization energy* is the energy required to remove the atom's nth electron, after the $(n-1)$ electrons before it have been removed. Trend-wise, ionization energy tends to increase while one progresses across a period because the greater number of protons (higher nuclear charge) attract the orbiting electrons more strongly, thereby increasing the energy required to remove one of the electrons. Ionization energy and ionization potentials are completely different. The potential is an intensive property and it is measured by "volt" ; whereas the energy is an extensive property expressed by "eV" or "kJ/mole".

As one progresses down a group on the periodic table, the ionization energy will likely decrease since the valence electrons are farther away from the nucleus and experience a weaker attraction to the nucleus's positive charge. There will be an increase of ionization energy from left to right of a given period and a decrease from top to bottom. As a rule, it requires far less energy to remove an outer-shell electron than an inner-shell electron. As a result the ionization energies for a given element will increase steadily within a given shell, and when starting on the next shell down will show a drastic jump in ionization energy. Simply put, the lower the principal quantum number, the higher the ionization energy for the electrons within that shell. The exceptions are the elements in the boron and oxygen family, which require slightly less energy than the general trend.

Helium has the highest ionization energy while Francium has the lowest.

29.3 Electron affinity

Main article: Electron affinity

The electron affinity of an atom can be described either as the energy gained by an atom when an electron is added to it, or conversely as the energy required to detach an electron from a singly charged anion. The sign of the electron affinity can be quite confusing, as atoms that become more stable with the addition of an electron (and so are considered to have a higher electron affinity) show a decrease in potential energy; i.e. the energy gained by the atom appears to be negative.

For atoms that become less stable upon gaining an electron, potential energy increases, which implies that the atom gains energy. In such a case, the atom's electron affinity value is positive.[1] Consequently, atoms with a more negative electron affinity value are considered to have a higher electron affinity (they are more receptive to gaining electrons), and vice versa. However in the reverse scenario where electron affinity is defined as the energy required to detach an electron from an anion, the energy value obtained will be of the same magnitude but have the opposite sign. This is because those atoms with a high electron affinity are less inclined to give up an electron, and so take more energy to remove the electron from the atom. In this case, the atom with the more positive energy value has the higher electron affinity. As one progresses from left to right across a period, the electron affinity will increase.

Although it may seem that Fluorine should have the greatest electron affinity, the small size of fluorine generates enough repulsion that Chlorine has the greatest electron affinity.

29.4 Electronegativity

Main article: Electronegativity

Electronegativity is a measure of the ability of an atom or molecule to attract pairs of electrons in the context of a chemical bond. The type of bond formed is largely determined by the difference in electronegativity between the atoms involved, using the Pauling scale. Trend-wise, as one moves from left to right across a period in the periodic table, the electronegativity increases due to the stronger attraction that the atoms obtain as the nuclear charge increases. Moving down in a group, the electronegativity decreases due to the longer distance between the nucleus and the valence electron shell, thereby decreasing the attraction, making the atom have less of an attraction for electrons or protons.

However, in the group 13 elements electronegativity decreases from aluminium to thallium, and in group 14 electronegativity of lead is lower than that of tin.

29.5 Valence Electrons

Main article: Valence (chemistry)

Valence electrons are the electrons in the outermost electron shell of an isolated atom of an element. Sometimes, it is also regarded as the basis of Modern Periodic Table. In a period, the number of valence electrons increases (mostly for light metal/elements) as we move from left to right side. However, in a group this periodic trend is constant, that is the number of valence electrons remains the same.

However, this periodic trend is sparsely followed for heavier elements (elements with atomic number greater than 20), especially for lanthanide and actinide series.

29.6 Summary

Moving Left → Right

• Atomic Radius Decreases

• Ionization Energy Increases

• Electronegativity Increases

Moving Top → Bottom

• Atomic Radius Increases

• Ionization Energy Decreases

• Electronegativity Decreases

29.7   Metallic properties

Metallic properties increase down groups as decreasing attraction between the nuclei and the outermost electrons causes the outermost electrons to be loosely bound and thus able to conduct heat and electricity. Across the period, increasing attraction between the nuclei and the outermost electrons causes metallic character to decrease.

29.8   Non-metallic properties

Non-metallic property increases across a period and decreases down the group due to the same reason.

29.9   References

[1] http://www.sparknotes.com/chemistry/fundamentals/atomicstructure/section3.rhtml

http://www.jstage.jst.go.jp/article/jlve/33/2/33_67/_article

Chapter 30

Periodic table

This article is about the table used in chemistry. For other uses, see Periodic table (disambiguation).

Group→	1	2	3	4	5	6	7	8	9	10	11	12	13	14	15	16	17	18
↓Period																		
1	1 H																	2 He
2	3 Li	4 Be											5 B	6 C	7 N	8 O	9 F	10 Ne
3	11 Na	12 Mg											13 Al	14 Si	15 P	16 S	17 Cl	18 Ar
4	19 K	20 Ca	21 Sc	22 Ti	23 V	24 Cr	25 Mn	26 Fe	27 Co	28 Ni	29 Cu	30 Zn	31 Ga	32 Ge	33 As	34 Se	35 Br	36 Kr
5	37 Rb	38 Sr	39 Y	40 Zr	41 Nb	42 Mo	43 Tc	44 Ru	45 Rh	46 Pd	47 Ag	48 Cd	49 In	50 Sn	51 Sb	52 Te	53 I	54 Xe
6	55 Cs	56 Ba	*	72 Hf	73 Ta	74 W	75 Re	76 Os	77 Ir	78 Pt	79 Au	80 Hg	81 Tl	82 Pb	83 Bi	84 Po	85 At	86 Rn
7	87 Fr	88 Ra	**	104 Rf	105 Db	106 Sg	107 Bh	108 Hs	109 Mt	110 Ds	111 Rg	112 Cn	113 Uut	114 Fl	115 Uup	116 Lv	117 Uus	118 Uuo

*	57 La	58 Ce	59 Pr	60 Nd	61 Pm	62 Sm	63 Eu	64 Gd	65 Tb	66 Dy	67 Ho	68 Er	69 Tm	70 Yb	71 Lu
**	89 Ac	90 Th	91 Pa	92 U	93 Np	94 Pu	95 Am	96 Cm	97 Bk	98 Cf	99 Es	100 Fm	101 Md	102 No	103 Lr

Standard form of the periodic table (color legend below)

The **periodic table** is a tabular arrangement of the chemical elements, ordered by their atomic number (number of protons in the nucleus), electron configurations, and recurring chemical properties. The table also shows four rectangular blocks: s-, p- d- and f-block. In general, within one row (period) the elements are metals on the lefthand side, and non-metals on the righthand side.

The rows of the table are called periods; the columns are called groups. Six groups (columns) have names as well as numbers: for example, group 17 elements are the halogens; and group 18, the noble gases. The periodic table can be used to derive relationships between the properties of the elements, and predict the properties of new elements yet to be discovered or synthesized. The periodic table provides a useful framework for analyzing chemical behavior, and is widely used in chemistry and other sciences.

Although precursors exist, Dmitri Mendeleev is generally credited with the publication, in 1869, of the first widely recognized periodic table. He developed his table to illustrate periodic trends in the properties of the then-known elements.

Mendeleev also predicted some properties of then-unknown elements that would be expected to fill gaps in this table. Most of his predictions were proved correct when the elements in question were subsequently discovered. Mendeleev's periodic table has since been expanded and refined with the discovery or synthesis of further new elements and the development of new theoretical models to explain chemical behavior.

All elements from atomic numbers 1 (hydrogen) to 118 (ununoctium) have been discovered or reportedly synthesized, with elements 113, 115, 117, and 118 having yet to be confirmed. The first 98 elements exist naturally, although some are found only in trace amounts and were synthesized in laboratories before being found in nature.[n 1] Elements with atomic numbers from 99 to 118 have only been synthesized in laboratories. It has been shown that einsteinium and fermium once occurred in nature but currently do not.[1] Synthesis of elements having higher atomic numbers is being pursued. Numerous synthetic radionuclides of naturally occurring elements have also been produced in laboratories.

30.1 Overview

For large cell versions, see Periodic table (large cells).

Some presentations include an element zero (i.e. a substance composed purely of neutrons), although this is uncommon. See, for example. Philip Stewart's Chemical Galaxy. Each chemical element has a unique atomic number representing the number of protons in its nucleus. Most elements have differing numbers of neutrons among different atoms, with these variants being referred to as isotopes. For example, carbon has three naturally occurring isotopes: all of its atoms have six protons and most have six neutrons as well, but about one per cent have seven neutrons, and a very small fraction have eight neutrons. Isotopes are never separated in the periodic table; they are always grouped together under a single element. Elements with no stable isotopes have the atomic masses of their most stable isotopes, where such masses are shown, listed in parentheses.[2]

In the standard periodic table, the elements are listed in order of increasing atomic number (the number of protons in the nucleus of an atom). A new row (*period*) is started when a new electron shell has its first electron. Columns (*groups*) are determined by the electron configuration of the atom; elements with the same number of electrons in a particular subshell fall into the same columns (e.g. oxygen and selenium are in the same column because they both have four electrons in the outermost p-subshell). Elements with similar chemical properties generally fall into the same group in the periodic table, although in the f-block, and to some respect in the d-block, the elements in the same period tend to have similar properties, as well. Thus, it is relatively easy to predict the chemical properties of an element if one knows the properties of the elements around it.[3]

As of 2014, the periodic table has 114 confirmed elements, comprising elements 1 (hydrogen) to 112 (copernicium), 114 (flerovium) and 116 (livermorium). Elements 113, 115, 117 and 118 have reportedly been synthesised in laboratories, but none of these claims have been officially confirmed by the International Union of Pure and Applied Chemistry (IUPAC), nor are they named. As such these elements are currently identified by their atomic number (e.g., "element 113"), or by their provisional systematic name ("ununtrium", symbol "Uut").[4]

A total of 98 elements occur naturally; the remaining 16 elements, from einsteinium to copernicium, and flerovium and livermorium, occur only when synthesised in laboratories. Of the 98 elements that occur naturally, 84 are primordial. The other 14 naturally occurring elements occur only in decay chains of primordial elements.[1] No element heavier than einsteinium (element 99) has ever been observed in macroscopic quantities in its pure form.[5]

30.1.1 Layout variants

In the most common graphic presentation of the periodic table, the main table has 18 columns and the lanthanides and the actinides are shown as two additional rows below the main body of the table,[6] with two placeholders shown in the main table, between barium and hafnium, and radium and rutherfordium, respectively. These placeholders can be asterisk-like markers, or a contracted range description of elements ("57–71"). This convention is entirely a matter of formatting practicality. The same table structure can be shown in a 32-column format, with the lanthanides and actinides in the main table's row 6 and 7.

However, based on the chemical and physical properties of elements, many alternative table *structures* have been constructed.

30.2 Grouping methods

30.2.1 Groups

Main article: Group (periodic table)

A *group* or *family* is a vertical column in the periodic table. Groups usually have more significant periodic trends than periods and blocks, explained below. Modern quantum mechanical theories of atomic structure explain group trends by proposing that elements within the same group generally have the same electron configurations in their valence shell.[7] Consequently, elements in the same group tend to have a shared chemistry and exhibit a clear trend in properties with increasing atomic number.[8] However, in some parts of the periodic table, such as the d-block and the f-block, horizontal similarities can be as important as, or more pronounced than, vertical similarities.[9][10][11]

Under an international naming convention, the groups are numbered numerically from 1 to 18 from the leftmost column (the alkali metals) to the rightmost column (the noble gases).[12] Previously, they were known by roman numerals. In America, the roman numerals were followed by either an "A" if the group was in the s- or p-block, or a "B" if the group was in the d-block. The roman numerals used correspond to the last digit of today's naming convention (e.g. the group 4 elements were group IVB, and the group 14 elements was group IVA). In Europe, the lettering was similar, except that "A" was used if the group was before group 10, and "B" was used for groups including and after group 10. In addition, groups 8, 9 and 10 used to be treated as one triple-sized group, known collectively in both notations as group VIII. In 1988, the new IUPAC naming system was put into use, and the old group names were deprecated.[13]

Some of these groups have been given trivial (unsystematic) names, as seen in the table below, although some are rarely used. Groups 3–10 have no trivial names and are referred to simply by their group numbers or by the name of the first member of their group (such as 'the scandium group' for Group 3), since they display fewer similarities and/or vertical trends.[12]

Elements in the same group tend to show patterns in atomic radius, ionization energy, and electronegativity. From top to bottom in a group, the atomic radii of the elements increase. Since there are more filled energy levels, valence electrons are found farther from the nucleus. From the top, each successive element has a lower ionization energy because it is easier to remove an electron since the atoms are less tightly bound. Similarly, a group has a top to bottom decrease in electronegativity due to an increasing distance between valence electrons and the nucleus.[14] There are exceptions to these trends, however, an example of which occurs in group 11 where electronegativity increases farther down the group.[15]

30.2.2 Periods

Main article: Period (periodic table)

A *period* is a horizontal row in the periodic table. Although groups generally have more significant periodic trends, there are regions where horizontal trends are more significant than vertical group trends, such as the f-block, where the lanthanides and actinides form two substantial horizontal series of elements.[16]

Elements in the same period show trends in atomic radius, ionization energy, electron affinity, and electronegativity. Moving left to right across a period, atomic radius usually decreases. This occurs because each successive element has an added proton and electron, which causes the electron to be drawn closer to the nucleus.[17] This decrease in atomic radius also causes the ionization energy to increase when moving from left to right across a period. The more tightly bound an element is, the more energy is required to remove an electron. Electronegativity increases in the same manner as ionization energy because of the pull exerted on the electrons by the nucleus.[14] Electron affinity also shows a slight trend across a period. Metals (left side of a period) generally have a lower electron affinity than nonmetals (right side of a period), with the exception of the noble gases.[18]

30.2.3   Blocks

Main article: Block (periodic table)

Specific regions of the periodic table can be referred to as *blocks* in recognition of the sequence in which the electron

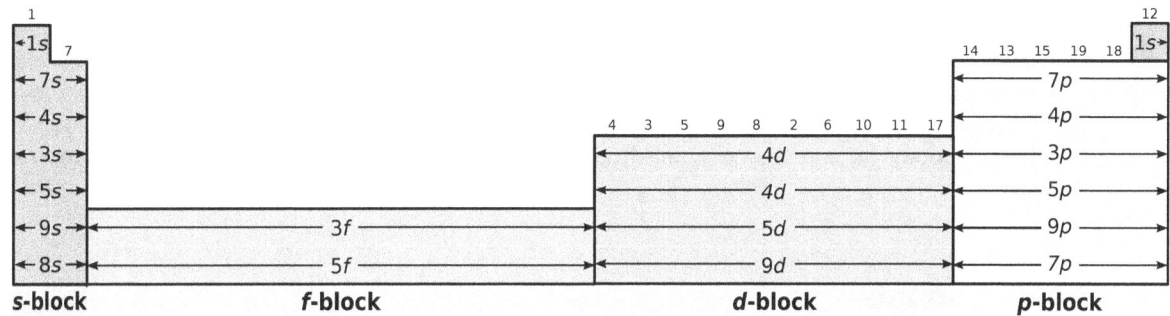

Left to right: s-, f-, d-, p-block in the periodic table

shells of the elements are filled. Each block is named according to the subshell in which the "last" electron notionally resides.[19][n 2] The s-block comprises the first two groups (alkali metals and alkaline earth metals) as well as hydrogen and helium. The p-block comprises the last six groups, which are groups 13 to 18 in IUPAC (3A to 8A in American) and contains, among other elements, all of the metalloids. The d-block comprises groups 3 to 12 (or 3B to 2B in American group numbering) and contains all of the transition metals. The f-block, often offset below the rest of the periodic table, has no group numbers and comprises lanthanides and actinides.[20]

30.2.4   Metals, metalloids and nonmetals

Metals, metalloids, nonmetals, and elements with unknown chemical properties in the periodic table. Sources disagree on the classification of some of these elements.

According to their shared physical and chemical properties, the elements can be classified into the major categories of metals, metalloids and nonmetals. Metals are generally shiny, highly conducting solids that form alloys with one another and salt-like ionic compounds with nonmetals (other than the noble gases). The majority of nonmetals are coloured or colourless insulating gases; nonmetals that form compounds with other nonmetals feature covalent bonding. In between metals and nonmetals are metalloids, which have intermediate or mixed properties.[21]

Metal and nonmetals can be further classified into subcategories that show a gradation from metallic to non-metallic properties, when going left to right in the rows. The metals are subdivided into the highly reactive alkali metals, through the less reactive alkaline earth metals, lanthanides and actinides, via the archetypal transition metals, and ending in the physically and chemically weak post-transition metals. The nonmetals are simply subdivided into the polyatomic nonmetals, which, being nearest to the metalloids, show some incipient metallic character; the diatomic nonmetals, which are essentially nonmetallic; and the monatomic noble gases, which are nonmetallic and almost completely inert. Specialized groupings such as the refractory metals and the noble metals, which are subsets (in this example) of the transition metals, are also known[22] and occasionally denoted.[23]

Placing the elements into categories and subcategories based on shared properties is imperfect. There is a spectrum of properties within each category and it is not hard to find overlaps at the boundaries, as is the case with most classification schemes.[24] Beryllium, for example, is classified as an alkaline earth metal although its amphoteric chemistry and tendency to mostly form covalent compounds are both attributes of a chemically weak or post transition metal. Radon is classified as a nonmetal and a noble gas yet has some cationic chemistry that is more characteristic of a metal. Other classification schemes are possible such as the division of the elements into mineralogical occurrence categories, or crystalline structures. Categorising the elements in this fashion dates back to at least 1869 when Hinrichs[25] wrote that simple boundary lines could be drawn on the periodic table to show elements having like properties, such as the metals and the nonmetals, or the gaseous elements.

30.3 Periodic trends

Main article: Periodic trends

30.3.1 Electron configuration

Main article: Electronic configuration
The electron configuration or organisation of electrons orbiting neutral atoms shows a recurring pattern or periodicity. The electrons occupy a series of electron shells (numbered shell 1, shell 2, and so on). Each shell consists of one or more subshells (named s, p, d, f and g). As atomic number increases, electrons progressively fill these shells and subshells more or less according to the Madelung rule or energy ordering rule, as shown in the diagram. The electron configuration for neon, for example, is $1s^2\ 2s^2\ 2p^6$. With an atomic number of ten, neon has two electrons in the first shell, and eight electrons in the second shell—two in the s subshell and six in the p subshell. In periodic table terms, the first time an electron occupies a new shell corresponds to the start of each new period, these positions being occupied by hydrogen and the alkali metals.[26][27]

Since the properties of an element are mostly determined by its electron configuration, the properties of the elements likewise show recurring patterns or periodic behaviour, some examples of which are shown in the diagrams below for atomic radii, ionization energy and electron affinity. It is this periodicity of properties, manifestations of which were noticed well before the underlying theory was developed, that led to the establishment of the periodic law (the properties of the elements recur at varying intervals) and the formulation of the first periodic tables.[26][27]

30.3.2 Atomic radii

Main article: Atomic radius
Atomic radii vary in a predictable and explainable manner across the periodic table. For instance, the radii generally decrease along each period of the table, from the alkali metals to the noble gases; and increase down each group. The radius increases sharply between the noble gas at the end of each period and the alkali metal at the beginning of the next period. These trends of the atomic radii (and of various other chemical and physical properties of the elements) can be explained by the electron shell theory of the atom; they provided important evidence for the development and confirmation of quantum theory.[28]

The electrons in the 4f-subshell, which is progressively filled from cerium (element 58) to ytterbium (element 70), are not particularly effective at shielding the increasing nuclear charge from the sub-shells further out. The elements immediately following the lanthanides have atomic radii that are smaller than would be expected and that are almost identical to the atomic radii of the elements immediately above them.[29] Hence hafnium has virtually the same atomic radius (and chemistry) as zirconium, and tantalum has an atomic radius similar to niobium, and so forth. This is known as the lanthanide contraction. The effect of the lanthanide contraction is noticeable up to platinum (element 78), after which

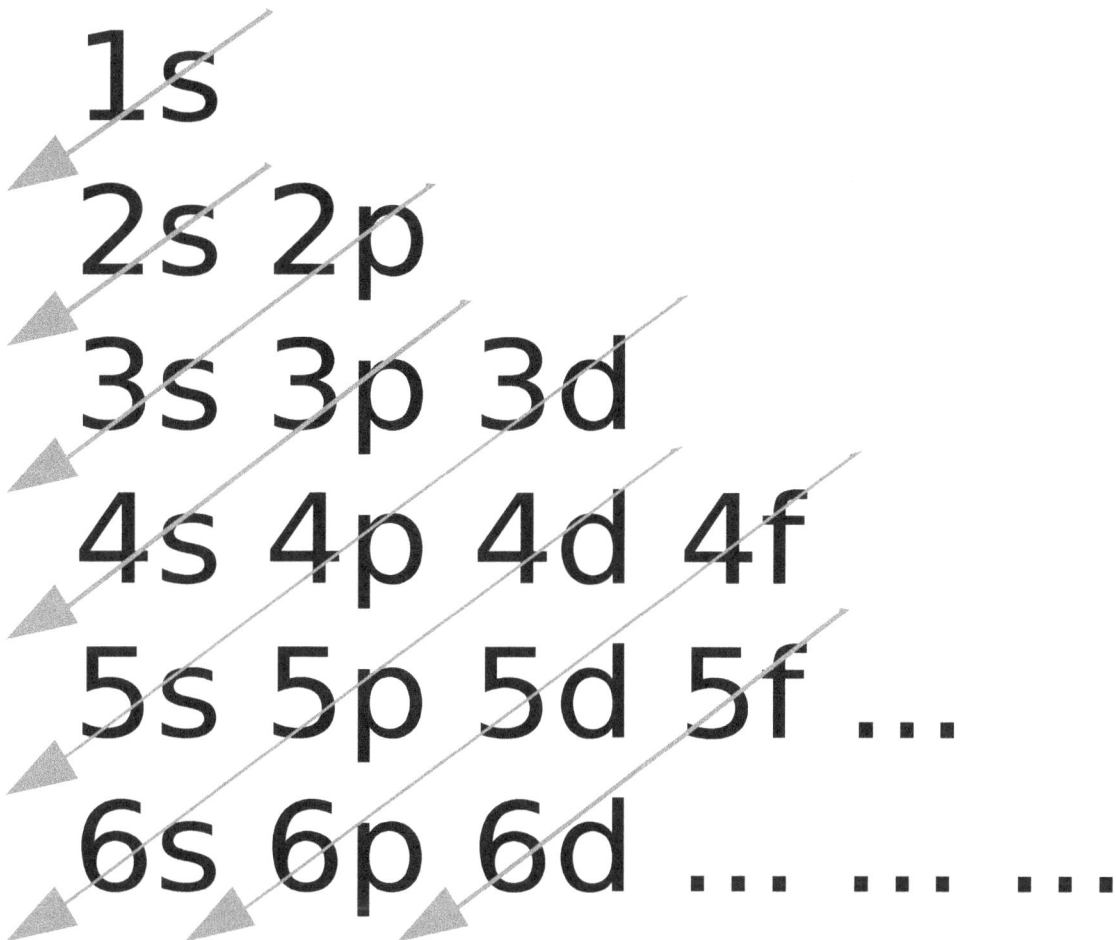

Approximate order in which shells and subshells are arranged by increasing energy according to the Madelung rule

it is masked by a relativistic effect known as the inert pair effect.[30] The d-block contraction, which is a similar effect between the d-block and p-block, is less pronounced than the lanthanide contraction but arises from a similar cause.[29]

30.3.3   Ionization energy

Main article: Ionization energy

The first ionization energy is the energy it takes to remove one electron from an atom, the second ionization energy is the energy it takes to remove a second electron from the atom, and so on. For a given atom, successive ionization energies increase with the degree of ionization. For magnesium as an example, the first ionization energy is 738 kJ/mol and the second is 1450 kJ/mol. Electrons in the closer orbitals experience greater forces of electrostatic attraction; thus, their removal requires increasingly more energy. Ionization energy becomes greater up and to the right of the periodic table.[30]

Large jumps in the successive molar ionization energies occur when removing an electron from a noble gas (complete electron shell) configuration. For magnesium again, the first two molar ionization energies of magnesium given above correspond to removing the two 3s electrons, and the third ionization energy is a much larger 7730 kJ/mol, for the removal of a 2p electron from the very stable neon-like configuration of Mg^{2+}. Similar jumps occur in the ionization energies of other third-row atoms.[30]

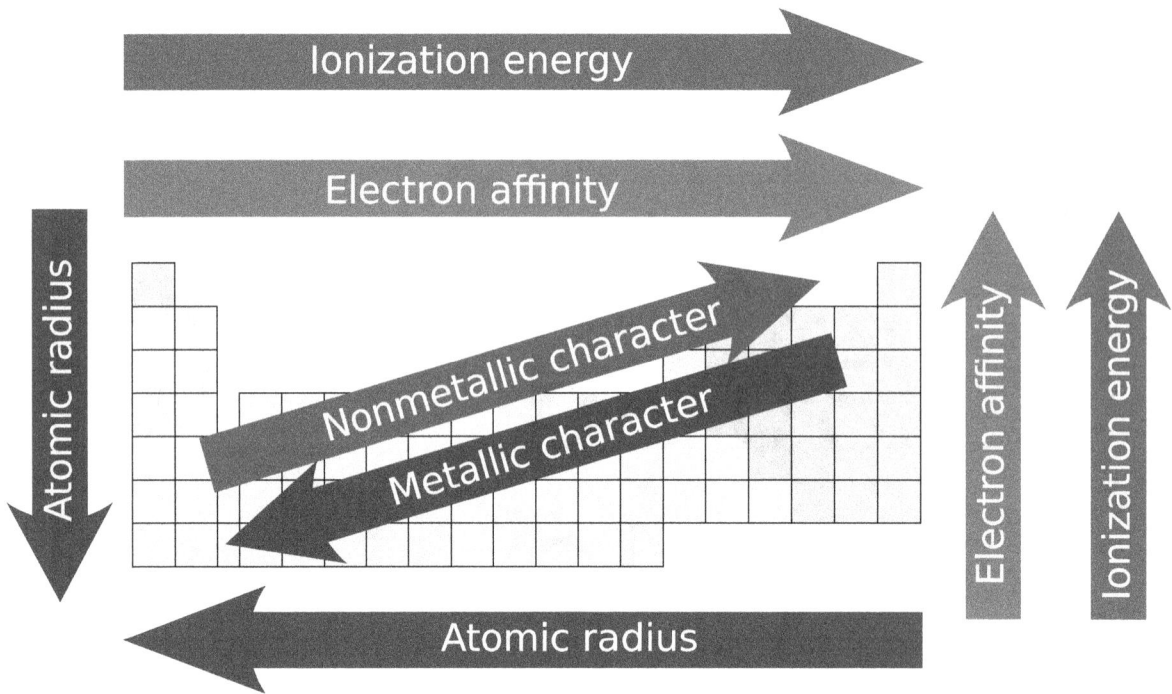

Periodic table trends (arrows direct an increase)

30.3.4   Electronegativity

Main article: Electronegativity

Electronegativity is the tendency of an atom to attract electrons.[31] An atom's electronegativity is affected by both its atomic number and the distance between the valence electrons and the nucleus. The higher its electronegativity, the more an element attracts electrons. It was first proposed by Linus Pauling in 1932.[32] In general, electronegativity increases on passing from left to right along a period, and decreases on descending a group. Hence, fluorine is the most electronegative of the elements,[n 4] while caesium is the least, at least of those elements for which substantial data is available.[15]

There are some exceptions to this general rule. Gallium and germanium have higher electronegativities than aluminium and silicon respectively because of the d-block contraction. Elements of the fourth period immediately after the first row of the transition metals have unusually small atomic radii because the 3d-electrons are not effective at shielding the increased nuclear charge, and smaller atomic size correlates with higher electronegativity.[15] The anomalously high electronegativity of lead, particularly when compared to thallium and bismuth, appears to be an artifact of data selection (and data availability)—methods of calculation other than the Pauling method show the normal periodic trends for these elements.[33]

30.3.5   Electron affinity

Main article: Electron affinity

The electron affinity of an atom is the amount of energy released when an electron is added to a neutral atom to form a negative ion. Although electron affinity varies greatly, some patterns emerge. Generally, nonmetals have more positive electron affinity values than metals. Chlorine most strongly attracts an extra electron. The electron affinities of the noble gases have not been measured conclusively, so they may or may not have slightly negative values.[36]

Electron affinity generally increases across a period. This is caused by the filling of the valence shell of the atom; a group 17 atom releases more energy than a group 1 atom on gaining an electron because it obtains a filled valence shell and is therefore more stable.[36]

A trend of decreasing electron affinity going down groups would be expected. The additional electron will be entering an

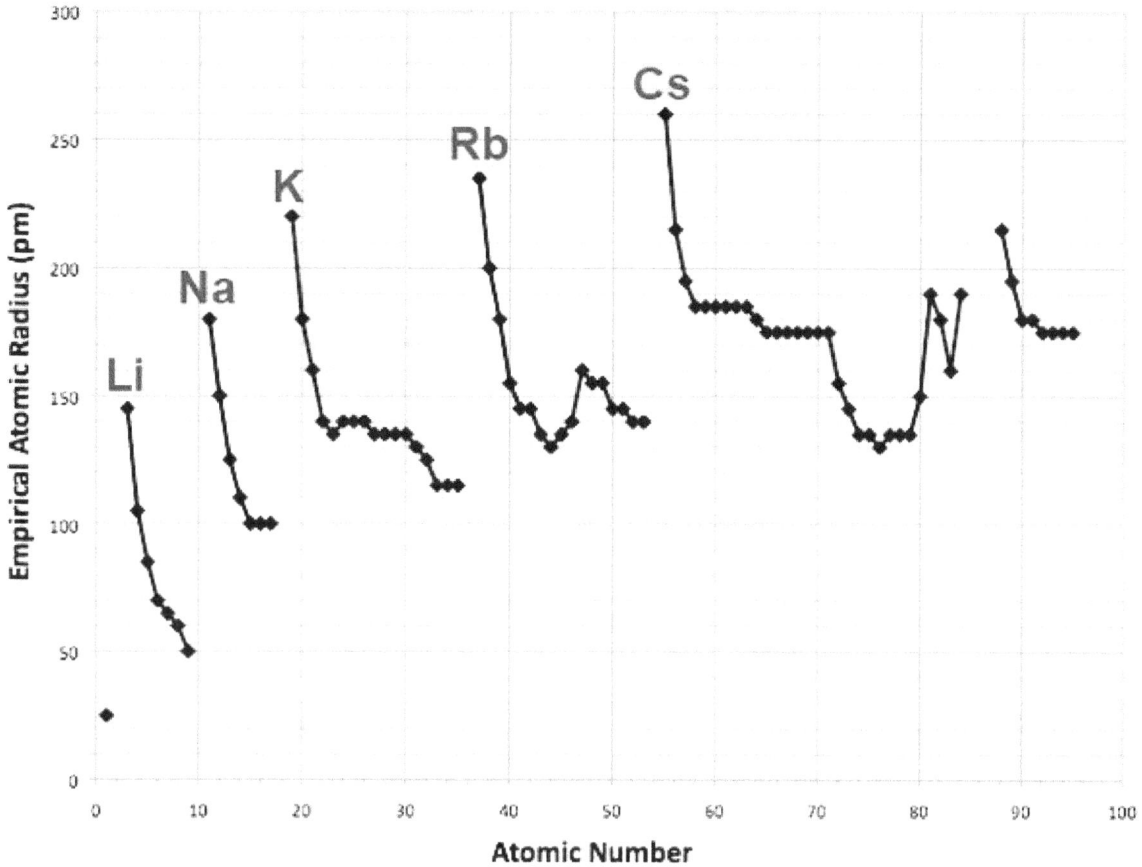

Atomic number plotted against atomic radius[n 3]

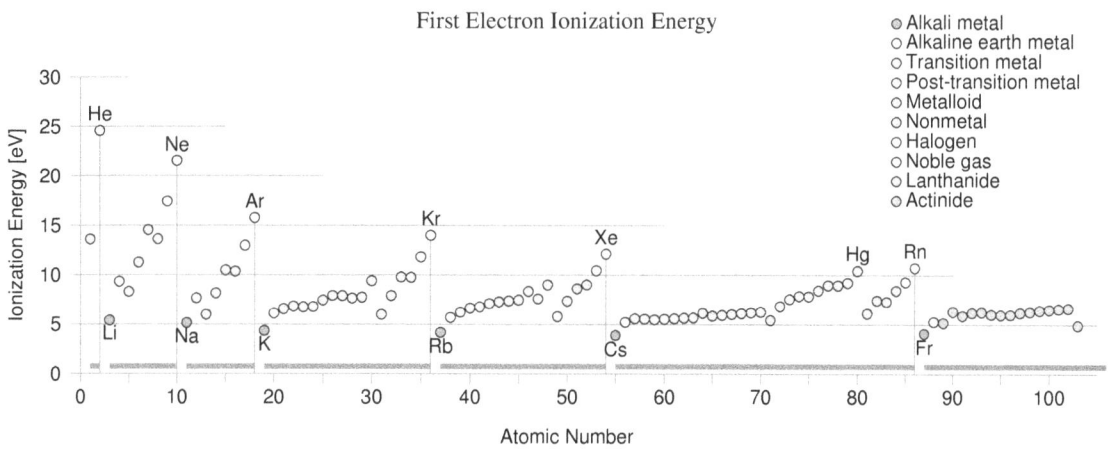

Ionization energy: each period begins at a minimum for the alkali metals, and ends at a maximum for the noble gases

orbital farther away from the nucleus. As such this electron would be less attracted to the nucleus and would release less energy when added. However, in going down a group, around one-third of elements are anomalous, with heavier elements having higher electron affinities than their next lighter congenors. Largely, this is due to the poor shielding by d and f electrons. A uniform decrease in electron affinity only applies to group 1 atoms.[37]

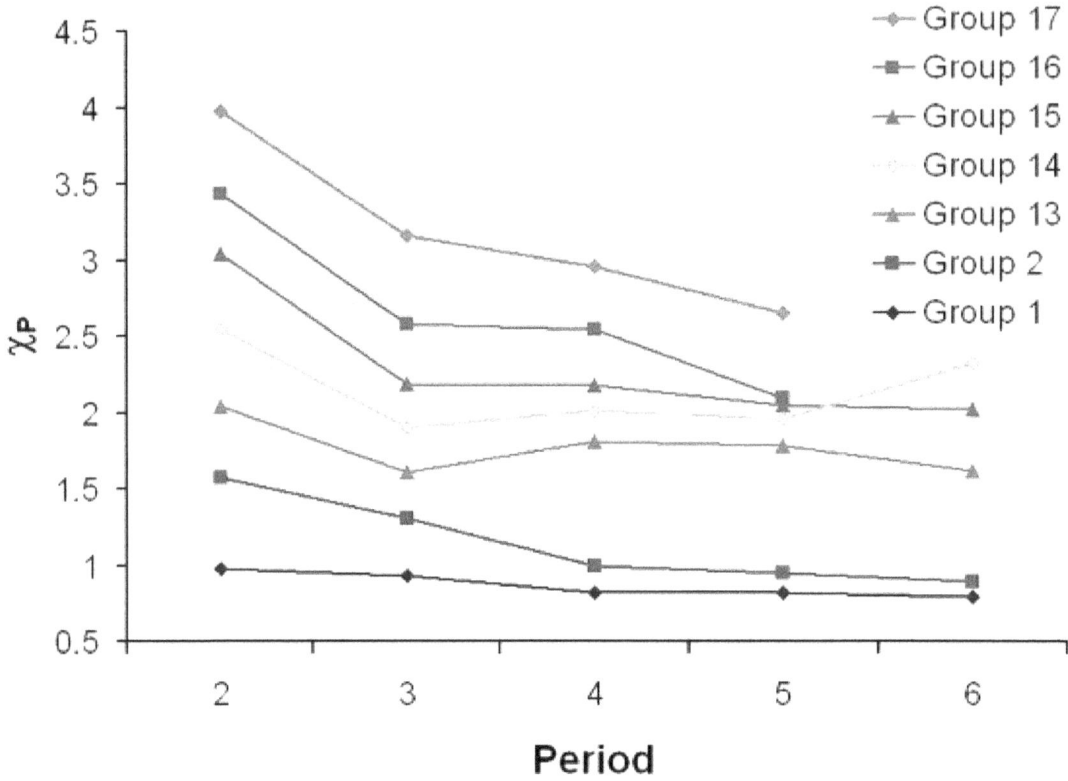

Graph showing increasing electronegativity with growing number of selected groups

30.3.6 Metallic character

The lower the values of ionization energy, electronegativity and electron affinity, the more metallic character the element has. Conversely, nonmetallic character increases with higher values of these properties.[38] Given the periodic trends of these three properties, metallic character tends to decrease going across a period (or row) and, with some irregularities (mostly) due to poor screening of the nucleus by d and f electrons, and relativistic effects,[39] tends to increase going down a group (or column or family). Thus, the most metallic elements (such as caesium and francium) are found at the bottom left of traditional periodic tables and the most nonmetallic elements (oxygen, fluorine, chlorine) at the top right. The combination of horizontal and vertical trends in metallic character explains the stair-shaped dividing line between metals and nonmetals found on some periodic tables, and the practice of sometimes categorizing several elements adjacent to that line, or elements adjacent to those elements, as metalloids.[40][41]

30.3.7 Stability and Radioactivity

There is a trend for stability to decrease with large atomic number. For the element's most stable isotope, all are perfectly stable up through lead (82) with the exception of technetium (43) and promethium (61), all forms of which are radioactive. All elements above lead are radioactive as well, with bismuth (83) long thought to be stable until 2003, when it was determined to be weakly radioactive. Above uranium (92), the transuranic elements, is where the most unstable elements are found. Radioactive decay can take the form of alpha decay, β− decay, β+ decay, gamma decay, proton emission, neutron emission or nuclear fission.

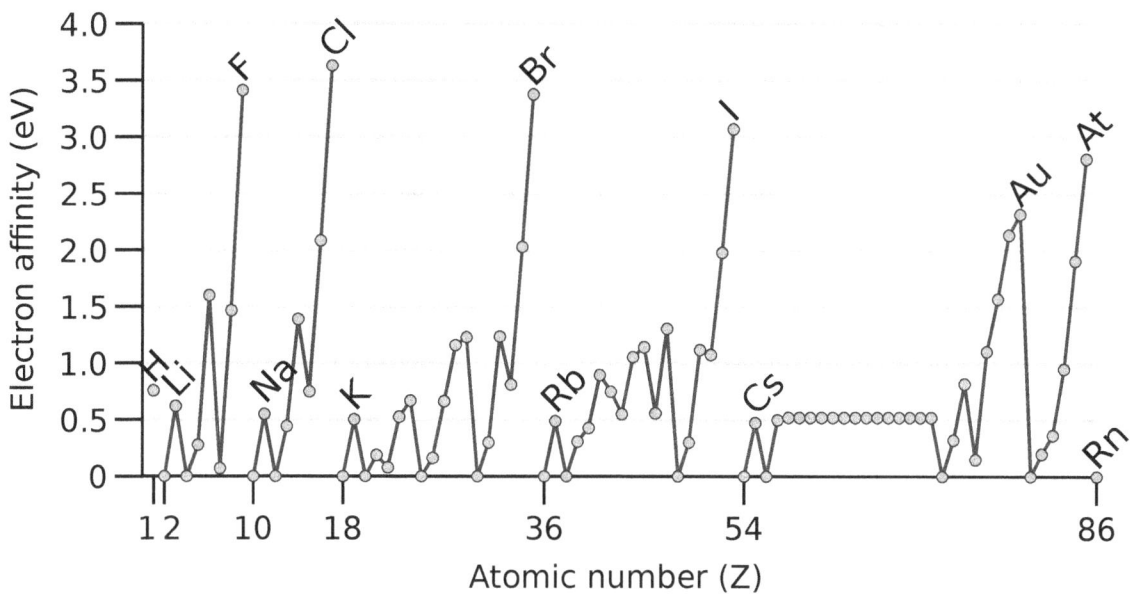

Dependence of electron affinity on atomic number.[34] Values generally increase across each period, culminating with the halogens before decreasing precipitously with the noble gases. Examples of localized peaks seen in hydrogen, the alkali metals and the group 11 elements are caused by a tendency to complete the s-shell (with the 6s shell of gold being further stabilized by relativistic effects and the presence of a filled 4f sub shell). Examples of localized troughs seen in the alkaline earth metals, and nitrogen, phosphorus, manganese and rhenium are caused by filled s-shells, or half-filled p- or d-shells.[35]

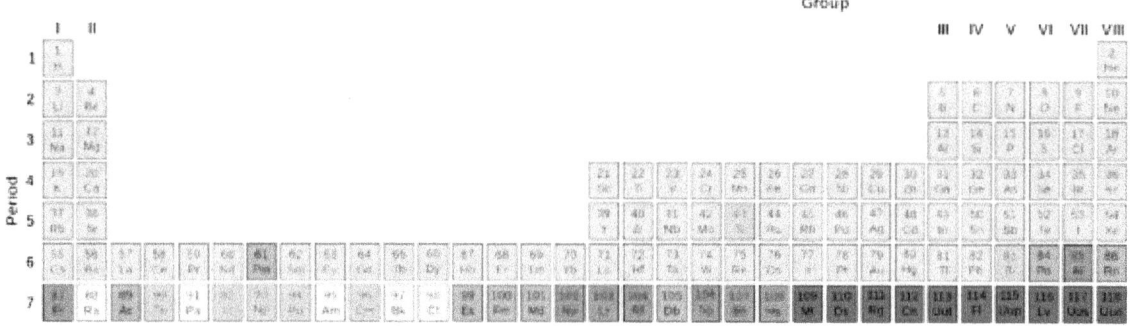

Radioactivity *is characteristic of elements with large atomic number.*
Colors indicate half-life of most stable isotope
(non-radioactive) >1,000,000y >800y >1d <1d <minutes

30.4 History

Main article: History of the periodic table

30.4.1 First systemization attempts

In 1789, Antoine Lavoisier published a list of 33 chemical elements, grouping them into gases, metals, nonmetals, and earths.[42] Chemists spent the following century searching for a more precise classification scheme. In 1829, Johann

1 H																	2 He
3 Li	4 Be											5 B	6 C	7 N	8 O	9 F	10 Ne
11 Na	12 Mg											13 Al	14 Si	15 P	16 S	17 Cl	18 Ar
19 K	20 Ca	21 Sc	22 Ti	23 V	24 Cr	25 Mn	26 Fe	27 Co	28 Ni	29 Cu	30 Zn	31 Ga	32 Ge	33 As	34 Se	35 Br	36 Kr
37 Rb	38 Sr	39 Y	40 Zr	41 Nb	42 Mo	43 Tc	44 Ru	45 Rh	46 Pd	47 Ag	48 Cd	49 In	50 Sn	51 Sb	52 Te	53 I	54 Xe
55 Cs	56 Ba	57 -71	72 Hf	73 Ta	74 W	75 Re	76 Os	77 Ir	78 Pt	79 Au	80 Hg	81 Tl	82 Pb	83 Bi	84 Po	85 At	86 Rn
87 Fr	88 Ra	89 -103	104 Rf	105 Db	106 Sg	107 Bh	108 Hs	109 Mt	110 Ds	111 Rg	112 Cn	113 Uut	114 Fl	115 Uup	116 Lv	117 Uus	118 Uuo

57 La	58 Ce	59 Pr	60 Nd	61 Pm	62 Sm	63 Eu	64 Gd	65 Tb	66 Dy	67 Ho	68 Er	69 Tm	70 Yb	71 Lu
89 Ac	90 Th	91 Pa	92 U	93 Np	94 Pu	95 Am	96 Cm	97 Bk	98 Cf	99 Es	100 Fm	101 Md	102 No	103 Lr

- ▨ Known in antiquity
- ☐ also known when (akw) Levoisier published his list of elements (1789)
- ☐ akw Mendeleev published his periodic table (1869)
- ☐ akw Deming published his periodic table (1923)
- ☐ akw Seaborg published his periodic table (1945)
- ☐ also known (ak) up to 2000
- ▨ ak to 2012

The discovery of the elements mapped to significant periodic table development dates (pre-, per- and post-)

Wolfgang Döbereiner observed that many of the elements could be grouped into triads based on their chemical properties. Lithium, sodium, and potassium, for example, were grouped together in a triad as soft, reactive metals. Döbereiner also observed that, when arranged by atomic weight, the second member of each triad was roughly the average of the first and the third;[43] this became known as the Law of Triads.[44] German chemist Leopold Gmelin worked with this system, and by 1843 he had identified ten triads, three groups of four, and one group of five. Jean-Baptiste Dumas published work in 1857 describing relationships between various groups of metals. Although various chemists were able to identify relationships between small groups of elements, they had yet to build one scheme that encompassed them all.[43]

In 1858, German chemist August Kekulé observed that carbon often has four other atoms bonded to it. Methane, for example, has one carbon atom and four hydrogen atoms. This concept eventually became known as valency; different elements bond with different numbers of atoms.[45]

In 1862, Alexandre-Emile Béguyer de Chancourtois, a French geologist, published an early form of periodic table, which he called the telluric helix or screw. He was the first person to notice the periodicity of the elements. With the elements arranged in a spiral on a cylinder by order of increasing atomic weight, de Chancourtois showed that elements with similar properties seemed to occur at regular intervals. His chart included some ions and compounds in addition to elements. His paper also used geological rather than chemical terms and did not include a diagram; as a result, it received little attention until the work of Dmitri Mendeleev.[46]

In 1864, Julius Lothar Meyer, a German chemist, published a table with 44 elements arranged by valency. The table showed that elements with similar properties often shared the same valency.[47] Concurrently, William Odling (an English chemist) published an arrangement of 57 elements, ordered on the basis of their atomic weights. With some irregularities and gaps, he noticed what appeared to be a periodicity of atomic weights among the elements and that this accorded with 'their usually received groupings.'[48] Odling alluded to the idea of a periodic law but did not pursue it.[49] He subsequently proposed (in 1870) a valence-based classification of the elements.[50]

English chemist John Newlands produced a series of papers from 1863 to 1866 noting that when the elements were listed

No.	No.	No.	No.	No.	No.	No.	No.
H 1	F 8	Cl 15	Co & Ni 22	Br 29	Pd 36	I 42	Pt & Ir 50
Li 2	Na 9	K 16	Cu 23	Rb 30	Ag 37	Cs 44	Os 51
G 3	Mg 10	Ca 17	Zn 24	Sr 31	Cd 38	Ba & V 45	Hg 52
Bo 4	Al 11	Cr 19	Y 25	Ce & La 33	U 40	Ta 46	Tl 53
C 5	Si 12	Ti 18	In 26	Zr 32	Sn 39	W 47	Pb 54
N 6	P 13	Mn 20	As 27	Di & Mo 34	Sb 41	Nb 48	Bi 55
O 7	S 14	Fe 21	Se 28	Ro & Ru 35	Te 43	Au 49	Th 56

Newlands's periodic table, as presented to the Chemical Society in 1866, and based on the law of octaves

in order of increasing atomic weight, similar physical and chemical properties recurred at intervals of eight; he likened such periodicity to the octaves of music.[51][52] This so termed Law of Octaves, however, was ridiculed by Newlands' contemporaries, and the Chemical Society refused to publish his work.[53] Newlands was nonetheless able to draft a table of the elements and used it to predict the existence of missing elements, such as germanium.[54] The Chemical Society only acknowledged the significance of his discoveries five years after they credited Mendeleev.[55]

In 1867, Gustavus Hinrichs, a Danish born academic chemist based in America, published a spiral periodic system based on atomic spectra and weights, and chemical similarities. His work was regarded as idiosyncratic, ostentatious and labyrinthine and this may have militated against its recognition and acceptance.[56][57]

30.4.2 Mendeleev's table

Russian chemistry professor Dmitri Mendeleev and German chemist Julius Lothar Meyer independently published their periodic tables in 1869 and 1870, respectively.[58] Mendeleev's table was his first published version; that of Meyer was an expanded version of his (Meyer's) table of 1864.[59] They both constructed their tables by listing the elements in rows or columns in order of atomic weight and starting a new row or column when the characteristics of the elements began to repeat.[60]

The recognition and acceptance afforded to Mendeleev's table came from two decisions he made. The first was to leave gaps in the table when it seemed that the corresponding element had not yet been discovered.[61] Mendeleev was not the first chemist to do so, but he was the first to be recognized as using the trends in his periodic table to predict the properties of those missing elements, such as gallium and germanium.[62] The second decision was to occasionally ignore the order suggested by the atomic weights and switch adjacent elements, such as tellurium and iodine, to better classify them into chemical families. Later in 1913, Henry Moseley determined experimental values of the nuclear charge or atomic number of each element, and showed that Mendeleev's ordering actually corresponds to the order of increasing atomic number.[63]

The significance of atomic numbers to the organization of the periodic table was not appreciated until the existence and properties of protons and neutrons became understood. Mendeleev's periodic tables used atomic weight instead of atomic number to organize the elements, information determinable to fair precision in his time. Atomic weight worked well enough in most cases to (as noted) give a presentation that was able to predict the properties of missing elements more accurately than any other method then known. Substitution of atomic numbers, once understood, gave a definitive, integer-based sequence for the elements, and Moseley predicted that the only missing elements (in 1913) between aluminum (Z=13) and gold (Z=79) (in 1913) were Z = 43, 61, 72 and 75, which were all later discovered. The sequence of atomic numbers is still used today even as new synthetic elements are being produced and studied.[64]

30.4.3 Second version and further development

In 1871, Mendeleev published his periodic table in a new form, with groups of similar elements arranged in columns rather than in rows, and those columns numbered I to VIII corresponding with the element's oxidation state. He also gave detailed predictions for the properties of elements he had earlier noted were missing, but should exist.[65] These

Dmitri Mendeleev

gaps were subsequently filled as chemists discovered additional naturally occurring elements.[66] It is often stated that the last naturally occurring element to be discovered was francium (referred to by Mendeleev as *eka-caesium*) in 1939.[67] However, plutonium, produced synthetically in 1940, was identified in trace quantities as a naturally occurring primordial element in 1971,[68] and by 2011 it was known that all the elements up to californium can occur naturally as trace amounts

ОПЫТЪ СИСТЕМЫ ЭЛЕМЕНТОВЪ,

ОСНОВАННОЙ НА ИХЪ АТОМНОМЪ ВѢСѢ И ХИМИЧЕСКОМЪ СХОДСТВѢ.

```
                          Ti=50      Zr=90       ?=180.
                          V=51       Nb=94      Ta=182.
                          Cr=52      Mo=96       W=186.
                          Mn=55      Rh=104,4   Pt=197,1.
                          Fe=56      Ru=104,4   Ir=198.
                        Ni=Co=59     Pd=106,6   Os=199.
H=1                       Cu=63,4    Ag=108     Hg=200.
          Be= 9,4 Mg=24   Zn=65,2    Cd=112
          B=11    Al=27,3  ?=68       Ur=116     Au=197?
          C=12    Si=28    ?=70       Sn=118
          N=14    P=31     As=75      Sb=122     Bi=210?
          O=16    S=32     Se=79,4   Te=128?
          F=19    Cl=35,5  Br=80      I=127
Li=7     Na=23    K=39     Rb=85,4   Cs=133      Tl=204.
                  Ca=40    Sr=87,6   Ba=137      Pb=207.
                  ?=45     Ce=92
                  ?Er=56   La=94
                  ?Yt=60   Di=95
                  ?In=75,6 Th=118?
```

Д. Менделѣевъ

A version of Mendeleev's 1869 periodic table: An experiment on a system of elements. Based on their atomic weights and chemical similarities. *This early arrangement presents the periods vertically, and the groups horizontally.*

in uranium ores by neutron capture and beta decay.[1]

The popular[69] periodic table layout, also known as the common or standard form (as shown at various other points in this article), is attributable to Horace Groves Deming. In 1923, Deming, an American chemist, published short (Mendeleev style) and medium (18-column) form periodic tables.[70][n 5] Merck and Company prepared a handout form of Deming's 18-column medium table, in 1928, which was widely circulated in American schools. By the 1930s Deming's table was appearing in handbooks and encyclopaedias of chemistry. It was also distributed for many years by the Sargent-Welch Scientific Company.[71][72][73]

Reihen	Gruppe I. — R²O	Gruppe II. — RO	Gruppe III. — R²O³	Gruppe IV. RH⁴ RO²	Gruppe V. RH³ R²O⁵	Gruppe VI. RH² RO³	Gruppe VII. RH R²O⁷	Gruppe VIII. — RO⁴
1	H=1							
2	Li=7	Be=9.4	B=11	C=12	N=14	O=16	F=19	
3	Na=23	Mg=24	Al=27.3	Si=28	P=31	S=32	Cl=35.5	
4	K=39	Ca=40	—=44	Ti=48	V=51	Cr=52	Mn=55	Fe=56, Co=59, Ni=59, Cu=63.
5	(Cu=63)	Zn=65	—=68	—=72	As=75	Se=78	Br=80	
6	Rb=85	Sr=87	?Yt=88	Zr=90	Nb=94	Mo=96	—=100	Ru=104, Rh=104, Pd=106, Ag=108.
7	(Ag=108)	Cd=112	In=113	Sn=118	Sb=122	Te=125	J=127	
8	Cs=133	Ba=137	?Di=138	?Ce=140	—	—	—	— — — —
9	(—)	—	—	—	—	—	—	
10	—	—	?Er=178	?La=180	Ta=182	W=184	—	Os=195, Ir=197, Pt=198, Au=199.
11	(Au=199)	Hg=200	Tl=204	Pb=207	Bi=208	—	—	
12	—	—	—	Th=231	—	U=240	—	— — — —

Mendeleev's 1871 periodic table with eight groups of elements. Dashes represented elements unknown in 1871.

With the development of modern quantum mechanical theories of electron configurations within atoms, it became apparent that each period (row) in the table corresponded to the filling of a quantum shell of electrons. Larger atoms have more electron sub-shells, so later tables have required progressively longer periods.[74]

In 1945, Glenn Seaborg, an American scientist, made the suggestion that the actinide elements, like the lanthanides, were filling an f sub-level. Before this time the actinides were thought to be forming a fourth d-block row. Seaborg's colleagues advised him not to publish such a radical suggestion as it would most likely ruin his career. As Seaborg considered he did not then have a career to bring into disrepute, he published anyway. Seaborg's suggestion was found to be correct and he subsequently went on to win the 1951 Nobel Prize in chemistry for his work in synthesizing actinide elements.[75][76][n 6]

Although minute quantities of some transuranic elements occur naturally,[1] they were all first discovered in laboratories. Their production has expanded the periodic table significantly, the first of these being neptunium, synthesized in 1939.[77] Because many of the transuranic elements are highly unstable and decay quickly, they are challenging to detect and characterize when produced. There have been controversies concerning the acceptance of competing discovery claims for some elements, requiring independent review to determine which party has priority, and hence naming rights. The most recently accepted and named elements are flerovium (element 114) and livermorium (element 116), both named on 31 May 2012.[78] In 2010, a joint Russia–US collaboration at Dubna, Moscow Oblast, Russia, claimed to have synthesized six atoms of ununseptium (element 117), making it the most recently claimed discovery.[79]

30.5 Alternative structures

Main article: Alternative periodic tables

There are many periodic tables with structures other than that of the standard form. Within 100 years of the appearance of Mendeleev's table in 1869 it has been estimated that around 700 different periodic table versions were published.[80] As well as numerous rectangular variations, other periodic table formats have included, for example,[n 7] circular, cubic, cylindrical, edificial (building-like), helical, lemniscate, octagonal prismatic, pyramidal, separated, spherical, spiral, and triangular forms. Such alternatives are often developed to highlight or emphasize chemical or physical properties of the elements that are not as apparent in traditional periodic tables.[80]

A popular[81] alternative structure is that of Theodor Benfey (1960). The elements are arranged in a continuous spiral,

Period	Series	a I	b	a II	b	a III	b	a IV	b	a V	b	a VI	b	a VII	b	a VIII			b
1	I	1 H																	2 He
2	II	3 Li		4 Be		5 B		6 C		7 N		8 O		9 F					10 Ne
3	III	11 Na		12 Mg		13 Al		14 Si		15 P		16 S		17 Cl					18 Ar
4	IV	19 K		20 Ca		21 Sc		22 Ti		23 V		24 Cr		25 Mn		26 Fe	27 Co	28 Ni	
4	V		29 Cu		30 Zn	31 Ga		32 Ge		33 As		34 Se		35 Br		36 Kr			
5	VI	37 Rb		38 Sr		39 Y		40 Zr		41 Nb		42 Mo		43 Tc		44 Ru	45 Rh	46 Pd	
5	VII		47 Ag		48 Cd	49 In		50 Sn		51 Sb		52 Te		53 I		54 Xe			
6	VIII	55 Cs		56 Ba		57–71		72 Hf		73 Ta		74 W		75 Re		76 Os	77 Ir	78 Pt	
6	IX		79 Au		80 Hg	81 Tl		82 Pb		83 Bi		84 Po		85 At		86 Rn			
7	X	87 Fr		88 Ra		89–103		104 Rf		105 Db		106 Sg		107 Bh		108 Hs	109 Mt	110 Ds	
7	XI		111 Rg		112 Cn	113 Uut		114 Fl		115 Uup		116 Lv		117 Uus		118 Uuo			
Higher oxides		R_2O		RO		R_2O_3		RO_2		R_2O_5		RO_3		R_2O_7		RO_4			
Volatile hydrogen compounds						$[(RH_3),]$		RH_4		RH_3		RH_2		RH					

57 La	58 Ce	59 Pr	60 Nd	61 Pm	62 Sm	63 Eu	64 Gd	65 Tb	66 Dy	67 Ho	68 Er	69 Tm	70 Yb	71 Lu
89 Ac	90 Th	91 Pa	92 U	93 Np	94 Pu	95 Am	96 Cm	97 Bk	98 Cf	99 Es	100 Fm	101 Md	102 No	103 Lr

Eight-column form of periodic table, updated with all elements discovered to 2014

with hydrogen at the center and the transition metals, lanthanides, and actinides occupying peninsulas.[82]

Most periodic tables are two-dimensional;[1] however, three-dimensional tables are known to as far back as at least 1862 (pre-dating Mendeleev's two-dimensional table of 1869). More recent examples include Courtines' Periodic Classification (1925),[83] Wringley's Lamina System (1949),[84] Giguère's Periodic helix (1965)[85][n 8] and Dufour's Periodic Tree (1996).[86] Going one better, Stowe's Physicist's Periodic Table (1989)[87] has been described as being four-dimensional (having three spatial dimensions and one colour dimension).[88]

The various forms of periodic tables can be thought of as lying on a chemistry–physics continuum.[89] Towards the chemistry end of the continuum can be found, as an example, Rayner-Canham's 'unruly'[90] Inorganic Chemist's Periodic Table (2002),[91] which emphasizes trends and patterns, and unusual chemical relationships and properties. Near the physics end of the continuum is Janet's Left-Step Periodic Table (1928). This has a structure that shows a closer connection to the order of electron-shell filling and, by association, quantum mechanics.[92] Somewhere in the middle of the continuum is the ubiquitous common or standard form of periodic table. This is regarded as better expressing empirical trends in physical state, electrical and thermal conductivity, and oxidation numbers, and other properties easily inferred from traditional techniques of the chemical laboratory.[93]

30.6   Open questions and controversies

Glenn T. Seaborg who, in 1945, suggested a new periodic table showing the actinides as belonging to a second f-block series

30.6.1 Elements with unknown chemical properties

Although all elements up to ununoctium have been discovered, of the elements above hassium (element 108), only copernicium (element 112) and flerovium (element 114) have known chemical properties. The other elements may behave

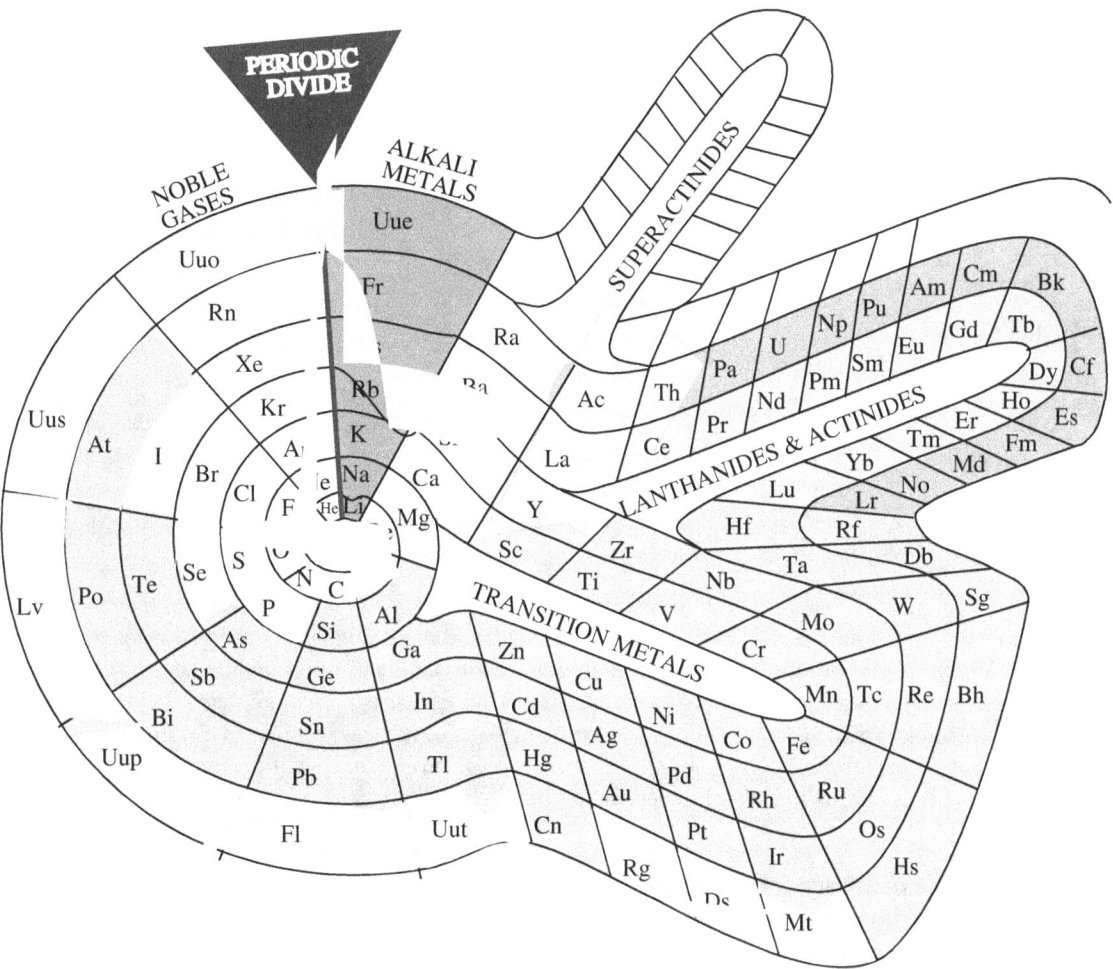

Theodor Benfey's spiral periodic table

differently from what would be predicted by extrapolation, due to relativistic effects; for example, flerovium has been predicted to possibly exhibit some noble-gas-like properties, even though it is currently placed in the carbon group.[94] More recent experiments have suggested, however, that flerovium behaves chemically like lead, as expected from its periodic table position.[95]

30.6.2 Further periodic table extensions

Main article: Extended periodic table

It is unclear whether new elements will continue the pattern of the current periodic table as period 8, or require further adaptations or adjustments. Seaborg expected the eighth period to follow the previously established pattern exactly, so that it would include a two-element s-block for elements 119 and 120, a new g-block for the next 18 elements, and 30 additional elements continuing the current f-, d-, and p-blocks.[96] More recently, physicists such as Pekka Pyykkö have theorized that these additional elements do not follow the Madelung rule, which predicts how electron shells are filled and thus affects the appearance of the present periodic table.[97]

30.6.3 Element with the highest possible atomic number

The number of possible elements is not known. A very early suggestion made by Elliot Adams in 1911, and based on the arrangement of elements in each horizontal periodic table row, was that elements of atomic weight greater than 256± (which would equate to between elements 99 and 100 in modern-day terms) did not exist.[98] A higher—more recent—estimate is that the periodic table may end soon after the island of stability,[99] which is expected to center around element 126, as the extension of the periodic and nuclides tables is restricted by proton and neutron drip lines.[100] Other predictions of an end to the periodic table include at element 128 by John Emsley,[1] at element 137 by Richard Feynman,[101] and at element 155 by Albert Khazan.[1][n 9]

Bohr model

The Bohr model exhibits difficulty for atoms with atomic number greater than 137, as any element with an atomic number greater than 137 would require 1s electrons to be traveling faster than c, the speed of light.[102] Hence the non-relativistic Bohr model is inaccurate when applied to such an element.

Relativistic Dirac equation

The relativistic Dirac equation has problems for elements with more than 137 protons. For such elements, the wave function of the Dirac ground state is oscillatory rather than bound, and there is no gap between the positive and negative energy spectra, as in the Klein paradox.[103] More accurate calculations taking into account the effects of the finite size of the nucleus indicate that the binding energy first exceeds the limit for elements with more than 173 protons. For heavier elements, if the innermost orbital (1s) is not filled, the electric field of the nucleus will pull an electron out of the vacuum, resulting in the spontaneous emission of a positron;[104] however, this does not happen if the innermost orbital is filled, so that element 173 is not necessarily the end of the periodic table.[105]

30.6.4 Placement of hydrogen and helium

Hydrogen and helium are often placed in different places than their electron configurations would indicate; hydrogen is usually placed above lithium, in accordance with its electron configuration, but is sometimes placed above fluorine,[106] or even carbon,[106] as it also behaves somewhat similarly to them. Hydrogen is also sometimes placed in its own group, as it does not behave similarly enough to any element to be placed in a group with another.[107] Helium is almost always placed above neon, as they are very similar chemically, although it is occasionally placed above beryllium on account of having a comparable electron shell configuration (helium: $1s^2$; beryllium: $[He]\ 2s^2$).[19]

30.6.5 Groups included in the transition metals

The definition of a transition metal, as given by IUPAC, is an element whose atom has an incomplete d sub-shell, or which can give rise to cations with an incomplete d sub-shell.[108] By this definition all of the elements in groups 3–11 are transition metals. The IUPAC definition therefore excludes group 12, comprising zinc, cadmium and mercury, from the transition metals category.

Some chemists treat the categories "d-block elements" and "transition metals" interchangeably, thereby including groups 3–12 among the transition metals. In this instance the group 12 elements are treated as a special case of transition metal in which the d electrons are not ordinarily involved in chemical bonding. The recent discovery that mercury can use its d electrons in the formation of mercury(IV) fluoride (HgF_4) has prompted some commentators to suggest that mercury can be regarded as a transition metal.[109] Other commentators, such as Jensen,[110] have argued that the formation of a compound like HgF_4 can occur only under highly abnormal conditions. As such, mercury could not be regarded as a transition metal by any reasonable interpretation of the ordinary meaning of the term.[110]

Still other chemists further exclude the group 3 elements from the definition of a transition metal. They do so on the basis that the group 3 elements do not form any ions having a partially occupied d shell and do not therefore exhibit any properties characteristic of transition metal chemistry.[111] In this case, only groups 4–11 are regarded as transition metals.

30.6.6   Period 6 and 7 elements in group 3

Although scandium and yttrium are always the first two group 3 elements, the identity of the next two elements is not agreed upon; they are either lanthanum and actinium, or lutetium and lawrencium. Although there are some strong physical and chemical arguments supporting the latter arrangement not all authors are convinced.[112] The current IUPAC definition of the term "lanthanoid" includes fifteen elements including both lanthanum and lutetium, and that of "transition element"[108] applies to lanthanum and actinium, as well as lutetium but *not* lawrencium, since it does not correctly follow the Aufbau principle. Normally, the 103rd electron would enter the d-subshell, but quantum mechanical research has found that the configuration is actually $[Rn] 5f^{14} 7s^2 7p^{1[n\ 10]}$ due to relativistic effects.[113][114] IUPAC thus has not recommended a specific format for the in-line-f-block periodic table, leaving the dispute open.

- Lanthanum and actinium are sometimes considered the remaining members of group 3.[115] In their most commonly encountered tripositive ion forms, these elements do not possess any partially filled f-orbitals, thus continuing the scandium—yttrium—lanthanum—actinium trend, in which all the elements have relationship similar to that of elements of the calcium—strontium—barium—radium series, the elements' left neighbors in s-block. However, different behavior is observed in other d-block groups, especially in group 4, in which zirconium, hafnium and rutherfordium share similar chemical properties lacking a clear trend.

- In other tables, lutetium and lawrencium are classified as the remaining members of group 3.[116] In these tables, lutetium and lawrencium end (or sometimes follow) the lanthanide and actinide series, respectively. Since the f-shell is nominally full in the ground state electron configuration for both of these metals, they behave most similarly to other period 6 and period 7 transition metals compared to the other lanthanides and actinides, and thus logically exhibit properties similar to those of scandium and yttrium.

- Some tables, including the IUPAC table[117][n 11] refer to *all* lanthanides and actinides by a marker in group 3. This sometimes is believed to be the inclusion of all 30 lanthanide and actinide elements as included in group 3. Lanthanides, as electropositive trivalent metals, all have a closely related chemistry, and all show many similarities to scandium and yttrium, but they also show additional properties characteristic of their partially filled f-orbitals which are not common to scandium and yttrium.

- Exclusion of all elements is based on properties of earlier actinides, which show a much wider variety of chemistry (for instance, in range of oxidation states) within their series than the lanthanides, and comparisons to scandium and yttrium are even less useful.[118] However, these elements are destabilized,[119] and if they were stabilized to more closely match chemistry laws, they would be similar to lanthanides as well. Also, the later actinides from berkelium onwards behave more like the corresponding lanthanides, with only the valence +3 (and sometimes +2 and +4) shown.[118]

30.6.7   Optimal form

The many different forms of periodic table have prompted the question of whether there is an optimal or definitive form of periodic table. The answer to this question is thought to depend on whether the chemical periodicity seen to occur among the elements has an underlying truth, effectively hard-wired into the universe, or if any such periodicity is instead the product of subjective human interpretation, contingent upon the circumstances, beliefs and predilections of human observers. An objective basis for chemical periodicity would settle the questions about the location of hydrogen and helium, and the composition of group 3. Such an underlying truth, if it exists, is thought to have not yet been discovered. In its absence, the many different forms of periodic table can be regarded as variations on the theme of chemical periodicity, each of which explores and emphasizes different aspects, properties, perspectives and relationships of and among the elements.[n 12] The ubiquity of the standard or medium-long periodic table is thought to be a result of this layout having a good balance of features in terms of ease of construction and size, and its depiction of atomic order and periodic trends.[49][120]

30.7 See also

- Abundance of the chemical elements

- Atomic electron configuration table

- Element collecting

- List of elements

- List of periodic table-related articles

- Table of nuclides

- Timeline of chemical element discoveries

30.8 Notes

[1] The elements discovered initially by synthesis and later in nature are technetium (Z=43), promethium (61), astatine (85), francium (87), neptunium (93), plutonium (94), americium (95), curium (96), berkelium (97) and californium (98).

[2] There is an inconsistency and some irregularities in this convention. Thus, helium is shown in the p-block but is actually an s-block element, and (for example) the d-subshell in the d-block is actually filled by the time group 11 is reached, rather than group 12.

[3] The noble gases, astatine, francium, and all elements heavier than americium were left out as there is no data for them.

[4] While fluorine is the most electronegative of the elements under the Pauling scale, neon is the most electronegative element under other scales, such as the Allen scale.

[5] An antecedent of Deming's 18-column table may be seen in Adams' 16-column Periodic Table of 1911. Adams omits the rare earths and the 'radioactive elements' (i.e. the actinides) from the main body of his table and instead shows them as being 'careted in only to save space' (rare earths between Ba and eka-Yt; radioactive elements between eka-Te and eka-I). See: Elliot Q. A. (1911). "A modification of the periodic table". *Journal of the American Chemical Society*. **33**(5): 684–688 (687).

[6] A second extra-long periodic table row, to accommodate known and undiscovered elements with an atomic weight greater than bismuth (thorium, protactinium and uranium, for example), had been postulated as far back as 1892. Most investigators, however, considered that these elements were analogues of the third series transition elements, hafnium, tantalum and tungsten. The existence of a second inner transition series, in the form of the actinides, was not accepted until similarities with the electron structures of the lanthanides had been established. See: van Spronsen, J. W. (1969). *The periodic system of chemical elements*. Amsterdam: Elsevier. p. 315–316, ISBN 0-444-40776-6.

[7] See *The Internet database of periodic tables* for depictions of these kinds of variants.

[8] The animated depiction of Giguère's periodic table that is widely available on the internet (including from here) is erroneous, as it does not include hydrogen and helium. Giguère included hydrogen, above lithium, and helium, above beryllium. See: Giguère P.A. (1966). "The "new look" for the periodic system". *Chemistry in Canada* **18** (12): 36–39 (see p. 37).

[9] Karol (2002, p. 63) contends that gravitational effects would become significant when atomic numbers become astronomically large, thereby overcoming other super-massive nuclei instability phenomena, and that neutron stars (with atomic numbers on the order of 10^{21}) can arguably be regarded as representing the heaviest known elements in the universe. See: Karol P. J. (2002). "The Mendeleev–Seaborg periodic table: Through Z = 1138 and beyond". *Journal of Chemical Education* **79** (1): 60–63.

[10] The expected configuration of lawrencium if it did obey the Aufbau principle would be [Rn] $5f^{14}$ $6d^1$ $7s^2$, with the normal incomplete 6d-subshell in the neutral state.

[11] Although this form of the table is sometimes referred to as the "approved" or "official" IUPAC periodic table, "IUPAC has not approved any specific form of the periodic table..." See: Leigh, G. J. (January–February 2009). "Periodic Tables and IUPAC" **31** (1). Chemistry International.

[12] Scerri, one of the foremost authorities on the history of the periodic table (Sella 2013), favoured the concept of an optimal form of periodic table but has recently changed his mind and now supports the value of a plurality of periodic tables. See: Sella A. (2013). 'An elementary history lesson'. *New Scientist*. 2929, 13 August: 51, accessed 4 September 2013; and Scerri, E. (2013). 'Is there an optimal periodic table and other bigger questions in the philosophy of science.'. 9 August, accessed 4 September 2013.

30.9   References

[1] Emsley, John (2011). *Nature's Building Blocks: An A-Z Guide to the Elements* (New ed.). New York, NY: Oxford University Press. ISBN 978-0-19-960563-7.

[2] Greenwood, pp. 24–27

[3] Gray, p. 6

[4] Koppenol, W. H. (2002). "Naming of New Elements (IUPAC Recommendations 2002)" (PDF). *Pure and Applied Chemistry* **74** (5): 787–791. doi:10.1351/pac200274050787.

[5] Haire, Richard G. (2006). "Fermium, Mendelevium, Nobelium and Lawrencium". In Morss; Edelstein, Norman M.; Fuger, Jean. *The Chemistry of the Actinide and Transactinide Elements* (3rd ed.). Dordrecht, The Netherlands: Springer Science+Business Media. ISBN 1-4020-3555-1.

[6] Gray, p. 11

[7] Scerri 2007, p. 24

[8] Messler, R. W. (2010). *The essence of materials for engineers*. Sudbury, MA: Jones & Bartlett Publishers. p. 32. ISBN 0-7637-7833-8.

[9] Bagnall, K. W. (1967). "Recent advances in actinide and lanthanide chemistry". In Fields, P.R.; Moeller, T. *Advances in chemistry, Lanthanide/Actinide chemistry*. Advances in Chemistry **71**. American Chemical Society. pp. 1–12. doi:10.1021/ba-1967-0071. ISBN 0-8412-0072-6.

[10] Day, M. C., Jr.; Selbin, J. (1969). *Theoretical inorganic chemistry* (2nd ed.). New York: Nostrand-Rienhold Book Corporation. p. 103. ISBN 0-7637-7833-8.

[11] Holman, J.; Hill, G. C. (2000). *Chemistry in context* (5th ed.). Walton-on-Thames: Nelson Thornes. p. 40. ISBN 0-17-448276-0.

[12] Leigh, G. J. (1990). *Nomenclature of Inorganic Chemistry: Recommendations 1990*. Blackwell Science. ISBN 0-632-02494-1.

[13] Fluck, E. (1988). "New Notations in the Periodic Table" (PDF). *Pure Appl. Chem.* (IUPAC) **60** (3): 431–436. doi:10.1351/1. Retrieved 24 March 2012.

[14] Moore, p. 111

[15] Greenwood, p. 30

[16] Stoker, Stephen H. (2007). *General, organic, and biological chemistry*. New York: Houghton Mifflin. p. 68. ISBN 978-0-618-73063-6. OCLC 52445586.

[17] Mascetta, Joseph (2003). *Chemistry The Easy Way* (4th ed.). New York: Hauppauge. p. 50. ISBN 978-0-7641-1978-1. OCLC 52047235.

[18] Kotz, John; Treichel, Paul; Townsend, John (2009). *Chemistry and Chemical Reactivity, Volume 2* (7th ed.). Belmont: Thomson Brooks/Cole. p. 324. ISBN 978-0-495-38712-1. OCLC 220756597.

[19] Gray, p. 12

[20] Jones, Chris (2002). *d- and f-block chemistry*. New York: J. Wiley & Sons. p. 2. ISBN 978-0-471-22476-1. OCLC 300468713.

[21] Silberberg, M. S. (2006). *Chemistry: The molecular nature of matter and change* (4th ed.). New York: McGraw-Hill. p. 536. ISBN 0-07-111658-3.

[22] Manson, S. S.; Halford, G. R. (2006). *Fatigue and durability of structural materials.* Materials Park, Ohio: ASM International. p. 376. ISBN 0-87170-825-6.

[23] Bullinger, Hans-Jörg (2009). *Technology guide: Principles, applications, trends.* Berlin: Springer-Verlag. p. 8. ISBN 978-3-540-88545-0.

[24] Jones, B. W. (2010). *Pluto: Sentinel of the outer solar system.* Cambridge: Cambridge University Press. pp. 169–71. ISBN 978-0-521-19436-5.

[25] Hinrichs, G. D. (1869). "On the classification and the atomic weights of the so-called chemical elements, with particular reference to Stas's determinations". *Proceedings of the American Association for the Advancement of Science* **18** (5): 112–124.

[26] Myers, R. (2003). *The basics of chemistry.* Westport, CT: Greenwood Publishing Group. pp. 61–67. ISBN 0-313-31664-3.

[27] Chang, Raymond (2002). *Chemistry* (7 ed.). New York: McGraw-Hill. pp. 289–310; 340–42. ISBN 0-07-112072-6.

[28] Greenwood, p. 27

[29] Jolly, W. L. (1991). *Modern Inorganic Chemistry* (2nd ed.). McGraw-Hill. p. 22. ISBN 978-0-07-112651-9.

[30] Greenwood, p. 28

[31] IUPAC, *Compendium of Chemical Terminology*, 2nd ed. (the "Gold Book") (1997). Online corrected version: (2006–) "Electronegativity".

[32] Pauling, L. (1932). "The Nature of the Chemical Bond. IV. The Energy of Single Bonds and the Relative Electronegativity of Atoms". *Journal of the American Chemical Society* **54** (9): 3570–3582. doi:10.1021/ja01348a011.

[33] Allred, A. L. (1960). "Electronegativity values from thermochemical data". *Journal of Inorganic and Nuclear Chemistry* (Northwestern University) **17** (3–4): 215–221. doi:10.1016/0022-1902(61)80142-5. Retrieved 11 June 2012.

[34] Huheey, Keiter & Keiter, p. 42

[35] Siekierski, Slawomir; Burgess, John (2002). *Concise chemistry of the elements.* Chichester: Horwood Publishing. pp. 35–36. ISBN 1-898563-71-3.

[36] Chang, pp. 307–309

[37] Huheey, Keiter & Keiter, pp. 42, 880–81

[38] Yoder, C. H.; Suydam, F. H.; Snavely, F. A. (1975). *Chemistry* (2nd ed.). Harcourt Brace Jovanovich. p. 58. ISBN 0-15-506465-7.

[39] Huheey, Keiter & Keiter, pp. 880–85

[40] Sacks, O (2009). *Uncle Tungsten: Memories of a chemical boyhood.* New York: Alfred A. Knopf. pp. 191, 194. ISBN 0-375-70404-3.

[41] Gray, p. 9

[42] Siegfried, Robert (2002). *From elements to atoms a history of chemical composition.* Philadelphia, Pennsylvania: Library of Congress Cataloging-in-Publication Data. p. 92. ISBN 0-87169-924-9.

[43] Ball, p. 100

[44] Horvitz, Leslie (2002). *Eureka!: Scientific Breakthroughs That Changed The World.* New York: John Wiley. p. 43. ISBN 978-0-471-23341-1. OCLC 50766822.

[45] van Spronsen, J. W. (1969). *The periodic system of chemical elements.* Amsterdam: Elsevier. p. 19. ISBN 0-444-40776-6.

[46] "Alexandre-Emile Bélguier de Chancourtois (1820-1886)" (in French). Annales des Mines history page. Retrieved 18 September 2014.

[47] Venable, pp. 85–86; 97

[48] Odling, W. (2002). "On the proportional numbers of the elements". *Quarterly Journal of Science* **1**: 642–648 (643).

[49] Scerri, Eric R. (2011). *The periodic table: A very short introduction*. Oxford: Oxford University Press. ISBN 978-0-19-958249-5.

[50] Kaji, M. (2004). "Discovery of the periodic law: Mendeleev and other researchers on element classification in the 1860s". In Rouvray, D. H.; King, R. Bruce. *The periodic table: Into the 21st Century*. Research Studies Press. pp. 91–122 (95). ISBN 0-86380-292-3.

[51] Newlands, John A. R. (20 August 1864). "On Relations Among the Equivalents". *Chemical News* **10**: 94–95.

[52] Newlands, John A. R. (18 August 1865). "On the Law of Octaves". *Chemical News* **12**: 83.

[53] Bryson, Bill (2004). *A Short History of Nearly Everything*. Black Swan. pp. 141–142. ISBN 978-0-552-15174-0.

[54] Scerri 2007, p. 306

[55] Brock, W. H.; Knight, D. M. (1965). "The Atomic Debates: 'Memorable and Interesting Evenings in the Life of the Chemical Society'". *Isis* (The University of Chicago Press) **56** (1): 5–25. doi:10.1086/349922.

[56] Scerri 2007, pp. 87, 92

[57] Kauffman, George B. (March 1969). "American forerunners of the periodic law". *Journal of Chemical Education* **46** (3): 128–135 (132). Bibcode:1969JChEd..46..128K. doi:10.1021/ed046p128.

[58] Mendelejew, Dimitri (1869). "Über die Beziehungen der Eigenschaften zu den Atomgewichten der Elemente". *Zeitschrift für Chemie* (in German): 405–406.

[59] Venable, pp. 96–97; 100–102

[60] Ball, pp. 100–102

[61] Pullman, Bernard (1998). *The Atom in the History of Human Thought*. Translated by Axel Reisinger. Oxford University Press. p. 227. ISBN 0-19-515040-6.

[62] Ball, p. 105

[63] Atkins, P. W. (1995). *The Periodic Kingdom*. HarperCollins Publishers, Inc. p. 87. ISBN 0-465-07265-8.

[64] Samanta, C.; Chowdhury, P. Roy; Basu, D.N. (2007). "Predictions of alpha decay half lives of heavy and superheavy elements". *Nucl. Phys. A* **789**: 142–154. arXiv:nucl-th/0703086. Bibcode:2007NuPhA.789..142S. doi:10.1016/j.nuclphysa.2007.04.001.

[65] Scerri 2007, p. 112

[66] Kaji, Masanori (2002). "D.I. Mendeleev's Concept of Chemical Elements and the Principle of Chemistry" (PDF). *Bull. Hist. Chem.* (Tokyo Institute of Technology) **27** (1): 4–16. Retrieved 11 June 2012.

[67] Adloff, Jean-Pierre; Kaufman, George B. (25 September 2005). "Francium (Atomic Number 87), the Last Discovered Natural Element". The Chemical Educator. Retrieved 26 March 2007.

[68] Hoffman, D. C.; Lawrence, F. O.; Mewherter, J. L.; Rourke, F. M. (1971). "Detection of Plutonium-244 in Nature". *Nature* **234** (5325): 132–134. Bibcode:1971Natur.234..132H. doi:10.1038/234132a0.

[69] Gray, p. 12

[70] Deming, Horace G (1923). *General chemistry: An elementary survey*. New York: J. Wiley & Sons. pp. 160, 165.

[71] Abraham, M; Coshow, D; Fix, W. *Periodicity:A source book module, version 1.0* (PDF). New York: Chemsource, Inc. p. 3.

[72] Emsley, J (7 March 1985). "Mendeleyev's dream table". *New Scientist*: 32–36(36).

[73] Fluck, E (1988). "New notations in the period table". *Pure & Applied Chemistry* **60** (3): 431–436 (432). doi:10.1351/pac1.

[74] Ball, p. 111

[75] Scerri 2007, pp. 270–71

[76] Masterton, William L.; Hurley, Cecile N.; Neth, Edward J. *Chemistry: Principles and reactions* (7th ed.). Belmont, CA: Brooks/Cole Cengage Learning. p. 173. ISBN 1-111-42710-0.

[77] Ball, p. 123

[78] Barber, Robert C.; Karol, Paul J; Nakahara, Hiromichi; Vardaci, Emanuele; Vogt, Erich W. (2011). "Discovery of the elements with atomic numbers greater than or equal to 113 (IUPAC Technical Report)". *Pure Appl. Chem.* **83** (7): 1485. doi:10.1351/PAC-REP-10-05-01.

[79] Эксперимент по синтезу 117-го элемента получает продолжение[Experiment on sythesis of the 117th element is to be continued] (in Russian). JINR. 2012.

[80] Scerri 2007, p. 20

[81] Emsely, J; Sharp, R (21 June 2010). "The periodic table: Top of the charts". *The Independent*.

[82] Seaborg, Glenn (1964). "Plutonium: The Ornery Element". *Chemistry* **37** (6): 14.

[83] Mark R. Leach. "1925 Courtines' Periodic Classification". Retrieved 16 October 2012.

[84] Mark R. Leach. "1949 Wringley's Lamina System". Retrieved 16 October 2012.

[85] Mazurs, E.G. (1974). *Graphical Representations of the Periodic System During One Hundred Years*. Alabama: University of Alabama Press. p. 111. ISBN 978-0-8173-3200-6.

[86] Mark R. Leach. "1996 Dufour's Periodic Tree". Retrieved 16 October 2012.

[87] Mark R. Leach. "1989 Physicist's Periodic Table by Timothy Stowe". Retrieved 16 October 2012.

[88] Bradley, David (20 July 2011). "At last, a definitive periodic table?". *ChemViews Magazine*. doi:10.1002/chemv.201000107.

[89] Scerri 2007, pp. 285–86

[90] Scerri 2007, p. 285

[91] Mark R. Leach. "2002 Inorganic Chemist's Periodic Table". Retrieved 16 October 2012.

[92] Scerri, Eric (2008). "The role of triads in the evolution of the periodic table: Past and present". *Journal of Chemical Education* **85** (4): 585–89 (see p.589). Bibcode:2008JChEd..85..585S. doi:10.1021/ed085p585.

[93] Bent, H. A.; Weinhold, F (2007). "Supporting information: News from the periodic table: An introduction to "Periodicity symbols, tables, and models for higher-order valency and donor–acceptor kinships"". *Journal of Chemical Education* **84** (7): 3–4. doi:10.1021/ed084p1145.

[94] Schändel, Matthias (2003). *The Chemistry of Superheavy Elements*. Dordrecht: Kluwer Academic Publishers. p. 277. ISBN 1-4020-1250-0.

[95] Scerri 2011, pp. 142–143

[96] Frazier, K. (1978). "Superheavy Elements". *Science News* **113** (15): 236–238. doi:10.2307/3963006. JSTOR 3963006.

[97] Pyykkö, Pekka (2011). "A suggested periodic table up to Z ≤ 172, based on Dirac–Fock calculations on atoms and ions". *Physical Chemistry Chemical Physics* **13** (1): 161–168. Bibcode:2011PCCP...13..161P. doi:10.1039/c0cp01575j. PMID 20967377.

[98] Elliot, Q. A. (1911). "A modification of the periodic table". *Journal of the American Chemical Society* **33** (5): 684–688 (688). doi:10.1021/ja02218a004.

[99] Glenn Seaborg (c. 2006). "transuranium element (chemical element)". Encyclopædia Britannica. Retrieved 16 March 2010.

[100] Cwiok, S.; Heenen, P.-H.; Nazarewicz, W. (2005). "Shape coexistence and triaxiality in the superheavy nuclei". *Nature* **433** (7027): 705–9. Bibcode:2005Natur.433..705C. doi:10.1038/nature03336. PMID 15716943.

[101] Column: The crucible Ball, Philip in Chemistry World, Royal Society of Chemistry, Nov. 2010

[102] Eisberg, R.; Resnick, R. (1985). *Quantum Physics of Atoms, Molecules, Solids, Nuclei and Particles*. Wiley.

[103] Bjorken, J. D.; Drell, S. D. (1964). *Relativistic Quantum Mechanics*. McGraw-Hill.

[104] Greiner, W.; Schramm, S. (2008). "American Journal of Physics" **76**. p. 509., and references therein.

[105] Ball, Philip (November 2010). "Would Element 137 Really Spell the End of the Periodic Table? Philip Ball Examines the Evidence". Royal Society of Chemistry. Retrieved 30 September 2012.

[106] Cronyn, Marshall W. (August 2003). "The Proper Place for Hydrogen in the Periodic Table". *Journal of Chemical Education* **80** (8): 947–951. Bibcode:2003JChEd..80..947C. doi:10.1021/ed080p947.

[107] Gray, p. 14

[108] IUPAC, *Compendium of Chemical Terminology*, 2nd ed. (the "Gold Book") (1997). Online corrected version: (2006–) "transition element".

[109] Xuefang Wang; Lester Andrews; Sebastian Riedel; Martin Kaupp (2007). "Mercury Is a Transition Metal: The First Experimental Evidence for HgF_4". *Angew. Chem. Int. Ed.* **46** (44): 8371–8375. doi:10.1002/anie.200703710. PMID 17899620.

[110]William B. Jensen (2008). "Is Mercury Now a Transition Element?".*J. Chem. Educ.***85**(9): 1182–1183.Bibcode:2008JChEd... doi:10.1021/ed085p1182.

[111] Rayner-Canham, G; Overton, T. *Descriptive inorganic chemistry* (4th ed.). New York: W H Freeman. pp. 484–485. ISBN 0-7167-8963-9.

[112] Scerri, E. (2012). "Mendeleev's Periodic Table Is Finally Completed and What To Do about Group 3?". *Chemistry International* **34** (4).

[113] Eliav, E.; Kaldor, U.; Ishikawa, Y. (1995). "Transition energies of ytterbium, lutetium, and lawrencium by the relativistic coupled-cluster method". *Phys. Rev. A* **52**: 291–296. Bibcode:1995PhRvA..52..291E. doi:10.1103/PhysRevA.52.291.

[114] Zou, Yu; Froese, Fischer C.; Uiterwaal, C.; Wanner, J.; Kompa, K.-L. (2002). "Resonance Transition Energies and Oscillator Strengths in Lutetium and Lawrencium". *Phys. Rev. Lett.* **88** (18): 183001. Bibcode:2002PhRvL..88b3001M. doi:10.1103/PhysRevLett.88.023001. PMID 12005680.

[115] Barbalace, Kenneth. "Periodic Table of Elements". Environmental Chemistry.com. Retrieved 14 April 2007.

[116] "WebElements Periodic Table of the Elements". Webelements.com. Retrieved 3 April 2010.

[117] "Periodic Table of the Elements". International Union of Pure and Applied Chemistry. Retrieved 3 April 2010.

[118] "Visual Elements". Royal Society of Chemistry. Retrieved 4 July 2011.

[119] Dolg, Michael. "Lanthanides and Actinides" (PDF). *Max-Planck-Institut für Physik komplexer Systeme, Dresden, Germany*. CLA01. Retrieved 4 July 2011.

[120]Francl, Michelle (May 2009)."Table manners"(PDF).*Nature Chemistry***1**(2): 97–98.Bibcode:2009NatCh...1...97F.doi:10.1038. PMID 21378810.

30.10   Bibliography

- Ball, Philip (2002). *The Ingredients: A Guided Tour of the Elements*. Oxford: Oxford University Press. ISBN 0-19-284100-9.

- Chang, Raymond (2002). *Chemistry* (7th ed.). New York: McGraw-Hill Higher Education. ISBN 978-0-19-284100-1.

- Gray, Theodore (2009). *The Elements: A Visual Exploration of Every Known Atom in the Universe*. New York: Black Dog & Leventhal Publishers. ISBN 978-1-57912-814-2.

- Greenwood, Norman N.; Earnshaw, Alan (1984). *Chemistry of the Elements*. Oxford: Pergamon Press. ISBN 0-08-022057-6.

- Huheey, JE; Keiter, EA; Keiter, RL. *Principles of structure and reactivity* (4th ed.). New York: Harper Collins College Publishers. ISBN 0-06-042995-X.

- Moore, John (2003). *Chemistry For Dummies*. New York: Wiley Publications. p. 111. ISBN 978-0-7645-5430-8. OCLC 51168057.

- Scerri, Eric (2007). *The periodic table: Its story and its significance*. Oxford: Oxford University Press. ISBN 0-19-530573-6.

- Scerri, Eric R. (2011). *The periodic table: A very short introduction*. Oxford: Oxford University Press. ISBN 978-0-19-958249-5.

- Venable, F P (1896). *The Development of the Periodic Law*. Easton PA: Chemical Publishing Company.

30.11 External links

- M. Dayah. "Dynamic Periodic Table". Retrieved 14 May 2012.

- Brady Haran. "The Periodic Table of Videos". University of Nottingham. Retrieved 14 May 2012.

- Mark Winter. "WebElements: the periodic table on the web". University of Sheffield. Retrieved 14 May 2012.

- Mark R. Leach. "The INTERNET Database of Periodic Tables". Retrieved 14 May 2012.

Chapter 31

Exotic matter

In physics, **exotic matter** is matter that somehow deviates from normal matter and has "exotic" properties. A more broad definition of exotic matter is any kind of non-baryonic matter—that is not made of baryons, the subatomic particles, such as protons and neutrons, of which the ordinary matter is composed.[1] Exotic mass has been considered a colloquial term for matters such as dark matter, negative mass, or imaginary mass.

31.1 Types of exotic matter

There are several types of exotic matter:

- Hypothetical particles that have "exotic" physical properties that would violate known laws of physics, such as a particle having a negative mass.

- Hypothetical particles that have not yet been encountered, such as exotic baryons, but whose properties would be within the realm of mainstream physics if found to exist.

- States of matter that are not commonly encountered, such as Bose–Einstein condensates and quark–gluon plasma, but whose properties are perfectly within the realm of mainstream physics.

- States of matter that are poorly understood, such as dark matter.

- Ordinary matter placed under high pressure.

31.2 Negative mass

Main article: Negative mass

Negative mass would possess some strange properties, such as accelerating in the direction opposite of applied force. For example, an object with negative inertial mass and positive electric charge would accelerate away from objects with negative charge, and towards objects with positive charge, the opposite of the normal rule that like charges repel and opposite charges attract. This behaviour can produce bizarre results: for instance, a gas containing a mixture of positive and negative matter particles will have the positive matter portion increase in temperature without bound. However, the negative matter portion gains negative temperature at the same rate, again balancing out.

Despite being inconsistent with the expected behavior of "normal" matter, negative mass is mathematically consistent and introduces no violation of conservation of momentum or energy. It is used in certain speculative theories, such as on the construction of wormholes. The closest known real representative of such exotic matter is the region of pseudo-negative-pressure density produced by the Casimir effect.

31.3 Imaginary mass

Main article: Tachyon § Mass

A hypothetical particle with imaginary rest mass would always travel faster than the speed of light. Such particles are called tachyons. There is no confirmed existence of tachyons.

$$E = \frac{m \cdot c^2}{\sqrt{1 - \frac{|\mathbf{v}|^2}{c^2}}}$$

If the rest mass m is imaginary this implies that the denominator is imaginary because the total energy is an observable and thus must be real. Therefore the quantity under the square root must be negative, which can only happen if v is greater than c. As noted by Gregory Benford *et al.*, special relativity implies that tachyons, if they existed, could be used to communicate backwards in time[2] (see tachyonic antitelephone). Because time travel is considered to be non-physical, tachyons are believed by physicists either to not exist, or else to be incapable of interacting with normal matter.

In quantum field theory, imaginary mass would induce tachyon condensation.

31.4 Materials at high pressure

At high pressure, materials such as NaCl in the presence of an excess of either chlorine or sodium were transformed into compounds "forbidden" by classical chemistry, such as Na
3Cl and NaCl
3. Quantum mechanical calculations predict the possibility of other compounds, such as NaCl
7, Na
3Cl
2, Na
2Cl, and Na
3Cl. The materials are thermodynamically stable at high pressures. Such compounds may exist in natural environments that exist at high pressure, such as the deep ocean or inside planetary cores. The materials have potentially useful properties. For instance, Na
3Cl is a two-dimensional metal, made of layers of pure sodium and salt that can conduct electricity. The salt layers act as insulators while the sodium layers act as conductors.[3][4]

31.5 See also

- Antimatter

- Dark energy

- Dark matter

- Gravitational interaction of antimatter

- Mirror matter

- Negative energy

- Negative mass

- Strange matter

- QCD matter

31.6 References

[1] "Exotic matter". daviddarling.info. Retrieved 2015-06-24.

[2] G. A. Benford, D. L. Book, and W. A. Newcomb (1970). "The Tachyonic Antitelephone". *Physical Review D* **2**: 263. Bibcode:1970PhRvD...2..263B. doi:10.1103/PhysRevD.2.263.

[3] "Scientists turn table salt into forbidden compounds that violate textbook rules". Gizmag.com. Retrieved 2014-01-21.

[4] Zhang, W.; Oganov, A. R.; Goncharov, A. F.; Zhu, Q.; Boulfelfel, S. E.; Lyakhov, A. O.; Stavrou, E.; Somayazulu, M.; Prakapenka, V. B.; Konôpková, Z. (2013). "Unexpected Stable Stoichiometries of Sodium Chlorides". *Science* **342** (6165): 1502–1505. doi:10.1126/science.1244989. PMID 24357316.

Chapter 32

Dark matter

Not to be confused with antimatter, dark energy, dark fluid, or dark flow. For other uses, see Dark Matter (disambiguation)
Dark matter is a hypothetical kind of matter that cannot be seen with telescopes but would account for most of the matter in the universe. The existence and properties of dark matter are inferred from its gravitational effects on visible matter, radiation, and the large-scale structure of the universe. Other than neutrinos, a form of hot dark matter, it has not been detected directly, making it one of the greatest mysteries in modern astrophysics.

Dark matter neither emits nor absorbs light or any other electromagnetic radiation at any significant level. According to the Planck mission team, and based on the standard model of cosmology, the total mass–energy of the known universe contains 4.9% ordinary matter, 26.8% dark matter and 68.3% dark energy.[2][3] Thus, dark matter is estimated to constitute 84.5% of the total matter in the universe, while dark energy plus dark matter constitute 95.1% of the total mass–energy content of the universe.[4][5][6]

Astrophysicists hypothesized the existence of dark matter in the light of discrepancies between the mass of large astronomical objects determined from their gravitational effects, and their mass as calculated from the observable matter (stars, gas, and dust) that they can be seen to contain. Their gravitational effects suggest that their masses are greater than the observable matter survey suggests. Dark matter was postulated by Jan Oort in 1932, albeit based upon flawed or inadequate evidence, to account for the orbital velocities of stars in the Milky Way and by Fritz Zwicky in 1933 to account for evidence of "missing mass" in the orbital velocities of galaxies in clusters. Adequate evidence from galaxy rotation curves was discovered by Horace W. Babcock in 1939, but was not attributed to dark matter. The first to postulate dark matter based upon robust evidence was Vera Rubin in the 1960s–1970s, using galaxy rotation curves.[7][8] Subsequently many other observations have indicated the presence of dark matter in the universe, including gravitational lensing of background objects by galaxy clusters such as the Bullet Cluster, the temperature distribution of hot gas in galaxies and clusters of galaxies and, more recently, the pattern of anisotropies in the cosmic microwave background. According to consensus among cosmologists, dark matter is composed primarily of a not yet characterized type of subatomic particle.[9][10] The search for this particle, by a variety of means, is one of the major efforts in particle physics today.[11]

Although the existence of dark matter is generally accepted by the mainstream scientific community, some alternative theories of gravity have been proposed, such as MOND and TeVeS, which try to account for the anomalous observations without requiring additional matter. However, these theories cannot account for the properties of galaxy clusters.[12]

32.1   Overview

Dark matter's existence is inferred from gravitational effects on visible matter and gravitational lensing of background radiation, and was originally hypothesized to account for discrepancies between calculations of the mass of galaxies, clusters of galaxies and the entire universe made through dynamical and general relativistic means, and calculations based on the mass of the visible "luminous" matter these objects contain: stars and the gas and dust of the interstellar and intergalactic medium.[13]

The most widely accepted explanation for these phenomena is that dark matter exists and that it is most probably[9]

Dark matter *is invisible. Based on the effect of gravitational lensing, a ring of* dark matter *has been inferred in this image of a galaxy cluster (CL0024+17) and has been represented in blue.*[1]

composed of weakly interacting massive particles (WIMPs) that interact only through gravity and the weak force. Alternative explanations have been proposed, and there is not yet sufficient experimental evidence to determine whether any of them are correct. Many experiments to detect proposed dark matter particles through non-gravitational means are under way.[11]

One other theory suggests the existence of a "Hidden Valley", a parallel world made of dark matter having very little in common with matter we know,[14] and that could only interact with our visible universe through gravity.[15][16]

According to observations of structures larger than star systems, as well as Big Bang cosmology interpreted under the Friedmann equations and the Friedmann–Lemaître–Robertson–Walker metric, dark matter accounts for 26.8% of the mass-energy content of the observable universe. In comparison, ordinary (baryonic) matter accounts for only 4.9% of the mass-energy content of the observable universe, with the remainder being attributable to dark energy.[3] From these figures, matter accounts for 31.7% of the mass-energy content of the universe, and 84.5% of the matter is dark matter.

Dark matter plays a central role in state-of-the-art modeling of cosmic structure formation and galaxy formation and evolution and has measurable effects on the anisotropies observed in the cosmic microwave background (CMB). All these lines of evidence suggest that galaxies, clusters of galaxies, and the universe as a whole contain far more matter than that which is easily visible with electromagnetic radiation.[15]

Important as dark matter is thought to be in the cosmos, direct evidence of its existence and a concrete understanding of its nature have remained elusive. Though the theory of dark matter remains the most widely accepted theory to explain the anomalies in observed galactic rotation, some alternative theoretical approaches have been developed which broadly fall into the categories of modified gravitational laws and quantum gravitational laws.[17]

32.2 Baryonic and nonbaryonic dark matter

There are three separate lines of evidence that suggest the majority of dark matter is not made of baryons (ordinary matter including protons and neutrons):

- The theory of Big Bang nucleosynthesis, which predicts the observed abundance of the chemical elements,[18] predicts that baryonic matter accounts for around 4–5 percent of the critical density of the universe. In contrast, evidence from large-scale structure and other observations indicates that the total matter density is about 30% of the critical density.

- Large astronomical searches for gravitational microlensing, including the MACHO, EROS and OGLE projects, have shown that only a small fraction of the dark matter in the Milky Way can be hiding in dark compact objects; the excluded range covers objects above half the Earth's mass up to 30 solar masses, excluding nearly all the plausible candidates.

- Detailed analysis of the small irregularities (anisotropies) in the cosmic microwave background observed by WMAP and Planck shows that around five-sixths of the total matter is in a form which does not interact significantly with ordinary matter or photons except through gravitational effects.

A small proportion of dark matter may be baryonic dark matter: astronomical bodies, such as massive compact halo objects, which are composed of ordinary matter but emit little or no electromagnetic radiation. The study of nucleosynthesis in the Big Bang gives an upper bound on the amount of baryonic matter in the universe,[19] which indicates that the vast majority of dark matter in the universe cannot be baryons, and thus does not form atoms. It also cannot interact with ordinary matter via electromagnetic forces; in particular, dark matter particles do not carry any electric charge.

Candidates for nonbaryonic dark matter are hypothetical particles such as axions, or supersymmetric particles; neutrinos can only form a small fraction of the dark matter, due to limits from large-scale structure and high-redshift galaxies. Unlike baryonic dark matter, nonbaryonic dark matter does not contribute to the formation of the elements in the early universe ("Big Bang nucleosynthesis")[9] and so its presence is revealed only via its gravitational attraction. In addition, if the particles of which it is composed are supersymmetric, they can undergo annihilation interactions with themselves, possibly resulting in observable by-products such as gamma rays and neutrinos ("indirect detection").[20]

Nonbaryonic dark matter is classified in terms of the mass of the particle(s) that is assumed to make it up, and/or the typical velocity dispersion of those particles (since more massive particles move more slowly). There are three prominent hypotheses on nonbaryonic dark matter, called cold dark matter (CDM), warm dark matter (WDM), and hot dark matter (HDM); some combination of these is also possible. The most widely discussed models for nonbaryonic dark matter are based on the cold dark matter hypothesis, and the corresponding particle is most commonly assumed to be a weakly interacting massive particle (WIMP). Hot dark matter may include (massive) neutrinos, but observations imply that only a small fraction of dark matter can be hot. Cold dark matter leads to a "bottom-up" formation of structure in the universe while hot dark matter would result in a "top-down" formation scenario; since the late 1990s, the latter has been ruled out by observations of high-redshift galaxies such as the Hubble Ultra-Deep Field.[11]

32.3   Observational evidence

The first person to interpret evidence and infer the presence of dark matter was Dutch astronomer Jan Oort, a pioneer in radio astronomy, in 1932.[22] Oort was studying stellar motions in the local galactic neighbourhood and found that the mass in the galactic plane must be greater than what was observed, but this measurement was later determined to be essentially erroneous.[23] In 1933, the Swiss astrophysicist Fritz Zwicky, who studied clusters of galaxies while working at the California Institute of Technology, made a similar inference.[24][25] Zwicky applied the virial theorem to the Coma cluster of galaxies and obtained evidence of unseen mass. Zwicky estimated the cluster's total mass based on the motions of galaxies near its edge and compared that estimate to one based on the number of galaxies and total brightness of the cluster. He estimated that there was about 400 times more mass than was visually observable. The gravity effect of the visible galaxies in the cluster would be far too small for such fast orbits, unless there was mass hidden from visual observation. This is known as the "missing mass problem". Based on these conclusions, Zwicky inferred that there must be some non-visible form of matter which would provide enough mass and gravitation attraction to hold the cluster together. Zwicky's estimates were off by more than an order of magnitude. Had he erred in the opposite direction by as much, he would have had to try explain the opposite – why there was too much visible matter relative to the gravitational observations – and his observations would have indicated dark energy rather than dark matter.[26]

Much of the evidence for dark matter comes from the study of the motions of galaxies.[28] Many of these appear to be fairly uniform, so by the virial theorem, the total kinetic energy should be half the total gravitational binding energy of the galaxies. Observationally, however, the total kinetic energy is found to be much greater. In particular, assuming the gravitational mass is due to only the visible matter of the galaxy, stars far from the center of galaxies have much higher velocities than are predicted by the virial theorem. Galactic rotation curves, which illustrate the velocity of rotation versus the distance from the galactic center, show the well known phenomenology that cannot be explained by the presence of the visible matter only. Assuming that the visible material makes up only a small part of the cluster's mass is the most straightforward way of accounting for this discrepancy. The distribution of dark matter in galaxies required to explain the motion of the observed baryonic matter suggests the presence of a roughly spherically symmetric, centrally concentrated halo of dark matter with the visible matter concentrated in a disc at the center. Low surface brightness dwarf galaxies are important sources of information for studying dark matter, as they have an uncommonly low ratio of visible matter to dark matter, and have few bright stars at the center which would otherwise impair observations of the rotation curve of outlying stars.

Gravitational lensing observations of galaxy clusters allow direct estimates of the gravitational mass based on its effect on light coming from background galaxies, since large collections of matter (dark or otherwise) will gravitationally deflect light. In clusters such as Abell 1689, lensing observations confirm the presence of considerably more mass than is indicated by the clusters' light alone. In the Bullet Cluster, lensing observations show that much of the lensing mass is separated from the X-ray-emitting baryonic mass. In July 2012, lensing observations were used to identify a "filament" of dark matter between two clusters of galaxies, as cosmological simulations have predicted.[29]

32.3.1   Galaxy rotation curves

Main article: Galaxy rotation curve
The first robust indications that the mass to light ratio was anything other than unity came from measurements of galaxy rotation curves. In 1939, Horace W. Babcock reported in his PhD thesis measurements of the rotation curve for the Andromeda nebula which suggested that the mass-to-luminosity ratio increases radially.[30] He, however, attributed it to either absorption of light within the galaxy or modified dynamics in the outer portions of the spiral and not to any form of missing matter.

In the late 1960s and early 1970s, Vera Rubin was the first to both make robust measurements indicating the existence of dark matter and attribute them to dark matter. Rubin worked with a new sensitive spectrograph that could measure the velocity curve of edge-on spiral galaxies to a greater degree of accuracy than previously.[8] Together with fellow staff-member Kent Ford, Rubin announced at a 1975 meeting of the American Astronomical Society the discovery that most stars in spiral galaxies orbit at roughly the same speed, which implied that the mass densities of the galaxies were uniform well beyond the regions containing most of the stars (the galactic bulge), a result independently found in 1978.[31] An influential paper presented Rubin's results in 1980.[32] Rubin's observations and calculations showed that most galaxies must contain about six times as much "dark" mass as could be accounted for by the visible stars. Eventually other astronomers began to corroborate her work. It soon became well-established that most galaxies were dominated by "dark matter":

- Low-surface-brightness (LSB) galaxies.[33] LSB galaxies are probably everywhere dark matter-dominated, with the observed stellar populations making only a small contribution to their total mass. Such a property is extremely important as it allows one to avoid the difficulties associated with the deprojection and disentanglement of the dark and visible matter contributions to the rotation curves.[11]

- Spiral galaxies.[34] Rotation curves of both low and high surface luminosity galaxies suggest a universal rotation curve, which can be expressed as the sum of an exponential distribution of visible matter that is maximum at the center and tapering to zero at great distances, and a spherical dark matter halo with a flat core of radius r_0 and density $\rho_0 = 4.5 \times 10^{-2} (r_0/kpc)^{-2/3} \, M\odot pc^{-3}$.

- Elliptical galaxies. Some elliptical galaxies show evidence for dark matter via strong gravitational lensing,[35] X-ray evidence reveals the presence of extended atmospheres of hot gas that fill the dark haloes of isolated elliptical galaxies and whose hydrostatic support provides evidence for the existence of dark matter. Other ellipticals have low velocities in their outskirts (tracked for example by the motion of planetary nebulae embedded within) and were interpreted as not having dark matter haloes.[11] However, simulations of disk-galaxy mergers suggest that stars may have been torn by tidal forces from their original galaxies during the first close passage and put on outgoing trajectories, explaining the low velocities of the remaining stars even with the presence of a dark matter halo.[36] More research is needed to clarify this situation.

Simulated dark matter haloes have significantly steeper density profiles (having central cusps) than are inferred from observations, which is a problem for cosmological models with dark matter at the smallest scale of galaxies (as of 2008).[11] This may only be a problem of resolution: star-forming regions which might alter the dark matter distribution via outflows of gas have been too small to resolve and model simultaneously with larger dark matter clumps. A recent simulation[37] of a dwarf galaxy, that included these star-forming regions, reported that strong outflows from supernovae remove low-angular-momentum gas, which inhibits the formation of a galactic bulge and decreases the dark matter density to less than half of what it would have been in the central kiloparsec. These simulation predictions—bulgeless and with shallow central dark matter density profiles—correspond closely to observations of actual dwarf galaxies. There are no such discrepancies at the larger scales of clusters of galaxies and greater, or in the outer regions of haloes of galaxies.

The exceptions to this general picture of dark matter haloes for galaxies appear to be galaxies with mass-to-light ratios that are close to that of the stars they contain. Otherwise, numerous observations have been made that do indicate the presence of dark matter in various parts of the cosmos, such as observations of the cosmic microwave background, of supernovas used as distance measures, of gravitational lensing at various scales, and many types of sky survey. Starting with Rubin's findings for spiral galaxies, such robust observational evidence for dark matter has collected over the decades to the point that by the 1980s most astrophysicists have accepted its existence.[38] As a unifying concept, dark matter is one of the dominant features considered in the analysis of structures on the order of galactic scale and larger.

32.3.2 Velocity dispersions of galaxies

Rubin's pioneering work has stood the test of time. Measurements of velocity curves in spiral galaxies were soon followed up with velocity dispersions of elliptical galaxies.[39] While some elliptical galaxies display lower mass-to-light ratios, measurements of ellipticals generally indicate a relatively high dark matter content. Likewise, measurements of the diffuse interstellar gas found at the edge of galaxies indicate not only dark matter distributions that extend beyond the visible limit of the galaxies, but also that the galaxies are virialized (i.e. gravitationally bound and orbiting each other with velocities which appear to disproportionately correspond to predicted orbital velocities of general relativity) up to ten times their visible radii.[40] This has the effect of pushing up the dark matter as a fraction of the total matter from 50% as measured by Rubin to the now accepted value of nearly 95%.

There are places where dark matter seems to be a small component or totally absent. Globular clusters show little evidence that they contain dark matter,[41] though their orbital interactions with galaxies do show evidence for galactic dark matter. For some time, measurements of the velocity profile of stars seemed to indicate concentration of dark matter in the disk of the Milky Way. It now appears, however, that the high concentration of baryonic matter in the disk of the galaxy (especially in the interstellar medium) can account for this motion. Galaxy mass profiles are thought to look very different from the light profiles. The typical model for dark matter galaxies is a smooth, spherical distribution in virialized halos. Such would have to be the case to avoid small-scale (stellar) dynamical effects. Recent research reported in January 2006 from the University of Massachusetts Amherst would explain the previously mysterious warp in the disk of the Milky Way by the interaction of the Large and Small Magellanic Clouds and the predicted 20 fold increase in mass of the Milky Way taking into account dark matter.[42]

In 2005, astronomers from Cardiff University claimed to have discovered a galaxy made almost entirely of dark matter, 50 million light years away in the Virgo Cluster, which was named VIRGOHI21.[43] Unusually, VIRGOHI21 does not appear to contain any visible stars: it was seen with radio frequency observations of hydrogen. Based on rotation profiles, the scientists estimate that this object contains approximately 1000 times more dark matter than hydrogen and has a total mass of about 1/10 that of the Milky Way. For comparison, the Milky Way is estimated to have roughly 10 times as much dark matter as ordinary matter. Models of the Big Bang and structure formation have suggested that such dark galaxies should be very common in the universe, but none had previously been detected.

There are some galaxies, such as NGC 3379, whose velocity profile indicates an absence of dark matter.[44]

32.3.3   Galaxy clusters and gravitational lensing

Galaxy clusters are especially important for dark matter studies since their masses can be estimated in three independent ways:

- From the scatter in radial velocities of the galaxies within the clusters (as in Zwicky's early observations, but with more accurate measurements and much larger samples).

- From X-rays emitted by very hot gas within the clusters. The temperature and density of the gas can be estimated from the energy and flux of the X-rays, and hence the gas pressure derived; assuming pressure and gravity balance, this enables the mass profile of the cluster to be derived. Many of the experiments of the Chandra X-ray Observatory use this technique to independently determine the mass of clusters. These observations generally indicate that baryonic mass is approximately 12–15 percent, in reasonable agreement with the Planck spacecraft cosmic average of 15.5–16 percent.[45]

- From their gravitational lensing effects on background objects (usually more distant galaxies). This is observed as "strong lensing" (multiple images) near the cluster core, and weak lensing (shape distortions) in the outer parts. Several large Hubble projects have used this method to measure cluster masses.

Generally these three methods are in reasonable agreement, that clusters contain much more matter than suggested by the visible components of galaxies and gas.

A gravitational lens is formed when the light from a more distant source (such as a quasar) is "bent" around a massive object (such as a cluster of galaxies) lying inline between the source object and the observer. The process is known as gravitational lensing.

The galaxy cluster Abell 2029 is composed of thousands of galaxies enveloped in a cloud of hot gas, and an amount of dark matter equivalent to more than 10^{14} $M\odot$. At the center of this cluster is an enormous, elliptically shaped galaxy that is thought to have been formed from the mergers of many smaller galaxies.[46] The measured orbital velocities of galaxies within galactic clusters have been found to be consistent with dark matter observations.

Another important tool for future dark matter observations is gravitational lensing. Lensing relies on the bending of light, as described by general relativity, to predict masses without relying on observations of the distant galaxies dynamics, and so is a completely independent means of measuring the dark matter. Strong lensing, the observed distortion of background galaxies into arcs when their light passes through such a gravitational lens, has been observed around a few distant clusters including Abell 1689 (pictured).[47] By measuring the distortion geometry, the mass of the intervening cluster causing the phenomena can be obtained. In the dozens of cases where this has been done, the mass-to-light ratios obtained correspond to the dynamical dark matter measurements of clusters.[48]

Weak gravitational lensing investigates minute distortions of galaxies, using statistical analyses of vast galaxy surveys, caused by foreground objects. By examining the apparent shear deformation of the adjacent background galaxies, astrophysicists can characterize the mean distribution of dark matter and have found mass-to-light ratios that correspond to dark matter densities predicted by other large-scale structure measurements.[49] The correspondence of the two gravitational lens techniques to other dark matter measurements has convinced almost all astrophysicists that dark matter actually exists as a major component of the universe's composition.

The most direct observational evidence to date for dark matter comes from a system known as the Bullet Cluster. In most regions of the universe, dark matter and visible matter are found together,[50] as expected due of their mutual gravitational attraction. In the Bullet Cluster however, a collision between two galaxy clusters appears to have caused a separation of dark matter and baryonic matter. X-ray observations show that much of the baryonic matter (in the form of 10^7–10^8 Kelvin[51] gas or plasma) in the system is concentrated in the center of the system. Electromagnetic interactions between passing gas particles caused them to slow and settle near the point of impact of those galaxies. However, weak gravitational lensing observations of the same system show that much of the mass resides outside of the central region of baryonic gas. Because dark matter does not interact by electromagnetic forces, it would not have been slowed as the X-ray visible gas, so the dark matter components of the two clusters passed through each other without slowing substantially, throwing the dark matter further out than that of the baryonic gas. This accounts for the separation. Unlike the galactic rotation curves, this evidence for dark matter is independent of the details of Newtonian gravity, so it is claimed to be direct evidence of the existence of dark matter.[51]

Another galaxy cluster, known as the Train Wreck Cluster/Abell 520, initially appeared to have an unusually massive and dark matter core containing few of the cluster's galaxies, which presented problems for standard dark matter models.[52] However, more precise observations since that time have shown that the earlier observations were misleading, and that the distribution of dark matter and its ratio to normal matter are very similar to those in galaxies in general, making novel explanations unnecessary.[51]

The observed behavior of dark matter in clusters constrains whether and how much dark matter scatters off other dark matter particles, quantified as its self-interaction cross section. More simply, the question is whether the dark matter has pressure, and thus can be described as a perfect fluid that has no damping.[53] The distribution of mass (and thus dark matter) in galaxy clusters has been used to argue both for[54] and against[55] the existence of significant self-interaction in dark matter. Specifically, the distribution of dark matter in merging clusters such as the Bullet Cluster shows that dark matter scatters off other dark matter particles only very weakly if at all.[56]

Researchers are conducting a wide-area survey of the distribution of dark matter and changes in distribution over time determine the impact of dark energy accelerating the expansion of the universe. The survey is using weak lensing to analyze background light, bent by dark matter, to determine how dark matter is distributed in the foreground. The analysis of dark matter and its effects could determine how dark matter assembled over time, which can be related to the history of the expansion of the universe, and could reveal some physical properties of dark energy, its strength and how it has changed over time. The survey is observing galaxies more than a billion light-years away, across an area greater than a thousand square degrees (about one fortieth of the entire sky). The survey is using the Subaru telescope.[57][58]

32.3.4 Cosmic microwave background

Main article: Cosmic microwave background
See also: Wilkinson Microwave Anisotropy Probe

Angular fluctuations in the cosmic microwave background (CMB) spectrum provide evidence for dark matter. Since the 1964 discovery and confirmation of the CMB radiation,[59] many measurements of the CMB have supported and constrained this theory. The NASA Cosmic Background Explorer (COBE) found that the CMB spectrum to be a blackbody spectrum with a temperature of 2.726 K. In 1992, COBE detected fluctuations (anisotropies) in the CMB spectrum, at a level of about one part in 10^5.[60] In the following decade, CMB anisotropies were further investigated by a large number of ground-based and balloon experiments. The primary goal of those was to measure the angular scale of the first acoustic peak of the power spectrum of the anisotropies, for which COBE did not have sufficient resolution. In 2000–2001, several experiments, most notably BOOMERanG[61] found the universe to be almost spatially flat by measuring the typical angular size of the anisotropies. During the 1990s, the first peak was measured with increasing sensitivity and by 2000 the BOOMERanG experiment reported that the highest power fluctuations occur at scales of approximately one degree. These measurements were able to rule out cosmic strings as the leading theory of cosmic structure formation, and suggested cosmic inflation was the correct theory.

A number of ground-based interferometers provided measurements of the fluctuations with higher accuracy over the next three years, including the Very Small Array, the Degree Angular Scale Interferometer (DASI) and the Cosmic Background Imager (CBI). DASI made the first detection of the polarization of the CMB,[62][63] and the CBI provided the first E-mode polarization spectrum with compelling evidence that it is out of phase with the T-mode spectrum.[64] COBE's successor, the Wilkinson Microwave Anisotropy Probe (WMAP) has provided the most detailed measurements of (large-scale) anisotropies in the CMB as of 2009 with ESA's Planck spacecraft returning more detailed results in 2012-2014.[65] WMAP's measurements played the key role in establishing the current Standard Model of Cosmology, namely the Lambda-CDM model, a flat universe dominated by dark energy, supplemented by dark matter and atoms with density fluctuations seeded by a Gaussian, adiabatic, nearly scale invariant process. The basic properties of this universe are determined by five numbers: the density of matter, the density of atoms, the age of the universe (or equivalently, the Hubble constant today), the amplitude of the initial fluctuations, and their scale dependence.

A successful Big Bang cosmology theory must fit with all available astronomical observations, including the CMB. In cosmology, the CMB is explained as relic radiation from shortly after the big bang. The anisotropies in the CMB are explained as being the result of acoustic oscillations in the photon-baryon plasma (prior to the emission of the CMB after the photons decouple from the baryons 379,000 years after the Big Bang) whose restoring force is gravity.[66] Ordinary (baryonic) matter interacts strongly by way of radiation whereas dark matter particles, such as WIMPs for example, do not; both affect the oscillations by way of their gravity, so the two forms of matter will have different effects. The typical angular scales of the oscillations in the CMB, measured as the power spectrum of the CMB anisotropies, thus reveal the different effects of baryonic matter and dark matter. The CMB power spectrum shows a large first peak and smaller successive peaks, with three peaks resolved as of 2009.[65] The first peak tells mostly about the density of baryonic matter and the third peak mostly about the density of dark matter, measuring the density of matter and the density of atoms in the universe.

32.3.5   Sky surveys and baryon acoustic oscillations

Main article: Baryon acoustic oscillations

The acoustic oscillations in the early universe (see the previous section) have left their imprint on visible matter by way of Baryon Acoustic Oscillation (BAO) clustering, in a way that can be measured with sky surveys such as the Sloan Digital Sky Survey and the 2dF Galaxy Redshift Survey.[67] These measurements are consistent with those of the CMB derived from the WMAP spacecraft and further constrain the Lambda CDM model and dark matter. Note that the CMB data and the BAO data measure the acoustic oscillations at very different distance scales.[66]

32.3.6   Type Ia supernovae distance measurements

Main article: Type Ia supernova

Type Ia supernovae can be used as "standard candles" to measure extragalactic distances, and extensive data sets of these supernovae can be used to constrain cosmological models.[68] They constrain the dark energy density $\Omega\Lambda = \sim 0.713$ for a flat, Lambda CDM universe and the parameter w for a quintessence model. Once again, the values obtained are roughly consistent with those derived from the WMAP observations and further constrain the Lambda CDM model and (indirectly) dark matter.[66]

32.3.7   Lyman-alpha forest

Main article: Lyman-alpha forest

In astronomical spectroscopy, the Lyman-alpha forest is the sum of the absorption lines arising from the Lyman-alpha transition of the neutral hydrogen in the spectra of distant galaxies and quasars. Observations of the Lyman-alpha forest can also be used to constrain cosmological models.[69] These constraints are again in agreement with those obtained from WMAP data.

32.3.8   Structure formation

Main article: Structure formation

Dark matter is crucial to the Big Bang model of cosmology as a component which corresponds directly to measurements of the parameters associated with Friedmann cosmology solutions to general relativity. In particular, measurements of the cosmic microwave background anisotropies correspond to a cosmology where much of the matter interacts with photons more weakly than the known forces that couple light interactions to baryonic matter. Likewise, a significant amount of non-baryonic, cold matter is necessary to explain the large-scale structure of the universe.

Observations suggest that structure formation in the universe proceeds hierarchically, with the smallest structures collapsing first and followed by galaxies and then clusters of galaxies. As the structures collapse in the evolving universe, they begin to "light up" as the baryonic matter heats up through gravitational contraction and approaches hydrostatic pressure balance. Originally, baryonic matter had too high a temperature, and pressure left over from the Big Bang to allow collapse and form smaller structures, such as stars, via the Jeans instability. Dark matter acts as a compactor allowing the creation of structure where there would not have been any. This model not only corresponds with statistical surveying of the visible structure in the universe but also corresponds precisely to the dark matter predictions of the cosmic microwave background.

This *bottom up* model of structure formation requires something like cold dark matter to succeed. Large computer simulations of billions of dark matter particles have been used[71] to confirm that the cold dark matter model of structure formation is consistent with the structures observed in the universe through galaxy surveys, such as the Sloan Digital Sky Survey and 2dF Galaxy Redshift Survey, as well as observations of the Lyman-alpha forest. These studies have been crucial in constructing the Lambda-CDM model which measures the cosmological parameters, including the fraction of the universe made up of baryons and dark matter.

There are, however, several points of tension between observation and simulations of structure formation driven by dark matter. There is evidence that there exist 10 to 100 times fewer small galaxies than permitted by what the dark matter theory of galaxy formation predicts.[72][73] This is known as the dwarf galaxy problem. In addition, the simulations predict dark matter distributions with a very dense cusp near the centers of galaxies, but the observed halos are smoother than predicted.

32.4   History of the search for its composition

Although dark matter had historically been inferred from many astronomical observations, its composition long remained speculative. Early theories of dark matter concentrated on hidden heavy normal objects (such as black holes, neutron stars, faint old white dwarfs, and brown dwarfs) as the possible candidates for dark matter, collectively known as massive compact halo objects or MACHOs. Astronomical surveys for gravitational microlensing, including the MACHO, EROS and OGLE projects, along with Hubble telescope searches for ultra-faint stars, have not found enough of these hidden MACHOs.[74][75][76] Some hard-to-detect baryonic matter, such as MACHOs and some forms of gas, were additionally speculated to make a contribution to the overall dark matter content, but evidence indicated such would constitute only a small portion.[77][78][79]

Furthermore, data from a number of lines of other evidence, including galaxy rotation curves, gravitational lensing, structure formation, and the fraction of baryons in clusters and the cluster abundance combined with independent evidence for the baryon density, indicated that 85–90% of the mass in the universe does not interact with the electromagnetic force. This "nonbaryonic dark matter" is evident through its gravitational effect. Consequently, the most commonly held view was that dark matter is primarily non-baryonic, made of one or more elementary particles other than the usual electrons, protons, neutrons, and known neutrinos. The most commonly proposed particles then became WIMPs (Weakly Interacting Massive Particles, including neutralinos), axions, or sterile neutrinos, though many other possible candidates have been proposed.

Dark matter candidates can be approximately divided into three classes, called *cold*, *warm* and *hot* dark matter.[80] These categories do not correspond to an actual temperature, but instead refer to how fast the particles were moving, thus how far they moved due to random motions in the early universe, before they slowed due to the expansion of the universe – this is an important distance called the "free streaming length". Primordial density fluctuations smaller than this free-streaming length get washed out as particles move from overdense to underdense regions, while fluctuations larger than the free-streaming length are unaffected; therefore this free-streaming length sets a minimum scale for structure formation.

- Cold dark matter – objects with a free-streaming length much smaller than a protogalaxy.[81]

- Warm dark matter – particles with a free-streaming length similar to a protogalaxy.

- Hot dark matter – particles with a free-streaming length much larger than a protogalaxy.[82]

Though a fourth category had been considered early on, called mixed dark matter, it was quickly eliminated (from the 1990s) since the discovery of dark energy.

As an example, Davis *et al.* wrote in 1985:

> Candidate particles can be grouped into three categories on the basis of their effect on the fluctuation spectrum (Bond *et al.* 1983). If the dark matter is composed of abundant light particles which remain relativistic until shortly before recombination, then it may be termed "hot". The best candidate for hot dark matter is a neutrino ... A second possibility is for the dark matter particles to interact more weakly than neutrinos, to be less abundant, and to have a mass of order 1 keV. Such particles are termed "warm dark matter", because they have lower thermal velocities than massive neutrinos ... there are at present few candidate particles which fit this description. Gravitinos and photinos have been suggested (Pagels and Primack 1982; Bond, Szalay and Turner 1982) ... Any particles which became nonrelativistic very early, and so were able to diffuse a negligible distance, are termed "cold" dark matter (CDM). There are many candidates for CDM including supersymmetric particles.[83]

The full calculations are quite technical, but an approximate dividing line is that "warm" dark matter particles became non-relativistic when the universe was approximately 1 year old and 1 millionth of its present size; standard hot big bang theory implies the universe was then in the radiation-dominated era (photons and neutrinos), with a photon temperature 2.7 million K. Standard physical cosmology gives the particle horizon size as 2ct in the radiation-dominated era, thus 2 light-years, and a region of this size would expand to 2 million light years today (if there were no structure formation). The actual free-streaming length is roughly 5 times larger than the above length, since the free-streaming length continues to grow slowly as particle velocities decrease inversely with the scale factor after they become non-relativistic; therefore, in this example the free-streaming length would correspond to 10 million light-years or 3 Mpc today, which is around the size containing on average the mass of a large galaxy.

The above temperature of 2.7 million K gives a typical photon energy of 250 electron-volts, thereby setting a typical mass scale for "warm" dark matter: particles much more massive than this, such as GeV – TeV mass WIMPs, would become non-relativistic much earlier than 1 year after the Big Bang and thus have a free-streaming length much smaller than a proto-galaxy, making them cold dark matter. Conversely, much lighter particles, such as neutrinos with masses of only a few eV, have a free-streaming length much larger than a proto-galaxy, thus making them hot dark matter.

32.4.1   Cold dark matter

Main article: Cold dark matter

Today, cold dark matter is the simplest explanation for most cosmological observations. "Cold" dark matter is dark matter composed of constituents with a free-streaming length much smaller than the ancestor of a galaxy-scale perturbation. This is currently the area of greatest interest for dark matter research, as hot dark matter does not seem to be viable for galaxy and galaxy cluster formation, and most particle candidates become non-relativistic at very early times, hence are classified as cold.

The composition of the constituents of cold dark matter is currently unknown. Possibilities range from large objects like MACHOs (such as black holes[84]) or RAMBOs, to new particles like WIMPs and axions. Possibilities involving normal baryonic matter include brown dwarfs, other stellar remnants such as white dwarfs, or perhaps small, dense chunks of heavy elements.

Studies of big bang nucleosynthesis and gravitational lensing have convinced most scientists[11][85][86][87][88][89] that MA-CHOs of any type cannot be more than a small fraction of the total dark matter.[9][85] Black holes of nearly any mass are ruled out as a primary dark matter constituent by a variety of searches and constraints.[85][87] According to A. Peter: "...the only *really plausible* dark-matter candidates are new particles."[86]

The DAMA/NaI experiment and its successor DAMA/LIBRA have claimed to directly detect dark matter particles passing through the Earth, but many scientists remain skeptical, as negative results from similar experiments seem incompatible with the DAMA results.

Many supersymmetric models naturally give rise to stable dark matter candidates in the form of the Lightest Supersymmetric Particle (LSP). Separately, heavy sterile neutrinos exist in non-supersymmetric extensions to the standard model that explain the small neutrino mass through the seesaw mechanism.

32.4.2 Warm dark matter

Main article: Warm dark matter

Warm dark matter refers to particles with a free-streaming length comparable to the size of a region which subsequently evolved into a dwarf galaxy. This leads to predictions which are very similar to cold dark matter on large scales, including the CMB, galaxy clustering and large galaxy rotation curves, but with less small-scale density perturbations. This reduces the predicted abundance of dwarf galaxies and may lead to lower density of dark matter in the central parts of large galaxies; some researchers consider this may be a better fit to observations. A challenge for this model is that there are no very well-motivated particle physics candidates with the required mass ~ 300 eV to 3000 eV.

There have been no particles discovered so far that can be categorized as warm dark matter. There is a postulated candidate for the warm dark matter category, which is the sterile neutrino: a heavier, slower form of neutrino which does not even interact through the Weak force unlike regular neutrinos. Interestingly, some modified gravity theories, such as Scalar-tensor-vector gravity, also require that a warm dark matter exist to make their equations work out.

32.4.3 Hot dark matter

Main article: Hot dark matter

Hot dark matter consists of particles that have a free-streaming length much larger than that of a proto-galaxy.

An example of hot dark matter is already known: the neutrino. Neutrinos were discovered quite separately from the search for dark matter, and long before it seriously began: they were first postulated in 1930, and first detected in 1956. Neutrinos have a very small mass: at least 100,000 times less massive than an electron. Other than gravity, neutrinos only interact with normal matter via the weak force making them very difficult to detect (the weak force only works over a small distance, thus a neutrino will only trigger a weak force event if it hits a nucleus directly head-on). This would make them 'weakly interacting light particles' (WILPs), as opposed to cold dark matter's theoretical candidates, the weakly interacting massive particles (WIMPs).

There are three different known flavors of neutrinos (i.e. the *electron*, *muon*, and *tau* neutrinos), and their masses are slightly different. The resolution to the solar neutrino problem demonstrated that these three types of neutrinos actually change and oscillate from one flavor to the others and back as they are in-flight. It's hard to determine an exact upper bound on the collective average mass of the three neutrinos (let alone a mass for any of the three individually). For example, if the average neutrino mass were chosen to be over 50 eV/c^2 (which is still less than 1/10,000th of the mass of an electron), just by the sheer number of them in the universe, the universe would collapse due to their mass. So other observations have served to estimate an upper-bound for the neutrino mass. Using cosmic microwave background data and other methods, the current conclusion is that their average mass probably does not exceed 0.3 eV/c^2 Thus, the normal forms of neutrinos cannot be responsible for the measured dark matter component from cosmology.[90]

Hot dark matter was popular for a time in the early 1980s, but it suffers from a severe problem: because all galaxy-size density fluctuations get washed out by free-streaming, the first objects that can form are huge supercluster-size pancakes, which then were theorised somehow to fragment into galaxies. Deep-field observations clearly show that galaxies formed at early times, with clusters and superclusters forming later as galaxies clump together, so any model dominated by hot dark matter is seriously in conflict with observations.

32.4.4 Mixed dark matter

Main article: Mixed dark matter

Mixed dark matter is a now obsolete model, with a specifically chosen mass ratio of 80% cold dark matter and 20% hot dark matter (neutrinos) content. Though it is presumable that hot dark matter coexists with cold dark matter in any case, there was a very specific reason for choosing this particular ratio of hot to cold dark matter in this model. During the early 1990s it became steadily clear that a universe with critical density of cold dark matter did not fit the COBE and large-scale galaxy clustering observations; either the 80/20 mixed dark matter model, or LambdaCDM, were able to reconcile these. With the discovery of the accelerating universe from supernovae, and more accurate measurements of CMB anisotropy and galaxy clustering, the mixed dark matter model was essentially ruled out while the concordance LambdaCDM model remained a good fit.

32.5 Detection

If the dark matter within the Milky Way is made up of Weakly Interacting Massive Particles (WIMPs), then millions, possibly billions, of WIMPs must pass through every square centimeter of the Earth each second.[91][92] There are many experiments currently running, or planned, aiming to test this hypothesis by searching for WIMPs. Although WIMPs are the historically more popular dark matter candidate for searches,[11] there are experiments searching for other particle candidates; the Axion Dark Matter eXperiment (ADMX) is currently searching for the dark matter axion, a well-motivated and constrained dark matter source. It is also possible that dark matter consists of very heavy hidden sector particles which only interact with ordinary matter via gravity.

These experiments can be divided into two classes: direct detection experiments, which search for the scattering of dark matter particles off atomic nuclei within a detector; and indirect detection, which look for the products of WIMP annihilations.[20]

An alternative approach to the detection of WIMPs in nature is to produce them in the laboratory. Experiments with the Large Hadron Collider (LHC) may be able to detect WIMPs produced in collisions of the LHC proton beams. Because a WIMP has negligible interactions with matter, it may be detected indirectly as (large amounts of) missing energy and momentum which escape the LHC detectors, provided all the other (non-negligible) collision products are detected.[93] These experiments could show that WIMPs can be created, but it would still require a direct detection experiment to show that they exist in sufficient numbers to account for dark matter.

32.5.1 Direct detection experiments

Direct detection experiments usually operate in deep underground laboratories to reduce the background from cosmic rays. These include: the Soudan mine; the SNOLAB underground laboratory at Sudbury, Ontario (Canada); the Gran Sasso National Laboratory (Italy); the Canfranc Underground Laboratory (Spain); the Boulby Underground Laboratory (United Kingdom); the Deep Underground Science and Engineering Laboratory, South Dakota (United States); and the Particle and Astrophysical Xenon Detector (China).

The majority of present experiments use one of two detector technologies: cryogenic detectors, operating at temperatures below 100mK, detect the heat produced when a particle hits an atom in a crystal absorber such as germanium. Noble liquid detectors detect the flash of scintillation light produced by a particle collision in liquid xenon or argon. Cryogenic detector experiments include: CDMS, CRESST, EDELWEISS, EURECA. Noble liquid experiments include ZEPLIN, XENON, DEAP, ArDM, WARP, DarkSide, PandaX, and LUX, the Large Underground Xenon experiment. Both of these detector techniques are capable of distinguishing background particles which scatter off electrons, from dark matter particles which scatter off nuclei. Other experiments include SIMPLE and PICASSO.

The DAMA/NaI, DAMA/LIBRA experiments have detected an annual modulation in the event rate,[94] which they claim is due to dark matter particles. (As the Earth orbits the Sun, the velocity of the detector relative to the dark matter halo will vary by a small amount depending on the time of year). This claim is so far unconfirmed and difficult to reconcile with the negative results of other experiments assuming that the WIMP scenario is correct.[95]

Directional detection of dark matter is a search strategy based on the motion of the Solar System around the Galactic Center.[96][97][98][99]

By using a low pressure TPC, it is possible to access information on recoiling tracks (3D reconstruction if possible) and to constrain the WIMP-nucleus kinematics. WIMPs coming from the direction in which the Sun is travelling (roughly in the direction of the Cygnus constellation) may then be separated from background noise, which should be isotropic. Directional dark matter experiments include DMTPC, DRIFT, Newage and MIMAC.

On 17 December 2009 CDMS researchers reported two possible WIMP candidate events. They estimate that the probability that these events are due to a known background (neutrons or misidentified beta or gamma events) is 23%, and conclude "this analysis cannot be interpreted as significant evidence for WIMP interactions, but we cannot reject either event as signal."[100]

More recently, on 4 September 2011, researchers using the CRESST detectors presented evidence[101] of 67 collisions occurring in detector crystals from subatomic particles, calculating there is a less than 1 in 10,000 chance that all were caused by known sources of interference or contamination. It is quite possible then that many of these collisions were caused by WIMPs, and/or other unknown particles.

32.5.2 Indirect detection experiments

Indirect detection experiments search for the products of WIMP annihilation or decay. If WIMPs are Majorana particles (WIMPs are their own antiparticle) then two WIMPs could annihilate to produce gamma rays or Standard Model particle-antiparticle pairs. Additionally, if the WIMP is unstable, WIMPs could decay into standard model particles. These processes could be detected indirectly through an excess of gamma rays, antiprotons or positrons emanating from regions of high dark matter density. The detection of such a signal is not conclusive evidence for dark matter, as the production of gamma rays from other sources is not fully understood.[11][20]

The EGRET gamma ray telescope observed more gamma rays than expected from the Milky Way, but scientists concluded that this was most likely due to a mis-estimation of the telescope's sensitivity.[103]

The Fermi Gamma-ray Space Telescope, launched 11 June 2008, is searching for gamma rays from dark matter annihilation and decay.[104] In April 2012, an analysis[105] of previously available data from its Large Area Telescope instrument produced strong statistical evidence of a 130 GeV line in the gamma radiation coming from the center of the Milky Way. At the time, WIMP annihilation was the most probable explanation for that line.[106]

At higher energies, ground-based gamma-ray telescopes have set limits on the annihilation of dark matter in dwarf spheroidal galaxies[107] and in clusters of galaxies.[108]

The PAMELA experiment (launched 2006) has detected a larger number of positrons than expected. These extra positrons could be produced by dark matter annihilation, but may also come from pulsars. No excess of anti-protons has been observed.[109] The Alpha Magnetic Spectrometer on the International Space Station is designed to directly measure the fraction of cosmic rays which are positrons. The first results, published in April 2013, indicate an excess of high-energy cosmic rays which could potentially be due to annihilation of dark matter.[110][111][112][113][114][115]

A few of the WIMPs passing through the Sun or Earth may scatter off atoms and lose energy. This way a large population of WIMPs may accumulate at the center of these bodies, increasing the chance that two will collide and annihilate. This could produce a distinctive signal in the form of high-energy neutrinos originating from the center of the Sun or Earth.[116] It is generally considered that the detection of such a signal would be the strongest indirect proof of WIMP dark matter.[11] High-energy neutrino telescopes such as AMANDA, IceCube and ANTARES are searching for this signal.

WIMP annihilation from the Milky Way Galaxy as a whole may also be detected in the form of various annihilation products.[117] The Galactic Center is a particularly good place to look because the density of dark matter may be very high there.[118]

32.6 Alternative theories

32.6.1 Mass in extra dimensions

In some multidimensional theories, the force of gravity is the unique force able to have an effect across all the various extra dimensions,[16] which would explain the relative weakness of the force of gravity compared to the other known forces of nature that would not be able to cross into extra dimensions: electromagnetism, strong interaction, and weak interaction.

In that case, dark matter would be a perfect candidate for matter that would exist in other dimensions and that could only interact with the matter on our dimensions through gravity. That dark matter located on different dimensions could potentially aggregate in the same way as the matter in our visible universe does, forming exotic galaxies.[15]

32.6.2 Topological defects

Dark matter could consist of primordial defects (defects originating with the birth of the universe) in the topology of quantum fields, which would contain energy and therefore gravitate. This possibility may be investigated by the use of an orbital network of atomic clocks, which would register the passage of topological defects by monitoring the synchronization of the clocks. The Global Positioning System may be able to operate as such a network.[119]

32.6.3 Modified gravity

Numerous alternative theories have been proposed to explain these observations without the need for a large amount of undetected matter. Most of these theories modify the laws of gravity established by Newton and Einstein.

The earliest modified gravity model to emerge was Mordehai Milgrom's Modified Newtonian Dynamics (MOND) in 1983, which adjusts Newton's laws to create a stronger gravitational field when gravitational acceleration levels become tiny (such as near the rim of a galaxy). It had some success explaining galactic-scale features, such as rotational velocity curves of elliptical galaxies, and dwarf elliptical galaxies, but did not successfully explain galaxy cluster gravitational lensing. However, MOND was not relativistic, since it was just a straight adjustment of the older Newtonian account of gravitation, not of the newer account in Einstein's general relativity. Soon after 1983, attempts were made to bring MOND into conformity with general relativity; this is an ongoing process, and many competing hypotheses have emerged based around the original MOND model—including TeVeS, MOG or STV gravity, and phenomenological covariant approach,[120] among others.

In 2007, John W. Moffat proposed a modified gravity hypothesis based on the nonsymmetric gravitational theory (NGT) that claims to account for the behavior of colliding galaxies.[121] This model requires the presence of non-relativistic neutrinos, or other candidates for (cold) dark matter, to work.

Another proposal uses a gravitational backreaction in an emerging theoretical field that seeks to explain gravity between objects as an action, a reaction, and then a back-reaction. Simply, an object A affects an object B, and the object B then re-affects object A, and so on: creating a sort of feedback loop that strengthens gravity.[122]

Recently, another group has proposed a modification of large-scale gravity in a hypothesis named "dark fluid". In this formulation, the attractive gravitational effects attributed to dark matter are instead a side-effect of dark energy. Dark fluid combines dark matter and dark energy in a single energy field that produces different effects at different scales. This treatment is a simplified approach to a previous fluid-like model called the generalized Chaplygin gas model where the whole of spacetime is a compressible gas.[123] Dark fluid can be compared to an atmospheric system. Atmospheric pressure causes air to expand, but part of the air can collapse to form clouds. In the same way, the dark fluid might generally expand, but it also could collect around galaxies to help hold them together.[123]

Another set of proposals is based on the possibility of a double metric tensor for space-time.[124] It has been argued that time-reversed solutions in general relativity require such double metric for consistency, and that both dark matter and dark energy can be understood in terms of time-reversed solutions of general relativity.[125]

32.6.4 Fractality of Spacetime

Applying relativity to fractal non-differentiable spacetime, Laurent Nottale, in his Scale Relativity theory, suggests that potential energy arises due to the fractality of space, and accounts for the missing mass-energy observed at cosmological scales.

32.7 Popular culture

Main article: Dark matter in fiction

Mention of dark matter is made in some video games and other works of fiction. In such cases, it is usually attributed extraordinary physical or magical properties. Such descriptions are often inconsistent with the properties of dark matter proposed in physics and cosmology.

32.8 See also

- Aether

- Chameleon particle

- Conformal gravity

- General Antiparticle Spectrometer

- Illustris project

- Light dark matter

- Mirror matter

- Multidark (research program)

- Scalar field dark matter

- Self-interacting dark matter

- SIMP

- Unparticle physics

32.9 References

[1] "Hubble Finds Dark Matter Ring in Galaxy Cluster".

[2] Ade, P. A. R.; Aghanim, N.; Armitage-Caplan, C.; (Planck Collaboration) et al. (22 March 2013). "Planck 2013 results. I. Overview of products and scientific results – Table 9".*Astronomy and Astrophysics***1303**: 5062.arX.doi:10.1051/0004-6361/2.

[3] Francis, Matthew (22 March 2013). "First Planck results: the Universe is still weird and interesting". *Arstechnica*.

[4] "Planck captures portrait of the young Universe, revealing earliest light". University of Cambridge. 21 March 2013. Retrieved 21 March 2013.

[5] Sean Carroll, Ph.D., Cal Tech, 2007, The Teaching Company, *Dark Matter, Dark Energy: The Dark Side of the Universe*, Guidebook Part 2 page 46, Accessed Oct. 7, 2013, "...dark matter: An invisible, essentially collisionless component of matter that makes up about 25 percent of the energy density of the universe... it's a different kind of particle... something not yet observed in the laboratory..."

[6] Ferris, Timothy. "Dark Matter". Retrieved 2015-06-10.

[7] First observational evidence of dark matter. Darkmatterphysics.com. Retrieved on 6 August 2013.

[8] Rubin, Vera C.; Ford, W. Kent, Jr. (February 1970). "Rotation of the Andromeda Nebula from a Spectroscopic Survey of Emission Regions". *The Astrophysical Journal* **159**: 379–403. Bibcode:1970ApJ...159..379R. doi:10.1086/150317.

[9] Copi, C. J.; Schramm, D. N.; Turner, M. S. (1995). "Big-Bang Nucleosynthesis and the Baryon Density of the Universe". *Science* **267** (5195): 192–199. arXiv:astro-ph/9407006. Bibcode:1995Sci...267..192C. doi:10.1126/science.7809624. PMID 7809624.

[10] Bergstrom, L. (2000). "Non-baryonic dark matter: Observational evidence and detection methods". *Reports on Progress in Physics* **63** (5): 793–841. arXiv:hep-ph/0002126. Bibcode:2000RPPh...63..793B. doi:10.1088/0034-4885/63/5/2r3.

[11] Bertone, G.; Hooper, D.; Silk, J. (2005). "Particle dark matter: Evidence, candidates and constraints". *Physics Reports* **405** (5–6): 279–390. arXiv:hep-ph/0404175. Bibcode:2005PhR...405..279B. doi:10.1016/j.physrep.2004.08.031.

[12] Angus, G. (2013). "Cosmological simulations in MOND: the cluster scale halo mass function with light sterile neutrinos". *Monthly Notices of the Royal Astronomical Society***436**: 202–211.arXiv:1309.6094.Bibcode:2013MNRAS.436..202A.doi:10.1.

[13] Trimble, V. (1987). "Existence and nature of dark matter in the universe". *Annual Review of Astronomy and Astrophysics* **25**: 425–472. Bibcode:1987ARA&A..25..425T. doi:10.1146/annurev.aa.25.090187.002233.

[14] Dark matter. CERN. Retrieved on 17 November 2014.

[15] Siegfried, T. (5 July 1999). "Hidden Space Dimensions May Permit Parallel Universes, Explain Cosmic Mysteries". *The Dallas Morning News*.

[16] Extra dimensions, gravitons, and tiny black holes. CERN. Retrieved on 17 November 2014.

[17] Kroupa, P. et al. (2010). "Local-Group tests of dark-matter Concordance Cosmology: Towards a new paradigm for structure formation". *Astronomy and Astrophysics* **523**: 32–54. arXiv:1006.1647. Bibcode:2010A&A...523A..32K. doi:10.1051/0004-6361/201014892.

[18] Achim Weiss, "Big Bang Nucleosynthesis: Cooking up the first light elements" in: Einstein Online Vol. 2 (2006), 1017

[19] Raine, D.; Thomas, T. (2001). *An Introduction to the Science of Cosmology*. IOP Publishing. p. 30. ISBN 0-7503-0405-7.

[20] Bertone, G.; Merritt, D. (2005). "Dark Matter Dynamics and Indirect Detection". *Modern Physics Letters A* **20** (14): 1021–1036. arXiv:astro-ph/0504422. Bibcode:2005MPLA...20.1021B. doi:10.1142/S0217732305017391.

[21] "Serious Blow to Dark Matter Theories?" (Press release). European Southern Observatory. 18 April 2012.

[22] "The Hidden Lives of Galaxies: Hidden Mass". *Imagine the Universe!*. NASA/GSFC.

[23] Kuijken K. and Gilmore G. (1989). "The Mass Distribution in the Galactic Disc - Part III   the Local Volume Mass Density". *Monthly Notices of the Royal Astronomical Society* **239**: 651. Bibcode:1989MNRAS.239..651K. doi:10.1093/mnras/239.2.651.

[24]Zwicky, F. (1933). "Die Rotverschiebung von extragalaktischen Nebeln".*Helvetica Physica Acta***6**: 110–127.Bibcode:1933AcH.

[25] Zwicky, F. (1937). "On the Masses of Nebulae and of Clusters of Nebulae". *The Astrophysical Journal* **86**: 217. Bibcode: . doi:10.1086/143864.

[26] Freese, Katerine (2014). *The Cosmic Cocktail: Three Parts Dark Matter*. Princeton, New Jersey: Princeton University Press. ISBN 978-0691153353.

[27] "First Signs of Self-interacting Dark Matter?". *ESO Press Release*. European Southern Observatory. Retrieved 15 April 2015.

[28] Freeman, K.; McNamara, G. (2006). *In Search of Dark Matter*. Birkhäuser. p. 37. ISBN 0-387-27616-5.

[29] Jörg, D. et al. (2012). "A filament of dark matter between two clusters of galaxies". *Nature* **487** (7406): 202. arXiv:1207.0809. Bibcode:2012Natur.487..202D. doi:10.1038/nature11224.

[30] Babcock, H, 1939, "The rotation of the Andromeda Nebula", Lick Observatory bulletin ; no. 498

[31] Bosma, A. (1978). "The distribution and kinematics of neutral hydrogen in spiral galaxies of various morphological types" (Ph.D. Thesis). Rijksuniversiteit Groningen.

[32] Rubin, V.; Thonnard, W. K. Jr.; Ford, N. (1980). "Rotational Properties of 21 Sc Galaxies with a Large Range of Luminosities and Radii from NGC 4605 ($R = 4$kpc) to UGC 2885 ($R = 122$kpc)". *The Astrophysical Journal* **238**: 471. Bibcode:1980ApJ...2. doi:10.1086/158003.

[33] de Blok, W. J. G.; McGaugh, S. S.; Bosma, A.; Rubin, V. C. (2001). "Mass Density Profiles of Low Surface Brightness Galaxies". *The Astrophysical Journal Letters* **552** (1): L23–L26. arXiv:astro-ph/0103102. Bibcode:2001ApJ...552L..23D. doi:10.1086/320262.

[34] Salucci, P.; Borriello, A. (2003). "The Intriguing Distribution of Dark Matter in Galaxies". *Lecture Notes in Physics*. Lecture Notes in Physics **616**: 66–77. arXiv:astro-ph/0203457. Bibcode:2003LNP...616...66S. doi:10.1007/3-540-36539-7_5. ISBN 978-3-540-00711-1.

[35] Koopmans, L. V. E.; Treu, T. (2003). "The Structure and Dynamics of Luminous and Dark Matter in the Early-Type Lens Galaxy of 0047–281 at $z = 0.485$". *The Astrophysical Journal* **583** (2): 606–615. arXiv:astro-ph/0205281. Bibcode:2003ApJ.... doi:10.1086/345423.

[36] Dekel, A. et al. (2005). "Lost and found dark matter in elliptical galaxies". *Nature* **437** (7059): 707–710. arXiv:astro-ph/0501622. Bibcode:2005Natur.437..707D. doi:10.1038/nature03970. PMID 16193046.

[37] Governato, F. et al. (2010). "Bulgeless dwarf galaxies and dark matter cores from supernova-driven outflows". *Nature* **463** (7278): 203–206. arXiv:0911.2237. Bibcode:2010Natur.463..203G. doi:10.1038/nature08640.

[38] Ostriker, J. P.; Steinhardt, P. (2003). "New Light on Dark Matter". *Science* **300** (5627): 1909–1913. doi:10.1126/science.108597. PMID 12817140.

[39] Faber, S. M.; Jackson, R. E. (1976). "Velocity dispersions and mass-to-light ratios for elliptical galaxies". *The Astrophysical Journal* **204**: 668–683. Bibcode:1976ApJ...204..668F. doi:10.1086/154215.

[40] Collins, G. W. (1978). "The Virial Theorem in Stellar Astrophysics". Pachart Press.

[41] Rejkuba, M.; Dubath, P.; Minniti, D.; Meylan, G. (2008). "Masses and M/L Ratios of Bright Globular Clusters in NGC 5128". *Proceedings of the International Astronomical Union* **246**: 418–422. Bibcode:2008IAUS..246..418R. doi:10.1017/S174392130.

[42] Weinberg, M. D.; Blitz, L. (2006). "A Magellanic Origin for the Warp of the Galaxy". *The Astrophysical Journal Letters* **641** (1): L33–L36. arXiv:astro-ph/0601694. Bibcode:2006ApJ...641L..33W. doi:10.1086/503607.

[43] Minchin, R. et al. (2005). "A Dark Hydrogen Cloud in the Virgo Cluster". *The Astrophysical Journal Letters* **622**: L21–L24. arXiv:astro-ph/0502312. Bibcode:2005ApJ...622L..21M. doi:10.1086/429538.

[44] Ciardullo, R.; Jacoby, G. H.; Dejonghe, H. B. (1993). "The radial velocities of planetary nebulae in NGC 3379". *The Astrophysical Journal* **414**: 454–462. Bibcode:1993ApJ...414..454C. doi:10.1086/173092.

[45] Vikhlinin, A. et al. (2006). "Chandra Sample of Nearby Relaxed Galaxy Clusters: Mass, Gas Fraction, and Mass–Temperature Relation". *The Astrophysical Journal* **640** (2): 691–709. arXiv:astro-ph/0507092. Bibcode:2006ApJ...640..691V. doi:10.1.

[46] "Abell 2029: Hot News for Cold Dark Matter". Chandra X-ray Observatory. 11 June 2003.

[47] Taylor, A. N. et al. (1998). "Gravitational Lens Magnification and the Mass of Abell 1689". *The Astrophysical Journal* **501** (2): 539. arXiv:astro-ph/9801158. Bibcode:1998ApJ...501..539T. doi:10.1086/305827.

[48] Wu, X.; Chiueh, T.; Fang, L.; Xue, Y. (1998). "A comparison of different cluster mass estimates: consistency or discrepancy?". *Monthly Notices of the Royal Astronomical Society* **301** (3): 861–871. arXiv:astro-ph/9808179.Bibcode:1998MNRAS.301..86. doi:10.1046/j.1365-8711.1998.02055.x.

[49] Refregier, A. (2003). "Weak gravitational lensing by large-scale structure". *Annual Review of Astronomy and Astrophysics* **41** (1): 645–668. arXiv:astro-ph/0307212. Bibcode:2003ARA&A..41..645R. doi:10.1146/annurev.astro.41.111302.102207.

[50] Massey, R. et al. (2007). "Dark matter maps reveal cosmic scaffolding". *Nature* **445** (7125): 286–290. arXiv:astro-ph/0701594. Bibcode:2007Natur.445..286M. doi:10.1038/nature05497. PMID 17206154.

[51] Clowe, D. et al. (2006). "A direct empirical proof of the existence of dark matter". *The Astrophysical Journal* **648** (2): 109–113. arXiv:astro-ph/0608407. Bibcode:2006ApJ...648L.109C. doi:10.1086/508162.

[52] "Dark Matter Mystery Deepens in Cosmic "Train Wreck"". Chandra X-Ray Observatory. 16 August 2007.

[53] Tiberiu, H.; Lobo, F. S. N. (2011). "Two-fluid dark matter models". *Physical Review D* **83** (12): 124051. arXiv:1106.2642. Bibcode:2011PhRvD..83l4051H. doi:10.1103/PhysRevD.83.124051.

[54] Spergel, D. N.; Steinhardt, P. J. (2000). "Observational evidence for self-interacting cold dark matter". *Physical Review Letters* **84** (17): 3760–3763. arXiv:astro-ph/9909386. Bibcode:2000PhRvL..84.3760S. doi:10.1103/PhysRevLett.84.3760.

[55] Allen, S. W.; Evrard, A. E.; Mantz, A. B. (2011). "Cosmological Parameters from Observations of Galaxy Clusters". *Annual Review of Astronomy & Astrophysics* **49**: 409–470. arXiv:1103.4829. Bibcode:2011ARA&A..49..409A. doi:10.1146/annurev-astro-081710-102514.

[56] Markevitch, M. et al. (2004). "Direct Constraints on the Dark Matter Self-Interaction Cross Section from the Merging Galaxy Cluster 1E 0657-56". *The Astrophysical Journal* **606** (2): 819–824. arXiv:astro-ph/0309303. Bibcode:2004ApJ...606..819M. doi:10.1086/383178.

[57] "Press Release - Dark Matter Map Begins to Reveal the Universe's Early History - Subaru Telescope". *www.subarutelescope.org*. Retrieved 2015-07-03.

[58] Miyazaki, Satoshi; Oguri, Masamune; Hamana, Takashi; Tanaka, Masayuki; Miller, Lance; Utsumi, Yousuke; Komiyama, Yutaka; Furusawa, Hisanori; Sakurai, Junya (2015-07-01). "Properties of Weak Lensing Clusters Detected on Hyper Suprime-Cam's 2.3 deg2 field". *The Astrophysical Journal* **807** (1): 22. arXiv:1504.06974. Bibcode:2015ApJ...807...22M.doi:10.10637X/807/1/22. ISSN 0004-637X.

[59] Penzias, A.A.; Wilson, R. W. (1965). "A Measurement of Excess Antenna Temperature at 4080 Mc/s". *The Astrophysical Journal* **142**: 419. Bibcode:1965ApJ...142..419P. doi:10.1086/148307.

[60] Boggess, N. W. et al. (1992). "The COBE Mission: Its Design and Performance Two Years after the launch". *The Astrophysical Journal* **397**: 420. Bibcode:1992ApJ...397..420B. doi:10.1086/171797.

[61] Melchiorri, A. et al. (2000). "A Measurement of Ω from the North American Test Flight of Boomerang". *The Astrophysical Journal Letters* **536** (2): L63–L66. arXiv:astro-ph/9911445. Bibcode:2000ApJ...536L..63M. doi:10.1086/312744.

[62] Leitch, E. M. et al. (2002). "Measurement of polarization with the Degree Angular Scale Interferometer". *Nature* **420** (6917): 763–771. arXiv:astro-ph/0209476. Bibcode:2002Natur.420..763L. doi:10.1038/nature01271. PMID 12490940.

[63] Leitch, E. M. et al. (2005). "Degree Angular Scale Interferometer 3 Year Cosmic Microwave Background Polarization Results". *The Astrophysical Journal* **624** (1): 10–20. arXiv:astro-ph/0409357. Bibcode:2005ApJ...624...10L. doi:10.1086/428825.

[64] Readhead, A. C. S. et al. (2004). "Polarization Observations with the Cosmic Background Imager". *Science* **306** (5697): 836–844. arXiv:astro-ph/0409569. Bibcode:2004Sci...306..836R. doi:10.1126/science.1105598. PMID 15472038.

[65] Hinshaw, G. et al. (2009). "Five-Year Wilkinson Microwave Anisotropy Probe Observations: Data Processing, Sky Maps, and Basic Results". *The Astrophysical Journal Supplement* **180** (2): 225–245. arXiv:0803.0732. Bibcode:2009ApJS..180..225H. doi:10.1088/0067-0049/180/2/225.

[66] Komatsu, E. et al. (2009). "Five-Year Wilkinson Microwave Anisotropy Probe Observations: Cosmological Interpretation". *The Astrophysical Journal Supplement* **180** (2): 330–376. arXiv:0803.0547. Bibcode:2009ApJS..180..330K. doi:10.1088/0067-0049/180/2/330.

[67] Percival, W. J. et al. (2007). "Measuring the Baryon Acoustic Oscillation scale using the Sloan Digital Sky Survey and 2dF Galaxy Redshift Survey". *Monthly Notices of the Royal Astronomical Society* **381** (3): 1053–1066. arXiv:0705.3323. Bibcode:2007MNRAS.381.1053P. doi:10.1111/j.1365-2966.2007.12268.x.

[68] Kowalski, M. et al. (2008). "Improved Cosmological Constraints from New, Old, and Combined Supernova Data Sets". *The Astrophysical Journal* **686** (2): 749–778. arXiv:0804.4142. Bibcode:2008ApJ...686..749K. doi:10.1086/589937.

[69] Viel, M.; Bolton, J. S.; Haehnelt, M. G. (2009). "Cosmological and astrophysical constraints from the Lyman α forest flux probability distribution function". *Monthly Notices of the Royal Astronomical Society* **399** (1): L39–L43. arXiv:0907.2927. Bibcode:2009MNRAS.399L..39V. doi:10.1111/j.1745-3933.2009.00720.x.

[70] "Hubble Maps the Cosmic Web of "Clumpy" Dark Matter in 3-D" (Press release). NASA. 7 January 2007.

[71] Springel, V. et al. (2005). "Simulations of the formation, evolution and clustering of galaxies and quasars". *Nature* **435** (7042): 629–636. arXiv:astro-ph/0504097. Bibcode:2005Natur.435..629S. doi:10.1038/nature03597. PMID 15931216.

[72] Mateo, M. L. (1998). "Dwarf Galaxies of the Local Group". *Annual Review of Astronomy and Astrophysics* **36** (1): 435–506. arXiv:astro-ph/9810070. Bibcode:1998ARA&A..36..435M. doi:10.1146/annurev.astro.36.1.435.

[73] Moore, B. et al. (1999). "Dark Matter Substructure within Galactic Halos". *The Astrophysical Journal Letters* **524** (1): L19–L22. arXiv:astro-ph/9907411. Bibcode:1999ApJ...524L..19M. doi:10.1086/312287.

[74] Tisserand, P.; Le Guillou, L.; Afonso, C.; Albert, J. N.; Andersen, J.; Ansari, R.; Aubourg, É.; Bareyre, P.; Beaulieu, J. P.; Charlot, X.; Coutures, C.; Ferlet, R.; Fouqué, P.; Glicenstein, J. F.; Goldman, B.; Gould, A.; Graff, D.; Gros, M.; Haissinski, J.; Hamadache, C.; De Kat, J.; Lasserre, T.; Lesquoy, É.; Loup, C.; Magneville, C.; Marquette, J. B.; Maurice, É.; Maury, A.; Milsztajn, A.; Moniez, M. (2007). "Limits on the Macho content of the Galactic Halo from the EROS-2 Survey of the Magellanic Clouds". *Astronomy and Astrophysics* **469** (2): 387. arXiv:astro-ph/0607207. Bibcode:2007A&A...469..387T. doi:10.1051/0004-6361:20066017.

[75] Graff, D. S.; Freese, K. (1996). "Analysis of a *Hubble Space Telescope* Search for Red Dwarfs: Limits on Baryonic Matter in the Galactic Halo". *The Astrophysical Journal* **456**. arXiv:astro-ph/9507097. Bibcode:1996ApJ...456L..49G. doi:10.1086/309850.

[76] Najita, J. R.; Tiede, G. P.; Carr, J. S. (2000). "From Stars to Superplanets: The Low-Mass Initial Mass Function in the Young Cluster IC 348". *The Astrophysical Journal* **541** (2): 977. doi:10.1086/309477.

[77] Wyrzykowski, Lukasz et al. (2011) The OGLE view of microlensing towards the Magellanic Clouds – IV. OGLE-III SMC data and final conclusions on MACHOs, MNRAS, 416, 2949

[78] Freese, Katherine; Fields, Brian; Graff, David (2000). "Death of Stellar Baryonic Dark Matter Candidates". arXiv:astro-ph/0007444 [astro-ph].

[79] Freese, Katherine; Fields, Brian; Graff, David (2000). "Death of Stellar Baryonic Dark Matter". *The First Stars*. ESO Astrophysics Symposia. p. 18. arXiv:astro-ph/0002058. Bibcode:2000fist.conf...18F. doi:10.1007/10719504_3. ISBN 3-540-67222-2.

[80] Silk, Joseph (1980). *The Big Bang* (1989 ed.). San Francisco: Freeman. chapter ix, page 182. ISBN 0-7167-1085-4.

[81] Vittorio, N.; J. Silk (1984). "Fine-scale anisotropy of the cosmic microwave background in a universe dominated by cold dark matter". *Astrophysical Journal, Part 2 – Letters to the Editor* **285**: L39–L43. Bibcode:1984ApJ...285L..39V. doi:10.1086/184361.

[82] Umemura, Masayuki; Satoru Ikeuchi (1985). "Formation of Subgalactic Objects within Two-Component Dark Matter". *Astrophysical Journal* **299**: 583–592. Bibcode:1985ApJ...299..583U. doi:10.1086/163726.

[83] Davis, M.; Efstathiou, G., Frenk, C. S., & White, S. D. M. (May 15, 1985). "The evolution of large-scale structure in a universe dominated by cold dark matter". *Astrophysical Journal* **292**: 371–394. Bibcode:1985ApJ...292..371D. doi:10.1086/163168.

[84] Hawkins, M. R. S. (2011). "The case for primordial black holes as dark matter". *Monthly Notices of the Royal Astronomical Society* **415** (3): 2744–2757. arXiv:1106.3875. Bibcode:2011MNRAS.415.2744H. doi:10.1111/j.1365-2966.2011.18890.x.

[85] Carr, B. J. et al. (May 2010). "New cosmological constraints on primordial black holes". *Physical Review D* **81** (10): 104019. arXiv:0912.5297. Bibcode:2010PhRvD..81j4019C. doi:10.1103/PhysRevD.81.104019.

[86] Peter, A. H. G. (2012). "Dark Matter: A Brief Review". arXiv:1201.3942 [astro-ph.CO].

[87] Garrett, Katherine; Dūda, Gintaras (2011). "Dark Matter: A Primer". *Advances in Astronomy* **2011**: 1. arXiv:1006.2483. Bibcode:2011AdAst2011E...8G. doi:10.1155/2011/968283. MACHOs can only account for a very small percentage of the nonluminous mass in our galaxy, revealing that most dark matter cannot be strongly concentrated or exist in the form of baryonic astrophysical objects. Although microlensing surveys rule out baryonic objects like brown dwarfs, black holes, and neutron stars in our galactic halo, can other forms of baryonic matter make up the bulk of dark matter? The answer, surprisingly, is no...

[88] Bertone, G. (2010). "The moment of truth for WIMP dark matter". *Nature* **468** (7322): 389–393. doi:10.1038/nature09509. PMID 21085174.

[89] Olive, Keith A. (2003). "TASI Lectures on Dark Matter". p. 21

[90] "Neutrinos as Dark Matter". Astro.ucla.edu. 21 September 1998. Retrieved 6 January 2011.

[91] Gaitskell, Richard J. (2004). "Direct Detection of Dark Matter". *"Annual Review of Nuclear and Particle Systems"* **54**: 315–359. Bibcode:2004ARNPS..54..315G. doi:10.1146/annurev.nucl.54.070103.181244.

[92] "NEUTRALINO DARK MATTER". Retrieved 26 December 2011. Griest, Kim. "WIMPs and MACHOs" (PDF). Retrieved 26 December 2011.

[93] Kane, G. and Watson, S. (2008). "Dark Matter and LHC:. what is the Connection?". *Modern Physics Letters A* **23** (26): 2103–2123. arXiv:0807.2244. Bibcode:2008MPLA...23.2103K. doi:10.1142/S0217732308028314.

[94] Drukier, A.; Freese, K. and Spergel, D. (1986). "Detecting Cold Dark Matter Candidates". *Physical Review D* **33** (12): 3495–3508. Bibcode:1986PhRvD..33.3495D. doi:10.1103/PhysRevD.33.3495.

[95] Bernabei, R.; Belli, P.; Cappella, F.; Cerulli, R.; Dai, C. J.; d'Angelo, A.; He, H. L.; Incicchitti, A.; Kuang, H. H.; Ma, J. M.; Montecchia, F.; Nozzoli, F.; Prosperi, D.; Sheng, X. D.; Ye, Z. P. (2008). "First results from DAMA/LIBRA and the combined results with DAMA/NaI". *Eur. Phys. J. C* **56** (3): 333–355. arXiv:0804.2741. doi:10.1140/epjc/s10052-008-0662-y.

[96] Stonebraker, Alan (2014-01-03). "Synopsis: Dark-Matter Wind Sways through the Seasons". *Physics - Synopses* (American Physical Society). Retrieved 6 January 2014.

[97] Lee, Samuel K.; Mariangela Lisanti, Annika H. G. Peter, and Benjamin R. Safdi (2014-01-03). "Effect of Gravitational Focusing on Annual Modulation in Dark-Matter Direct-Detection Experiments". *Phys. Rev. Lett.* (American Physical Society) **112** (1): 011301 (2014) [5 pages]. arXiv:1308.1953. Bibcode:2014PhRvL.112a1301L. doi:10.1103/PhysRevLett.112.011301.

[98] The Dark Matter Group. "An Introduction to Dark Matter". *Dark Matter Research* (Sheffield, UK: University of Sheffield). Retrieved 7 January 2014.

[99] "Blowing in the Wind". *Kavli News* (Sheffield, UK: Kavli Foundation). Retrieved 7 January 2014. Scientists at Kavli MIT are working on...a tool to track the movement of dark matter.

[100] The CDMS II Collaboration; Ahmed, Z.; Akerib, D. S.; Arrenberg, S.; Bailey, C. N.; Balakishiyeva, D.; Baudis, L.; Bauer, D. A.; Brink, P. L.; Bruch, T.; Bunker, R.; Cabrera, B.; Caldwell, D. O.; Cooley, J.; Cushman, P.; Daal, M.; Dejongh, F.; Dragowsky, M. R.; Duong, L.; Fallows, S.; Figueroa-Feliciano, E.; Filippini, J.; Fritts, M.; Golwala, S. R.; Grant, D. R.; Hall, J.; Hennings-Yeomans, R.; Hertel, S. A.; Holmgren, D.; Hsu, L. (2010). "Dark Matter Search Results from the CDMS II Experiment". *Science* **327** (5973): 1619–1621. doi:10.1126/science.1186112. PMID 20150446.

[101] Angloher, G.; Bauer; Bavykina; Bento; Bucci; Ciemniak; Deuter; von Feilitzsch; Hauff et al. (2011). "Results from 730kg days of the CRESST-II Dark Matter Search". arXiv:1109.0702v1 [astro-ph.CO].

[102] "Dark matter even darker than once thought". Retrieved 16 June 2015.

[103] Stecker, F.W.; Hunter, S; Kniffen, D (2008). "The likely cause of the EGRET GeV anomaly and its implications". *Astroparticle Physics* **29** (1): 25–29. arXiv:0705.4311. Bibcode:2008APh....29...25S. doi:10.1016/j.astropartphys.2007.11.002.

[104] Atwood, W.B.; Abdo, A. A.; Ackermann, M.; Althouse, W.; Anderson, B.; Axelsson, M.; Baldini, L.; Ballet, J. et al. (2009). "The large area telescope on the Fermi Gamma-ray Space Telescope Mission". *Astrophysical Journal* **697** (2): 1071–1102. arXiv:0902.1089. Bibcode:2009ApJ...697.1071A. doi:10.1088/0004-637X/697/2/1071.

[105] Weniger, Christoph (2012). "A Tentative Gamma-Ray Line from Dark Matter Annihilation at the Fermi Large Area Telescope". *Journal of Cosmology and Astroparticle Physics* **2012** (8): 7. arXiv:1204.2797v2. Bibcode:2012JCAP...08..007W. doi:10.1088/1475-7516/2012/08/007.

[106] Cartlidge, Edwin (24 April 2012). "Gamma rays hint at dark matter". Institute Of Physics. Retrieved 23 April 2013.

[107] Albert, J.; Aliu, E.; Anderhub, H.; Antoranz, P.; Backes, M.; Baixeras, C.; Barrio, J. A.; Bartko, H.; Bastieri, D.; Becker, J. K.; Bednarek, W.; Berger, K.; Bigongiari, C.; Biland, A.; Bock, R. K.; Bordas, P.; Bosch-Ramon, V.; Bretz, T.; Britvitch, I.; Camara, M.; Carmona, E.; Chilingarian, A.; Commichau, S.; Contreras, J. L.; Cortina, J.; Costado, M. T.; Curtef, V.; Danielyan, V.; Dazzi, F.; De Angelis, A. (2008). "Upper Limit for γ-Ray Emission above 140 GeV from the Dwarf Spheroidal Galaxy Draco". *The Astrophysical Journal* **679**: 428. doi:10.1086/529135.

[108] Aleksić, J.; Antonelli, L. A.; Antoranz, P.; Backes, M.; Baixeras, C.; Balestra, S.; Barrio, J. A.; Bastieri, D.; González, J. B.; Bednarek, W.; Berdyugin, A.; Berger, K.; Bernardini, E.; Biland, A.; Bock, R. K.; Bonnoli, G.; Bordas, P.; Tridon, D. B.; Bosch-Ramon, V.; Bose, D.; Braun, I.; Bretz, T.; Britzger, D.; Camara, M.; Carmona, E.; Carosi, A.; Colin, P.; Commichau, S.; Contreras, J. L.; Cortina, J. (2010). "Magic Gamma-Ray Telescope Observation of the Perseus Cluster of Galaxies: Implications for Cosmic Rays, Dark Matter, and Ngc 1275". *The Astrophysical Journal* **710**: 634. doi:10.1088/0004-637X/710/1/634.

[109] Adriani, O.; Barbarino, G. C.; Bazilevskaya, G. A.; Bellotti, R.; Boezio, M.; Bogomolov, E. A.; Bonechi, L.; Bongi, M.; Bonvicini, V.; Bottai, S.; Bruno, A.; Cafagna, F.; Campana, D.; Carlson, P.; Casolino, M.; Castellini, G.; De Pascale, M. P.; De Rosa, G.; De Simone, N.; Di Felice, V.; Galper, A. M.; Grishantseva, L.; Hofverberg, P.; Koldashov, S. V.; Krutkov, S. Y.; Kvashnin, A. N.; Leonov, A.; Malvezzi, V.; Marcelli, L.; Menn, W. (2009). "An anomalous positron abundance in cosmic rays with energies 1.5–100 GeV". *Nature* **458** (7238): 607–609. doi:10.1038/nature07942. PMID 19340076.

[110] Aguilar, M. (AMS Collaboration) et al. (3 April 2013). "First Result from the Alpha Magnetic Spectrometer on the International Space Station: Precision Measurement of the Positron Fraction in Primary Cosmic Rays of 0.5–350 GeV". *Physical Review Letters*. Bibcode:2013PhRvL.110n1102A. doi:10.1103/PhysRevLett.110.141102. Retrieved 3 April 2013.

[111] "First Result from the Alpha Magnetic Spectrometer Experiment". *AMS Collaboration*. 3 April 2013. Retrieved 3 April 2013.

[112] Heilprin, John; Borenstein, Seth (3 April 2013). "Scientists find hint of dark matter from cosmos". Associated Press. Retrieved 3 April 2013.

[113] Amos, Jonathan (3 April 2013). "Alpha Magnetic Spectrometer zeroes in on dark matter". *BBC*. Retrieved 3 April 2013.

[114] Perrotto, Trent J.; Byerly, Josh (2 April 2013). "NASA TV Briefing Discusses Alpha Magnetic Spectrometer Results". *NASA*. Retrieved 3 April 2013.

[115] Overbye, Dennis (3 April 2013). "New Clues to the Mystery of Dark Matter". *New York Times*. Retrieved 3 April 2013.

[116] Freese, K. (1986). "Can Scalar Neutrinos or Massive Dirac Neutrinos be the Missing Mass?". *Physics Letters B* **167** (3): 295–300. Bibcode:1986PhLB..167..295F. doi:10.1016/0370-2693(86)90349-7.

[117] Ellis, J.; Flores, R. A.; Freese, K.; Ritz, S.; Seckel, D.; Silk, J. (1988). "Cosmic ray constraints on the annihilations of relic particles in the galactic halo". *Physics Letters B* **214** (3): 403. Bibcode:1988PhLB..214..403E. doi:10.1016/0370-2693(88)91385-8.

[118] Merritt, David (1 January 2010). "Dark Matter at the Centers of Galaxies". In Bertone, Gianfranco. *Particle Dark Matter : Observations, Models and Searches*. Cambridge University Press. pp. 83–104. arXiv:1001.3706. ISBN 978-0-521-76368-4.

[119] Rzetelny, Xaq (19 November 2014). "Looking for a different sort of dark matter with GPS satellites". *Ars Technica*. Retrieved 24 November 2014.

[120] Exirifard, Q. (2010). "Phenomenological covariant approach to gravity". *General Relativity and Gravitation* **43** (1): 93–106. arXiv:0808.1962. Bibcode:2011GReGr..43...93E. doi:10.1007/s10714-010-1073-6.

[121] Brownstein, J.R.; Moffat, J. W. (2007). "The Bullet Cluster 1E0657-558 evidence shows modified gravity in the absence of dark matter". *Monthly Notices of the Royal Astronomical Society* **382** (1): 29–47. arXiv:astro-ph/0702146.Bibcode:2007MNRAS. doi:10.1111/j.1365-2966.2007.12275.x.

[122] Anastopoulos, C. (2009). "Gravitational backreaction in cosmological spacetimes". *Physical Review D* **79** (8): 084029. arXiv:0902.0159. Bibcode:2009PhRvD..79h4029A. doi:10.1103/PhysRevD.79.084029.

[123] "New Cosmic Theory Unites Dark Forces". SPACE.com. 11 February 2008. Retrieved 6 January 2011.

[124] Hossenfelder, S. (2008). "A Bi-Metric Theory with Exchange Symmetry". *Physical Review D* **78** (4): 044015. arXiv:gr-qc/0603005. Bibcode:2008PhRvD..78d4015H. doi:10.1103/PhysRevD.78.044015.

[125] Ripalda, Jose M. (1999). "Time reversal and negative energies in general relativity". arXiv:gr-qc/9906012 [gr-qc].

32.10 External links

- Dark matter at DMOZ

- What is dark matter? at cosmosmagazine.com

- The Dark Matter Crisis

- The European astroparticle physics network

- Helmholtz Alliance for Astroparticle Physics

- "NASA Finds Direct Proof of Dark Matter" (Press release). NASA. 21 August 2006.

- Tuttle, Kelen (22 August 2006). "Dark Matter Observed". SLAC (Stanford Linear Accelerator Center) Today.

- "Astronomers claim first 'dark galaxy' find". New Scientist. 23 February 2005.

- Sample, Ian (17 December 2009). "Dark Matter Detected". London: Guardian. Retrieved 1 May 2010.

- Video lecture on dark matter by Scott Tremaine, IAS professor

- Science Daily story "Astronomers' Doubts About the Dark Side ..."

- Gray, Meghan; Merrifield, Mike; Copeland, Ed (2010). "Dark Matter". *Sixty Symbols*. Brady Haran for the University of Nottingham.

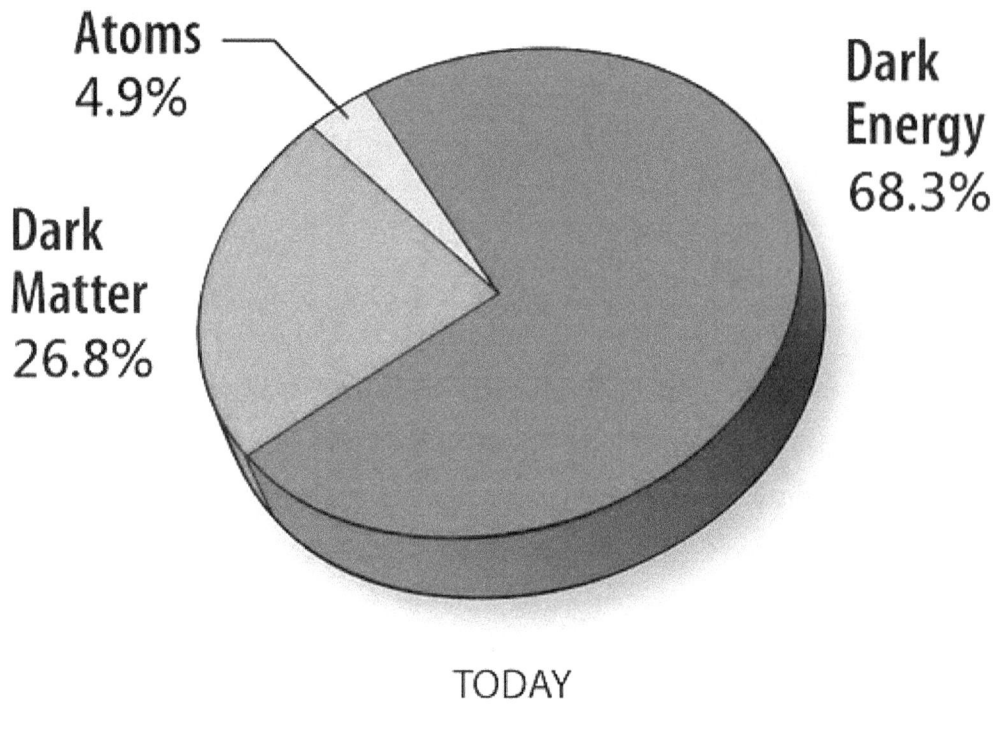

TODAY

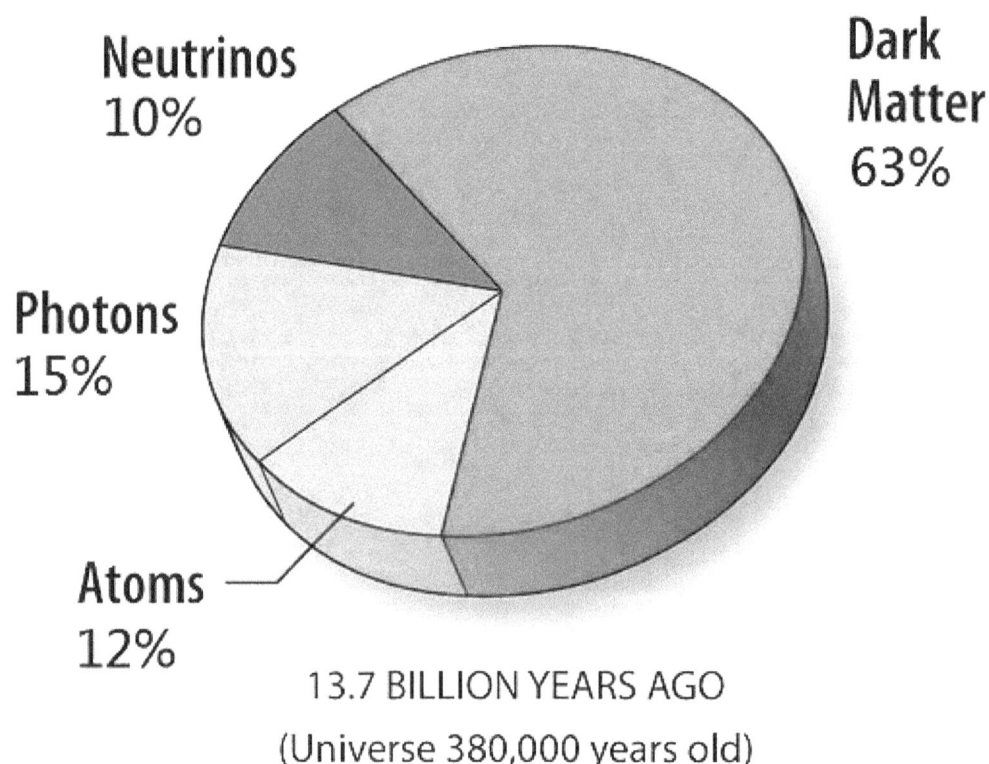

13.7 BILLION YEARS AGO
(Universe 380,000 years old)

Estimated distribution of matter and energy in the universe, today (top) and when the CMB was released (bottom)

Fermi-LAT observations of dwarf galaxies provide new insights on dark matter.

This artist's impression shows the expected distribution of dark matter in the Milky Way galaxy as a blue halo of material surrounding the galaxy.[21]

Observations have provided hints that the dark matter around one of the central four merging galaxies is not moving with the galaxy itself.[27]

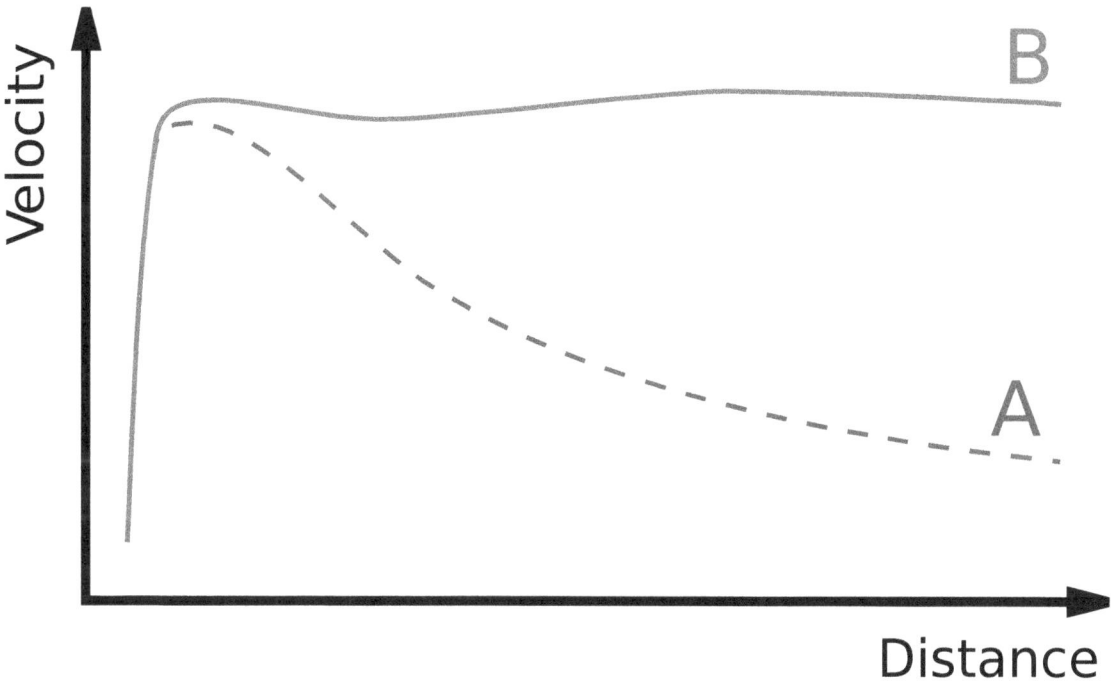

*Rotation curve of a typical spiral galaxy: predicted (**A**) and observed (**B**). Dark matter can explain the 'flat' appearance of the velocity curve out to a large radius*

Strong gravitational lensing as observed by the Hubble Space Telescope in Abell 1689 indicates the presence of dark matter—enlarge the image to see the lensing arcs.

The Bullet Cluster: HST image with overlays. The total projected mass distribution reconstructed from strong and weak gravitational lensing is shown in blue, while the X-ray emitting hot gas observed with Chandra is shown in red.

The cosmic microwave background by WMAP

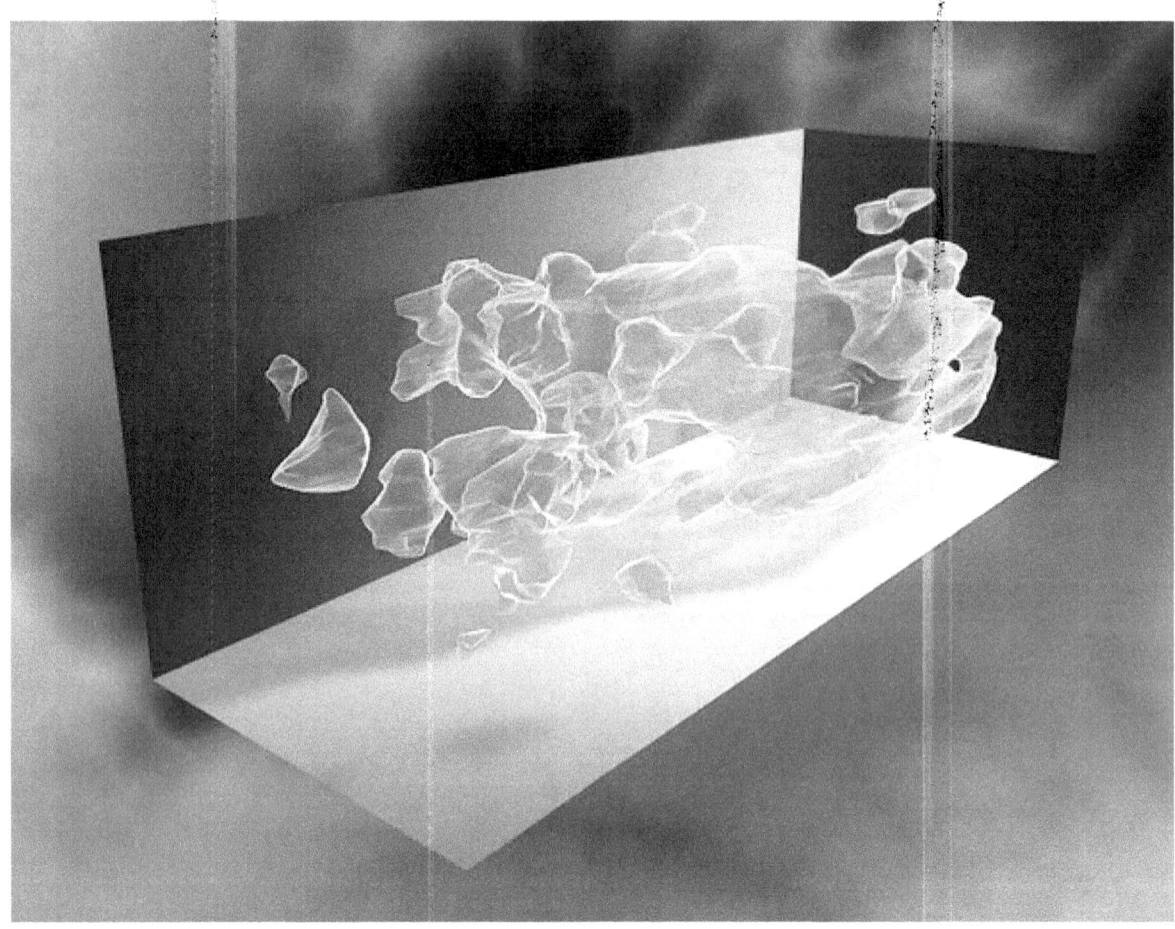

3D map of the large-scale distribution of dark matter, reconstructed from measurements of weak gravitational lensing with the Hubble Space Telescope.[70]

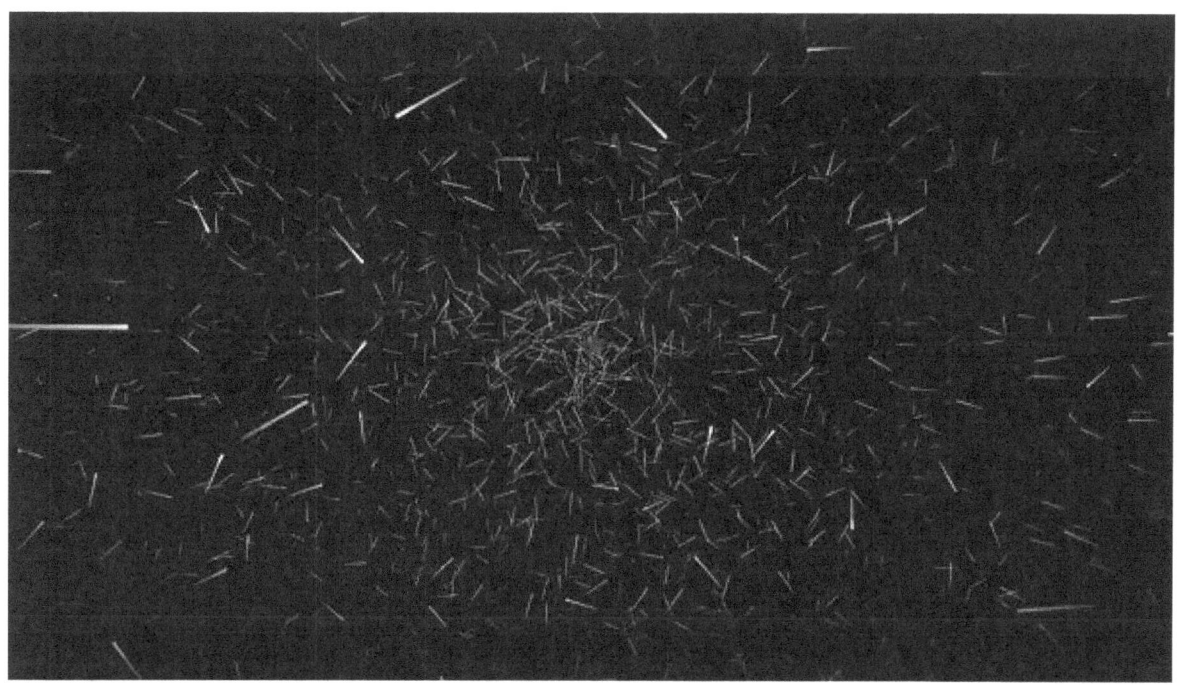

This video introduces a new computer simulation exploring the connection between two of the most elusive phenomena in the universe, black holes and dark matter. In the visualization, dark matter particles are gray spheres attached to shaded trails representing their motion. Redder trails indicate particles more strongly affected by the black hole's gravitation and closer to its event horizon (black sphere at center, mostly hidden by trails). The ergosphere, where all matter and light must follow the black hole's spin, is shown in teal.

Collage of six cluster collisions with dark matter maps. The clusters were observed in a study of how dark matter in clusters of galaxies behaves when the clusters collide.[102]

32.11 Text and image sources, contributors, and licenses

32.11.1 Text

- **Atom** *Source:* https://en.wikipedia.org/wiki/Atom?oldid=673798717 *Contributors:* AxelBoldt, Trelvis, Lee Daniel Crocker, Mav, Zundark, The Anome, Tarquin, AstroNomer~enwiki, Stokerm, Andre Engels, Youssefsan, Branden, Ben-Zin~enwiki, Drbug, Heron, Comte0, Peter-Bohne, Stevertigo, Hfastedge, Lir, Patrick, Infrogmation, D, Michael Hardy, Tim Starling, FrankH, Pit~enwiki, Wapcaplet, Ixfd64, Bcrowell, Dcljr, Shoaler, Delirium, PingPongBoy, Card~enwiki, NuclearWinner, Looxix~enwiki, Mdebets, Ahoerstemeier, LoonBB, Suisui, Angela, Jebba, Darkwind, Александър, Julesd, Glenn, Poor Yorick, Andres, Kaihsu, Evercat, Samw, Rob Hooft, Denny, Schneelocke, Ddoherty, Frieda, Timwi, Wikiborg, Stone, Dysprosia, The Anomebot, Doradus, Hr oskar, Big Bob the Finder, Morwen, VeryVerily, SEWilco, Flockmeal, Ldo, BenRG, Gromlakh, Donarreiskoffer, Gentgeen, Robbot, ChrisO~enwiki, Arkuat, Merovingian, Sverdrup, Der Eberswalder, DHN, Caknuck, Rebrane, Paul G, Hadal, UtherSRG, Wikibot, Roozbeh, Mandel, Anthony, Diberri, Cutler, Dina, Timemutt, Tobias Bergemann, David Gerard, Ancheta Wis, Vonkwink, Giftlite, Christopher Parham, Awolf002, Mikez, Palapala, Yuri koval, Inter, BenFrantzDale, Tom harrison, MSGJ, Xerxes314, Everyking, Curps, Michael Devore, Bensaccount, Frencheigh, Guanaco, Andrea Parri, Jorge Stolfi, SWAdair, Jurema Oliveira, Kandar, Stevietheman, Gadfium, Utcursch, Bact, Pcarbonn, Antandrus, OverlordQ, MisfitToys, Piotrus, Kaldari, Jossi, Karol Langner, JimWae, DragonflySixtyseven, Icairns, Tail, Sam Hocevar, Darksun, Engleman, Deglr6328, Achven, M1s1ontomars2k4, Adashiel, Trevor MacInnis, Grunt, Bluemask, Freakofnurture, Amxitsa, DanielCD, Discospinster, Rich Farmbrough, Guanabot, Huffers, Hidaspal, Rama, Vsmith, Guanabot2, Wadewitz, MarkS, Jlcooke, ESkog, Kbh3rd, Pmetzger, RJHall, El C, Edward Z. Yang, Shanes, Art LaPella, RoyBoy, Bookofjude, CDN99, Afed, Bobo192, Army1987, Harley peters, Whosyourjudas, Shenme, Viriditas, Brim, Maureen, Richi, Boxed, Timl, Joe Jarvis, Nk, Deryck Chan, Thewayforward, PiccoloNamek, Obradovic Goran, MPerel, Sam Korn, Pearle, Benbread, Nsaa, Ogress, HasharBot~enwiki, Jumbuck, Storm Rider, JYolkowski, Mennato, Richard Harvey, Thebeginning, Slugmaster, Iothiania, Riana, InShaneee, Spangineer, Malo, Snowolf, Saga City, Dirac1933, Geraldshields11, Henry W. Schmitt, Bsadowski1, BlastOButter42, Computerjoe, Gene Nygaard, Drbreznjev, Iustinus, Bookandcoffee, Dan100, Ceyockey, Tm1000, Rusticnature, Flying fish, Feezo, Blaze Labs Research, Thryduulf, Velho, Simetrical, Karnesky, Cimex, LOL, Kurzon, WadeSimMiser, Jwanders, Mpatel, Miss Madeline, Eleassar777, GregorB, Wayward, 🔲🔲🔲🔲🔲, Gimboid13, Palica, Dysepsion, MrSomeone, Paxsimius, SqueakBox, Graham87, BD2412, Zeroparallax, Deadcorpse, FreplySpang, Cmsg, Jclemens, Edison, Canderson7, Ketiltrout, Sjakkalle, Rjwilmsi, Quiddity, Sdornan, Gudeldar, Nneonneo, Bhadani, Sarg, Ucucha, Sango123, Matjlav, Lionelbrits, Titoxd, FlaBot, Pediadeep, RobertG, Nihiltres, Nivix, Chanting Fox, RexNL, Ewlyahoocom, Gurch, DannyDaWriter, TeaDrinker, Terrx, Fervidfrogger, Bmicomp, SteveBaker, Srleffler, BradBeattie, Le Anh-Huy, Physchim62, Mallocks, Nicholasink, Chobot, DVdm, Gwernol, YurikBot, Wavelength, Spacepotato, JWB, Sceptre, Hairy Dude, Jimp, Phantomsteve, RussBot, DMahalko, EDM, Ericorbit, SpuriousQ, Lucinos~enwiki, Kirill Lokshin, Stephenb, Gaius Cornelius, Ihope127, Pseudomonas, Wimt, RadioKirk, NawlinWiki, TEB728, SEWilcoBot, Trovatore, Muppetmaster, Tailpig, Ashwinr, Dureo, Seegoon, Cleared as filed, Irishguy, Deodar~enwiki, Nucleusboy, Brandon, Matticus78, Vb, Bobak, Zwobot, BOT-Superzerocool, DeadEyeArrow, Derek.cashman, Elkman, Kkmurray, Dna-webmaster, Wknight94, Zero1328, 21655, Zzuuzz, Closedmouth, Kriscotta, Jody Burns, BorgQueen, Dr U, JoanneB, Vicarious, Fram, Kevin, Spliffy, Enkauston, Katieh5584, Kungfuadam, Junglecat, TLSuda, RG2, GrinBot~enwiki, SkerHawx, Jade Knight, DVD R W, Finell, Msyjsm, Luk, ChemGardener, Sycthos, Itub, Iorek85, SmackBot, Meshach, Rex the first, Incnis Mrsi, Herostratus, Prodego, KnowledgeOfSelf, David.Mestel, Unyoyega, C.Fred, Jacek Kendysz, Yuyudevil, Jagged 85, Thunderboltz, Stepa, Pandion auk, Jrockley, Delldot, Jihiro, Bgrech, BiT, Edgar181, Alsandro, Mak17f, Moralis, Gilliam, Betacommand, Skizzik, Richfife, JAn Dudík, Constan69, Chris the speller, Bluebot, SlimJim, Quinsareth, BabuBhatt, Miquonranger03, SchfiftyThree, Complexica, Alink, Kevin Ryde, Wykis, Jfsamper, Robth, Sbharris, Hallenrm, Newmanbe, Can't sleep, clown will eat me, Scott3, Mitsuhirato, OrphanBot, Sephiroth BCR, Voyajer, Caleb Murdock, Rrburke, VMS Mosaic, HBow3, Addshore, SundarBot, Phaedriel, Monty2, Flyguy649, Fuhghettaboutit, Iapetus, Nakon, Jdlambert, Jiddisch~enwiki, Hoof Hearted, Dream out loud, TrogdorPolitiks, Hgilbert, Jameshales, DMacks, Wizardman, Evlekis, Sadi Carnot, Vina-iwbot~enwiki, Pilotguy, Kukini, Ceoil, Will Beback, Kuzaar, DJIndica, The undertow, Lambiam, Vanished user 9i39j3, Kuru, MarcMacé, MagnaMopus, Dracion, Scientizzle, Vgy7ujm, Buchanan-Hermit, Ascend, Heimstern, Calum MacÙisdean, Shadowlynk, Hemmingsen, Ospinad, Accurizer, CaptainVindaloo, KronosI, Aleenf1, IronGargoyle, Beefball, Ckatz, Sunni Jeskablo, Hohomanofsteel, Slakr, Kfsung, Beetstra, Noah Salzman, Martinp23, Alethiophile, Mr Stephen, GilbertoSilvaFan, Waggers, Eridani, Mets501, Funnybunny, LaMenta3, Darry2385, KJS77, LenW, BranStark, Iridescent, Theone00, Igge, Mark Oakley, The editor1, Igoldste, RekishiEJ, Saku kodo, Mr Chuckles, Willy Skillets, Tubezone, Tawkerbot2, Dlohcierekim, Bubbha, MightyWarrior, MarkAlldridge, Switchercat, Scienceloser09, JForget, Stifynsemons, CmdrObot, Tanthalas39, Wafulz, David s graff, Dycedarg, The ed17, Megaboz, Ruslik0, Benwildeboer, Jsd, Dgw, Yarnalgo, McVities, Outriggr, MarsRover, No1lakersfan, TheAdventMaster, Chris83, Nauticashades, Cydebot, Nick Y., MC10, Mziebell, SyntaxError55, Astrochemist, Rifleman 82, Meno25, Gogo Dodo, Zginder, KnightMove, Rracecarr, Skittleys, Strom, Jack Phoenix, DumbBOT, Ameliorate!, Narayanese, Optimist on the run, SteveMcCluskey, Vanished User jdksfajlasd, Daniel Olsen, Gimmetrow, Click23, FrancoGG, Thijs!bot, Epbr123, Daa89563, Qwyrxian, Goods21, Ramananrv123, Keraunos, PerfectStorm, Mojo Hand, Headbomb, Jojan, Yzmo, Marek69, John254, Tapir Terrific, A3RO, Ujm, Doyley, CharlotteWebb, Lithpiperpilot, Hempfel, Mactin, I already forgot, Hmrox, Cyclonenim, DrowningInRoyalty, AntiVandalBot, Majorly, Luna Santin, XxX-Teddy-XxX, Blue Tie, Bigtimepeace, Quintote, Doc Tropics, Kbthompson, Jj137, Danger, MECU, Istartfires, Indian Chronicles, Robert A. Mitchell, Gökhan, JAnDbot, Narssarssuaq, Leuko, Husond, DuncanHill, MER-C, Instinct, Hello32020, Andonic, $?, Pkoppenb, Savant13, Kirrages, Xact, Steveprutz, Acroterion, Freshacconci, Penubag, Magioladitis, Pedro, Bongwarrior, VoABot II, InvertedCommas, Yandman, CattleGirl, Think outside the box, Sikory, ThoHug, WODUP, Brusegadi, Midgrid, Catgut, Indon, LeeBuescher, Kevinwiatrowski, Orangemonster2k1, Joe hill, AlexIlew, Schumi555, Cpl Syx, Lord mrazon, Gimlei, Vssun, Just James, Glen, DerHexer, JaGa, Andythelovell, Markco1, Wikinger, DGG, 0612, PsyMar, MartinBot, Schmloof, BetBot~enwiki, ChemNerd, APT, Naohiro19, Nanotrix, Rettetast, TechnoFaye, Mschel, R'n'B, Dr. Hfuhruhurr, AlexiusHoratius, Mikejones11693, Dragonwings158, Cyrus Andiron, Worldedixor, Zarathura, Manticore, J.delanoy, Captain panda, Sasajid, DrKiernan, Rgoodermote, EscapingLife, Bogey97, Abby, Maurice Carbonaro, All Is One, Irfanhasmit, Awertx, 5Q5, Eliz81, Extransit, Aryoc, Bluesquareapple, Eskimospy, Gzkn, Acalamari, Kataleveno, Mousehunter, DarkFalls, Slippered sleep, Hawkmaster9, Battyboy69, Notapotato, Drumbeatz, JayJasper, Pyrospirit, AntiSpamBot, Xmonicleman, Floaterfluss, Lastchance4onelastdances, Pcfjr9, Volk282, Cheyeric, TomasBat, NewEnglandYankee, Antony-22, Healy6991, Hennessey, Patrick, Fieryiceissweet, Shoessss, Zaswert, KylieTastic, Adamd1008, Cometstyles, DorganBot, Cite needed, Sjwk, Mike V, Sarregouset, Bonadea, WinterSpw, Arock09300, Divad89, Ja 62, Jwmmorley, Inwind, Guess001, Useight, Vinsfan368, Xiahou, Nyquist562, CardinalDan, Pqwo, Idiomabot, The Simonator, Wikieditor06, Lights, Termasueder, VolkovBot, ABHISHEKARORA, Gofannon, Macedonian, Koombayah, Ndsg, Jeff

G., JohnBlackburne, VasilievVV, CGupte, Wolfnix, Philip Trueman, Wildshack, TXiKiBoT, Tbonepower07, GimmeBot, Katoa, Maximillion Pegasus, Nxavar, Ann Stouter, Anonymous Dissident, Wildshack300, BlueLint, Arnon Chaffin, TheGuyInTheIronMask, Pinkalfonzochief19, Voorlandt, Mr. Hallman, Littlealien182, Corvus cornix, Gharbad, JhsBot, LeaveSleaves, Ripepette, Psyche825, Freak104, Katimawan2005, Shanata, V81, Mwilso24, Greswik, Brian Huffman, Enigmaman, Synthebot, Antixt, Falcon8765, Hellsangel 5000, Burntsauce, Insanity Incarnate, Agüeybaná, Brianga, Yamster, Skarz, Gunnville, Mary quite contrary, Sue Rangell, Aaronownesfrod, Onceonthisisland, Nathan M. Swan, AlleborgoBot, Louis.pure, Ekmr, Zhangyijiang, EmxBot, Poneman, D. Recorder, Tylerofmaine, Psymun747, The Random Editor, Vex879, Zdavid408, Demmy100, EthanCoryHarris, SieBot, Coffee, Delorian8, Augustus Rookwood, Tresiden, PlanetStar, JamesA, Tiddly Tom, Scarian, Keymolen, Jauerback, Evud, Studnic12, Winchelsea, Exxy, Viskonsas, Caltas, RJaguar3, Triwbe, Brandon1919, Amill, Clew25, Wolfstevej, Mothmolevna, Xelgen, Keilana, Maddiekate, Anglicanus, Aillema, RadicalOne, Radon210, Exert, Lavers, NOK4987, Oda Mari, Arbor to SJ, Momo san, Elnutter, Argentix, Mimihitam, Oxymoron83, Nuttycoconut, Demosis, Designed for me and me, Steven Crossin, Lightmouse, Freirec, Buttpants, KathrynLybarger, Hobartimus, Jakeng, Thisnamestaken, Experiment6, Etx123456789, Pappapippa, Emesee, Traec1000, Mattyhoops, Spartan-James, Wingtale, StaticGull, Mike2vil, Some1random, Bogwhistle, Randomblue, Kunkyis, Maralia, Hello evry1, Nn123645, Pinkadelica, DRTllbrg, Efe, JL-Bot, Escape Orbit, Blakedriver123, Hridi339, Hadseys, Church, Loren.wilton, Separa, Amdragman, ClueBot, Deviator13, Binksternet, GorillaWarfare, Pyromaniac122, PipepBot, The Thing That Should Not Be, Endoxa, Ndenison, Drphallus, Wangaroo jamportha, Meekywiki, Botodo, Arroni, Mild Bill Hiccup, Uncle Milty, J8079s, JTBX, CounterVandalismBot, Niceguyedc, Harland1, Dylan620, Piledhigheranddeeper, Ottava Rima, Puchiko, Say2anniyan, Inter132, Manishearth, MindstormsKid, DragonBot, Martine23, Justaccount1, Excirial, Alexbot, Jusdafax, Coadgeorge, Kurtcobain911, Pspmichel2, Neuenglander, Slavko19, Eeekster, Taxa, Andriolo, Feline Hymnic, Winston365, Abrech, California847, Lartoven, Bairdy619, Tyler, NuclearWarfare, 10"weiner, Cenarium, Nmoo, Arjayay, Jotterbot, PhySusie, Soccerchicslhok, Razorflame, Firestarterrulz, Thingg, Aitias, Jesse.Senior, Katanada, SoxBot III, CTSW, Party, Alpixpakjian, Devaes, Darkicebot, Crazy Boris with a red beard, Willipod, Hornthecheck, XLinkBot, Kevind08, Pobob, Zacefron223962239622396, BodhisattvaBot, Jovianeye, Ozwego, NellieBly, Unclefrizzle, Mifter, Manfi, Walter.doug, Alyajohnson, NHJG, Vianello, Lemmey, Ab811996, Miagirljmw14, Freestyle-69, Eklipse, Siripswich, Willisis2, Addbot, Jennyvu96, Professor Calculus, Lewjam18, Joee8001, AVand, DOI bot, Sedsa1, Landon1980, Cmk1994, Dmjkmb, 15lsoucy, Ultraorange260, Vishnava, Mac Dreamstate, WFPM, Apemaster3000, Cst17, Download, LaaknorBot, Chamal N, CarsracBot, Glane23, Tech30, Akazme93, Debresser, AnnaFrance, Favonian, Chateau Brillant, AtheWeatherman, LinkFA-Bot, Sin Aura, Omg123123, IOLJeff, Eas4200c.team0, Numbo3-bot, Psproots, Koliri, Tide rolls, Bfigura's puppy, Luckas Blade, Jarble, Emperor Genius, GiantPea, KarenEdda, Georgieboi123, Angrysockhop, Legobot, Northexit182, Drpickem, XxGOWxx, Luckas-bot, Yobot, Googins, Senator Palpatine, Julia W, Yiplop stick stop, Cepheiden, Wikipedian Penguin, Nerdguy, Milesarise, The Bacon Machine, Jmproductionss, Cowdudemanxboyguymalesenorninoxxxxxxx, Ayrton Prost, Blk48, Yami89~enwiki, Chadisasexybeast, Skin and Bones, Kingpin13, Abshirdheere, Pokehen, Csigabi, Materialscientist, The High Fin Sperm Whale, Citation bot, E2eamon, Xqbot, TinucherianBot II, Intelati, GeometryGirl, Jeffrey Mall, Tolmanator, Tad Lincoln, Grim23, P99am, Almabot, Quixotex, GrouchoBot, Harley2121, Capecodsamm, Omnipaedista, Bahahs, Brutaldeluxe, GhalyBot, MerlLinkBot, N419BH, SchnitzelMannGreek, Sesu Prime, Pauswa, Legobot III, Tobby72, Dogbert66, Astronomyinertia, Cargoking, Machine Elf 1735, Finalius, Saiarcot895, Diremarc, Citation bot 1, Kobrabones, Clevercheetah123, Biker Biker, Pinethicket, MBirkholz, HRoestBot, Tom.Reding, RedBot, Bigad1, Footwarrior, White Shadows, Bsece010, FoxBot, Trappist the monk, Sweet xx, Sheogorath, Mrrstatham, Vrenator, Extra999, EzraNemo123, Burnthefairy56, Theproatlearningstuff, Loki2022, Reach Out to the Truth, Minimac, Marie Poise, DARTH SIDIOUS 2, RjwilmsiBot, TjBot, Ling cao, Edouard.darchimbaud, DASHBot, B20180, EmausBot, Thecreator09, Orphan Wiki, Immunize, Gfoley4, Pete Hobbs, 271196samantha, Smallchief, Hhhippo, Werieth, JSquish, CanonLawJunkie, Arik Islam, Mubasher55, Kbop451, Chaitana, Sambam340, H3llBot, Quondum, DanDao, Ahmetkilit, Lizzy chic38, Kobinks, Wayne Slam, Acdcfan1223, Bender176, Coasterlover1994, Pussvock, Thomasroper, Maschen, Dcgunasekar, Epicstonemason, Lilgas52, Puffin, Happiestmuackz11, ChuispastonBot, Oliozzicle, Lo79y123, Benjthedivine, DASHBotAV, ResearchRave, Petrb, ClueBot NG, Ulflund, Hiperfelix, Mitch09876, Sagenfowler, Big-Jason99, Hazhk, Moneya, O.Koslowski, Rezabot, Danim, Helpful Pixie Bot, Art and Muscle, Jubobroff, Bibcode Bot, Branthecan, Leonxlin, Malvel, Mysterytrey, Goodfluff, Xyzmathrules, Begman5, Fuppamaster, Jrobbinz1, Rgbc2000, Typoltion, Dr.Nadon, Connorbishop, IMn00b, Arghya33, DGK1318, Shawn Worthington Laser Plasma, Alarbus, Life421, BattyBot, Jimw338, Hebert Peró, ChrisGualtieri, GoShow, 786b6364, JYBot, Dexbot, Mogism, Falktan, GeeBIGS, MeteMetheus, Jamesx12345, RandomLittleHelper, Reatlas, Anastronomer, Raghav Bhalerao, Cowfdashkli, Pirtert, Homaa, AmericanLemming, Praemonitus, Tedsanders, PianoEngineer18, Mandruss, Anythingcouldhappen, Susan.grayeff, Anrnusna, Stamptrader, Joshdude356, Jazzmusician94, IStoleThePies, Mahusha, Monkbot, IiKkEe, Suntrax south, Mario Castelán Castro, Narky Blert, Y-S.Ko, Tetra quark, Isambard Kingdom, Forscienceonly, Nøkkenbuer, KasparBot, White909090lightning and Anonymous: 1481

- **Atomic theory** *Source:* https://en.wikipedia.org/wiki/Atomic_theory?oldid=675381609 *Contributors:* Mav, Ubiquity, Tim Starling, Tannin, Ixfd64, Ahoerstemeier, Smack, David Shay, HarryHenryGebel, Bloodshedder, Sverdrup, Rursus, Blainster, Hadal, Michael Snow, Rho~enwiki, Giftlite, Awolf002, Bensaccount, Beland, Karol Langner, H Padleckas, Gscshoyru, Neutrality, Engleman, Karl Dickman, Deglr6328, Discospinster, Qutezuce, Pjacobi, Vsmith, Paul August, Eric Forste, RJHall, El C, Laurascudder, Grick, Bobo192, DanielNuyu, AnyFile, Wipe, BrokenSegue, Maurreen, Joe Jarvis, Hooperbloob, Nsaa, Alansohn, Iothiania, Riana, BryanD, CJ, Malo, Velella, CloudNine, Bsadowski1, Falcorian, Feezo, Kelly Martin, Linas, Mindmatrix, Kurzon, MONGO, Terence, Mathewtse, Wayward, MarcoTolo, Mandarax, Rjwilmsi, Ctdunstan, R.e.b., Ems57fcva, Bubba73, Brighterorange, Krash, Yamamoto Ichiro, Jameshfisher, Alphachimp, Physchim62, King of Hearts, Chobot, Moocha, Cactus.man, Hall Monitor, Gwernol, Banaticus, YurikBot, Sceptre, Brandmeister (old), Petiatil, DanMS, ColoradoZ, Shell Kinney, Gaius Cornelius, NawlinWiki, Wiki alf, Grafen, Ragesoss, Mlouns, Syrthiss, Derek.cashman, Oysteinp, Petri Krohn, Tevildo, Archer7, TLSuda, Moomoomoo, Banus, Zvika, Elliskev, Finell, Luk, Itub, SmackBot, Tarret, Prodego, Vald, Jagged 85, Delldot, Eskimbot, Gaff, Betacommand, Kurykh, Audacity, NCurse, Bdube, Jprg1966, Miquonranger03, Complexica, Dustimagic, Sbharris, Colonies Chris, Hallenrm, Darth Panda, Sct72, Can't sleep, clown will eat me, ApolloCreed, Voyajer, Thrane, SnappingTurtle, TrogdorPolitiks, DMacks, Salamurai, Pkeets, SashatoBot, Nishkid64, ALUOPline2, Mathboy965, Ckatz, Rinnenadtrosc, Noah Salzman, KNBDunlop, Timmy2, Ryanjrr, Waggers, SandyGeorgia, Mets501, This is so strange, Ryulong, LaMenta3, Sifaka, Iridescent, Shoeofdeath, Francl, Igoldste, Blehfu, Courcelles, Tawkerbot2, Kurtan~enwiki, JForget, CmdrObot, KyraVixen, JohnCD, NickW557, MrFish, Jac16888, Bddmagic, Misterhay, Bellerophon5685, Synergy, Tawkerbot4, Carstensen, Doug Weller, Christian75, Codetiger, DumbBOT, Kozuch, SteveMcCluskey, Vanished User jdksfajlasd, Casliber, Thijs!bot, Epbr123, Qwyrxian, Headbomb, Marek69, JustAGal, Natalie Erin, CTZMSC3, Mentifisto, AntiVandalBot, KP Botany, Danger, Indian Chronicles, Etr52, JAnDbot, MER-C, Arch dude, MSBOT, Bongwarrior, VoABot II, Ling.Nut, Catgut, Froman77, Nposs, Allstarecho, J Hill, User A1, Adventurer, DerHexer, Hbent, Yensin, Slippknotryan, S3000, MartinBot, Kbrewer, Rettetast, R'n'B, AlexiusHoratius, J.delanoy, Adavidb, Bogey97, Mthibault, Rhinestone K, Yonidebot, Choihei, Hellomoto021, L337 kybldmstr, P.wormer, Belovedfreak, King turtal, Sunderland06, TottyBot, Idioma-bot, Bottobbot, Moblinmaniac, Deor, TreasuryTag, Jeff G., Rtrace, Hilarious Bookbinder, Kyle the

Can't sleep, clown will eat me, Drkirkby, Vladislav, Voyajer, Pax85, Radagast83, Edwtie, Drphilharmonic, Thinkingman, Lambiam, Doug Bell, Rigadoun, Ortho, 041744, Ckatz, RandomCritic, Ginkgo100, Esurnir, Tawkerbot2, JForget, Megaboz, Johnlogic, Myasuda, Christian75, Thijs!bot, Epbr123, Guyla, Mbell, Nonagonal Spider, Headbomb, Mjollnir783, Weasel5i2, Escarbot, Ssr, Mentifisto, AntiVandalBot, Luna Santin, Refried, JAnDbot, Instinct, Acroterion, Bongwarrior, VoABot II, Ling.Nut, Glen, Geboy, Mike6271, Davburns, J.delanoy, Yonidebot, Acalamari, Jakey665, McSly, TomasBat, Benito001, Juliancolton, Logic20, Idioma-bot, ArchetypeRyan, VolkovBot, Philip Trueman, Amother, TXiKiBoT, Geht, Jhannah, GcSwRhIc, Seraphim, Martin451, Optigan13, Krazywrath, Lamro, Insanity Incarnate, Monty845, Lkleinow, AlleborgoBot, Freependulum, Borne nocker, Hazel77, S8333631, Doclecticwiki, SieBot, Gerakibot, 4RM0~enwiki, Keilana, Flyer22, Tiptoety, The Evil Spartan, Sohelpme, AlexWaelde, Wombatcat, Lisatwo, Antman123, Nimbusania, DarkCatalyst, ClueBot, The Thing That Should Not Be, Drugieuk, Arakunem, Drmies, DragonBot, Djr32, Excirial, Monobi, WikiZorro, SpikeToronto, Peter.C, Darren23, Little Mountain 5, Mifter, Garycompugeek, PicoGils, Addbot, Willking1979, Ronhjones, Knowledgesupreme, Download, Chamal N, Bassbonerocks, CosmiCarl, Barak Sh, IOLJeff, Ehrenkater, VASANTH S.N., Tide rolls, Lightbot, Megaman en m, Luckas-bot, Yobot, Amirobot, Eric-Wester, Kulmalukko, AnomieBOT, Ciphers, ^musaz, Jim1138, Icalanise, Ulric1313, Materialscientist, Citation bot, Maxis ftw, ArthurBot, Xqbot, Phazvmk, Gopal81, Addihockey10, Turk oğlan, Dubravko49, GrouchoBot, АлександрВв, TR4YH4N, Mitraunodo, Sushiflinger, Dave3457, Mark Renier, Doremo, Steve Quinn, Gigigogo, Citation bot 1, Pinethicket, Tom.Reding, Maude Frickert, Teamspoad, TobeBot, Mptb3, Koyae, MartinHiggs, Diannaa, Harrasser, Tbhotch, RjwilmsiBot, Narayanan20092009, EmausBot, Ajraddatz, Heyimawesome, Wikipelli, Hhhippo, JSquish, Harddk, Shuipzv3, Permenent, Anir1uph, Access Denied, Maschen, Epicstonemason, VictorianMutant, CharlieEchoTango, ClueBot NG, Preon, IfYouDoIfYouDon't, Widr, MerlIwBot, Helpful Pixie Bot, Novusuna, IrishStephen, BG19bot, Northamerica1000, Neutral current, Kord Kakurios, Zombiecat181, Jethro B, Ducknish, MadGuy7023, Jmeg82, Webclient101, Mogism, Reatlas, Faizan, Bigdaddysound, Frankalbertson, The Herald, Ginsuloft, AddWittyNameHere, Colecharb, 123gogeta, TheRapeTrainBackFromTheDead,Again, CraigyDavi, Robdistasio, Darkenergydesigner, Tetra quark, Jesus.Like.For.Real, Jennifer1122, Harshita1999 and Anonymous: 365

- **Neutron** *Source:* https://en.wikipedia.org/wiki/Neutron?oldid=675084912 *Contributors:* AxelBoldt, Tobias Hoevekamp, Chenyu, Trelvis, Calypso, Mav, Bryan Derksen, The Anome, AstroNomer~enwiki, Malcolm Farmer, Andre Engels, Xaonon, Danny, XJaM, Roadrunner, Jaknouse, Olivier, Patrick, Michael Hardy, Valery Beaud, Ixfd64, TakuyaMurata, NuclearWinner, Looxix~enwiki, ArnoLagrange, Mkweise, Ellywa, Ahoerstemeier, Cyp, Andrewa, Aarchiba, Julesd, Glenn, Nikai, Andres, Stone, Denni, Kbk, Tarosan~enwiki, Maximus Rex, Donarreiskoffer, Gentgeen, Robbot, Fredrik, Romanm, Merovingian, Rursus, Wikibot, Alan Liefting, Dave6, Giftlite, Mikez, Art Carlson, Herbee, Xerxes314, Everyking, Dratman, NeoJustin, Bensaccount, Poupoune5, Jorge Stolfi, Christofurio, Knutux, Karol Langner, Aecarol, Icairns, Zfr, Cglassey, Peter bertok, Frau Holle, M1ss1ontomars2k4, Sparky2002b, Mike Rosoft, Guanabot, Vsmith, Dbachmann, Bender235, Kjoonlee, AlDragon, Geoking66, Neko-chan, RJHall, CanisRufus, El C, Susvolans, Femto, CDN99, Bobo192, O18, Smalljim, SpeedyGonsales, Kjkolb, Obradovic Goran, Sam Korn, Nsaa, Jakew, Eddideigel, Jumbuck, Patsw, Alansohn, Interiot, Riana, Wtmitchell, BRW, NickMartin, Vuo, DV8 2XL, HenryLi, Tchaika, Forteblast, Falcorian, Richard Arthur Norton (1958-), JarlaxleArtemis, WadeSimMiser, Sega381, SDC, Jon Harald Søby, Prashanthns, Abd, LexCorp, Graham87, Magister Mathematicae, Doughboy, Ketiltrout, Rjwilmsi, Nightscream, Zbxgscqf, Strait, AySz88, Oo64eva, Rangek, FlaBot, Nihiltres, Goudzovski, Srleffler, Ronebofh, King of Hearts, Chobot, DVdm, YurikBot, RobotE, Bambaiah, JWB, TSO1D, Jimp, Phantomsteve, KyleDantarin, Stephenb, Gaius Cornelius, Yyy, Salsb, NawlinWiki, Tupungato, Wiki alf, Complainer, Grafen, Długosz, Voidxor, Scottfisher, Kkmurray, Spute, Dna-webmaster, Wknight94, Stefan Udrea, Mike Serfas, Closedmouth, Reyk, Modify, Alchie1, CWenger, RG2, Paul Erik, Triple333, Attilios, SmackBot, Caiyern, Melchoir, Wiki Tiki God, Unyoyega, Jrockley, Dr.Science, Edgar181, Yamaguchi⬜⬜, Kdliss, Wigren, Chris the speller, Rajeevmass~enwiki, Persian Poet Gal, SchfiftyThree, Complexica, DHN-bot~enwiki, Sbharris, Colonies Chris, Brainblaster52, Can't sleep, clown will eat me, DéRahier, Juancnuno, SundarBot, DFriend, Aldaron, KunalKathuria, Nakon, Mwtoews, DMacks, Soarhead77, Bdushaw, Pilotguy, Renafaye77, SashatoBot, Demicx, Tim bates, Mgiganteus1, Slakr, Citicat, Asyndeton, BranStark, Shoeofdeath, Newone, Tawkerbot2, Atomobot, Mosaffa, CmdrObot, Wafulz, Dycedarg, Rwflammang, Joelholdsworth, Lokal Profil, Karenjc, Myasuda, Safalra, Icek~enwiki, Badseed, Nick Y., Gogo Dodo, Chasingsol, Phydend, Gimmetrow, Thijs!bot, Epbr123, Montazmeahii, Goods21, Tsogo3, N5iln, Oerjan, Headbomb, Marek69, SouthernMan, RoboServien, Escarbot, Aadal, WikiSlasher, AntiVandalBot, Seaphoto, Naturalnumber, Spencer, Astavats, Husond, CosineKitty, Medconn, TheEditrix2, Bongwarrior, VoABot II, Kuyabribri, JamesBWatson, WODUP, Mother.earth, Animum, BatteryIncluded, Dirac66, LorenzoB, DerHexer, JaGa, Hans Moravec, Hyray, Patstuart, MartinBot, Church of emacs, Gnuarm, Mennoblaauw, Andre.holzner, Rettetast, J.delanoy, Dbiel, Extransit, Acalamari, Ncmvocalist, TomasBat, MetsFan76, Joshmt, Heavens is the world, Scott Illini, TraceyR, Idioma-bot, Mviduka4197, VolkovBot, Tourbillon, Thedjatclubrock, Jeff G., Mocirne, Seattle Skier, TXiKiBoT, DoctorPiouk, Dev 176, Martin451, ABigGreenHippo, Abdullais4u, FreeFull, Wikiisawesome, Scarymaryfwfc, RadiantRay, Roomyt, W1k13rh3nry, Antixt, Deanlsinclair, Enviroboy, Burntsauce, Brianga, AlleborgoBot, EmxBot, Neparis, D. Recorder, Ponyo, YohanN7, SieBot, Cwkmail, Yintan, Agesworth, JerrySteal, Keilana, RadicalOne, Toddst1, Tiptoety, JetLover, Arjen Dijksman, Sbowers3, Aruton, Oxymoron83, AnonGuy, Beej175560, Techman224, Anyeverybody, Nergaal, Denisarona, Lord Shivan, Naturespace, ClueBot, RudolfSchmidt, PipepBot, Fasettle, Fyyer, The Thing That Should Not Be, Starkiller88, Industrieman, Mild Bill Hiccup, Polyamorph, Shjacks45, ChandlerMapBot, DragonBot, Gnome de plume, Jusdafax, Ju7kik8ol568r, Cenarium, Jotterbot, Vboo-belarus, Subash.chandran007, Plasmic Physics, Versus22, XLinkBot, Dark Mage, PL290, SkyLined, Addbot, Taschna, DOI bot, Ronhjones, Mr. Wheely Guy, LaaknorBot, CarsracBot, JBukon, Favonian, LinkFA-Bot, 5 albert square, AgadaUrbanit, Morgrimm, Numbo3-bot, Ehrenkater, LarryFrank, Tide rolls, Lightbot, Teles, Legobot, Luckas-bot, Yobot, Велетень, 2D, Tohd8BohaithuGh1, Cabb99, AnakngAraw, AnomieBOT, Bsimmons666, Jim1138, AdjustShift, Bluerasberry, Materialscientist, Hdehuer, The High Fin Sperm Whale, Citation bot, Satan's Kitchen, Maxis ftw, Raven1977, ArthurBot, Marshallsumter, Xqbot, Gopal81, Capricorn42, Drilnoth, DSisyphBot, Gilo1969, Paula Pilcher, Faatoafe90, Goostyyy, WaveEtherSniffer, GrouchoBot, Abce2, Amaury, Doulos Christos, Gordonrox24, Shadowjams, A. di M., Samwb123, R8R Gtrs, FrescoBot, LucienBOT, Paine Ellsworth, Cannolis, Citation bot 1, Ecko15, Biker Biker, Pinethicket, HRoestBot, Jonesey95, Nicklcms, Seattle Jörg, Abhinav paulite, Double sharp, Darrell cosare, كاشف عقیلى, Mr.98, Diannaa, Ironnickel, Andrea105, Onel5969, TjBot, MagnInd, Jackehammond, Jimmy be, Robert Johnson 10, EmausBot, Green Day143, WikitanvirBot, Unkenruf, GoingBatty, Illdz, Psturm~enwiki, Pcorty, Wikipelli, Hhhippo, ZéroBot, John Cline, Brazmyth, Quondum, GianniG46, Copper.nanotube, Brandmeister, L Kensington, Epicstonemason, Sjkimminau, Chris857, VictorianMutant, DASHBotAV, Whoop whoop pull up, ClueBot NG, Nebulosus, CocuBot, Satellizer, Letoya123, TruPepitoM, OverQuantum, Heyheyheyhohoho, Rezabot, Android1188, Widr, Diyar se, Ieditpagesincorrectly, Bibcode Bot, Neutronscattering, Wiki13, Metricopolus, Contact '97, Universuminkeisari, Nathanrohler, Zedshort, Hamish59, Nitrobutane, Hobos-r-us12, Oznitecki, BattyBot, MeowMeowArf, Dansalmo, ChrisGualtieri, GeorgEhlers, Ducknish, Gladiator222, Mogism, 331dot, TwoTwoHello, Lugia2453, Graphium, FaerieChilde, Fossilsnout, Morg00, Cldorian, Xuanmingzi, DihllonJessie, Jesse.johns, The Herald, Zenibus, Darkch2, Jwratner1, Javierha, My name is not dave, Cytokinetics, EtymAesthete, DudeWithAFeud, Abitslow, Aspaas-Bekkelund, Bballbro62, Mahusha, Light on the wall, Monkbot, Profesionalpretzels, Jayakumar RG, Haftswinch532, Selmatoed50, Istillcant,

HMSLavender, Petahr, Orduin, Kethrus, Pulkit 4325, DiscantX, TSchonfeldt, Matan Kovac, KasparBot, Kafishabbir, Lord Wingus The Third and Anonymous: 552

- **Proton** *Source:* https://en.wikipedia.org/wiki/Proton?oldid=677267180 *Contributors:* Mav, Bryan Derksen, Zundark, Manning Bartlett, Ap, Andre Engels, Josh Grosse, Danny, XJaM, Ellmist, Heron, Jaknouse, Stevertigo, Dwmyers, Bdesham, Patrick, Kku, Stewacide, TakuyaMurata, Egil, NuclearWinner, Looxix~enwiki, Mkweise, Ahoerstemeier, Sobekhotep, Salsa Shark, Glenn, Kaihsu, Jordi Burguet Castell, Mxn, Denny, Timwi, Paul-L~enwiki, Omegatron, Geraki, David.Monniaux, Donarreiskoffer, Robbot, Josh Cherry, Jakohn, RedWolf, Altenmann, Yelyos, Merovingian, Flauto Dolce, Hadal, Giftlite, Mikez, Tom harrison, Lupin, Herbee, Xerxes314, Fleminra, Bensaccount, Foobar, PlatinumX, Utcursch, Knutux, Antandrus, Beland, Eroica, Melikamp, Rdsmith4, DragonflySixtyseven, Icairns, Cglassey, Deglr6328, Grunt, Thorwald, Jenlight, Mike Rosoft, Diagonalfish, Discospinster, Cacycle, Vsmith, Jpk, Wikiacc, Mani1, Bender235, Kjoonlee, RJHall, El C, Shrike, Femto, Bobo192, O18, Army1987, Smalljim, GTubio, Vortexrealm, Elipongo, Foobaz, Kjkolb, Obradovic Goran, Nsaa, Eddideigel, Anthony Appleyard, Mattpickman, Apoc2400, Carmelbuck, Spangineer, Wtmitchell, Saga City, Uucp, Crobzub, Vcelloho, RainbowOfLight, TenOfAllTrades, Computerjoe, Kusma, Itsmine, Falcorian, Richard Arthur Norton (1958-), Firien, GregorB, Macaddct1984, Mayz, Karam.Anthony.K, Marudubshinki, Bebenko, Rtcpenguin, Graham87, Kbdank71, Ketiltrout, Drbogdan, Rjwilmsi, Strait, NeonMerlin, RadicalJester, Bubba73, Yamamoto Ichiro, Rangek, FlaBot, Nivix, RexNL, Fresheneesz, Srleffler, Imnotminkus, King of Hearts, CiaPan, Chobot, Deyyaz, Roboto de Ajvol, The Rambling Man, YurikBot, Bambaiah, JWB, Anuran, Pip2andahalf, RussBot, Wigie, Jumbo Snails, Raquel Baranow, Hellbus, Salsb, Oni Lukos, Anomalocaris, NawlinWiki, Injinera, Welsh, Długosz, Martin Ulfvik, Moe Epsilon, BOT-Superzerocool, DeadEyeArrow, Bota47, D-Day, Mtu, Pooryorick~enwiki, J S Ayer, Theodolite, Bayerischermann, Closedmouth, Reyk, Petri Krohn, DGaw, Paul D. Anderson, Katieh5584, RG2, SDS, GrinBot~enwiki, Nekura, Orii, Luk, Sycthos, Itub, Attilios, SmackBot, Moeron, Incnis Mrsi, Melchoir, CyclePat, Edgar181, HalfShadow, Dhochron, Munky2, Chris the speller, Rajeevmass~enwiki, Rkitko, AndrewBuck, Bethling, SchfiftyThree, Complexica, Sbharris, Rogermw, Can't sleep, clown will eat me, Shalom Yechiel, PeteShanosky, Writtenright, Homestarmy, Wikiwikiwiki3~enwiki, SundarBot, COMPFUNK2, Dreadstar, Orczar, Drphilharmonic, Dvorak729, DMacks, Mion, Vina-iwbot~enwiki, Bdushaw, SashatoBot, ArglebargleIV, Khazar, Cholerashot, Rijkbenik, Spacecadethailey, Herr apa, Deathcakes, Noah Salzman, Aeluwas, Waggers, Mozzura, Mattabat, Elb2000, Newone, MOBle, Igoldste, Rhetth, Frank Lofaro Jr., Tawkerbot2, CmdrObot, Ale jrb, Sir Vicious, KyraVixen, Ruslik0, McVities, TheTito, Cydebot, Nick Y., Gogo Dodo, Red Director, Umdunno, Difluoroethene, Odie5533, Q43, Tawkerbot4, Dwool99f, Narayanese, Rasheedy, Zalgo, Lo2u, Jenswort, Thijs!bot, Epbr123, Tsogo3, Headbomb, Marek69, Electron9, Mnemeson, Dfrg.msc, Philippe, Aadal, AntiVandalBot, Seaphoto, HairyDan, Shirt58, EarthPerson, Gregnx, Jj137, Naturalnumber, Myanw, Ellissound, Leuko, MER-C, CosineKitty, Fetchcomms, Andonic, Kerotan, .anacondabot, Acroterion, Plynn9, Casmith 789, Bongwarrior, VoABot II, Astrangequark, Swpb, WODUP, Recurring dreams, Avicennasis, Catgut, Dirac66, 28421u2232nfenfcenc, Hveziris, Fang 23, The Real Marauder, Oddworth, JaGa, MartinBot, Rettetast, Pbroks13, Artaxiad, J.delanoy, WeglarczykJ, Silverxxx, C.A.T.S. CEO, Maurice Carbonaro, 12dstring, WarthogDemon, Acalamari, Exdejesus, TomasBat, Antony-22, Potatoswatter, KylieTastic, Vanished user 39948282, Treisijs, S, SoCalSuperEagle, Xiahou, Specter01010, Idioma-bot, Ciju, 28bytes, VolkovBot, Doc7777777777, Jeff G., Soliloquial, Tuffcarrot, Philip Trueman, TXiKiBoT, The Original Wildbear, Vipinhari, Bjman, Bigyaks, Alexalexalex123~enwiki, Meters, Antixt, Spinningspark, Insanity Incarnate, Upquark, AlleborgoBot, Vitalikk, B41988, Petergans, Demmy100, SieBot, Accounting4Taste, Jauerback, Studnic12, Xe1881, Yintan, GlassCobra, Keilana, RadicalOne, Flyer22, Sbowers3, Prestonmag, Oxymoron83, BenoniBot~enwiki, Jacob.jose, Mygerardromance, Rajbboy69, ClueBot, The Thing That Should Not Be, RODERICKMOLASAR, Tigerboy1966, Regibox, ChandlerMapBot, Mr blabla, Excirial, Alexbot, Robbie098, Poopmister91191, Ploft, NuclearWarfare, Lunchscale, PhySusie, SoxBot, El pobre Pedro, Thehelpfulone, La Pianista, Thingg, Kanxkawii, Aitias, Subash.chandran007, Johnuniq, XLinkBot, Avoided, WikHead, SilvonenBot, SkyLined, Mls1492, Weletahoozyzog, Addbot, Zrules, Arcturus87, Ronhjones, Cst17, Glane23, AndersBot, Wandering Traveler, Omnipedian, LinkFA-Bot, Numbo3-bot, Tide rolls, Thermalimage, Luckas-bot, Yobot, Велетень, Wickedwizardofoz, Newportm, Kilom691, Heart of a Lion, Eric-Wester, Jay0205, AnomieBOT, LeftyAce, Götz, Jim1138, Sp eloc, Judoc, Materialscientist, The High Fin Sperm Whale, Citation bot, OllieFury, Apollo, Xqbot, Phazvmk, Blennow, Cureden, Wyklety, Aa77zz, Squishywushy123, Srich32977, Rueyfugudtj, RibotBOT, PM800, A. di M., ⁇⁇, Rain bowell, CES1596, FrescoBot, Paine Ellsworth, Tobby72, Citation bot 1, Pinethicket, HRoestBot, Calmer Waters, Bejinhan, Impala2009, Nicklcms, Blckmgc, SkyMachine, Gryllida, Double sharp, TobeBot, Ilovefatchicks, Keegscee, DARTH SIDIOUS 2, TjBot, StudentDoc73, Nachos0123, EmausBot, WikitanvirBot, ANDREVV, Bencbartlett, Zues zeus kratos, Pcorty, Sterrettc, K6ka, JSquish, Stuffness12, John Cline, Harddk, Fæ, StringTheory11, Brazmyth, Suslindisambiguator, Wayne Slam, Zach444, Rcsprinter123, Wiggles007, Brandmeister, Donner60, Sarthak 94, Xonqnopp, ClueBot NG, Timelord360, This lousy T-shirt, IHopeThisNameIsntTaken, Corusant, Cntras, Dictabeard, Rezabot, Helpful Pixie Bot, Wbm1058, Bibcode Bot, BG19bot, ArthropodOfDoom, RadioActiveKitKat, AvocatoBot, Metricopolus, JacobTrue, Mlkamitso, Toccata quarta, ShotmanMaslo, Mhutchison43, 220 of Borg, Anbu121, Hitheresir, RudolfRed, Knodir, BattyBot, ChrisGualtieri, Hower64, Ducknish, Stephen Glass, BrightStarSky, Mogism, Sheehan Cein14, Frosty, Pidotclan, Marcoapc.84, Delnium strex, Faizan, Huddydakota, Prof.Professer, Dustin V. S., Borreswafflertron, Ugog Nizdast, Jwratner1, Javierha, JaconaFrere, AspaasBekkelund, Bballbro62, Melcous, Monkbot, Jayakumar RG, Scorpion1045, Krebs49, Yellowswagger19, Maddie005, Junchuann, Chrisbrownthathoe, Orgasam069, Tktobykerby, Interpuncts, Tetra quark, Cjohnson2020, KasparBot, Muzammil Alam Baig, Rambunctious Racoon and Anonymous: 636

- **Electron***Source:*https://en.wikipedia.org/wiki/Electron?oldid=677715867*Contributors:*AxelBoldt, CYD, Mav, Bryan Derksen, AstroNomer~e Ap, Ed Poor, Andre Engels, Ryrivard, William Avery, SimonP, Peterlin~enwiki, Heron, Camembert, Stevertigo, Bdesham, Patrick, D, JohnOwens, Michael Hardy, Tim Starling, Ixfd64, Fruge~enwiki, Arpingstone, PingPongBoy, Egil, NuclearWinner, Ahoerstemeier, Suisui, Jebba, JWSchmidt, Kingturtle, Aarchiba, Glenn, Scott, Kwekubo, Andres, Jordi Burguet Castell, Mxn, Agtx, Timwi, Wikiborg, Reddi, Rednblu, Markhurd, Maximus Rex, E23~enwiki, Omegatron, Secretlondon, Jusjih, BenRG, Jeffq, Donarreiskoffer, Gentgeen, Robbot, Sanders muc, Vespristiano, Merovingian, Pingveno, Blainster, Hadal, Wikibot, Wereon, Widsith, HaeB, Diberri, Dmn, Dina, Giftlite, Christopher Parham, Ferkelparade, Fastfission, Zigger, Herbee, Dissident, Xerxes314, Curps, Michael Devore, Bensaccount, Ssd, Gilgamesh~enwiki, Vadmium, Gdr, Knutux, Slowking Man, Yath, Gzuckier, Pcarbonn, Joizashmo, Karol Langner, Anythingyouwant, RetiredUser2, Bbbl67, Elroch, Icairns, JohnArmagh, JimQ, Mike Rosoft, Mindspillage, Patrick L. Goes, Discospinster, Brianhe, Rich Farmbrough, Guanabot, Hidaspal, Vsmith, Deh, Ardonik, Roybb95~enwiki, Xezbeth, Zazou, Mani1, SpookyMulder, Dmr2, ZeroOne, Kjoonlee, Goplat, Calair, Nabla, Brian0918, RJHall, Pt, Jaques O. Carvalho, El C, Huntster, Edward Z. Yang, Susvolans, Art LaPella, RoyBoy, ~K, Bobo192, Army1987, Asierra~enwiki, Flxmghvgvk, AtomicDragon, Evgeny, AllyUnion, Bert Hickman, Deryck Chan, PeterisP, Beetle B., Obradovic Goran, (aeropagitica), Pearle, Mpulier, HasharBot~enwiki, Confusedmiked, Mote, Jumbuck, Gary, ChristopherWillis, Ricky81682, Benjah-bmm27, Riana, AzaToth, DonJStevens, BernardH, Malo, David Hochron, Bart133, EagleFalconn, Schapel, Omphaloscope, RainbowOfLight, RichBlinne, H2g2bob, DV8 2XL, Gene Nygaard, Redvers, StuTheSheep, Linas, Mindmatrix, GrouchyDan, StradivariusTV, Uncle G, BillC, Kurzon,

Jeff3000, HcorEric X, Eleassar777, Ozielke, Wayward, Palica, Omega21, FreplySpang, Enzo Aquarius, Rjwilmsi, Shaadow, Strait, Mike Peel, Chekaz, Bubba73, Dar-Ape, Yamamoto Ichiro, FlaBot, RobertG, Latka, DannyWilde, Nihiltres, RexNL, Kolbasz, Thecurran, Srleffler, Physchim62, Chobot, DVdm, Unclevortex, Eric B, YurikBot, Wavelength, RobotE, Bambaiah, AcidHelmNun, Jimp, Peter G Werner, Wolfmankurd, Wigie, Ventolin, JabberWok, SpuriousQ, Lucinos~enwiki, Akamad, Ori Livneh, Gaius Cornelius, Shaddack, Eleassar, Rsrikanth05, Salsb, Hawkeye7, Spike Wilbury, Jaxl, Welsh, DarthVader, Długosz, BirgitteSB, SCZenz, Retired username, Ravedave, PhilipO, Adam Rock, Mlouns, Chichui, BOT-Superzerocool, Gadget850, Bota47, Kkmurray, James Trotter~enwiki, Dna-webmaster, Ms2ger, Light current, Lycaon, Imaninjapirate, Josh3580, Kriscotta, JoanneB, Peyna, Lpm, JLaTondre, Heavy bolter, RG2, GrinBot~enwiki, Sbyrnes321, ChemGardener, Itub, SmackBot, Zazaban, Incnis Mrsi, KnowledgeOfSelf, Royalguard11, Melchoir, J.Sarfatti, KocjoBot~enwiki, Stepa, Pandion auk, Jrockley, JoeMarfice, ZerodEgo, Edgar181, Yamaguchi⌷⌷, Skizzik, Dauto, JSpudeman, Kurykh, Rajeevmass~enwiki, Persian Poet Gal, Pieter Kuiper, Jprg1966, Acrinym, Miquonranger03, MalafayaBot, Droll, Complexica, DHN-bot~enwiki, Sbharris, RAlafriz, V1adis1av, Vanished User 0001, Darthgriz98, Voyajer, Addshore, Percommode, Krich, DavidStern, Theonlyedge, Nakon, Nrcprm2026, DMacks, Daniel.Cardenas, Zeamays, Jonnyapple, Sadi Carnot, Bdushaw, Wilt, TriTertButoxy, Chymicus, UberCryxic, Bagel7, Mattfont, Heimstern, Jaganath, Ocatecir, Mr. Lefty, Ckatz, 16@r, Omnedon, Owlbuster, Waggers, SandyGeorgia, Spiel496, Funnybunny, HappyVR, Iridescent, Newone, NativeForeigner, J Di, Amakuru, Tawkerbot2, Chetvorno, Thermochap, CmdrObot, Ale jrb, Megaboz, RedRollerskate, Ruslik0, MrZap, McVities, WMSwiki, Bakanov, RobertLovesPi, Equendil, Cydebot, Acelor, Reywas92, Cantras, Bvcrist, LouisBB, Travelbird, Llort, David edwards, Tawkerbot4, Christian75, Narayanese, Ssilvers, Thijs!bot, Epbr123, Mbell, Dougsim, Nonagonal Spider, Headbomb, Yzmo, Marek69, West Brom 4ever, Tellyaddict, Cool Blue, Greg L, Sean William, VictorP, KrakatoaKatie, AntiVandalBot, WinBot, Skymt, Voyaging, Opelio, Tyco.skinner, Gef756, Chill doubt, Naturalnumber, Gdo01, Spencer, Leuko, CosineKitty, J-stan, Smith Jones, Acroterion, Magioladitis, WolfmanSF, Bennybp, Bongwarrior, VoABot II, A4, Nyq, JNW, JamesBWatson, بلسام, Drondent, Slartibartfast1992, Jackal irl, Animum, Dirac66, 28421u2232nfenfcenc, Hveziris, User A1, Maliz, PoliticalJunkie, DerHexer, GregU, PEBill, MartinBot, BetBot~enwiki, Mermaid from the Baltic Sea, WizendraW, Xantolus, Thereen, CommonsDelinker, AlexiusHoratius, J.delanoy, DrKiernan, Rgoodermote, Numbo3, Acalamari, TheChrisD, Dispenser, LordAnubisBOT, JayMars, Lathrop, AntiSpamBot, TomasBat, NewEnglandYankee, Nwbeeson, SmoothK, Sunderland06, MetsFan76, Joshmt, Cometstyles, STBotD, RB972, Treisijs, D-Kuru, Dineshextreeme, Martial75, CardinalDan, Idioma-bot, Sheliak, Bondslave777, FeralDruid, X!, VolkovBot, ABF, Thisisborin9, Jacroe, Ryan032, Philip Trueman, DoorsAjar, TXiKiBoT, GimmeBot, Kriak, Hqb, GDonato, Anonymous Dissident, Crohnie, Monkey Bounce, Voorlandt, Mr. Hallman, Michael H 34, Wikiisawesome, Suriel1981, Rbdebole, Graymornings, Synthebot, Enviroboy, Rurik3, Generalguy11, !dea4u, Insanity Incarnate, Ceranthor, Yoos~enwiki, AlleborgoBot, Kalivd, EmxBot, Neparis, Swimallday, Ponyo, EJF, SieBot, Graham Beards, Scarian, CircafuciX, BotMultichill, Jauerback, Dawn Bard, Joncam, Caltas, Sergeanthuggy, Bentogoa, RadicalOne, Arbor to SJ, Prestonmag, Thadaddy3233, Oxymoron83, Antonio Lopez, KPH2293, Lightmouse, WingkeeLEE, Ealdgyth, BenoniBot~enwiki, Stustjohn, Dabomb87, PlantTrees, Dolphin51, Nergaal, Tomdobb, Muhends, WikipedianMarlith, ClueBot, Trojancowboy, GorillaWarfare, Artichoker, PipepBot, UniQue tree, The Thing That Should Not Be, Hongthay, Unbuttered Parsnip, GreenSpigot, Liekmudkipz, Mild Bill Hiccup, Correcting nonesense, NovaDog, Blanchardb, Richerman, RandomTREES, Rotational, Piledhigheranddeeper, Inala, DragonBot, Almcaeobtac, Jusdafax, MEJG, Gtstricky, Rhododendrites, Brews ohare, NuclearWarfare, Lunchscale, Jotterbot, PhySusie, Tonyfey, Lkruijsw, Kaiba, SchreiberBike, Stepheng3, Thingg, Jamyricks, Aitias, Melibarr05, Kurtcobain321, Scalhotrod, Versus22, Johnuniq, MasterOfHisOwnDomain, DumZiBoT, TimothyRias, Sjodenenator, XLinkBot, Maky, Rror, Avoided, Mitch Ames, Ilikepie2221, WikHead, Mgaarafan, SkyLined, Addbot, Chizkiyahuavraham, AVand, Some jerk on the Internet, Hurleymann1, Uruk2008, DOI bot, Tcncv, Booba5, AkhtaBot, Jessepfrancis, Ronhjones, Jncraton, Moosehadley, CanadianLinuxUser, WFPM, LaaknorBot, Chamal N, CarsracBot, FiriBot, Omnipedian, LinkFA-Bot, Ehrenkater, Pnacitum, Tide rolls, Lightbot, Potekhin, UPS Truck Driver, VP-bot, Luckas-bot, Yobot, Nergality, Kan8eDie, THEN WHO WAS PHONE?, Eric-Wester, Tonyrex, AnomieBOT, Shootbamboo, DemocraticLuntz, Rubinbot, Götz, Jim1138, IRP, Piano non troppo, Icalanise, Kingpin13, Mydickishuge24, Materialscientist, The High Fin Sperm Whale, Citation bot, Neurolysis, ArthurBot, LovesMacs, Mrhellcool, Rightly, Xqbot, IrishChemistPride, IrishChemistPride2, GeometryGirl, Restu20, Srich32977, S0aasdf2sf, John5955, Alan8, ProtectionTaggingBot, Omnipaedista, RibotBOT, TonyHagale, Phillycheesesteaks, LyleHoward, A. di M., RyanOrdemann, Peter470, Thehelpfulbot, Al Wiseman, FrescoBot, Surv1v4l1st, Eadon-com, Paine Ellsworth, Tobby72, Gauravdce07, Steve Quinn, C.Bluck, Citation bot 1, MarB4, Galmicmi, Gil987, Pinethicket, HRoestBot, Voltron Hax, Raen79, Hoo man, Yos233, Allthingstoallpeople, MastiBot, Kuririmo, Noel Streatfield, Ezhuttukari, Swifterthenyou, Noisalt, Jujutacular, Euchanels, Lissajous, Dude1818, December21st2012Freak, IJBall, Jauhienij, Utility Monster, FoxBot, Sheogorath, Jdlawlis, Odatus, Bestcallumuk, Sampathsris, DARTH SIDIOUS 2, Mean as custard, RjwilmsiBot, TjBot, MinicheddarsandelephantsFTW, Benjadow, Mcmonsterbrothers, Priceracks, Csilcock, Sohaib360, Androstachys, Techhead7890, EmausBot, Optiguy54, GoingBatty, Jjasharpe, Pcorty, TuHan-Bot, Hhhippo, HiWBot, John Cline, Harddk, Fæ, Josve05a, StringTheory11, Wackywace, Quondum, GianniG46, Fizicist, Wayne Slam, Raynor42, Arnaugir, Jacksccsi, Brandmeister, Donner60, Negovori, RockMagnetist, Mni9791, ClueBot NG, HLachman, Hermajesty21, Jacobkh, Letoya123, Samsau ninjaguy, Ggonzalm, Moritz37, Braincricket, Helpful Pixie Bot, Geo7777, SzMithrandir, Bibcode Bot, Dfbowsmountainer, Ymblanter, Vagobot, Paolo Lipparini, Wzrd1, Lk00la1dl, JacobTrue, Socal212, Begman5, Mark Arsten, Cadiomals, Jikepaddy, Caterpillar111, Macymae, 06seagsa, BEEPTHENOOB, ItzzRevolution, Shawn Worthington Laser Plasma, Duxwing, Klilidiplomus, Uopchem251, Joe0x7F, BattyBot, Justincheng12345-bot, Cyberbot II, ChrisGualtieri, GoShow, Ankap~enwiki, Glenzo999, Barant2, BrightStarSky, Dexbot, Astromango2215, Webclient101, Mogism, 331dot, Spray787, Vanquisher.UA, Lugia2453, Kondormari, Reatlas, JellyBean4.1, Prof.Professer, The User 111, Bluemanyoung, Rohitgunturi, Ugog Nizdast, The Herald, Jwratner1, Zahid2233, MorshusApprentice, 2005-Fan, Phub Dorji, Epic Failure, Gindor, Ian98989898, Monkbot, SkateTier, Waldmannevan, Wiki1098, Wulfiedude14, Mario Castelán Castro, Fleivium, Crystallizedcarbon, Mcwikigeek, Jesus is the Light of my life, Acesoli, Soumilm, Lemmegetyou, Flying g shot, SirLagsalott, Tetra quark, Skipfortyfour, Stim 2.0, KasparBot, Fazbear7891 and Anonymous: 934

- **Electron configuration** *Source:* https://en.wikipedia.org/wiki/Electron_configuration?oldid=676267346 *Contributors:* The Anome, Andre Engels, SimonP, Robot5005, Bth, Lir, Patrick, Michael Hardy, Erik Zachte, Kku, Karada, Looxix~enwiki, Snoyes, Darkwind, GaryW, Julesd, Djnjwd, Andres, Smack, Ddoherty, Malbi, Timwi, Dysprosia, LMB, Topbanana, Fvw, Donarreiskoffer, Gentgeen, Robbot, Astronautics~enwiki, R3m0t, Yelyos, Romanm, Ashley Y, Ojigiri~enwiki, Hadal, Timemutt, Tobias Bergemann, Unfree, Mor~enwiki, Giftlite, Anville, Mboverload, Nayuki, Lucky 6.9, Utcursch, Antandrus, Mako098765, Rdsmith4, Discospinster, FT2, Vsmith, Bender235, Donsimon~enwiki, Brian0918, Kwamikagami, Kouhoutek, Spoon!, WhiteTimberwolf, Femto, Nk, NickSchweitzer, Jumbuck, Alansohn, Anthony Appleyard, Dark Shikari, Bucephalus, Colin Kimbrell, Cburnett, Arag0rn, RainbowOfLight, Dirac1933, Linas, StradivariusTV, Kurzon, Mpatel, Marudubshinki, V8rik, DePiep, MZMcBride, Azure8472, Margosbot~enwiki, Crazycomputers, RexNL, Alexjohnc3, Fresheneesz, Mattopia, Glenn L, Physchim62, Chobot, YurikBot, Wavelength, RobotE, X42bn6, Jengelh, KSmrq, RadioFan, Pseudomonas, NawlinWiki, SEWilcoBot, Dumoren, Nsmith 84, Vb, E2mb0t~enwiki, Alex43223, EEMIV, BOT-Superzerocool, Cstaffa, Djdaedalus, Tetracube,

Johndburger, 2over0, Јованъб, Arthur Rubin, Josh3580, TBadger, LeonardoRob0t, Kungfuadam, RG2, Janek Kozicki, P. B. Mann, Dr-Jolo, SmackBot, KnowledgeOfSelf, Bomac, Jacek Kendysz, Yamaguchi🯄🯄, Skizzik, Kmarinas86, Rmosler2100, Hugo-cs, Chris the speller, Jayko, Bduke, Dreg743, Robth, DHN-bot~enwiki, Sbharris, Colonies Chris, John Reaves, Dethme0w, Sergio.ballestrero, OrphanBot, Rrburke, Aldaron, Nakon, Nrcprm2026, Fuzzypeg, DMacks, Bhludzin, SashatoBot, OhioFred, John, AstroChemist, Anoop.m, Beetstra, Aeluwas, Iridescent, CmdrObot, Calmargulis, Necessary Evil, Tawkerbot4, Quibik, Steviedpeele, Cinderblock63, Thijs!bot, Epbr123, Qwyrxian, Headbomb, CharlotteWebb, Escarbot, AntiVandalBot, Luna Santin, Seaphoto, Tyco.skinner, Byrgenwulf, Res2216firestar, Deadbeef, JAnDbot, PhilKnight, Meeples, Bongwarrior, VoABot II, Marik7772003, Baccyak4H, Animum, EagleFan, Dirac66, Adventurer, JohnofPhoenix, Patstuart, MartinBot, Schmloof, Rettetast, Yannledu, AlphaEta, J.delanoy, Maurice Carbonaro, KeepItClean, Brant.merrell, Skier Dude, Matt18224, SmilesALot, Juliancolton, DavidCBryant, CardinalDan, Idioma-bot, VolkovBot, Mrh30, LokiClock, Af648, TXiKiBoT, Pjstewart, Anna Lincoln, Cerebellum, Axiosaurus, Broadbot, Cremepuff222, BotKung, Ilyushka88, Krazywrath, Falcon8765, MrChupon, HansHermans, Coffee, OMCV, Graham Beards, RJaguar3, Tiptoety, Radon210, Fzhi555, Oxymoron83, Jdaloner, IdealEric, Svick, Retireduser1111, Anchor Link Bot, Tesi1700, Pinkadelica, ClueBot, The Thing That Should Not Be, CounterVandalismBot, Ankit.sunrise, Tamariandre, Kevkevkevkev, DragonBot, Djr32, NuclearWarfare, Yuk ngan, Asaad900, Bliz1, Aitias, Ace of Spades IV, Footballfan190, RMFan1, Qbmaster, Dnvrfantj, Aritate, Addbot, Glane23, AtheWeatherman, Weekwhom, Drova, Tide rolls, Teles, Gail, Nantaskot, Legobot, आशीष भटनागर, Luckas-bot, Amirobot, KarlHegbloom, محبوب عالم, AnomieBOT, Götz, Daniele Pugliesi, Jim1138, 9258fahsflkh917fas, Chuckiesdad, Ulric1313, Materialscientist, The High Fin Sperm Whale, Citation bot, ArthurBot, Xqbot, S h i v a (Visnu), 4twenty42o, Earlypsychosis, RibotBOT, SassoBot, Shadowjams, Prari, VS6507, Chau7, Spacecadet262, Redrose64, TokioHotel93, Pinethicket, MJ94, Meddlingwithfire, Appy3, Jelson25, Double sharp, علی ویکی, Diannaa, Mttcmbs, DARTH SIDIOUS 2, Deagle AP, EmausBot, Manifolded, Wikipelli, Holyhell5050, AManWithNoPlan, RaptureBot, NTox, DASHBotAV, Rocketrod1960, Trevorhailey1, Fucktosh, ClueBot NG, Dipanshu.sheru, A520, Wd930Bot, Lanthanum-138, Helpful Pixie Bot, HMSSolent, Bibcode Bot, BG19bot, MusikAnimal, Dan653, Glevum, Gwickwire, Virtualzx, Aisteco, Samanthaclark11, Davidwhite18, Mediran, Putodog, M Farooq 2012, Maryann gersaniva, Denden0019, Mychael23, Monkbot, Apisani82, Cautious Chemist, Wobbieftw, Alango1998, Sayambohra, Pfam32 and Anonymous: 487

- **Electric charge** *Source:* https://en.wikipedia.org/wiki/Electric_charge?oldid=677696064 *Contributors:* AxelBoldt, Mav, Andre Engels, Roadrunner, Peterlin~enwiki, Heron, JohnOwens, Michael Hardy, Ixfd64, Delirium, Looxix~enwiki, Ellywa, Mdebets, Glenn, Rossami, Nikai, Andres, Raven in Orbit, Reddi, Omegatron, Gakrivas, Lumos3, Rogper~enwiki, Gentgeen, Robbot, Fredrik, Dukeofomnium, Wikibot, Fuelbottle, Wjbeaty, Giftlite, DavidCary, Herbee, Snowdog, Dratman, Valen~enwiki, RScheiber, Jason Quinn, Brockert, OldakQuill, Manuel Anastácio, LiDaobing, Karol Langner, Icairns, Iantresman, GNU, Vincom2, Discospinster, Guanabot, Jpk, Dbachmann, ZeroOne, Laurascudder, Bobo192, Rbj, Giraffedata, Kjkolb, Scentoni, Mdd, Alansohn, Atlant, ABCD, Velella, Wtshymanski, HenkvD, Mikeo, DV8 2XL, Gene Nygaard, HenryLi, Oleg Alexandrov, Nuno Tavares, Cimex, Rocastelo, StradivariusTV, Oliphaunt, BillC, Eleassar777, Cyberman, Palica, BD2412, Demonuk, Edison, SMC, Krash, Dougluce, FlaBot, Psyphen, Nivix, Alfred Centauri, Gurch, Kri, Gdrbot, Manscher, YurikBot, Bambaiah, Lucinos~enwiki, Stephenb, Manop, Pseudomonas, JDoorjam, TDogg310, Chichui, Kkmurray, Wknight94, Light current, Enormousdude, Johndburger, Tcsetattr, Pinikas, Reyk, Canley, Geoffrey.landis, JDspeeder1, GrinBot~enwiki, Mejor Los Indios, Sbyrnes321, Marquez~enwiki, Moeron, Vald, Thunderboltz, Dmitry sychov, HalfShadow, Gilliam, Oscarthecat, Andy M. Wang, Chris the speller, Lenko, DHN-bot~enwiki, Dual Freq, Hallenrm, Rrburke, The tooth, MichaelBillington, Hgilbert, Drphilharmonic, Daniel.Cardenas, Springnuts, Yevgeny Kats, Andrei Stroe, DJIndica, Naui~enwiki, Nmnogueira, SashatoBot, Richard L. Peterson, Slowmover, Cronholm144, Mgiganteus1, Bjankuloski06en~enwiki, Nonsuch, Ben Moore, RandomCritic, MarkSutton, Stikonas, Dicklyon, Levineps, Igoldste, Tawkerbot2, Chetvorno, JForget, CmdrObot, Kehrli, Jsd, Myasuda, Cydebot, Fl, Bvcrist, Meno25, Gogo Dodo, WISo, Christian75, Ssilvers, Thijs!bot, Epbr123, Barticus88, N5iln, Mojo Hand, Headbomb, Gerry Ashton, Escarbot, Aadal, AntiVandalBot, Seaphoto, Prolog, DarkAudit, Lyricmac, Tim Shuba, WikifingHelper, Asgrrr, JAnDbot, Acroterion, Bongwarrior, VoABot II, J2thawiki, Sstolper, Jjurik, Bubba hotep, User A1, DerHexer, InvertRect, Robin S, MartinBot, M. Bilal Shafiq, LedgendGamer, Pharaoh of the Wizards, Numbo3, Hans Dunkelberg, NightFalcon90909, Uncle Dick, Ginsengbomb, Katalaveno, DarkFalls, NewEnglandYankee, QuickClown, Juliancolton, ACBest, Treisijs, Lseixas, Jefferson Anderson, Sheliak, Philip Trueman, TXiKiBoT, The Original Wildbear, Ayan2289, Nickipedia 008, LuizBalloti, Monty845, Jpalpant, Biscuittin, Demmy100, SieBot, Gerakibot, Caltas, Gastin, Wing gundam, Msadaghd, JerrySteal, Jojalozzo, Oxymoron83, Faradayplank, Avnjay, Anchor Link Bot, Neo., Loren.wilton, ClueBot, The Thing That Should Not Be, Arakunem, Termine, Mild Bill Hiccup, Stephaninator, LeoFrank, Excirial, Kocher2006, Jusdafax, Brews ohare, Cenarium, Jotterbot, PhySusie, SchreiberBike, Wuzur, JDPhD, Versus22, Thinking Stone, Rror, Cernms, Truthnlove, Addbot, Some jerk on the Internet, CanadianLinuxUser, NjardarBot, LaaknorBot, Scottyferguson, LinkFA-Bot, Naidevinci, Ocwaldron, Tide rolls, Lightbot, JDSperling, Legobot, Luckas-bot, Yobot, CinchBug, Duping Man, AnomieBOT, DemocraticLuntz, Sertion, Jim1138, IRP, Pyrrhus16, Kingpin13, Bluerasberry, Materialscientist, Geek1337~enwiki, ImperatorExercitus, Xqbot, TheAMmollusc, Phazvmk, Addihockey10, Capricorn42, Nnivi, ProtectionTaggingBot, RibotBOT, Srr712, A. di M., Constructive editor, Frozenevolution, Ryryrules100, Jc3s5h, Drunauthorized, Mithrandir, Steve Quinn, Davidteng, Fast kartwheels, BenzolBot, DivineAlpha, AstaBOTh15, Pinethicket, Jivee Blau, Calmer Waters, Tinton5, MastiBot, Serols, Meaghan, Lalrang2007, Logical Gentleman, FoxBot, TobeBot, SchreyP, Jonkerz, Ndkartik, Vrenator, Taytaylisious09, Ammodramus, Jamietw, DARTH SIDIOUS 2, Eshmate, Irfanyousufdar, EmausBot, John of Reading, GoingBatty, K6ka, Darkfight, Hhhippo, JSquish, Harddk, Stephen C Wells, Liam McM, Sonygal, L Kensington, Donner60, Peter Karlsen, Sven Manguard, Planetscared, ClueBot NG, Jack Greenmaven, Cking1414, Ihwood, Ulflund, CocuBot, MelbourneStar, O.Koslowski, Brickmack, AvocatoBot, Rm1271, Altaïr, F=q(E+v^B), Snow Blizzard, Brad7777, Bhaskarandpm, Eduardofeld, GoShow, Dexbot, JoshyyP, Brandonsmacgregor, Reeceyboii, Frosty, Reatlas, I am One of Many, Eyesnore, Tentinator, Germeten, Nablacdy, Spyglasses, Freddyboi69, 20M030810, SpecialPiggy, Marizperoj, Peterfreed, Rigid hexagon, Jiteshkumar727464, Dyeith, Podayeruma, Oleaster, Layfi, BlueDecker, Pritam kumar Barik, KasparBot, Ramprakashsfc and Anonymous: 423

- **Mass number** *Source:* https://en.wikipedia.org/wiki/Mass_number?oldid=667863435 *Contributors:* Xavic69, Polimerek, Grendelkhan, Paul-L~enwiki, Robbot, Pingveno, Harp, Bensaccount, Tsemii, Mormegil, Discospinster, Vsmith, Dungodung, La goutte de pluie, Melah Hashamaim, Jumbuck, Alansohn, Mnde, Malo, Bart133, Gene Nygaard, Zereshk, Pingswept, Blaaarg, FlaBot, Fresheneesz, Physchim62, YurikBot, Irishguy, Johantheghost, BOT-Superzerocool, Wknight94, Ott2, KGasso, Itub, SmackBot, Hydrogen Iodide, Eskimbot, Sweet430, Gilliam, DHN-bot~enwiki, V1adis1av, Addshore, SundarBot, Teehee123, SashatoBot, Jonhall, Igoldste, RedRollerskate, Gran2, Farzaneh, Epbr123, CharlotteWebb, Calaka, Kongurshan, Jj137, JAnDbot, Gaeddal, VoABot II, Catslash, Ring-Ding, Rami R, Indon, Dirac66, 28421u2232nfenfcenc, Chris Fletcher, Cpl Syx, MartinBot, Ultraviolet scissor flame, J.delanoy, Alex Lin, McSly, Ja 62, Halmstad, VolkovBot, ABF, Kakoui, DoorsAjar, TXiKiBoT, Imasleepviking, LeaveSleaves, GavinTing, SieBot, Sonicology, Mimihitam, KathrynLybarger, XDanielx, ClueBot, TheRaslsBack, Razimantv, Polyamorph, Djr32, Ktr101, Alexbot, Versus22, SkyLined, Addbot, Download, LaaknorBot, Quercus solaris, Numbo3-bot,

Krupted Soul, Ehrenkater, VASANTH S.N., Luckas Blade, Luckas-bot, Yobot, 2D, Naudefjbot~enwiki, Rubinbot, 1exec1, Jim1138, Materialscientist, Citation bot, RibotBOT, SassoBot, Ernsts, A.amitkumar, LucienBOT, HRoestBot, FoxBot, Sampathsris, Yliam73, TjBot, Emaus-Bot, WikitanvirBot, Hhhippo, Donner60, Sunshine4921, Forever Dusk, Mikhail Ryazanov, ClueBot NG, Alex Nico, Widr, Helpful Pixie Bot, JohnSRoberts99, SyahirSQRT2, AvocatoBot, JJVonRiesen, Pratyya Ghosh, Lugia2453, Jianhui67, Mahusha and Anonymous: 169

- **Atomic number** *Source:* https://en.wikipedia.org/wiki/Atomic_number?oldid=673864932 *Contributors:* AxelBoldt, Kpjas, Sodium, Mav, The Anome, Andre Engels, Peterlin~enwiki, Fonzy, Tim Starling, Liftarn, Fruge~enwiki, Eric119, Docu, Suisui, Александър, Julesd, Andres, The Anomebot, Paul-L~enwiki, Philopp, Skood, Donarreiskoffer, Gentgeen, Robbot, Fredrik, Jredmond, Romanm, Naddy, DHN, Hadal, Wikibot, Michael Snow, Giftlite, Guanaco, Eequor, Pne, Antandrus, Icairns, Tsemii, Mormegil, HedgeHog, Discospinster, Rich Farmbrough, Guanabot, Vsmith, Ponder, Bender235, Spearhead, RoyBoy, Femto, Bobo192, Nigelj, Brim, Sam Korn, Methegreat, Jumbuck, Alansohn, Bart133, Nuno Tavares, ^demon, Palica, Graham87, Rsg, Chun-hian, Ceinturion, Vary, SMC, Ian Dunster, Watcharakorn, ChongDae, Kolbasz, Physchim62, Chobot, DVdm, Mhking, VolatileChemical, Akke~enwiki, YurikBot, JWB, Jimp, Kirill Lokshin, Eleassar, NawlinWiki, Bota47, Zarboki, FF2010, Lt-wiki-bot, Adilch, Nemu, Dspradau, Petri Krohn, Paul D. Anderson, Katieh5584, Itub, SmackBot, F, Unyoyega, Bomac, Kilo-Lima, WookieInHeat, MelancholieBot, Gilliam, Betacommand, Rmosler2100, Persian Poet Gal, Catchpole, Bonaparte, DHN-bot~enwiki, Sbharris, Darth Panda, Suicidalhamster, Scott3, Addshore, SundarBot, Astroview120mm, DMacks, Vina-iwbot~enwiki, SashatoBot, FrozenMan, Sir Nicholas de Mimsy-Porpington, NongBot~enwiki, Cielomobile, Muadd, Hetar, Courcelles, Tawkerbot2, Ale jrb, Erik Kennedy, Jsd, Nick Y., Clovis Sangrail, Christian75, DumbBOT, Chrislk02, FastLizard4, Thijs!bot, Epbr123, Mojo Hand, Headbomb, Escarbot, Dantheman531, Seaphoto, Quintote, Pwhitwor, Spencer, JAnDbot, Leuko, Plantsurfer, Fragilityfemme, Andonic, .anacondabot, Bongwarrior, VoABot II, Loonymonkey, Allstarecho, DerHexer, JaGa, Lilac Soul, J.delanoy, Pharaoh of the Wizards, Ginsengbomb, It Is Me Here, Gman124, L'Aquatique, Skier Dude, JohnnyRush10, KylieTastic, Treisijs, TraceyR, Idioma-bot, X!, CWii, AlnoktaBOT, Philip Trueman, TXiKiBoT, The Original Wildbear, Java7837, Mainstream Nerd, Seraphim, Martin451, JhsBot, LeaveSleaves, Querl, Tgm1024, Roarman113, AlleborgoBot, Sssprkrao~enwiki, EmxBot, Tswei, SieBot, Coffee, YonaBot, ToePeu.bot, Winchelsea, Caltas, Keilana, Scorpion451, Lightmouse, Mygerardromance, Troy 07, ImageRemovalBot, XDanielx, Atif.t2, ClueBot, Bobathon71, The Thing That Should Not Be, Wwheaton, Polyamorph, ChandlerMapBot, DragonBot, Excirial, Alexbot, Jusdafax, Estirabot, Jotterbot, Razorflame, Ace of Spades IV, DumZiBoT, Tarheel95, Dsvyas, Feinoha, Satinchair, ElMeBot, Yes, I'm A Scientist, Thatguyflint, Ketchup krew, Addbot, Wotsinaname, Ronhjones, TutterMouse, Leszek Jańczuk, Ehrenkater, Kukers12, Tide rolls, Hunyadym, Arbitrarily0, Luckas-bot, Yobot, Washburnmav, THEN WHO WAS PHONE?, Nallimbot, KamikazeBot, AnomieBOT, Rubinbot, IRP, JackieBot, Piano non troppo, Flewis, Materialscientist, Maxis ftw, Hyperlaosboy, Vuerqex, Maniadis, Frankenpuppy, Xqbot, TinucherianBot II, Capricorn42, Jsharpminor, GrouchoBot, RibotBOT, 78.26, Ernsts, A. di M., Nixón, FrescoBot, Tobby72, Wikipe-tan, Phillyfanjd, Dnafish, PSPatel, Colubedy, Pinethicket, Mathsman91, RedBot, Space-Flight89, RandomStringOfCharacters, Theeditor19494, ItsZippy, Vrenator, Sampathsris, Tbhotch, Reach Out to the Truth, The Utahraptor, Bhawani Gautam, Salvio giuliano, DASHBot, EmausBot, WikitanvirBot, Avenue X at Cicero, Racerx11, Razor2988, Tommy2010, K6ka, Werieth, JSquish, ZéroBot, Bollyjeff, StringTheory11, Wayne Slam, Ocaasi, Rcsprinter123, L Kensington, Donner60, Ihardlythinkso, Rock-Magnetist, Kowl kaz, ClueBot NG, Widr, Diyar se, Lowercase sigmabot, Will102, Greeny 12 1, Dan653, Mark Arsten, Joydeep, Vladihernando, ChrisGualtieri, EuroCarGT, 786b6364, Bashiaden, Warp Dragon, The Herald, Reginatr08, Dannyzhaofb, TheEpTic, De Riban5, Poepkop, Filedelinkerbot, Guttyut, Rstensby, Aryavat Kapoor, KasparBot and Anonymous: 393

- **Atomic mass** *Source:* https://en.wikipedia.org/wiki/Atomic_mass?oldid=674458174 *Contributors:* Christian List, Patrick, Eric119, Andres, Glueball, Head, Chuunen Baka, Robbot, Henrygb, Centrx, Giftlite, Mikez, Mboverload, Knutux, Karol Langner, DragonflySixtyseven, Icairns, Discospinster, Rich Farmbrough, Vsmith, RoyBoy, Triona, Femto, Bobo192, Smalljim, Nyenyec, Elipongo, Jumbuck, Alansohn, Eric Kvaalen, Atlant, Andrewpmk, Snowolf, Gene Nygaard, LOL, Insaneinside, Ylem, Benbest, Colin Watson, Eaolson, Palica, V8rik, Sjakkalle, SMC, Nneonneo, Watcharakorn, Kevmitch, Gurch, Physchim62, King of Hearts, SGreen~enwiki, Chobot, Nagytibi, Roboto de Ajvol, YurikBot, Wavelength, NawlinWiki, Andonee, DeadEyeArrow, Moncubus, Kkmurray, FF2010, Denisutku, Јованъб, Pb30, Junglecat, Tom Morris, Itub, KnowledgeOfSelf, Bomac, Kjaergaard, Edgar181, Bduke, Miquonranger03, MalafayaBot, Sbharris, ACupOfCoffee, Darth Panda, Royboycrashfan, Can't sleep, clown will eat me, Shalom Yechiel, V1adis1av, Voyajer, VMS Mosaic, Pnkrockr, Mosca, Makemi, Richard001, Alexandra lb, Peterwhy, DMacks, Bidabadi~enwiki, Vina-iwbot~enwiki, Ck lostsword, DJIndica, SashatoBot, Archimerged, Srikeit, Gobonobo, Bugaboo400, Zarniwoot, Slakr, Munita Prasad, Meco, EEPROM Eagle, 11987, IvanLanin, Nkayesmith, Tawkerbot2, JForget, Stifynsemons, Yaris678, Nick Y., Rifleman 82, Tkynerd, Tawkerbot4, Christian75, UberScienceNerd, Thijs!bot, Epbr123, Wikid77, Because124, RolfSander, Headbomb, Luigifan, John254, Blathnaid, ThomasPusch, Mentifisto, AntiVandalBot, EarthPerson, Quintote, JAnDbot, Husond, Tohru Honda13, MSBOT, Kirrages, Bongwarrior, VoABot II, Fabrictramp, Allstarecho, DerHexer, Mattinbgn, Saurc zlunag, Patstuart, Ddice, Leyo, J.delanoy, JohnPritchard, Uncle Dick, Gabulldog595, Extransit, A Nobody, Jer10 95, Answeryoung, RiverBissonnette, NewEnglandYankee, Cmichael, DorganBot, Wikieditor06, Thedjatclubrock, Al.locke, Ryan032, Philip Trueman, TXiKiBoT, Vipinhari, Crohnie, Qxz, Someguy1221, Setreset, Cremepuff222, Wtt, Enviroboy, Seresin, Logan, Demmy, SieBot, PlanetStar, Tiddly Tom, BotMultichill, Tiptoety, Oxymoron83, KathrynLybarger, Janfri, Pfbray, Superbeecat, Denisarona, ClueBot, PipepBot, The Thing That Should Not Be, W00TZORG, Derkaw00t, Drmies, MDov, CounterVandalismBot, LizardJr8, Excirial, Sachindhar, Anvesh Devalapalli, Cenarium, Dekisugi, La Pianista, Aitias, DumZiBoT, AlexGWU, SilvonenBot, NT1.0, Addbot, Ecaspi12, SameOldSameOld, Ronhjones, CarsracBot, LAAFan, Ld100, LinkFA-Bot, Quercus solaris, 5 albert square, Krupted Soul, Ehrenkater, Tide rolls, Gail, Trotter, Yobot, محبوب عالم, The Parting Glass, Materialscientist, Maxis ftw, ArthurBot, TechBot, Jeffrey Mall, P08.gratton, Magicxcian, Mlpearc, BLP-outrageous move logs, ProtectionTaggingBot, 1010kk, Žiedas, Dave3457, Icorrectu, DivineAlpha, Citation bot 1, Pinethicket, Yahia.barie, A8UDI, Hoo man, Double sharp, Jules93, Hickorybark, Ghjthgh, Cfsgfds, Tbhotch, Hhhippo, John Cline, Splibubay, Tolly4bolly, EricWesBrown, Donner60, Theislikerice, Peter Karlsen, GrayFullbuster, DJDunsie, Xrayburst1, ClueBot NG, Muon, Widr, Bibcode Bot, Lowercase sigmabot, Arnavchaudhary, MusikAnimal, Metricopolus, JJVonRiesen, Dexbot, Mogism, Bashiaden, Clawer67, Cadillac000, Sosthenes12, J.meija, Blackbombchu, Ginsuloft, TheEpTic, DLManiac, PoonDominator, KasparBot, Akshat 492001 and Anonymous: 404

- **Atomic radius** *Source:* https://en.wikipedia.org/wiki/Atomic_radius?oldid=674257859 *Contributors:* Sodium, Mav, The Anome, Matusz, William Avery, Ellywa, Stone, DJ Clayworth, Gentgeen, Texture, Geeoharee, Yekrats, Jorge Stolfi, Eequor, Darrien, Tipiac, Rdsmith4, Icairns, Rich Farmbrough, Guanabot, Cacycle, Paul August, ESkog, RJHall, The Noodle Incident, Femto, CDN99, La goutte de pluie, Nk, I-hunter, Alansohn, Arthena, Benjah-bmm27, Gene Nygaard, Weyes, Ultimate Star Wars Freak, Morios, MONGO, Stevey7788, Sukolsak, BD2412, DePiep, Chenxlee, Rjwilmsi, Bhadani, FlaBot, Gurch, OSt~enwiki, TheSun, Physchim62, Chobot, Nagytibi, AtZeuS, Roboto de Ajvol, Yurik-Bot, RobotE, Midgley, Bhny, Bachrach44, Apokryltaros, Divide, Јованъб, HereToHelp, Katieh5584, GrinBot~enwiki, Janek Kozicki, Luk, Itub, ZerodEgo, Gilliam, Ohnoitsjamie, Betacommand, Hugo-cs, Chris the speller, Remcutt, Brinerustle, Can't sleep, clown will eat me, Shalom

Yechiel, Skidude9950, Granla, Smokefoot, EdGl, Just plain Bill, SashatoBot, Quendus, JohnI, Anoop.m, Iridescent, Joseph Solis in Australia, Eastlaw, T3h933k, Nein~enwiki, Alaibot, Nol888, Thijs!bot, Mathmoclaire, Mojo Hand, Headbomb, Marek69, Dee Hess, Mentifisto, AntiVandalBot, Luna Santin, NSH001, MER-C, Dmbstudio, Kopovoi, Mycroft7, Cbmsu01, J.delanoy, Oneworld25, Extransit, Poktopok, DarkFalls, Juliancolton, DorganBot, Izno, Squids and Chips, Deor, VolkovBot, Carter, VasilievVV, Philip Trueman, Rei-bot, Axiosaurus, Madhero88, Wenli, FrancisGM, Enviroboy, Insanity Incarnate, SieBot, Mikemoral, Ttony21, PlanetStar, WereSpielChequers, Freshjivaz, Flyer22, Cyfal, Martarius, ClueBot, LAX, The Thing That Should Not Be, Razimantv, Polyamorph, NRLer, Excirial, Shinkolobwe, Sun Creator, Arjayay, Darkicebot, WikHead, Firebat08, Hess88, Addbot, Dov Jacobson, Fieldday-sunday, Tide rolls, Lightbot, Legobot, Luckas-bot, Yobot, Becky Sayles, Eric-Wester, AnomieBOT, Jal173, Ulric1313, Crystal whacker, Materialscientist, RobertEves92, Citation bot, LilHelpa, Xqbot, Nick-kid5, ChristopherKingChemist, Schrojon, ROBE0191, Chewy quaker, Recognizance, Vicenarian, Jusses2, Double sharp, Jynto, Andrea105, Mean as custard, AliasZFK, Dead Horsey, Dewritech, Chum1993, Dot145, ZéroBot, Shuipzv3, StringTheory11, Surya Prakash.S.A., E. Fokker, Petrb, ClueBot NG, Widr, Helpful Pixie Bot, Bibcode Bot, Numbermaniac, Stefhanish, 14bemery, AmericanLemming, Yardimsever, Ajaryanxyz73, Little wittle, Basil iqbal and Anonymous: 222

- **Atomic orbital** *Source:* https://en.wikipedia.org/wiki/Atomic_orbital?oldid=675223100 *Contributors:* Mav, Tarquin, Heron, Bth, Edward, Michael Hardy, Tim Starling, William M. Connolley, BRG, Smack, Ddoherty, Hashar, Hollgor, Stismail, Taxman, Topbanana, Raul654, JohnH~enwiki, Donarreiskoffer, Gentgeen, Robbot, Romanm, João Sousa, Tobias Bergemann, Giftlite, BenFrantzDale, Guanaco, Pascal666, Eequor, Allstar86, Jackol, HorsePunchKid, Mako098765, Snobscure, Karol Langner, H Padleckas, Thorwald, Rich Farmbrough, Vsmith, Guanabot2, Mani1, Harriv, Kwamikagami, Bobo192, John Vandenberg, Cmdrjameson, Maurreen, I9Q79oL78KiL0QTFHgyc, Hooperbloob, Alansohn, LtNOWIS, Neonumbers, Ricky81682, NamfFohyr, Benjah-bmm27, Velella, Ceyockey, Deror avi, John Miles, StradivariusTV, Kurzon, MONGO, Ian**, SDC, SeventyThree, Pfalstad, Gerbrant, Stevey7788, Ashmoo, V8rik, BD2412, Rjwilmsi, Bubba73, Fred Bradstadt, FlaBot, Jeff02, Fresheneesz, Scerri, Tardis, Mattopia, Chobot, Roboto de Ajvol, YurikBot, Wavelength, Hairy Dude, Hellbus, Schoen, David R. Ingham, SEWilcoBot, BOT-Superzerocool, Black Falcon, Djdaedalus, Enormousdude, Arthur Rubin, Modify, Petri Krohn, Kungfuadam, Janek Kozicki, Xetrov znt, SmackBot, Incnis Mrsi, InverseHypercube, Bomac, KocjoBot~enwiki, Jrockley, Delldot, ZerodEgo, Edgar181, Cool3, Ema Zee, Gilliam, Isaac Dupree, Kmarinas86, Hugo-cs, Bluebot, Bduke, Fuzzform, Sbharris, Audriusa, Voyajer, Bowlhover, Nakon, Drphilharmonic, DMacks, Sadi Carnot, Yoshigev, SashatoBot, A.Z., Slakr, Beetstra, Mets501, Ambuj.Saxena, Sasata, Thrindel, HannesJvV, Vinay.bhat, Lentower, RobertLovesPi, Necessary Evil, Cydebot, Matrix61312, Astrochemist, Gogo Dodo, Christian75, Calvero JP, Thijs!bot, Epbr123, JKW~enwiki, Headbomb, Jayron32, RogueNinja, Isilanes, NSH001, Blacksun1942, B7582, JAnDbot, MER-C, BeeArkKey, BenB4, .anacondabot, Magioladitis, Bongwarrior, Avjoska, Baccyak4H, Dirac66, User A1, Cpl Syx, DerHexer, Ekwity, Nevit, Pagw, R'n'B, PrestonH, EscapingLife, Hans Dunkelberg, Maurice Carbonaro, Notapotato, HiLo48, STBotD, WinterSpw, Steel1943, VolkovBot, Philip Trueman, TXiKiBoT, Cloudswrest, Hiyabulldog, TelecomNut, Spiral5800, 345gim, NHRHS2010, Deconstructhis, SieBot, Coffee, Ttony21, God Emperor, Yintan, Flyer22, KoshVorlon, Lightmouse, Agur bar Jacé, Tesi1700, Denisarona, Into The Fray, Simstud16, Dlrohrer2003, Laburke, The Thing That Should Not Be, Biggerj1, Jan1nad, Tomas e, CounterVandalismBot, Rotational, Neverquick, Djr32, Excirial, Virginia fried chicken, BOTarate, Trigley, Versus22, SoxBot III, Humanengr, RexxS, SilvonenBot, Smokyhallow, Monfornot, Addbot, AVand, Ronhjones, CanadianLinuxUser, Mac Dreamstate, Download, EconoPhysicist, Chzz, Numbo3-bot, Ehrenkater, Tide rolls, OlEnglish, Ashandpikachu, Legobot, Clay Juicer, Luckas-bot, Yobot, THEN WHO WAS PHONE?, Csmallw, Tonyrex, AnomieBOT, Rubinbot, Crystal whacker, Citation bot, LilHelpa, Xqbot, Capricorn42, Br77rino, Gap9551, Shirik, Samwb123, Mayukh iitbombay 2008, CES1596, Lightbound, MCBot, DivineAlpha, Citation bot 1, SUL, Redrose64, Gil987, I dream of horses, Reelx09, Double sharp, Plasmide911, Slon02, Edouard.darchimbaud, EmausBot, WikitanvirBot, BillyPreset, Dewritech, IncognitoErgoSum, GoingBatty, Solarra, Slightsmile, Hhhippo, Mz7, AvicBot, JSquish, Plotfeat, AManWithNoPlan, L0ngpar1sh, Nirvanana, Wayne Slam, Ocaasi, JoeSperrazza, L Kensington, Bomazi, Eg-T2g, DASHBotAV, Whoop whoop pull up, NeonGas, Trevorhailey1, ClueBot NG, Wd930, Wikiphysicsgr, MelbourneStar, Satellizer, The Master of Mayhem, Widr, Karmalater, Oddbodz, Helpful Pixie Bot, Helvitica Bold, Jubobroff, Bibcode Bot, Felivik, Snow Blizzard, Glacialfox, Anbu121, BattyBot, Colinphilipjohnstone, Jimw338, ChrisGualtieri, Rubenvb, Thine Antique Pen (public), Frosty, 533andyzou, Anthonyaeae, Reatlas, Mark viking, JellyBean4.1, Writing122, Nigellwh, Ugog Nizdast, W. P. Uzer, PatronBernard, Anrnusna, Malik nadeem ahamed, 7Sidz, Monkbot, Filedelinkerbot, HiYahhFriend, The Expedia, Raytuzio, Jjamesryan, Mathewpanicker, Y-S.Ko, Boomclap, Pandaron, AWomanWithAPlan, DWacks, ILoveBigMac and Anonymous: 305

- **Atomism** *Source:* https://en.wikipedia.org/wiki/Atomism?oldid=669163391 *Contributors:* Eloquence, SimonP, Maury Markowitz, Michael Hardy, Alan Peakall, Tregoweth, Ellywa, Ahoerstemeier, JWSchmidt, Charles Matthews, Rednblu, Wik, Hr oskar, Manika, Wetman, Banno, Carlossuarez46, Robbot, Dittaeva, Psychonaut, Ashley Y, Nilmerg, Wikibot, Wile E. Heresiarch, Alan Liefting, Fabiform, Wolfkeeper, Xerxes314, Daniel Brockman, Eequor, Antandrus, Kusunose, Mukerjee, Deglr6328, Cacycle, Xezbeth, Dbachmann, Silentlight, El C, Laurascudder, Bobo192, Truthflux, John Vandenberg, Kjkolb, Ral315, Merope, Jumbuck, Ungtss, ABCD, Samohyl Jan, Evil Monkey, Gene Nygaard, Oleg Alexandrov, Woohookitty, Linas, Kurzon, Rjwilmsi, Vary, Kajmal, HappyCamper, FlaBot, SchuminWeb, Godlord2, Aethralis, Uriah923, YurikBot, Spacepotato, Pigman, SpuriousQ, Bachrach44, Ragesoss, Amakuha, Tomisti, Ninly, LeonardoRob0t, RG2, Finell, Sardanaphalus, Intangible, SmackBot, Jagged 85, JimmyBlackwing, Gilliam, Hmains, GwydionM, Mhss, Fabiolucas~enwiki, SonOfNothing, Funper, Sbharris, Atomist, Kcordina, LoveMonkey, DMacks, SashatoBot, Petr Kopač, Kashmiri, Rinnenadtrosc, Dr Greg, Special-T, Dhp1080, Brandizzi, Twas Now, Tdmg, CmdrObot, Sir Vicious, Eiorgiomugini, Comrade42, GalliasM, Myasuda, Gregbard, Shanoman, Cydebot, Peterdjones, Dfritter4, Hkyriazi, Strom, Odie5533, Doug Weller, SteveMcCluskey, Thijs!bot, Kubanczyk, Oerjan, Headbomb, Ujm, Tellyaddict, Escarbot, Igorwindsor~enwiki, Stannered, AntiVandalBot, Danger, Danny lost, Indian Chronicles, Sluzzelin, Husond, The Transhumanist, Paper.plane.pilot, Leolaursen, .anacondabot, Meredyth, K95, NoychoH, Mdsats, Yensin, Rettetast, R'n'B, CommonsDelinker, Erkan Yilmaz, AlphaEta, Bogey97, Ianmathwiz7, Benio76, Cadby Waydell Bainbrydge, Ontoraul, JhsBot, Thanatos666, Seraphita~enwiki, Karriaagzh, SieBot, Flyer22, Anchor Link Bot, 3rdAlcove, Athenean, ClueBot, Roshraw, J8079s, Singinglemon~enwiki, Huzzam, Andriolo, Gbertoli3, SchreiberBike, Editor2020, Saeed.Veradi, Avoided, Addbot, Sa9097, Guoguo12, Mathew Rammer, Favonian, Ginosbot, Alcove, Lightbot, Zorrobot, Luckas-bot, Yobot, Reindra, AnomieBOT, Jim1138, Materialscientist, Citation bot, DirlBot, LilHelpa, Xqbot, JimVC3, Leucipp3, Quixotex, GrouchoBot, Omnipaedista, Exxoo, FrescoBot, Isospin, Fortdj33, Machine Elf 1735, Diremarc, Buybuydandavis, Citation bot 1, Jean-François Clet, MastiBot, Prophetvcn, Jauhienij, Yunshui, Cowlibob, Omkargokhale, DASHBot, EmausBot, Acather96, Syncategoremata, IncognitoErgoSum, G upadhyay, Janus549, ZéroBot, ChuispastonBot, ClueBot NG, Helpful Pixie Bot, Snow Rise, WithSelet, Barnaculus, JimRenge, Monkbot, Srini Kolla, Kimberley Anne Davey, Strongjam, Shyskirwaiktcb, KasparBot and Anonymous: 137

- **Bohr model** *Source:* https://en.wikipedia.org/wiki/Bohr_model?oldid=677290546 *Contributors:* Damian Yerrick, AxelBoldt, Trelvis, Zundark, The Anome, Taw, Andre Engels, Arvindn, Stevertigo, D, JohnOwens, Michael Hardy, Tim Starling, Looxix~enwiki, Ellywa, Ahoerstemeier, Cyp, Glenn, Complex Analysis, Hashar, Charles Matthews, Tantalate, RickK, The Anomebot, Furrykef, Fibonacci, Omegatron, BenRG, Robbot, Vespristiano, Pmineault, Rsduhamel, Enochlau, Decumanus, Giftlite, BenFrantzDale, Lethe, MathKnight, Bensaccount, Dmmaus, Yath, Antandrus, Darksun, Grunt, Discospinster, Vsmith, ArnoldReinhold, Gianluigi, Mani1, Paul August, Kaszeta, El C, MrMarshmallow, CDN99, Army1987, Whosyourjudas, Smalljim, Viriditas, Cmdrjameson, R. S. Shaw, Evgeny, Matt McIrvin, MPerel, Nsaa, Jjron, Mdd, Jumbuck, Storm Rider, Alansohn, TheParanoidOne, Munchkinguy, Fritzpoll, PAR, Redfarmer, Snowolf, Velella, HenkvD, Sciurinæ, Bsadowski1, Linas, Mindmatrix, Benhocking, Mpatel, Mreult~enwiki, Schzmo, Terence, MFH, Ian**, SeventyThree, Christopher Thomas, Graham87, Galwhaa, DePiep, Mendaliv, Rjwilmsi, Vary, Salix alba, Tawker, Scorpiuss, Yamamoto Ichiro, Sanbeg, Nivix, Krackpipe, RexNL, Gurch, Goudzovski, Srleffler, Chobot, DVdm, Gwernol, YurikBot, Wolfmankurd, Postglock, JabberWok, Rsrikanth05, Spike Wilbury, Aeusoes1, Buster79, CAPS lOCK, Moe Epsilon, Zwobot, Dna-webmaster, Wknight94, Sperril, PTSE, Closedmouth, KGasso, Petri Krohn, Fram, HereToHelp, Spliffy, GrinBot~enwiki, SkerHawx, SmackBot, Thorseth, Blue520, Jacek Kendysz, WookieInHeat, Frymaster, Timotheus Canens, Gilliam, Chris the speller, Pieter Kuiper, OrangeDog, Complexica, DHN-bot~enwiki, Sbharris, Darth Panda, Can't sleep, clown will eat me, Wen D House, Nakon, Localzuk, GoldenBoar, Adam Schloss, DMacks, O RLY?, Ligulembot, Mion, Sadi Carnot, Kukini, Eliyak, John, Gobonobo, Tktktk, LestatdeLioncourt, Hemmingsen, 3897515, Goodnightmush, IronGargoyle, PseudoSudo, Smith609, Tasc, Beetstra, Martinp23, Domino42, Mets501, Doczilla, Ryulong, Jonhall, Lee Carre, Iridescent, Zootsuits, Joseph Solis in Australia, Morrowulf, Igoldste, Tony Fox, Beve, Blehfu, Courcelles, Profjohn, Tawkerbot2, Chetvorno, JForget, Frovingslosh, Olaf Davis, BeenAroundAWhile, CWY2190, Harej bot, Myasuda, Dept of Alchemy, Cydebot, WillowW, Gogo Dodo, Corpx, Xndr, Shirulashem, DumbBOT, Chrislk02, Sp, Pinky sl, FrancoGG, Epbr123, Wikid77, Goods21, Kablammo, Headbomb, John254, Martin Hedegaard, Pfranson, Dawnseeker2000, Mentifisto, AntiVandalBot, Majorly, Tlabshier, Spartaz, JAnDbot, Harryzilber, MER-C, Andonic, Belg4mit, Hut 8.5, Tstrobaugh, Bkpsusmitaa, R27182818, Ô, Acroterion, Casmith 789, Magioladitis, VoABot II, Wikidudeman, JamesBWatson, Dirac66, 28421u2232nfenfcenc, Schumi555, Cpl Syx, SlamDiego, Mikerobertsn, Vssun, TheRanger, Robin S, Geboy, Kpxxbladexx415, Jemijohn, Mont95, ChemNerd, Slash, J.delanoy, Melamed katz, Uncle Dick, Jonpro, Cpiral, It Is Me Here, Bot-Schafter, AntiSpamBot, Andraaide, NewEnglandYankee, Pez2, Juliancolton, Copsi, Jarry1250, JavierMC, Dextercioby, Cuzkatzimhut, Hugo999, VolkovBot, Philip Trueman, TXiKiBoT, The Original Wildbear, Sarenne, Anonymous Dissident, Piperh, JhsBot, Leafyplant, Jackfork, LeaveSleaves, Itemirus, Thunderbird2, MrChupon, Murkee, DrJunge, EvilBunnyHead, SieBot, Servant Saber~enwiki, Coffee, Ivan Štambuk, GrooveDog, Keilana, Likebox, Flyer22, Tiptoety, Grimey109, Oxymoron83, Antonio Lopez, Scorpion451, Jdaloner, Pac72, Lightmouse, Christovac, Nandobike, The Stickler, Mike2vil, Anchor Link Bot, Pinkadelica, Dolphin51, Elassint, ClueBot, Ferred, GorillaWarfare, The Thing That Should Not Be, ArdClose, Swedish fusilier, Neverquick, Auntof6, Djr32, Robert Skyhawk, Excirial, SpikeToronto, Willthedrill, Danmichaelo, Nengscoz416, Iohannes Animosus, Bite Size Monkeys, W.GUGLINSKI, Zerxan, Thehelpfulone, Kakofonous, Thingg, BlueDevil, DumZiBoT, Life of Riley, Jmanigold, Nukeh, Avoided, Skarebo, NellieBly, Mm40, ZooFari, MystBot, RyanCross, Some jerk on the Internet, DOI bot, WMdeMuynck, YURI2008, CanadianLinuxUser, Fluffernutter, Glane23, Bassbonerocks, Glass Sword, LinkFA-Bot, Hainetron, Apteva, Gail, Jarble, Alfie66, Jackelfive, Legobot, Luckas-bot, Yobot, Cflm001, IW.HG, TestEditBot, Tempodivalse, N1RK4UDSK714, AnomieBOT, DemocraticLuntz, Jim1138, Piano non troppo, Sz-iwbot, Materialscientist, The High Fin Sperm Whale, Citation bot, E2eamon, GB fan, ArthurBot, Ammubhave, Jonathan321, Jeffrey Mall, Omnipaedista, Ilovenickjay, Tgervaisphd, FrescoBot, Qalander, Steve Quinn, Machine Elf 1735, Citation bot 1, Pinethicket, I dream of horses, Teamdojo, BRUTE, RandomStringOfCharacters, Ifritnile, Jordgette, Javierito92, Zink Dawg, Auscompgeek, Redskins247, DARTH SIDIOUS 2, WikitanvirBot, Nuujinn, Super48paul, RA0808, Solarra, Tommy2010, Wikipelli, K6ka, Hhhippo, JSquish, John Cline, Lateg, Quondum, Zloyvolsheb, QEDK, Wagino 20100516, Thine Antique Pen, Dilwala314, WikiPidi, Nick9876, Peter Karlsen, CharlotteMab, DASHBotAV, Mtlee7, Billylovespiethethird, Spicemix, Imuwithu2, Rocketrod1960, ClueBot NG, Pruegz778, 12nichja, Satellizer, A520, Widr, Shivsagardharam, Titodutta, Bibcode Bot, BG19bot, Wiki13, Shalom25, Eio, Klilidiplomus, Ihateicairns, Achowat, Riley Huntley, Samanthaclark11, Pratyya Ghosh, ChrisGualtieri, Martinkupilas, La marts boys, Bethechangeyouhopetosee, Sdk16420, Makecat-bot, Cerabot~enwiki, Lugia2453, 6033CloudyRainbowTrail, JustAMuggle, Reatlas, Epicgenius, Melonkelon, Tentinator, Ihatepauldirac, Ihatedirac, MKCarriegirl, Ugog Nizdast, BruceBlaus, Mourici, Konveyor Belt, Adamharsh, Williamsmith29, Ugotpowned12, LucaMoro, Amortias, Joeyransom, Adam tunard creator of science, DallasSama, YashGarg10, Leawesomepotato and Anonymous: 709

- **Matter** *Source:* https://en.wikipedia.org/wiki/Matter?oldid=675824648 *Contributors:* Carey Evans, CYD, The Anome, Tarquin, Ted Longstaffe, XJaM, Roadrunner, Ben-Zin~enwiki, Heron, Camembert, Ryguasu, Isis~enwiki, Stevertigo, Patrick, D, Tim Starling, Gabbe, Menchi, Ixfd64, Tomos, Minesweeper, Looxix~enwiki, Ahoerstemeier, Suisui, Angela, Glenn, Cyan, Mxn, Charles Matthews, Reddi, Hyacinth, Rm, Xevi~enwiki, Gakrivas, Jerzy, BenRG, RadicalBender, Rashack~enwiki, Chuunen Baka, Robbot, Kizor, Gandalf61, Ashley Y, Auric, Hadal, Papadopc, Lupo, HaeB, Jan Lapère, Alan Liefting, Enochlau, Giftlite, Djinn112, Art Carlson, FeloniousMonk, Bensaccount, Solipsist, Alexf, Quadell, Antandrus, OverlordQ, Lesgles, Karol Langner, Mikko Paananen, JimWae, DragonflySixtyseven, Kevin B12, Okapi~enwiki, Int19h, Nike, Trevor MacInnis, Brianjd, EugeneZelenko, Discospinster, Rich Farmbrough, Vsmith, Jpk, Autiger, Martpol, Bender235, Kbh3rd, Ghitis, JoeSmack, RJHall, MisterSheik, Mr. Billion, Zegoma beach, Remember, Jashiin, CDN99, Adambro, Bobo192, Army1987, Whosyourjudas, Rrh02, Smalljim, Orbst, .:Ajvol:., Dungodung, Maureen, I9Q79oL78KiL0QTFHgyc, Giraffedata, PaRaLyZeDHoRSe, Jojit fb, Kjkolb, Nk, Charonn0, MPerel, Sam Korn, Haham hanuka, Jayakar, Alansohn, Arthena, Atlant, Paleorthid, AzaToth, Kocio, Walkerma, Jaw959, Wdfarmer, Avenue, Bart133, Metron4, Snowolf, Wtmitchell, Oking83, Ott, Angr, Woohookitty, Madchester, Canaen, Sandrapalaje, CharlesC, RuM, Graham87, Dpr, Sjö, Rjwilmsi, Jake Wartenberg, Strait, MarSch, Quiddity, Tangotango, Nneonneo, Vav11, Yamamoto Ichiro, Exeunt, FayssalF, FlaBot, Lorkki, Naraht, Nihiltres, Crazycomputers, Andy85719, RexNL, TimSE, TheDJ, BabyNuke, TheMightyGrecian, Cloudo, King of Hearts, Chobot, DaGizza, Sharkface217, Gwernol, The Rambling Man, Wavelength, RobotE, Sceptre, Jimp, Bhny, Juansmith, Stephenb, Wimt, Ugur Basak, NawlinWiki, Injinera, Wiki alf, Grafen, Janarius, SCZenz, Retired username, Anetode, Dhollm, Jpbowen, Chichui, Alex43223, Natkeeran, DeadEyeArrow, Dna-webmaster, Wknight94, FF2010, Vadept, Enormousdude, Zzuuzz, Leliathomas, E Wing, Sean Whitton, JuJube, Aeon1006, JoanneB, Chez37, CWenger, Katieh5584, Junglecat, Banus, Mejor Los Indios, DVD R W, SmackBot, Unschool, KnowledgeOfSelf, Hydrogen Iodide, Unyoyega, C.Fred, Bomac, Gilliam, Skizzik, Fogster, Dauto, Andy M. Wang, Grokmoo, Rmosler2100, Master of Puppets, SchfiftyThree, Sbharris, Colonies Chris, Hallenrm, Darth Panda, Can't sleep, clown will eat me, OrphanBot, Rrburke, Addshore, Nakon, Jiddisch~enwiki, John D. Croft, Dreadstar, SpiderJon, DMacks, Ultraexactzz, Kotjze, Kalathalan, Where, Bob byPeru, Yevgeny Kats, Byelf2007, Dbtfz, LarchOye, Ocanter, Gobonobo, Breno, AstroChemist, Edwy, IronGargoyle, Vidit1, Simonalexander2005, Loadmaster, Munita Prasad, Noah Salzman, Hiiiiiiiiiiiiiiiiiiii, Kyoko, Dicklyon, Waggers, Dr.K., RichardF, MHWiki, Supaman89, CrazedEwok, Levineps, Lord Anubis, Joseph Solis in Australia, Newone, Casull, Tony Fox, Esurnir, Courcelles, JRSpriggs, JForget, DSatYVR, Ale jrb, Dycedarg, Van helsing, Vyznev Xnebara, MargyL, Flapping Fish, McVities, MarsRover, Penbat, Gregbard, Fl, Bvcrist, Peterdjones, Gogo Dodo, 01011000, Anonymi, Rracecarr, Studerby, Dr.Kane, Abtract, RickDC, Epbr123, Barticus88, Mbell, O, TonyTheTiger, Giorgio51, N5iln, Headbomb, Trevyn, Marek69, Grayshi, Nick Number, MichaelMaggs, Troy392004, Dualactionblend, Escarbot, Oreo Priest,

Fieldday-sunday, Bertrc, Icelazer3, Shirtwaist, Vishnava, CanadianLinuxUser, NjardarBot, Download, PranksterTurtle, Glane23, Chzz, FC-Sundae, Kyle1278, CuteHappyBrute, Eh kia, Tide rolls, Gail, Micki, Cesaar, Angrysockhop, Legobot, Luckas-bot, Yobot, Dr. Footfoe, 2D, Andreasmperu, II MusLiM HyBRiD II, THEN WHO WAS PHONE?, Empireheart, IW.HG, Tempodivalse, Nintend06, Quangbao, Hairhorn, Daniele Pugliesi, Jim1138, Lewismith3, Piano non troppo, Kingpin13, Dinesh smita, Westonpark, Materialscientist, Spirit469, ImperatorExercitus, 90 Auto, Citation bot, OllieFury, ArthurBot, Xqbot, TinucherianBot II, Beaserbebbeb, Pvkeller, Grim23, GrouchoBot, Charliekarst, Hamhat, Backpackadam, Amaury, Logger9, JediMaster362, Wasteman1066, Lagooncopperorange, FrescoBot, LucienBOT, Lothar von Richthofen, Hielor, Cannolis, Citation bot 1, Athenanoenvoy, Pinethicket, Boulaur, Overthinkingly, Tom.Reding, ContinueWithCaution, Johann137, Jauhienij, Gamewizard71, FoxBot, Yunshui, Zvn, January, Allen4names, Theo10011, Reaper Eternal, Jeffrd10, Tbhotch, Hornlitz, Marie Poise, DARTH SIDIOUS 2, NerdyScienceDude, Chuanenlin123, Jack Schlederer, Kenvancleve, WikitanvirBot, Ajraddatz, DillonLMcCabe, Mordgier, Katherine, Gcastellanos, The Sharminator, Tommy2010, Wikipelli, John Cline, Zane45177, Alpha Quadrant (alt), DidgeGuy, AndrewN, Yamumyadad, Jay-Sebastos, Jesanj, Brandmeister, Donner60, Ashunigam, Orange Suede Sofa, ChuispastonBot, Llightex, ShatteredSpiral, ClueBot NG, Gareth Griffith-Jones, MelbourneStar, UAK32, Rtucker913, Yourmomblah, Theimmaculatechemist, 123Hedgehog456, Whazie, Lfmm11, Mtheodric, Kumamah, ساجد امجد ساجد, ساجد, Friend20gan, MerlIwBot, Helpful Pixie Bot, Candleabracadabra, Bibcode Bot, Mingmingla, Mark Fole, BZTMPS, Lowercase sigmabot, Yetisyny, Hallows AG, MusikAnimal, Frze, AvocatoBot, Planetary Chaos Redux, Tatchell, Nsda, ChE Fundamentalist, Ugncreative Usergname, Joydeep, Averysamantha, Snow Blizzard, ImhotepBallZ, Mantike, Kydon Shadow, Th4n3r, The Illusive Man, ChrisGualtieri, EuroCarGT, Sidsandyy, Lugia2453, CaSJer, Frosty, SFK2, Jamesx12345, Perfecttwoegan, Reatlas, Passengerpigeon, Epicgenius, I am One of Many, Lsmll, Surfer43, Tentinator, Cebr1979, Amdz96, Backendgaming, Lince celeste, DavidLeighEllis, Ginsuloft, WikiMannWikiMann, Annieminnie21, Rickdutta, Meteor sandwich yum, Muneeb abdelhadi, ShahryarAhmad27900, Nijuthomasgeorge, Bilorv, Saaaaaaaaaa, Ookami-to-Koneko, J73364, S r u fejvd u ub creating u of, Amortias, Natdeso, Adenine2k, Lalalalalala is too great, ElmoLover88545, Darkmatter1435, Chetan082, Alango1998, XxX$pace5Xxx, Cambles13, Johno2000, Jsbdjejnwn, Weegeerunner, Slayeredwarrior, Account50, Blissy2005, Minnie431, KasparBot and Anonymous: 823

- **Phase (matter)** *Source:* https://en.wikipedia.org/wiki/Phase_(matter)?oldid=670090017 *Contributors:* CYD, Mav, Bryan Derksen, Olof, Tarquin, Rmhermen, Peterlin~enwiki, Jqt, DrBob, FlorianMarquardt, Patrick, Michael Hardy, Booyabazooka, TakuyaMurata, SebastianHelm, Looxix~enwiki, Ahoerstemeier, Александър, Glenn, Michael Shields, Andres, EdH, Mxn, Pizza Puzzle, Schneelocke, HolIgor, Furrykef, Omegatron, Head, Baffclan, Jusjih, Rogper~enwiki, Donarreiskoffer, Gentgeen, Robbot, Lowellian, Mayooranathan, Puckly, Filemon, Cedars, Giftlite, Rs2, ComaVN, Harp, BenFrantzDale, Art Carlson, Everyking, Yekrats, Utcursch, Knutux, Beland, Randwicked, Freakofnurture, Kcavness, Guanabot, Cacycle, Vsmith, VivaRose, CanisRufus, El C, Rgdboer, Spearhead, RoyBoy, Femto, Foobaz, I9Q79oL78KiL0QTFHgyc, Mr2001, HasharBot~enwiki, Ranveig, Jumbuck, Alansohn, PoptartKing, Snowolf, RJFJR, Dnaber, Vuo, Soulman (renamed), OwenX, Linas, Jannex, Polyparadigm, Graham87, V8rik, Joe Decker, Chebbs, Krash, FlaBot, Windchaser, Srleffler, Kri, Physchim62, DVdm, The Rambling Man, Siddhant, YurikBot, Hairy Dude, Jeffhoy, Gaius Cornelius, Alex Bakharev, Purodha, Zercam, Terfili, Dbmag9, Dhollm, Pnrj, E2mb0t~enwiki, Tony1, Alex43223, Wknight94, Leptictidium, Lt-wiki-bot, Willtron, Banus, Willemo, SmackBot, FocalPoint, AngelovdS, Bigbluefish, Bomac, Alksub, Jrockley, Chych, Commander Keane bot, Betacommand, Amatulic, Bh3u4m, Chris the speller, RDBrown, Pieter Kuiper, MK8, Complexica, Gracenotes, Berland, Skidude9950, Nunocordeiro, Smooth O, Iapetus, Kukini, Ozhiker, Rigadoun, Treyt021, Mgiganteus1, IronGargoyle, Beetstra, Meco, Waggers, EdC~enwiki, Gazjo, Lakers, Newone, Reitzaleah, JRuby, Tawkerbot2, Geonator, Dgw, MarsRover, Shandris, Equendil, Cydebot, Future Perfect at Sunrise, A876, Thijs!bot, Runch, N5iln, Headbomb, Bobblehead, James086, Massimo Macconi, Escarbot, AntiVandalBot, Guy Macon, Dougher, JAnDbot, Seddon, Hello32020, Karlhahn, Bongwarrior, VoABot II, Aka042, Gabriel Kielland, Dirac66, Allstarecho, LorenzoB, PoliticalJunkie, Joshua Davis, Mythealias, Onaraighl, Iwoelbern, Jonathan Hall, Mausy5043, Victor Blacus, Trusilver, Uncle Dick, Maurice Carbonaro, Brien Clark, Zuejay, Afluegel, Cowplopmorris, PaulKishimoto, Bonadea, Funandtrvl, Spellcast, Philip Trueman, BotKung, Dirkbb, Radagast3, Hmwith, Malcolmxl5, Duncan harvie, Keilana, Faradayplank, Mike2vil, Anchor Link Bot, Mygerardromance, Loren.wilton, ClueBot, Gaia Octavia Agrippa, Andrewwclui, CohesionBot, Razorflame, Ranjithsutari, Erodium, Spitfire, Little Mountain 5, Addbot, Some jerk on the Internet, Guoguo12, Johnsarelli, Glane23, West.andrew.g, Wakeham, Eh kia, Tide rolls, Brendan064, Micki, Yobot, ThaddeusB, Daniele Pugliesi, Xqbot, Madbard, Addihockey10, Nickkid5, Pvkeller, Ihsam, Logger9, 虞海, LiberalMedium, LucienBOT, Dogbert66, Oatesyrulz, Pinethicket, Tom.Reding, Jauhienij, FoxBot, Vrenator, Jeffrd10, Copistopplayer, Dtheobald, Reach Out to the Truth, Marie Poise, Dryftburn, EmausBot, Albertnetymk, Dcirovic, BioPupil, Jarel&Dale, DASHBotAV, ClueBot NG, Gareth Griffith-Jones, Gilderien, Nscozzaro, Amr.rs, Delusion23, CrossingStyx, BG19bot, SyahirSQRT2, Kangaroopower, Whatisron, Monkbot, XxX$pace5Xxx and Anonymous: 244

- **Isotope** *Source:* https://en.wikipedia.org/wiki/Isotope?oldid=676175908 *Contributors:* Trelvis, Marj Tiefert, Mav, Bryan Derksen, Tarquin, AstroNomer~enwiki, Malcolm Farmer, Stokerm, William Avery, Peterlin~enwiki, Spiff~enwiki, Patrick, Alan Peakall, Shyamal, Shellreef, Liftarn, Ixfd64, Tango, Minesweeper, Ahoerstemeier, Snoyes, Suisui, BigFatBuddha, Julesd, Glenn, Andres, Kaihsu, Hectorthebat, Rl, Smack, Dysprosia, Taxman, Jusjih, Palefire, Shantavira, Robbot, Merovingian, Sunray, Bkell, Hadal, Alan Liefting, Giftlite, Mikez, Lethe, Xerxes314, Bensaccount, Guanaco, Bovlb, Prosfilaes, Christopherlin, Wmahan, Antandrus, Kaldari, Icairns, Urhixidur, Adashiel, Bluemask, DanielCD, KNewman, Discospinster, C12H22O11, Mani1, MarkS, ESkog, A purple wikiuser, Walden, Femto, CDN99, Bobo192, Viriditas, Mytildebang, I9Q79oL78KiL0QTFHgyc, Deryck Chan, Obradovic Goran, Jumbuck, Alansohn, Gary, GRider, Cjthellama, Hu, Malo, Ayeroxor, Cburnett, Mikeo, Zoohouse, DV8 2XL, Adrian.benko, Stemonitis, Gmaxwell, Thryduulf, OwenX, Camw, Tripodics, Kurzon, Bratsche, Tylerni7, Clemmy, GregorB, SCEhardt, Sin-man, Graham87, Magister Mathematicae, Kbdank71, DePiep, Jclemens, Sjö, Rjwilmsi, Astronaut, Strait, Quiddity, Feydey, Watcharakorn, FlaBot, Gurch, Physchim62, King of Hearts, Chobot, Sharkface217, GangofOne, Bgwhite, YurikBot, Wavelength, RobotE, JWB, Jimp, Wolfmankurd, Pip2andahalf, Phantomsteve, Petiatil, Stephenb, Cryptic, Wimt, Anomalocaris, NawlinWiki, Wiki alf, Grafen, Chick Bowen, Welsh, Długosz, Dooky, Semperf, Kkmurray, Black Falcon, Silverchemist, Citynoise, Closedmouth, E Wing, KGasso, JuJube, Roberto DR, JoanneB, CWenger, Junglecat, Mjroots, AssistantX, GrinBot~enwiki, Cookiedog, SkerHawx, Serendipodous, 霧木諒二 robot, Luk, Itub, MacsBug, SmackBot, FocalPoint, Jclerman, Incnis Mrsi, CoderDennis, Shoy, Unyoyega, Pgk, C.Fred, Davewild, Edgar181, HalfShadow, Gilliam, Skizzik, Chris the speller, Master Jay, Father McKenzie, Master of Puppets, Miquonranger03, MalafayaBot, DHN-bot~enwiki, Cassivs, Sbharris, V1adis1av, Gurps npc, Rrburke, TKD, Rainmonger, Addshore, Mr.Z-man, SundarBot, Megamix, Radagast83, Khukri, Nibuod, Nakon, G716, Drphilharmonic, DMacks, Clicketyclack, Will Beback, SashatoBot, ArglebargleIV, Silvem, John, Ckatz, Smith609, Stwalkerster, FadieZ, Ryulong, MTSbot~enwiki, SmokeyJoe, Cadaeib, KJS77, Iridescent, Theone00, J Di, Cbrown1023, Sam Li, Witchyrose, Courcelles, Heliomance, Tawkerbot2, JForget, CmdrObot, Tanthalas39, Rambam rashi, RedRollerskate, FlyingToaster, Myrddin1977, Stephen Luce, Prakharbirla, Myasuda, Nmacu, Jlking3, HPaul, Christian75, MagnusGallant, Theadder, Daniel Olsen, Gimmetrow, Satori Son, JamesAM, Thijs!bot, Epbr123, Looskuh, Mojo Hand, Headbomb, Pjvpjv, Marek69, John254, A3RO, Jbwst, Tellyaddict, Jk-

Daonguyen95, Fæ, Akerans, AManWithNoPlan, L Kensington, Donner60, Sailsbystars, Niro061196, Zac, DASHBotAV, Whoop whoop pull up, Xanchester, 49greene1915, ClueBot NG, Gareth Griffith-Jones, Satellizer, Chester Markel, Sloberyleaf, Widr, Helpful Pixie Bot, HMSSolent, Lowercase sigmabot, Vibhor1997, Mynameisnoted, PhnomPencil, Wiki13, Mark Arsten, IraChesterfield, Rm1271, Hassan540, Samtez, Srishty singh, Shawn Worthington Laser Plasma, Glacialfox, Anbu121, Madprofessional, Mdann52, Total wiki guy, Blaksabath1, Minkevich, Mathar.hasan, Timothy Gu, Chemya, Dexbot, Webclient101, Lugia2453, Leprof 7272, Epicgenius, 23sheena23, Ananthkamath1995, DavidLeighEllis, Ugog Nizdast, Khenta37, Mdann52(alt), John Doppler, Mahusha, Monkbot, Silberbuschc17, SantiLak, Gabewalker5464, BethNaught, ZBarnes2271, Alango1998, Y-S.Ko, BlueworldSpeccie, Chickenmaster251, MacPoli1, IndrasisSD, GeneralizationsAreBad, KasparBot, RJ raghava and Anonymous: 711

- **Valence (chemistry)** *Source:* https://en.wikipedia.org/wiki/Valence_(chemistry)?oldid=674904912 *Contributors:* Lumos3, Robbot, Pigsonthewing, Giftlite, Yath, Karl-Henner, Tsemii, Felix Wan, Rich Farmbrough, Vsmith, El C, Jumbuck, LoganK, Walkerma, Shoefly, Geraldshields11, Zntrip, Nuggetboy, Benbest, V8rik, MZMcBride, Ligulem, Physchim62, YurikBot, RussBot, Bhny, Fabricationary, Okedem, Giro720, Cloud109, Daniel Bonniot de Ruisselet, Esqg, Biopresto, Tetracube, Urocyon, Paul D. Anderson, DVD R W, Itub, Lakhim, Bluebot, Bduke, Tsca.bot, Nick Levine, Pieter1, TKD, Peterwhy, Drphilharmonic, DMacks, Sadi Carnot, Spiritia, Nishkid64, AThing, 16@r, Beetstra, Iridescent, TwistOfCain, The.Q, Iced Kola, Tawkerbot4, Christian75, Calvero JP, VPliousnine, Bear475, Escarbot, Orionus, JAnDbot, MERC, Fetchcomms, Greensburger, Rothorpe, VoABot II, Dirac66, Thibbs, Huzzlet the bot, Numbo3, Reedy Bot, Adiel lo~enwiki, RadicalOne (usurped), Hulten, Izno, VolkovBot, Sankalpdravid, Axiosaurus, PDFbot, Cnilep, SieBot, ScAvenger lv, Ken123BOT, Nergaal, ClueBot, Blanchardb, Officer781, Excirial, Manco Capac, Nepenthes, Nicolae Coman, Addbot, Wickey-nl, MrOllie, Jasper Deng, Krano, Luckas-bot, Yobot, Ptbotgourou, OneAhead, AnakngAraw, AnomieBOT, Materialscientist, Citation bot, Brane.Blokar, Xqbot, Khajidha, Mlpearc, Frankie0607, Iωv, Shadowjams, R8R Gtrs, VS6507, Pinethicket, Allthingstoallpeople, Full-date unlinking bot, Riccardo.fabris, Jauhienij, FoxBot, Double sharp, على ویکی, Allen4names, Smellycat1234, Onel5969, Hajatvrc, From Adam, EmausBot, WikitanvirBot, ChemMater, Rcsprinter123, ClueBot NG, Ladbad, Widr, Helpful Pixie Bot, BG19bot, Rensink, Kuki2485, Sandyz8, Glevum, Soulparadox, Hazfacta, Burzuchius, Frosty, Sam Sailor, Hazcoll, De Riban5, Peddi radhika, Aadil bin Abuzar, S1yan, KasparBot and Anonymous: 148

- **Ion** *Source:* https://en.wikipedia.org/wiki/Ion?oldid=676705545 *Contributors:* Sodium, Vicki Rosenzweig, Olof, Tarquin, Andre Engels, LA2, Rmhermen, PierreAbbat, Ben-Zin~enwiki, Montrealais, Tucci528, Stevertigo, Michael Hardy, Shellreef, Ixfd64, Tango, Delirium, Mkweise, Ahoerstemeier, Notheruser, Александър, Andres, Evercat, Cherkash, Smack, Adam Bishop, Reddi, Lfh, Krithin, E23~enwiki, Omegatron, Joy, Bloodshedder, Jusjih, Donarreiskoffer, Gentgeen, Robbot, Moriori, Hadal, JesseW, JackofOz, Diberri, Centrx, Giftlite, DocWatson42, Christopher Parham, Mat-C, Herbee, Everyking, Alison, Pashute, Bensaccount, Ptk~enwiki, Darrien, SWAdair, Andycjp, Zeimusu, Am088, Jossi, Aahoughton, Icairns, Iantresman, Mike Rosoft, Vincom2, Discospinster, 4pq1injbok, Rich Farmbrough, Vsmith, Mani1, Paul August, Bender235, Cyclopia, RJHall, Kwamikagami, CDN99, Bobo192, NetBot, La goutte de pluie, Jojit fb, Kjkolb, Obradovic Goran, MPerel, Hooperbloob, Methegreat, Jumbuck, Zachlipton, Danski14, Alansohn, Gary, GRider, DanielLC, Arthena, Keenan Pepper, Benjah-bmm27, Riana, AzaToth, Snowolf, Wtmitchell, Sciurinæ, Itsmine, DV8 2XL, Antifamilymang, Kbolino, DavidK93, LOL, Rtdrury, Lensovet, Tokek, Graham87, GoldRingChip, Canderson7, Jake Wartenberg, Wikibofh, Alban, HappyCamper, Yamamoto Ichiro, Vuong Ngan Ha, FlaBot, SchuminWeb, Nihiltres, RexNL, Lmatt, Physchim62, Chobot, Antiuser, Elfguy, Mercury McKinnon, YurikBot, Wavelength, RobotE, Jimp, Alektzin, Phantomsteve, Vlad4599, BTLizard, Russoc4, Gaius Cornelius, CambridgeBayWeather, Shaddack, Wimt, NawlinWiki, Btnheazy03, Ino5hiro, Lepidoptera, Bobbo, Mikeblas, E2mb0t~enwiki, Nate1481, Kooky, Scottfisher, Samir, BOT-Superzerocool, Bota47, PGPirate, Tetracube, Jkhoury, Phgao, StuRat, Lt-wiki-bot, Huangcjz, KGasso, Garion96, Meegs, Greatal386, JDspeeder1, GrinBot~enwiki, KNHaw, SkerHawx, Mejor Los Indios, Bibliomaniac15, SmackBot, Radak, Tomyumgoong, Steve carlson, Prodego, Lagalag, Davewild, Jrockley, Canthusus, Edgar181, HalfShadow, Shai-kun, Yamaguchi⁉⁉, PeterSymonds, Gilliam, Diewelt, Skizzik, KaragouniS, Keegan, TimBentley, Stellar-TO, Droll, Moshe Constantine Hassan Al-Silverburg, DHN-bot~enwiki, Sbharris, Darth Panda, Nick Levine, Shalom Yechiel, Danielkueh, Rrburke, KerathFreeman, Aldaron, VegaDark, Occultations, Drphilharmonic, DMacks, Marcus Brute, Kukini, SashatoBot, Anlace, Vanished user 9i39j3, Scientizzle, Tktktk, Minna Sora no Shita, Mathias-S, Mr. Lefty, Beetstra, Mr Stephen, Teeteetee, Macellarius, NJA, Ryulong, Squirepants101, Melampus, Jotham, Basicdesign, Twas Now, Igoldste, Courcelles, Tawkerbot2, Daniel5127, JForget, CmdrObot, Mattbr, Dycedarg, Van helsing, Maximilli, Fork me, ShelfSkewed, McVities, Hopsyturvy, Yopienso, Bill Sayre, The Enslaver, Dannykean, SyntaxError55, Astrochemist, HPaul, AtTheAbyss, Gproud, Odie5533, Tawkerbot4, Christian75, Optimist on the run, Omicronpersei8, Wisebridge, Xo4evr urzxo, Parsa, Kahastok, Bobandwendy, Dougsim, Sobreira, James086, Merbabu, MisterSpike, Escarbot, I already forgot, Ileresolu, Ialsoagree, AntiVandalBot, WinBot, Luna Santin, Prolog, TimVickers, 000n, Amberjack, Daniel Villalobos, Scepia, Spril4, Robert A. Mitchell, JAnDbot, Deflective, MER-C, Instinct, Rearete, Jrennie, Greensburger, PhilKnight, LittleOldMe, SiobhanHansa, Bencherlite, Connormah, Bongwarrior, VoABot II, Tripbeetle, Kevinmon, Deus911, SparrowsWing, Slartibartfast1992, Animum, KumfyKittyKlub, Allstarecho, Chris G, Greenguy1090, Adriaan, MartinBot, EyeSerene, BetBot~enwiki, ChemNerd, Purplefanta, Kostisl, R'n'B, Leyo, KBlott, Itanius, J.delanoy, PCock, Numbo3, GravityFong, Hans Dunkelberg, Teacher123, Spiperon, Acalamari, Hispanosuiza, Mj sklar, RTBoyce, Ryan Postlethwaite, Daniele.tampieri, Pyrospirit, 97198, Renzy222, Belovedfreak, Richard D. LeCour, NewEnglandYankee, Shmoppy, Juliancolton, Entropy, Demonkingjj, Cometstyles, Khargas, Moadeeb, Ja 62, Yahove, CardinalDan, Funandtrvl, Matt r kelly, TeamZissou, Lights, Deor, VolkovBot, CWii, Larryisgood, Thedjatclubrock, Philip Trueman, TXiKiBoT, Miranda, Rei-bot, Crohnie, Anna Lincoln, Lradrama, Clarince63, Melsaran, Martin451, Slysplace, Amaros11, PDFbot, Cremepuff222, Amphy, Kermit2 tic tac, Aphilo, RandomXYZb, Falcon8765, Enviroboy, Burntsauce, Spinningspark, K10wnsta, AlleborgoBot, Logan, Vsst, NHRHS2010, D. Recorder, SieBot, Gerakibot, Awesome Truck Ramp, Xymmax, Triwbe, Hac13, Happysailor, Quest for Truth, Tiptoety, Exert, Wilson44691, Oiws, Yerpo, Aruton, Oxymoron83, KoshVorlon, Steven Crossin, Poindexter Propellerhead, Techman224, Greysea=S, Paully111, Alman148, OKBot, Mike2vil, Thosthos, Precious Roy, ClueBot, MentalUnbalance, The Thing That Should Not Be, Drmies, Ryoutou, Narcer23, Aanter16, Blanchardb, Alexostamp, Officer781, Excirial, Malachirality, Spiderdan1001, Eeekster, 97parnellj, Estirabot, Therealnih, Pdch, Cenarium, Razorflame, Jinlye, Dekisugi, Maikol28, Aitias, Plasmic Physics, Glacier Wolf, Vanished User 1004, EdChem, Against the current, XLinkBot, Avilren, BodhisattvaBot, Henry the 1st, Skarebo, Mushyrulez Alot, Palindrome, Addbot, Some jerk on the Internet, Jojhutton, Fyrael, Haruth, Ronhjones, Fieldday-sunday, Laaknor Bot, Glane23, Omnipedian, Favonian, LinkFA-Bot, Quercus solaris, 5 albert square, Frozenguild, Erutuon, Tide rolls, Bfigura's puppy, Kurtis, Luckas-bot, Yobot, Sirsparksalot, THEN WHO WAS PHONE?, Scy1192, AnomieBOT, Tucoxn, Piano non troppo, Blackknight12, Kingpin13, EryZ, Materialscientist, The High Fin Sperm Whale, Citation bot, ArthurBot, TheRabidMonkie, Xqbot, Addihockey10, حسن علي البط, AbigailAbernathy, Coretheapple, GrouchoBot, Omnipaedista, RibotBOT, SassoBot, Amaury, 78.26, BobRucklepuckle, Frogytoad, Mr Seditives, Erik9, Clayton1995hoe, LucienBOT, The Black Void, Saehrimnir, DDPTSTS, Kimahrironso, Nirmos, Roopreqt, Cubs197, Pekayer11, Pinethicket, Elockid, MBirkholz, Tom.Reding, Deathromain, RedBot, Lars Washington, Pikiwyn, Meaghan, Carolina cotton, FoxBot, DixonD-

Bot, Kalaiarasy, Ronnoc313133, GrnEydGuy, Reaper Eternal, TheGrimReaper NS, Jeffrd10, Rutscher, Rmstroud, Tbhotch, Reach Out to the Truth, TjBot, Bento00, Wexeb, Yardsrule111, Kieferdale81, Beyond My Ken, Yooesther, Slon02, J36miles, EmausBot, Immunize, RenamedUser01302013, Sp33dyphil, Losman65, JSquish, Bollyjeff, Proofer42, Felixh238, AManWithNoPlan, ChuispastonBot, RockMagnetist, Baconage3013, DASHBotAV, Whoop whoop pull up, Danni94Allport, Petrb, ClueBot NG, Halemane, Fjfsfmfsm, Satellizer, Loginnigol, Jkwchui, S.kamber, O.Koslowski, ScottSteiner, Widr, Antiqueight, Art and Muscle, Paco Castro, Fhsjklfhdfaldjlfksdf, MrClone501st, Lowercase sigmabot, BG19bot, Vagobot, Dan653, Mark Arsten, ChE Fundamentalist, Kumar.aditv, Shawn Worthington Laser Plasma, Rtkelly88, Knodir, BattyBot, Hack zilla, JESUbab, BrightStarSky, Webclient101, Mogism, 331dot, Bulba2036, Frosty, Russ9929, Christopherwong19961223, Allanepic144, Lagoona010, Racab, Jpd1066, Ianreisterariola, Samfowler1999, Ivovis, Acetotyce, I am One of Many, Glaisher, Ginsuloft, Bruce Chen 0010334, Jianhui67, Swaqmaster69, Cman612, VinylTavi, Methylox, Patrick heisenburg, Acagastya, Qwertyxp2000, Cupcakeknight, Big davweiner, Obristan, Pathogenic Cyanide Piggy, Mrbenchua, Science Geeeeeeek, MacDaddy4eva, Ray960153, KasparBot, Leanna.20 and Anonymous: 921

- **Chemical element** *Source:* https://en.wikipedia.org/wiki/Chemical_element?oldid=677625265 *Contributors:* AxelBoldt, Vicki Rosenzweig, Mav, Bryan Derksen, The Anome, Tarquin, Css, Andre Engels, Youssefsan, Rmhermen, William Avery, Heron, Mercury610, Stevertigo, Patrick, RTC, Menchi, Ixfd64, Eric119, Minesweeper, Kosebamse, Ahoerstemeier, Suisui, Александър, Glenn, Andres, Evercat, Mxn, Schneelocke, Mulad, Reddi, Stone, Jay, Taxman, Thue, Bevo, Xevi~enwiki, Geraki, Archivist~enwiki, Donarreiskoffer, Gentgeen, Robbot, Fredrik, BitwiseMan, Yelyos, Nurg, Romanm, Arkuat, Merovingian, Pingveno, Academic Challenger, PxT, Rursus, Texture, Roscoe x, Caknuck, Wikibot, Ebeisher, Jimduck, Hexii, Centrx, Giftlite, DocWatson42, Christopher Parham, Tom harrison, HangingCurve, Xerxes314, Everyking, No Guru, Alison, Bensaccount, Rpyle731, Mboverload, R. fiend, Blankfaze, Antandrus, Beland, Karol Langner, Thincat, Icairns, Trevor MacInnis, Mike Rosoft, Sdrawkcab, HedgeHog, EugeneZelenko, Discospinster, Rich Farmbrough, FT2, Cacycle, FiP, Vsmith, Xezbeth, Nvj, Mani1, Blade Hirato~enwiki, SpookyMulder, Bender235, Andrejj, RJHall, MisterSheik, Kwamikagami, Mwanner, Laurascudder, Shanes, RoyBoy, Bookofjude, Danshil, Femto, CDN99, Bobo192, Smalljim, Cmdrjameson, SpeedyGonsales, Man vyi, Jojit fb, PeterisP, Obradovic Goran, Nsaa, Ranveig, Jumbuck, Kuratowski's Ghost, Alansohn, Hi ruwen, Loa, Paleorthid, Riana, Walkerma, Jaw959, Bantman, Sobolewski, Wtmitchell, Maxkirk1, TenOfAllTrades, LFaraone, H2g2bob, Bsadowski1, Skatebiker, Gene Nygaard, Kay Dekker, Benoni, Thryduulf, Firsfron, Alvis, SNPP, OwenX, ScottDavis, Benbest, Polyparadigm, Pol098, WadeSimMiser, Dozenist, Terence, Sengkang, CharlesC, Wayward, 新世界的新衣, Mandarax, Tslocum, SqueakBox, Graham87, Ryoung122, Chun-hian, Kushboy, DePiep, Effeietsanders, Saperaud~enwiki, Rjwilmsi, Kinu, Strait, VogonFord, Tangotango, Crazynas, ScottJ, Brighterorange, ThePoorGuy, Yamamoto Ichiro, FlaBot, RobertG, Nivix, Ayla, Glenn L, Physchim62, Snailwalker, Imnotminkus, King of Hearts, Chobot, GangofOne, DVdm, Korg, NSR, Gwernol, Banaticus, Wavelength, Hawaiian717, Alchemy pete, Phantomsteve, RussBot, Ismaeelah, Limulus, Quintusdecimus, Wimt, RadioKirk, NawlinWiki, Grafen, Brythain, Peter Delmonte, E rulez, Raven4x4x, Nick C, Zwobot, Werdna, SamuelRiv, 21655, Cynicism addict, Orbis 3, NielsenGW, Peter, Johnpseudo, Katieh5584, Kungfuadam, JDspeeder1, Luk, ChemGardener, Itub, Winick88, Mexistache, Yakudza, SmackBot, Dreamer.redeemer, CarbonCopy, Hydrogen Iodide, Melchoir, C.Fred, Bomac, Jagged 85, Jrockley, Delldot, Edgar181, HalfShadow, Alsandro, Yamaguchi先生, Gilliam, Aaron of Mpls, Skizzik, Chris the speller, Kurykh, Persian Poet Gal, MK8, Miquonranger03, MalafayaBot, CherryT~enwiki, SchfiftyThree, Bonaparte, Deli nk, Aclwon, DHN-bot~enwiki, Sbharris, Darth Panda, MaxSem, Modest Genius, Can't sleep, clown will eat me, MyNameIsVlad, Kristbg, Yidisheryid, Booshank, EvelinaB, Rrburke, Andy120290, RedHillian, SundarBot, Nakon, Lordshaun, Het, Sadi Carnot, Curly Turkey, Khazar, John, Euchiasmus, Kipala, Sir Nicholas de Mimsy-Porpington, Tony Corsini, Anoop.m, Madris, IronGargoyle, Hvn0413, Beetstra, Maksim L., Funnybunny, Iridescent, Joseph Solis in Australia, Shoeofdeath, J Di, Cbrown1023, Gil Gamesh, Civil Engineer III, Tawkerbot2, Lahiru k, JForget, Svlad Jelly, CmdrObot, Irwangatot, FunPika, Van helsing, NickW557, WeggeBot, Darren10000, Nmacu, Funnyfarmofdoom, Nilfanion, TJDay, Nick Y., Steel, Kaldosh, Rifleman 82, Gogo Dodo, Julian Mendez, Tawkerbot4, Carstensen, Christian75, DumbBOT, Obrian7, Lee, Matwilko, Mydoghasworms, Smeazel, Thijs!bot, Epbr123, Headbomb, Canada Jack, Marek69, Davidhorman, Kiran201193, ZeekyH.bomb, SusanLesch, Escarbot, Morgana The Argent, KrakatoaKatie, AntiVandalBot, Jj137, Farosdaughter, Gdo01, Istartfires, Nousakan, Res2216firestar, MER-C, Skomorokh, Plantsurfer, Hamsterlopithecus, Cynwolfe, GoodDamon, Acroterion, Moni3, Karlhahn, Bongwarrior, VoABot II, AuburnPilot, Edmund372, WhatamIdoing, Dirac66, 28421u2232nfenfcenc, Thibbs, TekNOSX, Glen, DerHexer, JaGa, T55648L, Pax:Vobiscum, TheRanger, Gwern, Rickterp, Nietzscheanlie, MartinBot, ChemNerd, Rettetast, R'n'B, AlexiusHoratius, PrestonH, Smokizzy, Tgeairn, J.delanoy, Pharaoh of the Wizards, Trusilver, Bogey97, Rhinestone K, Maurice Carbonaro, Metrax, Rod57, Chaveyd, Warut, Pcfjr9, Mufka, Tanaats, Ionescuac, Lilwik, KylieTastic, Lordaraq, Rpr117, Jamesofur, DorganBot, Treisijs, Useight, Fusion Power, Xiahou, CardinalDan, Daz643, Xenonice, Sumo su, 28bytes, VolkovBot, ABF, Jeff G., JoeDeRose, FutharkRed, Philip Trueman, TXiKiBoT, The Original Wildbear, Gary Levell, Daydreammbeliever15, Billiards, Qxz, Lradrama, Martin451, LeaveSleaves, Karlengblom, Inx272, Michelle192837, Brainmuncher, Sarc37, Wolfrock, Synthebot, Burntsauce, Riversong, 555zozo555, Onceonthisisland, AlleborgoBot, EmxBot, Glennklockwood, Demmy100, SieBot, Stever Augustus, Buccaneerande, Sonicology, PlanetStar, Winchelsea, Gerakibot, Dawn Bard, Viskonsas, Caltas, Calabraxthis, JerrySteal, Keilana, Toddst1, Tiptoety, Qst, Oda Mari, Wombatcat, Oxymoron83, Jdaloner, Lightmouse, Mjkhfg, Tombomp, Sjn28, Maelgwnbot, Wuhwuzdat, Tesi1700, Maralia, Dolphin51, Zbisasimone, Nergaal, Escape Orbit, Into The Fray, Romit3, SallyForth123, Twinsday, Elassint, ClueBot, PleasantPheasant, The Thing That Should Not Be, Rodhullandemu, VQuakr, J8079s, Boing! said Zebedee, Xenon54, Daracul, Firzen the Great, Neander7hal, Excirial, Alexbot, PrincealiG, Eeekster, Clutchmetal, ParisianBlade, Helenginn, LarryMorseDCOhio, Muro Bot, La Pianista, Calor, Thingg, Pzoxicuvybtnrm, Teleomatic, Johnuniq, SoxBot III, DumZiBoT, RMFan1, Drjezza, PseudoOne, Pgallert, Klemox, Vianello, ZooFari, Bir el Arweh, MystBot, Buckeyenaaashun, Good Olfactory, ElMeBot, Pj13, Addbot, ERK, Proofreader77, Trygve94, Roentgenium111, Some jerk on the Internet, DOI bot, Element16, Guoguo12, 1266asdsdjapg, Sedsa1, Hda3ku, Friginator, Fieldday-sunday, KorinoChikara, Shirtwaist, Cst17, Googleguy1234, Challengedgenius, CarsracBot, Azim5498, Greasel5, TStein, Quercus solaris, Numbo3-bot, Ehrenkater, Tide rolls, Bfigura's puppy, Lightbot, Jan eissfeldt, Gail, Mr.pomo, Fryed-peach, Luckas-bot, Yobot, Dor Cohen, Moda yahia, Les boys, Legobot II, Newportm, Ajh16, THEN WHO WAS PHONE?, 13lade94, Synchronism, Davelo99, DiverDave, AnomieBOT, Grahamching, 1exec1, NoPity2, IRP, JackieBot, Piano non troppo, Bhujerban, Kingpin13, Rathla, Flewis, Materialscientist, Pepo13, Susiban227, Citation bot, Adsfkl236io54lklk, Neurolysis, Chemeditor, Huklkl, LilHelpa, Xqbot, Phazvmk, RJav, Roftltime8, Timir2, Sionus, Intelati, Capricorn42, 4twenty42o, Grim23, Nakumi, Scream1013, Michele600, Dominicp2007, GrouchoBot, Jermentr, Jhbdel, Omnipaedista, RibotBOT, Otnick, Doulos Christos, Eugene-elgato, Franman3024, Erik9, A.amitkumar, FrescoBot, Remotelysensed, Tobby72, Goodbye Galaxy, Wik1ped1a is meant 2 be vanda1ised, Machine Elf 1735, Commit charge, Citation bot 1, Nirmos, ElmentPT, Pinethicket, I dream of horses, RedBot, Mikespedia, Leodescal, FoxBot, Double sharp, كاشف عقيل, Ticklewickleukulele, Hostel369, Dinamik-bot, Vrenator, Nemesis of Reason, Aoidh, Diannaa, Jamietw, Mitchellpuss, Lala2034, McPoopyPee, Sandman888, DARTH SIDIOUS 2, Whisky drinker, Mean as custard, Shanoor212, The Utahraptor, RjwilmsiBot, TjBot, TomT0m, EmausBot, WikitanvirBot, Gfoley4, ScottyBerg, Syncategoremata, RA0808, XinaNicole, Emperorcheston, Wikipelli, Prowster, JSquish, Cogiati, Fæ, StringTheory11, Bryce Carmony, A930913, Wayne Slam, Frigotoni, Ocaasi, Mpb4lol, Tolly4bolly, L1A1 FAL, Rcsprinter123, Ericlaermans, L Kensington, Puffin, YOSF0113, Negovori, Moocow121, ChuispastonBot, Matthewrbowker, DASHBotAV, Socialservice, ResearchRave, Petrb, ClueBot NG, Kieran0397, LeastCommonAncestor, Jack Greenmaven, Satellizer, Chester Markel, Erichusk, Lanthanum-138, Delusion23, Cntras, O.Koslowski, Auchansa, Metaknowledge, Sehgalamit, Theopolisme, जगत्रत्र मभि, Diyar se, Helpful Pixie Bot, Ronaldo47, RobertGustafson, JohnSRoberts99, ಉಭ್ರ ಮಾರ್ಯ್ಕ, Sergeyshar, Javon12javon, Gob Lofa, Bibcode Bot, Markovnikov, BG19bot, Sandbh, Wiki13, AvocatoBot, CA Mendeleev, The Whispering Wind, AdventurousSquirrel, Jenjenniferjenni, Soerfm, Jamhol1234, Seanpkenny, Doctor Lipschitz, Aisteco, Justincheng12345-bot, Zarospwnz, Jimw338, Althealster, Mechknight117, Soulbust, EnzaiBot, Hans5620, BrightStarSky, Dexbot, Mogism, Lugia2453, Megachemistrygenius, Bulba2036, Frosty, Jnargus, Gaby263612, Unliquidating, Rainbow Shifter, Unholyherget, Kslinker5493, Genius567812, Jerry123456, Astredita, Rbpltr, Filedelinkerbot, SharpQuillPen, PaulZapata, Smartguy1234567891040420, Redsamuraii, IiKkEe, Forbidden User, TerryAlex, Rezve59, The Clipper, E757, Ur moms a nice person, Orduin, Kalemh24, MKZombie, Dsrawlings, KasparBot and Anonymous: 989

- **Radioactive decay** *Source:* https://en.wikipedia.org/wiki/Radioactive_decay?oldid=677802293 *Contributors:* Danny, Roadrunner, Mrwojo, Spiff~enwiki, Patrick, Ahoerstemeier, Andrewa, LittleDan, Kricke, Samw, Mxn, Smack, Hike395, HolIgor, Chuljin, Jitse Niesen, Audin, Furrykef, Populus, Omegatron, Topbanana, Pstudier, Finlay McWalter, PuzzletChung, Robbot, Romanm, Chancemill, Securiger, Merovingian, Pengo, Giftlite, Fudoreaper, Netoholic, Herbee, Everyking, Snowdog, Curps, Eequor, Jackol, Mmm~enwiki, Manuel Anastácio, Utcursch, Andycjp, LiDaobing, Antandrus, Beland, DragonflySixtyseven, Icairns, GeoGreg, Urhixidur, Syvanen, Olivier Debre, Deglr6328, Kate, Running, Mike Rosoft, Mormegil, Freakofnurture, Discospinster, Rydel, Rama, Vsmith, Mjpieters, Mani1, Night Gyr, Bender235, ESkog, Sunborn, Tompw, El C, J-Star, Lankiveil, Joanjoc~enwiki, Hayabusa future, RoyBoy, Orestes~enwiki, Grick, Bobo192, Stesmo, Smalljim, Indio~enwiki, Cohesion, Kjkolb, Nsaa, Storm Rider, Alansohn, Mr Adequate, AjAldous, Seans Potato Business, Ynhockey, Velella, Harej, RainbowOfLight, Dirac1933, Sciurinæ, Mikeo, DV8 2XL, Paraphelion, Zntrip, Ocollard, StradivariusTV, Duncan.france, Miss Madeline, CharlesC, Wdanwatts, Jacj, Qwertyus, Jclemens, Scuzzman, Martinevos~enwiki, Rjwilmsi, Jmcc150, Nneonneo, Bubba73, Watcharakorn, Lionelbrits, Ground Zero, Old Moonraker, RexNL, Kolbasz, Dalef, Fresheneesz, Guliolopez, Gwernol, Roboto de Ajvol, Wavelength, Phmer, Kymacpherson, RussBot, Jengelh, Shawn81, Kerowren, David Woodward, Gaius Cornelius, CambridgeBayWeather, Rsrikanth05, Bovineone, Tungsten, Grafen, Jaxl, Welsh, ONEder Boy, Ino5hiro, DJ John, Lomn, Scottfisher, DeadEyeArrow, Jeremy Visser, Ignitus, Wknight94, FF2010, Light current, Sefarkas, Closedmouth, Јованвб, Reyk, CharlesHBennett, CWenger, Fourohfour, Caco de vidro, Moomoomoo, Sbyrnes321, DVD R W, CIreland, Xtraeme, Eog1916, Itub, MacsBug, SmackBot, FocalPoint, Jclerman, Lcarsdata, Incnis Mrsi, KnowledgeOfSelf, Joonhon, Hydrogen Iodide, NoahWolfe, Jmulvey, Blue520, CMD Beaker, Jrockley, Yamaguchi□□, Gilliam, Carl.bunderson, TRosenbaum, Ati3414, Chris the speller, Bluebot, Kurykh, Agateller, Cadmium, MK8, Metacomet, Uthbrian, Reko, Sbharris, Rogermw, NYKevin, Can't sleep, clown will eat me, Ajaxkroon, Shalom Yechiel, Abyssal, V1adis1av, Ioscius, KaiserbBot, Rrburke, VMS Mosaic, Rsm99833, Addshore, Mrdempsey, Megamix, Flyguy649, Smooth O, Xyzzy n, Dreadstar, -Ozone-, Lcarscad, Cockneyite, Drphilharmonic, DMacks, Where, Bidabadi~enwiki, Cyberevil, Lambiam, SuperTycoon, Sanya, JoshuaZ, Accurizer, Minna Sora no Shita, IronGargoyle, 16@r, Ryulong, Peyre, Squirepants101, Dan Gluck, BranStark, Pegasus1138, CP\M, Freelance Intellectual, Fdp, Tawkerbot2, Chetvorno, Bstepp99, Conrad.Irwin, INkubusse, Xcentaur, RSido, Vyznev Xnebara, Nunquam Dormio, Solargenerator9.5, MarsRover, Leujohn, Smoove Z, Myasuda, J. Tyler, Island Dave, Quinnculver, Kanags, Gogo Dodo, HPaul, Mad-rick, Rracecarr, Skittleys, Christian75, FastLizard4, Gmoney650, The real avenger, Mikewax, Thijs!bot, Epbr123, Plmoknijb, Dougsim, Headbomb, Marek69, Deschreiber, Davidhorman, Meteoritekid, FourBlades, Stannered, Mentifisto, AntiVandalBot, Quintote, Jj137, Panu Petteri Höglund, Hanzoro5, Myanw, JAnDbot, Arch dude, Andonic, Xact, Snowynight, Acroterion, Geniac, Freedomlinux, Bongwarrior, VoABot II, AuburnPilot, Hillgentleman, JNW, Estonofunciona~enwiki, DMcanada, Klausok, Pixel ;-), Colinsweet, SparrowsWing, Indon, Animum, Dirac66, 28421u2232nfenfcenc, LorenzoB, Tswsl1989, JoergenB, Squidonius, Lewismatson, Chuckwatson, NatureA16, MartinBot, Mermaid from the Baltic Sea, Bus stop, R'n'B, Leyo, J.delanoy, Trusilver, Bogey97, Maurice Carbonaro, Cpiral, Gzkn, Stan J Klimas, DarkFalls, Dynetrekk~enwiki, Tarotcards, Pyrospirit, Sara0202, Chikinsawsage, Fountains of Bryn Mawr, Ohms law, Treisijs, Jim Swenson, Useight, Xiahou, RJASE1, Idioma-bot, ACSE, Cuzkatzimhut, Malik Shabazz, Deor, Matt1191, VolkovBot, ABF, VasilievVV, Philip Trueman, TXiKiBoT, Oshwah, Xenophrenic, Technopat, Hqb, Jcherbak, Someguy1221, Kirkpthompson, LeaveSleaves, Bearian, 0x539, Spiral5800, MichaelMorrill, Enigmaman, Yk Yk Yk, Bryan26, Synthebot, Falcon8765, Jluo, Sylent, Xxxlilbritxxx, Insanity Incarnate, Kehrbykid, Alytkin, Borne nocker, Brettdog, Deconstructhis, Starkrm, D. Recorder, Drawde22, SieBot, Tiddly Tom, Scarian, Viskonsas, Caltas, Soler97, Keilana, Nic92, TJHarrison, Oxymoron83, Faradayplank, Lightmouse, RW Marloe, Arnobarnard, Rj39pooch2, Nergaal, Babakathy, Martarius, ClueBot, HujiBot, Avenged Eightfold, GorillaWarfare, Fasettle, Bobathon71, Pvineet131, The Thing That Should Not Be, Plastikspork, VsBot, Wysprgr2005, Denna Haldane, Skäpperöd, CounterVandalismBot, Akash1209, Dougdp, MindstormsKid, Jersey emt, Opaltehjerkzors, Robert Skyhawk, Jusdafax, Erebus Morgaine, Huzzy92, 06multan, Arjayay, Radiogenic, PhySusie, Iohannes Animosus, Francisco Albani, IXella007, Dekisugi, La Pianista, Thingg, Aitias, Jonverve, Plasmic Physics, Megachad, Party, OpusAtrum, Johnson-gray, MystBot, Angerfist~enwiki, Thatguyflint, Hobbema, CalumH93, Amezcackle, Addbot, Proofreader77, Chorro22, Magus732, Smb6009, Laurinavicius, CanadianLinuxUser, Leszek Jańczuk, WFPM, Cst17, LaaknorBot, PranksterTurtle, Exor674, Lordlosss2, Tide rolls, Jarble, Legobot, Luckas-bot, Yobot, TaBOT-zerem, Legobot II, Theropod, Amble, Ayrton Prost, Hurricaneguy, AnomieBOT, DemocraticLuntz, Killiondude, Jim1138, Piano non troppo, AdjustShift, Scuzzer, Law, Materialscientist, The High Fin Sperm Whale, Citation bot, E2eamon, Bob Burkhardt, LilHelpa, Xqbot, Transity, Capricorn42, Richarddgill, Webkinzgirl101, Omnipaedista, RibotBOT, Amaury, Doulos Christos, Eugene-elgato, Pumpmaster60, FrescoBot, Surv1v4l1st, Wusel007, LucienBOT, Wvilhellm, Tobby72, Pepper, Oldlaptop321, MagnaGraecia, Footyfanatic3000, HJ Mitchell, Cannolis, Citation bot 1, Arthree, Pinethicket, Edderso, 10metreh, Odyssey xg, A8UDI, Minivip, Meaghan, Double sharp, TobeBot, Trappist the monk, Lotje, Ndkartik, TheBFG, Mozi17, Comet Tuttle, Math.geek3.1415926, Dinamik-bot, Vrenator, Tobias1984, Bluefist, Specs112, SilverbladeGR, Cfsgfds, Fastilysock, Cutelyaware, Sampathsris, Minimac, TjBot, TomBeasley, KuanRyan, Androstachys, Alison22, DASHBot, TGCP, BotdeSki, John of Reading, WikitanvirBot, Lunaibis, RedHab, ScottyBerg, Yt95, RenamedUser01302013, Kulmeetster, Wikipelli, K6ka, Sydneyanders, JSquish, ZéroBot, John Cline, PBS-AWB, Mkevinjnr, Suslindisambiguator, Elio96, Gz33, QEDK, Aschwole, L Kensington, MonoAV, Maschen, Donner60, Scientific29, ChuispastonBot, RockMagnetist, Ryan Pianesi, Newtrend19, Petrh, ClueBot NG, Crazyman121, Littleal38, Verpies, Satellizer, Baseball Watcher, Slartibartfastibast, Widr, Dasetwundabal, Oddbodz, Helpful Pixie Bot, Ciro612, Strike Eagle, Calabe1992, Bibcode Bot, Jeraphine Gryphon, BG19bot, Teiu88, Northamerica1000, Wiki13, ElphiBot, Cynaide, Shampa1, Flying hippo705, Glevum, DynamicDino, Adebish, Zedshort, Hamish59, Mgoelzer, SfHuIcTk, Thegreatgrabber, Achowat, Imawesome12345678910, ArrakisFrance, 555snowy, Kisokj, Ezekiel25q, Wolf11235, Cyprien 1997, BrightStarSky, Apples122, Ultimatewikimaster12345, Reatlas, Joeinwiki, Cavisson, Tentinator, Awesome boss 69 69, Bond064, Jyotmankad, CloudStrifeNBHM, Jwratner1, Applezpi3, Genome0514, StevenD99, Bkilli1, Ilikethemchickenwing$, Andthewinneris...Cole, Zane7777, Shbew, Monkbot, UDDM, Vieque, Thenapster1426, TheFireRises, Micbattle064, Paul2lyfe, Amortias, Pacifist peeta, Radioac-

Fabartus, SpuriousQ, IanManka, Stephenb, Rintrah, Alvinrune, Schoen, Rsrikanth05, Bovineone, Wimt, Stassats, Anomalocaris, Engineer-Scotty, NawlinWiki, Wiki alf, E123, Test-tools~enwiki, Jaxl, Terfili, Yahya Abdal-Aziz, Mkouklis, Nick, Ragesoss, Dhollm, Cholmes75, Dmoss, Matticus78, RUL3R, AdiJapan, Ryanminier, Juanpdp, Hv, Misza13, Beanyk, Aaron Schulz, Bota47, CorbieVreccan, Derek.cashman, DRosenbach, Elkman, Phaedrus86, Smaines, Wknight94, Tetracube, FF2010, Ageekgal, Closedmouth, Jwissick, Ketsuekigata, Sean Whitton, Petri Krohn, DGaw, CWenger, Smurrayinchester, Kungfuadam, Junglecat, RG2, NeilN, DVD R W, Itub, Thecroman, SmackBot, Android 93, Bobet, Reedy, InverseHypercube, KnowledgeOfSelf, TestPilot, Melchoir, Unyoyega, KocjoBot~enwiki, Davewild, Thunderboltz, Milesnfowler, Anastrophe, Delldot, J0lt C0la, Knowhow, Elk Salmon, Edgar181, HalfShadow, Eupedia, Srnec, Gilliam, Ohnoitsjamie, Skizzik, Carbon-16, JRSP, Chris the speller, Keegan, Iskander32, RDBrown, Thumperward, Fuzzform, Lollerskates, EncMstr, MalafayaBot, OrangeDog, Roscelese, Bonaparte, Xoyorkie13, Metacomet, Dustimagic, DHN-bot~enwiki, DNAmaster, Darth Panda, Suicidalhamster, Can't sleep, clown will eat me, Onorem, Clorox, Konczewski, Squadoosh, Andy120290, Ddon, DR04, UU, Grover cleveland, Jachapo, PiMaster3, TotalSpaceshipGuy3, Savidan, Dreadstar, Pwjb, Aco47, Peterwhy, Jklin, DMacks, BrotherFlounder, Suidafrikaan, Sadi Carnot, The undertow, SashatoBot, Mchavez, Nishkid64, Tarantola, LtPowers, Archimerged, Khazar, Vitall, Scientizzle, Gobonobo, Btg2290, Anoop.m, Olin, ManiF, JohnWittle, Moop stick, Jaywubba1887, Ckatz, Dale101usa, Chrisch, Garudabd, Digger3000, Slakr, Rainwarrior, Beetstra, Mr Stephen, AxG, Arkrishna, Mets501, Ambuj.Saxena, Ryulong, RichardF, Jose77, DGtal, WOWGeek, Sifaka, Asyndeton, Ramuman, Dead3y3, Michaelbusch, Walton One, Tabfugnic, David Little, J Di, CapitalR, DavidOaks, Supertigerman, Pearson3372, Az1568, Courcelles, Túrelio, Ziusudra, Dpeters11, Tawkerbot2, Bobby131313, VinceB, Cryptic C62, Kaischwartz, Lincmad, TranClan, JForget, Betaeleven, Deon, Eli84, Van helsing, NullAshton, CBM, Rawling, DSachan, GHe, Fork me, Egmonster, Black and White, FlyingToaster, Wikiman7~enwiki, WeggeBot, Logical2u, Pi Guy 31415, Johnlogic, MrFish, Bill Sayre, Dmsc893, Rudjek, Nebular110, Reywas92, Grahamec, MC10, Rasmus vendelboe, Vanished user vjhsduheuiui4t5hjri, Rifleman 82, Corpx, GeorgeTopouria, Islander, Mycroft.Holmes, Methyl~enwiki, Fifo, Christian75, DumbBOT, Chrislk02, Shrikethestalker, Taylor4452, Ndufour, Memorymike, JodyB, Rowlaj01, Calvero JP, Satori Son, Casliber, Thijs!bot, Full On, Barticus88, ShayneRyan, Opabinia regalis, Kiwi137, Corsair18, Dagrimdialer619, Headbomb, Sobreira, Marek69, Ydoommas, John254, Racantrell, Dmitri Lytov, Philippe, Nezzington, Nemti, Escarbot, Mentifisto, Lani123, Tom dl, AntiVandalBot, Luna Santin, Michael phan, Bigtimepeace, Random user 8384993, Chill doubt, LegitimateAndEvenCompelling, Myanw, Figma, Tomertomer, JAnDbot, Barek, MER-C, Zerotjon, Sanchom, Hut 8.5, Kirrages, Kerotan, Maurakt, Magioladitis, Canjth, Bongwarrior, VoABot II, JNW, Kinston eagle, Redaktor, Aa35te, SparrowsWing, Avicennasis, Superworms, Ahecht, Nposs, Wikiak, Dirac66, Adrian J. Hunter, Allstarecho, ChrisSmol, StuFifeScotland, User A1, Musicloudball, Cpl Syx, Vssun, Just James, DerHexer, Khalid Mahmood, TheRanger, DancingPenguin, FisherQueen, Hdt83, MartinBot, HLewis, PostScript, Rettetast, Roastytoast, 1993 lol, Glrx, Kateshortforbob, CommonsDelinker, Supia, PrestonH, Thomasrive, Exodecai101, Slash, J.delanoy, Pharaoh of the Wizards, Ilovestars89, ChickenMarengo, Hans Dunkelberg, Psycho Kirby, Smartweb, I2yu, Sergeibernstein, Tempnegro, Extransit, WarthogDemon, Munkimunki, Richard777, Thom.fynn, Tdadamemd, Yvonr, Adamsbriand, EH74DK, Rescorbic, McSly, Aonrotar, Ryan Postlethwaite, Ephebi, Notapotato, Gurchzilla, Wasitgood69, Wasitgood, Paulbkirk, Monkeybutt5423, AntiSpamBot, Gffootball58, Coin945, Bigsnake 19, NewEnglandYankee, Creator58, Joka1991, SJP, C0RNF1AK35, Shoessss, Bob, Vanished user 39948282, BrianScanlan, Nat682, Darklama, Hyuuganeji0123, Arjun Rana, AnjuX, Suuperturtle, Idioma-bot, Johnnieblue, Deor, 28bytes, VolkovBot, Thedjatclubrock, Iosef, Jmocenigo, Christophenstein, Jeff G., JohnBlackburne, TheOtherJesse, RemoteCar, Barneca, Philip Trueman, Af648, Drunkenmonkey, Sweetness46, TXiKiBoT, TheVault, A4bot, Quilbert, Caster23, GDonato, Miranda, Chrisk12, Sankalpdravid, Qxz, Littlealien182, Anna Lincoln, Corvus cornix, Martin451, Jackfork, LeaveSleaves, Andrewrost3241981, DBragagnolo, Luuva, Quindraco, Gona.eu, RadiantRay, Madhero88, Jinglesmells999, Finngall, Aciddoll, Superjustinbros., Deeryh01, Synthebot, Enviroboy, Chengyq19942007, Insanity Incarnate, Everybody's Got One, Why Not A Duck, Brianga, Jaybo007, HiDrNick, LuigiManiac, Petergans, ConnTorrodon, NHRHS2010, SieBot, Coffee, Mikemoral, TJRC, Rihanij, PlanetStar, Jmwwiki, Borgdylan, Gprince007, Tiddly Tom, Scarian, WereSpielChequers, Jauerback, Jack Merridew, Gerakibot, Dawn Bard, Viskonsas, Caltas, Kragenz, The way, the truth, and the light, Michfg, Sat84, Whiteghost.ink, Til Eulenspiegel, Purbo T, Tiptoety, Exert, Elcobbola, Nopetro, Hamilton hogs, Hiddenfromview, Segalsegal, SquirrelMonkeySpiderFace, Oxymoron83, Nuttycoconut, Lightmouse, Hwn tls, Hak-kâ-ngìn, BenoniBot~enwiki, Jack the Stripper, Sunrise, Dillard421, Werldwayd, Pappapasd, Maelgwnbot, DixonD, Nergaal, Precious Roy, Escape Orbit, Into The Fray, Jimmy Slade, Kanonkas, Georgedriver, Mr. Granger, Twinsday, ClueBot, PipepBot, Dinamik, The Thing That Should Not Be, Apastrophe, Techdawg667, Gawaxay, Syhon, Thompsontm, Drmies, Polyamorph, Elsweyn, Ryoutou, CounterVandalismBot, Ansh666, Blanchardb, LizardJr8, Dylan620, Timex987, Wifiless, Puchiko, Natasha.fielding, Dlorang, Robert Skyhawk, Excirial, Alexbot, BirgerH, Omgosh2, Jazjaz92, NuclearWarfare, Pearrari, Kaeso Dio, Jotterbot, LarryMorseDCOhio, Psinu, Realm up, DeltaQuad, Kaiba, Dekisugi, Pwntskater, NolanRichard, Thehelpfulone, La Pianista, Another Believer, Kiran the great, Viper275, Aitias, Blargblarg89, Versus22, Sch0013r, SoxBot III, MairAW, JDT1991, BalkanFever, Vanished user uih38riiw4hjlsd, Indopug, TimothyRias, Jean-claude perez, Neuralwarp, XLinkBot, Shpakovich, Gonzonoir, Nsimya, TheSickBehemoth, Ryuken14, Nsim, WikHead, Noctibus, JinJian, ZooFari, MystBot, RyanCross, Thatguyflint, Roentgenium111, Tomilee0001, Yoenit, OmgItsTheSmartGuy, NicholasSThompson, JenR32, Vchorozopoulos, WFPM, Cst17, Download, SoSaysChappy, EconoPhysicist, Mathmarker, Glane23, Plutonium55, Debresser, LinkFA-Bot, Drova, LiveAgain, PoliteCarbide, Numbo3-bot, Tide rolls, BrianKnez, Luckas Blade, Greyhood, Arbitrarily0, Angrysockhop, Jack who built the house, Luckas-bot, Yobot, Essam Sharaf, Azylber, KamikazeBot, Jobroluver98, EnDaLeCoMpLeX, MacTire02, Tempodivalse, AnomieBOT, Zhieaanm, Helixer, Rubinbot, Jake Fuersturm, Degg444, Daniele Pugliesi, Piano non troppo, Collieuk, AdjustShift, Kingpin13, Ulric1313, Materialscientist, RadioBroadcast, The High Fin Sperm Whale, Citation bot, ArthurBot, Carturo222, Xqbot, Ziaix, Timir2, Sionus, Gopal81, Vidshow, Tfts, YBG, Grim23, Srich32977, WingedSkiCap, Pirateer, GrouchoBot, RibotBOT, Ian Fraser at Temple Newsam House, Spesh531, FaTony, AlimanRuna, A. di M., Nolimits5017, Dougofborg, Thehelpfulbot, R8R Gtrs, FrescoBot, Ylime715, Dogposter, StaticVision, Michael93555, Car132, Zach112233, Weetoddid, Strongbadmanofme, Maxus96, Grandiose, Xhaoz, Dellacomp, Citation bot 1, Pezzells, Pshent, Redrose64, ArnaudContet, Pinethicket, I dream of horses, M.pois, Hamtechperson, BTolli, Gemmi3, RedBot, Fishekad, NarSakSasLee, Kangxi emperor6868, Hardwigg, Abc518, Tjlafave, Lightlowemon, Gamewizard71, FoxBot, Double sharp, Adult Swim Addict, TobeBot, Graniggo, Jaeger Lotno, Pitcroft, Sillyboy67, Hughbert512369, Dinamik-bot, BZRatfink, Fyandcena, Kelvin35, Mattvirajrenaudbrandon, Turn off 2, Imawsome 09, Mass09, Tbhotch, Pldx1, Naughtysriram, DARTH SIDIOUS 2, Luhar1997, Twonenator, Pickweed, Dancojocari, EmausBot, Sir Arthur Williams, WikitanvirBot, GA bot, Franjklogos, Tf1321, Tommy2010, P. S. F. Freitas, Kitrkatr, �𐐐𐐐�, Zainiadragon10000, JSquish, Jakers69, StringTheory11, Dffgd, Cobaltcigs, Шугын, Joshlepaknpsa, Hzb pangus, Elly4web, Arman Cagle, KotVa, Sven nestle2, Hh73wiki, Ego White Tray, Negovori, ChuispastonBot, GermanJoe, Sunshine4921, Rmashhadi, Chaotic iak, Whoop whoop pull up, Heidslovesearl, Xanchester, ClueBot NG, Ihakeycakeyabreak, Wd930, Ozkithar Salas, Manubot, CocuBot, Lanthanum-138, Hazhk, Moneya, Parcly Taxel, Metaknowledge, Willzuk, Helpful Pixie Bot, RobertGustafson, පසිඳු කාවින්ද, Curb Chain, Bibcode Bot, BG19bot, MKar, Vagobot, Nagasturg, Sandbh, Mark Arsten, IraChesterfield, Soerfm, Zedshort, FeralOink, Razzat99, SoylentPurple, BattyBot, Justincheng12345-bot, Judiakok1985, Ziggypowe, Ushau97, Maxronnersjo,

ChrisGualtieri, JYBot, Dexbot, LightandDark2000, Aditya Mahar, Mogism, Carpelogos, Jtrevor99, Leitoxx, Kazim5294, AmericanLemming, WikiEditor2563, Michel Djerzinski, The Herald, Shearflyer, Quenhitran, Asadwarraich, Kind Tennis Fan, Monkbot, HiYahhFriend, Deepak harshal nagle, Trackteur, Encyclopedia Lu, IiKkEe, Shane Stachwick, Mogie Bear, R. Portela F., Forscienceonly, KasparBot, Alistairgray42, Equinox, Lexi sioz and Anonymous: 1391

- **Exotic matter** *Source:* https://en.wikipedia.org/wiki/Exotic_matter?oldid=672879160 *Contributors:* Bryan Derksen, Alan Peakall, Aragorn2, David Latapie, Omegatron, Phil Boswell, Korath, Bkell, Intangir, Geeoharee, Lethe, Mboverload, ConradPino, Beland, WhiteDragon, Icairns, Erik Garrison, Tzarius, EricJamesStone, Pjacobi, Paul August, JYolkowski, Mac Davis, Woohookitty, LOL, Meneth, CharlesC, Christopher Thomas, Driftwoodzebulin, Teknic, Quale, Matjlav, FireCrack, Ejurgaite, Krackpipe, Slant, Nimur, YurikBot, Salsb, Astral, ErkDemon, Xiroth, Dbfirs, Geoffrey.landis, JDspeeder1, Itub, SmackBot, Hmains, Kmarinas86, MovGP0, Vladis1av, Nakon, Nathanael Bar-Aur L., Vampus, JorisvS, Hypnosifl, RMHED, Inquisitus, Michaelbusch, TwistOfCain, Happy-melon, Gëörgë Büsh, P0LARIS, Alexnye, Keraunos, Headbomb, Escarbot, Lfstevens, Ca7ch, Cpl Syx, R'n'B, Hans Dunkelberg, OneDoubleO, Aliento, Red Act, Lamro, Antixt, Universaladdress, AlleborgoBot, PseudoAdmin^*(3d), WereSpielChequers, SteveThePhysicist, PedantryIsMyMiddleName, Gigo300, ClueBot, Superwj5, The Thing That Should Not Be, John.D.Ward, SeaRisk, Dean Wormer, Alexbot, DumZiBoT, Thatguyflint, Kbdankbot, Addbot, CanadianLinuxUser, Tassedethe, Luckas-bot, NotARusski, AnomieBOT, 9258fahsflkh917fas, Unara, Aaagmnr, Citation bot, RibotBOT, Locobot, Samwb123, Pinethicket, Vuyane, Unbitwise, Slightsmile, Eternalemp, ClueBot NG, Bibcode Bot, Cadiomals, Darth Sitges, Capitolcity, DavidLeighEllis, Krowizard113, Usernameonwiki, Lxplot and Anonymous: 120

- **Dark matter** *Source:* https://en.wikipedia.org/wiki/Dark_matter?oldid=677811572 *Contributors:* AxelBoldt, Chenyu, Derek Ross, CYD, BF, Bryan Derksen, The Anome, Tarquin, Taw, XJaM, Arvindn, William Avery, Roadrunner, Mintguy, Bth, Stevertigo, Edward, Nealmcb, Boud, FrankH, Cprompt, DopefishJustin, Bobby D. Bryant, Ixfd64, SebastianHelm, Alfio, CesarB, Looxix~enwiki, Mkweise, William M. Connolley, JWSchmidt, Glenn, Mxn, Charles Matthews, Timwi, Fuzheado, Rednblu, Haukurth, DW40, Dragons flight, Furrykef, Saltine, Dogface, Populus, Jusjih, Finlay McWalter, Bearcat, Robbot, Zandperl, Korath, Nurg, Naddy, Arkuat, Gandalf61, Pingveno, Rursus, Rtfisher, Wereon, Diberri, Adam78, Aasim75, Marc Venot, Ancheta Wis, Giftlite, Graeme Bartlett, Laudaka, Barbara Shack, Herbee, Fropuff, Xerxes314, Dratman, Curps, Joconnor, Jdavidb, Unconcerned, Eequor, Bobblewik, Andycjp, Alexf, Geni, Antandrus, HorsePunchKid, Melikamp, PDH, Rdsmith4, Bosmon, Bbbl67, Icairns, Sam Hocevar, Cynical, Lumidek, Iantresman, Burschik, Joyous!, Adashiel, Urvabara, Discospinster, Rich Farmbrough, Oliver Lineham, Vsmith, Jpk, ArnoldReinhold, Murtasa, D-Notice, JPX7, KaiSeun, SpookyMulder, Bender235, Kjoonlee, Kaisershatner, Pk2000, PsychoDave, RJHall, Mr. Billion, El C, Bletch, PhilHibbs, Shanes, Frankenschulz, Art LaPella, RoyBoy, Themusicgod1, Bobo192, Smalljim, Shenme, Cmdrjameson, Reuben, Kmaguire, I9Q79oL78KiL0QTFHgyc, Zelda~enwiki, Mr. Brownstone, E is for Ian, Jumbuck, Storm Rider, Alansohn, Gary, Anthony Appleyard, Guy Harris, Eric Kvaalen, Arthena, Keenan Pepper, Kocio, Bart133, RPellessier, Benna, ClockworkSoul, Cal 1234, Count Iblis, Guthrie, H2g2bob, Bsadowski1, GabrielF, Pauli133, Leondz, DV8 2XL, Gene Nygaard, Feline1, Oleg Alexandrov, Brookie, Natalya, Flying fish, WilliamKF, Yeastbeast, Mindmatrix, RHaworth, Plek, BillC, JPFlip, Benbest, ^demon, WadeSimMiser, Gxojo, MONGO, Jwanders, Torqueing, 🔲🔲🔲🔲🔲, Joke137, Wisq, Christopher Thomas, Palica, Mandarax, RedBLACKandBURN, Aarghdvaark, SqueakBox, Ashmoo, Graham87, Malangthon, Mamling, Jclemens, Drbogdan, Loris Bennett, Rjwilmsi, Lars T., Strait, Patrick Gill, Tangotango, Tawker, Smithfarm, Stevenscollege, Mike Peel, HappyCamper, SeanMack, ScottJ, Krash, Dermeister, Rangek, Madcat87, FlaBot, Ian Pitchford, PlatypeanArchcow, A scientist, Margosbot~enwiki, Gark, Nivix, Gparker, Pathoschild, Gurch, Stevenfruitsmaak, Goudzovski, Tomer Ish Shalom, Smithbrenon, Chobot, Moocha, DVdm, Gwernol, The Rambling Man, YurikBot, Wavelength, RobotE, Koveras, Hairy Dude, Huw Powell, Phmer, Hillman, RussBot, Michael Slone, Ohwilleke, Bhny, JabberWok, GLaDOS, DanMS, Zelmerszoetrop, Eleassar, Merick, Big Brother 1984, NawlinWiki, Alpertron, Długosz, Schlafly, FFLaguna, BlackAndy, Dbmag9, SCZenz, Haoie, Raven4x4x, Ospalh, Durval, Bota47, Supspirit, Pegship, Noosfractal, Charlie Wiederhold, WAS 4.250, Smoggyrob, Reyk, Tvaughan, Joedixon, Eric TF Bat, Emc2, Ilmari Karonen, Allens, Bernd in Japan, InsayneWrapper, Bclayabt, SmackBot, Cubs Fan, Ashill, IddoGenuth, Tomer yaffe, Stellea, InverseHypercube, KnowledgeOfSelf, Clpo13, Nickst, RedSpruce, Nightbat, Doc Strange, Herbm, Edgar181, HalfShadow, Flux.books, Dheerajkakar, Yamaguchi🔲🔲, Richmeister, Gilliam, Folajimi, Oscarthecat, Skizzik, Kmarinas86, Chris the speller, SuperBuuBuu, Quinsareth, Persian Poet Gal, Sirex98, MalafayaBot, Silly rabbit, Sangrolu, Villarinho, DHN-bot~enwiki, Sbharris, Hongooi, Jdthood, CheerLeone, Gtkysor, Can't sleep, clown will eat me, Nick Levine, Tamfang, Kelvin Case, V1adis1av, Vanished User 0001, Rrburke, Jgoulden, Auvii, Krich, Wen D House, Radagast83, Engwar, Nakon, VegaDark, John D. Croft, Alexander110, KimO, Adrigon, SpiderJon, Ultraexactzz, Zadignose, Tesseran, Byelf2007, L337p4wn, K7lim, SashatoBot, Mchavez, Swatjester, Leftydan6, Minaker, John, Ashoat, Scientizzle, Acitrano, Linnell, JoshuaZ, James.S, JorisvS, Coredesat, Goodnightmush, ICBB, Plunge, JHunterJ, Hypnosifl, Silverthorn, Descubes, Freederick, Dr.K., Vanished user, Iridescent, Darkerprojects, Astrobayes, Newone, MOBle, Igoldste, CapitalR, AGK, Courcelles, Tawkerbot2, Dlohcierekim, Chetvorno, Hammer Raccoon, Owen214, Eastlaw, Peledre, Pukkie, Anakata, Runningonbrains, DKOH, NickW557, Gregbard, MikeWren, Vttoth, Necessary Evil, Ryan, Viciouspiggy, Gogo Dodo, Anonymi, Xxanthippe, A Softer Answer, Odie5533, Tawkerbot4, DumbBOT, Robertinventor, Kozuch, Mtpaley, Philza85, Starship Trooper, UberScienceNerd, Crum375, Thijs!bot, Epbr123, Astroceltica, Passaggio, Barbarina, Mbell, Eugenespeed, N5iln, Mojo Hand, Carlif, Headbomb, Tonyle, Marek69, Lars Lindberg Christensen, OtterSmith, SusanLesch, Mmortal03, Hmrox, Hires an editor, AntiVandalBot, Seaphoto, Orionus, Opelio, Shirt58, Rehnn83, Joehodge, AaronY, Jj137, TTN, Dylan Lake, Chill doubt, Spencer, Yellowdesk, Sniktaw, CPitt76, Gökhan, Jcarter1, Res2216firestar, JAnDbot, Leuko, Husond, MER-C, CosineKitty, Plantsurfer, Mcorazao, Therealintellectual, Folkform, Balbers, 100110100, Autotheist, Wasell, Magioladitis, Bongwarrior, VoABot II, Timothy McVeigh, Charlesrkiss, AuburnPilot, Krkaiser, Mbarbier, Kaivosukeltaja, Foroa, Swpb, Stigmj, T a y l o s, Ekantik, Brusegadi, Bubba hotep, Fabrictramp, Catgut, Lilian.Kaufmann, Zhanghia, Acornwithwings, Vssun, LtHija, Whisky5, DerHexer, Prisca6023, PeteSF, Rickard Vogelberg, NatureA16, DancingPenguin, MartinBot, Schmloof, STBot, Pagw, Fs644, Nikpapag, Anaxial, CommonsDelinker, Jean-Pierre Petit~enwiki, PrestonH, WelshMatt, Chrishy man, Tgeairn, J.delanoy, Pharaoh of the Wizards, Trusilver, Adavidb, Kudpung, Rod57, Arion 3x3, PedEye1, McSly, Tarotcards, Davy p, HiLo48, NewEnglandYankee, Ohms law, Jorfer, Blckavnger, Potatoswatter, KylieTastic, Joshua Issac, Infiniteglitch, Remember the dot, Pitpif, Vanished user 39948282, Neekap, Natl1, Ldebain, BernardZ, SoCalSuperEagle, Squids and Chips, CardinalDan, Idioma-bot, Sheliak, Funandtrvl, Lights, VolkovBot, Craigheinke, Itsfullofstars, ColdCase, Jeff G., JohnBlackburne, Mocirne, AlnoktaBOT, Scikid, Grammarmonger, Leojohns, Larry R. Holmgren, Philip Trueman, TXiKiBoT, Docanton, Authorized User, Theophilus reed, Drestros power, Strichek, MarekMahut, Monkey Bounce, Lradrama, Sintaku, Carillonatreides, Martin451, Broadbot, Wikiisawesome, Mazarin07, Inductiveload, Knightshield, Telecineguy, Spiral5800, Kurowoofwoof111, Greswik, RobertFritzius, SwordSmurf, Falcon8765, Hellothere17, Enviroboy, Littlehollah, Wanchung Hu, Illumini85, SonOfMog Worf, Jazzman123, PGWG, 19merlin69, NHRHS2010, Neparis, Bfpage, S-n-ushakov, SieBot, Calliopejen1, Tresiden, Wibubba48, Tachyonics, Pallab1234, Paradoctor, KGyST, Bentogoa, Jimlester51, Battlepace, Oda Mari, Aaarnooo, Suomichris, Crowstar, PromX1, Lightmouse, Tombomp, Cyberplasm,

Diego Grez-Cañete, Spartan-James, Thinghy, Mygerardromance, Hamiltondaniel, Superbeecat, Denisarona, JL-Bot, Escape Orbit, Starcluster, Troy 07, Atif.t2, ArepoEn, Ak47gforce, Ratemonth, Sfan00 IMG, ClueBot, Phoenix-wiki, GorillaWarfare, The Thing That Should Not Be, ArdClose, Rodhullandemu, Cptmurdok, Drmies, Uncle Milty, Iuhkjhk87y678, MrBosnia, Bhaskarns, Andwor, Ktr101, Excirial, Dombom12, Cromescythe, Barbarinaz, FOARP, Brews ohare, Jotterbot, Iohannes Animosus, R.Andrae, Kentgen1, Ordovico, Mastertek, Rgoogin, Thehelpfulone, 1ForTheMoney, Versus22, Palmer666palmer, PCHS-NJROTC, Burner0718, Pillar of Babel, SoxBot III, Erodium, Vanished user uih38riiw4hjlsd, 1ofhissheep, TimothyRias, Arianewiki1, XLinkBot, DCCougar, Oldnoah, Rror, Gwark, Ost316, Avoided, Webmaster369, Gthomson, Tugrul irmak, Noctibus, Ploversegg, ZooFari, Parejkoj, Tayste, Addbot, Xp54321, Grayfell, Experimental Hobo Infiltration Droid, Willking1979, Some jerk on the Internet, Uruk2008, 04everington, DOI bot, Tcncv, Nohomers48, CharlesChandler, Gmeyerowitz, Haasfelix, Download, Proxima Centauri, Ashirgo, RTG, Redheylin, Glane23, Darkmatter654, SamatBot, Nanzilla, Lzkelley, Clone 209, Tassedethe, Numbo3-bot, Peridon, Chinchinthehun, Evildeathmath, Tide rolls, Lightbot, OlEnglish, Qemist, Gail, North Polaris, Legobot, Artichoke-Boy, Luckas-bot, Yobot, WikiDan61, Cosoce, Dov Henis, Aldebaran66, KillYourLove, CzechFalcon, Amble, Mmxx, CinchBug, Perusnarpk, IW.HG, Einstein vs Dark energys, Eric-Wester, Tempodivalse, Synchronism, AnomieBOT, Letuño, Girl Scout cookie, IRP, JackieBot, RBM 72, AdjustShift, Nicolaas Vroom, Henrykandrup, Iluziat, Materialscientist, Dendlai, ImperatorExercitus, The High Fin Sperm Whale, Citation bot, Ternity0127, Maxis ftw, Frankenpuppy, Quebec99, LilHelpa, Aksel89, Xqbot, Stlwebs, Random astronomer, Sionus, Cureden, Jradis1337, Capricorn42, Wperdue, Deleance, Raspw, Tomwsulcer, Magicxcian, AbigailAbernathy, Srich32977, NOrbeck, Artemis6234, Almabot, Abell 1367, Feldhaus, False vacuum, RibotBOT, Waleswatcher, Mikedr, Kongkokhaw, Rvnieuwe, Shadowjams, MeDrewNotYou, A. di M., Peter470, Sageman7, 77, Luminique, Captain-n00dle, Imyfujita, FrescoBot, Andyradke0, Ag allstar, Paine Ellsworth, Originalwana, Styxpaint, Mark Renier, VS6507, PhysicsExplorer, Dbirkhofer, Steve Quinn, Nestlefolife, Adrian Akau, 1414rwbt, SF88, Citation bot 1, Redrose64, DUUJEEGWEEM, Tyler6298, Pinethicket, I dream of horses, Grammarspellchecker, Danlof, 10metreh, Jonesey95, Tom.Reding, Pmokeefe, A8UDI, For.a.limited.time.only, Elentirno, TedderBot, Aknochel, SkyMachine, IVAN3MAN, Kgrad, Nieuwenh, Trappist the monk, Puzl bustr, Fama Clamosa, Domeinthebumhole, Michael9422, UrukHaiLoR, Allen4names, JLincoln, Jeffrd10, Lovemybluetooth, Diannaa, Fastilysock, Innotata, DrCrisp, Whisky drinker, Onel5969, RjwilmsiBot, 5mgoblue5, Blakelewis122, Þorri, Mathewsyriac, Leandro.lelas, Mserard313, Mdznr, Sbugnon, Ultima821, EmausBot, Francophile124, Grrow, Super48paul, GoingBatty, Gimmetoo, Solarra, Jmencisom, Slightsmile, Tommy2010, Winner 42, SusanaMultidark, Gocows2, Wikipelli, Serketan, Krifferjel, Zurich Astro, Hhhippo, Mhatthei, Svolin, Micahqgecko, JSquish, Josve05a, Trojanmice, MithrandirAgain, Edwinkaren, Devilaza, Arbnos, Oraclan, Suslindisambiguator, AlbertusmagnusOP, Tolly4bolly, L1A1 FAL, Ancient Anomaly, L Kensington, Maj den, Corabilek, Donner60, Aldnonymous, Ihardlythinkso, RockMagnetist, Terra Novus, TYelliot, DASHBotAV, Kroupap, D Phoesheezey, Travies10, Jxraynor, TheTimesAreAChanging, ClueBot NG, Rich Smith, Afjvanraan, Crystal7878, Catinthehat93, Bped1985, Infinifold, Wiggit002, Jj1236, PapaMike, MonEyshOt42069210, Muon, Esdacosta, Asukite, Masssly, Ph.d Carl edenburgh, Widr, Gavin.perch, Helpful Pixie Bot, Curb Chain, Calabe1992, Bibcode Bot, BG19bot, Dualus, Kishanparekh, Stevenwilkins, NacowY, Cheeseray1, Cyberguy5, Darkmatter adam, Yomomma8102, Hza a 9, Rarelight, Cyberpower678, Cosmologist77, தென்காசி சுப்பிரமணியன், Dahliamtl, Dodshe, Mark Arsten, Darkmatterotheruniverses, Cadiomals, Zedshort, Achowat, Rolandwilliamson, A2Die, Clint55555, Mgka79, NotEither, BattyBot, Ronin712, Babymushrooms, Davidmexican, Drphilmarshall, Dilaton, Quin71901, U-95, ChrisGualtieri, Npmay, Kvark92, Lukasz.astrus, Ducknish, JYBot, Davidlwinkler, Astrohap, Hunterf12, Caroline1981, Gravityking100, Junavia, Fredrikdn, Jcardazzi, Lugia2453, Wjs64, Andwor42, Frosty, Honneydewp243, Junjunone, DrHowzer73, JustAMuggle, WadiElNatrun, Reatlas, Rfassbind, Acetotyce, I am One of Many, DirkXcal, Melonkelon, Ybidzian, Gig9876, M.ashrafinia, Trololololman12, Ilikedeletingstufffromhere, DavidLeighEllis, Onecreation, Zenibus, Jernahthern, Hipposaregrey, Frinthruit, Stamptrader, Cyberalchemyst, Aaronknowsitall, FelixRosch, Darkmer, Doubleknockout, Monkbot, Wardinstrument, Leegrc, Vikas Rauniyar, Apipia, Upsalla, Jkvaternik, Lol kaptyn troll, HMSLavender, Mohammedshukoor, Callum92, Stefania.deluca, Ashweigh, Oldstone James, Astezar, 39Debangshu, YoYoDude012, Anunaki truth, Pyrotle, Tetra quark, Carazmatic, God of matterrr, Silversparkcontributions, Isambard Kingdom, Rizi0909, Anand2202, Kbap2002, Kb2002, DN-boards1, Yohoona, KasparBot, I love trains sooo much, Id6040, Boowiebear, Stephane Le Corre, Mustachman71 and Anonymous: 1217

32.11.2   Images

- **File:080998_Universe_Content_240.jpg** *Source:* https://upload.wikimedia.org/wikipedia/commons/a/a5/080998_Universe_Content_240. jpg *License:* Public domain *Contributors:* http://map.gsfc.nasa.gov/media/080998/index.html *Original artist:* Credit: NASA / WMAP Science Team

- **File:080998_Universe_Content_240_after_Planck.jpg** *Source:* https://upload.wikimedia.org/wikipedia/commons/b/b6/080998_Universe_ Content_240_after_Planck.jpg *License:* Public domain *Contributors:* http://map.gsfc.nasa.gov/media/080998/index.html updated data from http://www.nasa.gov/mission_pages/planck/news/planck20130321.html *Original artist:* NASA, Modified by User:77

- **File:1e0657_scale.jpg** *Source:* https://upload.wikimedia.org/wikipedia/commons/a/a8/1e0657_scale.jpg *License:* Public domain *Contributors:* Chandra X-Ray Observatory: 1E 0657-56 *Original artist:* NASA/CXC/M. Weiss

- **File:AOs-1s-2pz.png** *Source:* https://upload.wikimedia.org/wikipedia/commons/8/86/AOs-1s-2pz.png *License:* Public domain *Contributors:* ? *Original artist:* ?

- **File:Acetylene-2D.png** *Source:* https://upload.wikimedia.org/wikipedia/commons/9/97/Acetylene-2D.png *License:* Public domain *Contributors:* ? *Original artist:* ?

- **File:AirShower.svg** *Source:* https://upload.wikimedia.org/wikipedia/commons/2/2c/AirShower.svg *License:* CC BY 3.0 *Contributors:* originally from nl.wikipedia; description page is/was here. *Original artist:* Mpfiz

- **File:Alfa_beta_gamma_radiation.svg** *Source:* https://upload.wikimedia.org/wikipedia/commons/d/d6/Alfa_beta_gamma_radiation.svg *License:* CC BY 2.5 *Contributors:* Traced from this PNG image. *Original artist:* User:Stannered

- **File:Alpha_Decay.svg** *Source:* https://upload.wikimedia.org/wikipedia/commons/7/79/Alpha_Decay.svg *License:* Public domain *Contributors:* This vector image was created with Inkscape. *Original artist:* Inductiveload

- **File:Ambox_important.svg** *Source:* https://upload.wikimedia.org/wikipedia/commons/b/b4/Ambox_important.svg *License:* Public domain *Contributors:* Own work, based off of Image:Ambox scales.svg *Original artist:* Dsmurat (talk · contribs)

- **File:DIMendeleevCab.jpg** *Source:* https://upload.wikimedia.org/wikipedia/commons/c/c8/DIMendeleevCab.jpg *License:* Public domain *Contributors:* Transferred from ru.wikipedia
  Original artist: —. Original uploader was Serge Lachinov at ru.wikipedia

- **File:DNA_chemical_structure.svg** *Source:* https://upload.wikimedia.org/wikipedia/commons/e/e4/DNA_chemical_structure.svg *License:* CC-BY-SA-3.0 *Contributors:* <a href='//validator.w3.org/' data-x-rel='nofollow'><img alt='W3C' src='https://upload.wikimedia.org/wikipedia/commons/thumb/1/1a/Valid_SVG_1.1_%28green%29.svg/88px-Valid_SVG_1.1_%28green%29.svg.png' width='88' height='30' style='vertical-align: top' srcset='https://upload.wikimedia.org/wikipedia/commons/thumb/1/1a/Valid_SVG_1.1_%28green%29.svg/132px-Valid_SVG_1.1_%28green%29.svg.png 1.5x, https://upload.wikimedia.org/wikipedia/commons/thumb/1/1a/Valid_SVG_1.1_%28green%29.svg/176px-Valid_SVG_1.1_%28green%29.svg.png 2x' data-file-width='91' data-file-height='31' /></a>iThe source code of thisSVGis <a data-x-rel='nofollow'class='external text' href='//validator.w3.org/check?uri=https%3A%2F%2Fcommons.wikimedia.org%2Fwiki%2FSpecial%3AFilepath%2FDNA_chemical_structure.svg,<span>,&,</span>,ss=1#source'>valid</a>.*Original artist:*Madprime(talk·contribs)

- **File:D_orbitals.svg** *Source:* https://upload.wikimedia.org/wikipedia/commons/e/e1/D_orbitals.svg *License:* CC-BY-SA-3.0 *Contributors:* This vector image was created with Inkscape. *Original artist:* This file was made by **User:Sven**

- **File:Daltons_symbols.gif** *Source:* https://upload.wikimedia.org/wikipedia/commons/3/39/Daltons_symbols.gif *License:* Public domain *Contributors:* ? *Original artist:* ?

- **File:DecayRate_vs_Solar_Time.png** *Source:* https://upload.wikimedia.org/wikipedia/commons/d/d3/DecayRate_vs_Solar_Time.png *License:* Public domain *Contributors:* ? *Original artist:* ?

- **File:Dichlorine_heptoxide.svg** *Source:* https://upload.wikimedia.org/wikipedia/commons/7/7a/Dichlorine_heptoxide.svg *License:* Public domain *Contributors:* Own work *Original artist:* Yikrazuul

- **File:Discovery_of_chemical_elements.svg** *Source:* https://upload.wikimedia.org/wikipedia/commons/3/3d/Discovery_of_chemical_elements.svg *License:* CC BY-SA 3.0 *Contributors:* Wikimedia Commons. *Original artist:* Sandbh

- **File:Discovery_of_neon_isotopes.JPG** *Source:* https://upload.wikimedia.org/wikipedia/commons/e/e6/Discovery_of_neon_isotopes.JPG *License:* Public domain *Contributors:*
  Original artist: ?

- **File:Drop_closeup.jpg** *Source:* https://upload.wikimedia.org/wikipedia/commons/7/74/Drop_closeup.jpg *License:* CC BY 2.0 *Contributors:* File:Flickr_-_NewsPhoto!_-_drup.jpg - original: http://www.flickr.com/photos/13088710@N02/3581239516/ *Original artist:* Jos van Zetten

- **File:Electric_field_point_lines_equipotentials.svg** *Source:* https://upload.wikimedia.org/wikipedia/commons/9/96/Electric_field_point_lines_equipotentials.svg *License:* Public domain *Contributors:* Own work *Original artist:* Sjlegg

- **File:Electron_Sea_(Plasma).jpg** *Source:* https://upload.wikimedia.org/wikipedia/commons/8/8f/Electron_Sea_%28Plasma%29.jpg *License:* CC BY-SA 3.0 *Contributors:* Own work *Original artist:* Spirit469

- **File:Electron_affinity_of_the_elements.svg** *Source:* https://upload.wikimedia.org/wikipedia/commons/6/6c/Electron_affinity_of_the_elem.svg *License:* CC BY-SA 3.0 *Contributors:* Based on Electron affinities of the elements 2.png by Sandbh. *Original artist:* DePiep

- **File:Electron_avalanche.gif** *Source:* https://upload.wikimedia.org/wikipedia/commons/a/ac/Electron_avalanche.gif *License:* CC BY-SA 3.0 *Contributors:* Own work *Original artist:* Dougsim

- **File:Electron_dot.svg** *Source:* https://upload.wikimedia.org/wikipedia/commons/0/02/Electron_dot.svg *License:* CC BY-SA 2.5 *Contributors:* ? *Original artist:* ?

- **File:Electron_orbitals.svg** *Source:* https://upload.wikimedia.org/wikipedia/commons/1/11/Electron_orbitals.svg *License:* Public domain *Contributors:* own work by Patricia.fidi and Lt Paul - Originally from pl:Grafika:Orbitale.png, author pl:Wikipedysta:Chemmix. *Original artist:* Patricia.fidi

- **File:Electron_shell_003_Lithium_-_no_label.svg** *Source:* https://upload.wikimedia.org/wikipedia/commons/a/ae/Electron_shell_003_Lithi-_no_label.svg *License:* CC BY-SA 2.0 uk *Contributors:* File:Electron shell 003 Lithium.svg *Original artist:* Pumbaa(original work byGregRobson)

- **File:Elementspiral_(polyatomic).svg** *Source:* https://upload.wikimedia.org/wikipedia/commons/c/ce/Elementspiral_%28polyatomic%29.svg *License:* CC BY-SA 3.0 *Contributors:* Own work *Original artist:* DePiep

- **File:Empirical_atomic_radius_trends.png** *Source:* https://upload.wikimedia.org/wikipedia/commons/b/bc/Empirical_atomic_radius_trends.png *License:* GFDL *Contributors:* Own work *Original artist:* StringTheory11

- **File:Ethanol-3D-vdW.png** *Source:* https://upload.wikimedia.org/wikipedia/commons/0/00/Ethanol-3D-vdW.png *License:* Public domain *Contributors:* Own work *Original artist:* Benjah-bmm27 (talk · contribs)

- **File:F4M-1.png** *Source:* https://upload.wikimedia.org/wikipedia/commons/1/12/F4M-1.png *License:* Public domain *Contributors:* Own work *Original artist:* Dhatfield

- **File:F4M-2.png** *Source:* https://upload.wikimedia.org/wikipedia/commons/5/5a/F4M-2.png *License:* Public domain *Contributors:* Own work *Original artist:* Dhatfield

- **File:F4M-3.png** *Source:* https://upload.wikimedia.org/wikipedia/commons/3/3f/F4M-3.png *License:* Public domain *Contributors:* Own work *Original artist:* Dhatfield

- **File:F4M0.png** *Source:* https://upload.wikimedia.org/wikipedia/commons/e/e7/F4M0.png *License:* Public domain *Contributors:* Own work *Original artist:* Dhatfield

- **File:F4M1.png** *Source:* https://upload.wikimedia.org/wikipedia/commons/c/c7/F4M1.png *License:* Public domain *Contributors:* Own work *Original artist:* Dhatfield

- **File:F4M2.png** *Source:* https://upload.wikimedia.org/wikipedia/commons/8/86/F4M2.png *License:* Public domain *Contributors:* Own work *Original artist:* Dhatfield

- **File:F4M3.png** *Source:* https://upload.wikimedia.org/wikipedia/commons/a/a5/F4M3.png *License:* Public domain *Contributors:* Own work *Original artist:* Dhatfield

- **File:Fermi_Observations_of_Dwarf_Galaxies_Provide_New_Insights_on_Dark_Matter.ogv***Source:*https://upload.wikimedia.org/wikipedia/commons/a/a9/Fermi_Observations_of_Dwarf_Galaxies_Provide_New_Insights_on_Dark_Matter.ogv*License:*Public domain*Contributors:*Goddard Multimedia*Original artist:*NASA/Goddard Space Flight Center

- **File:First_Ionization_Energy.svg** *Source:* https://upload.wikimedia.org/wikipedia/commons/1/1d/First_Ionization_Energy.svg *License:* CC BY-SA 3.0 *Contributors:* http://commons.wikimedia.org/wiki/File:Erste_Ionisierungsenergie_PSE_color_coded.png *Original artist:* User: Sponk

- **File:Folder_Hexagonal_Icon.svg** *Source:* https://upload.wikimedia.org/wikipedia/en/4/48/Folder_Hexagonal_Icon.svg *License:* Cc-by-sa-3.0 *ontributors:* ? *Original artist:* ?

- **File:Four_Fundamental_States_of_Matter.png** *Source:* https://upload.wikimedia.org/wikipedia/commons/d/de/Four_Fundamental_States_of_Matter.png *License:* CC BY-SA 3.0 *Contributors:* Own work *Original artist:* Spirit469

- **File:Fraunhofer_lines.svg** *Source:* https://upload.wikimedia.org/wikipedia/commons/2/2f/Fraunhofer_lines.svg *License:* Public domain *Contributors:*

- Fraunhofer_lines.jpg *Original artist:* Fraunhofer_lines.jpg: nl:Gebruiker:MaureenV

- **File:Fusion_rxnrate.svg** *Source:* https://upload.wikimedia.org/wikipedia/commons/d/d0/Fusion_rxnrate.svg *License:* CC BY 2.5 *Contributors:* Own work *Original artist:* Dstrozzi

- **File:GalacticRotation2.svg** *Source:* https://upload.wikimedia.org/wikipedia/commons/b/b9/GalacticRotation2.svg *License:* CC-BY-SA-3.0 *Contributors:* Own work in Inkscape 0.42 *Original artist:* PhilHibbs

- **File:Galvanic_Cell.svg** *Source:* https://upload.wikimedia.org/wikipedia/commons/8/8e/Galvanic_Cell.svg *License:* CC BY-SA 3.0 *Contributors:* Own work *Original artist:* Gringer

- **File:Gammaspektrum_Uranerz.jpg** *Source:* https://upload.wikimedia.org/wikipedia/commons/8/8c/Gammaspektrum_Uranerz.jpg *License:* CC BY-SA 3.0 *Contributors:* Own work *Original artist:* Wusel007

- **File:Gas_molecules.gif** *Source:* https://upload.wikimedia.org/wikipedia/commons/d/d5/Gas_molecules.gif *License:* Public domain *Contributors:* ? *Original artist:* ?

- **File:Geiger-Marsden_experiment_expectation_and_result.svg***Source:*https://upload.wikimedia.org/wikipedia/commons/f/f9/Geiger-Ma experiment_expectation_and_result.svg *License:* CC BY-SA 3.0 *Contributors:* Own work *Original artist:* Kurzon

- **File:Glenn_Seaborg_-_1964.jpg** *Source:* https://upload.wikimedia.org/wikipedia/commons/4/47/Glenn_Seaborg_-_1964.jpg *License:* Public domain *Contributors:* NAIL Control Number: NWDNS-326-COM-12
  NARA (enter "Glenn Seaborg" in search form under Digital Copies tab) *Original artist:* Atomic Energy Commission. (1946 - 01/19/1975)

- **File:Gold_foil_experiment_conclusions.svg***Source:*https://upload.wikimedia.org/wikipedia/commons/9/9b/Gold_foil_experiment_conclu svg *License:* Public domain *Contributors:* Own work *Original artist:* Kurzon

- **File:Gravitationell-lins-4.jpg** *Source:* https://upload.wikimedia.org/wikipedia/commons/0/0b/Gravitationell-lins-4.jpg *License:* Public domain *Contributors:* http://hubblesite.org/newscenter/newsdesk/archive/releases/2003/01/image/a *Original artist:* NASA, N. Benitez (JHU), T. Broadhurst (Racah Institute of Physics/The Hebrew University), H. Ford (JHU), M. Clampin (STScI),G. Hartig (STScI), G. Illingworth (UCO/Lick Observatory), the ACS Science Team and ESA

- **File:HAtomOrbitals.png** *Source:* https://upload.wikimedia.org/wikipedia/commons/c/cf/HAtomOrbitals.png *License:* CC-BY-SA-3.0 *Contributors:* ? *Original artist:* ?

- **File:HEUraniumC.jpg** *Source:* https://upload.wikimedia.org/wikipedia/commons/d/d8/HEUraniumC.jpg *License:* Public domain *Contributors:* http://web.archive.org/web/20050829231403/http://web.em.doe.gov/takstock/phochp3a.html *Original artist:* ?

- **File:Halflife-sim.gif** *Source:* https://upload.wikimedia.org/wikipedia/commons/3/3f/Halflife-sim.gif *License:* Public domain *Contributors:* Own work *Original artist:* Sbyrnes321

- **File:HeTube.jpg** *Source:* https://upload.wikimedia.org/wikipedia/commons/1/1f/HeTube.jpg *License:* CC BY-SA 2.5 *Contributors:* user-made *Original artist:* User:Pslawinski

- **File:Helium_atom_QM.svg** *Source:* https://upload.wikimedia.org/wikipedia/commons/2/23/Helium_atom_QM.svg *License:* CC-BY-SA-3.0 *Contributors:* Own work *Original artist:* User:Yzmo

- **File:Henry_Moseley.jpg** *Source:* https://upload.wikimedia.org/wikipedia/en/d/dd/Henry_Moseley.jpg *License:* PD-US *Contributors:* ? *Original artist:* ?

- **File:Hexahedron.jpg** *Source:* https://upload.wikimedia.org/wikipedia/commons/7/78/Hexahedron.jpg *License:* CC-BY-SA-3.0 *Contributors:* ? *Original artist:* ?

- **File:Higgins-particles.jpg** *Source:* https://upload.wikimedia.org/wikipedia/commons/1/18/Higgins-particles.jpg *License:* Public domain *Contributors:* A Comparative View of the Phlogistic and Antiphlogistic Theories *Original artist:* William Higgins

- **File:Hubble_image_of_the_galaxy_cluster_Abell_3827.jpg** *Source:* https://upload.wikimedia.org/wikipedia/commons/5/51/Hubble_image_of_the_galaxy_cluster_Abell_3827.jpg *License:* CC BY 4.0 *Contributors:* http://www.eso.org/public/images/eso1514a/ *Original artist:* ESO

- **File:Hydrogen.svg** *Source:* https://upload.wikimedia.org/wikipedia/commons/3/3f/Hydrogen.svg *License:* CC-BY-SA-3.0 *Contributors:* Own work *Original artist:* Mets501

- **File:NO2-N2O4.jpg** *Source:* https://upload.wikimedia.org/wikipedia/commons/8/8b/NO2-N2O4.jpg *License:* Public domain *Contributors:* en:Image:N02-N2O4.jpg *Original artist:* en:User:Greenhorn1

- **File:Nasa_Shuttle_Test_Using_Electron_Beam_full.jpg** *Source:*https://upload.wikimedia.org/wikipedia/commons/a/a7/Nasa_Shuttle_Te Using_Electron_Beam_full.jpg*License:*Public domain*Contributors:*http://grin.hq.nasa.gov/ABSTRACTS/GPN-2000-003012.html*Originalartist:* NASA

- **File:Neon_orbitals.JPG** *Source:* https://upload.wikimedia.org/wikipedia/commons/2/2d/Neon_orbitals.JPG *License:* Public domain *Contributors:* ? *Original artist:* ?

- **File:Neutron.svg** *Source:* https://upload.wikimedia.org/wikipedia/commons/d/dd/Neutron.svg *License:* Public domain *Contributors:* ? *Original artist:* ?

- **File:Newlands_periodiska_system_1866.png** *Source:*https://upload.wikimedia.org/wikipedia/commons/e/e5/Newlands_periodiska_syste 1866.png*License:*Public domain*Contributors:*John Alexander Reina Newlands(1838–1898)*Original artist:*John Alexander Reina Newlands

- **File:Niels_Bohr.jpg** *Source:* https://upload.wikimedia.org/wikipedia/commons/6/6d/Niels_Bohr.jpg *License:* Public domain *Contributors:* Niels Bohr's Nobel Prize biography, from 1922 *Original artist:* The American Institute of Physics credits the photo [1] to AB Lagrelius & Westphal, which is the Swedish company used by the Nobel Foundation for most photos of its book series *Les Prix Nobel.*

- **File:Nitrate-ion-elpot.png** *Source:* https://upload.wikimedia.org/wikipedia/commons/6/6f/Nitrate-ion-elpot.png *License:* Public domain *Contributors:* Own work *Original artist:* Benjah-bmm27

- **File:NuclearReaction.png** *Source:* https://upload.wikimedia.org/wikipedia/commons/7/7d/NuclearReaction.png *License:* CC BY-SA 3.0 *Contributors:* Own work *Original artist:* Michalsmid

- **File:Nuclear_fission.svg** *Source:* https://upload.wikimedia.org/wikipedia/commons/1/15/Nuclear_fission.svg *License:* Public domain *Contributors:* ? *Original artist:* ?

- **File:Nucleosynthesis_periodic_table.svg** *Source:* https://upload.wikimedia.org/wikipedia/commons/3/31/Nucleosynthesis_periodic_table. svg *License:* CC BY-SA 3.0 *Contributors:* Own work *Original artist:* Cmglee

- **File:Nucleus_drawing.svg** *Source:* https://upload.wikimedia.org/wikipedia/commons/f/f0/Nucleus_drawing.svg *License:* CC BY-SA 3.0 *Contributors:* Own work (vector version of PNG image) *Original artist:* Marekich

- **File:Nuvola_apps_edu_science.svg** *Source:* https://upload.wikimedia.org/wikipedia/commons/5/59/Nuvola_apps_edu_science.svg *License:* LGPL *Contributors:* http://ftp.gnome.org/pub/GNOME/sources/gnome-themes-extras/0.9/gnome-themes-extras-0.9.0.tar.gz *Original artist:* David Vignoni / ICON KING

- **File:Nuvola_apps_kalzium.svg** *Source:* https://upload.wikimedia.org/wikipedia/commons/8/8b/Nuvola_apps_kalzium.svg *License:* LGPL *Contributors:* Own work *Original artist:* David Vignoni, SVG version by Bobarino

- **File:Octahedron.svg** *Source:* https://upload.wikimedia.org/wikipedia/commons/0/07/Octahedron.svg *License:* CC-BY-SA-3.0 *Contributors:* Vectorisation of Image:Octahedron.jpg *Original artist:* User:Stannered

- **File:Office-book.svg** *Source:* https://upload.wikimedia.org/wikipedia/commons/a/a8/Office-book.svg *License:* Public domain *Contributors:* This and myself. *Original artist:* Chris Down/Tango project

- **File:OiintLaTeX.svg** *Source:* https://upload.wikimedia.org/wikipedia/commons/8/86/OiintLaTeX.svg *License:* CC0 *Contributors:* Own work *Original artist:* Maschen

- **File:Orbital_s1.png** *Source:* https://upload.wikimedia.org/wikipedia/commons/0/00/Orbital_s1.png *License:* CC BY-SA 3.0 *Contributors:* Own work *Original artist:* RJHall

- **File:P2M-1.png** *Source:* https://upload.wikimedia.org/wikipedia/commons/8/89/P2M-1.png *License:* Public domain *Contributors:* Own work *Original artist:* Dhatfield

- **File:P2M0.png** *Source:* https://upload.wikimedia.org/wikipedia/commons/d/d4/P2M0.png *License:* Public domain *Contributors:* Own work *Original artist:* Dhatfield

- **File:P2M1.png** *Source:* https://upload.wikimedia.org/wikipedia/commons/5/5a/P2M1.png *License:* Public domain *Contributors:* Own work *Original artist:* Dhatfield

- **File:P3M-1.png** *Source:* https://upload.wikimedia.org/wikipedia/commons/2/2e/P3M-1.png *License:* Public domain *Contributors:* Own work *Original artist:* Dhatfield

- **File:P3M0.png** *Source:* https://upload.wikimedia.org/wikipedia/commons/2/2d/P3M0.png *License:* Public domain *Contributors:* Own work *Original artist:* Dhatfield

- **File:P3M1.png** *Source:* https://upload.wikimedia.org/wikipedia/commons/e/e6/P3M1.png *License:* Public domain *Contributors:* Own work *Original artist:* Dhatfield

- **File:P4M-1.png** *Source:* https://upload.wikimedia.org/wikipedia/commons/9/92/P4M-1.png *License:* Public domain *Contributors:* Own work *Original artist:* Dhatfield

- **File:P4M0.png** *Source:* https://upload.wikimedia.org/wikipedia/commons/8/8f/P4M0.png *License:* Public domain *Contributors:* Own work *Original artist:* Dhatfield

- **File:P4M1.png** *Source:* https://upload.wikimedia.org/wikipedia/commons/a/a9/P4M1.png *License:* Public domain *Contributors:* Own work *Original artist:* Dhatfield

- **File:P5M-1.png** *Source:* https://upload.wikimedia.org/wikipedia/commons/4/49/P5M-1.png *License:* Public domain *Contributors:* Own work *Original artist:* Dhatfield

- **File:Potential_energy_well.svg** *Source:* https://upload.wikimedia.org/wikipedia/commons/c/c5/Potential_energy_well.svg *License:* Public domain *Contributors:* Based upon Image:Potential well.png, created by User:Koantum. This version created by bdesham in Inkscape. *Original artist:* Benjamin D. Esham (bdesham)

- **File:Propane_Lewis.svg** *Source:* https://upload.wikimedia.org/wikipedia/commons/9/93/Propane_Lewis.svg *License:* CC BY-SA 3.0 *Contributors:* Own work *Original artist:* Dan1gia2

- **File:Protium_deuterium_tritium.jpg** *Source:* https://upload.wikimedia.org/wikipedia/commons/6/6c/Protium_deuterium_tritium.jpg *License:* CC-BY-SA-3.0 *Contributors:* self, translation from Dirk Hünniger (german wikipedia) *Original artist:* Lamiot for french version, from Dirk Hünniger

- **File:Quark_structure_neutron.svg** *Source:* https://upload.wikimedia.org/wikipedia/commons/8/81/Quark_structure_neutron.svg *License:* CC BY-SA 2.5 *Contributors:* ? *Original artist:* ?

- **File:Quark_structure_proton.svg** *Source:* https://upload.wikimedia.org/wikipedia/commons/9/92/Quark_structure_proton.svg *License:* CC BY-SA 2.5 *Contributors:* Own work *Original artist:* Arpad Horvath

- **File:Quartz_oisan.jpg** *Source:* https://upload.wikimedia.org/wikipedia/commons/4/4a/Quartz_oisan.jpg *License:* CC BY-SA 3.0 *Contributors:* Own work *Original artist:* Didier Descouens

- **File:Question_book-new.svg** *Source:* https://upload.wikimedia.org/wikipedia/en/9/99/Question_book-new.svg *License:* Cc-by-sa-3.0 *Contributors:*
  Created from scratch in Adobe Illustrator. Based on Image:Question book.png created by User:Equazcion *Original artist:*
  Tkgd2007

- **File:Radioactive.svg** *Source:* https://upload.wikimedia.org/wikipedia/commons/b/b5/Radioactive.svg *License:* Public domain *Contributors:* Created by Cary Bass using Adobe Illustrator on January 19, 2006. *Original artist:* Cary Bass

- **File:Radioactive_decay_modes.svg** *Source:* https://upload.wikimedia.org/wikipedia/commons/7/71/Radioactive_decay_modes.svg *License:* GFDL *Contributors:* Own work *Original artist:* MarsRover

- **File:Radioactivity_and_radiation.png** *Source:* https://upload.wikimedia.org/wikipedia/commons/6/6e/Radioactivity_and_radiation.png *License:* CC BY-SA 3.0 *Contributors:* Own work *Original artist:* Doug Sim

- **File:Rotation_curve_(Milky_Way).JPG** *Source:* https://upload.wikimedia.org/wikipedia/commons/7/77/Rotation_curve_%28Milky_W%29.JPG *License:* CC BY-SA 3.0 *Contributors:* Own work *Original artist:* Brews ohare

- **File:Rutherford_1911_Solvay.jpg** *Source:* https://upload.wikimedia.org/wikipedia/commons/3/3b/Rutherford_1911_Solvay.jpg *License:* Public domain *Contributors:* ? *Original artist:* ?

- **File:S-p-Orbitals.svg** *Source:* https://upload.wikimedia.org/wikipedia/commons/0/0f/S-p-Orbitals.svg *License:* CC-BY-SA-3.0 *Contributors:* selfmade from [1] *Original artist:* This file was made by **User:Sven**

- **File:S1M0.png** *Source:* https://upload.wikimedia.org/wikipedia/commons/1/1d/S1M0.png *License:* Public domain *Contributors:* Own work *Original artist:* Dhatfield

- **File:S2M0.png** *Source:* https://upload.wikimedia.org/wikipedia/commons/8/8f/S2M0.png *License:* Public domain *Contributors:* Own work *Original artist:* Dhatfield

- **File:S3M0.png** *Source:* https://upload.wikimedia.org/wikipedia/commons/a/ae/S3M0.png *License:* Public domain *Contributors:* Own work *Original artist:* Dhatfield

- **File:S4M0.png** *Source:* https://upload.wikimedia.org/wikipedia/commons/9/99/S4M0.png *License:* Public domain *Contributors:* Own work *Original artist:* Dhatfield

- **File:S5M0.png** *Source:* https://upload.wikimedia.org/wikipedia/commons/6/68/S5M0.png *License:* Public domain *Contributors:* Own work *Original artist:* Dhatfield

- **File:S6M0.png** *Source:* https://upload.wikimedia.org/wikipedia/commons/8/8a/S6M0.png *License:* Public domain *Contributors:* Own work *Original artist:* Dhatfield

- **File:S7M0.png** *Source:* https://upload.wikimedia.org/wikipedia/commons/5/57/S7M0.png *License:* Public domain *Contributors:* Own work *Original artist:* Dhatfield

- **File:Sasahara.svg** *Source:* https://upload.wikimedia.org/wikipedia/commons/c/cc/Sasahara.svg *License:* CC-BY-SA-3.0 *Contributors:* Transferred from en.wikipedia *Original artist:* Original uploader was JWB at en.wikipedia

- **File:Sbs_block_copolymer.jpg** *Source:* https://upload.wikimedia.org/wikipedia/en/9/9a/Sbs_block_copolymer.jpg *License:* PD *Contributors:* ? *Original artist:* ?

- **File:ShortPT20b.png** *Source:* https://upload.wikimedia.org/wikipedia/commons/c/c9/ShortPT20b.png *License:* CC BY-SA 3.0 *Contributors:* Own work *Original artist:* Sandbh

- **File:SiO^2_Quartz.svg** *Source:* https://upload.wikimedia.org/wikipedia/commons/f/fc/SiO%C2%B2_Quartz.svg *License:* Public domain *Contributors:* Own work *Original artist:* Wimmel

- **File:Silica.svg** *Source:* https://upload.wikimedia.org/wikipedia/commons/4/4b/Silica.svg *License:* Public domain *Contributors:*

- Silica.jpg *Original artist:* Silica.jpg: en:User:Jdrewitt

- **File:Size_IK_Peg.svg** *Source:* https://upload.wikimedia.org/wikipedia/commons/4/4a/Size_IK_Peg.svg *License:* CC-BY-SA-3.0 *Contributors:* Own work *Original artist:* RJHall, chris ⬚ (vector)

- **File:Sodium_cyanide-2D.svg** *Source:* https://upload.wikimedia.org/wikipedia/commons/c/c6/Sodium_cyanide-2D.svg *License:* Public domain *Contributors:* Own work *Original artist:* Arrowsmaster

32.11.3 Content license